Microwave Remote Sensing of Sea Ice

Geophysical Monograph Series

Including

IUGG Volumes
Maurice Ewing Volumes
Mineral Physics Volumes

GEOPHYSICAL MONOGRAPH SERIES

Mineral Physics Volumes

Geophysical Monograph 68

Microwave Remote Sensing of Sea Ice

Frank D. Carsey

Editor

American Geophysical Union

Published under the aegis of the AGU Books Board.

Library of Congress Cataloging-in-Publication Data

Microwave remote sensing of sea ice / Frank D. Carsey, editor.
    p.   cm. — (Geophysical monograph ; 68)
    Includes bibliographical references and index.
    ISBN 0-87590-033-X
    1. Sea ice—Remote sensing.   2. Microwave remote sensing.
I. Carsey, Frank D.   II. Series.
GB2401.72.R42M53   1992
551.3'43—dc20   92-41032
                                                                      CIP

ISSN:0065-8448
ISBN 0-87590-033-X

Contents

Tables

ACKNOWLEDGMENTS

This manuscript was prepared by the Jet Propulsion Laboratory, California Institute of Technology, and supported by the National Aeronautics and Space Administration and the Office of Naval Research.

In generating this book we have relied extensively upon reviewers from the polar geosciences community. We solicited and received from engineers and scientists over 100 reviews of the chapters and even of the entire book. We are aware of the very valuable time that all these reviewers have contributed to the quality of the book, and we thank them. In particular, we wish to point out that Charles Livingstone of the Canada Centre for Remote Sensing, Alex Stogryn of Aerojet, and Norbert Untersteiner of the University of Washington did a splendid job of reviewing and commenting on the whole book. We also thank the Jet Propulsion Laboratory staff for the production work, and we point out especially the very great skills and remarkably long hours of Jeanne Holm and Robin Dumas (and we thank Kaitlyn Holm for the fine timing of her arrival). The summary appendix of archived satellite data was assembled for this book by R. Weaver of the National Snow and Ice Data Center in Boulder, and the glossary was provided by W. F. Weeks of the University of Alaska at Fairbanks. We appreciate the support of ONR and NASA through the respective program managers, Charles Luther and Robert Thomas.

We recognize the many individuals, from agency scientists to ordinary citizens, who make satellite-supported science possible.

We also take time to remember the pioneers who understood, some decades ago, the high value of satellite-observation technology to the polar sciences and lent their great energy to advancing it. We dedicate this book to the memory of one of these pioneers, Bill Campbell. He was a friend, colleague, co-author, and boundlessly enthusiastic spokesman for our field. His career spanned more than 20 productive years of international projects beginning with the U.S.–U.S.S.R. Bering Sea work in the early 1970s and extending through the U.S.–Norway Barents Sea work earlier this year. We will miss him.

William Joseph Campbell, 1930–1992

PREFACE

Human activities in the polar regions have undergone incredible changes in this century. Among these changes is the revolution that satellites have brought about in obtaining information concerning polar geophysical processes. Satellites have flown for about three decades, and the polar regions have been the subject of their routine surveillance for more than half that time. Our observations of polar regions have evolved from happenstance ship sightings and isolated harbor icing records to routine global records obtained by those satellites. Thanks to such abundant data, we now know a great deal about the ice-covered seas, which constitute about 10% of the Earth's surface. This explosion of information about sea ice has fascinated scientists for some 20 years. We are now at a point of transition in sea ice studies; we are concerned less about ice itself and more about its role in the climate system. This change in emphasis has been the prime stimulus for this book.

A number of us who have worked in this developmental phase have collaborated on this book, which seeks to describe the current science and technology of the observation of ice with satellite-borne microwave instruments. Our intent is to render the use of satellite ice data into a working tool for the satellite data sets of the recent past and immediate future.

The book is more than a collection of papers. It is structured in integrated chapters. The authors represent institutions around the world at the forefront of these studies. While the chapters are free-standing, they are also linked into what we intend to be an integrated whole. We outlined and discussed the contents of the book in a series of meetings sponsored by the Office of Naval Research (ONR) and the National Aeronautics and Space Adminis-

tration (NASA) and held in the last part of 1990 and through 1991. In setting out the chapters, we sought to provide an orderly progression from basic observations, through modeling, to geophysical interpretation, and on to the use of the data in simulations of the roles and responses of ice in the global climate system.

We have attempted to comprehensively and uniformly cover the issues related to data interpretation. It will come as no surprise that we have not had the space here to cover all the issues to our satisfaction; that might not even be possible. Areas where our forced brevity left significant topics unaddressed include the use of satellite microwave data in polar operations, the remote sensing of processes of the ice margin, and the host of effects arising from seasonal change. It may be that each of these topics should be dealt with in its own book. Our result is a book of consensus, for the most part, but we hope that some of our occasional and appropriate disagreements are still visible; this subject is not yet finished.

Frank Carsey, Editor in Chief,
   Jet Propulsion Laboratory, Pasadena
Roger Barry, University of Colorado, Boulder
Josefino Comiso, Goddard Space Flight Center, Greenbelt
D. Andrew Rothrock, University of Washington, Seattle
Robert Shuchman, Environmental Research Institute of Michigan, Ann Arbor
W. Terry Tucker, Cold Regions Research and Engineering Laboratory, Hanover
Wilford Weeks, University of Alaska, Fairbanks
Dale Winebrenner, University of Washington, Seattle

Editorial Board, July 1992

Chapter 1. Introduction

FRANK D. CARSEY

Jet Propulsion Laboratory, California Institute of Technology, 4800 Oak Grove Drive, Pasadena, California 91109

ROGER G. BARRY

Cooperative Institute for Research in Environmental Sciences, University of Colorado, Boulder, Colorado 80309-0449

WILFORD F. WEEKS

Alaska SAR Facility, Geophysical Institute, University of Alaska, Fairbanks, Alaska 99775-0800

1.1 SEA ICE

The polar oceans and their adjacent seas are, at least seasonally, covered by a thin, uneven sheet of sea ice formed by the freezing of ocean surface water. The area of ocean covered by this ice varies strongly with the season. During the peak coverage, in the spring of each polar region, the ice extends into the mid-latitudes where its presence impacts human activities. During minimum coverage, in late summer and early fall, the ice is confined to the most remote regions of the oceans—the Arctic Basin and the Antarctic margins. Typical extents are shown in the Arctic and Antarctic location map, Figure 1-1.

Figure 1-1 also suggests why the satellite is becoming an opportune platform for polar geophysical information. In particular, in the polar regions, the spatial scales are vast, transportation is difficult and hazardous, operations are expensive, and the climate is hostile. Satellite instruments are thus a natural tool for environmental observations, and numerous visible and infrared spaceborne systems have made, and will continue to make, valuable contributions to ice surveillance [Massom, 1991]. To overcome these sensors'

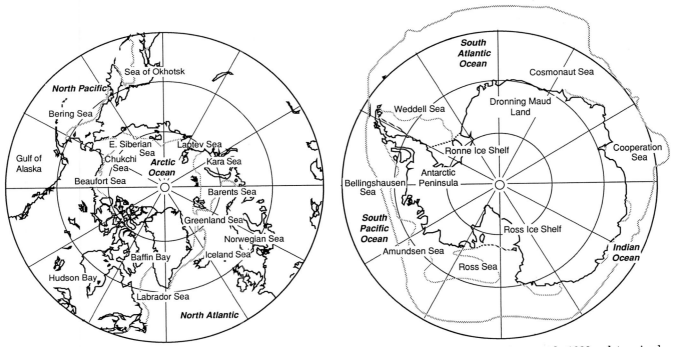

Fig. 1-1. Place maps for the north and south polar regions. The dotted lines show the maxima and minima in extent for 1989 as determined by the Special Sensor Microwave/Imager (SSM/I).

Microwave Remote Sensing of Sea Ice
Geophysical Monograph 68
©1992 American Geophysical Union

limitations to weather and light levels, microwave systems have been designed and flown since the 1970's. This book addresses the current technology of the microwave observation of sea ice.

Navigators, explorers, geographers, climatologists, and other researchers have long been interested in sea ice; the history of sea ice observation is extensive. Our interest in sea ice properties and behavior derives from its roles in the climate system and in polar operations. In the climate system, the sea ice cover has been placed in a unique position as the key harbinger of the global warming pre-dicted as the greenhouse-gas concentration increases [Stouffer et al., 1989]. This position makes the interpretation of sea ice data very important; these interpretations are presently the central task of the sea ice community.

The sea ice cover, in fact, vigorously interacts with the ocean and atmosphere (Figure 1-2) [Carmack, 1986]. The spatial and temporal scales of these interactions are in response to similar scales in the atmospheric and oceanic forcing terms, ranging spatially from meters to thousands of kilometers and temporally from diurnal to decadal.

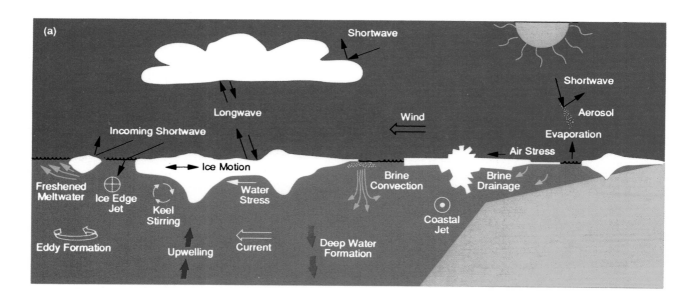

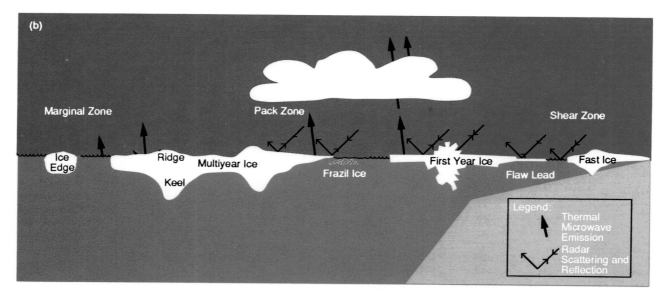

Fig. 1-2. (a) The physical processes of air–sea–ice interaction and (b) microwave emission and scattering from various surface elements are shown in schematic form.

1.2 SEA ICE IN CLIMATE AND OPERATIONS

Sea ice participates in the key large-scale processes of the Earth's climate system, the absorption and emission of radiant energy, and the poleward flux of heat. This participation comes about through processes involving the atmosphere, the ocean, and the radiation field. A few of these can be usefully outlined.

• Radiation Balance. The polar climate is essentially the result of radiative heat loss and compensatory poleward oceanic and atmospheric heat transport [Barry, 1989]. In winter the negative radiation balance results in strong cooling of the polar climate, and in summer solar radiation is the major source of heat input to the ice. The snow-covered ice also has a very high albedo relative to that of the open ocean. Therefore, changes in sea ice extent cause drastic changes in the surface albedo of the high-latitude seas.

• Surface Heat and Brine Fluxes. While the regional climate is radiatively controlled, the local processes of air–sea–ice interaction are governed by turbulent heat fluxes. Although the fully developed ice cover is an effective insulator between the cold air and the relatively warm ocean, areas of open water and thin ice lose heat rapidly during the cold seasons [Maykut, 1986]. The resulting ice growth injects the cold brine rejected during ice formation into the upper ocean. These fluxes of heat and brine result in the most significant source of dense water for the world Ocean [Carsey, 1991].

• Freshwater Fluxes. In the reverse of the brine-generation process, melting ice supplies fresh water to the upper ocean. This low-density fresh water can have a significant role in North Atlantic circulation [Aagaard and Carmack, 1989], and this meltwater contains significant biological material [Smith and Sakshaug, 1990].

• Ice Margin Processes. At the margin of the sea ice cover, the abrupt transition to open water or a coastline gives rise to unique processes, including water mass formation, oceanic upwelling, eddy formation, and atmospheric instability generation (e.g., [Muench et al., 1987]).

• Operations. Sea ice operations, including navigation and trafficability on and below the surface, drill ship operations in the marginal seas, and harbor operations, are concerned with locating areas of thin ice, identifying hazards such as very thick ice, and forecasting ice conditions.

1.3 THE SEA ICE VARIABLES

The list of sea ice variables of primary interest to both the operational and scientific data user can be summarized in a fairly short list, even though these users pursue very different ends:

• The ice extent and thickness distribution and their changes, ice growth and melt, with emphasis on ice covers less than 50 cm and greater than 2 m thick
• The snow depth to an accuracy of about 5 cm, and the snow wetness to an accuracy of about 1%
• The summer melt, including melt-pond coverage to an accuracy of 3%
• The ice motion and deformation on a scale of 5 km and to an accuracy of 2 cm/s
• The winds and currents that push the ice, and the cloud cover, air, and ocean temperatures that control growth and melt

1.4 ICE EXTENT

The ice extent, the latitude of the ice edge as a function of longitude, is of both climatological and practical importance. A key prediction of the greenhouse-gas-induced climate change is that the sea ice extent will respond early to the altered conditions (e.g., [Stouffer et al., 1989]). Satellite microwave data have succeeded in generating an exceptionally complete and accurate record of the ice extent for both polar regions over the past 18 years (e.g., [Zwally et al., 1983; Parkinson et al., 1987]), and this record continues today (Appendix B). This type of data is shown in Figure 1-3, in which the seasonal pattern is derived from the SSM/I instrument. Analyses of the trends in these data have been performed (e.g., [Gloersen and Campbell, 1991]), and the hemispheric ice extents (Figure 1-4) have been determined to be quite steady, with a slight decrease in the Arctic and a very slight increase in the Antarctic over approximately the decade of the eighties. Ice extent is an interesting variable; while we see it to be remarkably stationary at present, it is often predicted to be a sensitive indicator of climate change over the next decades. Ice extent and other properties are observed on much finer scales with synthetic aperture radar (SAR) images, examples of which are shown in Figure 1-5.

1.5 THE MICROWAVE PROPERTIES OF ICE

Approximately 10% of the world Ocean is coverd by sea ice during some portion of the year. Compositionally, this vast expanse is fairly uniform and is largely composed of the most common polymorph of ice, Ice Ih. In winter, a tiny fraction of this region is open ocean. In summer, another fraction is fresh meltwater wetting the snow or ponded on the ice's surface. Finally the ice (and snow) contain a trace, usually less than 1%, of brine. Outside of the world's "wet ocean," only the ice-covered regions of Earth exhibit this kind of uniform composition. However, the scattering and emission of microwaves by sea ice are markedly sensitive to even these small variations in composition and structure. As a result, microwaves offer the prospect of effective monitoring of the sea ice variables in which we are interested.

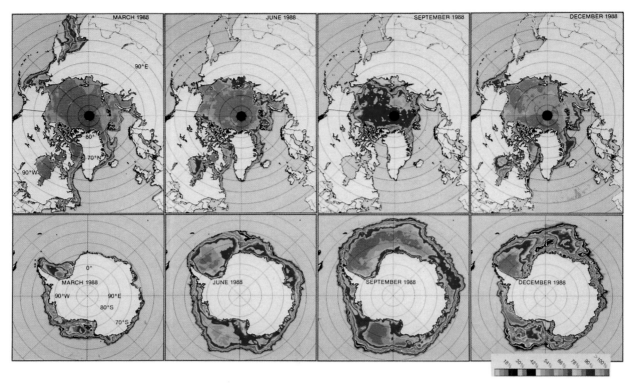

Fig. 1-3. Microwave radiometer data from the SSM/I instrument interpreted with the bootstrap algorithm (described in Chapter 10) to yield the seasonal cycle of ice concentration for both polar regions. (Figure courtesy of J. Comiso.)

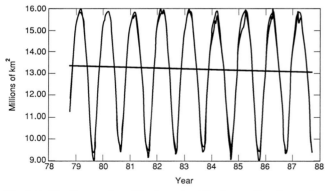

Fig. 1-4. The time series of the ice extent for the north polar region as determined by Scanning Multichannel Microwave Radiometer (SMMR) data [Gloersen and Campbell, 1991]. The smooth line represents the long-term mean cycle.

1.6 TECHNOLOGY DEVELOPMENT

Our approach to generating geophysical information on the sea ice variables is shown schematically in Figure 1-6. This approach involves using satellites to acquire observations of ice conditions, processing systems to run algorithms that estimate geophysical variables, and forecasting sys-

tems to ingest the derived geophysical variables, along with general environmental data and analyses, to obtain forecasts of ice conditions and to clarify the role of the ice cover in the global climate. We will address all the elements of Figure 1-6 except the specifics of climate change, which have been treated extensively in recent literature (e.g., [Barry, 1989]).

Developing techniques for monitoring ice conditions with microwave instruments has required carefully conducted field campaigns involving simultaneous observations at the surface and from aircraft and spacecraft. The first major campaign was the 1973 US–USSR Bering Sea Experiment (BESEX), which set the pattern for future studies combining ice geophysics and remote sensing (e.g., [Ramseier et al., 1982]).

The development approach to remote sensing of sea ice is summarized in Figure 1-7. This figure shows the progressive development in which data and models work in complement to define the signatures expressed in the algorithms that identify and evaluate ice processes and conditions. In this formalism, models are central to understanding our database. While we have argued that the ice-covered seas are compositionally simple, there are both local diversity and daily changes in ice thickness, snow structure and wetness, brine and air distribution, and physical temperature. The consequence of this diversity in

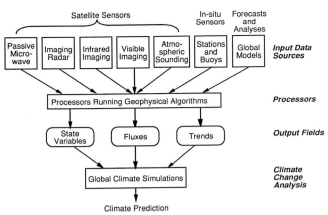

Fig. 1-6. Applying satellite data of the ice-covered seas to the task of climate change analysis.

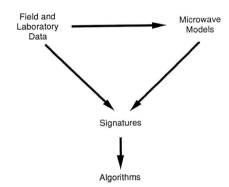

Fig. 1-7. Developing geophysical algorithms involves examining in-situ, aircraft, and satellite data; integrating those data into models that extend our signature into a broader "parameter space"; and inverting the signature into a processing algorithm.

Fig. 1-5. The first European Remote Sensing Satellite (ERS-1) SAR data for the ice of the Bering Sea. (a) On February 9, 1992, winds compact the ice, and (b) on February 27, 1992, polynyas are seen below St. Lawrence and St. Matthew Islands. Light areas are wind-roughened open water and very dark areas are new ice. (Copyright ESA, 1992.)

our empirical databases can be overwhelming. Often what seems to be noise can later be shown to be useful information. A careful framework of physical understanding can sort through nature's bewildering variation and find essential order.

Quantitative signature modeling and the corresponding interplay between modeling and experiment are keys to this understanding. With validated models in hand we can investigate the response of the idealized ice–snow–air–water–brine system to conditions where a purely empirical approach to characterization would be expensive, impractical, or hazardous. Also, we can efficiently design new systems of measurement or analysis to more effectively probe particular geophysical parameters of interest.

A prime example of the essential benefits of modeling can be appreciated in the case of new or young ice. The surface of newly formed ice is wet with highly saline brine rejected from the growing ice, and is nearly certain to acquire a snow cover that will soak up this brine. The resulting three-layer system (the sea ice, the briny snow, and the pure snow) can exhibit great spatial and temporal variability. Thus, the field investigator has a difficult puzzle to solve. The appropriate model, however, can predict the interactions of the microwaves with the system components for all cases once we have the key observations required to establish model parameters.

In the early years of microwave observations, the in-situ data sets were quite sparse, the models were heuristic, the satellite data-processing algorithms were interim in nature, and the geophysical products were noisy and contained gaps in time. At present, real progress has been made in developing algorithms for geophysical products: We can reliably estimate some ice variables, and there are time series spanning nearly two decades. We are by no means finished: Algorithms for estimating many key variables remain to be developed; validating algorithm products has defied our efforts to date; and the mathematical sophistication of our models has outpaced their validation so the modeling and observations do not yet optimally complement each other.

1.7 INSTRUMENTS IN ORBIT, DATA SETS IN HAND

The quantity and quality of microwave data over the Earth's sea ice is quite exciting. There are data sets in hand (Appendix B) [Massom, 1991, Part II] addressing ice extent, type, concentration, motion, and reflectance. There are instruments in orbit and other instruments planned for flight to extend and improve these observations. The overall timetable for instruments collecting sea ice data is shown in Figure 1-8. The data set being collected now is impressive (see Appendix A): a five-channel radiometer called the SSM/I, synthetic aperture radars [Curlander and McDonough, 1991] on the European Remote Sensing Satellite and the Japanese Earth Resources Satellite, visible-light and infrared sensors with a range of resolutions, a microwave altimeter, and various atmospheric sounders. It is up to the scientific community to exploit these data sets using, in part, the information in this book, which discusses a technology that has been evolving over the past 20 years.

1.8 THE BOOK

This book covers several areas of interest in microwave remote sensing of sea ice. The individual chapters cover the sea ice, its behavior, predictive models and algorithms, products, and technology.

Chapters 2 and 3 discuss the ice itself and the general connection between its microwave and physical properties. Sea ice is described as a material and the basic connection between its characteristics and the microwave properties controlling remote sensing signals are reviewed. Far more detail is available on the subject of ice (e.g., [Weeks and Ackley, 1986]).

The observed and modeled microwave behavior of important ice forms in the context of spaceborne instruments is

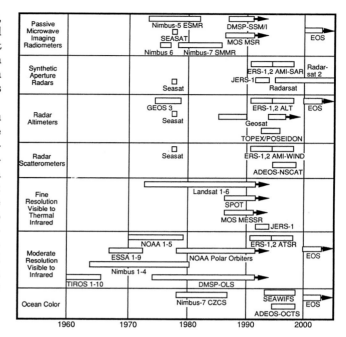

Fig. 1-8. Planned and historical satellite missions with data valuable to sea ice research. Soviet satellites have acquired data in some categories, but these have been omitted since access to these data is uncertain.

covered in Chapters 4 through 9. Data acquired in situ and in the laboratory are tabulated and discussed in the context of season, region, and the ice's developmental stage. The theoretical behavior of microwaves in the basic ice types found on the sea is examined through a number of approaches.

The algorithms now in use to produce estimates of geophysical variables and the status of research to develop methods for generating additional sea ice information are discussed in Chapters 10 through 22. Some algorithms that are essentially mature, such as for ice type, and others that are highly developmental, such as for snow cover properties, are shown.

The application of satellite-derived ice information in simulations of the behavior of ice in the climate system is covered in Chapters 23 and 24. The use of ice cover data in initiating and updating ice dynamics and thermodynamics in predictive models is reviewed, and the application of Kalman smoothing to the refinement of interpreting the sea-ice seasonal cycle is discussed.

Chapter 25 discusses new approaches to microwave observations of sea ice. The application of data analysis technology from other disciplines and a new radar technology are discussed.

Chapter 26 is a summary including anticipated future activities and opportunities.

We hope you find this book useful and interesting.

REFERENCES

Aagaard, K. and E. Carmack, The role of sea ice and other fresh water in the Arctic circulation, *Journal of Geophysical Research, 94*, pp. 14,485–14,498, 1989.

Barry, R. G., The present climate of the Arctic Ocean and possible past and future states, *The Arctic Seas. Climatology, Oceanography, Geology and Biology*, edited by Y. Herman, pp. 1–46, van Nostrand Reinhold, New York, 1989.

Carmack, E., Circulation and mixing in ice-covered waters, *The Geophysics of Sea Ice*, edited by N. Untersteiner, pp. 641–712, NATO ASI Series B: Physics vol. 146, Plenum Press, New York, 1986.

Carsey, F. D., An approach to brine and freshwater fluxes interpreted from radar and microwave radiometer data, *Deep Convection and Deep Water Formation in the Oceans*, edited by P. C. Chu and J. S. Gascard, pp. 123–133, Elsevier Science Publishers, New York, 1991.

Curlander, J. C. and R. N. McDonough, *Synthetic Aperture Radar, Systems and Signal Processing*, 647 pp., Wiley Interscience, New York, 1991.

Gloersen, P. and W. Campbell, Recent variations in Arctic and Antarctic sea ice covers, *Nature, 352*, pp. 33–36, 1991.

Massom, R., *Satellite Remote Sensing of Polar Regions*, 307 pp., Lewis Publications, Boca Raton, Florida, 1991.

Maykut, G. A., The surface heat and mass flux, *The Geophysics of Sea Ice*, edited by N. Untersteiner, pp. 395–464, NATO ASI Series B: Physics vol. 146, Plenum Press, New York, 1986.

Muench, R. D., S. Martin, and J. Overland, Preface: second marginal ice zone research collection, *Journal of Geophysical Research, 92*, entire issue, 1987.

Parkinson, C. L., J. C. Comiso, H. J. Zwally, D. J. Cavalieri, P. Gloersen, and W. J. Campbell, *Arctic Sea Ice, 1973–1976: Satellite Passive-Microwave Observations*, 296 pp., NASA SP-489, National Aeronautics and Space Administration, Washington, DC, 1987.

Ramseier, R., P. Gloersen, W. Campbell, and T. Chang, Mesoscale descriptions for the principal Bering Sea Experiment, *Proceedings of the Final Symposium on the Results of the Joint Soviet–American Expedition*, edited by Y. Kondratyev et al., pp. 231–236, A. A. Balkema, Rotterdam, 1982.

Smith, W. and E. Sakshaug, Polar phytoplankton, *Polar Oceanography, Part B, Chemistry, Biology, and Geology*, edited by W. Smith, pp. 477–517, Academic Press, San Diego, California, 1990.

Stouffer, R., S. Manabe, and K. Bryan, Interhemispheric asymmetry in climate response to a gradual increase of atmospheric CO_2, *Nature, 342*, pp. 660–682, 1989.

Weeks, W. F. and S. F. Ackley, The growth, structure, and properties of sea ice, *The Geophysics of Sea Ice*, edited by N. Untersteiner, pp. 9–164, NATO ASI Series B: Physics vol. 146, Plenum Press, New York, 1986.

Zwally, H. J., J. C. Comiso, C. L. Parkinson, W. J. Campbell, F. D. Carsey, and P. Gloersen, *Antarctic Sea Ice, 1973–1976: Satellite Passive Microwave Observations*, 206 pp., NASA SP-459, National Aeronautics and Space Administration, Washington, DC, 1983.

Chapter 2. Physical Properties of Sea Ice Relevant to Remote Sensing

W. B. TUCKER III, DONALD K. PEROVICH, AND ANTHONY J. GOW

Cold Regions Research and Engineering Laboratory, 72 Lyme Road, Hanover, New Hampshire 03755-1290

WILFORD F. WEEKS

Alaska SAR Facility, Geophysical Institute, University of Alaska, Fairbanks, Alaska 99775-0800

MARK R. DRINKWATER

Jet Propulsion Laboratory, California Institute of Technology, 4800 Oak Grove Drive, Pasadena, California 91109

An understanding of the physical processes and properties of sea ice is critical in interpreting microwave signatures. The incorporation of salt in the form of brine inclusions in the ice makes sea ice a vastly different material than freshwater ice. The amount of brine incorporated is largely growth-rate dependent. In columnar ice brine is trapped in inclusions within ice crystals in a platelet substructure, while in frazil ice the inclusions are located between the crystal boundaries. Brine drainage begins immediately after ice formation, occurring slowly during the growth season, but increasing considerably during summer. In the Arctic, enhanced surface melting, coupled with increased interconnectivity of the brine inclusions, almost completely flushes the salt from the ice in the upper layers, leaving air voids and channels; a process that greatly increases the ice's porosity.

The crystal texture of the upper meter of the ice also shows the effects of protracted warming, with rounding of crystal boundaries and virtual elimination of the brine platelet substructure. Such effects are not evident in Antarctic sea ice where pronounced surface melting apparently does not occur, resulting in an absence of both surface melt ponds and the melt-pond–hummock relief so characteristic of similar floes in the Arctic.

Pressure ridges are evident in both polar regions, although they are larger and more numerous in the Arctic. Small-scale surface properties, such as ice surface roughness, are very important to microwave remote sensing. Frost flowers and snow on recently formed ice absorb brine, greatly affecting the surface dielectric properties. Excessive snow cover can depress both multiyear and first-year floes sufficiently to cause sea water to flood the surface, resulting in the misidentification of ice types in microwave algorithms.

2.1 INTRODUCTION

Remote sensing in the polar regions has significantly increased our understanding of sea ice extent and variability. The capability of all-weather day and night surveillance of the ice packs has both enhanced research capabilities and helped to accomplish operational objectives. For instance, continuous monitoring of the ice extent and concentration, facts that are routinely available from satellite passive microwave systems, is proving invaluable to scientific investigations involving sea ice climatology. This capability has also improved navigation and assisted petroleum development operations. Other useful products, such as the automated characterization of ice types and velocities, are currently being developed. In addition, field studies associated with these remote sensing developments have contributed to our understanding of the physical properties of the ice.

Early sea-ice remote-sensing experiments consisted of aircraft missions that carried sensors designed for either land, oceanic, or atmospheric applications. Microwave sensors have evolved considerably since those early missions, and some are now designed with sea ice as one of the primary media of interest. Both active and passive microwave sensors play major roles in these missions. Useful geophysical products can be derived from each type of sensor by using a range of frequencies, polarizations, and look angles. Ice properties exhibit tremendous variability in both vertical and horizontal dimensions. Ice types can range from essentially freshwater ice in refrozen melt ponds to very saline new ice to extremely porous ancient multiyear ice.

In this chapter, we address the sea ice properties that are important to microwave remote sensing. For passive microwave measurements, the ice's emissivity is the electromagnetic characteristic of interest. Emissivity is a function of the effective surface reflectance and of both surface and volume scattering. The reflectivity depends upon the bulk dielectric properties, which are functions of the distribution of the brine, air, and solid salt within the ice. Scattering

Microwave Remote Sensing of Sea Ice
Geophysical Monograph 68

depends upon the surface roughness and inhomogeneities within the ice, as well as the bulk dielectric constant. Active microwave instruments measure the backscatter, which is dominated by surface and volume scattering. Again, roughness and spatial variations in the dielectric properties contribute to the magnitude of the scattering.

Physical properties of the ice that will be discussed are the crystal structure, brine and air content, and surface properties. The properties and evolution of both first-year and multiyear ice, defined here as ice that has survived one or more summers, are described. This description of ice properties and processes is intended to set the stage for the chapters that follow, describing the nature of microwave–sea-ice interactions in some detail.

2.2 FIRST-YEAR ICE

2.2.1 Growth of First-Year Ice

The freezing of seawater differs from that of freshwater in two ways due to the inclusion of salt. First, salt depresses the freezing point of water (in degrees Centigrade) according to the approximation $T_f=-0.003-0.0527S_w-0.00004S_w^2$ [Neumann and Pierson, 1966], where S_w is the salinity of the water in parts per thousand. Second, the temperature of the maximum density of seawater for salinities greater than 24.7‰ is less than the freezing point. The result is that continued cooling of seawater in this salinity range results in an unstable vertical density distribution. This leads to convective overturning that continues until the water reaches the freezing point. Because of the oceanic density structure, this convection is limited to a relatively shallow layer. This well mixed layer, which presumably ranges from 10- to 40-m thick [Doronin and Kheisin, 1975], must cool to the freezing point before freezing can begin. Subsequent brine rejection by the growing ice causes additional convective overturning that can further deepen the mixed layer.

Initial ice formation begins at the water surface, where the heat loss is greatest, providing the small amount of supercooling necessary for ice growth. Ice growth begins with the formation of small platelets and needles, called frazil. As frazil crystals continue forming, a soupy mixture of unconsolidated crystals and seawater is created, commonly referred to as grease ice. With continued freezing under quiet conditions, the frazil crystals begin to coalesce, freezing together to form a solid cover up to 10-cm thick. This thin cover, which behaves elastically, is called nilas. Often, however, the action of wind and waves influences the configuration of the solid cover that forms. In leads, wind frequently herds the frazil to the downwind side, forming accumulations that can reach 1 m in thickness. Wind-driven circulation in leads can also cause frazil to accumulate in bands. In large open areas, such as occur in the marginal ice zones of the Arctic and Antarctic, wind and wave action cause the formation of pancake ice, which consists of circular masses of semiconsolidated frazil. Pancakes commonly grow to diameters of 0.3 to 3.0 m, and in many cases develop several-centimeter-high raised rims from bumping together and from sloshing of newly formed crystals onto existing pancakes. The slush forming in the open areas between the pancakes eventually consolidates to form either a solid field of pancakes or a combination of pancake and sheet ice. In the Antarctic, the rafting of pancake ice is thought to be largely responsible for the thick masses of ice with a frazil texture. Such ice floes have been extensively observed in the Weddell Sea [Lange et al., 1989].

An important aspect of the occurrence of frazil is that it forms the initial ice cover and, once it has consolidated, insulates the underlying ocean from the cold atmosphere. Further growth of ice must take place beneath this initial skim. Most often, this occurs as congelation growth, which is the freezing of seawater directly to the bottom of the existing sheet as the result of heat conduction upward through the overlying ice. In these cases, the growth rate is determined by the temperature gradient in the existing sheet and its effective thermal conductivity.

Once the ice sheet forms, the crystals at the ice–water interface lose one degree of growth freedom. As a result, noncompetitive crystal growth can only occur when the grain boundaries are exactly perpendicular to the freezing interface. Through a process referred to as geometric selection, crystals in unfavored orientations are eliminated by crystals oriented in the favored growth direction. This selection process occurs in a transition layer that is usually 5- to 10-cm thick [Weeks and Ackley, 1986]. The final crystal structure that develops contains vertically elongated columnar crystals aligned parallel to the direction of heat flow and having their c-axes oriented generally parallel to the ice–water interface.

2.2.2 Crystal Structure

Exclusive of the effects of snow cover, typical Arctic sea ice consists texturally of three layers resulting from the conditions during growth described above. The first two layers are generally relatively thin and consist of the initial frazil layer underlain by the transition layer. Columnar ice generally comprises the remainder of the ice sheet. There are, however, ample exceptions to this typical structure. Large accumulations of frazil-textured ice have been reported in Antarctic sea ice [Gow et al., 1987a; Lange et al., 1989], and significant amounts of frazil have been observed in samples from the marginal ice zones of the Arctic. Mixtures of the two crystal types are also not uncommon.

The elongated crystals characteristic of columnar ice contain a predominant crystal substructure. Within each grain, pure ice plates are separated by parallel layers of brine inclusions. These inclusions are the result of constitutional supercooling that allows the systematic entrapment of brine. Constitutional supercooling is caused by the more rapid diffusion of heat than salt at the freezing interface, resulting in a supercooled condition in the fluid ahead of the interface. As a result a nonplanar, dendritic interface develops with the associated entrapment of brine between dendritic plates, which are composed of pure ice.

This aggregation of plates at the base of the sea ice sheet is called the skeleton layer and is generally 1- to 3-cm thick (Figure 2-1). The spacing between the plates ranges from a few tenths to about 1 mm, and is largely a function of the growth rate [Lofgren and Weeks, 1969; Nakawo and Sinha, 1984]. Therefore, each columnar ice crystal contains a number of brine inclusions located in the grooves between the pure ice plates. Figure 2-2 shows a horizontal thin-section photograph depicting the ice-plate–brine-inclusion substructure within individual crystals of sea ice. The salinity of the ice sheet is ultimately determined by the abundance of brine inclusions and the amount of solid salt and brine within the inclusions.

The grain size of these columnar crystals varies considerably. The horizontal diameter of the crystals can be a few tenths to several centimeters, while the vertical length of

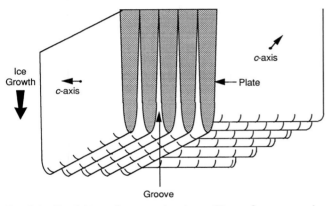

Fig. 2-1. Dendrite and groove structure of two columnar sea ice crystals. Residual brine is trapped in grooves between dendrites (plates), which remain essentially salt free.

Fig. 2-2. Horizontal cross-section of a thin section of sea ice showing multiplate and brine layer substructure of individual crystals. Arrow indicates c-axis orientation in three of the crystals. Scale subdivisions measure 1 mm.

the crystals can be tens of centimeters. Crystal sizes generally increase with depth [Weeks and Hamilton, 1962; Paige, 1966]. Typical diameters range from 0.5 cm near the top of the sheet to 2 to 3 cm near the bottom of the sheet.

As noted, typical congelation growth results in columnar crystals that have their c-axes oriented substantially in the horizontal plane. The c-axes are normal to the dendritic plates (Figure 2-1). In some cases, the crystals show no preferred alignment within the horizontal plane. However, strong alignments in this plane are quite common. Weeks and Gow [1978, 1980], Langhorne [1983], Langhorne and Robinson [1986], and Stander and Michel [1989] found that the direction of preferred c-axis alignment was coincident with the direct of current motion at the ice–water interface. It is hypothesized that crystal growth is enhanced in this direction by more effective removal of the solute buildup at the growing interface by increased turbulence. Of significance to remote sensing is that the preferred alignment of the c-axes has been found to polarize signals of impulse radar systems [Campbell and Orange, 1974; Kovacs and Morey, 1978]. Using a radar with a center frequency of 625 MHz, which is typically used to determine ice thickness, no bottom reflection was obtained when the radar antenna E-field was oriented normal to the alignment of the c-axes in first-year fast ice. Observations indicate that aligned c-axes can develop in the bulk of the congelation component of first-year ice [Weeks and Gow, 1978, 1980] and that such alignments may extend over areas of many tens of square kilometers [Cherepanov, 1971; Weeks and Gow, 1978].

Although we have thus far concentrated almost entirely on columnar ice, granular ice can also be an important constituent of the ice cover. As mentioned earlier, the initial consolidated skim usually consists of granular ice. Snow ice, another form of granular ice, may also form on the ice surface. This ice type results from the freezing of the snow cover that has become saturated with water whose source may be seawater, rainwater, or meltwater. Snow ice can generally be distinguished from other granular ice types because it is somewhat coarser grained and contains more air bubbles. Such ice can be fresh or saline depending on the conditions of saturation.

Frazil ice may also be observed at some depth in the ice sheet. Under certain circumstances, frazil is believed to form at depth in the water column and aggregate onto the bottom of the ice sheet. The large amounts of frazil observed in the Weddell Sea of Antarctica may form in this manner [Gow et al., 1982, 1987a]. The rafting of pancake ice, which is extensive in the Weddell Sea, may also be responsible for enhancing frazil ice thickness [Wadhams et al., 1987; Lange et al., 1989]. In the Arctic, the occurrence of frazil appears to be much more restricted. Most often it appears to be associated with turbulent hydrodynamic conditions associated with the earliest stages of ice growth. Granular ice and mixed columnar and granular ice have also been observed in substantial proportions in cores obtained from ridges [Tucker et al., 1987]. In this case, frazil is believed to originate in the voids between the blocks of ice incorporated when the ridge was formed. It has also been found on the flanks and keels of ridges [Tucker et al., 1987], suggesting that freely suspended frazil crystals in the water column are collected by the ridge as it moves relative to the seawater. The crystals then accumulate on the sides of ridge keels, where they eventually consolidate. Lange and Eicken [1991b] believe that deformation processes can cause frazil ice to be incorporated both between and beneath columnar ice sections.

Granular ice occurs in a wide variety of crystal sizes. Snow ice has crystal sizes ranging from 3 to 5 mm. Nilas, the thin ice cover resulting from the consolidation of grease ice, can have grains as small as a few tenths of a millimeter up to 2 mm. Likewise, pancake ice typically has small grains, measuring a few millimeters or less. Frazil crystals found deep within the ice sheet range from millimeters to several centimeters in diameter. In many cases fine-grained columnar crystals may be confused with granular ice, particularly in a horizontal thin section. Usually, however, the brine platelet substructure is sufficient to distinguish between the two types of ice.

2.2.3 Incorporation of Brine and Air

The incorporation of salt and air into the ice sheet is extremely important with regard to remote sensing in that the electromagnetic properties of the sheet, including the dielectric constant, the reflectivity, and volume and surface scattering, are heavily dependent on the distribution of liquid brine and gas within the ice. Also crucial to remote sensing is that the incorporation of brine gives rise to strong vertical anisotropy in a congelation ice sheet. This unique structure results from the extended columnar crystals, the vertical elongation of brine pockets, and the vertical orientation of brine drainage systems.

The temperature regime in the ice, both at the time of observation and at the time of initial formation of each layer, controls many of the characteristics of the sheet. The brine inclusion process is temperature dependent in that the plate spacing is almost solely determined by the growth rate. With more rapid freezing, the plate spacing is narrower and more brine is entrapped, resulting in higher salinities. The amount of entrapped salt is correspondingly less at lower growth rates. The plate spacing has been found to be generally inversely proportional to the square root of the growth rate [Weeks and Ackley, 1986]. Brine is also entrapped as pockets in consolidated granular ice. However, as these crystals do not have the characteristic substructure, the inclusions are found along grain boundaries and at the intersections of grains.

The amount of salt entrapped as brine in the ice also depends on the salinity of the water. Laboratory measurements by Weeks and Lofgren [1967] and Cox and Weeks [1975, 1988] indicate that the relationship between the initial salinity of the ice (S_i) and the salinity of the water from which the ice formed (S_w) can be expressed as $S_i = K_{eff}S_w$, where K_{eff} is the effective distribution coefficient, which is growth-rate dependent.

The amount of liquid brine in the ice sheet at any time after initial entrapment is dictated by the requirements of phase equilibrium [Weeks and Ackley, 1986], which fix the salinity of the brine that coexists with ice at a given temperature. That is, changes in the temperature of the ice will be accompanied by changes in the size of the brine pockets resulting from freezing or melting on the walls of the pockets until the salinity of the brine reaches the new equilibrium composition. The result is that, due to temperature changes, constantly varying proportions of brine, air, and ice exist in the brine inclusions so that the brine salinity at every level in the ice sheet is exactly the equilibrium value at that temperature.

Because the most rapid freezing invariably occurs at the top of the ice sheet, it might be assumed that the highest salinities would be found there. Although this is generally true, it must be noted that the process of brine drainage begins immediately after initial entrapment. The dominant mechanisms for brine loss from growing ice sheets are gravity drainage and brine expulsion. Brine expulsion results from pressure increases in the brine pockets due to freezing. Any decrease in the temperature of the ice will cause ice to form on the walls of the brine cavities, thereby increasing the salinity of the remaining brine to maintain phase equilibrium. Since the new ice on the cavity wall has about a 10% greater volume than the water from which it froze, some brine is expelled through cracks around the pocket caused by the pressure increase associated with the volume change. While most of the expelled brine would be expected to drain downward into the warmer and therefore more porous underlying ice, in thin ice, brine is also expelled upward onto the ice surface. This produces a thin layer of highly concentrated brine on the surface of the ice [Martin, 1979]. This process appears to be enhanced by capillary effects leading to the "wicking up" of brine by snow and frost flowers [Drinkwater and Crocker, 1988]. Present data suggest that brine expulsion is primarily of importance in the upper layers of sea ice sheets.

The major desalination mechanism of a growing ice sheet is gravity drainage. Because the top of the ice is colder, the brine higher in the ice sheet is denser and will drain downward. This effect is enhanced by a downward increase in the permeability of the ice. The formation of brine channels, vertical tubes centimeters to tens of centimeters long created by the interconnection of brine pockets, appears to be particularly important in increasing permeability. Although some aspects of the formation and behavior of these features are well understood [Eide and Martin, 1975; Lake and Lewis, 1970], many of the details have, as yet, not been thoroughly investigated. As a result of the continuing desalination process, the bulk salinity of the ice sheet decreases as it thickens and ages. This effect is demonstrated by the idealized profiles shown in Figure 2-3. These show the typical C-shaped profiles found in growing first-year ice. The upper portion of the ice has a higher salinity because it initially entraps more salt as a result of its more rapid growth rate. The high salinities in the lower portion of the ice result from the fact that although the

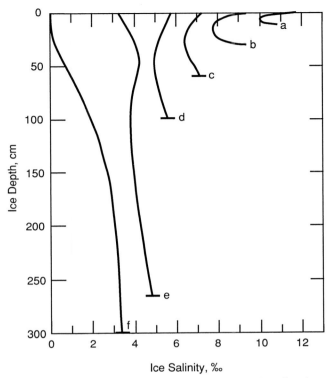

Fig. 2-3. Idealized salinity profiles in Arctic sea ice of various thicknesses. Curves (a) through (d) represent changes observed in first-year ice. The remaining curves are for two types of profiles found in multiyear ice. Curve (e) typifies the salinity distribution beneath low areas where the surface is close to the freeboard level, while curve (f) represents the salinity profile beneath hummocks and other elevated areas. (Reprinted with permission from Maykut [1985]. Copyright CRC Press, Inc.)

cating the brine cavities. Total porosities (air plus brine) can reach 200‰, particularly at the bottom of the ice sheet. More often, measured total porosities are found to be 20‰ to 60‰. Normally densities of first-year ice are about 0.91 to 0.92 mg/m³, with a full range from 0.89 to 0.925 mg/m³ [Cold Regions Research and Engineering Laboratory (CRREL), unpublished document, Hanover, New Hampshire].

2.2.4 Polar Contrasts

The differences in the properties of Arctic and Antarctic first-year ice exist because of the very different growth conditions that are directly related to contrasts in the land and sea relationships in the two polar regions. The Antarctic continent lies substantially poleward of 70° S, and sea ice forms from the coast outwards. In general, ice that forms in the south is advected generally northward, where it melts [Ackley and Holt, 1984]. Therefore, Antarctic sea ice essentially constitutes a marginal ice zone that undergoes very large changes in ice extent and whose northward progression is hindered by the warmer ocean, warmer atmosphere, and ocean swell. The Arctic, on the other hand, contains an ocean basin surrounded by land. The ice extent here undergoes less seasonal change than in the Antarctic and much less ice is advected from the basin because there are few passages through which ice can exit the basin. Also, the land-locked nature of the basin plus the large amount of multiyear ice that survives the summer combine to limit the effective open water fetch at the southern limits of the pack. As a result, both wave effects and the contribution of frazil ice are less in the Arctic than in the Antarctic.

Besides major differences in the ice extent and concentration, profound differences exist between ice thicknesses in the two regions. First-year ice thickness in the Antarctic varies greatly with location, measuring on the order of 0.50 to 0.75 m in parts of the Weddell Sea [Wadhams et al., 1987; Lange and Eicken, 1991a] to as much as 3 m in coastal embayments, where air temperatures are colder and the oceanic heat flux is negligible or in some cases even negative. In the Arctic, it is not uncommon to have first-year ice in excess of 1 m, frequently reaching 2 m [Bilello, 1980], though ice in the peripheral seas of the Arctic is substantially thinner. For example, in the Bering and Labrador Seas, where air temperatures are more moderate than in the Arctic Basin, thicknesses of undeformed ice seldom exceed 1.0 m. However, rafting occurs frequently in the thin ice of these areas, increasing the mean thicknesses to 1.25 to 1.50 m [Drinkwater and Squire, 1989; Carsey et al., 1989].

Major differences are apparent in the crystal structures of Arctic and Antarctic ice. In the Antarctic, far more frazil or granular ice has been observed than in the Arctic. Recent studies of Arctic Basin sea ice have reported granular ice contents of undeformed ice to be about 10% to 15% [Gow et al., 1987b]. In contrast, ice sampled in the Weddell Sea contained about 50% to 60% frazil ice. Recent investigations by Jeffries and W. F. Weeks [unpublished data] confirm that frazil is also an important component of sea ice

initial salt entrapment in this ice is commonly less than in the overlying ice, because this ice has just formed, it has lost little brine. Also, the lower layers of the ice receive brine that has drained from above. At mid-levels, slower growth rates and sufficient time to allow drainage to begin combine to produce lower salinities. These effects also act to reduce the bulk salinities as the ice becomes thicker.

The desalination process results in small-scale variations in salinity. For example, Weeks and Lee [1962], Tucker et al. [1984a], and Eicken et al. [1991] demonstrated that salinity profiles of cores taken within 1.0 m of each other from the same ice sheet can be significantly different. They observed salinities of the upper 10 cm differing by as much as 2‰ for ice that was structurally identical. These variations were attributed to the location of vertical brine drainage channels, with ice in the vicinity of a drainage channel being less saline.

Gas (air) is also entrapped in the ice as it forms. The gas volume subsequently increases as brine drainage takes place, leaving empty cavities and drainage channels. As might be expected, the volume of gas trapped at initial freezing is quite small, generally ranging from 0‰ to 15‰. Air volumes increase significantly as drainage occurs, va-in the Western Ross Sea. Figure 2-4 compares the granular and columnar contents of recent measurements made in the Arctic and Antarctic. A contributing factor to the widespread occurrence of frazil in the Antarctic is increased wind and wave turbulence, which enhances frazil production. Associated with these more dynamic conditions is the rafting of frazil-rich pancake ice and ice sheets. Platelet ice, which consists of large plate- and wafer-like crystals, has also been observed in the Antarctic [Gow et al., 1987a; Dieckmann et al., 1986; Lange et al., 1988]. This type of ice has yet to be identified in the Arctic. Unfortunately, other large areas of Antarctic sea ice remain virtually unstudied. Thus, extrapolation from the Weddell Sea regarding structure and other physical characteristics could lead to errors.

The amount of multiyear ice also differs between the Arctic and Antarctic. Because Antarctic ice is so efficiently advected from the coastal regions north to the marginal areas where it melts, multiyear ice appears to be scarce around most of Antarctica. Large oceanic heat fluxes efficiently melt the thin first-year ice. Ice in the Antarctic regions does not appear to last longer than 1 to 2 years [Gow and Tucker, 1990], except in isolated bays where fast ice may remain in place for several years. In sharp contrast, Koerner [1973] found that multiyear ice in the Arctic comprised at least 70% of the ice cover during a surface transect of the Central Arctic. Comiso [1990] also found large concentrations of Arctic multiyear ice using passive microwave data. Stochastic model studies by Colony and Thorndike [1985] indicated that ice-floe residence times of 5 to 7 years in the Arctic Basin are not unusual.

Antarctic first-year ice appears to be only slightly more saline (0.3‰ to 0.5‰) than its Arctic counterpart [Gow and Tucker, 1990]. The higher Antarctic salinities probably result from more rapid growth rates associated with generally colder temperatures. It may also be possible that more brine is retained in consolidated frazil because its structure impedes the development of brine drainage mechanisms. Gow et al. [1987a] found no compelling evidence that such was the case for the sea ice they examined in the Weddell Sea. To date, observations do not verify this hypothesis.

2.3 Evolution of Multiyear Ice

2.3.1 Desalination

Ice undergoes dramatic changes during the melt season. The most obvious change is the large decrease in salinity. This desalination occurs primarily in response to temperature changes within the ice. Since the largest increases in temperature occur near the surface, the greatest desalination is also observed in the upper layers of the ice. Two effects are important in decreasing the salinity. First, the warming causes significant changes in the brine inclusions. As temperatures rise, ice goes into solution in the brine to produce the new lower equilibrium salinities. Associated with this brine volume increase, the brine pockets enlarge and tend to coalesce into vertical channels that facilitate gravity drainage of the brine and desalination of the ice. Martin [1979] showed photographic evidence of these channels.

In the Arctic, the improved system of brine drainage is further enhanced by the second major effect of summer—flushing of the ice with surface meltwater. Melting snow and ice provide a source of fresh water that can percolate through the ice, greatly reducing the salinity of the upper 50 to 100 cm. The largest changes occur during the late spring and early summer, when significant melting of the snow cover takes place. Salinities of the upper 50 to 100 cm are reduced to less than 1‰, down from previous values of 4‰ to 7‰. Although desalination occurs in Antarctic sea ice as it ages, the desalination is much less than in the Arctic, mainly because surface melting is uncommon [Andreas and Ackley, 1982]; the snow cover remains intact, melt ponds do not form, and albedos remain high. Therefore, in the Southern Hemisphere, flushing is not an important mechanism in draining brine from the ice.

Plots of bulk ice salinities versus thickness for Arctic and Antarctic sea ice are shown in Figure 2-5. For Arctic first-year ice there is a substantial decrease in salinity with thickness due to less entrapment as the growth rate decreases. Brine drainage activity is probably also manifested in the decreased salinities of thicker (1 to 2 m) first-year ice. The salinities of the thicker multiyear ice also decrease with increasing thickness, although the gradient is much less. The overall bulk salinity of Arctic multiyear ice is 3‰ to 4‰. Comparison of Figures 2-5(a) and (b) indicates that Antarctic ice has slightly higher salinities than Arctic ice of comparable thickness and does not appear to be as strongly dependent on thickness. Certainly the absence of flushing contributes to higher salinities in the Antarctic ice.

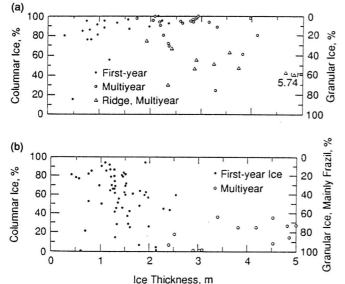

Fig. 2-4. Percentage of columnar and granular ice versus ice thickness in floes from (a) the Fram Strait [Gow et al., 1987b] and (b) the Weddell Sea [Gow et al., 1987a].

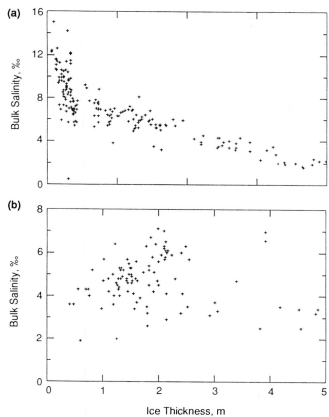

Fig. 2-5. Scattergrams of the bulk ice salinity versus ice thickness for (a) Arctic sea ice during the growth season and (b) Antarctic sea ice. The data were obtained from a variety of published and unpublished sources.

2.3.2 Crystal Retexturing

The crystalline texture of the ice is also affected by the summer melt cycle. This effect appears especially true of Arctic ice, which is apparently enhanced by the protracted warming and melting of the ice surface. In effect, the warming causes the retexturing of the crystals. The originally sharp crystal boundaries become rounded and the well-defined brine platelet layer structure is substantially modified or even eliminated. Figure 2-6 shows an example of this retexturing. The depth of the retexturing may exceed 100 cm. While still columnar in texture, the remaining structure bears little resemblance to that of the original ice. It is brine poor with salinities near zero, and glacial-like in appearance. Such thermally activated retexturing has to date not been reported in Antarctic multiyear ice.

2.3.3 Formation and Modification of the Surface Layer

In the previous two sections, we have described effects that modify the properties of the ice on its way to becoming multiyear ice. These changes are most pronounced in the uppermost layers of the ice, primarily the top meter, the portion of the ice that is most important to high-frequency remote-sensing instruments [Tucker et al., 1991]. The two major changes are the large reduction in salinity and crystal retexturing. These alterations combine to produce ice with properties that resemble those of freshwater ice.

Another major effect is an increase in porosity of the upper layers of the ice. As the brine drains from the upper layers of the ice, the air volume in the upper 10 to 30 cm of the multiyear ice is greatly increased. This is accompanied by a decrease in density, to values as low as 0.6 mg/m^3 [CRREL, unpublished data, Hanover, New Hampshire]. Higher porosities and lower densities appear to be especially likely on weathered ridges where the increased absorption of solar radiation may play a large role. There is also less snow to melt off of the ridge peaks. The decrease in salinity is also greater in elevated, drained areas as opposed to the ice at or below sea level.

Other modifications of the surface include the effects of fresh water or snow saturated with water freezing on the surface. Occasionally, layers of fresh ice have been observed on the surface of multiyear ice [Tucker et al., 1987]. This effect might be expected if only part of the snow melted off, but, in the Arctic, melting of the snow cover is usually quite complete. We expect that this formation of a fresh ice layer occurs in late summer or fall from rain that freezes on the surface or recent snow that melts and refreezes. This effect as well as the increased porosities discussed above is not observed in Antarctic sea ice.

In the Fram Strait region of the Eastern Arctic, the surfaces of multiyear floes can be flooded with seawater. Snow covers in this region are thicker than those observed in the Western Arctic [Tucker et al., 1987]. On some floes, the snow becomes deep enough to depress the floe and cause surface flooding. In late spring, this was observed on roughly 30% of the floes sampled in the Fram Strait [Perovich et al., 1988]. The seawater then infiltrates the upper layers of the multiyear ice creating a high salinity. Microwave signatures from this flooded ice can easily be confused for first-year ice [Tucker et al., 1991], with consequent errors in the estimation of multiyear ice concentrations [Comiso, 1990]. Such flooding has been observed in the Antarctic on both first-year and multiyear ice [Lange et al., 1990; Wadhams et al., 1987].

2.3.4 Melt Ponds and Hummocks

Melt ponds are a significant feature of Arctic multiyear sea ice. In fact, the characteristic appearance of multiyear ice largely results from the presence of melt ponds. The surface of multiyear ice has a characteristic hummocky appearance produced by alternating low refrozen melt ponds and adjacent hummocks. Small hummocks may have originally been level ice, whereas larger hummocks probably are weathered remnants of deformed ice.

In early summer, meltwater covers a large percentage of the surface of the floe as a result of snow melting on the level ice surface. When the melt cycle becomes more advanced, the coverage of the surface by melt ponds will be reduced to

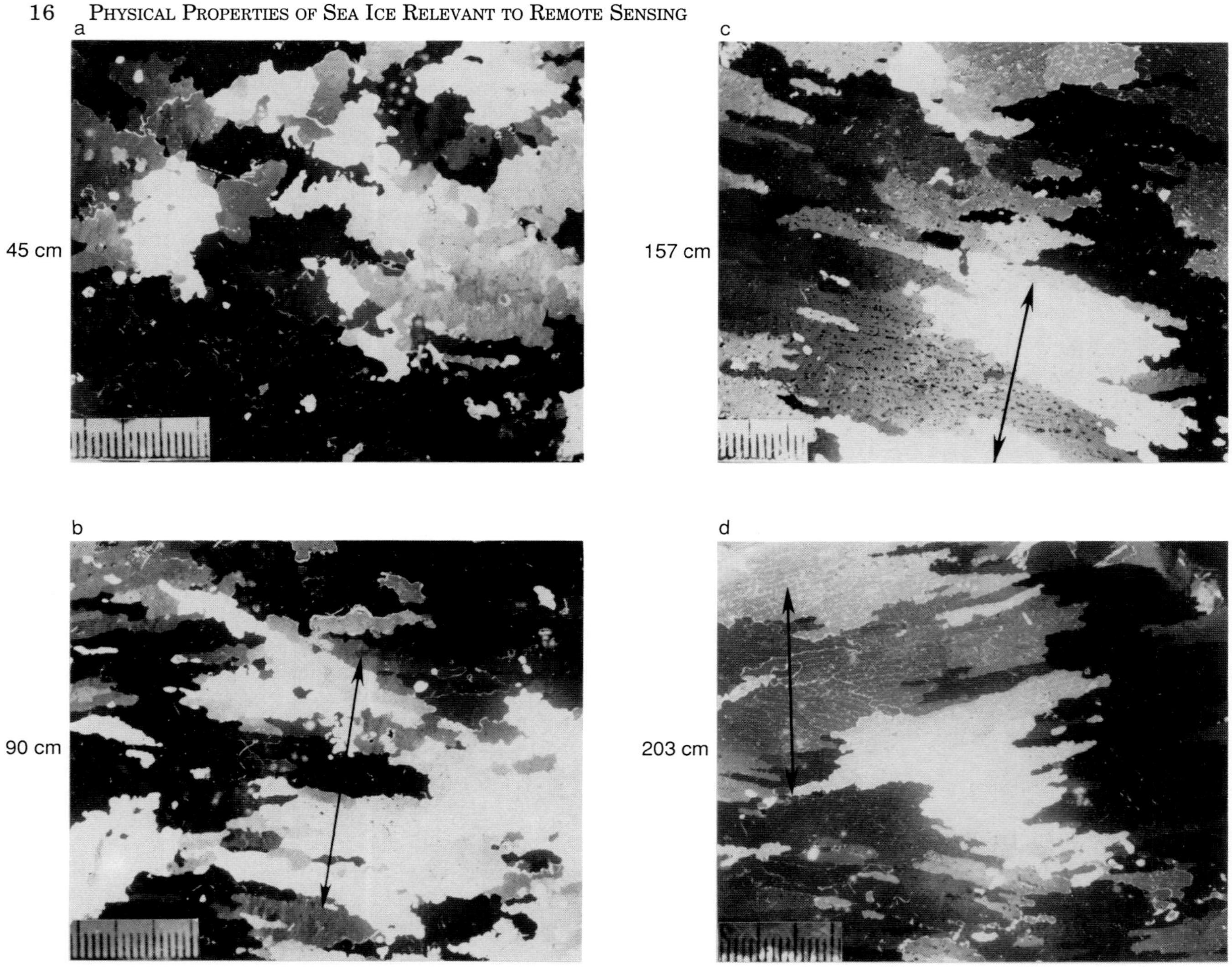

Fig. 2-6. Horizontal thin sections demonstrating retexturing of ice in the upper layers of an Arctic multiyear floe for (a) 45 cm, (b) 90 cm, (c) 157 cm, and (d) 203 cm. Notice that in (b) the original c-axis alignment (arrowed) is still preserved despite nearly complete loss of brine pocket substructures, in contrast to substructure still maintained in deeper unretextured ice [Gow et al., 1987b]. Small-scale subdivisions measure 1 mm.

about 30% [Maykut, 1985]. Surface drainage into adjacent leads is also effective in reducing aerial coverage. The melt ponds decrease the surface albedo and absorb more radiation, gradually deepening and decreasing the area of surface coverage. Ponds are commonly 50 cm to 1-m deep. They occasionally melt completely through a floe, in which case the pond may subsequently become filled with seawater. When such ponds freeze, they produce ice with essentially first-year characteristics.

Ponds begin to freeze during the late summer and early fall. At this time approximately 10% to 30% of the ice's surface is covered by ponds. With the exception just mentioned, refrozen melt ponds consist of fresh ice very similar to lake ice. A core through a pond nearly 1-m thick underlain by retextured columnar ice is shown in Figure 2-7. The top of this core profile shows the zero salinity characteristic of fresh pond ice. Another interesting struc-

tural feature of the pond ice is the very coarse-grained nature of the crystals exhibiting substantially vertical c-axes. This is a feature often observed in lake ice [Gow, 1986]. Grain sizes in melt ponds can range from very small (millimeter) to quite large (several centimeters). The melt pond ice often contains long tubular bubbles a few millimeters in diameter, but tens of centimeters long. Densities of melt pond ice range from 0.85 to 0.92 mg/m^3, depending upon the amount of air entrapped. Air volumes can be as large as 80‰ to 90‰ in the case of ice containing a high concentration of bubbles, but more often are 20‰ to 40‰ [CRREL, unpublished data, Hanover, New Hampshire].

Melt ponds are uncommon in the Antarctic because the extensive surface melting common to the Arctic seldom occurs. Andreas and Ackley [1982] speculated that lower relative humidities and stronger winds enhance turbulent heat losses. Under these conditions, surface melting will

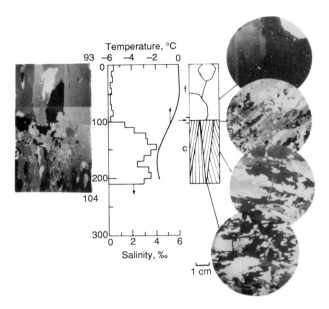

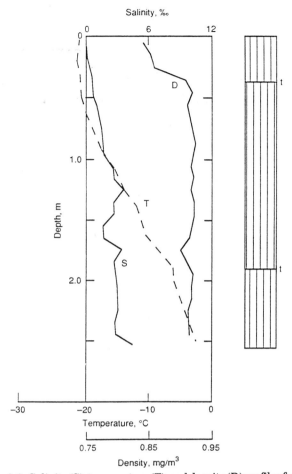

Fig. 2-7. Salinity, temperature, and structure profiles from a ponded multiyear ice floe in the Fram Strait. The vertical thin section shows the transition between the ponded and columnar ice at around 100 cm in depth. The c-axis alignments in horizontal thin sections are indicated by arrows [Gow et al., 1987b].

Fig. 2-8. Salinity (S), temperature (T), and density (D) profiles from an ice core drilled through a hummock on a multiyear floe. The ice at this location on the floe consisted entirely of columnar ice, with Tr referring to an obvious transition [A. J. Gow, CRREL, unpublished data, Hanover, New Hampshire].

only occur when the air temperature is significantly above 0°C, a situation only rarely observed in the Antarctic.

Hummocks also have low salinities because their raised relief enhances brine drainage during summer. The upper 10 to 20 cm in particular may retain very little brine. An example of a core profile from a hummock on a floe composed entirely of columnar ice is shown in Figure 2-8. The density of the upper 10 cm is quite low, caused mainly by the presence of evacuated brine channels with some additional porosity due to melting enhanced by the absorption of shortwave radiation. The air volume is very large (300‰) for the upper 10 cm. As expected, salinities are very low. It is often difficult to obtain accurate quantitative measurements of density for the near-surface ice of multiyear hummocks and ridges because of the brittle degraded nature of the ice.

On mid-relief ice (areas between melt ponds and hummocks), upper layer salinities are also quite low, often close to zero. Densities are also low, typically 0.75 to 0.85 mg/m³, although they are not as low as the near surface of the well-drained hummocks. Air volumes in the upper layers range from 80‰ to 95‰, decreasing with depth to 5‰ to 25‰. Likewise, densities and salinities increase with depth.

2.3.5 Properties of Multiyear Versus First-Year Ice

In comparing multiyear and first-year ice properties, it is useful to consider the Arctic and Antarctic separately, since the ice properties are so different in these two areas.

The Arctic shall be addressed first, since the greater variability in properties occurs there. Large differences are apparent in the thickness and appearance of first-year and multiyear ice. Arctic first-year ice grows to a maximum thickness of about 2.0 m in its single season. The equilibrium thickness of undeformed multiyear ice is 3.0 to 4.0 m [Maykut and Untersteiner, 1971], though the thickness of multiyear ice varies greatly. On a single floe, its thickness may range from less than a meter to tens of meters. The thinner ice results from pronounced ablation during the summer, probably beneath a melt pond, while the thick ice features are typically weathered first-year ridges.

To illustrate this variability, Figure 2-9 shows the ice thickness of three 100-m grids taken from the same floe in the Eastern Arctic [CRREL, unpublished data, Hanover, New Hampshire]. The grids were separated by several hundred meters. The figure implies that floes can be made of morphological regimes having totally different characteristics. For instance, Figure 2-9(a) shows thick very

Fig. 2-9. Ice thickness variations on a multiyear floe in the Beaufort Sea for (a) mean thickness = 7.13 m, standard deviation = 3.47 m; (b) mean thickness = 3.63 m, standard deviation = 0.48 m; and (c) mean thickness = 5.33 m, standard deviation = 1.36 m. Data are from three 100 × 100-m areas located several hundred meters apart on a single multiyear floe [W. B. Tucker, CRREL, unpublished data, Hanover, New Hampshire].

deformed ice, while Figure 2-9(b) is essentially undeformed, and Figure 2-9(c) contains components of both deformed and undeformed ice. Mean thicknesses can also be very different between individual floes. In measuring thicknesses at 5 and 10 m spacings, W. B. Tucker [document in preparation, Cold Regions Research and Engineering Laboratory, Hanover, New Hampshire] found mean thicknesses ranging from 2.62 to 5.60 m for eight different multiyear floes. The total range of ice thicknesses on these floes varied from 0.93 to 18.30 m. Figure 2-10 shows the probability density function of the 3770 drill-hole thickness measurements made on these eight floes. The figure shows that the most likely multiyear ice thickness is 3.5 to 4.0 m, and also shows the long positive tail indicative of the presence of ridges.

The surface appearances of first-year and multiyear ice are quite different. Undeformed first-year ice is flat with little freeboard. When it has been deformed, the ridges consist of very distinct collections of angular blocks. Multiyear ice typically has a rolling hummocky appearance. Even undeformed multiyear floes have a surface relief of 10 to 20 cm as a result of differential melting. The undulating surface is easily identified from an aircraft if there is little snow cover. Multiyear ridges are distinct from first-year ridges because they are weathered and rounded, with little or no sign of the original block structure evident.

There are also major property differences in salinity, crystal texture, and porosity between first-year and multiyear ice. Figure 2-11 shows property profiles of multiyear and first-year ice sampled in early winter. Rather than porosity, the density is shown, but the differences in porosity are inherent in the density. The major differences are clarified by comparing these profiles. The salinity of first-year ice is considerably higher, particularly near the surface. For the upper 30 cm, first-year ice has a salinity of 6‰ to 7‰, while in the multiyear ice it is less than 1‰. For the remainder of the cores, salinities of first-year ice are consistently higher, except at the very bottom of the cores. First-year ice is characterized by well-defined platelet substructure in its crystal texture, whereas multiyear ice shows none. The grain boundaries in the upper 1.0 m of multiyear ice are not as sharp as those of first-year ice—a clear indication of crystal retexturing. Finally, the densi-

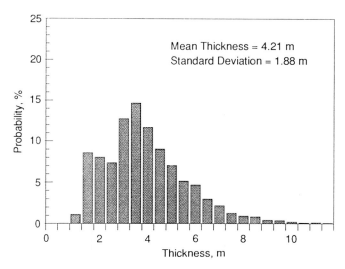

Fig. 2-10. Probability density function of 3770 ice thickness measurements made on eight Arctic multiyear floes.

ties of multiyear ice are lower than those of first-year ice in the upper layers. This reflects the increase in porosity due to the evacuation of brine inclusions and the enlargement of brine cavities and drainage channels during the melt season. Densities below 1.0 m are comparable in both cores.

A similar examination of the thickness and properties of first-year and multiyear ice in the Antarctic is instructive. Thickness variations are also found on Antarctic multiyear floes. For example, Eicken et al. [1991] found that the thickness of three cores spaced at 20 m on a single floe ranged from 1.27 to 2.15 m. A major difference from Arctic multiyear ice is that the thicknesses are considerably less due to the fact that ridges are much smaller than those found in the Arctic [Weeks et al., 1988]. Lange and Eicken [1991a] found a mean thickness of 1.53 m for apparently undeformed second-year ice and a mean thickness of 2.53 m for deformed second-year ice. However, Gow et al. [1987a] found thicknesses on different multiyear floes in the Weddell Sea ranging from 2.4 m to greater than 5.0 m. Many of the thicker floes showed no signs of having been deformed, indicating that thickness and the extent of deformation in

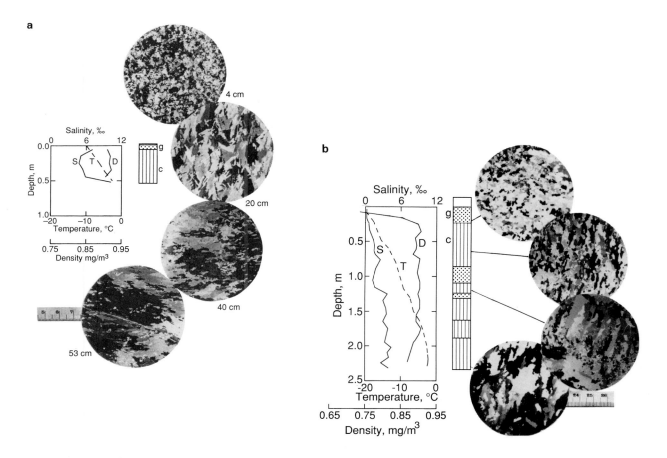

Fig. 2-11. Physical property and structure profiles of (a) lead ice 53 days after freeze-up [Gow et al., 1990] and (b) multiyear ice [CRREL, unpublished data, Hanover, New Hampshire].

the Weddell Sea pack ice can vary greatly with location in the pack.

Figure 2-12 shows core profiles of first-year and multiyear ice collected from the Weddell Sea, Antarctica. Only salinity and structural profiles were measured for these cores. The salinity profiles show that multiyear ice is somewhat less saline in the upper 40 cm. Below that, however, salinities of both ice types appear to hover around 4‰ to 5‰. The one exception to this is in the multiyear ice at a depth of 140 to 170 cm, where salinity falls to 2‰ to 3‰. This likely coincides with the transition from first- to second-year growth as evidenced by a change in the structure of the ice from congelation to frazil. It appears that frazil can be incorporated into the ice structure after the first year's growth as easily as during the first year. This implies that at least some frazil is forming in the water column beneath the existing ice sheet or being formed in leads and subsequently swept under adjacent ice [Lange and Eicken, 1991b]. Densities differ little between first-year and multiyear ice. Both the densities and the salinities indicate that minimal brine drainage and expansion of brine channels take place

during the transformation of first- to second-year ice in the Weddell Sea.

Antarctic ice also exhibits large variations within a single floe. Examining the stratigraphy and salinities from closely spaced cores (with a separation of approximately 20 m) of two-year-old ice, Eicken et al. [1991] concluded that the variations in mean properties of cores taken within a few meters can vary as much as properties within an entire geographic region. Thicknesses of this relatively thin multiyear ice ranged from 1.27 to 1.54 m and the mean salinities varied from 4.1‰ to 7.0‰. The largest variability in the properties exists in the upper layers, which, as mentioned previously, are extremely important to microwave remote sensing. These variations appeared to be related to formation of the floe and deformation. Thus, it appears that large property variations can occur in Antarctic multiyear ice floes, although probably not to the extent observed in Arctic multiyear ice floes with their characteristic hummock–melt-pond morphology.

While we know that the variability of ice properties on multiyear ice is quite large, it has not been systematically quantified. It would be useful to conduct further surveys of properties on single multiyear floes to assist in quantifying these rather large property variations. This is particularly critical for the upper layers that most significantly affect microwave signatures. Knowledge is also needed of the range over which property variations occur relative to the size of the microwave footprint.

2.4 SURFACE PROPERTIES

Any discussion of remote sensing of sea ice would be incomplete without some attention being paid to the large variety of roughness features apparent on the surface of the ice cover. This is particularly important in the present context in that the surface roughness itself causes large variations in remote sensing signatures. Here we will dwell on roughness at two basic spatial scales. The first will be the large scale that we designate as that horizontal scale ranging from meters to kilometers and primarily includes features associated with deformation, although topography caused by melt ponds may be included. The other scale focuses on the small-scale roughness of the ice and snow surface ranging from millimeters to meters. There is a definite overlap between these two scales; for instance there are both small- and large-scale roughnesses associated with a brash ice field.

2.4.1 Deformation Features

Winds and currents keep the ice pack in nearly continuous motion, resulting in significant deformation of the ice cover. This can lead to the formation of a pack consisting of ice floes separated by water openings, which in itself is a roughness element since the freeboards of floes can vary from millimeters to several meters, as is the case of a ridge located on the edge of a floe. In the central pack, floes may be several kilometers in diameter. In the marginal ice zone

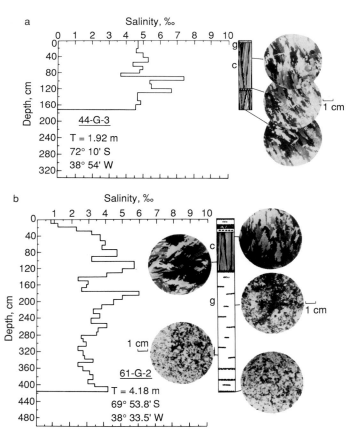

Fig. 2-12. Salinity and structure profiles of (a) first-year ice and (b) multiyear ice from the Weddell Sea, Antarctica. First-year ice consists of 94% congelation ice, whereas the multiyear ice is composed of 75% granular frazil formed during the second year of growth [Gow et al., 1987a].

(MIZ), on the other hand, wave and wind action may thoroughly pulverize the ice, reducing floe sizes to less than a meter. Also characteristic of the MIZ and over large parts of the Antarctic is the formation of pancake ice.

Figure 2-13 shows small floes in the Labrador Sea MIZ incorporated into a brash ice field. Here the brash is made up of pulverized floes and newly forming frazil and is interspersed with broken blocks and floe pans.

Deformed ice constitutes a significant portion of the ice cover. Estimates made from submarine ice draft profiles indicate that the mass of ice contained in ridges is 25% to 50% of the total mass of ice in the Arctic [Wadhams, 1984; Williams et al., 1975]. Recent analysis of ice thickness measurements of Arctic multiyear floes [W. B. Tucker, document in preparation, CRREL, Hanover, New Hampshire] indicated that as much as 50% of multiyear ice may be deformed. That is, these floes contain a large component of ice that was deformed in its first-year state and that has subsequently aged and become incorporated into multiyear floes. Indeed, this deformed ice may be a major reason why that particular floe survived, having its thickness and strength increased by incorporation of the deformed ice.

The most notable roughness features caused by the deformation of sea ice are ridges and rubble fields. The ice sheet deforms under compression and shear, and blocks are piled both above and below the surface. Measurements of block sizes in a number of ridge sails [Tucker et al., 1984b] indicate that while most ridges are formed of ice less than 1-m thick, a considerable number are of thicker ice. Tucker et al. [1984b] also found a relationship between the ridge heights and the thickness of the ice contained within the ridges, which apparently results from the relation of ice strength to thickness. Freely floating ridges may have sails that exceed 10 m in height, while grounded ridges near shore can approach 20 m. Submarine ice draft measure-

Fig. 2-13. Incorporation of floe pans into a brash ice field in the Labrador Sea marginal ice zone during March 1989. (Photograph by M. R. Drinkwater.)

ments have found ridge keels extending to depths greater than 40 m [Wadhams, 1984].

Ridges can be long linear features extending for several kilometers. More often, however, they are sinuous and extend for only several hundred meters. Repeated ridging causes rubble fields. Examples of typical Arctic first-year and multiyear ridges are shown in Figure 2-14. Because ridges are significant obstacles to surface transportation over pack ice, their statistics have been extensively studied. Surface ridge statistics have been obtained with airborne laser profilometers for many years [Wadhams, 1984; Hibler et al., 1974]. While estimating the mass of deformed ice from ridge counts extracted from laser terrain profiles may lead to errors [Rothrock, 1986], useful statistics on the surface roughness due to ridges can be obtained. Although much more information is available on ridging statistics in the Arctic than the Antarctic, Weeks et al. [1988] used the little available data to contrast ridging in various Arctic regions to that of the Ross Sea (Table 2-1). Lytle and Ackley [1991] found ridging in the Eastern Weddell Sea to be similar to that in the Ross Sea. Granberg and Leppäranta [1990], however, found evidence of increased ridging in the western sector of the Weddell Sea. These limited results, as well as impressions based on extensive visual observations, indicate that ridging is much more intense in the Arctic, with more and higher ridges than in the Antarctic. This difference presumably results from a less compact and less constrained pack being deformed in the Antarctic. Table 2-1 points out that near-coastal regions in both areas contain more ridges, a manifestation of the deformation occurring when the ice is forced towards the coast.

First-year ridges are composed of piles of very angular ice blocks. Voids between the blocks are frequently filled with wind-compacted snow. Tucker et al. [1984b] found that the sizes of the blocks generally scale linearly with the ice thickness, the largest dimension being about twice the thickness of the block. The sharp-edged blocks are ablated during summer to the point where there is little evidence of the original structure when the freezing season commences. Multiyear ridges are typically well rounded, hummocky features with few, if any, large voids. Their heights are also significantly less than first-year ridges. A typical multiyear ridge is shown in Figure 2-14(d). The porosities of the near-surface ice are high, while the brine volume is nearly zero, both resulting from the enhanced ablation and drainage occurring on the ridges during summer. There is a dearth of observations of multiyear ridges in the Antarctic. In the absence of surface melting, it is expected that these ridges would maintain their original blocky structure.

On a somewhat smaller roughness scale are broken ice fields and rafting. Frequently, sheets of thin ice are broken by wind, waves, or deformation, and the blocks often raft. When the water containing these blocks freezes, the newly formed sheet can have a vertical roughness ranging from a few centimeters up to a meter. Horizontal scales vary depending upon the original formation conditions, but can vary from centimeters to tens of meters. Rafting of ice sheets is a common occurrence. These may be long, linear,

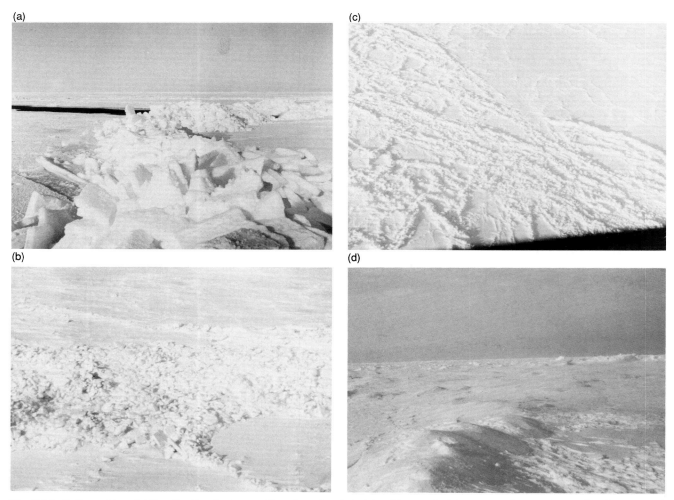

Fig. 2-14. Deformed ice: (a) free floating first-year ridge about 2 m in height; (b) grounded first-year ridge over 6 m in height; (c) aerial view of a 400-m-wide rubble field in the near-shore Beaufort Sea; and (d) multiyear ridge about 3 m in height.

Table 2-1. Ridging parameters from the Ross Sea compared to data from various Arctic locations for winter and early spring. Parameters were obtained from laser profilometer data from Weeks et al. [1988].

Location	Ridges per km	Mean ridge height, m
Ross Sea coastal	6.8	1.25
Ross Sea offshore	1.9	1.21
Central Beaufort Sea	2.6	1.47
Beaufort–Chukchi shear zone	4.4	1.51
West Eurasian Basin	4.7	1.50
North Greenland	8.2	1.68

or sinuous features extending several kilometers with vertical reliefs of less than a meter. Frequently, however, finger rafts occur, which have a distinctive square wave appearance. Like other rafts, horizontal and vertical dimensions are extremely variable.

While not a result of ice deformation, another mesoscale surface roughness feature is caused by the snow cover. The snow cover in both the Arctic and Antarctic is nonuniform, and significant topographic variations are created primarily by wind. Most noticeable are snow drifts, which, when compacted and hardened, are known as sastrugi. They are often associated with a morphological feature of the ice surface, such as an ice block, and they taper away from such a feature with the direction of the wind at the time they were created. These features are often several hundred meters in length with heights above the level surface of up to 1 m. Frequently observed on floes are rows of sastrugi, trending in the same general direction. Snow that drifts in around the bases of pressure ridges frequently reaches depths greater than 1 m.

2.4.2 *Small-Scale Surface Roughness*

Knowledge of the small-scale sea ice surface roughness is critical for the interpretation of microwave remote-sensing data. Over recent years, the development of a variety of instruments for this purpose has led to quantitative measurements of the roughness properties of a number of ice types. Most of these measurements have been made in the millimeter or centimeter range using profilometers employing either mechanical, photographic, or laser technology. Four principal methods are currently being used: (a) a mechanical roughness comb or rill meter [Drinkwater, 1989]; (b) a camera-based system with an automated digitizing and analysis system [Paterson et al., 1991]; (c) a tracked laser instrument for precise ranging along a meter baseline [Johansson, 1988; Farmer et al., 1991]; and (d) a manual method where a vertical slab is retrieved and photographed against a grid background [R. G. Onstott, personal communication, 1991].

Despite the numerous techniques, only a few measurements of small-scale ice roughness have been made. Drinkwater [1989] reported that the centimeter-scale roughness of first-year ice in the Labrador Sea may conform to either an exponential or Gaussian correlation function depending upon the degree of deformation of the surface. The surface was composed of flat ice pans separated by levees built from deformed brash ice or ridged blocks. The smooth ice had a standard deviation of 1.74 cm with a correlation length of 12.8 cm, while the standard deviation of the deformed ice was 4.43 cm and the correlation length was 18.6 cm. These measurements agree with those of Paterson et al. [1991], who found standard deviations of 0.78 and 5.26 cm for smooth and rough ice in the Labrador Sea. First-year ice in other Arctic areas appears to be smoother. This is probably because it has not been subjected to the harsher environment that characterizes the Labrador Sea MIZ. Kim et al. [1985] reported standard deviations of smooth and somewhat rougher ice to be 0.15 and 0.37 cm, presumably for samples from the Beaufort Sea. Likewise, in Chapter 5, R. G. Onstott reports a range of first-year ice roughnesses from 0.05 to 0.49 cm with correlation lengths of 0.5 to 3.7 cm for ice in the Eastern and Western Arctic. Farmer et al. [1991] also observed generally smaller roughnesses in the Beaufort Sea. They found standard deviations ranged from 0.18 to 0.40 cm and correlation lengths of 2.0 to 12.0 cm.

Observations of small-scale roughnesses of multiyear ice are sparse. Kim et al. [1985] reported standard deviations of 0.15 and 0.81 for smooth and rough multiyear ice, and correlation lengths of 8.2 and 8.9 cm, respectively. R. G. Onstott in Chapter 5 reports roughnesses of multiyear hummocks ranging from 0.26 to 0.89 cm, with correlation lengths of 2.8 to 6.0 cm. The standard deviations of frozen melt-pond surfaces were found to be much less, 0.08 to 0.21 cm. Correlation lengths ranged between 2.4 and 3.5 cm for these much smoother surfaces.

2.5 SNOW COVER

For much of the year, sea ice has another important surface feature—a snow cover. The areal extent and thickness of the snow cover is important for a number of reasons, particularly those concerning the thermodynamics of sea ice. During winter the snow acts as an insulating blanket, decreasing heat exchange with the ocean and retarding freezing, while in the summer the high albedo of the snow reduces the shortwave radiation input impeding surface melting [Maykut and Untersteiner, 1971]. How a snow cover impacts microwave signatures of sea ice and whether microwave data can be used to determine snow cover properties are important issues.

To date there has not been a large-scale effort to systematically measure the snow cover thickness on sea ice in the polar regions. However, there have been a few isolated studies. The results of these studies can be used to make several salient points regarding the temporal evolution of a snow cover, the geographical variability of snow depth, and the variations in snow depth between ice types. The thickness of the snow cover can exhibit tremendous variability, both temporally and spatially. Temporally, there are dramatic changes in snow cover during the summer melt cycle and the early snowfalls of fall freeze-up. In addition, there are the almost continual changes due to the wind redistributing the snow. On a large scale, very little is known regarding how snowfall varies within and between the different regions of the Arctic. On a small scale, snow depths are heavily influenced by ice surface topography and can vary from the bare ice of a frozen melt pond to a meter-thick snow cover in the lee of a pressure ridge over horizontal distances of only a few meters.

Field observations from the Central Arctic [Untersteiner, 1961] showed a steady increase in snow depth from August to October of about 0.15 m, followed by a gradual increase of 0.05 m until the onset of the melt season. Once melt began, the snow cover melted completely in only a few weeks. Based on a survey of prior measurements, Maykut and Untersteiner [1971] estimated a total snow depth of 0.40 m for ice in the Arctic Basin, 0.30 m of which was assumed to fall during September and October.

Table 2-2 summarizes snow survey results for three different regions in the Arctic. Each snow survey consisted of point measurements of snow depth made on a grid encompassing several thousand square meters. There is a large geographical disparity in mean snow depth with ice in the Greenland Sea MIZ having an average snow cover two and a half times as thick as the Eastern Arctic and three and a half times as thick as the Southern Beaufort Sea. Small-scale variations in snow depth are evident in the range of point measurements of snow depth, which vary from zero to over a meter.

Wadhams et al. [1987] made extensive measurements of snow depth on sea ice in the Weddell Sea. They found the mean snow depth to be quite variable with a general trend towards deeper snow at higher latitudes. The snow depth on a given floe depended primarily on the age of the floe and

Table 2-2. Summary of snow depths for Arctic multiyear ice. Beaufort Sea data are from W. B. Tucker [CRREL, unpublished data, Hanover, New Hampshire] and MIZ are from Perovich et al. [1988].

Location	Date	Latitude	Longitude	Number of points sampled	Mean	Standard deviation	Min. snow depth, cm	Max. snow depth, cm
Southern Beaufort Sea	April 1986	70° N	150° W	225	12.8	11.1	0	58
Eastern Arctic	November 1988	82° N	30° E	484	18.8	19.5	0	104
Greenland Sea MIZ	March 1987	78° N	2° W	570	46.5	24.1	3	120

on recent meteorological events, such as snowstorms and windstorms. Mean snow depths ranged from 0.00 to 0.40 m, with values typically falling between 0.10 and 0.15 m.

Studies in the Greenland Sea MIZ indicate that snow depth is significantly greater on multiyear ice than on first-year ice. Tucker et al. [1987] found that early summer snow depths on multiyear ice in the Greenland Sea MIZ ranged from 0.03 to 0.65 m, with an average of 0.28 m, while the snow on first-year ice never exceeded 0.20 m and averaged 0.08 m. Perovich et al. [1988] confirmed this during a winter experiment in the same region. They found a statistically significant difference in mean snow depth between multiyear floes (0.47 m) and first-year floes (0.11 m). As mentioned earlier, a heavy snow cover can impact the properties of the underlying sea ice by causing surface flooding with seawater.

The snow cover over Arctic sea ice is fairly simple. It consists primarily of wind-blown, hard-packed snow composed of well-rounded grains with diameters of 0.25 to 0.50 mm and a density between 0.3 and 0.4 mg/m³. Ice layers, resulting from warm periods, are quite common in the snowpack. Occasionally, depth hoar can be found at the base of the snow cover. During most of the year the snow remains dry, usually having significant free water only during the summer melt period.

Frost flowers (Figure 2-15) can significantly impact the microwave remote-sensing properties of young ice [Drinkwater, 1988; Drinkwater and Crocker, 1988]. These flowers are formed by ice deposited on the surface directly from the vapor phase. The extent and form of the flowers depend primarily on the air temperature and level of super-saturation near the surface. Frost flower conditions can change rapidly, forming a dense mat in only a few hours and disappearing with only a small amount of surface flooding. In that frost flowers draw brine upwards from the new ice surface by capillary action as the freezing ice surface rejects brine, salinities as high as 110‰ have been measured in frost flowers [Drinkwater and Crocker, 1988]. The brine-enriched frost flowers form a rough dielectric interface in contrast to the smooth underlying ice surface. This interface strongly impacts the microwave signature and can lead to a misidentification of the ice type as discussed in later chapters.

The small-scale roughness of snow or frost flowers can also be important to microwave sensors. Unfortunately, there is a dearth of such measurements. Recent observations of the snow roughness on first-year ice in the Beaufort Sea found standard deviations to range from 0.18 to 0.40 cm,

with correlation lengths of 2.0 to 12.0 cm [Farmer et al., 1991]. According to A. Carlström [personal communication, 1991], the mean root mean square roughness of young ice surfaces covered by frost flowers is 0.72 cm with a correlation length of 1.0 cm. These few measurements were obtained from the Beaufort Sea and the Central Arctic.

2.6 ICE PROPERTY STATISTICS

As interest in using microwave remote-sensing observations to monitor geophysical processes in sea ice has in-

(a)

(b)

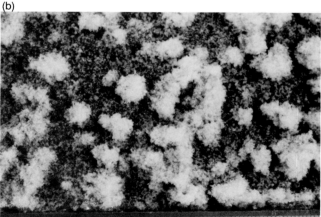

Fig. 2-15. Frost flowers: (a) enlarged view of individual frost flowers and (b) dense cover of clustered frost flowers on young sea ice.

creased, there has also been a growing awareness that in order to interpret these observations, an understanding is necessary of how variations in microwave signatures relate to differences in the physical properties and crystalline structure of the ice cover. To this point, only qualitative descriptions of the ice have been related to microwave signatures. Nonetheless, these characterizations, consisting of ice surface conditions and vertical profiles of ice crystal structure, temperature, salinity, brine volume, and density, have been shown to play a vital role in interpreting both passive and active microwave data [Arcone et al., 1986; Tucker et al., 1987, 1991].

However, for all their utility, such descriptive characterizations are insufficient for theoretical microwave models. Winebrenner et al. [1989] discussed the importance of a statistical description of ice structure for microwave modeling and have enumerated several useful statistical measures. Two such prominent microwave scattering models are the strong fluctuation theory [Tsang and Kong, 1981; Stogryn, 1984] and the dense medium radiative transfer model [Winebrenner et al., 1989].

The strong fluctuation theory requires characterization of the ice using the two-dimensional correlation function. Correlation functions for sea ice were investigated by Lin et al. [1988] in a combined experimental and theoretical effort. They analyzed a single horizontal thin-section photograph and found that the correlation function was exponential with correlation lengths comparable to the average size of the brine inclusions. Vallese and Kong [1981] examined correlation lengths for snow and determined that there was a correspondence to the average grain size of the snow.

The most extensive correlation function study for sea ice was performed by Perovich and Gow [1991] on saline ice grown in an outdoor pond. In this investigation, the correlation functions were determined for more than 50 thin-section photographs of this saline ice using a personal-computer-based image-processing system [Perovich and Hirai, 1988]. Figure 2-16 features a representative example of a correlation function for thin saline ice. A horizontal thin section was taken from the bottom of a 95-mm-thick ice sheet that had been growing for 48 hours, Figure 2-16(a). The ice was columnar with the c-axes of the component crystals being substantially horizontal. The sample was photographed at $-10°C$ and had a salinity of 11‰, giving a brine volume of 6.1%. The two-dimensional correlation function, Figure 2-16(b), is peaked with a sharp drop-off, small regular sidelobes, and a shape that is more elliptical than circular. The correlation lengths in the major and minor directions are 0.15 and 0.12 mm, respectively, lengths that are closely associated with the physical size of the brine inclusions.

Perovich and Gow [1991] established that correlation lengths for young sea ice are directly related to physical features of the ice microstructure, such as the size of brine pockets in salty ice (typically tenths of a millimeter) and vapor inclusions in desalinated ice (typically millimeters). Like the ice microstructure, the correlation function of sea ice can be quite variable, depending on such parameters as

(a)
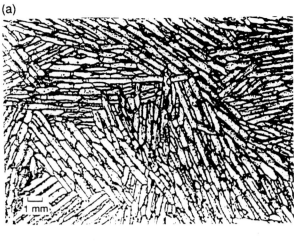

(b)
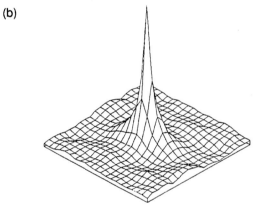

(c)
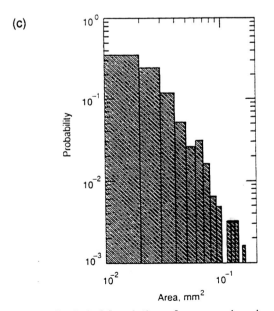

Fig. 2-16. Statistical descriptions of young sea ice microstructure: (a) horizontal thin section of young ice; (b) correlation function of this thin section; and (c) brine-pocket size distribution of the thin section [Perovich and Gow, 1991].

growth conditions, ice temperature, and brine volume. A growing ice sheet has the coldest temperatures at the surface and the warmest at the ice–water interface, with the brine volume increasing with depth. This produces a general trend of correlation lengths increasing with depth. They found that this trend can be overridden by structural changes in the ice microstructure, such as variations in the ice platelet diameter or a switch from columnar to granular crystals.

There are other possible statistical descriptors besides the correlation function. The brine pocket geometry and size distribution are important for the dense medium microwave models [Winebrenner et al., 1989]. Arcone et al. [1986] examined brine pocket geometry, using both horizontal and vertical thin sections of young columnar ice. They found that brine pockets were ellipsoidal with the long axis aligned within 15° of vertical. The vertical length of the brine pockets was typically 1 to 2 mm. In the horizontal, the major and minor axes of individual brine pockets varied considerably from 0.01 to 1.7 mm. A representative ratio of the three axes of the brine inclusion ellipsoids (length:major:minor) was 10:3:1.

As an example of brine pocket size distribution, Figure 2-16(c) shows the brine inclusion size distribution for the thin section presented in Figure 2-16(a). There were 615 inclusions in this 270 mm^2 section, with a total area of 17 mm^2 and a mean area of 0.028 mm^2. If we assume that the inclusions were elliptical with an aspect ratio of 0.8, similar to the correlation lengths for this thin section, then the mean values of the major and minor axes were 0.10 and 0.08 mm, respectively. These values are somewhat smaller but roughly comparable to the correlation lengths (0.15 and 0.12) calculated for this image. The brine-inclusion size distribution shows a sharp drop-off of probability (P) as the area (A) increases. This drop-off is well fit ($R^2 = 0.96$) by a power law with $P = 1.9 \times 10^{-5} A^{-2.5}$.

2.7 SUMMARY

In this chapter, we have attempted to illuminate aspects of sea ice that are believed to affect microwave remote sensing. In doing so, it was necessary to delve in some detail into certain processes, such as ice growth and its dynamic and thermal modifications. This was done primarily to provide a brief background useful in understanding the state of the ice at various stages in its history.

Although the physical properties of sea ice have been studied for many years, they have recently taken on new significance due largely to increased remote sensing of the polar regions. While the emphasis has been to characterize and understand properties important to remote sensing, the increased attention to ice properties has enabled us to better understand properties and processes in their own right. This process is expected to continue as sensors are continually refined.

There is a need to provide better quantification of properties. This process has begun as described in the last

section and will continue under the guidance of the needs of microwave models. Techniques of quantitatively describing small- and large-scale properties, both in vertical and horizontal dimensions, will be necessary to provide input to microwave models. Coupled with quantification of properties will be the need to model the ice properties themselves. One possibility is to develop empirical models to predict the temporal evolution of ice properties. As our understanding of complex processes increases, the models can become more physically based, and should aim to predict the spatial and temporal evolution of properties from climatic or thermodynamic input. Needless to say, ice properties and processes must continue to be studied in laboratory and field experiments to enhance the theoretical development of process models, which in turn can provide a firm base for microwave models.

REFERENCES

Ackley, S., On the differences in ablation seasons of the Arctic and Antarctic sea ice, *Journal of Atmospheric Science, 39(3),* pp. 440–447, 1982.

Andreas, E. L. and S. F. Ackley, On the differences in ablation seasons of the Arctic and Antarctic sea ice, *Journal of Atmospheric Science, 39(3),* pp. 440–447, 1982.

Arcone, S. A., A. J. Gow, and S. McGrew, Structure and dielectric properties at 4.8 and 9.5 GHz of saline ice, *Journal of Geophysical Research, 91,* pp. 14,281–14,303, 1986.

Bilello, M. A., Decay patterns of fast sea ice in Canada and Alaska, *Sea Ice Processes and Models,* edited by R. S. Pritchard, pp. 313–326, University of Washington Press, Seattle, Washington, 1980.

Campbell, K. J. and A. S. Orange, The electrical anisotropy of sea ice in the horizontal plane, *Journal of Geophysical Research, 79,* pp. 5059–5063, 1974.

Carsey, F. D., S. A. Digby, S. Argus, M. J. Collins, B. Holt, C. E. Livingstone, and C. L. Tang, Overview of LIMEX'87 ice observations, *IEEE Transactions on Geoscience and Remote Sensing, 27(5),* pp. 468–484, 1989.

Cherepanov, N. V., Spatial arrangement of sea ice crystal structure, *Problemy Arktiki i Antarktiki, 38,* pp. 176–181, 1971.

Colony, R. and A. S. Thorndike, Sea ice motion as a drunkard's walk, *Journal of Geophysical Research, 90,* pp. 965–974, 1985.

Comiso, J. C., Arctic multiyear ice classification and summer ice cover using passive microwave satellite data, *Journal of Geophysical Research, 95(C8),* pp. 13,411–13,422, 1990.

Cox, G. F. and W. F. Weeks, *Brine Drainage and Initial Salt Entrapment in Sodium Chloride Ice,* CRREL Research Report 345, 85 pp., Cold Regions Research and Engineering Laboratory, Hanover, New Hampshire, 1975.

Cox, G. F. N. and W. F. Weeks, Numerical simulations of the profile properties of undeformed first-year sea ice during the growth season, *Journal of Geophysical Research, 93,* pp. 12,449–12,460, 1988.

Dieckmann, G., G. Rohardt, H. Hellimer, and J. Kipfstul, The occurrence of ice platelets at 250-m depth near the Filchner Ice Shelf and its significance for sea ice biology, *Deep Sea Research, 33,* pp. 141–148, 1986.

Doronin, Y. P. and D. E. Kheisin, *Sea Ice,* 323 pp., Girdrometeoizdat Publishers, Leningrad, 1975 (NSF Trans. TT75-52088, 1977).

Drinkwater, M. R., Important changes in microwave scattering properties of young snow-covered sea ice as indicated from dielectric modeling, *Proceedings of the 1988 International Geoscience and Remote Sensing Symposium (IGARSS'88),* pp. 793–797, University of Edinburgh, Edinburgh, UK, 1988.

Drinkwater, M. R., LIMEX'87 ice surface characteristics; implications for C-band SAR backscatter signatures, *IEEE Transactions on Geoscience and Remote Sensing, 27(5),* pp. 501–513, 1989.

Drinkwater, M. R. and G. B. Crocker, Modelling changes in the dielectric and scattering properties of young snow-covered sea ice at GHz frequencies, *Journal of Glaciology, 34(118),* pp. 274–282, 1988.

Drinkwater, M. R. and V. A. Squire, C-band SAR observations of marginal ice zone rheology in the Labrador Sea, *IEEE Transactions on Geoscience and Remote Sensing, GE-27,* pp. 522–534, 1989.

Eicken, H., M. A. Lange, and G. S. Dieckmann, Spatial variability of sea-ice properties in the Northwestern Weddell Sea, *Journal of Geophysical Research, 96,* pp. 10,603–10,615, 1991.

Eide, L. I. and S. Martin, The formation of brine drainage features in young sea ice, *Journal of Glaciology, 14,* pp. 137–154, 1975.

Farmer, L. D., M. R. Drinkwater, T. I. Lukowski, L. D. Arsenault, C. E. Livingstone, A. Kovacs, D. L. Bell, F. Fetterer, S. A. Digby–Carsey, L. M. H. Ulander, A. Carlström, W. Stringer, and L. Shapiro, *Beaufort Sea Ice— 1, Field Data Report,* 102 pp., Naval Oceanographic and Atmospheric Research Laboratory, Branch Office, Hanover, New Hampshire, 1991.

Gow, A. J., Orientation textures in ice sheets of quietly frozen lakes, *Journal of Crystal Growth, 74,* pp. 247–258, 1986.

Gow, A. J. and W. B. Tucker III, Sea ice in the polar regions, *Polar Oceanography, Part A: Physical Science,* edited by Walker O. Smith, pp. 47–122, Academic Press, San Diego, California, 1990.

Gow, A. J., S. F. Ackley, W. F. Weeks, and J. W. Govoni, Physical and structural characteristics of Antarctic sea ice, *Annals of Glaciology, 3,* pp. 113–117, 1982.

Gow, A. F., S. F. Ackley, K. R. Buck, and K. M. Golden, *Physical and Structural Characteristics of Weddell Sea Pack Ice,* CRREL Report 87-14, 70 pp., Cold Regions Research and Engineering Laboratory, Hanover, New Hampshire, 1987a.

Gow, A. J., W. B. Tucker, and W. F. Weeks, *Physical Properties of Summer Sea Ice in the Fram Strait, June– July,* CRREL Report 87-16, 81 pp., Cold Regions Research and Engineering Laboratory, Hanover, New Hampshire, 1987b.

Gow, A. J., D. A. Meese, D. K. Perovich, and W. B. Tucker III, The anatomy of a freezing lead, *Journal of Geophysical Research, 95,* pp. 18,221–18,232, 1990.

Granberg, H. B. and M. Lepparänta, *Helicopter-Borne Remote Sensing of Antarctic Sea Ice Using a Laser Profiler, Synchronized Video and 70 mm Camera During Finnarp-89,* paper presented at International Association for Hydraulic Research (IAHR) Ice Symposium 1990, IAHR, Espoo, Finland, 1990.

Hibler, W. D., A. J. Mock, and W. B. Tucker, Classification and variation of sea ice in the western Arctic Basin, *Journal of Geophysical Research, 79,* pp. 2735–2743, 1974.

Johansson, R., *Laser-Based Surface Roughness Measurements of Snow and Sea Ice on the Centimeter Scale,* Research Report 162, 34 pp., Department of Radio and Space Science, Chalmers University of Technology, Göteborg, Sweden, 1988.

Kim, Y. S., R. K. Moore, R. G. Onstott, and S. Gogineni, Towards identification of optimum radar parameters for sea ice monitoring, *Journal of Glaciology, 31(109),* pp. 214–219, 1985.

Koerner, R. M., The mass balance of sea ice of the Arctic Ocean, *Journal of Glaciology, 12,* pp. 173–185, 1973.

Kovacs, A. and R. Morey, Radar anisotropy of sea ice due to preferred azimuthal orientation of the horizontal c-axes of ice crystals, *Journal of Geophysical Research, 83,* pp. 6037–6046, 1978.

Lake, R. A. and E. L. Lewis, Salt rejection by sea ice during growth, *Journal of Geophysical Research, 75(30),* pp. 583–597, 1970.

Lange, M. A. and H. Eicken, The sea ice thickness distribution in the Northwestern Weddell Sea, *Journal of Geophysical Research, 96,* pp. 4821–4837, 1991a.

Lange, M. A. and H. Eicken, Texture characteristics of sea ice and the major mechanisms of ice growth in the Weddell Sea, *Annals of Glaciology, 15,* pp. 210–215, 1991b.

Lange, M. A., S. F. Ackley, P. Wadhams, G. S. Dieckmann, and H. Eicken, Development of sea ice in the Weddell Sea, Antarctica, *Annals of Glaciology, 12,* pp. 92–96, 1989.

Lange, M. A., P. Schlosser, S. F. Ackley, P. Wadhams, and G. S. Dieckmann, O^{18} concentrations in the sea ice of the Weddell Sea, Antarctica, *Journal of Glaciology, 36,* pp. 315–323, 1990.

Langhorne, P. J., Laboratory experiments on crystal orientation in NaCl ice, *Annals of Glaciology, 4,* pp. 163–169, 1983.

Langhorne, P. J. and W. H. Robinson, Alignment of crystals in sea ice due to fluid motion, *Cold Regions Science and Technology, 12,* pp. 197–214, 1986.

Lin, F. C., J. A. Kong, R. T. Shin, A. J. Gow, and S. A. Arcone, Correlation function study for sea ice, *Journal of Geophysical Research, 93,* pp. 14,055–14,063, 1988.

Lofgren, G. and W. F. Weeks, Effect of growth parameters on substructure spacing in NaCl ice crystals, *Journal of Glaciology, 7,* pp. 153–163, 1969.

Lytle, V. I. and S. F. Ackley, Sea ice ridging in the Eastern Weddell Sea, *Journal of Geophysical Research, 96(10),* pp. 18,411–18,416, 1991.

Martin, S., A field study of brine drainage and oil entrainment in first-year sea ice, *Journal of Glaciology, 22,* pp. 473–502, 1979.

Maykut, G. A., The ice environment, *Sea Ice Biota,* edited by R. A. Horner, pp. 21–82, CRC Press, Boca Raton, Florida, 1985.

Maykut, G. A. and N. Untersteiner, Some results from a time-dependent thermodynamic model of sea ice, *Journal of Geophysical Research, 76,* pp. 1550–1575, 1971.

Nakawo, M. and N. K. Sinha, A note on the brine layer spacing of first-year sea ice, *Atmosphere–Ocean, 22(2),* pp. 193–206, 1984.

Neumann, G. and W. J. Pierson, Jr., *Principles of Physical Oceanography,* 545 pp., Prentice–Hall, Englewood Cliffs, New Jersey, 1966.

Paige, R. A., *Crystallographic Studies of Sea Ice in McMurdo Sound, Antarctica,* Technical Report R494, 31 pp., Naval Civil Engineering Laboratory, Port Hueneme, California, 1966.

Paterson, J. S., B. Brisco, S. Argus, and G. Jones, In-situ measurements of micro-scale surface roughness of sea ice, *Arctic, 44(1), Supplement 1,* pp. 140–146, 1991.

Perovich, D. K. and A. J. Gow, A statistical description of the microstructure of young sea ice, *Journal of Geophysical Research, 96,* pp. 16,943–16,953, 1991.

Perovich, D. K. and A. Hirai, Microcomputer-based image-processing systems, *Journal of Glaciology, 34,* pp. 249–252, 1988.

Perovich, D. K., A. J. Gow, and W. B. Tucker III, Physical properties of snow and ice in the winter marginal ice zone of Fram Strait, *Proceedings of the IGARSS'88 Symposium,* pp. 1119–1123, Edinburgh, Scotland, ESA SP-284, European Space Agency, Noordwijk, Netherlands, 1988.

Rothrock, D. A., Ice thickness distribution—measurement and theory, *The Geophysics of Sea Ice,* edited by N. Untersteiner, pp. 551–575, NATO ASI Series B: Physics vol. 146, Plenum Press, New York, 1986.

Stander, E. and B. Michel, The effects of fluid flow on the development of preferred orientations in sea ice: laboratory experiments, *Cold Regions Science and Technology, 17,* pp. 153–161, 1989.

Stogryn, A., Correlation functions for random granular media in strong fluctuation theory, *IEEE Transactions on Geoscience and Remote Sensing, GE-22,* pp. 150–154, 1984.

Tsang, L. and J. A. Kong, Scattering of electromagnetic waves from random media with strong permittivity fluctuations, *Radio Science, 16,* pp. 303–320, 1981.

Tucker, W. B., A. J. Gow, and J. A. Richter, On small-scale variations of salinity in first-year sea ice, *Journal of Geophysical Research, 89,* pp. 6505–6514, 1984a.

Tucker, W. B., D. S. Sodhi, and J. W. Govoni, Structure of first-year pressure ridge sails in the Alaskan Beaufort Sea, *The Alaskan Beaufort Sea: Ecosystems and Environment,* edited by P. W. Barnes et al., pp. 115–135, Academic Press, New York, 1984b.

Tucker, W. B., III, A. J. Gow, and W. F. Weeks, Physical properties of summer sea ice in Fram Strait, *Journal of Geophysical Research, 92(C7),* pp. 6787–6803, 1987.

Tucker, W. B., III, T. C. Grenfell, R. G. Onstott, D. K. Perovich, A. J. Gow, R. A. Shuchman, and L. L. Sutherland, Microwave and physical properties of sea ice in the winter marginal ice zone, *Journal of Geophysical Research, 96,* pp. 4573–4587, 1991.

Untersteiner, N., On the mass and heat budget of the Arctic sea ice, *Archiv fuer Meteorologie, Geophysik, und Bioklimatologie, A(12),* pp. 151–182, 1961.

Vallese, F. and J. A. Kong, Correlation function studies for snow and ice, *Journal of Applied Physics, 52,* pp. 4921–4925, 1981.

Wadhams, P., Sea ice morphology and its measurement, *Arctic Technology and Policy,* edited by I. Dyer and C. Chryssostomidis, pp. 179–195, Hemisphere Publishing Corporation, Washington, DC, 1984.

Wadhams, P., M. A. Lange, and S. F. Ackley, The ice thickness distribution across the Atlantic section of the Antarctic Ocean in midwinter, *Journal of Geophysical Research, 92(C13),* pp. 14,535–14,552, 1987.

Weeks, W. F. and S. F. Ackley, The growth, structure and properties of sea ice, *The Geophysics of Sea Ice,* edited by N. Untersteiner, pp. 9–164, NATO ASI Series B: Physics vol. 146, Plenum Press, New York, 1986.

Weeks, W. F. and A. J. Gow, Preferred crystal orientations in the fast ice along the margins of the Arctic Ocean, *Journal of Geophysical Research, 83,* pp. 5105–5121, 1978.

Weeks, W. F. and A. J. Gow, Crystal alignments in the fast ice of Arctic Alaska, *Journal of Geophysical Research, 85,* pp. 1137–1146, 1980.

Weeks, W. F. and W. L. Hamilton, Petrographic characteristics of young sea ice, Point Barrow, Alaska, *American Mineralogy, 47,* pp. 945–961, 1962.

Weeks, W. F. and O. S. Lee, The salinity distribution in young sea ice, *Arctic, 15,* pp. 92–108, 1962.

Weeks, W. F. and G. Lofgren, The effective solute distribution during the freezing of NaCl solutions, *Proceedings of the International Conference on Low Temperature Science, Physics of Snow and Ice, I(1),* pp. 579–597, edited by H. Oura, Institute of Low Temperature Science, Hokkaido University, Sapporo, Japan, 1967.

Weeks, W. F., S. F. Ackley, and J. W. Govoni, Sea ice ridging in the Ross Sea, Antarctica, as compared with sites in the Arctic, *Journal of Geophysical Research, 94,* pp. 4984–4988, 1988.

Williams, E., C. W. M. Swithinbank, and G. deQ Robin, A submarine study of Arctic pack ice, *Journal of Glaciology, 15,* pp. 349–362, 1975.

Winebrenner, D. P., L. Tsang, B. Wen, and R. West, Sea ice characterization measurements needed for testing of microwave remote sensing models, *IEEE Journal of Oceanic Engineering, 14(2),* pp. 149–158, 1989.

Chapter 3. The Physical Basis for Sea Ice Remote Sensing

Martti Hallikainen

Laboratory of Space Technology, Helsinki University of Technology, Otakaari 5A, 02150 Espoo, Finland

Dale P. Winebrenner

Polar Science Center, Applied Physics Laboratory, University of Washington, 1013 NE 40th Street, Seattle, Washington 98105

3.1 Definition of Electromagnetic Quantities

The electromagnetic frequency (f) and free-space wavelength (λ_0) are related by

$$\lambda_0 = c / f \tag{1}$$

where c is the speed of light. The frequencies of interest for this book, i.e., the microwave range, cover the region from 1 to 300 GHz; the corresponding wavelength ranges from 30 cm to 1 mm. Note that this broad definition of the microwave range also includes millimeter waves, frequencies of 30 to 300 GHz, or wavelengths of 1 cm to 1 mm.

The physical basis for microwave remote sensing of sea ice is the interaction of microwaves with the sea ice layer. The two main instruments used for microwave remote sensing are the radiometer and radar. A radiometer is a sensitive receiver used to measure the intensity of the noise-like radiation emitted by the target, whereas a radar, sensing the reflection or backscatter of its own radiation, includes both a transmitter and a receiver.

In microwave radiometry, the measured emission intensity is converted to an equivalent brightness temperature

$$T_B (f, \theta_i) = e (f, \theta_i) T \tag{2}$$

where e is the dimensionless emission coefficient ($0 < e \le 1$), T is the physical temperature of the target in kelvins, and θ_i is the local incidence angle off nadir. The emission coefficient is determined by the dielectric properties and the surface roughness of the target. In radiometry of sea ice, radiation emitted by each layer of the water–ice–snow system contributes to the observed brightness temperature. Due to complexities within the snow and sea ice media, the relation in Equation (2) describes only approximately the observed brightness temperature. (See Chapters 4 and 8 for a more detailed analysis of the brightness temperature and its dependence on surface roughness and other characteristics.) For a general analysis of the behavior of the brightness temperature, the reader is referred to Ulaby et al. [1982, 1986] and Colwell [1983].

The microwaves transmitted by a radar are reflected and scattered by a target. The backscattered intensity is measured by the receiver, and the result is expressed as the differential backscattering coefficient σ_0 (usually referred to as the backscattering coefficient).

The backscattering coefficient is defined in terms of the incident and scattered electric fields $E_t{}^i$ and $E_r{}^s$, as follows:

$$\sigma_{tr}^0 \left(\theta_i, \phi_i\right) = \frac{4\pi R^2 \left| E_r^s \right|^2}{A \left| E_t^i \right|^2} \tag{3}$$

where t and r are the polarizations of the transmitted and received fields, respectively; R is the distance from the radar to the target; and A is the area illuminated by the radar. Angles θ_i and ϕ_i define the incident direction of the transmitted power. The backscattering coefficient is usually expressed in decibels,

$$\sigma^0 \,(\text{dB}) = 10 \log_{10} \sigma^0 \tag{4}$$

For a general analysis of the behavior of the backscattering coefficient, the reader is referred to Ulaby et al. [1982, 1986] and Colwell [1983]. In order to characterize the microwave interaction with the snow–ice–water system, we must define several basic electromagnetic quantities.

3.1.1 Relative Permittivity

The relative permittivity, denoted $\varepsilon = \varepsilon' - j\varepsilon''$ (where $j = \sqrt{-1}$) is a complex number that characterizes the electrical properties of media. The real part of the relative permittivity ε' gives the contrast with respect to free space ($\varepsilon'_{\text{air}} = 1$), whereas the imaginary part of the permittivity ε'' gives the electromagnetic loss of the material. Here, ε' and ε'' are often referred to as the dielectric constant and the dielectric loss factor, respectively.

3.1.2 Propagation, Absorption, and Phase Constant

For an electromagnetic plane wave propagating in the z-direction, the intensity of the electric field at point z can be expressed as

$$E(z) = E_0 e^{-\gamma z} \tag{5}$$

Microwave Remote Sensing of Sea Ice
Geophysical Monograph 68
©1992 American Geophysical Union

where E_0 is the field intensity at $z = 0$. The complex propagation constant of the medium is denoted by γ and is given by

$$\gamma = \alpha + j\beta \qquad (6)$$

where α is the absorption constant and β is the phase constant. The absorption constant describes the transformation of energy into other forms of energy, such as heat. The phase constant is equal to the wave number $k = 2\pi/\lambda$ in a lossless medium. They are related to the complex permittivity by

$$\alpha = k_0 \left| Im\left\{ \sqrt{\varepsilon} \right\} \right| \qquad (7a)$$

$$\beta = k_0 Re\left\{ \sqrt{\varepsilon} \right\} \qquad (7b)$$

where k_0 is the wave number in free space. The power absorption coefficient κ_a is defined as

$$\kappa_a = 2\alpha \qquad (8)$$

It is often expressed in dB/m through the relation

$$\kappa_a \,(dB/m) = 8.686\alpha \,(Np/m) \qquad (9)$$

3.1.3 Extinction and Scattering Coefficient

The total electromagnetic loss in a scattering medium consists of absorption loss (electromagnetic power transformed into other forms of energy, such as heat) and scattering loss (energy is caused to travel in directions other than that of the incident radiation). Scattering loss is caused by particles of different ε embedded in a host medium. The extinction coefficient (total loss) is thus

$$\kappa_e = \kappa_a + \kappa_s \qquad (10)$$

where κ_s denotes the scattering loss.

3.1.4 Penetration Depth

Part of a wave incident upon the surface of a medium from the air in the z-direction is transmitted across the boundary into the medium. The penetration depth is defined as the depth δ_p at which

$$\frac{P(\delta_p)}{P(0+)} = \frac{1}{e} \qquad (11a)$$

or

$$\int_0^{\delta_p} \kappa_e(z)\,dz = 1 \qquad (11b)$$

where z is the direction normal to the surface, $P(\delta_p)$ is the transmitted power at δ_p, and $P(0+)$ is the transmitted power just beneath the surface [Ulaby et al., 1986]. If scattering in the medium is ignored, $\kappa_e \equiv \kappa_a = 2\alpha$, and if, additionally, $\alpha(z)$ does not depend on z,

$$\delta_p = \frac{1}{\kappa_a}$$
$$= \frac{\sqrt{\varepsilon'}}{k_0 \varepsilon''}; \quad \varepsilon'' \ll \varepsilon' \qquad (11c)$$

The penetration depth indicates the maximum depth of the medium that contributes to the backscattering coefficient and brightness temperature.

3.2 DIELECTRIC AND EXTINCTION PROPERTIES OF THE WATER–ICE–SNOW SYSTEM

In remote sensing, materials of interest are generally classified into one of the following groups: (a) homogeneous substances, (b) electrolytic solutions, or (c) heterogeneous mixtures [Ulaby et al., 1986]. Pure water and ice belong to group (a). Seawater and brine belong to group (b). Sea ice (a mixture of ice crystals, air pockets, and liquid brine inclusions) and snow (a mixture of air, ice particles, and possibly liquid water inclusions) belong to group (c).

For a medium consisting of a host material that contains a concentration of particles of another material, the relative permittivity of the mixture consists of two components,

$$\varepsilon(x,y,z;\hat{p}) = \varepsilon_m(\hat{p}) + \varepsilon_f(x,y,z;\hat{p}) \qquad (12)$$

where $\varepsilon_m(\hat{p}) \equiv \,<\varepsilon(x,y,z;\hat{p})>$ is the average value of the permittivity of the medium and $\varepsilon_f(\hat{p})$ is the fluctuating component. Thus, $\varepsilon_m(\hat{p})$ is independent of the position within the medium, but it may be a function of the polarization unit vector $\hat{p}$, which denotes the direction of the electric field of the incident wave with respect to the inclusions.

3.2.1 Pure Water, Seawater, and Brine

As compared with other natural media, the complex permittivity of water is exceptionally high. This is due to its polar molecular structure [Debye, 1929]. Dissolved salts in seawater (salinity up to 35‰ by weight) increase the ionic conductivity. Consequently, the dielectric loss factor of seawater is higher than that of pure water at frequencies below 10 GHz. The salinity of liquid brine in sea ice is governed by its temperature and can exceed 200‰ at low temperatures.

The dielectric behavior of pure water follows the Debye equation [Debye, 1929],

$$\varepsilon_w = \varepsilon_{w\infty} + \frac{\varepsilon_{w0} - \varepsilon_{w\infty}}{1 + j2\pi f\,\tau_w} \tag{13}$$

where ε_{w0} is the static dielectric constant of pure water, $\varepsilon_{w\infty}$ is the high-frequency limit of ε_w, τ_w is the relaxation time of pure water, and f is the frequency.

Both real and imaginary parts of ε_w of pure water depend on frequency and temperature. The temperature dependence of τ_w, ε_{w0}, and $\varepsilon_{w\infty}$ has been discussed by several investigators, and numerical expressions for these quantities are available in the literature [Lane and Saxton, 1953; Stogryn, 1971; Klein and Swift, 1977].

Seawater contains dissolved salts, including $NaCl$, $MgCl_2$, Na_2SO_4, $CaCl_2$, and KCl, and the average salinity in the oceans is 32.54‰ [Anderson, 1960]. Salts increase the dielectric loss of water by adding free charge carriers. Hence, the loss factor of seawater includes an additional term due to ionic conductivity, and the complex permittivity of seawater follows the modified Debye equation,

$$\varepsilon_{sw} = \varepsilon_{sw\infty} + \frac{\varepsilon_{sw0} - \varepsilon_{sw\infty}}{1 + j2\pi f\,\tau_{sw}} - j\frac{\sigma_{sw}}{\varepsilon_0} \tag{14}$$

where the subscript sw refers to seawater and σ_{sw} is the ionic conductivity of seawater. Numerical expressions for the terms of Equation (14) are available in the literature [Klein and Swift, 1977; Stogryn, 1971; Weyl, 1964].

Figure 3-1 shows the complex permittivity of pure and seawater (salinity 32.54‰) at 0° and 20°C. The real part is practically independent of salinity. The imaginary part is governed by the Debye relaxation process at frequencies above 10 GHz and by the ionic conductivity at frequencies below 3 GHz.

Liquid brine is concentrated in the form of elongated inclusions in sea ice [Assur, 1960]. The permittivity of liquid brine basically follows Equation (14). However, the salinity of liquid brine is determined by its temperature; when the ice temperature decreases, additional water freezes out, thereby increasing the salt concentration in the brine. Empirical expressions for the brine salinity have been derived by Assur [1960] and Poe et al. [1972]. Chapter 2 describes the salinity of liquid brine in sea ice.

Since the salinity of brine may be much higher than that of seawater, some of the empirical expressions derived for the terms of Equation (14) are not valid for liquid brine. Stogryn and Desargeant [1985] measured the dc conductivity and the dielectric properties of seawater liquid brine in equilibrium with sea ice. Their empirical equation for the dc conductivity is

$$\sigma_b = -T\exp\left(0.5193 + 0.8755\times 10^{-1}T\right)$$
$$\text{if } T \geq -22.9°C$$
$$= -T\exp\left(1.0334 + 0.1100T\right) \tag{15}$$
$$\text{if } T \leq -22.9°C$$

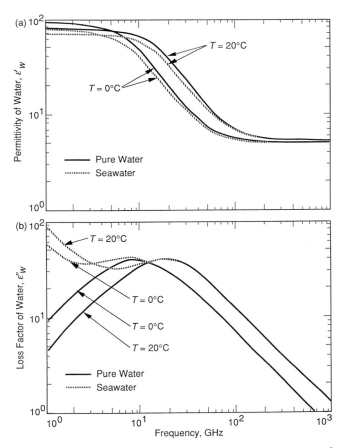

Fig. 3-1. The (a) permittivity and (b) loss factor of pure and seawater at 0° and 20°C [Ulaby et al., 1986].

where T is the temperature in degrees centigrade and σ_b is in S/m. From the values of ε_b, they determined expressions for the Debye parameters ε_{b0}, $\varepsilon_{b\infty}$, and τ_b:

$$\varepsilon_{b0} = \frac{939.66 - 19.068T}{10.737 - T} \tag{16a}$$

$$\varepsilon_{b\infty} = \frac{82.79 + 8.19T^2}{15.68 + T^2} \tag{16b}$$

$$\begin{aligned}2\pi\tau_b = {}& 0.10990 + 0.13603\times 10^{-2}\,T \\ & + 0.20894\times 10^{-3}\,T^2 \\ & + 0.28167\times 10^{-5}\,T^3\end{aligned} \tag{16c}$$

where T is the temperature in degrees centigrade and $2\pi\tau_b$ is in nanoseconds. Note that Equations (16a) to (16c) are valid for temperatures between −2.8° and −25.0°C.

Figure 3-2 shows the dielectric properties of liquid brine, based on the empirical results of Stogryn and Desargeant [1985]. In general, ε'_b and ε''_b decrease with decreasing temperature. The rapid increase of ε''_b with decreasing

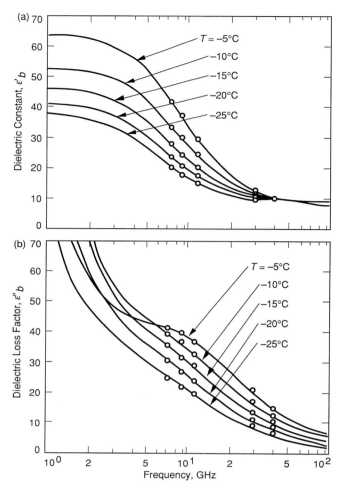

Fig. 3-2. The dielectric (a) constant and (b) loss factor of brine as a function of frequency with temperature as a parameter [Stogryn and Desargeant, 1985].

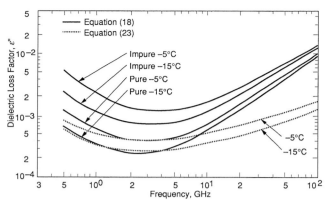

Fig. 3-3. Dielectric loss factor of freshwater ice by Mätzler and Wegmüller [1987], Equation (18), and by Tiuri et al. [1984], Equation (23).

temperature. The rapid increase of ε''_b with decreasing frequency below 3 GHz is caused by the ionic conductivity. The Debye relaxation is just barely visible at $T = -5°C$; at lower temperatures, it is masked by the ionic conductivity.

3.2.2 Freshwater Ice

Freshwater ice is free from salt, but usually it includes air bubbles, impurities, and cracks. Pure ice (frozen distilled water) is an idealization of freshwater ice. The dielectric properties of freshwater ice at microwave frequencies are determined by two physical processes, namely the high-frequency tail of a relaxation spectrum (relaxation frequency in the kilohertz range) and the low-frequency tail of far-infrared absorption bands [Evans, 1965; Mätzler and Wegmüller, 1987]. These processes result in a practically constant value for the dielectric constant ε'_i at microwave

frequencies and in a minimum of the dielectric loss factor ε''_i centered around 1 to 5 GHz.

Based on experimental data, the relative dielectric constant of pure and freshwater ice may be assigned the constant value

$$\varepsilon'_i = 3.17 \tag{17}$$

between 10 MHz and 1000 GHz [Mätzler and Wegmüller, 1987; Cumming, 1952].

For the imaginary part of the permittivity, results of various investigators show substantial scatter. The scatter is caused by varying amounts of impurities in the ice samples and by measurement inaccuracies (partly due to small values of ε''_i). Reviews of the results have been given by Warren [1984] and Ulaby et al. [1986].

Mätzler and Wegmüller [1987] developed a numerical equation for the dielectric loss factor of freshwater ice, based on the two physical processes that influence ε''_i,

$$\varepsilon''_i = A/f + Bf^C \tag{18}$$

where f is the frequency in GHz. The values of constants A, B, and C were determined for the temperatures of $-5°$ and $-15°C$. For $-5°C$, the values are A = 0.0026, B = 0.00023, and C = 0.87. For $-15°C$, the values are A = 0.0013, B = 0.00012, and C = 1.0. The behavior of ε''_i from Equation (18) is depicted in Figure 3-3, along with another empirical expression, Equation (23).

According to Equation (18), ε''_i is between 0.001 and 0.01 in the 1- to 100-GHz range. Thus, freshwater ice is a low-loss medium and its power absorption coefficient at 1 GHz (and $-5°C$) is 0.09 dB/m and penetration depth is 100 m. At 10 GHz, the corresponding numbers are 0.9 dB/m and 10.0 m. This means that contributions to the brightness temperature and the backscattering coefficient are not limited to the topmost ice layers.

3.2.3 Sea Ice

3.2.3.1. Experimental observations. Sea ice consists of freshwater ice, liquid brine, and air. In addition to the dielectric properties of these constituents, several additional parameters influence the complex permittivity of sea ice, including the volume fraction of each constituent and geometry (shape, size, and orientation) of brine pockets with respect to the propagation direction of the electromagnetic wave. Most of these parameters depend on ice temperature.

The pure ice occurs in the form of thin platelets with the crystallographic *c*-axes perpendicular to the plane of the platelets. These platelets form grains whose *c*-axes can have various orientations. For frazil and multiyear ice the axes may be randomly oriented, while for columnar ice the polar angle of the *c*-axes tends to increase from 0° (vertical) near the ice surface to 90° as the depth increases. Due to these differences, the dielectric properties of sea ice depend on ice type. For a more detailed description of the structure of sea ice, the reader is referred to Chapter 2.

Experimental investigations of the microwave dielectric behavior of sea ice, simulated sea ice (frozen seawater), and NaCl ice (frozen NaCl solution) have been carried out at frequencies below 40 GHz (Table 3-1). An extremely limited amount of data is available on the dielectric properties of multiyear sea ice. In most studies, temperatures below −5°C were used, although rapid changes in the permittivity occur at temperatures close to 0°C. The two most useful sets of experimental Arctic sea ice data were obtained by Vant [1976] and Vant et al. [1974, 1978]. Vant [1976] and Vant et al. [1978] reported the same results, but the former included detailed results for each ice sample.

Measurements of simulated sea ice and/or NaCl ice have been made by several investigators (Table 3-1). Preparation of artificial sea ice samples offers a convenient way to control the salinity and thus study the basic behavior of saline ice. The structure of artificial samples may differ from that of natural sea ice, depending on the sample size and freezing conditions. Hoekstra and Cappillino [1971] made their ice samples by flash freezing seawater in a small sample holder, which resulted in high values of ε_{si}. However, Arcone et al. [1986] found no difference in structure for samples grown naturally in a large tank.

Figures 3-4 to 3-6 show experimental ε'_{si} and ε''_{si} values obtained by various investigators as a function of temperature. Data are shown for 1 GHz, 4 to 5 GHz, and 10 to 16 GHz, because these frequency bands are widely used for both active and passive remote sensing of sea ice. Note that the temperature scale is logarithmic to show the rapid changes occurring at temperatures above −5°C. For the permittivity of sea ice at frequencies above 16 GHz, the reader is referred to Sackinger and Byrd [1972] (experimental data), Vant et al. [1978] (modeling), and Stogryn [1987] (modeling). Due to substantial scatter in most of the original data sets, only curves fitted to data points are shown in Figures 3-4 to 3-6.

Experimental values by Vant et al. [1978] at 1 GHz and by Hallikainen [1983] at 0.9 GHz are shown in Figure 3-4. The values of Vant et al. [1978] show that ε'_{si} and ε''_{si} for multiyear ice are lower than those for first-year ice; this is due to the lower salinity. Since the relative brine volume of low-salinity sea ice near 0°C may be higher than that of high-salinity sea ice at lower temperatures (the relative brine volume of sea ice of 1‰ salinity is 103‰ at −0.5°C, whereas the corresponding number for sea ice of 4‰ salinity is 40‰ at −5°C; see Chapter 2), the maximum values obtained by Hallikainen [1983] are of the same order of magnitude as those obtained by Vant et al. [1978] for high-salinity Arctic sea ice.

At C-band, few measurements have been made at temperatures above −5°C. The results of Vant et al. [1978] for Arctic sea ice at 4 GHz are compared with the values of Arcone et al. [1986] for simulated sea ice at 4.75 GHz in Figure 3-5. The abrupt changes in the slopes of the ε''_{si} curves of Vant et al. [1978] may be due to crystallization of NaCl at −22.9°C. The unusual decrease of ε''_{si} with increasing salinity in the results for simulated sea ice may be due to differences in the size and length of brine pockets and, thus, conductivity [Arcone et al., 1986].

Experimental values at 10 GHz cover frazil, columnar, multiyear, and artificial sea ice (see Figure 3-6). The dielectric loss is higher for frazil than for columnar sea ice with the electromagnetic wave propagating in the vertical direction. The values of ε''_{si} obtained by Hoekstra and Cappillino [1971] are higher than those obtained by other investigators. This may be due to their sample-preparation techniques (flash freezing seawater in the sample holder), which result in an orientation of brine pockets different from that in natural sea ice.

The effect of the orientation of brine pockets with respect to the electric field of an electromagnetic wave propagating in the sea ice medium has been investigated by Sackinger and Byrd [1972] (Arctic sea ice), Vant et al. [1978] (artificial ice), and Hallikainen [1983] (low-salinity sea ice). The results of Sackinger and Byrd [1972] for Arctic sea ice in the 26- to 40-GHz range indicate that at −7°C, ε''_{si} is about 300% higher for an orientation of 90° (horizontal) than that for 0° (vertical). The difference decreases with decreasing temperature and is about 150% at −21.5°C. The results of Vant et al. [1978] for artificial ice suggest that there is no difference in ε'_{si} and ε''_{si} for angles of 45°, 60°, and 90°.

Experimental values for the absorption coefficient κ_a of natural sea ice (orientation 0°) are shown in Figure 3-7. In general, κ_a increases with increasing frequency and temperature. The loss of multiyear ice is considerably lower than that of first-year ice, mainly due to much lower salinity. The values for first-year low-salinity sea ice show that the temperature dependence of the absorption coefficient is high at temperatures near 0°C. At 10 GHz, the differences among the results for columnar, frazil, and multiyear ice are substantial.

Figure 3-8 shows experimental values for the penetration depth of natural sea ice, which were computed from the results in Figure 3-7. In the 1- to 10-GHz range, the

TABLE 3-1. Experimental investigations of the dielectric properties of natural and simulated sea ice in the 0.5- to 40-GHz range. The orientation from vertical is written in parentheses if not given explicitly in the reference source.

Reference	Frequency, GHz	Ice type	Salinity, ‰	Temp., °C	Orientation from vertical, deg	Number of samples	Comments
Hoekstra and Cappillino [1971]	0.1 to 24 (5 frequencies)	Simulated	4, 6, 8, 10, 12	−80 to −2 (maximum)	(0)	8	Seawater flash frozen in holder. One to three samples per frequency, one per salinity. Temperature range varies depending on sample. Open-circuited coaxial technique up to 4 GHz. Short circuited waveguide technique at 10/24 GHz. Sample size (reported only for 24 GHz): 0.15 cm^3.
Hoekstra and Cappillino [1971]	9.8	NaCl	7.3, 14.6	−33 to −4	(0)	2	NaCl solution flash frozen in sample holder. Short circuited waveguide technique. Sample size not reported.
Sackinger and Byrd [1972]	26 to 40	First year, Arctic Sea	2.8, 3.4, 7.2	−32 to −7	0 and 90	Not reported	Short-circuited waveguide technique. Frequency sweep from 26 to 40 GHz. Sample length range: 1.2 to 3.5 cm.
Vant et al. [1974]	10	First year, Bering Sea	3.2 to 4.6	−60 to −3	(0)	4	Two columnar and two frazil ice samples. Waveguide transmission technique. Sample length not reported.
Vant et al. [1974]	10	Multiyear, Bering Sea	0.61, 0.70	−60 to −3	(0)	2	Waveguide transmission technique. Sample length not reported.
Vant et al. [1974]	31.4, 34.0	NaCl	8.0	−7	(0)	2	Free-space transmission technique. Sample: large slab.
Bogorodskii and Khoklov [1975]	10	First year, Bering Sea	1.9 to 21.6	−15 to −1	0	130	Results presented for two temperature ranges (−8° to −15°C and −5° to −1°C) only. Free-space transmission technique. Sample diameter, 28 cm; thickness range, 4 to 20 cm.
Vant et al. [1978]	0.1 to 4.0 (7 frequencies)	First year, Beaufort Sea	5.1 to 10.5	−40 to −5	0	7	Samples represent winter–spring ice. Coaxial cage measurement technique. Sample diameter, 7.6 cm; length, 8 cm.
Vant et al. [1978]	4.0, 7.5	First year, Beaufort Sea	4.1 to 7.5	−40 to −4	0	4	Sample diameter, 7.6 cm; length, 2 cm. See above for additional information.
Vant et al. [1978]	0.1 to 7.5 (8 frequencies)	Multiyear, Beaufort Sea	1.3, 1.7	−40 to −5	0	2	See above for measurement technique and sample size.
Vant et al. [1978]	0.1 to 4.0 (7 frequencies)	NaCl	3.0 to 3.9	−40 to −6	0, 30, 45, 60, 90	10	See above for measurement technique and sample size.
Hallikainen [1977]	16	NaCl	0.9 to 3.8	−15 to −5	0	4	Free-space transmission technique. Sample size: 33 cm square. Each sample measured as a function of thickness.

TABLE 3-1. Experimental investigations of the dielectric properties of natural and simulated sea ice in the 0.5- to 40-GHz range. The orientation from vertical is written in parentheses if not given explicitly in the reference source. (Continued)

Reference	Frequency, GHz	Ice type	Salinity, ‰	Temp., °C	Orientation from vertical, deg	Number of samples	Comments
Hallikainen [1983]	0.6, 0.9	First year, Gulf of Finland	0.0 to 1.6	−18 to −0.05	0 (189 samples) 30 (100 samples) 90 (84 samples)	373	Covers all conditions for two winters. Each sample measured only at original temperature. Emphasis at temperatures above −2°C (200 samples). Doubly reentrant cavity technique. Sample diameter, 4.7 cm; length, 1.6 cm.
Arcone et al. [1986]	4.75, 9.8	Simulated	2.4 to 5.5 (2.3 to 4.4)	−32 to −2	0	7	Ice grown in an outdoor pool (sea salt + water). Salinity measured before and after testing. Detailed structure analysis. Free-space transmission technique. Sample, 46-cm square; thickness range, 8.5 to 23 cm.
Hallikainen et al. [1988]	10	First year, Gulf of Bothnia	0.0 to 1.0	−6.0 to −0.3	0	28	Each sample measured only at original temperature. Waveguide transmission technique. Sample length, 12 cm; volume, 28 cm³.

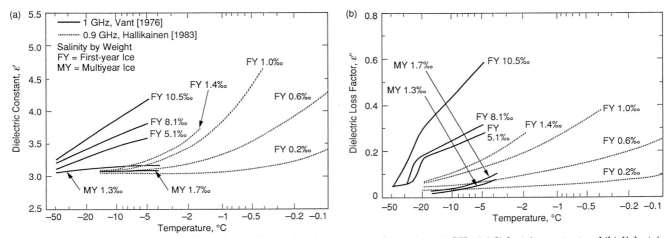

Fig. 3-4. Experimental permittivity values obtained by various investigators for sea ice at 1 GHz: (a) dielectric constant and (b) dielectric loss factor.

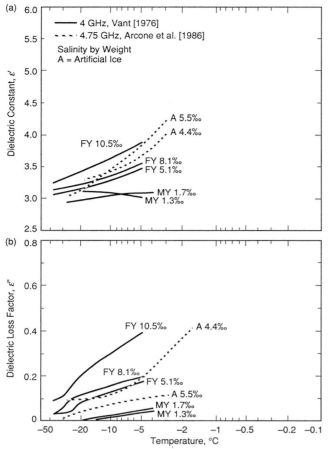

Fig. 3-5. Experimental permittivity values obtained by various investigators for sea ice in the 4- to 5-GHz range: (a) dielectric constant and (b) dielectric loss factor.

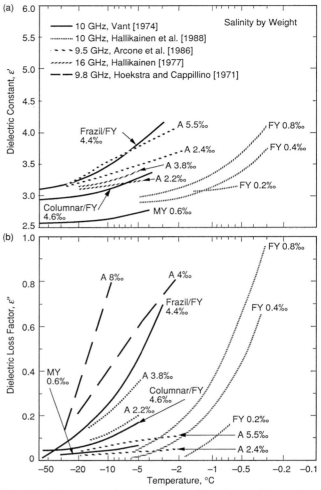

Fig. 3-6. Experimental permittivity values obtained by various investigators for sea ice in the 10- to 16-GHz range: (a) dielectric constant and (b) dielectric loss factor.

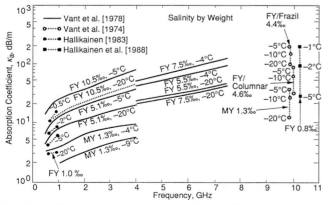

Fig. 3-7. Experimental power absorption coefficient of sea ice, obtained by various investigators.

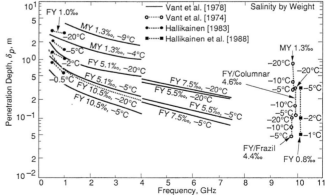

Fig. 3-8. Experimental penetration depth for sea ice, computed from the values in Figure 3-7.

penetration depth is between 5 and 100 cm for first-year ice and 30 and 500 cm for multiyear ice. Figure 3-8 indicates that sensors operating at X-band provide information on sea ice mainly from the topmost 5 to 80 cm, depending on ice type, salinity, and temperature; whereas the corresponding numbers for L-band sensors are 40 to 500 cm.

3.2.3.2. Modeling approaches. Numerous attempts have been made to relate the dielectric properties of sea ice to the relative brine volume V_b [Vant et al., 1974, 1978; Arcone et al., 1986]. Linear equations of the form $\varepsilon_{si} = f(V_b)$ were tested extensively by Vant et al. [1978]. Although the relationship is not basically linear [Hallikainen, 1983], simple relations can be developed separately for each frequency of interest. Numerical relations for 1 GHz (based on data sets obtained by Vant et al. [1978], Arcone et al. [1986], and Hallikainen et al. [1988]); 4 GHz [Vant, 1976]; and 10 GHz [Vant, 1976; Arcone et al., 1986; Hallikainen et al., 1988], are

1 GHz:

$$\varepsilon'_{si} = 3.12 + 0.009V_b \qquad (19a)$$

$$\varepsilon''_{si} = 0.04 + 0.005V_b \qquad (19b)$$

4 GHz:

$$\varepsilon'_{si} = 3.05 + 0.0072V_b \qquad (19c)$$

$$\varepsilon''_{si} = 0.02 + 0.0033V_b \qquad (19d)$$

10 GHz:

$$\varepsilon'_{si} = 3.0 + 0.012V_b \qquad (19e)$$

$$\varepsilon''_{si} = 0.0 + 0.010V_b \qquad (19f)$$

where V_b is the relative brine volume and $V_b \leq 70‰$. Equations (19a) through (19f) do not explicitly take into account the effect of ice density upon the real part of the permittivity. The correlation coefficient for Equations (19a) through (19f) is, in general, between 0.7 and 0.8.

Theoretical dielectric models of sea ice have been developed and/or evaluated by Vant et al. [1978], Gloersen and Larabee [1981], Stogryn [1987], and Sihvola and Kong [1988, 1989]. Vant et al. [1978] tested the applicability of the confocal model by Tinga et al. [1973]. The confocal model provides a satisfactory agreement with most of their experimental observations on the dielectric behavior of sea ice in the 0.1- to 7.5-GHz range. In the model, the brine inclusions are assumed to be symmetrical prolate spheroids with an axial ratio of 20. Additionally, the brine inclusions are assumed to be oriented at an angle of about 40° relative to the direction of propagation.

Stogryn [1987] used the bilocal approximation of the strong fluctuation theory in order to develop a model for the tensor dielectric constant of sea ice that is treated as a random anisotropic medium. Comparisons with measurements on several ice types over the temperature range –2° to –32°C and the frequency range of 0.1 to 40 GHz were made, which resulted in a reasonably good agreement between the theoretical and experimental values. Figure 3-9 shows a comparison for frazil ice at 10 GHz.

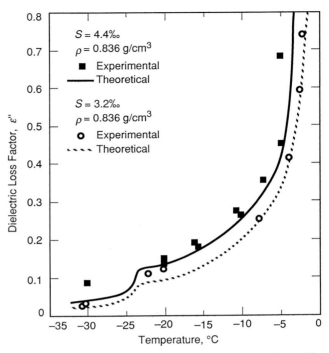

Fig. 3-9. Comparison of theoretical and experimental values of the complex permittivity of frazil sea ice at 10 GHz [Stogryn, 1987].

Sihvola and Kong [1988, 1989] derived general mixing formulas for discrete scatterers immersed in a host medium. The inclusion particles are assumed to be ellipsoidal and their diameter is assumed to be much smaller than the wavelength. The electric field inside the scatterers is determined by quasistatic analysis. The mixing equations were applied to sea ice; however, the feasibility of the method has not been tested by comparing the theoretical values with experimental data.

3.2.4 Snow

3.2.4.1. Dry snow. Dry snow consists of ice crystals and air voids. After a snowfall, the shapes of ice particles are naturally modified. Thermodynamically, the ice crystals seek equilibrium, for which the ratio of surface area to volume is minimum [Gray and Male, 1981]. Any water in the snow medium collects at points of contact between the ice grains. The metamorphism caused by melting and freezing changes the microstructure of the snow. The ice grains become rounded during the melting process, and some of the smaller grains disappear completely. Snow that has undergone several melt–freeze cycles tends to form multiple clusters. In general, the density of snow slowly increases with time due to metamorphism and melt–freeze cycles. The density of dry snow varies from 0.1 g/cm³ (newly fallen snow) to 0.5 g/cm³ (refrozen snow). The interface between snow and sea ice may be gradual, consisting of a high-density snow–ice layer. Due to flooding or wicking, this layer may be saline.

Electromagnetically, dry snow is a dielectric mixture of ice and air and, therefore, its complex permittivity is governed by the dielectric properties of ice, snow density, and ice-particle shape. Since the real part of the permittivity of ice is $\varepsilon'_i = 3.17$ at frequencies between 10 MHz and 1000 GHz (Equation 17) and is practically independent of temperature, the dielectric constant of dry snow, ε'_{ds}, is only a function of density.

Several empirical expressions for ε'_{ds} at microwave frequencies have been obtained by various investigators [Hallikainen et al., 1986; Tiuri et al., 1984; Mätzler and Wegmüller, 1987; Cumming, 1952]. The results agree well with one another. For the most extensive data set used in these measurements, the result between 3 and 37 GHz [Hallikainen et al., 1986] is

$$\varepsilon'_{ds} = 1 + 1.9\rho_{ds} \text{ for } \rho_{ds} \le 0.5 \text{ g/cm}^3$$

$$= 0.51 + 2.88\rho_{ds} \text{ for } \rho_{ds} \ge 0.5 \text{ g/cm}^3 \quad (20)$$

where ρ_{ds} is the density of dry snow in g/cm^3. Based on Equation (20), $\varepsilon'_{ds} = 1.57$ for snow density that is 0.3 g/cm^3. The density of snow is mostly below 0.5 g/cm^3. The behavior of the dielectric constant of dry snow can be explained with dielectric mixing models. A review of various models is given by Ulaby et al. [1986].

Using the two-phase dielectric mixing formula for spherical ice inclusions [Tinga et al., 1973], the imaginary part of the permittivity of dry snow ε''_{ds} results in a straightforward manner from the dielectric behavior of freshwater ice [Ulaby et al., 1986],

$$\varepsilon''_{ds} = \frac{0.34 V_i \varepsilon''_i}{(1 - 0.417 V_i)^2} \quad (21)$$

where V_i is the volume fraction of ice in the mixture, and ε''_i is the loss factor of ice. Also, V_i is related to the snow density ρ_{ds} through

$$V_i = \frac{\rho_{ds}}{0.916} \quad (22)$$

where 0.916 g/cm^3 is the density of pure ice. According to Equation (21), the dielectric loss factor of normal dry snow (density of 0.3 g/cm^3) is about 15% of the value for pure ice.

Experimental data for the loss factor of dry snow below 13 GHz are given by Tiuri et al. [1984] and Cumming [1952]. Figure 3-10 shows experimental ε''_{ds} values obtained by Tiuri et al. [1984] and Cumming [1952]. The best fit to the observed values of Tiuri et al. [1984] was provided by the following expression:

$$\varepsilon''_{ds} = 1.59 \times 10^6 \left[0.52\rho_{ds} + 0.62\rho_{ds}^2 \right]$$

$$\times \left[f^{-1} + 1.23 \times 10^{-14} \sqrt{f} \right] e^{0.036T} \quad (23)$$

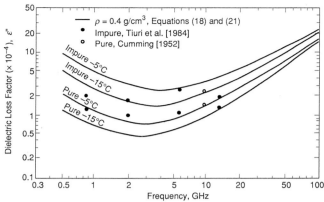

Fig. 3-10. Experimental data for the dielectric loss factor of dry snow by various investigators and a comparison with an empirical expression.

where T is the temperature in degrees centigrade and f is the frequency in hertz. Figure 3-10 also shows a comparison of the ε''_{ds} values obtained from Equations (18) and (21). The model combining the results of Ulaby et al. [1986], the mixing model of Equation (18), and those of Mätzler and Wegmüller [1987], the empirical model of Equation (21), gives results that are higher by a factor of 2.5 than those given by the empirical model of Tiuri et al. [1984]. More data are needed in order to resolve the difference.

Based on theoretical calculations using the Mie theory (assuming independent scattering), scattering dominates absorption in dry snow (ice grain size 1 mm) at frequencies above 15 GHz [Ulaby et al., 1986]. In order to estimate the total loss at millimeter-wave frequencies, the extinction properties of several dry-snow types were measured in the 18- to 90-GHz range by Hallikainen et al. [1987]. The following empirical expressions were developed to relate the extinction coefficient of dry snow, denoted κ_{eds}, to the observed snow particle diameter d_0:

$$\kappa_{eds} = 1.5 + 7.4 d_0^{2.3} \text{ dB/m at 18 GHz} \quad (24a)$$

$$= 30 d_0^{2.1} \text{ dB/m at 35 GHz} \quad (24b)$$

$$= 180 d_0^{2.0} \text{ dB/m at 60 GHz} \quad (24c)$$

$$= 300 d_0^{1.9} \text{ dB/m at 90 GHz} \quad (24d)$$

The particle diameter (observed by photography) is in millimeters. Equations (24a) to (24c) can be combined into a single equation of the form

$$\kappa_{eds} = 0.0018 f^{2.8} d_0^{2.0} \text{ dB/m for 18 to 60 GHz} \quad (25)$$

In Equation (25), f is in gigahertz and d_0 is in millimeters. Equations (24) and (25) hold for particle sizes below 1.6 mm.

The empirical expressions in Equation (24) are depicted in Figure 3-11. They indicate that the extinction coefficient

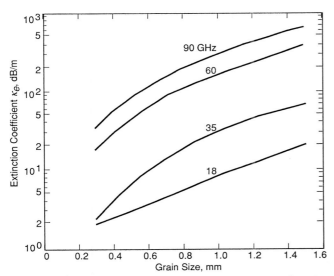

Fig. 3-11. Empirical extinction coefficient of dry snow as a function of snow particle diameter [Hallikainen et al., 1987].

at 90 GHz may exceed 500 dB/m for large grain sizes. At 18 GHz, the corresponding result is about 20 dB/m. Figure 3-11 implies that at 35, 60, and 90 GHz, the scattering coefficient ($\kappa_s = \kappa_e - \kappa_a$) is much larger than the absorption coefficient for all realistic grain sizes. Note that at 18 GHz, the absorption and scattering coefficients are comparable in magnitude for small-grained snow.

The strong fluctuation theory provides results that agree reasonably well with the empirical extinction coefficient values in the 18- to 60-GHz range for all realistic grain sizes, and also at 90 GHz for grain sizes smaller than 0.9 mm [Hallikainen et al., 1987].

3.2.4.2. Wet snow. Wet snow is a mixture of ice crystals, liquid water, and air. For snow on top of the sea ice layer, liquid water may be saline. The geometry and porosity of wet snow depend on its liquid water content. It has been concluded that snow has two distinct regimes of liquid saturation [Colbeck, 1982]. In the lower range (pendular regime), air is continuous throughout the pore space, and liquid water occurs in the form of isolated inclusions. In the higher range of liquid saturation (funicular regime), liquid water is continuous throughout the pore space and air occurs as distinct bubbles trapped by narrow constrictions in the pores. There is a sharp transition between the two regimes.

Electromagnetically, wet snow is a three-component dielectric mixture consisting of ice particles, air, and liquid water. As discussed in Sections 3.2.1 and 3.2.2, both water and ice exhibit Debye-type relaxation spectra. The relaxation frequency (the frequency at which the maximum dielectric loss occurs) of ice is in the kilohertz range, whereas that for water at 0°C is 9 GHz. The complex permittivities of ice and water depend on frequency and temperature. Consequently, the permittivity of wet snow is a function of frequency, temperature, volumetric water content, snow density, the shape of ice particles, and the shape of water inclusions.

Since the permittivity of water is substantially higher than that of ice and air, the dielectric behavior of wet snow is governed by the volume fraction of water. Basically, ε'_{ws} and ε''_{ws} are compressed versions of those for water at 0°C. The dielectric behavior of wet snow has been investigated by Linlor [1980] between 4 and 12 GHz, by Tiuri et al. [1984] at 1 and 4 GHz, and by Hallikainen et al. [1986] between 3 and 37 GHz. Two of the models, the Debye-like semiempirical model, and the theoretical two-phase Polder–Van Santen mixing model, were found to describe adequately the dielectric behavior of wet snow.

The argument for a Debye-like model is as follows: Since the water inclusions in snow have an ε_w about 40 times larger than ε'_{ds}, the spectral behavior of wet snow is dominated by the dispersion behavior of water. The developed Debye-like model is given by

$$\varepsilon'_{ws} = A + \frac{Bm_v^x}{1 + (f/f_0)^2} \tag{26a}$$

$$\varepsilon''_{ws} = \frac{C(f/f_0)m_v^x}{1 + (f/f_0)^2} \tag{26b}$$

where f_0 is the relaxation frequency, f is the frequency, and A, B, C, and x are all constants determined by fitting the model to the measured data in the 3- to 37-GHz range. The approach gives

$$A = 1.0 + 1.83\rho_{ds} + 0.02A_1 m_v^{1.015} + B_1 \tag{27a}$$

$$B = 0.073A_1 \tag{27b}$$

$$C = 0.073A_2 \tag{27c}$$

$$x = 1.31 \tag{27d}$$

$$f_0 = 9.07 \text{ GHz} \tag{27e}$$

where constants A_1, A_2, and B_1 are given by

$$A_1 = 0.78 + 0.03f - 0.58 \times 10^{-3}f^2 \tag{28a}$$

$$A_2 = 0.97 - 0.39 \times 10^{-2}f + 0.39 \times 10^{-3}f^2 \tag{28b}$$

$$B_1 = 0.31 - 0.05f + 0.87 \times 10^{-3}f^2 \tag{28c}$$

where f is in gigahertz. The values of constants A_1, A_2, and B_1 in Equation (28) should be applied at frequencies above 15 GHz. At lower frequencies, we may set

$$A_1 = A_2 = 1.0 \tag{29a}$$

$$B_1 = 0 \tag{29b}$$

Figure 3-12 shows the effect of liquid water content on the real and imaginary parts of the permittivity of wet snow according to Equations (26) to (29). The dry snow density in Figure 3-12 is defined [Ambach and Denoth, 1972]

$$\rho_{ds} = \frac{\rho_{ws} - m_v}{1.0 - m_v} \qquad (30)$$

where ρ_{ds} is the density of dry snow in g/cm^3, ρ_{ws} is the density of wet snow in g/cm^3, and m_v is the volumetric liquid water content within the snow sample. In Figure 3-12, the dry snow density is 0.25 g/cm^3, which was the average observed value in the study of Hallikainen et al. [1986].

The results in Figure 3-12 confirm that the dielectric behavior of wet snow follows that of water at 0°C; the level of ε_{ws} is determined by the liquid water content of snow.

The modified Debye-like equation presented above does not take into account the geometry of the wet snow medium (the shape of water inclusions, funicular and pendular regimes). The two-phase Polder–Van Santen dielectric mixing model [Polder and Van Santen, 1946] can be applied to wet snow by assuming that wet snow consists of dry snow as the host material with water inclusions embedded in it. The water droplets are randomly distributed and randomly oriented ellipsoids with depolarization factors A_{w1}, A_{w2}, and A_{w3}. The mixing formula is given by

$$\varepsilon_{ws} = \varepsilon_{ds} + \frac{m_v \varepsilon_{ws}}{3} (\varepsilon_w - \varepsilon_{ds}) \sum_{j=1}^{n} \left[\varepsilon_{ws} + (\varepsilon_w - \varepsilon_{ws})A_{wj} \right]^{-1} \qquad (31)$$

where m_v is the liquid water content.

According to Colbeck [1982], liquid water inclusions in snow are approximately needle-shaped ($A_{w1} = 0.0$, $A_{w2} = A_{w3} = 0.5$ for perfect needles) in the pendular regime (low values of m_v), but they become approximately disk-shaped ($A_{w1} = A_{w2} = 0.0$, $A_{w3} = 1.0$ for disks) in the funicular regime (high values of m_v). The best fit of Equation (31) to experimental data was obtained by assuming that the inclusion shapes depend on snow water content and, additionally, that the water inclusions are asymmetrical in shape ($A_{w1} \neq A_{w2} \neq A_{w3}$) [Hallikainen et al., 1986]. Transition from the pendular regime to the funicular regime was observed to take place around $m_v = 3\%$.

Figure 3-13 depicts the power absorption coefficient of wet snow, calculated using the modified Debye-like model. Even a volumetric water content of 2% results in absorption of tens of decibels per meter, depending on frequency. Figure 3-14 shows the corresponding penetration depth, indicating that in the case of wet snow, microwave radar and radiometer measurements of sea ice only provide information on the snowpack on top of the ice layer. This is discussed further in Chapters 4, 5, 16, and 17.

3.3 SCATTERING IN SEA ICE AND ITS SNOW COVER

Because sea ice and snow are physically and dielectrically inhomogeneous, i.e., members of group (c) in the nomenclature of Section 3.2, electromagnetic waves encountering these substances are not simply reflected and refracted; illuminating radiation is scattered into a range of directions. Scattering depends on the electromagnetic frequency, polarization, and direction of incidence, and may arise from irregular snow and ice boundaries, as well as interior dielectric inhomogeneities. Scattering from a given sample of sea ice also depends on the particular, random interface boundaries and arrangements of snow grains, brine pockets, etc., within the sample. Scattering is, therefore, best discussed in terms of statistical moments of scattered fields and their relationships to the statistics of properties of ice and snow.

Experimentally, we observe the fields scattered or emitted from a number of independent ice locations or samples, the properties of which are (assumed to be) identical in a statistical sense. We then average field quantities to estimate a particular moment of the scattered field for given values of frequency, incidence angle, and, in the case of radar, polarization of the illumination. For example, the backscattering cross section is proportional to the mean power scattered into a given direction, see Equation (3). It is sometimes helpful to think of the theoretical calculation

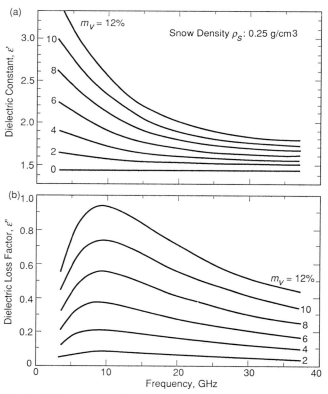

Fig. 3-12. Permittivity of snow, according to the modified Debye-like model plotted as a function of frequency and (a) dielectric constant and (b) dielectric loss factor, with liquid water content as a parameter [Hallikainen et al., 1986].

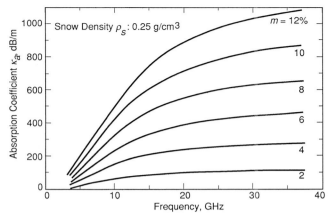

Fig. 3-13. Absorption coefficient of snow, calculated using the modified Debye-like model, Equations (26) to (28), as a function of frequency with liquid water content as a parameter.

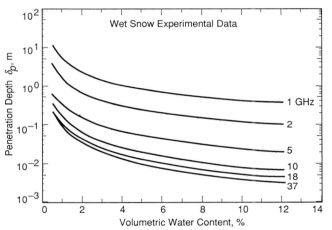

Fig. 3-14. Penetration depth for snow, computed from Figure 3-13, as a function of volumetric water content with frequency as a parameter.

of scattered field moments in terms of a thought experiment similar to actual experiments. We envision for each set of ice types and conditions an ensemble of ice samples from which we draw particular realizations, one after another. We place these in the path of an illuminating wave with specified parameters, and "observe" the resulting fields at points inside and outside the ice. The resulting sequence of fields at a given point, or propagating in a given direction, may be averaged taking into account their absolute phase (i.e., coherently), or not, to produce various theoretical field moments. Though this procedure is not actually used in theoretical computations, it facilitates useful conceptual divisions of the scattered fields into component parts (see Sections 3.3.1 and 3.3.2).

The scattered field moments, as functions of frequency, polarization, incidence angle, etc., constitute the signature of a given ice type. Recent research has provided some qualitative understanding of signatures of the main ice types and their seasonal changes. However, quantitative

relationships between ice and snow properties and signatures are complex and, at present, understood only roughly. The roles of layering in ice and snow, ice grain, brine-pocket and air-bubble size distributions, surface roughness, and other factors are topics of active research. Chapter 8 seeks to clarify the present capabilities of quantitative signature modeling, as well as the present degree of uncertainty regarding those capabilities. Here we survey the salient features of signatures for various ice types, discuss qualitatively the physics behind the observations, and indicate areas of current uncertainty and investigation. To begin, we review fundamental concepts in volume and rough surface scattering as they apply to the study of sea ice signatures.

3.3.1 Volume Scattering Fundamentals

Consider first scattering from spatial fluctuations in dielectric properties within the ice and/or snow, that is, volume scattering. It is useful to think of volume scattering as taking place in two stages. In the first stage, the incident wave excites a phase-coherent mean wave that is refracted and propagates into the snow and ice, as if the material were homogeneous and possessed an effective dielectric constant (see Figure 3-15). This mean field is that part of the transmitted radiation with a constant phase relationship to the illuminating wave, no matter what realization from the ensemble of statistically identical samples we put in its path. The effective dielectric constant is generally complex. Its imaginary part results both from absorption and from scattering of the mean field into waves that on (ensemble) average, have no consistent phase relation to that of the illumination. The terms "effective permittivity" and "dielectric constants" of snow and sea ice in Sections 3.2.3 and 3.2.4 properly refer to this effective dielectric constant.

The scattered waves that have no consistent phase relationship to the illuminating wave are collectively called the

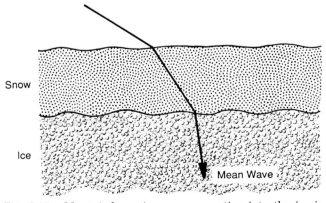

Fig. 3-15. Mean (coherent) wave propagation into the ice is governed by the effective permittivity, which in general includes forward-scattering effects. The strength of scattering depends on the lossiness of the medium and the size and dielectric constant of inhomogeneities. First-year ice is electromagnetically lossy, whereas old ice is relatively transparent and contains many effective scatterers (bubbles).

incoherent scattered field. This field typically sprays into a broad range of directions and is thus what we observe in the backscattering direction. When essentially all the incoherent field results from scattering directly out of the mean field, we say that single scattering dominates the incoherent field. When a significant part of the incoherent field is itself scattered, rescattered, and so on, we say that multiple scattering of the incoherent field is taking place. Multiple scattering occurs when there is little absorption to soak up energy and when individual scattering events redirect a significant amount of radiation. On the other hand, single scattering tends to dominate when the medium is lossy so that multiply scattered waves are strongly attenuated. Note that very different scattering regimes can result for a given effective dielectric constant, depending on whether the imaginary part is due primarily to absorption or to scattering "out of beam." In the former case, scattered radiation is quickly absorbed and volume scattering is weak, whereas in the latter case, volume scattering may be quite strong. It may be surprising that strong volume scattering does not require that a large fraction of the scattering volume be taken up by scatterers. Fractions of even a few percent can cause strong volume scattering; the level of absorption and strength of individual scattering events are the key factors.

3.3.2 Surface Scattering Fundamentals

If the surfaces separating snow, ice, and/or water from air are not plane, then radiation will also be scattered from those surfaces. In this case, incident radiation induces secondary, Huygens' sources on the surface that then reradiate. Coherent fields are reflected and refracted into the material according to Snell's law, while incoherent fields again spray off into a range of directions. Thus, as in volume scattering, backscattering from rough surfaces is incoherent (except in the case of smooth surfaces and vertical incidence). The coherent fields are diminished, compared with what they would be for plane surfaces, to make up for the incoherent energy that goes into other directions. Roughness must generally be comparable to the radiation wavelength to significantly alter the energy balance between coherent and incoherent fields. Except in certain limits, the classification of rough surface scattering as single or multiple scattering is less clear than that for volume scattering; however, the classical rough surface scattering models (first-order perturbation theory and the Kirchhoff approximation) are generally regarded as single-scattering approximations.

Scattering from surface roughness may be the dominant scattering mechanism in several circumstances. When the dielectric constant of the scattering material is large, there is little transmission into the material and thus little or no volume scattering can take place. Seawater is a good example. Water surfaces with roughness heights of even 5% to 10% of the radiation wavelength produce considerable backscattering (i.e., Bragg scattering) at microwave frequencies. Materials with much smaller dielectric con-

stants may be primarily surface scatterers if dielectric constant fluctuations within the material are weak, if fluctuations occur only on very short length scales as compared to the wavelength, or if the material is lossy.

3.3.3 Microwave Signatures and Scattering Physics

The physics underlying an active microwave signature may be quite different from the physics underlying a passive signature of the same piece of sea ice. For practical reasons, virtually all operational active observations are made in the special case of backscattering. As noted above, backscattering consists of incoherent scattered fields (except perhaps in some cases at vertical incidence, i.e., in some instances in altimetry, as discussed in Chapter 7). Thus, no matter how small the fraction of incident energy that is incoherently scattered, active signature observations depend only on such scattering. By contrast, passive signatures depend (in the case of an isothermal emitting medium) on both coherent and incoherent reflection and scattering via Kirchhoff's law [Peake, 1959]

$$\varepsilon_j\left(\theta_i\right) = 1 - \left|R_j\left(\theta_i,\ \phi_i\right)\right|^2 - \frac{1}{4\pi\cos\theta_i}\int_0^{\pi/2}\int_{-\pi}^{\pi}\sin\theta\,d\theta\,d\phi$$

$$\times\left[\sigma_{jh}^o\left(\theta,\ \phi;\ \theta_i,\ \phi_i\right) + \sigma_{jv}^o\left(\theta,\ \phi;\ \theta_i,\ \phi_i\right)\right] \qquad (32)$$

where $\varepsilon_j(\theta_i,\ \phi_i)$ is the emissivity for polarization j = H or V, at nadir and azimuthal angles of observation θ_i and ϕ_i, respectively; $|R_j|^2$ is the Fresnel power reflection coefficient for the same polarization; and $\sigma_{jv}^o\ (\theta,\ \phi;\ \theta_i,\ \phi_i)$ and $\sigma_{jh}^o\ (\theta,\ \phi;\ \theta_i,\ \phi_i)$ are the incoherent (bistatic) differential scattering cross sections in the scattering direction $(\theta,\ \phi)$ for H- and V-polarized radiation incident from the direction $(\theta_i,\ \phi_i,)$ respectively, see Equation (3). In cases where incoherent scattering is weak, the reflection coefficient, and thus the effective dielectric constant, determines the emissivity, and passive signatures show little effect of such scattering. As the strength of scattering increases, incoherently scattered intensity and thus active signatures tend to increase. However, a better scatterer typically makes a poorer absorber. According to Kirchhoff's law, the material therefore also makes a poorer emitter; as scattering increases, brightness temperatures for a given physical temperature typically decrease. In addition, multiple scattering tends to decrease the difference between vertically and horizontally polarized brightness temperatures because of the disorder it creates. Based on the physical and dielectric characterizations of sea ice in Chapter 2 and this chapter, we can now make some general comments on the microwave signatures and underlying scattering processes for seawater and several sea ice types.

3.3.3.1. Seawater. Seawater has a power reflection coefficient on the order of 0.5 in the microwave regions due to its large dielectric constant. Scattering from open water, therefore, takes place at the surface. The microwave intensity backscattered from water increases with increasing local wind because wind roughens the water surface. On the other hand, microwave transmission into seawater and the proportion of incident radiation that is absorbed are relatively small. Thus, water is a poor emitter in the microwave region of the spectrum. Typical brightness temperatures are on the order of half the physical temperature, and horizontally polarized brightness temperatures fall substantially below their vertically polarized counterparts [Ulaby et al., 1982; Onstott et al., 1987]. Although roughness must be considerable to cause a noticeable change in emission, wind does roughen water surfaces sufficiently to impact passive microwave estimation of sea ice concentration and type (see Chapter 10).

3.3.3.2. New, gray, and pancake ice types. Consider new ice, ranging from a few centimeters to 10–20 cm thick, forming on the ocean surface (we refer in this section only to new ice forming under cold conditions and defer discussion of melting ice to Section 3.3.3.6). As the ice forms, it initially retains a considerable amount of brine. Its temperature cannot decrease much below that of the seawater in which it floats. New ice is, therefore, characterized by a high-fractional brine volume and an effective dielectric constant that, while considerably smaller than that of seawater, is large relative to that of thicker ice. New ice transmits more microwave radiation across its upper surface than does seawater, but it also strongly absorbs that radiation.

Passive signatures of new ice show vertically polarized brightness temperatures within a few percent of physical temperatures, even for ice only a few centimeters thick. Horizontally polarized signatures are 10% to 15% lower, but are much higher than those for seawater. Investigations to date indicate a minimal effect of scattering on passive new ice signatures between 6 and 37 GHz, except in the case of pancake ice, which shows considerable surface roughness effects (see Chapter 14). However, at least some passive signatures of undeformed new ice at 6.7 and 10 GHz appear to be affected by ice thickness and the profile of brine volume versus depth within the ice (see Chapter 8). This is not a scattering effect, but rather it is due to a modulation of the reflection coefficient by the depth profile of the effective permittivity.

Backscattering from new ice ranges from strong for deformed and/or rough ice (i.e., ice with a rough upper surface) to weak for bare, undeformed new ice, at least in the absence of an effect attributed to frost flowers (and described in the next subsection). Strong backscattering is associated with rough surface scattering from the air–ice or snow–ice interface [Livingstone and Drinkwater, 1991; Bredow and Gogineni, 1990]. However, the mechanisms behind weak backscattering from bare, undeformed new ice remain unclear. In at least one well-characterized case, scattering from the air–ice interface alone is insufficient to explain the observed level of backscattering at 5 GHz ([Bredow and Gogineni, 1990]; see also Chapter 8). Volume scattering

models predict a different polarization dependence than is observed (again, see Chapter 8). Some additional scattering physics, such as scattering from internal ice layer boundaries or the ice–water interface, may be needed to explain these observations. The effect of snow cover on backscattering from new ice, as well as backscattering mechanisms at frequencies above 10 GHz, are also topics of current research.

3.3.3.3. Frost flower effects. A particularly puzzling phenomenon appears in connection with frost flowers on ice in leads. Backscattering at 5 and 10 GHz from ice with thicknesses of 10 to 30 cm increases sharply over values for thinner ice, nearly reaching the high values characteristic of old ice (see below). Backscattering at these frequencies then decreases by roughly 5 to 7 decibels with increasing thickness, stabilizing at typical values for thick first-year ice. This phenomenon has been observed by U.S., Canadian, and Soviet investigators [R. G. Onstott, personal communication, Environmental Research Institute of Michigan, Ann Arbor, 1990; L. Arsenault, personal communication, 1990; T. C. Grenfell, personal communication, University of Washington, Seattle, 1990], and all associate its occurrence with frost flowers in some way. Speculation on the mechanism behind this phenomenon includes scattering from a rough effective surface produced by brine wicking in the frost flowers and some kind of volume scattering effect by ice crystals in the frost flowers. The latter would be difficult to understand given the transparency of dry snow at 5 and 10 GHz. However, there have to date been no tests of any of the hypotheses, so physics of this important phenomenon remain obscure. A schematic plot of backscattering versus the stage of first-year ice development is given in Chapter 5.

3.3.3.4. First-year ice. As the ice grows through the stage known as gray ice and into first-year ice, it loses brine, its surface grows colder and usually rougher, and the ice often acquires a snow cover. The lower brine volume, the consequently lower electromagnetic absorption, and increased roughness tend to increase scattering. Brightness temperatures decrease only a few percent as the ice ages to become thick first-year ice, but the difference between vertically and horizontally polarized brightness temperatures decreases noticeably. Backscattering cross sections for first-year ice are roughly 5 decibels higher than for new ice at frequencies from 1 GHz up to at least 10 GHz [Onstott et al., 1987; Livingstone et al., 1987]. There is presently little agreement within the ice signature modeling community concerning the relative importance of surface versus volume scattering in causing these observed changes. Even a thin snow cover is sometimes observed to increase backscattering and lower brightness temperatures significantly ([Kim et al., 1984]; see also Chapter 4). Volume scattering in snow may be strong enough at 37 GHz and higher frequencies for passive signatures to respond directly to the snow cover [Hallikainen et al., 1987]. However, it appears that backscattering effects at typical SAR frequencies (5 and 10 GHz) must result from the formation of slush at the snow–ice interface rather than from the very weak volume scattering in snow at those frequencies.

3.3.3.5. *Old ice.*

Old ice that has survived the previous summer melt season is generally also present in the Arctic. This is true to a much lesser degree in the Antarctic, and old ice there is morphologically less distinct from Antarctic first-year ice (see Chapter 2). Thus our discussion here of old ice will concern only old ice in the Arctic. The upper part of old ice consists generally of fresh, raised areas (typically hummocks) with bubbly, low-density upper layers and lower lying, higher density areas that are evidently refrozen melt ponds. These melt pond areas are most often fresh, but may also be saline. Snow cover is generally thickest on old ice, but also highly variable. All these factors lead to great variability in old ice signatures ([Grenfell, 1992]; see also Chapters 5 and 8). The millimeter-sized bubbles and low absorption in the upper part of old ice lead to strong volume scattering. The average backscattering cross section of old ice is therefore much stronger than that of first-year ice at frequencies of 5 GHz and above. Passive signatures show only small polarization differences and decreasing brightness temperatures with increasing frequency, due to the increased strength of the scattering bubbles at shorter wavelengths [Onstott et al., 1987]. Melt pond areas tend to show lower backscattering and higher brightness temperatures more characteristic of first-year ice [Grenfell, 1992]. Observations of old ice that has been flooded because of its snow burden also show signatures like those of first-year ice (see Chapter 4). Thus, it appears that strong volume scattering in the bubbly, low-density upper layer of old ice is the most important single process in forming old ice signatures. The relative importance of scattering in snow versus the air bubble layer on old ice, the significance of surface scattering versus frequency, and impedance matching by snow cover (due to its intermediate permittivity between air and ice) are topics of current investigation (see Chapter 8).

3.3.3.6. *Spring melt onset and summer ice.*

The spring appearance of even small quantities of liquid water in snow and on the ice drastically alters microwave signatures. As discussed in Section 3.2.4.2 and Chapters 16 and 17, volumetric snow wetness of even 1% to 2% increases absorption sufficiently to make even 10 cm of snow virtually opaque in most of the microwave spectrum. A thin layer of wet ice can effectively mask the underlying ice. Thus, a strong scatterer, such as old ice, displays a marked drop in backscattering and a rise in brightness temperatures to near first-year ice levels. The change in first-year ice is less drastic and not necessarily monotonic [Livingstone et al., 1987; Onstott and Gogineni, 1985]. First-year ice passive signatures change little upon melting, although some increase in the polarization difference at 6 and 10 GHz is observed in advanced melt conditions [Onstott et al., 1987]. Signatures through the summer vary erratically, evidently because of

changes in melt pond coverage, local freeze–thaw cycles, and other effects, but show some trends [Carsey, 1985].

3.3.3.7. *Fall freeze-up.*

Finally, as colder weather in the fall freezes the upper surface of any remaining ice, the absorption due to liquid water disappears and volume scattering in porous layers appears (or reappears) [Carsey, 1985]. Diurnal freeze–thaw cycling is likely to take place. Snow is also likely to fall during this period. Its thermal insulation may slow or even reverse the freezing trend. However, signature data for the freeze-up period are sparse. Thus, little can presently be said concerning this period.

3.4 SUMMARY

Sea ice and snow are physically and dielectrically inhomogeneous media consisting of pure ice, brine, air, and during the melt season, fresh water. The large dielectric constants and loss factors of fresh water, seawater, and brine cause those materials to strongly affect the dielectric properties and signatures of sea ice.

The dielectric properties of seawater and brine are relatively well-known as functions of salinity, temperature, and microwave frequency. The real part of the permittivity of pure ice is exceptionally constant over the entire microwave range at a value of 3.17. The imaginary part of the permittivity of pure ice is small, but present uncertainties in its dependence on temperature and frequency are significant. Our knowledge of sea ice permittivities is imprecise, especially as concerns high microwave frequencies and the directional anisotropy (i.e., the tensor character) of congelation ice. In dry snow, the real part of the permittivity is controlled by snow density at all microwave frequencies, whereas the dominant influence on the imaginary part changes from absorption below 10 GHz to scattering at 15 GHz and above. The imaginary part of the permittivity of wet snow depends strongly on volumetric snow wetness and frequency due to the microwave properties of fresh water. Absorption in wet snow is large even for a volumetric snow wetness of 2%.

The major distinctions and changes in ice microwave signatures are generally understandable and involve strong versus weak scattering and the large difference between the dielectric properties of ice versus water. However, it is clear that in a number of cases we still have fundamental uncertainties about the roles and relative importance of various scattering mechanisms. Our ability to compute signatures quantitatively is limited; in fact, even the limits on precision and on the ice type/condition scenarios that we can treat remain unclear. Quantitative signature modeling and the current state of knowledge are addressed in more detail in Chapter 8, and special cases are discussed for thin ice (Chapter 14), snow (Chapter 16), ice in the warmer seasons (Chapter 17), and low salinity ice (Chapter 20).

REFERENCES

Ambach, W. and A. Denoth, Studies on the dielectric properties of snow, *Zeitschrift für Gletscherkunde und Glazialgeologie, bd. VIII, heft 1–2*, pp. 113–123, 1972.

Anderson, D. L., The physical constants of sea ice, *Research Applications in Industry, 13*, pp. 310–318, 1960.

Arcone, S. A., A. G. Gow, and S. McGrew, Structure and dielectric properties at 4.8 and 9.5 GHz of saline ice, *Journal of Geophysical Research, 91(C12)*, pp. 14,281–14,303, 1986.

Assur, A., *Composition of Sea Ice and Its Tensile Strength*, Research Report 44, U.S. Army Snow and Ice and Permafrost Research Establishment, Wilmette, Illinois, 1960.

Bogorodokii, V. V. and G. P. Khoklov, Electrical properties of ice in the ice edge zone of the Bering Sea at 10 GHz frequency, *Proceedings of the Final Symposium on the Results of the Joint Soviet–American Expedition, Leningrad, May 1974*, pp. 219–233, 1975.

Bredow, J. and S. P. Gogineni, Comparison of measurements and theory for backscatter from bare and snow-covered saline ice, *IEEE Transactions on Geoscience and Remote Sensing, 28(4)*, pp. 456–463, 1990.

Carsey, F. D., Summer Arctic sea ice character from satellite microwave data, *Journal of Geophysical Research, 90(C3)*, pp. 5015–5034, 1985.

Colbeck, S. C., The geometry and permittivity of snow at high frequencies, *Journal of Applied Physics, 53*, pp. 4495–4500, 1982.

Colwell, R. N., editor, *Manual of Remote Sensing, Vol. I*, American Society of Photogrammetry, Falls Church, Virginia, 1983.

Cumming, W., The dielectric properties of ice and snow at 3.2 cm, *Journal of Applied Physics, 23*, pp. 768–773, 1952.

Debye, P., *Polar Molecules*, Dover, New York, 1929.

Evans, S., Dielectric properties of ice and snow—A review, *Journal of Glaciology, 5*, pp. 773–792, 1965.

Gloersen, P. and J. K. Larabee, *An Optical Model for the Microwave Properties of Sea Ice*, NASA TM-83865, NASA/Goddard Space Flight Center, Greenbelt, Maryland, 1981.

Gray, D. M. and D. H. Male, editors, *Handbook of Snow: Principles, Processes, Management and Use*, Pergamon Press, Elmsford, New York, 1981.

Grenfell, T. C., Surface-based passive microwave studies of multiyear sea ice, *Journal of Geophysical Research—Oceans, 97(C3)*, pp. 3485–3501, 1992.

Hallikainen, M. T., *Dielectric Properties of NaCl Ice at 16 GHz*, Report S-107, Helsinki University of Technology, Radio Laboratory, 37 pp., Espoo, Finland, 1977.

Hallikainen, M. T., A new low-salinity sea ice model for UHF radiometry, *International Journal of Remote Sensing, 4*, pp. 655–681, 1983.

Hallikainen, M. T., F. T. Ulaby, and M. Abdelrazik, Dielectric properties of snow in the 3 to 37 GHz range, *IEEE Transactions on Antennas and Propagation, AP-34*, pp. 1329–1340, 1986.

Hallikainen, M. T., F. T. Ulaby, and T. E. Van Deventer, Extinction behavior of dry snow in the 18 to 90 GHz range, *IEEE Transactions on Geoscience and Remote Sensing, GE-25(6)*, pp. 737–745, 1987.

Hallikainen, M. T., M. Toikka, and J. Hyyppä, Microwave dielectric properties of low-salinity sea ice, *Proceedings of the IGARSS'88*, pp. 419–420, 1988.

Hoekstra, P. and P. Cappillino, Dielectric properties of sea and sodium chloride ice at UHF and microwave frequencies, *Journal of Geophysical Research, 76*, pp. 4922–4931, 1971.

Kim, Y.-S., R. G. Onstott, and R. K. Moore, The effect of a snow cover on microwave backscatter from sea ice, *IEEE Journal of Oceanic Engineering, OE-9(5)*, pp. 383–388, 1984.

Klein, L. A. and C. T. Swift, An improved model for the dielectric constant of seawater at microwave frequencies, *IEEE Transactions on Antennas and Propagation, AP-25*, pp. 104–111, 1977.

Lane, J. and J. Saxton, Dielectric dispersion in pure polar liquids at very high radio frequencies, II. The effect of electrolytes in solution, *Proceedings of the Royal Society, 214A*, pp. 531–545, 1953.

Linlor, W. I., Permittivity and attenuation of wet snow between 4 and 12 GHz, *Journal of Applied Physics, 51*, pp. 2811–2816, 1980.

Livingstone, C. E. and M. R. Drinkwater, Springtime C-band SAR backscatter signatures of Labrador Sea marginal ice: Measurements versus modelling predictions, *IEEE Transactions on Geoscience and Remote Sensing, GE-29(1)*, pp. 29–41, 1991. (Correction, *IEEE Transactions on Geoscience and Remote Sensing, GE-29(3)*, p. 472, 1991.)

Livingstone, C. E., K. P. Singh, and A. L. Gray, Seasonal and regional variations of active/passive microwave signatures of sea ice, *IEEE Transactions on Geoscience and Remote Sensing, GE-25(2)*, pp. 159–173, 1987.

Mätzler, C. and U. Wegmüller, Dielectric properties of freshwater ice at microwave frequencies, *Journal of Physics D: Applied Physics, 20*, pp. 1623–1630, 1987.

Onstott, R. G. and S. Gogineni, Active microwave measurements of Arctic sea ice under summer conditions, *Journal of Geophysical Research, 90(C3)*, pp. 5035–5044, 1985.

Onstott, R. G., T. C. Grenfell, C. Mätzler, C. A. Luther, and E. A. Svendsen, Evolution of microwave sea ice signatures during early summer and midsummer in the marginal ice zone, *Journal of Geophysical Research, 92(C7)*, pp. 6825–6835, 1987.

Peake, W. H., Interaction of electromagnetic waves with some natural surfaces, *IRE Transactions on Antennas and Propagation, AP-7*, pp. S324–S329, 1959.

Poe, G., A. Stogryn, and A. T. Edgerton, *A Study of the Microwave Emission Characteristics of Sea Ice*, Final Technical Report 1749R-2, Aerojet Electrosystems Company, Azusa, California, 1972.

Polder, D. and J. H. Van Santen, The effective permeability of mixtures of solids, *Physica, 12*, pp. 257–271, 1946.

Sackinger, W. M. and R. C. Byrd, *Reflection of Millimeter Waves From Snow and Sea Ice*, IAEE Report 7203, Institute of Arctic Environmental Engineering, University of Alaska, Fairbanks, Alaska, 1972.

Sihvola, A. and J. A. Kong, Effective permittivity of dielectric mixtures, *IEEE Transactions on Geoscience and Remote Sensing, GE-26*, pp. 420–429, 1988. (Correction *IEEE Transactions on Geoscience and Remote Sensing, GE-27*, pp. 101–102, 1989.)

Stogryn, A., Equations for calculating the dielectric constant of saline water, *IEEE Transactions on Microwave Theory and Techniques, MTT-19*, pp. 733–736, 1971.

Stogryn, A., An analysis of the tensor dielectric constant of sea ice at microwave frequencies, *IEEE Transactions on Geoscience and Remote Sensing, GE-25(2)*, pp. 147–158, 1987.

Stogryn, A. and G. J. Desargeant, The dielectric properties of brine in sea ice at microwave frequencies, *IEEE Transactions on Antennas and Propagation, AP-33(5)*, pp. 523–532, 1985.

Tinga, W. R., W. A. G. Voss, and D. F. Blossey, Generalized approach to multiphase dielectric mixture theory, *Journal of Applied Physics, 44*, pp. 3897–3902, 1973.

Tiuri, M., A. Sihvola, E. Nyfors, and M. T. Hallikainen, The complex dielectric constant of snow at microwave frequencies, *IEEE Journal of Oceanic Engineering, OE-9*, pp. 377–382, 1984.

Ulaby, F. T., R. K. Moore, and A. K. Fung, *Microwave Remote Sensing—Active and Passive, Vol. II: Radar Remote Sensing and Surface Scattering and Emission Theory*, Addison–Wesley Publishing Company, Reading, Massachusetts, 1982.

Ulaby, F. T., R. K. Moore, and A. K. Fung, *Microwave Remote Sensing—Active and Passive, Vol. III: From Theory to Applications*, Artech House, Dedham, Massachusetts, 1986.

Vant, M. R., *A Combined Empirical and Theoretical Study of the Dielectric Properties of Sea Ice Over the Frequency Range 100 MHz to 40 GHz*, Technical Report, Carleton University, Ontario, Ottawa, 1976.

Vant, M. R., R. Gray, R. Ramseier, and V. Makios, Dielectric properties of fresh and sea ice at 10 and 35 GHz, *Journal of Applied Physics, 45*, pp. 4712–4717, 1974.

Vant, M. R., R. O. Ramseier, and V. Makios, The complex-dielectric constant of sea ice at frequencies in the range 0.1 to 40 GHz, *Journal of Applied Physics, 49(3)*, pp. 1264–1280, 1978.

Warren, S. G., Optical constants of ice from the ultraviolet to the microwave, *Applied Optics, 23*, pp. 1206–1225, 1984.

Weyl, P., On the change in electrical conductance of seawater with temperature, *Limnology and Oceanography, 9*, pp. 75–78, 1964.

Chapter 4. Passive Microwave Signatures of Sea Ice

Duane T. Eppler, L. Dennis Farmer, and Alan W. Lohanick
Polar Oceanography Branch Office, Naval Research Laboratory, 72 Lyme Road, Hanover, New Hampshire 03755-1290

Mark R. Anderson
Department of Geography, 311 Avery Hall, University of Nebraska, Lincoln, Nebraska 68588-0135

Donald J. Cavalieri, Josefino Comiso, and Per Gloersen
Laboratory for Hydrospheric Processes, Goddard Space Flight Center, Greenbelt, Maryland 20771

Caren Garrity
Alfred-Wegener-Institut für Polar- und Meeresforschung, Postfach 120161, Columbusstrasse, D-2850 Bremerhaven, Germany

Thomas C. Grenfell
Department of Atmospheric Sciences, University of Washington, Seattle, Washington 98195

Martti Hallikainen
Laboratory of Space Technology, Helsinki University of Technology, Otakaari 5A, 02150 Espoo, Finland

James A. Maslanik
Cooperative Institute for Research in Environmental Sciences, University of Colorado, Boulder, Colorado 80309-0449

Christian Mätzler
Institute of Applied Physics, University of Bern, Sidlerstrasse 5, 3012 Bern, Switzerland

Rae A. Melloh
Cold Regions Research and Engineering Laboratory, 72 Lyme Road, Hanover, New Hampshire 03755-1290

Irene Rubinstein
Ice Research and Development, York University, 4700 Keele Street, North York, Ontario, M3J 1P3, Canada

Calvin T. Swift
Department of Electrical and Computer Engineering, University of Massachusetts, Amherst, Massachusetts 01003

4.1 Introduction

Passive microwave research conducted over the past two decades has been directed toward understanding signatures of sea ice on disparate scales afforded by sensors located on the ice surface, in aircraft flying at a range of altitudes, and onboard satellites in space. Early work, which used data obtained chiefly from aircraft and satellite sensors, established that the three dominant surfaces in the Arctic, open water, first-year ice, and multiyear ice, could be discriminated from each other on the basis of their contrasting radiometric signatures. These observations laid the groundwork for operational algorithms now used to derive ice concentration information from satellite data (Chapter 10). A significant role was played by the surface mea-surements from field experiments [Marginal Ice Zone Experiment (MIZEX), Coordinated Eastern Arctic Research Experiment (CEAREX), and Labrador Sea Ice Margin Experiment (LIMEX); Chapter 20] and laboratory investigations [Cold Regions Research and Engineering Laboratory Experiment (CRRELEX); Chapter 9]. This subsequent work demonstrated that although a given ice type is characterized by a general signature that dominates the microwave signal, the instantaneous signature is actually a sum of scattering and emission contributions from features such as ridges, hummocks, melt ponds, the snowpack, the snow–ice interface, and relict snow and ice from previous melt episodes. The state of these secondary parameters is controlled in large part by factors related to the sequence of processes to which a floe has been subjected. Deformational regimes and meteorological conditions unique to specific regions through which floes have passed, as well as freeze–thaw cycles on diurnal and seasonal scales, affect microwave signatures.

Microwave Remote Sensing of Sea Ice
Geophysical Monograph 68
©1992 American Geophysical Union

The material presented here generally follows the progression of research from local to regional. Section 4.2 reviews and compares the general utility of surface-, aircraft-, and satellite-based passive microwave sensors for research. Section 4.3 discusses passive microwave signatures observed for different polar surfaces with each class of sensors. A discussion of signatures observed for ice features follows in Section 4.4. Section 4.5 discusses phenomena, such as snow, weather, and the seasonal cycle, that alter or mediate underlying ice signatures. Section 4.6 describes passive microwave remote sensing of freshwater ice on terrestrial lakes. Section 4.7 summarizes the state of the art and outlines key areas where additional research is needed if we are to capitalize fully on the progress of the past 20 years.

4.2 SENSORS

4.2.1 Surface-Based Sensors

Surface-based sensors offer the opportunity to determine pure-type radiances and to investigate local variability in electromagnetic characteristics of ice that occur at extremely small spatial scales. Fundamental radiometric signatures of different Arctic surfaces and features that occur within them (e.g., ridges, fractures, and melt ponds) have been obtained with a growing suite of sensors (Appendix B). Most surface-based sensors are simple radiometers used to obtain either spot measurements from stationary tripods or profiles from sled mounts and from rail mounts on ships. In some cases, radiometers have been mated to computer-driven scanning mechanisms so that images can be constructed after the fact from assemblages of radiometer traces.

The advantage of surface-based measurements is three-fold. First, physical ice properties can be documented in detail when measurements are taken. This makes possible a data set of accurate radiometric signatures related to specific and known physical properties of the surface. Second, the relatively small area encompassed by the sensor footprint (commonly between 1 and 2 m square) reduces heterogeneity within the observed surface, further enhancing interpretability of results. Third, the same surface can be imaged repeatedly from a variety of incidence angles and over time so that radiometric signatures can be characterized more completely.

4.2.2 Aircraft-Based Sensors

Aircraft sensors provide broad-area coverage at useful spatial resolution (typically between 10 and 100 m, depending on sensor characteristics and aircraft altitude) and offer the opportunity to map areal variation in radiometric signatures of ice types and ice features. High-quality imagery is obtained in darkness and through overcast. Passive microwave images obtained from aircraft cover a sufficiently broad swath to show floes and leads, and to document the general fine-scale structure of the pack. At the same time, spatial resolution is sufficient to resolve individual features such as ridges, hummocks, fractures, and melt ponds. Further, aircraft images capture the radiometric texture of various surfaces, as defined by resolution-scale variation in brightness temperature, that is necessary to fully exploit the ice classification potential of microwave sensors.

Aircraft sensors have been constructed to cover a range of frequencies (Appendix B) and have received wide use in a series of applications. As reconnaissance tools, the sensors have been used to map ice conditions in support of polar operations. Imagery gathered in support of sensor validation programs has been used to assess the accuracy of ice concentration information retrieved from satellite sensors. In research applications, aircraft imagers and profilers have been used to investigate region-scale variability in radiometric characteristics of sea ice and ice features.

Passive microwave imagery presented in this chapter was acquired with the Ka-band Radiometric Mapping System (KRMS), an airborne imager that operates at 33.6 GHz (vertical) with a bandwidth of 1.3 GHz. The minimum signal detected by the sensor as measured in laboratory bench tests is 0.05 K/s; operational sensitivity is believed to be better than 0.5 K. Dynamic range is 370 K. Calibration is obtained using an internal hot load and an external cold source. Radiances obtained with this sensor are converted to brightness temperatures using methods described by Eppler et al. [1984], Eppler and Heydlauff [1990], and Farmer et al. [1989].

4.2.3 Satellite-Based Sensors

Satellite sensors provide the capability to monitor the entire polar ice cover on a daily basis without regard to light level or cloudiness. Since 1972, three generations of spaceborne passive microwave imagers have been launched by the United States: the Nimbus-5 Electrically Scanning Microwave Radiometer (ESMR) in December 1972 [Wilheit, 1972]; the Nimbus-7 Scanning Multichannel Microwave Radiometer (SMMR) in October 1978 [Gloersen and Barath, 1977; Gloersen et al., 1984]; the Defense Meteorological Satellite Program (DMSP) Special Sensor Microwave/Imager (SSM/I) I in June 1987 [Hollinger et al., 1987]; and the DMSP SSM/I II in December 1990. Appendix B summarizes technical characteristics of these sensors. The spatial resolution of these sensors, typically about 15 to 30 km, precludes the possibility of resolving individual ice floes and leads within the ice pack. Nonetheless, information in the signal integrated over this field of view allows for reasonably accurate determination of ice concentration and ice type [Cavalieri et al., 1991; Steffen and Schweiger, 1991; Chapters 10 and 11].

4.3 PASSIVE MICROWAVE SIGNATURES OF SEA ICE

Electromagnetic properties of sea ice, which determine its passive microwave signature, change with time as the open water freezes, as the new ice sheet thickens and ages, and then as the summer melt flushes brine from the ice column and fosters development of a porous surface layer. Accumulation and metamorphosis of snow, recrystallization of ice at the snow–ice interface, and deformation of the ice sheet mediate these underlying signatures and contribute to variance associated with radiometric measurements made for different ice types.

Signatures observed for different ice types vary with the spatial resolution of the sensor. For example, whereas surface sensors typically provide point measurements of radiometric characteristics within a floe, satellite sensors measure radiation that emanates from the range of different surfaces falling within the footprint and return an integrated picture of the bulk characteristics of the pack. The discussion that follows, in which we characterize radiometric signatures of different ice types, treats surface-, aircraft-, and space-based observations separately. We begin at the ice surface with high-resolution measurements, and then discuss signatures observed with coarser resolution aircraft and satellite sensors. For descriptions of physical characteristics of these ice types, refer to World Meteorological Organization [1970], Weeks and Ackley [1982], Steffen [1986], and Chapter 2. Table 4-1 summarizes emissivities of different ice types and surface conditions addressed in subsequent discussions.

4.3.1 Observations With Surface-Based Sensors

Results included in this section were derived from radiometric observations made from sleds or from ships where the ice was accessible for direct measurements of physical properties. Sensors were located between 1 and 20 m from the surface, giving spatial resolutions from about 0.25 to 5 m. In virtually all cases of natural sea ice, fluctuations in brightness temperature and emissivity have been found at the finest spatial resolutions measured (about 0.25 to 0.50 m) (see, for example, Grenfell [1992]). Where possible, error bars are used in figures to reflect observed fluctuations. Observational frequencies range from 4.9 to 94 GHz and have been chosen to avoid regions of high atmospheric opacity (20 to 24 GHz and 40 to 80 GHz). Frequency resolution ranges from 300 to 500 MHz, depending on frequency. Observations presented here were made with $R = 50°$ and at both vertical and horizontal polarizations (V-pol and H-pol, respectively), which includes the observational parameters of ESMR, SMMR, and SSM/I.

4.3.1.1. Open water and new ice. Development of an ice sheet in turbulent seas begins when water in the mixed layer becomes supercooled and ice nucleates throughout the upper water column in the form of dispersed disk- or needle-shaped crystals called frazil. In the absence of steady wind, frazil, which is less dense than sea water, floats to the surface where it coalesces in a mat as grease ice. In cases where wind persists, frazil is herded downwind, sometimes after becoming entrained in circulation cells that are induced by wind in the mixed layer, where it accumulates at the leeward edge of the polynya or lead. Development of an ice sheet under calm conditions may be initiated without formation of significant volumes of frazil, in which case congelation ice grows essentially from the outset. Additional discussion of these processes is provided in Chapters 2 and 14.

From a radiometric perspective, water and ice represent contrasting materials. The water surface is highly reflective (so cold sky is seen by the sensor) and emits little energy, and so appears cold radiometrically (Figure 4-1). The open water signature also is characterized by strong polarization, as indicated by the large difference in emissivity observed between measurements of vertically and horizontally polarized radiation. Saline first-year ice is strongly emissive and so appears warm radiometrically (Figure 4-1). First-year ice signatures show weak polarization. This difference in polarization provides a basis for estimating the concentration of open water (and by subtraction, total ice) in coarse-resolution satellite data from SMMR and SSM/I (Chapter 10).

The fact that emissivity peaks after the water surface is covered to some thickness with ice (Figure 4-2) reflects the fact that the energy emitted from the ice sheet comes not from the surface itself but from the volume of the medium within the skin depth, which is almost entirely submerged below sea level. Liquid water on the surface, trapped between crystals in the incompletely frozen ice sheet and beneath the ice, thus continues to contribute to the microwave signature until interstitial water within the skin depth has frozen, the sheet has consolidated, and the surface has become dry. Even when freezing is complete, however, some brine will remain until the physical temperature of the ice sheet falls significantly below minimum temperatures encountered in the polar regions (Chapter 2). Therefore, the emissivity of a freezing surface and the degree to which radiation is polarized should be interpreted as an indication of the degree to which freezing is complete within the skin depth of the new ice sheet.

4.3.1.2. Nilas. Emissivities measured for grease ice are only slightly greater than those of open water and show similar strong polarization because liquid water constitutes a significant part of the medium within the skin depth (Figure 4-3). As consolidation of frazil ice within the skin depth proceeds, emissivity increases and polarization decreases in complex ways as discussed in Chapter 14. It is important to recognize that changes in radiometric signature observed with growth of the ice sheet are not simply direct measures of ice thickness.

4.3.1.3. Young ice. Change in radiometric characteristics of young ice, which is intermediate in thickness and age between nilas and first-year ice, is poorly understood and depends on sea state. For calm conditions, the maximum emissivity is reached at some thickness near 1 cm and declines slightly to first-year values as shown in Figure 4-2 and discussed in Chapter 14.

4.3.1.4. First-year ice. Emissivities measured for dry, cold first-year ice (Figure 4-1) show great consistency.

Table 4-1. Microwave emissivities of sea ice.

Ice type	Pol.	4.9	6.7	10	18.7	21	37	90	94
Water	V	0.505 (0.015)	0.513 (0.015)	0.532 (0.015)	0.570 (0.033)	0.617 (0.015)	0.662 (0.029)	0.792 (0.019)	0.753 [0.026]
	H	0.253 (0.015)		0.295 (0.020)	0.332 (0.018)	0.332 (0.018)	0.392 (0.015)	0.528 (0.022)	0.488 [0.050]
New	V	0.560		0.568	0.623		0.703	0.850	
	H	0.280		0.315	0.368		0.417	0.573	
Nilas (dark)	V			0.705	0.760		0.810	0.885	
	H		0.580	0.613	0.678		0.769	0.846	
Nilas (gray)	V		0.775	0.813	0.837		0.880	0.915	
	H		0.720	0.765	0.800		0.840	0.880	
Nilas (light)	V			0.910	0.950		0.960	0.955	
	H			0.850	0.890		0.930	0.925	
Densely packed 3-cm-thick pancakes	V	0.740	0.750	0.748	0.811	0.826	0.868	0.893	
	H	0.595		0.638	0.700	0.715	0.761	0.780	
First-year ice	V	0.935 [0.010]	0.900 (0.020)	0.924 (0.036)	0.941 (0.019)	0.960 (0.019)	0.955 (0.015)	0.926 (0.045)	0.934 (0.015)
	H	0.850 [0.065]	0.840 (0.025)	0.876 (0.021)	0.888 (0.019)	0.910 (0.020)	0.913 (0.013)	0.886 (0.031)	0.895 (0.009)
Dry multiyear	V	0.926 [0.037]	0.925 (0.020)	0.890 (0.035)	0.850 (0.068)	0.787 [0.080]	0.764 (0.079)	0.680 (0.105)	0.566 [0.061]
	H	0.865 [0.049]	0.820 (0.030)	0.817 (0.075)	0.780 (0.080)	0.635 [0.125]	0.706 (0.115)	0.650 (0.011)	0.535 [0.061]
Flooded multiyear	V			0.910 (0.056)	0.942 (0.036)		0.953 (0.015)	0.902 (0.062)	
	H			0.858 (0.080)	0.912 (0.050)		0.927 (0.040)	0.867 (0.068)	
Frozen melt pond	V		0.910 [0.050]	0.913 [0.039]	0.969 [0.030]		0.970 [0.017]	0.876 [0.069]	
	H		0.813 [0.069]	0.814 [0.061]	0.877 [0.037]		0.885 [0.036]	0.818 [0.093]	
Summer melting	V	0.945 [0.018]	0.944 [0.020]	0.960 (0.018)	0.960 (0.020)	0.960 [0.006]	0.970 (0.010)	0.953 (0.010)	0.960 (0.010)
	H	0.820 [0.020]	0.870 [0.025]	0.890 (0.025)	0.890 (0.025)	0.910 [0.012]	0.933 (0.015)	0.920 (0.020)	0.933 (0.015)
Summer frozen surface crust (extreme case)	V	0.955		0.965	0.960	0.950	0.898	0.728	
	H	0.923		0.933	0.920	0.900	0.833	0.678	

( ) = standard deviation of pooled numbers extracted from the literature, rather than individual measurements.

[ ] = standard deviation when only one source was available; numbers are missing if tolerances were not given.

This table was compiled by using data from Comiso et al. [1989], Grenfell [1986, 1992], Grenfell and Comiso [1986], Grenfell and Lohanick [1985], Lohanick and Grenfell [1986], Mätzler [1986], Mätzler et al. [1985], The Norsex Group [1983], Onstott et al. [1987], and Svendsen et al. [1987].

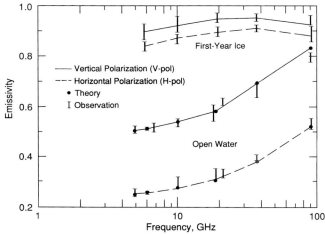

Fig. 4-1. Theoretical values for emissivity of calm water (derived using the permittivity data of Stogryn and Desargent [1985]) for field measurements of smooth to slightly rough open water made in leads and polynyas with shipboard and surface-based radiometers, and for field measurements of first-year ice plotted as a function of frequency ($R = 50°$). Error associated with open water measurements is interpreted to arise in large part from roughening of the water surface by wind, which represents the major source of variation in open water emissivity. Data were obtained in conjunction with the Norwegian Ocean Remote Sensing Experiment (NORSEX), MIZEX'83, MIZEX'84, MIZEX'87, during experiments in the Weddell and Bering Seas, and at Tuktoyaktuk and Mould Bay, Northwest Territories.

Winter-long measurements of the surface salinity of a first-year ice sheet indicate that salinity remains above 6‰ [Nakawo and Sinha, 1981; Steffen and Maslanik, 1988]. The winter brine drainage, characteristically a few parts per thousand, has no apparent influence on the radiance.

The dominant factor that contributes to variability of winter first-year ice signatures observed in aircraft and satellite data is snow. Snow-related changes in microwave signatures arise both from microwave scattering and emission unique to the snowpack and from physical changes to the first-year ice sheet induced by the snowpack (Sections 4.5.1 and 4.5.2 and Chapter 16).

4.3.1.5. Pancake ice. The emissivity of pancake ice is difficult to interpret as a pure ice type because throughout much of the growth stage, the pancakes are smaller than the sensor spatial resolution, even for surface radiometers. As discussed in Chapter 14, the net brightness of a field of pancakes can be a rough measure of ice thickness, but there is no working hypothesis as to the nature of the relationship.

4.3.1.6. Multiyear ice. Arctic multiyear floes typically are aggregates that include pieces of drained, saline, and freshwater ice. Each floe carries the record of a unique process history in the form of disparate assemblages of ice ridges, hummocks, fractures, melt ponds, and relict snow–ice surfaces. As a consequence, radiometric signatures of

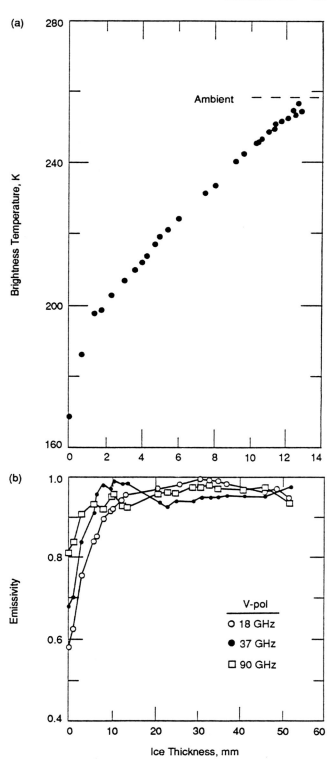

Fig. 4-2. (a) Brightness temperature (33.6 GHz, V-pol) plotted as a function of ice thickness for ice forming in a freezing lead near Barrow, Alaska [Eppler et al. 1986]. (b) Emissivity as a function of ice thickness at 18, 37, and 90 GHz (V-pol) for saline ice in a laboratory tank during CRRELEX [Grenfell et al., 1988].

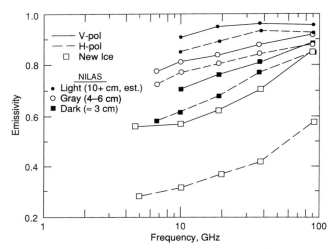

Fig. 4-3. Emissivity plotted as a function of frequency for grease ice and nilas of three different thicknesses ($R = 50°$). Data were obtained in conjunction with NORSEX and during experiments in the Weddell and Bering Seas.

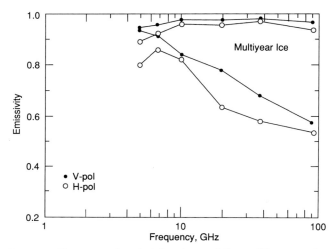

Fig. 4-4. The range of emissivity observed for multiyear sea ice plotted as a function of frequency ($R = 50°$). The envelope shown includes multiyear measurements taken over a period of years at a range of Arctic sites. Data were obtained in conjunction with MIZEX'83, MIZEX'84, MIZEX'87, NORSEX, and CEAREX'88.

multiyear ice are significantly more complex than signatures of first-season ice and show substantial variability (Figures 4-4 and 4-5) [Eppler et al., 1986; Grenfell, 1992]. Disregarding the contribution made by ridges, melt ponds, and other features to multiyear signatures (Section 4.4), multiyear ice itself has variable radiometric properties (e.g.,[Carsey, 1982, 1985]). A significant factor that contributes to this variance is the thickness and density of the highly porous surface layer that develops as sea ice undergoes melt–freeze cycling, both on a daily basis during summer months and on an annual basis. The low-density surface layer appears to develop by at least two different mechanisms, one in which melting is initiated along crystal boundaries and the other in which bubbles in the ice become melting centers. Scattering within this layer forces emissivity to decrease, an effect that is enhanced at higher frequencies [Grenfell, 1991]. In some instances unmelted snow may cap this zone, adding further complexity to the range of possible multiyear signatures.

4.3.2 Observations With Aircraft-Based Imagers

Ice signatures obtained from aircraft platforms generally are consistent both with surface measurements of ice signatures and with interpretations of evolving ice physics. Spatial resolution afforded by aircraft sensors is coarser than that of surface sensors, however, and fine-scale radiometric variability observed over short distances with surface radiometers commonly is lost as a result. Figure 4-6, for example, shows three images of the same floe of multiyear ice as observed from 300, 900, and 3000 m with a 33.6 GHz (V-pol) airborne sensor. Frozen melt ponds, which appear as radiometrically warm (dark) bodies, are clearly evident in the 300-m image. The largest, highest contrast melt ponds persist in the 900-m image, but the smaller ponds are

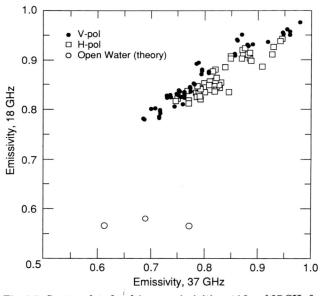

Fig. 4-5. Scatter plot of multiyear emissivities at 18 and 37 GHz for vertically and horizontally polarized radiation ($R = 50°$). Data come from measurements taken at over 200 Arctic sites. Data were obtained in conjunction with MIZEX'83, MIZEX'84, MIZEX'87, NORSEX, and CEAREX'88.

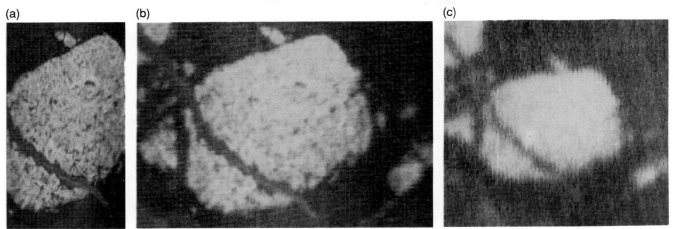

Fig. 4-6. Passive microwave images of a multiyear ice floe surrounded by first-year ice as observed from (a) 300 m (1000 ft), (b) 900 m (3000 ft), and (c) 3000 m (10,000 ft) with the Ka-band (33.6 GHz, V-pol) Radiometric Mapping System (KRMS). Light tones mark radiometrically cool areas of the floe, dark tones mark warm areas. Ground distance across the width of the scene imaged from 300 m (1000 ft) (a) is approximately 730 m (2400 ft).

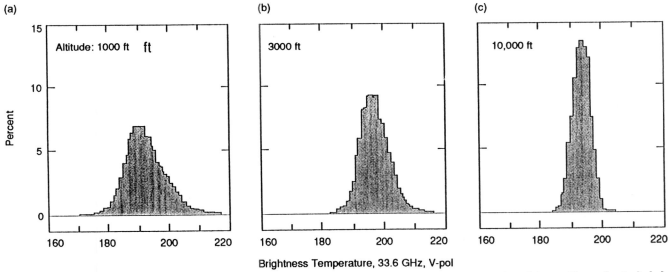

Fig. 4-7. Frequency distributions of brightness temperatures (33.6 GHz, V-pol) measured for the portion of the multiyear floe included in all three images of Figure 4-6 for (a) 300 m, (b) 900 m, and (c) 3000 m.

not resolved, although they contribute to the general mottled appearance of the floe. From 3000 m, none of the ponds appear as discrete features.

Frequency distributions of brightness temperatures measured within the portion of the floe observed at all three altitudes reflect the decrease in spatial resolution (Figure 4-7). Variance is greatest in the 300-m image, the histogram of which shows a broader, lower peak than the other distributions, and a tail toward warmer temperatures. The peak becomes progressively narrower, higher, and more symmetric as spatial resolution of the image decreases, and signatures of fine-scale features become integrated with signatures of adjacent ice across the ever larger antenna footprint.

The wide swath that aircraft imagers provide offers an enhanced view of broader scale radiometric relationships in the pack than can be observed with surface sensors. Signatures obtained with aircraft sensors, whether in the form of radiometric texture across individual ice floes, as discrete features such as ridges, fractures, and melt ponds, or as mesoscale variability in the overall character of the pack, are strongly dependent upon the resolving power of the instrument.

4.3.2.1. Open water and new ice. Figure 4-8 shows passive 33.6 GHz microwave signatures of new ice forming in a freezing polynya; Figure 4-9 is a coincident photograph of the central portion of the microwave image. Streamers and tadpole bodies of frazil, slush, and shuga mark wind-driven Langmuir circulation cells in the upper meters of exposed water. The brightness temperature of open water is 145 K, and its emissivity is 0.53, assuming a physical temperature of 1.8°C. Brightness temperatures measured for the frazil streamers range from 145 to approximately 180 K. If the physical temperature of the frazil bodies is assumed to be the same as that of the surrounding sea water, then this brightness temperature range corresponds to emissivities between 0.53 and 0.66, which is consistent with surface measurements (Figures 4-1 and 4-3).

Frazil blown downwind across the polynya has accumulated at the top left margin of the image and at right center where it has frozen into nilas. Mean brightness temperatures measured from the aircraft image for these surfaces fall between 180 and 235 K, which is intermediate between brightness temperatures typical of open water and first-year ice. Calculation of emissivity is less certain in this case because the physical temperature of neither the ice surface nor the emitting volume of ice seen by the sensor is known precisely. Inasmuch as a significant fraction of the material within the skin depth remains liquid, the mean physical temperature of the vertical section of ice seen by the sensor must be near that of freezing sea water. Using –1.8°C yields emissivities between 0.66 and 0.87 for nilas, which represent lower limits of the emissivity range.

4.3.2.2. Nilas. Aircraft observations of nilas radiances are generally consistent with surface measurements (Figures 4-2 and 4-3). Figures 4-10 and 4-11 show a lead, frozen with light nilas, in a sheet of thin first-year ice. A recently formed crack, frozen with dark nilas, crosses both the lead and adjacent first-year ice. Relatively slight differences in the visible-light image (Figure 4-11) correspond to significant changes in brightness temperature. The image shows that dark nilas is significantly cooler radiometrically than light nilas.

4.3.2.3. First-year ice. Images of first-year sea ice are characterized by signatures that display strikingly uniform radiometric textures, Figure 4-12(a), particularly when compared with those observed across multiyear floes (Figures 4-6, 4-8, and 4-10) [Eppler et al., 1986]. Figure 4-13(a) shows mean brightness temperatures measured for first-year ice in the Chukchi Sea along the 168th meridian between 73° and 76° N latitude; Figure 4-13(b) shows standard deviations associated with these means. Low values of standard deviation (typically between 1.0 and 2.0 K, and almost invariably less than 3.0 K) demonstrate the high degree of consistency and uniformity typical of radiometric signatures characteristic of individual slabs of first-year ice.

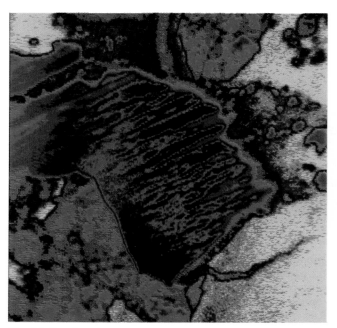

Fig. 4-8. Passive microwave image of a freezing polynya. (Figure 4-9 shows a coincident aerial mosaic of this scene.) Imaged area is approximately 3600 m across. The image was obtained from an altitude of 1525 m on March 20, 1983, with the KRMS (33.6 GHz, V-pol).

Fig. 4-9. Coincident aerial mosaic of the freezing polynya in Figure 4-14; the mosaic was obtained with an RC-10 camera. The camera was carried aboard the aircraft with the passive microwave sensor; the photograph was obtained simultaneously with the passive microwave image.

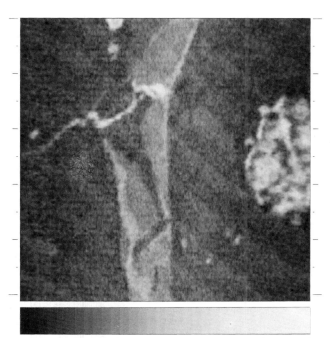

Fig. 4-10. Passive microwave image of nilas and young ice. (Figure 4-11 shows a coincident aerial mosaic of this scene.) Light tones mark radiometrically cool areas; dark tones mark warm areas. Ground distance across the scene, which was imaged from 300-m altitude with the (33.6 GHz, V-pol) KRMS sensor, is approximately 725 m.

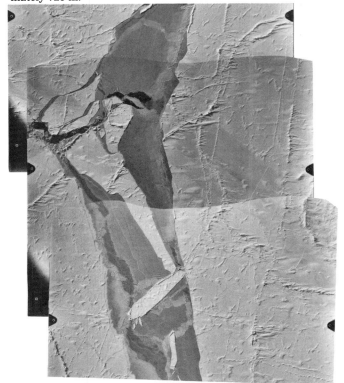

Fig. 4-11. Coincident aerial mosaic of the passive microwave image in Figure 4-10.

Brightness temperature differences observed between different slabs of ice, however, and between first-year ice in different regions can be significant. Mean brightness temperatures measured for first-year floes in the Chukchi Sea extend over a range of almost 20 K (Figure 4-13) in spite of low intrafloe variances. The mean temperature of first-year ice, measured over an area of approximately 900 km^2 near the northern coast of Alaska, was observed to be approximately 5 K warmer than that over an area of similar size farther offshore near the multiyear pack [Cavalieri et al., 1991] (Figure 4-12). Such variability probably results more from differences in scattering and emission characteristics of the snowpack and the snow–ice interface than from innate differences in radiometric properties of the underlying ice sheet [Eppler et al., 1990; see Sections 4.5.1 and 4.5.2].

4.3.2.4. Multiyear ice. Radiometric variability of multiyear ice observed in the course of surface studies (Figures 4-4 and 4-5) is manifest in aircraft data both as differences between mean brightness temperatures measured for different multiyear floes (Figures 4-14 and 4-15) and as the complex radiometric texture observed in images (Figures 4-16, 4-17, and 4-10). Figure 4-14 shows mean brightness temperatures and standard deviations measured for individual multiyear floes in the Chukchi Sea. These data come from the same image set used to derive first-year brightness temperatures plotted in Figure 4-13. The variance observed for multiyear brightness temperatures covers a greater range than that measured for first-year ice, and mean brightness temperatures of the multiyear data vary

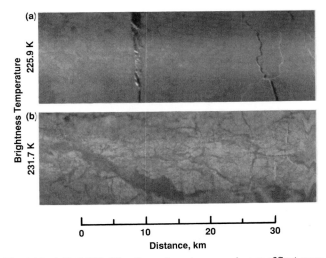

Fig. 4-12. A 33.6-GHz (V-pol) passive microwave image of first-year sea ice in the Beaufort Sea, north of Deadhorse, Alaska (taken on March 11, 1988). Radiometrically warm surfaces are dark; cool surfaces are light. The mean brightness temperature of first-year ice is shown to the right of each scene. (a) Ice closest to shore shows little spatial variability in brightness temperature and is radiometrically warm. Several recent rifts in the ice cover cross the scene from top to bottom. (b) First-year ice located farther from shore shows distinct floe structure and is radiometrically cooler than ice nearer to shore.

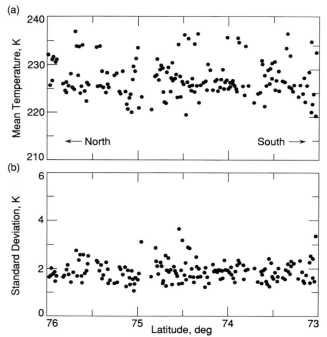

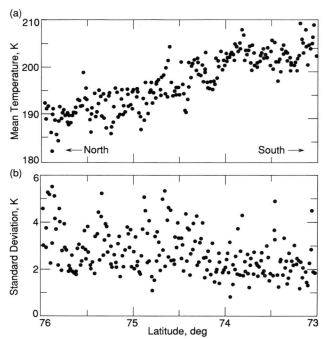

Fig. 4-13. Brightness temperature characteristics of first-year sea ice in the Chukchi Sea at 168° W longitude, measured at 33.6 GHz (V-pol) with the KRMS airborne imager. Each dot represents the mean brightness temperature measured across a different area of first-year ice. Each first-year ice area was selected to be adjacent to a multiyear floe, the mean temperatures of which are shown in Figure 4-21. (a) Mean first-year ice brightness temperature plotted as a function of latitude. (b) Standard deviation plotted as a function of latitude for these same data.

Fig. 4-14. Mean brightness-temperature (33.6 GHz, V-pol) characteristics of 214 multiyear sea ice floes in the Chukchi Sea along the 168th meridian between 73° and 76° N latitude. Each dot represents a different floe. Floes used in this analysis are adjacent to (and are paired with) first-year ice observations presented in Figure 4-13. (a) Brightness temperature plotted as a function of latitude. (b) Standard deviation plotted for these same data.

with latitude, increasing from north to south. First-year data show no such trend (Figure 4-13), which suggests that variation in multiyear brightness temperature arises not from a physical temperature gradient across the region (which should have a similar effect on first-year and multiyear ice), but from differences in the physical character of multiyear ice from one end of the imaged swath to the other.

Mesoscale variation in brightness temperature is observed elsewhere in the Arctic as well. Figure 4-15, which shows mean brightness temperature measured for 174 multiyear floes as a function of distance along a 2000-km flight track between Harrison Bay, Alaska, and Ellesmere Island, Northwest Territories, shows systematic variation in brightness temperature. Floes in some areas are radiometrically cooler than in other areas (compare points C, D, and G with points E, H, I, and J in Figure 4-15). Cavalieri et al. [1991] noted that in some instances zones in which floes are radiometrically warm coincide with areas in which the Arctic pack is highly deformed, under compressive stress, and lacks leads of significant width.

The increased range of variance associated with brightness temperatures measured for multiyear ice, Figure 4-14(b), (as compared with first-year ice) arises from

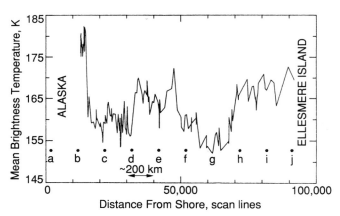

Fig. 4-15. Mean brightness temperature of 174 multiyear floes located along a 2000-km flight track between Harrison Bay, Alaska, and Ellesmere Island, Northwest Territories. Data were obtained with KRMS (33.6 GHz, V-pol).

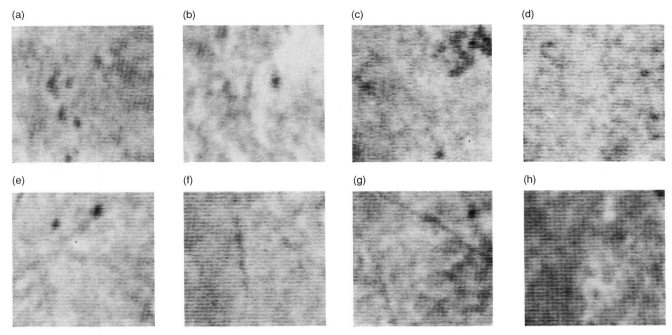

(a)   (b)   (c)   (d)

(e)   (f)   (g)   (h)

Fig. 4-16. Variation in the radiometric texture of multiyear ice at 33.6 GHz (V-pol). Light-toned areas are radiometrically cool; dark-toned areas are warm. Each image shows the central part of large multiyear floes. Floes were selected to represent as broad an area as available data would permit and show ice from different locations in the Chukchi Sea (a) 71.1° N, 194.9° E, (b) 75.5° N, 192.3° E, and (c) 73.1° N, 191.5° E; the Beaufort Sea (d) 72.5° N, 216.5° E and (e) 74.7° N, 220.2° E; and the Arctic Basin along the Canadian Archipelago (f) 77.1° N, 225.7° E, (g) 80.3° N, 238.9° E, and (h) 82.8° N, 261.7° E. All images were obtained from 6000 m between March 8 and 14, 1988. Each scene shows and area approximately 3.5 km².

the complexity of radiometric textures observed for multiyear floes. This is perhaps most evident in scenes obtained from low altitudes (900 m and below) [Figures 4-6(a) and (b) and 4-10], but can be observed in high altitude imagery as well (1500 m and higher) (Figures 4-16 and 4-17). Textures observed for multiyear ice in these high altitude scenes range from relatively featureless surfaces with low radiometric variance, Figures 4-16(d) and 4-17(d), to surfaces that show distinct features and high variance, Figures 4-16(b) and 4-17(b). The variety of textures reflect varied process histories of different floes and differences in the resulting assemblages of ridges, hummocks, melt ponds [dark, warm spots in Figures 4-6(a) and (b) and 4-16(a) and (e)], and apparent fractures [dark, warm lineaments in Figures 4-16(f) and (g)]. Variation in physical characteristics of the upper layer of partially decomposed (melted) snow and ice, which surface data show alter emissivity, also contributes to radiometric variability.

4.3.3 Observations From Satellite Sensors

Satellite passive microwave sensors provide nearly complete daily coverage of the Earth's polar regions and represent the best source of comprehensive data required for monitoring the behavior of sea ice on synoptic scales. The trade-off required to achieve complete polar coverage is that spatial resolution of these satellite sensors is limited to approximately 30 km. This places limitations on the extent to which ice characteristics can be derived from these data. At best we can determine the fractional coverage of three types of polar surfaces with current multichannel radiometers [Rothrock et al., 1988; Comiso, 1986], primarily because the sensor footprint fails to resolve individual objects (e.g., floes, leads, and ridges) required to recognize a full range of ice types and ice features.

Figure 4-18 shows microwave brightness temperatures of ice-free ocean (commonly referred to as open water), first-year ice, and multiyear ice as observed in the Arctic with the SSM/I on January 17, 1988. Each point represents the mean brightness temperature for a square area encompassed by 10×10 SSM/I grid map elements; each square grid element measures 25 km on a side. The short vertical bars represent ±1 standard deviation for the 100 map-element samples. In general, these data are consistent with surface and aircraft measurements (Figures 4-1 and 4-4), although some differences are evident. For example, at 37 GHz, satellite data show open water to be warmer radiometrically than multiyear ice, whereas both surface and aircraft data show the opposite (Figures 4-1, 4-4, and 4-8).

This discrepancy is interpreted as a function of three factors unique to the satellite observations. First, the surface within the large sensor footprint is likely to be inhomogeneous and will comprise several different surface types. Second, surface measurements of water typically are

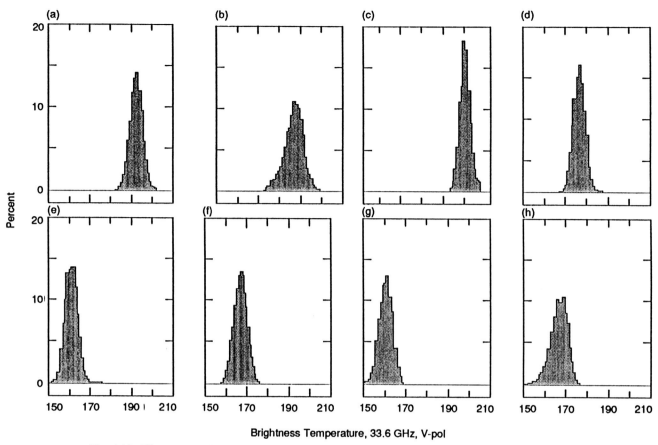

Fig. 4-17. Histograms of brightness temperatures within each of the images in Figure 4-16(a)–(h).

taken during calm periods whereas satellite measurements are likely to include foam or significant roughening of the water surface by wind and swell, all of which increase brightness temperature [Stogryn, 1972; Il'in et al., 1985; Smith, 1988]. Third, satellite sensors view the surface through the full thickness of the atmosphere, which contains more water vapor over open ocean than over ice.

Polarization and spectral characteristics of SSM/I Arctic data are shown as scatter plots in Figure 4-19. Each point represents the brightness temperature measured for one SSM/I pixel (approximately 25 km²). Polarizations of different surface types form an arrow with two distinct clusters, one at the tail (which is composed of ocean pixels) and the other at the head (which is composed of ice pixels). Good contrast between these clusters suggests that water can be discriminated from ice on the basis of polarization characteristics of the two surfaces. Spectral data for vertically polarized observations define a Z-shaped pattern that includes three clusters that correspond to open water, first-year ice, and multiyear ice, Figure 4-19(b). First-year ice and multiyear ice pixels plot as discrete clusters, suggesting that multiyear ice can be discriminated from first-year ice on the basis of spectral differences. These relationships

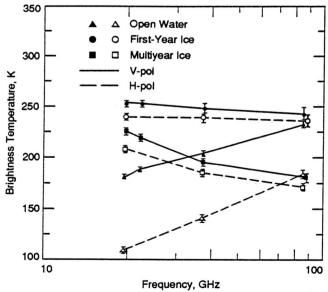

Fig. 4-18. Polarization and spectral characteristics of open water, first-year and multiyear sea ice as observed with the DMSP SSM/I on January 17, 1988.

(a)

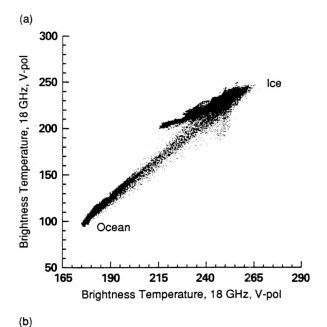

(b)

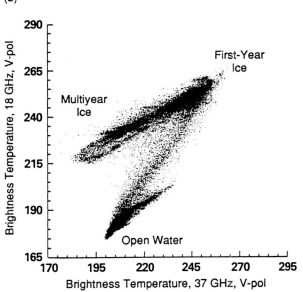

Fig. 4-19. DMSP SSM/I brightness-temperature plots for the Northern Hemisphere ocean areas poleward of 50° N for March 11, 1988: (a) 18 GHz, V-pol versus 18 GHz, H-pol; and (b) 37 GHz, V-pol versus 18 GHz, V-pol.

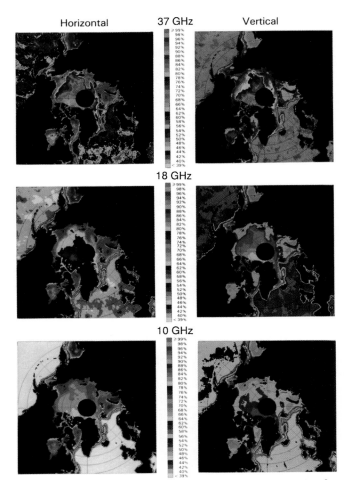

Fig. 4-20. Mean microwave emissivities of the Arctic region for March 1979, calculated for the SMMR 37-, 18-, and 10-GHz channels using methods described by Comiso et al. [1989].

from those observed near ice edges. This contrast increases with frequency, suggesting that part of the observed variability is due to enhanced scattering in snow at higher frequencies.

4.4 SEA ICE FEATURES

Within the relatively uniform background of surfaces composed of ice types discussed above occur features that are the products of deformational processes and melt. Signatures of these features, which include ridges, fractures, and melt ponds, add to the radiometric complexity and variability of the ice. Additionally, icebergs and ice islands represent features in their own right that form a small, but not insignificant, constituent in some regions. Most features that occur in the polar pack are composed of ice types discussed above. Their passive microwave signatures can be understood largely in terms of radiometric characteristics of their component parts.

form the basis of ice algorithms discussed in Chapters 10, 11, 12, and 14.

Considerable variability is observed in the signature of the pack over areas of the Central Arctic [Campbell et al., 1976, 1978; Carsey, 1982; Comiso, 1990; Cavalieri et al., 1991]. Figure 4-20 shows Arctic maps of mean effective emissivity in March 1979, for six of the 10 SMMR channels. Emissivities typical of the Central Arctic differ significantly

4.4.1 Leads and Fractures

Fracture and lead signatures can be divided into three general categories based on the stage of growth of ice that covers them and the nature of radiometric signatures that result. First, newly formed fractures that expose open water appear as radiometrically cold features with very low brightness temperatures that typically show medium to high contrast with adjacent ice (Figure 4-21). Second, fractures that are in the process of freezing and that are incompletely covered with ice are characterized by ephemeral signatures spanning a broad range of brightness temperatures that change rapidly with time as new ice accumulates (Figure 4-2). Ice in these features typically shows banded signatures oriented parallel to the lead trend (Figure 4-21) that reflect processes by which new ice accumulates in the presence of wind or currents [Eppler and Farmer, 1991]. Signatures of features in each of these first two categories are relatively short lived and display dynamic changes in both signature and radiometric contrast with their surroundings over periods of hours. The third cat-

egory, which consists of fractures that are frozen completely with ice greater than 10-cm thick, shows radiometrically warm, stable signatures that cool slowly with time. These features show high contrast with adjacent, radiometrically cool multiyear ice, but are not as easily detected where they occur adjacent to first-year ice that displays similar radiometric characteristics.

4.4.2 Ridges

Processes associated with the formation of pressure ridges determine the character of initial radiometric signatures associated with these deformational features. Signatures

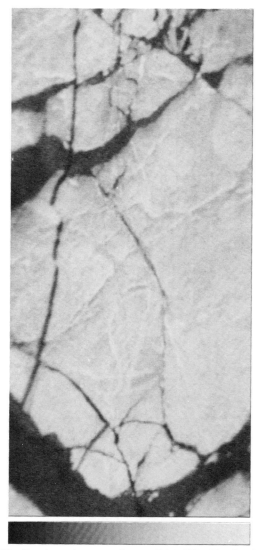

Fig. 4-22. Passive microwave image (33.6 GHz, V-pol) of radiometrically cool ridges in a multiyear floe. Image was obtained on March 20, 1983, in the Chukchi–Beaufort Sea region north of Point Barrow. Flight altitude was 1525 m. Ground distance across the scene is approximately 3600 m.

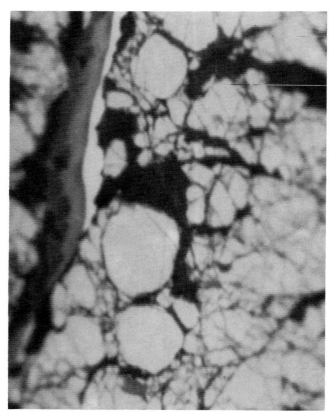

Fig. 4-21. Passive microwave image (33.6 GHz, V-pol) of a freezing lead showing banded structure created when frazil forms in exposed water (white band) and is herded to the downwind side of the lead where it accumulates as slush (gray structures that parallel the lead trend on the left side of the lead). Ground distance across the scene is approximately 14.5 km. Image obtained from an altitude of 3050 m on March 11, 1988.

of newly formed ridges can be understood in terms of ice from which ridge blocks are derived.

Ridge signatures evolve with time as the physical nature of the ice that constitutes each feature changes due to brine migration, metamorphosis, accumulation of snow, and change in radiometric characteristics of background ice. Numerous ridges are cooler than adjacent multiyear ice (Figure 4-22), which suggests that radiometric properties of ridge ice evolve differently than those of adjacent flat-lying multiyear ice by processes that are not well documented. In general, drainage of brine and changes in the ice surface and bubble distribution can be expected with concomitant radiance changes.

4.4.3 Melt Ponds

Frozen melt ponds typically are warmer radiometrically than the ice within which they occur and contribute to variance of emissivities measured for multiyear ice (Figures 4-6 and 4-7). Comparison between emissivities observed over topographically high surfaces on multiyear floes (which are probably relict hummocks or very old ridges) and topographically low surfaces (which are filled with frozen melt water) shows that the two types of surfaces have different signatures at all but the lowest frequencies (Figure 4-23).

4.4.4 Icebergs

Icebergs, which are composed of freshwater ice, are characterized by radiometrically cool signatures in passive microwave images (Figure 4-24). They form high-contrast targets when set in a field of radiometrically warm first-year ice, but are difficult to detect among multiyear floes if brightness temperature alone is examined. Because of

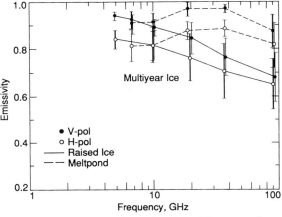

Fig. 4-23. Emissivity plotted as a function of frequency for raised ice on multiyear floes and for melt ponds ($R = 50°$). Note that the decline in emissivity for melt pond data at 90 GHz is attributed to the snow cover, which typically was between 60- and 80-mm deep. Emissivities rose significantly when the snow was removed. Melt pond data were obtained in conjunction with CEAREX'88; raised ice data are from NORSEX, MIZEX'87, and CEAREX'88.

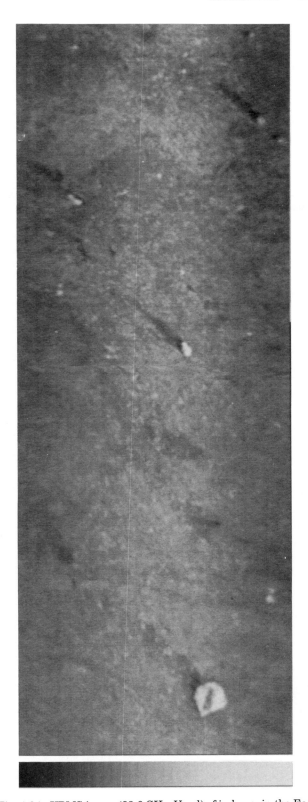

Fig. 4-24. KRMS image (33.6 GHz, V-pol) of icebergs in the East Greenland drift stream off the east coast of Greenland in March 1987.

their deep draft and high above-water profile, icebergs are subjected to different wind and current stresses than adjacent ice, and commonly drift at different rates. They also respond more slowly to changes in winds and currents due to their large mass. As a result, passive microwave images commonly show a radiometrically warm wake of newly formed ice mixed with brash and broken sea ice trailing behind each berg, created as the pack parts to flow around the obstruction, and then freezes behind it.

4.5 MEDIATING EFFECTS

Fundamental signatures of different ice types and signatures of the features they contain are affected by physical phenomena that occur between the ice surface and the passive microwave sensor. Snow and the interface between snow and the underlying sea ice sheet both have signatures in their own right and also mediate the signal emanating from underlying ice through processes of scattering and absorption. The atmosphere emits and absorbs radiation, the intensity of which depends on its temperature, thickness, and moisture content. Storms drop precipitation and alter the physical character of ice and snow. Ambient air temperature, which determines temperature gradients through snow and ice, not only varies with the passage of weather systems, but also fluctuates on diurnal and seasonal scales.

4.5.1 Snow

Lohanick [1990] measured brightness temperature as a function of distance on the surface of multiyear ice with a nonuniform snow cover and showed that snow could mask underlying ice features, in this case frozen melt ponds (Figure 4-25). Snow thus makes potentially significant contributions to microwave signatures. An assessment of the relative importance of a snow cover in a particular situation can be obtained using two quantities: the average thickness of snow cover on sea ice, and the penetration

depth of microwave radiation in snow. Chapter 2 discusses snow depth on sea ice in various regions.

Onstott et al. [1987] summarized measurements and calculated values of penetration depths in moist snow at frequencies up to 37 GHz, and Kuga et al. [1991] modeled scattering in moist snow at 95 GHz. Mätzler [1985] measured attenuation and scattering coefficients in dry snow. Results of the three studies are shown in Figure 4-26. Dry-snow values from Mätzler [1985] would not be significantly affected by including scattering at frequencies up to and including 37 GHz, but scattering would rapidly become important at higher frequencies.

Considering optical thicknesses greater than about 0.5, a snow cover will significantly affect the brightness temperature of sea ice for penetration depths less than the thickness of the snow. For typical sea ice snow covers in the range of hundreds of millimeters, dry snow affects the brightness temperature at frequencies above about 30 GHz, and moist snow with a water content of only 1% affects brightness temperature at virtually all microwave frequencies (Figure 4-26). Grenfell and Lohanick [1985] experimentally determined attenuation coefficients in moist snow at 10 GHz to be between about 3 and 4 dB/m. Comparison with the 10-GHz curve seen in Figure 4-26 shows snow-moisture values in the range of 1% to 2%.

Grenfell [1986] made surface radiometer observations of ice between 100- and 400-mm thick in the marginal ice zone and reported averaged brightness temperatures as a function of frequency for dry-snow cover in three thickness ranges. He found that thinner snow showed a generally flat spectrum, while thicker snow had a negative slope, consistent with more volume scattering expected in the snow at higher frequencies (Figure 4-27). Comiso et al. [1989] observed brightness temperatures of sea ice covered with dry snow as the snow was removed in layers. The 90-GHz brightness temperature of thicker snow rose by tens of kelvins as the snow was removed, but changed little at lower frequencies, a result that agrees with calculations of attenuation which

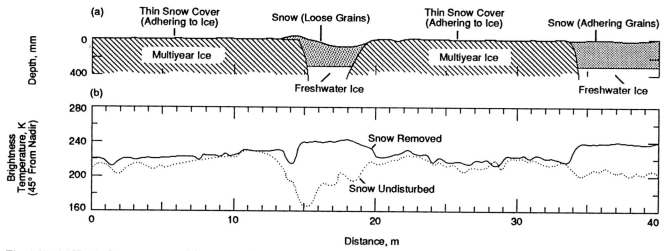

Fig. 4-25. (a) Vertical cross section of the 40-m multiyear ice study site with a nonuniform snow cover, and (b) brightness temperature at 37 GHz along the site, before and after snow removal [Lohanick, 1991].

include volume scattering in the snow at 90 GHz. In the laboratory, Grenfell and Comiso [1986] measured brightness temperature at several frequencies before and after a snowfall on thin saline ice. The brightness temperature did not change significantly at 10 GHz, but generally dropped at higher frequencies, probably due to volume scatter caused by the snow grains, which had maximum dimensions of 5 mm (in needles) near the top of the snow and about 2 mm (in more rounded grains) near the bottom (Figure 4-28). Some detailed studies of snow-cover effects are presented in Chapter 16.

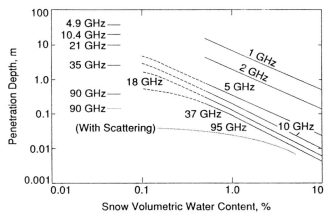

Fig. 4-26. Penetration depth in snow of microwave radiation as a function of snow volumetric water content for several frequencies. Solid portions of curves are replotted from Onstott et al. [1987]. Interpolated (dashed) portions of curves are based on the dry-snow values from Mätzler [1985], without scattering. Dotted lines are: 95 GHz from Kuga et al. [1991], and dry snow at 90 GHz from Mätzler [1985], both with scattering. Snow densities for the studies were 350 to 400 kg/m³. Kuga et al. [1991] used an average snow grain diameter of 1 mm.

4.5.2 Flood and Meltwater at the Snow–Ice Interface

When snow falls on bare sea ice, its interaction with the substrate is determined by snow type (which depends largely on air temperature and humidity), by its depth and density (snow limits the rate at which heat can leave the ice, in effect trapping heat near the snow–ice interface, and its weight can deform or submerge the ice), and by the condition of the upper layers of the ice (the availability of salts to affect the melting point). The interaction of snow with underlying ice and salts determines the dielectric constant (the mixing of the constituent dielectrics) and roughness of the interface layer (in the present context actually a dielectric roughness), and thus may significantly change the microwave signature of the ice, even if the snow cover is optically thin. Lohanick and Grenfell [1986] found that brightness temperature did not correlate well with snow depth for depths up to 300 mm on cold first-year ice (Figure 4-29).

The snow–ice interface typically is modeled as a surface, thus having only the property of roughness (a rough, mathematically continuous boundary, separating two materials having different dielectric constants). Microwave field observations, which are discussed below, suggest an expansion of this definition to include a relatively ill-defined volume that includes all physical effects due to the contact of saline ice with snow, which would not have occurred in either material under other circumstances. For example, the presence of salt in the ice determines the melting point of the slush that forms during flooding of the ice surface, and the snow thickness determines the amount of flooding. Likewise, the temperature of the slush, and thereby the rate at which the slush freezes, is also determined by the amount of snow (as a direct result of its thermal insulating properties). The slush layer, which forms at the contact between snow and ice, is therefore specific to the interaction of saline ice with its snow cover.

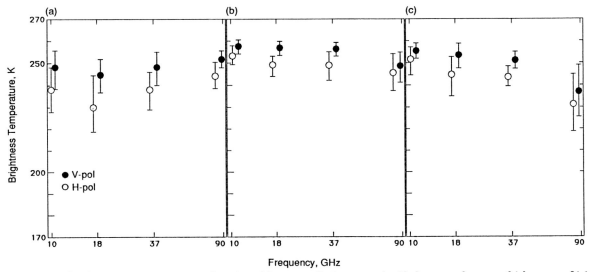

Fig. 4-27. Average brightness temperature as a function of frequency for ice covered with dry snow for snow thicknesses of (a) <3 mm, (b) 3 to 50 mm, and (c) >50 mm [Grenfell, 1986].

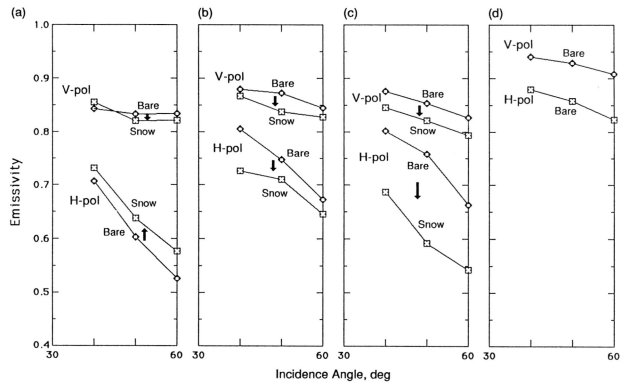

Fig. 4-28. Brightness temperature as a function of incidence angle at frequencies of (a) 10 GHz, (b) 18 GHz, (c) 37 GHz, and (d) 90 GHz, before and after a 45-mm snowfall on thin saline ice. Snow density increased from 120 kg/m^3 near the top to 350 kg/m^3 in the slushy 20-mm-thick bottom layer. Snow grains near the top were needles about 0.1 mm by 0.2 to 5 mm, and near the bottom were moist grains 1 to 2 mm in diameter [Grenfell and Comiso, 1986].

To isolate the contribution of the snow–ice interface to the observed brightness temperature, Lohanick [1990] studied a 10-m-long snow-covered section of multiyear ice at 33 GHz (Figure 4-30). Observations were made with a surface radiometer with the snow in place. Then all snow was removed except for the dense metamorphosed bottom layer in one segment of the experiment site. The brightness temperature of the segment in which the metamorphosed layer was left intact did not change until the layer was removed, indicating that it alone was contributing about 20 K to the brightness temperature. Removal of the meta-morphosed layer smoothed out the brightness temperature profile to that of bare ice.

Lohanick [1991] observed a 20-cm snowfall on undeformed 25-cm-thick saline ice, and the changes in the snow and interface layer that occurred over the subsequent 3-week period. Air temperature did not rise above about –5°C during this time. He obtained brightness temperatures at 10 and 85 GHz with surface-based radiometers. A slush layer, which formed at the base of the snow shortly after snowfall began, caused the 10-GHz brightness temperature to drop about 100 K from its bare-ice values (Figure 4-31). As the slush layer froze into an added frazil ice layer over a 7-day period, the 10-GHz brightness temperature rose to near blackbody values (higher than the original bare-ice ones). This laboratory situation apparently simulated

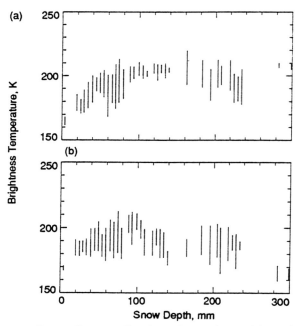

Fig. 4-29. Range of horizontally polarized (a) 10-GHz and (b) 37-GHz brightness temperatures (vertical lines) as a function of snow thickness along a test site on first-year ice in winter [Lohanick and Grenfell, 1986].

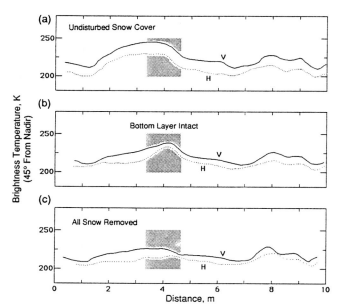

Fig. 4-30. Vertically polarized brightness temperature at 33.6 GHz and a 45° incidence angle, along a 10-m site on multiyear ice with a nonuniform dry snow cover. (a) Brightness temperature for undisturbed snow. The original snow thicknesses ranged from 50 to 270 mm. (b) All snow has been removed except for a 50-mm-thick layer in an area that was originally covered by 250 mm of snow. (c) The 50-mm-thick layer is also removed, showing that it contributed about 20 K to the original brightness temperature at that point in the site [Lohanick, 1990].

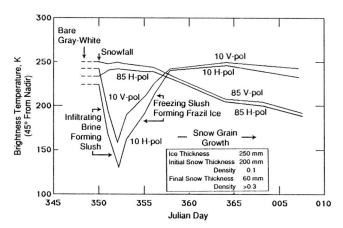

Fig. 4-31. Brightness temperature at 10 and 85 GHz as a function of time before (dashed lines) and after (solid lines) the first snowfall on undeformed gray-white ice. Snow thickness decreased from 200 to about 60 mm, and its density increased from 85 to 145 kg/m³, during the period shown. The 100 K decrease in 10-GHz brightness temperature coincided with the presence of a slush layer that formed when highly saline brine, forced up through the ice by the snow load, mixed with the grains in the lowest 10 mm of the snow layer. The slush layer froze into a superimposed frazil ice layer by JD 358 [Lohanick, 1991].

snowfall on undeformed gray-white ice, and the formation of an added frazil ice layer to the top ice surface as observed by Gow et al. [1990], on a freezing lead during the CEAREX drift phase.

4.5.3 Atmospheric Effects

Atmospheric conditions affect satellite-derived ice measurements [Gloersen and Cavalieri, 1986; Maslanik, 1991]. Several authors suggest that short-term changes in ice concentration and ice type retrieved from satellite data are related to disruption of the ice pack by seasonally varying synoptic weather systems [Anderson and Crane, 1984; Cahalan and Chiu, 1986; Zwally and Walsh, 1987; Serreze et al., 1990]. Ice concentration retrievals also are affected by attenuation and emission characteristics of the atmosphere itself.

Attenuation of the microwave signal by precipitation causes more severe problems, particularly at higher frequencies. As seen in Figure 4-19(b), which plots SSM/I 37 GHz V-pol against 19 GHz V-pol, the open water cluster is smeared, a phenomenon that is interpreted to arise from attenuation and emission by cloud liquid water. The spread is greater at 37 GHz than at 19 GHz due to greater influence of cloud liquid water on higher frequencies. In the marginal ice zone, ice-edge information can be retrieved for rain rates less than 10 mm/hour within acceptable limits of error. For precipitation rates higher than 10 mm/hour, atmospheric attenuation confounds surface-type interpretation and ice-concentration retrievals become unreliable.

4.5.4 Seasonal Variation

Brightness temperatures and emissivities derived from satellite sensors vary seasonally [Zwally et al., 1983; Comiso, 1983, 1990; Cavalieri et al., 1984; Parkinson et al., 1987]. Figure 4-32 shows patterns of change in the minimum and maximum SMMR 18- and 37-GHz brightness temperatures over one annual cycle for a 600×600 km area in the central Arctic north of the Canadian Archipelago in which multiyear ice is the dominant surface type. These satellite observations generally agree with observations made using surface instruments and aircraft sensors [Troy et al., 1981; Hollinger et al., 1984; Grenfell and Lohanick, 1985; Grenfell and Comiso, 1986; Livingstone et al., 1987a, 1987b; Onstott et al., 1987].

Livingstone et al. [1987b] suggested that the sea ice year can be divided into five stages, roughly equivalent to seasons, on the basis of surface processes and the character of microwave signatures that they produce: winter, initial warming (early melt), melt onset, advanced melt, and freeze-up. Surface wetness, ice melt, snow cover, new ice growth, and ice surface temperature are the main factors that combine to define these ice seasons. Carsey [1989] reviewed several of these stages relative to remote sensing.

In the following discussion, we describe the annual brightness-temperature cycle shown in Figure 4-32 in the context of Livingstone's ice seasons. Several points should

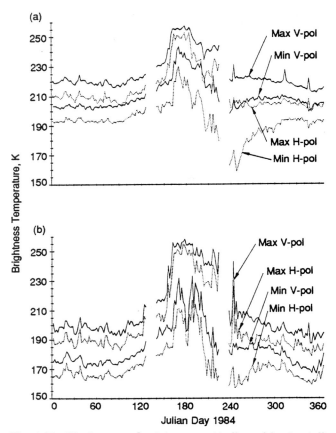

Fig. 4-32. Maximum and minimum vertically and horizontally polarized SMMR brightness temperatures for the (a) 18- and (b) 37-GHz channels. Data shown were derived from pixels within a 600×600-km region north of the Canadian Archipelago [Cavalieri et al., 1990].

be noted. First, whereas Livingstone et al. [1987b] based their classification on interpretation of active microwave data in which volume and surface scattering processes dominate, the passive microwave signal addressed here arises predominantly from emission and surface effects. Second, Julian days (JD's) referenced in the discussion pertain only to the data in Figure 4-32; transitions in other regions and in other years will occur on different days. Finally, inasmuch as seasonal variation has been addressed by relatively few studies, the analysis presented here is limited in scope.

4.5.4.1. Winter. During the winter, cold dry ice is overlain by cold dry snow. Between JD 0 (January 1) and 100 (April 9), Figure 4-32 shows that the 18- and 37-GHz radiances for both vertical and horizontal polarizations track along parallel but discrete traces throughout the winter. Fluctuations in radiance, which are represented by spikes and peaks in the brightness temperature data, are interpreted to be related to changes in scattering characteristics of the snow cover [Anderson et al., 1985; Comiso, 1986;

Grenfell, 1986; Hall et al., 1986; Anderson, 1987; Steffen and Maslanik, 1988].

4.5.4.2. Initial warming. Beginning about JD 100 (April 9), brightness temperatures in all channels begin to increase in response to general seasonal warming. This upturn of brightness temperature signals the beginning of the early melt period. Moisture is present in the snowpack and increases steadily during this period. Snow metamorphosis is enhanced as a result of diurnal freeze–thaw cycles [Livingstone et al., 1987b]. Signatures at the two frequencies continue to follow parallel tracks.

4.5.4.3. Melt onset. Between JD 150 (May 30) and 170 (June 19), brightness temperatures increase sharply in response to the onset of summer melt. Radiances at 18 and 37 GHz merge, and the ability to discriminate first-year ice from multiyear ice is lost. The snow cover becomes very nearly a perfect blackbody emitter as the free water content of snow increases. The end of the melt-onset period coincides with saturation of the upper snow cover with liquid water [Livingstone et al., 1987b]. At this point unmelted snow becomes slush, and voids in the porous surface layer of low-density ice become flooded. The skin is reduced to the upper surface of the snow, emissivity increases, and the radiometrically cool multiyear ice signature disappears [Grenfell and Lohanick, 1985; Foster et al., 1984].

4.5.4.4. Advanced melt. As the melt progresses, the surface becomes a complex mixture of granular layers, saturated and drained ice, and melt ponds [Onstott and Gogineni, 1985; Carsey, 1985; Onstott et al., 1987]. From a radiometric perspective, the ice surface cannot be described simply, but displays a range of brightness temperature and polarization characteristics [Grenfell and Lohanick, 1985]. Brightness temperatures remain at a high plateau during the first half of the melt period (JD 170 to 195) and then begin a gradual decline. Concomitant increase in polarization may be related to a progressive increase in the percent of the ice surface that is covered by pooled water. Melt pond formation clearly must affect the spatial and temporal variability of brightness temperatures as seen from satellites. Substantial changes in brightness temperatures during the late summer and autumn have been interpreted as reductions in ice concentration [Gloersen et al., 1978; Barry and Maslanik, 1989; Serreze et al., 1990] combined with melt effects, and as transient changes due to surface ponding and/or precipitation [Crane et al., 1982]. Crane and Anderson [1989] demonstrated how the transient effects of melt on microwave emission can be used to relate progression of surface melt to climatic conditions. Several investigators have suggested that analysis of this signature variability may provide information concerning melt processes [Grenfell and Lohanick, 1985; Barry and Maslanik, 1989; Maslanik, 1991], and lead to improved classification algorithms that assess the degree to which the ice surface is flooded [Comiso, 1990].

4.5.4.5. Freeze-up. Between JD 230 (August 18) and 290 (October 17), the slope of the brightness temperature traces changes, perhaps indicating that the area involved in melt ponds is no longer increasing. Over this same interval, the 18- and 37-GHz radiances begin to diverge, suggesting that the surface has stopped melting and has begun to freeze. By the end of the period, the signatures have assumed their characteristic winter values [Carsey, 1982].

4.6 LAKE ICE AND RIVER ICE

Perhaps because of limited availability of high-resolution data, passive microwave studies of river and lake ice are few [Schmugge et al., 1973; Hall et al., 1978, 1981; Swift et al., 1980a, b; Cameron et al., 1984; Melloh et al., 1991]. This section describes two independent field experiments that define river and lake ice signatures. First, a series of flights were made over Lake Erie in February 1978, and the Straits of Mackinac, Michigan, in March 1979, with a C-band radiometer [Swift et al., 1980a, b]. Second, flights were made over lakes and rivers in the vicinity of Fairbanks, Alaska, in March 1988 with a Ka-band (33.6 GHz) imager [Melloh et al., 1991]. Surface data were collected in conjunction with both experiments.

4.6.1 C-band Data (Great Lakes)

Figure 4-33 shows two brightness-temperature profiles that are representative of data obtained during the Great Lakes missions. Baseline brightness temperatures for smooth ice typically range from 195 to 210 K along these flight lines, with a slight increase from east to west (away from land). In-situ measurements show that ice thickness increases in this same direction.

Radiometric signatures of deformational features show clear contrast with respect to the signature of background ice. Excursions from the background temperature, which correspond to land, ridges, rough ice (rubble), and recently frozen ship channels, exceed 220 K in most instances. For pressure ridges and rubble, this contrast probably arises from an increase in mean ice thickness as a result of upheavals. In-situ documentation of one of the pressure ridges, 85°03′, Figure 4-33(a), showed extensive rubble with massive blocks of ice 30-cm thick thrust upwards as high as 2 m. The record portion marked "shoals," Figure 4-33(b), is a region of shallow water where ice rubble accumulates. Several closely spaced thickness measurements taken here and in the adjacent section of smooth ice show that smooth ice was 0.6-m thick where a constant 195 K brightness temperature was measured. Increasing brightness temperature at longitude 85°04′, Figure 4-33(b), corresponds to an active ship lead that had recently frozen.

4.6.2 Ka-band Data (Alaska)

Figure 4-34 shows 33.6-GHz passive microwave images of Harding Lake, Alaska, which is located approximately 60 km southeast of Fairbanks. The images were obtained

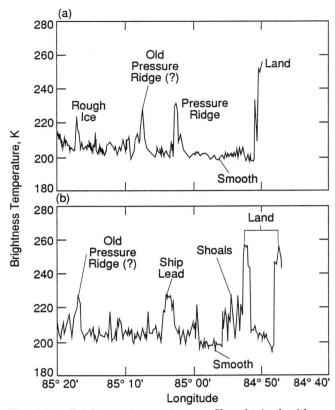

Fig. 4-33. Brightness temperature profiles obtained with an airborne C-band radiometer: (a) February 1978, over Lake Erie, and (b) March 1979, near the Straits of Mackinac.

during daylight hours on March 8, 1988, during a field experiment described by Melloh et al. [1991]. Surface measurements made between the dates of the two flights permit surface characteristics to be correlated with radiometric signatures evident in the images.

From a radiometric perspective, three surface types are evident in the March 8 image (Figure 4-34). The first type, making up the majority of the lake surface, is radiometrically cool (180 to 200 K, green and light green) and corresponds to clear ice which, although overlain in some places by fields of snow dunes, lacks a layer of bubbles. The second surface type, which is arrayed in a series of intersecting linear features, is warmer radiometrically (200 to 220 K, yellow and orange) for reasons that are not clear.

The third surface type is radiometrically cooler than the other two and coincides with a shallow shelf along the north rim of the lake (150 to 190 K, blue and green). Ice here was frozen to the lake bottom where a core was taken. Snow also was deeper and more continuous over this zone than that encountered elsewhere on the lake. It is not completely clear whether the radiometrically cold temperatures observed in this zone result from increased snow depth and continuity, or reflect the fact that ice is frozen to lake sediment at the bottom of the ice column.

(a)

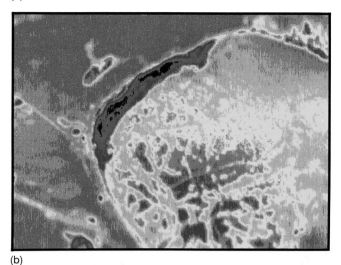

(b)

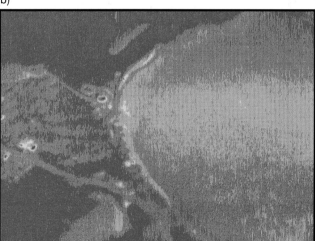

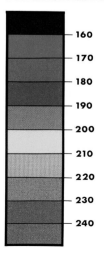

Fig. 4-34. Passive microwave imagery of Harding Lake near Fairbanks, Alaska, obtained in March 1988, with an airborne Ka-band passive microwave imaging system (KRMS). On March 8, radiometrically warm, linear features (orange and yellow) mark fracture patterns in the lake ice. The radiometrically cool band (blue and green) along the upper shore of the lake coincides with a shallow shelf.

Progress made through more than two decades of research now enables information concerning sea ice type and concentration to be derived globally on a routine, daily basis from satellite data. We can begin to build, for the first time, a long-term record of patterns of variation in the polar sea ice covers at synoptic scales. At finer spatial scales, ice signatures are defined in sufficient detail to permit real-time, automated monitoring of the ice pack from reconnaissance aircraft, should operational requirements demand this.

With the dominant aspects of ice signatures well understood and the tools to extract basic information from passive microwave observations in place, a host of technical issues critical to understanding and interpreting sometimes subtle aspects of ice and snow signatures remain unresolved. A clear gap exists between information now generated using passive microwave radiometry and that needed to improve global models of ice movement and oceanic and atmospheric circulation. Uncertainties associated with the following points are of particular importance.

- Characterization of ice types: Are there means to interpret the range of signatures of thin ice and associated features? Does formation of rafts, folds, and ridges significantly affect emissivity and scattering by altering brine drainage or accumulation and snow and slush accumulation? Is brightness temperature functionally related to the concentration of dispersed frazil concentration in water openings? Is the microwave signature of second-year ice unique or is it characterized by a range of signatures? What is the origin of multiyear ice signatures? To what extent does spatial variation in the relict snow–ice layer alter the radiometric signature, the variation in snow depth, or the melt pond coverage?

- Snow and the snow–ice interface: What contribution does snow make to microwave signatures of sea ice? Which characteristics of the snowpack have the most significant effect on signature? What snow characteristics (if any) can be measured/mapped remotely with passive microwave sensors? Does the character of the snow–ice interface follow a predictable evolutionary track with time? How does the character of the snow–ice interface in first-year ice differ from that in multiyear ice? How does this alter microwave signatures of underlying ice? Does the nature of the snow–ice interface exercise significant control on metamorphosis of the snowpack?

- Seasonal and diurnal freeze–thaw cycles: What do signatures observed during melt and freeze-up indicate with respect to physical changes in snow and ice properties? What variables govern the development of the snow–ice layer observed on multiyear ice? Does geographic position have a significant effect on radiometric signatures that arise from melt-related features?

- Sensor characteristics: To what extent can features that occur at subpixel scales be inferred from passive microwave data? Can lead evolution be monitored with coarse-resolution passive microwave sensors? What is the optimum suite of frequencies, polarizations, and spatial resolutions for sensing snow and sea ice characteristics, ice dynamics, or weather effects?

REFERENCES

Anderson, M. R., Snow melt on sea ice surfaces as determined from passive microwave satellite data, *Large Scale Effects of Seasonal Snow Cover,* edited by B. E. Goodison, R. G. Barry, and J. Dozier, International Association of Hydrological Sciences, IAHS Publication No. 166, Wallingford, England, pp. 329–342, 1987.

Anderson, M. R. and R. G. Crane, Arctic atmosphere-ice interaction studies using Nimbus-7 SMMR, presented at a Conference on Satellite Meteorology Remote Sensing and Applications, Clearwater, Florida, June 25–29, 1984, Preprint A85-37726-17-47, American Meteorology Society, Boston, pp. 132–136, 1984.

Anderson, M. R., R. G. Crane, and R. G. Barry, Characteristics of Arctic Ocean ice determined from SMMR data for 1979: Case studies in the seasonal sea ice zone, *Advances in Space Research, 5(6),* pp. 257–261, 1985.

Barry, R. G. and J. A. Maslanik, Arctic sea ice characteristics and associated atmosphere-ice interactions, *GeoJournal, 18(1),* pp. 35–44, 1989.

Cahalan, R. F. and L. S. Chiu, Large-scale short-period sea ice-atmosphere interactions, *Journal of Geophysical Research, 91(C9),* pp. 10,709–10,717, 1986.

Cameron, M. A., R. Gorman, M. Manore, A. Owens, G. Shirtliffe, and R. Ramseier, Passive microwave observation of St. Lawrence seaway ice and snow, *Ninth Canadian Symposium on Remote Sensing,* pp. 121–129, 1984.

Campbell, W. J., P. Gloersen, W. J. Webster, T. T. Wilheit, and R. O. Ramseier, Beaufort Sea ice zones as delineated by microwave imagery, *Journal of Geophysical Research, 81,* pp. 1103–1110, 1976.

Campbell, W. J., J. Wayenberg, J. B. Ramseyer, R. O. Ramseier, M. R. Vant, R. Weaver, A. Redmond, L. Arsenault, P. Gloersen, H. J. Zwally, T. T. Wilheit, T. C. Chang, D. Hall, L. Gray, D. C. Meeks, M. L. Bryan, F. T. Barath, C. Elachi, F. Leberl, and T. Fan, Microwave remote sensing of sea ice in the AIDJEX main experiment, *Boundary-Layer Meteorology, 13,* pp. 309–337, 1978.

Carsey, F. D., Arctic sea ice distribution at end of summer 1973–1976 from satellite microwave data, *Journal of Geophysical Research, 87,* pp. 5809–5835), 1982.

Carsey, F. D., Summer Arctic sea ice character from satellite microwave data, *Journal of Geophysical Research, 90(C3),* pp. 5015–5034, 1985.

Carsey, F., Review and status of remote sensing of sea ice, *IEEE Journal of Oceanic Engineering, 14(2),* pp. 127–138, 1989.

Cavalieri, D., P. Gloersen, and W. J. Campbell, Determination of sea ice parameters with the Nimbus 7 SMMR, *Journal of Geophysical Research, 89(D4),* pp. 5355–5369, 1984.

Cavalieri, D., B. Burns, and R. Onstott, Investigation of the effects of summer melt on the calculation of sea ice concentration using active and passive microwave data, *Journal of Geophysical Research, 95(C4),* pp. 5359–5369, 1990.

Cavalieri, D., J. Crawford, M. R. Drinkwater, D. T. Eppler, L. D. Farmer, R. R. Jentz, and C. C. Wackerman, Aircraft active and passive microwave validation of sea ice concentration from the Defense Meteorological Satellite Program special sensor microwave imager, *Journal of Geophysical Research, 96(C12),* pp. 21,989–22,008, 1991.

Comiso, J. C., Sea ice effective microwave emissivities from satellite passive microwave and infrared observations, *Journal of Geophysical Research, 88(C12),* pp. 7686–7704, 1983.

Comiso, J. C., Characteristics of Arctic winter sea ice from satellite multispectral microwave observations, *Journal of Geophysical Research, 91(C1),* pp. 975–994, 1986.

Comiso, J. C., Arctic multiyear ice classification and summer ice cover using passive microwave satellite data, *Journal of Geophysical Research, 95(C8),* pp. 13,411–14,422, 1990.

Comiso, J. C., T. C. Grenfell, D. L. Bell, M. A. Lange and S. F. Ackley, Passive microwave in situ observations of winter Weddell sea ice, *Journal of Geophysical Research, 94(C8),* pp. 10,891–10,905, 1989.

Crane, R. G. and M. Anderson, Spring melt patterns in the Kara/Barents Sea: 1984, *GeoJournal, 18,* pp. 25–33, 1989.

Crane, R. G., R. G. Barry, and H. J. Zwally, Analysis of atmosphere-sea ice interactions in the Arctic Basin using ESMR microwave data, *International Journal of Remote Sensing 3(3),* pp. 259–276, 1982.

Eppler, D. T. and L. D. Farmer, Texture analysis of radiometric signatures of new sea ice forming in Arctic leads, *IEEE Transactions on Geoscience and Remote Sensing, 29(2),* pp. 233–241, 1991.

Eppler, D. T. and B. M. Heydlauff, *Digitizing KRMS Analog Data on a Personal Computer,* NOARL Report 219, Naval Oceanographic and Atmospheric Research Laboratory, Stennis Space Center, Mississippi, 1990.

Eppler, D., L. Farmer, A. Lohanick, and M. Hoover, *Digital Processing of Ka-band Microwave Images for Sea Ice Classification,* NORDA Report 51, Naval Ocean Research and Development Activity, National Space Technology Labs, Bay St. Louis, Mississippi, 1984.

Eppler, D. T., L. D. Farmer, A. W. Lohanick, and M. Hoover, Classification of sea ice types with single-band (33.6 GHz) airborne passive microwave imagery, *Journal of Geophysical Research, 91(C9),* pp. 10,661–10,695, 1986.

Eppler, D. T., L. D. Farmer, and A. W. Lohanick, On the relationship between ice thickness and 33.6 GHz brightness temperature observed for first-season sea ice, *Sea Ice Properties and Processes,* edited by S. F. Ackley and W. W. Weeks, CRREL Monograph 90-1, Cold Regions Research and Engineering Laboratory, Hanover, New Hampshire, pp. 229–232, 1990.

Farmer, L. D., D. T. Eppler, and A. W. Lohanick, *Converting Digital KRMS Values to Units of Brightness Temperature,* NORDA Technical Note 427, Naval Ocean Research and Development Activity, National Space Technology Laboratories, Bay St. Louis, Mississippi, 1989.

Foster, J. L., D. K. Hall, A. T. C. Chang, and A. Rango, An overview of passive microwave snow research and results, *Reviews of Geophysics and Space Physics, 22(2),* pp. 195–208, 1984.

Gloersen, P. and F. Barath, A scanning multichannel microwave radiometer for Nimbus-G and Seasat-A, *IEEE Journal of Oceanic Engineering, 2(2),* pp. 172–178, 1977.

Gloersen, P. and D. J. Cavalieri, Reduction of weather effects in the calculation of sea ice concentration from microwave radiances, *Journal of Geophysical Research, 91(C3),* pp. 3913–3919, 1986.

Gloersen, P., H. J. Zwally, T. C. Chang, D. K. Hall, W. J. Campbell, and R. O. Ramseier, Time-dependence of sea ice concentration and multiyear ice fraction in the Arctic Basin, *Boundary-Layer Meteorology, 13,* pp. 339–359, 1978.

Gloersen, P., D. J. Cavalieri, A. T. C. Chang, T. T. Wilheit, W. J. Campbell, O. M. Johannessen, K. B. Katsaros, K. F. Künzi, D. B. Ross, D. Staelin, E. P. L. Windsor, F. T. Barath, P. Gudmundsen, and R. O. Ramseier, A summary of results from the first Nimbus-7 SMMR observations, *Journal of Geophysical Research, 89,* pp. 5335–5344, 1984.

Gow, A. J., D. A. Meese, D. K. Perovich, and W. B. Tucker, III, The anatomy of a freezing lead, *Journal of Geophysical Research, 95(C10),* pp. 18,221–18,232, 1990.

Grenfell, T. C., Surface-based passive microwave observations of sea ice in the Bering and Greenland Seas, *IEEE Transactions on Geoscience and Remote Sensing, GE-24(3),* pp. 378–382, 1986.

Grenfell, T. C., Surface-based passive microwave studies of multiyear sea ice, *Journal of Geophysical Research, 97(C3),* pp. 3485–3501, 1992.

Grenfell, T. C. and J. C. Comiso, Multifrequency passive microwave observations of first-year sea ice grown in a tank, *IEEE Transactions on Geoscience and Remote Sensing, GE-24(6),* pp. 826–831, 1986.

Grenfell, T. C. and A. W. Lohanick, Temporal variations of the microwave signatures of sea ice during the late spring and early summer near Mould Bay, Northwest Territories, *Journal of Geophysical Research, 90(C3),* pp. 5063–5074, 1985.

Grenfell, T., D. Bell, A. Lohanick, C. Swift, and K. St. Germain, Multifrequency passive microwave observations of saline ice grown in a tank, *Proceedings of the IGARSS'88 Symposium,* , pp. 1687–1690, 1988.

Hall, D. K., J. L. Foster, A. Rango, and A. T. C. Chang, Passive microwave studies of frozen lakes, *Proceedings of the American Society of Photogrammetry,* pp. 195–208, Fall Technical Meeting, Albuquerque, New Mexico, 1978.

Hall, D. K., J. L. Foster, A. T. C. Chang, and A. Rango, Freshwater ice thickness observations using passive microwave sensors, *IEEE Transactions on Geoscience and Remote Sensing, GE-19(4),* pp. 189–193, 1981.

Hall, D. K., A. T. C. Chang, and J. L. Foster, Detection of the depth-hoar layer in the snow-pack of the Arctic coastal plain of Alaska, U.S.A., using satellite data, *Journal of Glaciology, 32(110),* pp. 87–94, 1986.

Hollinger, J. P., B. E. Troy, Jr., R. O. Ramseier, K. W. Asmus, M. F. Hartman, and C. A. Luther, Microwave emission from high Arctic sea ice during freeze-up, *Journal of Geophysical Research, 89(C5),* pp. 8104–8122, 1984.

Hollinger, J. P., R. Lo, G. Poe, R. Savage, and J. Pierce, *Special Sensor Microwave/Imager User's Guide,* Naval Research Laboratory, Washington, DC, 1987.

Ketchum, R. D., L. D. Farmer, and J. P. Welsh, *K-Band Radiometric Mapping of Sea Ice,* 18 pp., NORDA Technical Note 179, Naval Ocean Research and Development Activity, National Space Technology Laboratories, Bay St. Louis, Mississippi, 1983.

Kuga, Y., F. T. Ulaby, T. F. Haddock, and R. D. DeRoo, Millimeter wave radar scattering from snow. I. Radiative transfer model, *Radio Science, 26(2),* pp. 329–341, 1991.

Livingstone, C. E., A. L. Gray, K. P. Singh, R. G. Onstott, and L. D. Arsenault, Microwave sea ice signatures near the onset of melt, *IEEE Transactions on Geoscience and Remote Sensing, GE-25(2),* pp. 174–187, 1987a.

Livingstone, C. E., A. L. Gray, and K. P. Singh, Seasonal and regional variations of active/passive microwave signatures of sea ice, *IEEE Transactions on Geoscience and Remote Sensing, GE-25,* pp. 159–173, 1987b.

Lohanick, A. W., Some observations of established snow cover on saline ice and their relevance to microwave remote sensing, in *Sea Ice Properties and Processes,* edited by S. F. Ackley and W. W. Weeks, CRREL Monograph 90-1, Cold Regions Research and Engineering Laboratory, Hanover, New Hampshire, pp. 61–67, 1990.

Lohanick, A. W., Microwave brightness temperatures of laboratory-grown undeformed first-year ice with an evolving snow cover, *Journal of Geophysical Research,* in press, 1991.

Lohanick, A. W. and T. C. Grenfell, Variations in brightness temperature over cold first-year sea ice near Tuktoyaktuk, Northwest Territories, *Journal of Geophysical Research, 91(C4),* pp. 5133–5144, 1986.

Maslanik, J. A., Effects of weather on the retrieval of sea ice concentration and ice type from passive microwave data, *International Journal of Remote Sensing, 13,* pp. 37–54, 1992.

Mätzler, C., *Interaction of Microwaves With the Natural Snow Cover,* a habilitation thesis, Institute of Applied Physics, University of Bern, Switzerland, 1985.

Mätzler, C., editor, Microwave signatures of Arctic Sea ice under summer melt conditions, *Proceedings of Workshop at Institute of Applied Physics*, University of Bern, Switzerland, 1986.

Mätzler, C., R. E. Ramseier, and E. Svendsen, Polarization effects in sea ice signatures, *IEEE Journal of Oceanic Engineering, OE-9*, pp. 333–338, 1985.

Melloh, R. A., D. T. Eppler, L. D. Farmer, L. W. Gatto, and E. F. Chacho, *Interpretation of Passive Microwave Imagery of Surface Snow and Ice, Harding Lake, Alaska*, CRREL Report 91-11, Cold Regions Research and Engineering Laboratory, Hanover, New Hampshire, 1991.

Nakawo, M. and N. K. Sinha, Growth rate and salinity profile of first-year ice in the high Arctic, *Journal of Glaciology, 27(96)*, pp. 315–330, 1981.

The Norsex Group, Norwegian remote sensing experiment in a marginal ice zone, *Science, 220(4599)*, pp. 781–787, 1983.

Onstott, R. G. and S. Gogineni, Active microwave measurements of Arctic sea ice under summer conditions, *Journal of Geophysical Research, 90(C3)*, pp. 5035–5044, 1985.

Onstott, R. G., T. C. Grenfell, C. Mätzler, C. A. Luther, and E. A. Svendsen, Evolution of microwave sea ice signatures during early summer and midsummer in the marginal ice zone, *Journal of Geophysical Research, 92(C7)*, pp. 6825–6835, 1987.

Parkinson, C. L., J. C. Comiso, H. J. Zwally, D. J. Cavalieri, P. Gloersen, and W. J. Campbell, *Arctic Sea Ice, 1973–1976: Satellite passive-microwave observations*, NASA SP-489, National Aeronautics and Space Administration, Washington, DC, 1987.

Rothrock, D. A., D. R. Thomas, and A. S. Thorndike, Principal component analysis of satellite passive microwave data over sea ice, *Journal of Geophysical Research, 93(C3)*, pp. 2321–2332, 1988.

Schmugge, T., T. T. Wilheit, P. Gloersen, M. F. Meir, D. Frank, and I. Dormhirn, Microwave signatures of snow and freshwater ice, *Advance Concepts and Techniques in the Study of Snow and Ice Resources*, National Academy of Sciences, Washington, DC, pp. 551–562, 1973.

Serreze, M. C., J. A. Maslanik, R. H. Preller, and R. G. Barry, Sea ice concentrations in the Canada Basin during 1988: Comparisons with other years and evidence of multiple forcing mechanisms, *Journal of Geophysical Research, 95(C12)*, pp. 22,253–22,267, 1990.

Smith, P. M., The emissivity of sea foam at 19 and 37 GHz, *IEEE Transactions on Geoscience and Remote Sensing, 26(5)*, pp. 541–547, 1988.

Steffen, K., *Atlas of the Sea Ice Types, Deformation Processes, and Openings in the Ice, North Water Project*, Zurcher Geographische Schriften Heft 20, Geographisches Institut, Zurich, Switzerland, 1986.

Steffen, K. and J. A. Maslanik, Comparison of Nimbus 7 scanning multichannel microwave radiometer radiance and derived sea ice concentrations with Landsat imagery for the North Water area of Baffin Bay, *Journal of Geophysical Research, 93(C9)*, pp. 10,769–10,781, 1988.

Steffen, K. and A. Schweiger, NASA Team algorithm for sea ice concentration retrieval from Defense Meteorological Satellite Program special sensor microwave imager: Comparison with Landsat satellite imagery, *Journal of Geophysical Research, 96(C12)*, pp. 21,971–21,987, 1991.

Stogryn, A., The emissivity of sea foam at microwave frequencies, *Journal of Geophysical Research, 77*, pp. 1658–1666, 1972.

Stogryn, A. and G. J. Desargent, The dielectric properties of brine in sea ice at microwave frequencies, *IEEE Transactions on Antennas and Propagation, AP-33*, pp. 523–532, 1985.

Svendsen, E., C. Mätzler, and T. C. Grenfell, A model for retrieving total sea ice concentration from a spaceborne dual-polarized passive microwave instrument operating near 90 GHz, *International Journal of Remote Sensing, 8*, pp. 1479–1487, 1987.

Swift, C. T., W. L. Jones, R. F. Harrington, J. C. Fedors, R. H. Couch, and B. L. Jackson, Microwave radar and radiometric remote sensing of lake ice, *Geophysical Research Letters, 7*, pp. 243–246, 1980a.

Swift, C. T., R. F. Harrington, and H. F. Thornton, Airborne microwave radiometer remote sensing of lake ice, *Proceedings of EASCON'80 Conference*, IEEE, New York, pp. 369–373, 1980b.

Tiuri, M., A. Sihvola, E. Nyfors, and M. Hallikainen, The complex dielectric constant of snow at microwave frequencies, *IEEE Journal of Oceanic Engineering, OE-9(5)*, pp. 377–382, 1984.

Troy, B. E., J. P. Hollinger, R. M. Lerner, and M. M. Wisler, Measurement of the microwave properties of sea ice at 90 GHz and lower frequencies, *Journal of Geophysical Research, 86(C5)*, pp. 4283–4289, 1981.

Weeks, W. F. and S. F. Ackley, *The Growth, Structure, and Properties of Sea Ice*, CRREL Monograph 82-1, Cold Regions Research and Engineering Laboratory, Hanover, New Hampshire, 1982.

Wilheit, T. T., The Electrically Scanned Microwave Radiometer (ESMR) experiment, *Nimbus-5 User's Guide*, NASA Goddard Space Flight Center, Greenbelt, Maryland, pp. 59–105, 1972.

World Meteorological Organization, *WMO Sea ice Nomenclature*, WMO Report 259, 155 pp., Secretariat of the World Meteorological Organization, Geneva, Switzerland, 1970.

Zwally, H. J. and J. E. Walsh, Comparison of observed and modeled ice motion in the Arctic Ocean, *Annals of Glaciology, 9*, pp. 136–144, 1987.

Zwally, H. J., J. C. Comiso, C. L. Parkinson, W. J. Campbell, F. D. Carsey, and P. Gloersen, *AntArctic Sea Ice, 1973–1976: Satellite Passive Microwave Observations*, NASA SP-459, National Aeronautics and Space Administration, Washington, DC, 1983.

Chapter 5. SAR and Scatterometer Signatures of Sea Ice

ROBERT G. ONSTOTT

Environmental Research Institute of Michigan, P.O. Box 8618, Ann Arbor, Michigan 48107

This chapter discusses the radar measurement of sea ice. The underlying theme is the physics that now make it possible to use active microwave techniques to obtain information about the frozen ocean instead of making direct in-situ measurements and that will ultimately make it possible to use those techniques for continuous, global monitoring. Basic remote-sensing concepts are presented and are followed by discussions of the interaction of electromagnetic energy with various ice types; the discussions include the effects of the environment and of physical-property perturbations. Critical links among ice type and form, physical and electrical properties, and backscatter signature will be highlighted and discussed. The examples used are primarily taken from Arctic research; however, the principles presented are equally applicable to the Antarctic region.

5.1 INTRODUCTION

Active remote sensors that operate in the microwave portion of the electromagnetic spectrum (i.e., radars operating at wavelengths from 2 to 24 cm) provide their own source of illumination and have been used to obtain information pertaining to ice-covered waters. Examples of such instruments include scatterometers and imaging radars.

Knowledge of ice type, thickness, age, and state are of particular interest in the study of the polar oceans. Of these parameters, the Holy Grail is the ability to estimate the distribution of ice thickness accurately, which is particularly critical in supporting climate studies. Figure 5-1 illustrates the key backscatter interactions for multiyear and first-year sea ice, and for open water without wind. Multiyear ice (i.e., ice that has survived a summer melt) can be distinguished from first-year ice by its greater thickness (1.5 m or greater), its lower salinity (2.5‰ versus 7.7‰ for first-year ice), and thicker snow cover (0.4 m versus 0.1 m for first-year ice). The backscatter from multiyear ice is a function of both surface and volume scattering (because of the very low upper-ice salinity, the increased radio wave penetration over time, and a dense population of gas bubbles); whereas backscatter from first-year ice and from the ocean is dominated, in large part, by the roughness of their surfaces (i.e., both are high-loss materials).

In-situ observations in the Arctic during the fall freeze-up, late winter, spring, and summer have been conducted to acquire empirical data to describe the microwave properties of sea ice. These data contribute to studies whose goal is to determine the ability of remote sensing to classify ice types and to characterize the physical and electrical parameters that control backscatter intensity. A wide range of frequencies, polarizations, and incidence angles have been employed to determine exactly how backscatter is influenced and to optimize the selection of sensor parameters so as to extract the geophysical parameters of greatest interest.

Results indicate that many features, including ice types, pressure ridges, and other large-scale topographical features, lead (polynya) formations, and icebergs, have unique microwave signatures that may be distinguished by using radar. Hence, it has been shown that backscatter is influenced by different aspects of the sea ice structure. Selection of wavelength, polarization, and viewing angle allows a measure of control over the depth at which the snow and ice are examined and helps in the determination of the dominant scattering mechanisms. These selection options are among the tools the scientist has to capture the desired geophysical information.

Active microwave research on solid oceans began in 1956 with a flight by the Naval Research Laboratory R5D Flying Laboratory. The goal was to determine the scattering characteristics of sea ice in the region near Thule, Greenland, when 0.4 to 10 GHz radars were used and the depression angles were less than 20°. The first attempt to map sea ice with the intent to retrieve information about ice properties was made in the early 1960's by the U.S. Army Cold Regions

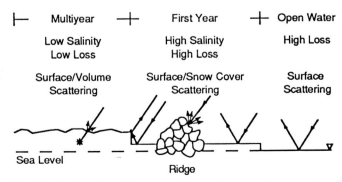

Fig. 5-1. Backscatter interactions for multiyear ice, first-year ice, and smooth open water.

Microwave Remote Sensing of Sea Ice
Geophysical Monograph 68
©1992 American Geophysical Union

Research and Engineering Laboratory (CRREL) [Anderson, 1966]. Interest in mapping sea ice began with the determination that during winter, first-year and old ice could be distinguished. Water was also found to produce a comparatively weak backscatter, an additional element of this study warranting further investigation. Even though sensors at that time provided little to support a quantitative assessment, these results have propelled research efforts to this day.

Historically, in the 1970's and 1980's, programs centered on determining what information was contained in radar image data of sea ice; often, interpretation of the imagery was based on the intercomparison of radar and optical imagery. The rigor of investigating the backscatter behavior and the mechanisms responsible for it gained momentum in the middle 1970's during a time when in-situ measurements complemented the aerial observation programs. The institution of air and in-situ campaigns was pioneered and promoted by Ramseier et al. [1982] and others, beginning with the 1973 US–USSR Bering Sea Experiment and the 1975–1976 Arctic Ice Dynamics Joint Experiment (AIDJEX) efforts [Campbell et al., 1978]. The addition of in-situ active microwave measurements and ice characterizations was instituted by R. K. Moore (University of Kansas) in 1976. The approaches developed in the 1970's continue today, with refinements in the sets of features sampled, the capabilities of the measurement sensors, and the missions of the experiments.

5.2 SCATTERING MEASUREMENTS

Until the late 1970's, knowledge of the active microwave signature of sea ice was more qualitative than quantitative. This is primarily attributed to difficulty with surface-truthing aircraft radars (which were flown over much greater distances than could be studied easily by surface parties) and to limitations on the size of an in-situ activity and its large logistics cost. Hence, the primary surface-truthing tool was aerial photography. Early radars were rarely designed to provide an absolute measurement, and the utilization of relative backscatter data was difficult even in making pass-to-pass comparisons; it was definitely difficult in making sensor-to-sensor comparisons.

Radar scatterometers were developed to provide accurate measurement of the scattering properties of terrain and ocean surfaces. The radar backscatter is defined in terms of a differential radar cross section per unit area σ°. This representation is used when backscatter is from an incoherent collection of a very large number of independent scatterers [Ulaby et al., 1982].

Aircraft campaigns began with observations at single frequencies and polarizations. Today, aircraft and in-situ systems have complete polarization diversification and operate at multiple frequencies. Sensors that are fully polarimetric allow the near-simultaneous retrieval of both orthogonal transmit and orthogonal receive polarizations (VV, VH, HV, and HH) and preserve the relative phase difference between channels (see Chapter 24). Side-looking

airborne radars (SLAR's) and synthetic aperture radar (SAR) observe a ground pixel at only one incidence angle. Near-surface scatterometers often allow for a full microwave characterization over a wide range of frequencies, polarizations, and incidence angles. Aircraft scatterometers have a similar capability, but have been limited to a few frequencies and a more difficult absolute calibration.

The measurement programs had two important tasks. The primary task was to provide for the optimum use of existing radar systems and support the development of future satellite and aircraft radars, both aims to be furthered by documenting the mean and variability of the radar backscatter for the various ice forms as functions of season, region, and formation conditions. This description allowed for selection of sensor frequency, angle, and polarization to get the information desired. The second aspect, the primary interest of today, was full characterization of the mechanisms responsible for an ice microwave signature and development of the understanding that would allow the optimum retrieval of geophysical information.

Progress toward this second goal began with the works of Anderson [1966], Dunbar [1969], Ketchum and Tooma [1973], Dunbar and Weeks [1975], and Johnson and Farmer [1971]. Quantitative measurements began with Rouse [1969] and the study of data acquired in May 1967, with the NASA 13.3-GHz (VV) scatterometer on sea ice in the Beaufort Sea. A correlation between ice type and the magnitude of the backscattering coefficient was made. This was extended by Parashar et al. [1977], describing 1970 measurements at 400 MHz and 13.3 GHz, compared with 15-GHz multipolarized images. Gray et al. [1977] suggested that seven ice types could be classified at 15 GHz by using combinations of the magnitude and the angular dependence of the scattering coefficient. Complementary work by Glushkov and Komarov [1971] and Loshchilov and Voyevodin [1972] in the USSR during this period led to the development in the early 1970's of an operational ice surveillance system based on a real aperture radar (RAR) system called Toros that operated at 16 GHz.

5.2.1 Description of In-Situ Data Collections

An increase in scientific and operational interests in the ice-covered regions of the Arctic Ocean resulted in an extensive set of experiments involving the use of radar for monitoring the properties of sea ice. These experiments used aircraft-borne scatterometers, SLAR's, and SAR's, as well as surface-based scatterometers. The goal of these efforts was to gain more information regarding the ice and the radar methods for measuring its properties. The result of the observations has been the documentation of scattering and scene properties for various seasons, regions, ice types, and ice formation conditions.

The in-situ microwave and physical ice property characterizations that serve as the primary basis of this chapter were initiated in 1977. Since that time, there have been observations conducted in the Beaufort Sea, Central Arctic, Greenland Sea, Eastern Arctic Ocean, Labrador Sea, and

Barents Sea. Seasons have included fall freeze-up, early winter, late winter, early spring, and summer. Experiments, locations, seasons, and observations are summarized in Table 5-1.

Sensors for the in-situ measurement programs have been varied. The transportable microwave active spectrometer (TRAMAS) was an important contributor in the late 1970's (see Figure 5-2). Sensor development has continued with the addition of rapid data-acquisition systems that allow efficient operation from both helicopter and ship.

5.2.2 Radar Integral Equation

A radar incorporates a transmitter and receiver. The transmitted energy that returns back to the radar is of great interest; its magnitude is determined by the scattering properties within the illuminated area. The average received power (P_r) may be obtained by the use of the integral radar equation for distributed targets

$$P_r = \frac{\lambda^2}{(4\pi)^3} \int_A \frac{P_t G^2 \sigma^\circ}{R^4} \tag{1}$$

where P_t is the power transmitted, G is the antenna gain, λ is the wavelength, R is the range from the radar to the terrain, A is the illuminated area, and σ° is the radar-scattering coefficient. The radar-scattering coefficient is a function of frequency, incidence angle, and polarization and is an absolute measure of scattering behavior.

TABLE 5-1. Near-surface-based scatterometer investigations.

Mission, year of data collection	Location	Month	Ice types	Frequencies, band
CEAREX,[a] 1989	Fram Strait	January–March	OW–MY	P, L, C, X, Ku, Ka, W
CRRELEX,[b] 1989	Ice tank	January and February	OW–ThFY	L, C, X, Ka (all are fully polarimetric)
CEAREX, 1988	Fram Strait	September–December	OW–ThFY–MY	P, L, C, X, Ku, Ka, W
CRRELEX, 1988	Ice tank	January and February	OW–ThFY	L, C, X (all are fully polarimetric) X, Ku, Ka, W (all are noncoherent)
BEPERS,[c] 1988	Bothnia	March	Brackish	X
MIZEX,[d] 1987	Fram Strait	March	OW–MY	P, L, C, X, Ku, Ka
CRRELEX, 1985	Ice tank	January–March	OW–ThFY	C, X, Ku
MIZEX, 1984	Fram Strait	June and July	OW ThFY–MY	L, C, X, Ku
MIZEX, 1983	Fram Strait	June and July	OW ThFY–MY	L, C, X, Ku
FIREX,[e] 1983	Mould Bay	April	FY SY	C, X, Ku
FIREX, 1982	Mould Bay	June and July	OW TFY MY	L, C, X, Ku
Lab Sea, 1982	Labrador Sea	February	OW NW–ThFY	L, C, X, Ku
Sursat,[f] 1981	Mould Bay	September and October	Gray MY	L, C, X, Ku
YMER,[g] 1980	East Greenland Sea	September	OW FY–MY	L, C, X, Ku
Sursat, 1979	Tuktoyaktuk	March	Brackish FY	L, C, X, Ku
UKansas,[h] 1978	Pt. Barrow	April	FY lake	L, C, X, Ku
UKansas, 1977	Pt. Barrow	May	TFY MY lake	L, X, Ku

[a] CEAREX = Coordinated Eastern Arctic Research Experiment
[b] CRRELEX = Cold Regions Research and Engineering Laboratory Experiment
[c] BEPERS = Bothnian Experiment in Preparation of ERS-1
[d] MIZEX = Marginal Ice Zone Experiment
[e] FIREX = Free-flying Imaging Radar Experiment
[f] Sursat = Surveillance Satellite Experiment
[g] YMER = a Swedish Experiment
[h] UKansas = University of Kansas Experiment

FY = first-year ice
MY = multiyear ice
ThFY = thin first-year ice
TFY = thick first-year ice
SY = second-year ice
OW = open water
NW = new water

Fig. 5-2. TRAMAS was the first in-situ active microwave system to obtain measurements of sea ice, and was developed in 1976 at the University of Kansas [Onstott et al., 1979].

5.2.3 Origin of Backscatter

What is the origin of the backscatter for sea ice? The basic hypothesis is that backscatter is influenced by the sea ice structure. Questions center on the importance of scatter from the surface of the snow layer, the interior of the snow, the snow–ice interface, the upper portion of the ice sheet, the interior of the ice sheet, and the ice–water interface. The way in which sea ice forms, its history, and its age are important in determining its microwave properties. When ice is young, its surface may be smooth (if it grows undisturbed) or rough (if it is agitated by wave action). Ice may exhibit a thin layer of brine on its surface, thereby limiting electromagnetic wave propagation to depths of a wavelength or less. With the high concentration of brine in its interior, new ice near its freezing point is extremely lossy and, again, penetration is limited.

With age, the volume of brine in first-year ice reduces (see Chapter 2), but the ice remains lossy. During summer, the ice that survives the process of melt has developed additional surface relief, has had the density of the upper portion of the ice sheet reduced, and has desalinated. These

three characteristics contribute to the difference in the microwave signatures of FY and MY ice. A cross-sectional view is provided in Figure 5-3. First-year ice has few internal scatterers, and these are small compared to a wavelength. The penetration depth in multiyear ice may be several wavelengths, and there is a significant number of gas bubbles with diameters of 1 to 3 mm found within this volume. The physical models shown in Figure 5-4 help focus attention on the important aspects of these two ice types.

5.2.3.1. Microwave signature for new, young, and thick first-year ice.
The dominant backscatter mechanism associated with first-year ice is surface scattering and/or scattering from the region very near the ice surface. The radar-scattering coefficient may be expressed as

$$\sigma°_{pp}\left(\theta_i\right) = K\Gamma^2(p, f, \theta_i, \varepsilon_r{}^*)\, SF\,(p, f, \sigma, l) \qquad (2)$$

which is a function of the Fresnel reflection coefficient (Γ), a shape function (SF), and a scaling constant (K). The reflection coefficient is a function of incidence angle (θ_i), the

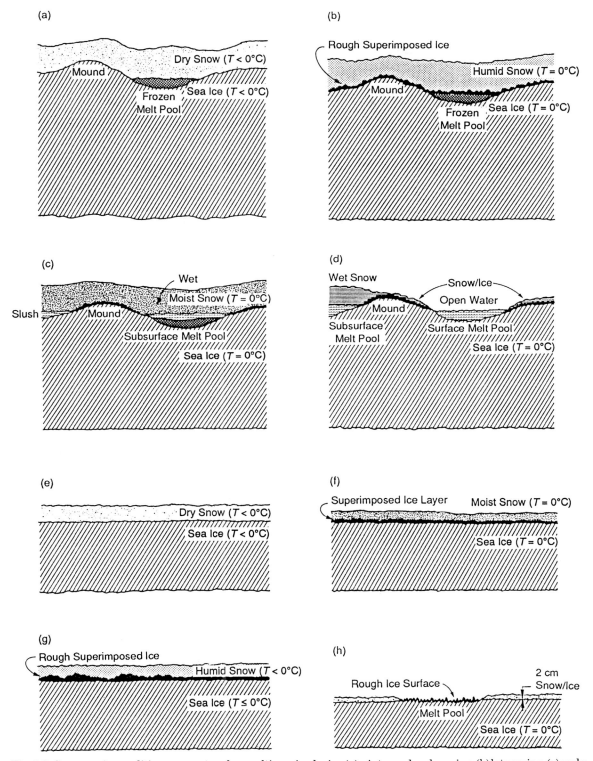

Fig. 5-3. Snow and ice conditions encountered on multiyear ice during (a) winter and early spring, (b) late spring, (c) early summer to midsummer, and (d) midsummer to late summer. Snow and ice conditions encountered on first-year ice during (e) winter and early summer, (f) late spring, (g) early summer to midsummer, and (h) midsummer to late summer [Onstott et al., 1987].

(a)

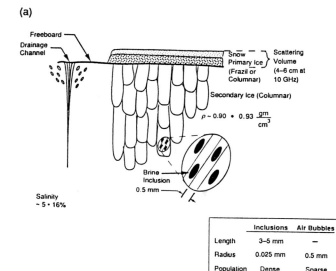

(b)

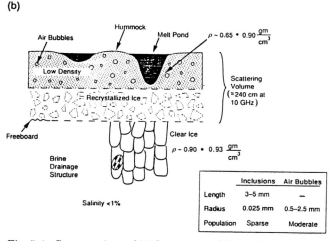

Fig. 5-4. Cross sections of (a) first-year and (b) multiyear sea ice.

transmit–receive antenna polarization *(p)*, frequency *(f)*, and the complex dielectric constant $(\varepsilon_r{}^*)$. The shape function is a function of polarization, frequency, the root mean square (rms) surface height (σ), and the surface correlation length (ℓ).

In summary, the important sea ice parameters (primary) are $\varepsilon_r{}^*$, because it acts to set the absolute backscatter level, and the surface-roughness statistics σ and ℓ, because they influence the shape function. They act independently of environmental forcing and have been decoupled to study their impact on the backscatter signature. See Chapters 8 and 9 for additional discussion on the depth of penetration and for comparison of measured and predicted data.

5.2.3.2. Microwave signature for multiyear ice. The backscatter of multiyear ice is attributed to gas bubbles in the upper portion of the ice sheet. (See Chapters 8 and 9.) Propagation through a volume of dielectric discontinuities

or discrete scatterers (i.e., gas bubbles in an ice matrix) may produce volume scattering. These results may be predicted (zero-order) by assuming that each bubble is characterized by a cross section proportional to its radius to the sixth power. The total backscatter is then proportional to the sum of all cross sections [Kim, 1984; Kim et al., 1984a] and may be expressed as the contribution from the surface and the ice volume as given by

$$\sigma°_{pp}(\theta_i) = \sigma°_s(\theta_i) + \gamma^2(\theta_i) \; \sigma°_v(\theta_i{}') \qquad (3)$$

The surface scattering contribution $\sigma°_s$ is identical to that described earlier. The intensity of the volume backscatter is weighted by the transmission coefficient γ, which is reasonably close to 1. The volume scattering coefficient may be expressed as

$$\sigma°_v(\theta_i{}') = N\sigma_b \cos(\theta_i{}')\,[1 - 1/L^2(\theta_i{}')]/(2k_e) \qquad (4)$$

where N is the number of particles, σ_b is the scattering cross section per particle, L^2 is the two-way loss factor, and k_e is the extinction coefficient. The number of particles is a function of the density of the layer (ρ) and the radius (r) of the particles cubed and given by

$$N = (1 - \rho/0.926)/(4\pi r^3/3) \qquad (5)$$

Hence in the strong volume scattering cases associated with multiyear ice, the backscatter will increase with increasing radii of the gas bubbles and as the total number of discrete scatterers increases. The number of scatterers increases with decreasing density and increasing thickness of the low-density ice (LDI) layer. A cross-sectional view of multiyear ice is shown in Figure 5-5, with ranges of expected thickness, density, salinity, and bubble diameter. In Figure 5-6, the relative contributions of smooth and rough surfaces and of a volume with a smooth and a rough surface are shown as a function of frequency for typical ice characterization values associated with multiyear ice. Note that backscatter intensity increases with increasing frequency in all cases.

5.2.3.3. Empirical observation of the sources of scatter. The measurements of first-year and multiyear ice illustrated in Figure 5-7 were made at 5.25 GHz, VH-polarization, with an incidence angle of 45° [R. G. Onstott, manuscript in preparation, 1992]. Backscatter intensity is plotted as a function of position in the ice and snow (1 m in ice for each 1.75 KHz), and the surface distance moved (40 m).

A multiyear ice sheet is typically composed of three scattering features: hummocks, melt pools, and ridges. The backscatter from the undulating part of the ice sheet (hummocks and melt pools) is considered here because melt pools and hummocks often dominate the response of a multiyear ice floe, and their similarities and differences serve to illustrate the importance of the upper portion of the ice sheet in determining the microwave signature response for

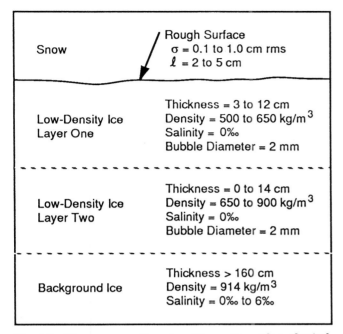

Snow	Rough Surface σ = 0.1 to 1.0 cm rms ℓ = 2 to 5 cm
Low-Density Ice Layer One	Thickness = 3 to 12 cm Density = 500 to 650 kg/m^3 Salinity = 0‰ Bubble Diameter = 2 mm
Low-Density Ice Layer Two	Thickness = 0 to 14 cm Density = 650 to 900 kg/m^3 Salinity = 0‰ Bubble Diameter = 2 mm
Background Ice	Thickness > 160 cm Density = 914 kg/m^3 Salinity = 0‰ to 6‰

Fig. 5-5. The upper portion of multiyear ice, with a physical-property description of the various layers.

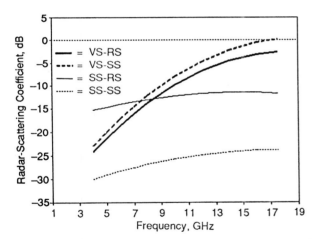

Fig. 5-6. Relative contributions of surface and volume scattering for MY ice [Kim, 1984]. The surface roughness parameters are the correlation length ℓ of 8.6 cm for both smooth and rough surfaces, and height standard deviation σ of 0.81 cm for the rough surface and 0.15 cm for the smooth surface. The ice density is 700 kg/m^3, the air bubble radius is 1 mm, and the complex dielectric constant is $3.15 - j0.01$. (VS = volume scatter, RS = rough surface, and SS = smooth surface.)

(a)

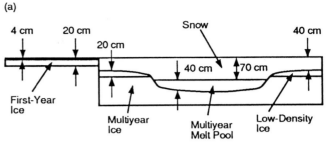

(b)

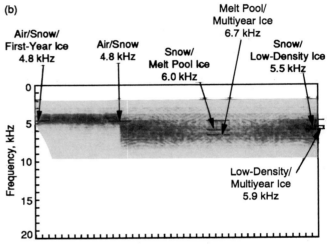

(c)

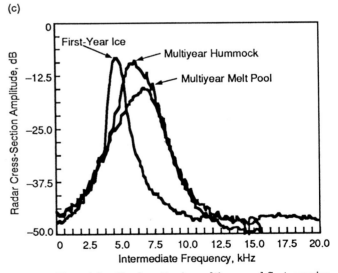

Fig. 5-7. The origin of backscatter in multiyear and first-year ice is illustrated with a probing radar scanned across a 40-m ground track. (a) The surface truth diagram is shown. (b) The backscatter response at 5.25 GHz, 45° incidence angle, and VH-polarization is shown as a function of ground position. (c) Average intensity responses for the first-year ice, hummock, and melt pool versus interior ice position are also shown. Intermediate frequency is proportional to the range at approximately 1750 KHz/m of ice.

multiyear ice. For this ice sheet, there is no distinction between the physical and electrical properties of a multiyear hummock and melt pool (fresh water), except for the ice in the top 20 to 40 cm. Their backscatter characteristics are quite distinct, however, and provide insight into the underlying electromagnetic processes.

These data serve to demonstrate empirically and visually that the physical-property differences in the uppermost portion of a multiyear ice sheet are responsible for the determination of its microwave signature (see Figure 5-7 and Table 5-2). A dense population of gas bubbles (1 to 2 mm in diameter is typical) is required to produce the strong backscatter response associated with multiyear ice. The roughness of the snow–ice interface plays a secondary role.

Melt pools typically have very smooth surfaces and produce weak backscatter. The important observation is that the population of air bubbles trapped in the melt pool ice is too small (the density is approximately that of pure ice) to produce significant backscatter. Penetration of the radar signal to the bottom of the melt pool is apparent, and the centroid of the backscatter response is associated with the pure-ice–sea-ice interface.

There are numerous air bubbles present in the upper portion of hummocked ice (their density is 825 kg/m^3), and these are of a size (2 mm in diameter) that may produce an enhanced backscatter return. The backscatter of this hummock and melt pool differs in intensity by a factor of about 32. In addition, the centroid of the hummock response is associated with the center of this low-density ice layer. Also, the response from the snow cover on either the melt pool or the hummock is relatively weak.

Results suggest that the backscatter for first-year ice is limited to the region immediately around the snow–ice interface and that there is no significant penetration in a 20-cm-thick ice sheet. The range extent of this response is explained by the radiation pattern of the radar.

TABLE 5-2. Intercomparison of key physical properties of the melt pool and hummock multiyear ice features.

Parameter	Multiyear melt pool	Multiyear hummock
Ice sheet thickness	3.15 m	3.25 m
Freeboard	5 cm	15 cm
Snow thickness	25 cm	20 cm
Feature thickness	30 cm	20 cm
Feature bulk density	914 kg/m^3	825 kg/m^3
Bubble layer description	Sparse	Abundant
Bubble diameter mean	3 mm	2 mm
Bubble, number/cm^3	0.1	3.3
Salinity, ‰	0	0

5.2.4 Water and Ice Backscatter Discussion

5.2.4.1. Open water. One of the issues of critical importance, especially in navigation and determination of ice concentration, is the discrimination of ice from water. The near-vertical (0° to 10°) backscatter response of water is typically higher than that of ice, whereas the backscatter response falls off quickly as the incidence angle increases beyond 10°. Backscatter for water in the open ocean can be written in a simplified form such as

$$\sigma°(u,\theta,f,p,\varphi) = C(f,\theta,p,\varphi)u^{\alpha(f,\theta,p,\varphi)} \qquad (6)$$

where $\sigma°$ is the normalized radar-scattering coefficient, C is the wind-speed scaling constant, u is the equivalent neutral-stability wind speed measured at a 10-m height, α is the wind-speed exponent, θ is the incidence angle, f is the operating frequency, p is the transmit–receive polarization, and φ is the angle between the radar-look direction and wind direction.

The backscatter response is a function of wind speed. Factors that include fetch, air–sea temperature difference (important for determining stability), wave slope, the orientation of the radar to the wind, and wave direction are important in determining the absolute backscatter intensity [Ulaby et al., 1986]. Examples of values for the upwind/downwind case and an angle of 50° are $\approx 6.76 \times 10^{-4}$ for the scaling constant C, and 1.18, 1.53, and 1.64 for the wind exponent α for L-HH, C-VV, and X-VV, respectively [Onstott and Shuchman, 1989].

Water between floes in the marginal ice zone (MIZ) and in the ice pack presents a response dissimilar to that of the open ocean; backscatter values are typically much lower. This has not been fully explored, but is attributed to wave dampening and wind shadowing by floes. Operation at L-VV presents a special situation that has been illustrated in Seasat SAR imagery—especially in cases including large bodies of water (i.e., polynyas). The backscatter intensity is similar to that of ice. Operation at HH-polarization, however, reduces the water response by 5 to 7 dB, enough to provide an improved separation of water and thick ice. This polarization response provides an opportunity to separate new ice from open water (see Chapter 24). Operation at higher frequencies gives more ice–water contrast. This is understandable, since open-water backscatter increases less quickly with increasing frequency than does backscatter from sea ice.

The backscatter responses for water in the open ocean (with wind speeds of 6 to 8 m/s), for moderate sea conditions, and in the MIZ are illustrated in Figures 5-8 and 5-9 for a frequency of 9.6 GHz, HH-polarization, and an angle of 25° [Onstott, 1990a]. The ocean and ice (when either is at 100% concentration) share a similar response level, but water between ice floes has an intensity that is dramatically less (by 13 dB). These data were acquired by a helicopter flying a scan beginning from a position about 16 km from the ice edge in the open ocean to the ice edge, then across the ocean–ice edge until large floes were encountered near the start of the pack ice, an additional 20 km. Illustrated in these radar data is a transition from the strong backscatter of the open ocean, to the backscatter of a region of calm water (5 km in extent) immediately adjacent to the ice edge, to that of a well-delineated and compact ice edge, into that of a region of 90% ice concentration and large multiyear floes. The weakest returns in the MIZ during summer and winter are associated with the water (or new ice) between ice floes. Ice and water are often difficult to distinguish due

to the system noise floor or to the setting of the operating point of the receiver in an attempt to prevent saturation by strong ice returns.

The liquid ocean produced a variety of microwave signatures that, in large part, were influenced by the position of the edge. Observed conditions likely reflect a complicated interaction between the winds, waves, and currents that is due to the effects of cold air from the pack crossing from the cold ocean near the ice edge to the warmer open ocean.

Additional ocean-signature variations were observed and attributed to the dampening of swell and gravity waves by the ice floes. In Figure 5-10, the liquid-ocean signatures are shown as a function of position from the ice edge. Signatures break into four major categories: (a) open ocean, (b) ocean prior to the ice edge and in the MIZ, which is immediately near the edge, (c) ocean in the MIZ, and (d) ocean in the MIZ and the shadows of ice floes.

Discrimination of ice and water may be problematic even at the higher frequencies at the ice edge when on-ice winds are involved. Even at X-band, the open ocean may produce a backscatter response with an intensity similar to that of

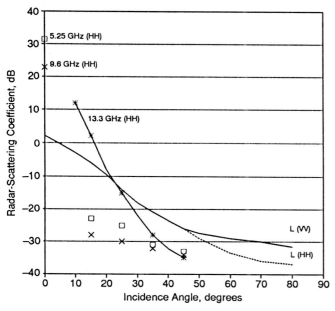

Fig. 5-8. Radar-scattering coefficient data at L-band for a medium sea [Wetzel, 1990], at 13.3 GHz (HH) for a calm ocean [Gray et al., 1982], and 5.3 GHz (HH) and 9.6 GHz (HH) for water between ice floes in the MIZ [Onstott et al., 1987].

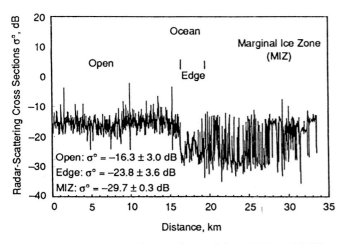

Fig. 5-9. Transect across the ice edge on July 5, 1984, at 9.6 GHz, HH-polarization, and 25° in the marginal ice zone in the Fram Strait [Onstott, 1990a].

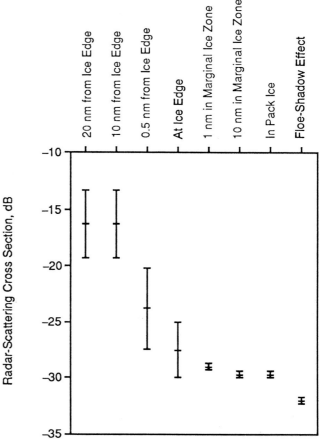

Fig. 5-10. Radar-scattering cross sections of the liquid ocean at 9.6 GHz, HH-polarization, and 25°, obtained as a function of position from the ice edge on July 5, 1984, in the marginal ice zone in the Fram Strait [Onstott, 1990a].

ice. A compacted ice edge prevents the detection of water between floes. Therefore, distinguishing the open ocean from the MIZ may require the exploitation of ice-feature shapes and differences in texture.

Examination of the polarization ratio ($\sigma°_{VV}/\sigma°_{HH}$) is instructive. Measurements show that water has a ratio greater than that of ice. Examination of the ratio of the Fresnel reflectivity coefficients (which represent power) supports this observation. This ratio is driven by the magnitudes of the dielectric constant of the material; the greater the dielectric constant, the larger the ratio at large incidence angles. Ice has a dielectric constant about an order of magnitude less than that of seawater.

5.2.4.2. New ice.

The key characteristic of the backscatter from new ice is its weakness. Wind produces ripples (capillary and short gravity waves) on water. The formation of ice crystals (grease) attenuates and impedes the formation of waves. The thicker the crystal layer, the greater the attenuation. The important effect of this process is the production of a surface that is smooth to the radar. In calm conditions, the ice sheet is effectively mirror-like, with only rafting events providing relief. Surface-roughness values are on the order of 0.05-cm rms—a factor of 20 smaller than a radar wavelength at 30 GHz. In addition, it has been observed (see Chapter 9) that the dielectric constant reduces by a factor of at least 4 to 6 from that of seawater. Hence, with the dramatic reduction in both the scale of surface roughness and in dielectric constant, backscatter from new ice is expected to be considerably smaller than that for open water. It may be small enough that it is difficult to measure accurately at the middle and large incidence angles.

5.2.4.3. Nilas and gray ice.

Two processes are important in determining the backscatter from nilas and gray ice. These include modification of the surface-roughness statistics and cooling of the upper ice sheet (the sheet cools with increasing ice thickness). New ice grown under calm conditions represents the limiting case of ice with a very smooth surface (except in the case of multiyear ice with melt pools). Any process, especially the formation of rime (e.g., frost flowers that act to wick up brine and then form roughness elements) on a new ice-sheet surface or the deposition of snow crystals, acts to increase surface roughness; there is no turning back! An increase in surface roughness at this stage translates into an increase in backscatter. The formation of frost flowers is the most important contributor in increasing small-scale surface roughness; hence, it is often the controlling backscatter mechanism for young ice. Intensities at microwave frequencies may reach levels almost approaching those of multiyear ice. By the time the gray-white stage is reached, snow has infiltrated and the backscatter intensity decays.

5.3 IMPORTANCE OF ENVIRONMENTAL AND PHYSICAL PROPERTIES

5.3.1 Temperature

5.3.1.1. Multiyear ice. Scattering for multiyear ice has been shown to be determined by the uppermost portion of the ice sheet. The ice there is relatively salt free and may contain large numbers of discrete scatterers. It is not anticipated that backscatter is impacted greatly by temperature variation (for the cases where $T_{ice} < -5°C$) because penetration through the top 10 to 20 cm is the minimum required to produce a multiyear-ice-like response. The dielectric properties of salt-free ice are relatively insensitive to temperature variations. Temporal observations during September to November have just recently been made (i.e., during CEAREX) and will provide a detailed examination of the temporal response of sea ice and the changes in ice-sheet and snow properties.

5.3.1.2. First-year ice. If the backscatter for first-year ice is largely determined by scattering from the ice surface, the impact of ambient temperature is to change the reflection coefficient of the ice sheet and possibly to promote the further formation of a slush layer. We confine the discussion here to changes in dielectric constant, and radar-scattering coefficients predicted for ice 1-m thick with a 2-cm snow layer for the temperature range of $-50°$ to $-2°C$ are compared with in-situ observations in Figure 5-11. It is

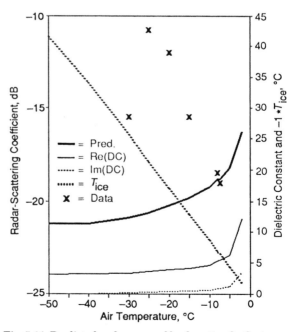

Fig. 5-11 Predicted and measured backscatter for first-year ice (1-m thickness) for air temperature from $-50°$ to $-2°C$. The frequency is 9.6 GHz and the incidence angle is $40°$. (Pred. = predicted, Re(DC) = real part of complex dielectric constant, IM(DC) = imaginary part of complex dielectric constant, and I_{ice} = ice surface temperature.)

shown that dielectric properties increase with rapidly increasing temperatures beginning at –10°C. Based on this knowledge, backscatter (as defined by the physical-optics model) would be expected to increase with an increasing dielectric constant. In addition, the backscatter response would be expected to change by less than 2 dB between the temperatures of –50° and –8°C. However, when compared with in-situ observations, a trend opposite to this is noted. These results suggest that the principal response to low temperature may be the freezing of the slush layer, even though there may be an increase in volume scattering because of an increased depth of penetration (i.e., 7 cm at $T = -50°C$ and 1 cm at $T = -8°C$).

5.3.1.3. Temperature dependence observed in airborne scatterometer data.

Livingstone et al. [1983] have compiled observations made over several years and include data obtained at various temperatures. In Figure 5-12, the temperature-dependence data at $\sigma°_{HH}$ and $\sigma°_{HV}$ for an angle of 25° are shown. These data suggest that as temperatures decrease, the backscatter for first-year ice decreases, whereas multiyear ice backscatter increases.

5.3.2 Snow Cover

The importance of snow on sea ice is determined by season (cold or warm) and ice type. The impact of a dry and

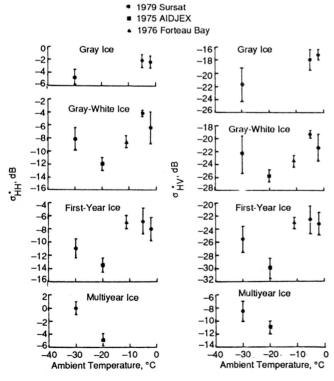

Fig. 5-12. Temperature dependence of $\sigma°_{HH}(\theta \simeq 25°)$ and $\sigma°_{HV}(\theta \simeq 25°)$ for gray, gray-white, first-year and multiyear ice [Livingstone et al., 1983].

moist snow layer on first-year and multiyear ice will be discussed.

5.3.2.1. Dry snow.

Snow on ice serves as a thermal blanket, since snow has a thermal conductivity much less than that of sea ice. In addition, it may provide an impedance-matching (electromagnetic) function. The scene reflectivity is determined by the combination of the dielectric and scattering properties of the snow and the ice sheet. The snow may improve the matching between the atmosphere and the ice sheet and also reduce the effect of surface roughness. The snow may also facilitate the transfer of brine from the ice sheet and produce a collection of slush at the ice–snow interface. The impact of a developing slush layer with a high liquid content may be (a) to reduce ice sheet surface roughness (i.e., through fill-in) when the surface is rough, (b) to create a dielectrically rough surface (i.e., a vertically variable boundary between dry and saturated snow), (c) to increase the dielectric constant contrast at the ice–snow interface, or (d) to provide a transitional dielectric layer between snow and ice. The dominant mechanism has not been fully defined, but, in all cases, observations suggest that the backscatter is weak in an absolute sense.

5.3.2.2. Dry snow on multiyear ice.

When compared with that of multiyear ice, the volume scattering of a thin layer of snow is small. In addition, the impact of raising the ice-surface temperature is small because the upper portion of the ice sheet is relatively brine free, hence absorption does not change greatly with increasing temperature. A calculation made to determine the impact of a 10-cm layer of dry snow with $T_{air} = -20°C$ and $T_{ice} = -13.7°C$ shows that the overall effect on the volume scatter coefficient of the multiyear ice is to lower $\sigma°$ by about 0.3 dB [Kim et al., 1984b].

5.3.2.3. Dry snow on first-year ice.

Observations show that snow plays an important role in determining the backscatter response for first-year ice [Onstott et al., 1982]. Dramatic increases in ice-surface temperature, especially when the ice is thin, may cause an increase in the complex dielectric constant and the production of brine slush at the snow–ice interface. If the snow burden is too great, the ice sheet may submerge and flood with seawater. All of these actions often produce very similar backscatter intensities.

Measurable signal differences have been observed between ice that is snow covered and ice that is barren. On the average, thick first-year ice with up to 8 cm of snow cover was reported to have backscattering coefficients 1 to 5 dB higher than thick first-year ice that was snow free. The effect was small at L-band (about 0.5 dB), and variable at frequencies from 9 to 17 GHz (a high of about 5 dB and a low of about 1 dB). Results from an examination of snow on smooth first-year ice are presented in Figure 5-13 for frequencies from 1.5 to 17 GHz. The data presented here represent an average derived from backscatter at 20° to 60°. It is interesting to note that the response for frequencies

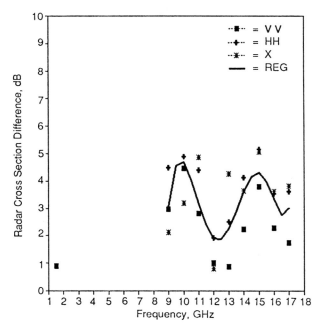

Fig. 5-13. The difference in radar-scattering coefficients for the case of thick first-year ice with and without an 8-cm snow layer. These data represent the average backscatter difference for angles between 20° and 60°. (X = cross-polarization.)

from 9 to 17 GHz is apparently oscillatory about a mean of 3 dB. This suggests that an interference process is responsible. Such a process can occur when a thin dielectric separates two semi-infinite dielectric media (e.g., air and ice). The amplitude of the response also appears to be polarization dependent. This result is of interest because it suggests that coherent radar responses may be observed in nature and are not necessarily dominated by physical-property variation (i.e., thickness of layers) and radar signal scintillation (i.e., fading). Other radar observations of this type have probably been all but ignored based on the above two premises. Coherence effects have been observed in the passive microwave observation of snow on terrain [J. Kong, personal communication, 1989]. The effect of snow on sea ice is shown as a function of snow depth in Figure 5-14 at 8.6 and 17.6 GHz for an angle of 50° [Kim et al., 1984b]. The backscatter predicted for a semi-infinite half space of snow, for both a snow layer on a smooth surface and snow on a rough surface, is also presented. These predictions, based on empirical measurements and a zero-order model, indicate that snow on ice with a smooth surface is an important contributor to backscatter (and is increasingly important with increasing frequency), whereas snow on a rough surface may contribute little, except at high frequencies. These predictions do not take into account potential interference effects, which may explain the close agreement between measurements and prediction seen in the case at 8.6 GHz and the much poorer agreement at 17.6 GHz. Figure 5-13 shows that the data at 8.6 and 17.6 GHz appear to represent data near a peak and a null.

5.3.2.4. Moist snow. During summer, snow is undergoing melt; water may be percolating through the snow and collecting into subsurface melt pools that are in transition and becoming surface melt pools. For the first half of the summer, the most important sea ice characteristic is the presence of a moist snowpack. Measurements of drained snow indicate that the bulk wetness of the snowpack is from 5% to 6% by volume during peak melt and that the majority of the snowpack is old and has a density of 400 to 500 kg/m^3 [Onstott et al., 1987]. The electrical properties of dry and wet snow are very different; the electrical properties of dry snow are nearly independent of frequency, whereas moist snow has properties that are very much a function of frequency. The penetration depths calculated for the peak melt conditions described above are about 4.5, 1.8, 1.5, and 1.3 cm for 5.25, 9.6, 13.6, and 16.6 GHz, respectively. Up to midsummer, the snowpack on multiyear ice may be at least 40 cm (i.e., there are many penetration depths between the air–snow and the snow–ice interfaces). It is very important to note that at the higher microwave frequencies, such as 13.6 GHz, only 1 to 3 cm of snow are required before the snow layer completely dominates the backscatter response, because for each penetration depth the incident signal experiences a 9-dB round-trip loss. It has been documented [Onstott et al., 1987] that operation at the longer wavelengths (e.g., 6 cm) allows ice type discrimination during the first two weeks of summer (i.e., to June 20) and that during peak melt (a period often centered around July 7), longer wavelengths (e.g., 24 cm) allow some ice type discrimination due to the ability to penetrate the snow cover and sense the evolving layer of superimposed ice at the snow–ice interface of the thinner first-year ice. In Figures 5-14(c) and (d), the expected behavior for moist snow is presented for small moisture values characteristic of snow in early summer.

5.3.3 Surface Roughness

The small-scale roughness of the air–ice or ice–snow interface is important in the determination of the backscatter intensity for both first-year and multiyear ice, as is shown in Figure 5-6. Since roughness is often the primary backscatter mechanism for first-year ice, except, possibly, at very cold temperatures and when the ice is thick, the first-year ice response to roughness changes is the most dramatic of the two major ice categories. Given that first-year ice may have very smooth to very rough surfaces, the range of backscatter responses may be large and variations of 10 to 15 dB may be observed. In Figure 5-15, the angular responses for smooth first-year ice (thickness of 1.37 m), thick first-year (TFY) ice of pancake-ice origin (pans of 2 to 3 m with 4-cm-high rims, and an ice thickness of 1.65 m), and pancake ice (1- to 2- m pans and a thickness of about 10 to 15 cm) are illustrated. The response shown for the TFY ice cases is typical. As roughness increases, the near-vertical response decreases and the middle- and large-angle backscatter increases. The pancake ice example is shown because it represents a case where the roughness and

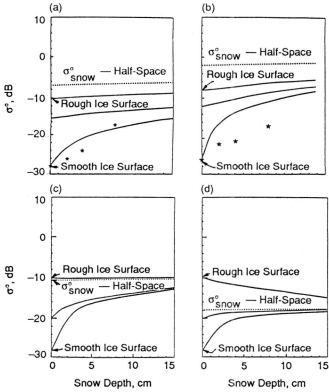

Fig. 5-14. The effect of a dry snow cover on sea ice is shown as a function of snow depth at (a) 8.6 GHz, and (b) 17.6 GHz for an angle of 50°. The solid lines are predicted $\sigma°$ for snow cover and can be very low (smooth ice) or close to that of a half space of dry snow (rough ice). The effect of wet snow on sea ice is shown for a liquid water content of (c) 1% and (d) 3% for 8.6 GHz for an angle of 50° [Kim, 1984; Kim et al., 1984a].

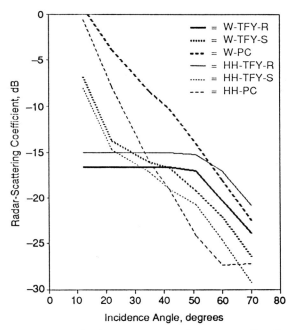

Fig. 5-15. The radar backscatter response is shown as a function of angle for smooth first-year ice of pancake origin (pans of 2- to 3-m diameters, 4-cm-high rims, and a thickness of 1.65 m), and of pancake ice (1- to 2-m pans and a thickness of 10 to 15 cm). These observations were made at 9.6 GHz. (W = water, R = rough, S = smooth, and PC = pancake ice.)

dielectric constant are great and produce responses with intensities that may be confused with the responses and intensities produced by multiyear ice. A further examination of the importance of surface roughness for first-year ice is provided in the discussion of laboratory measurements (see Chapter 9).

Additionally, in-situ observations by Onstott [1991] show that multiyear ice with a smooth snow–ice interface produces up to 4 dB greater backscatter than multiyear ice with a rough interface, which confirms theoretical predictions [Kim et al., 1984a] shown in Figure 5-6. It is also observed that a relatively smooth interface is typical. The importance of surface roughness is further examined in the sections discussing the properties of the low-density ice layer of multiyear ice (Sections 5.3.6 and 5.3.7). Measured roughness statistics are compiled in Table 5-3 to illustrate the variation in roughness observed for sea ice grown in the laboratory and for in-situ observations.

5.3.4 Dielectric Constant and Salinity

The dielectric constant of ice in the upper portion of the ice sheet is known to be one of the key parameters that set the absolute backscatter level. Observations suggest that once first-year ice attains an ice thickness greater than about 10 cm during cold conditions, the differences between the dielectric properties of first-year and multiyear ice become of secondary importance, since the ice types become more similar than different. It is also observed that surface roughness and the presence of a large number of discrete scatterers (of sizes within an order of magnitude of the radar wavelength in the medium) near the ice–snow interface begin to dominate and account for the differences in backscatter level. Additional discussions of the importance of the dielectric constant are found throughout this chapter and in the discussion of laboratory measurements (see Chapter 9).

Differences in dielectric constant may be exploited through the examination of the polarization ratio $P_r = \sigma°_{VV}/\sigma°_{HH}$. This ratio is expected to have a value of 1 (0 dB) at an incidence angle of 0° and to increase with increasing incidence angle. In addition, the larger the dielectric constant, the larger the possible polarization ratio for observations of calm open water, new ice (5.5 cm), thin first-year ice (24 and 42 cm), and multiyear ice during winter. (See Figure 5-16.) These data were obtained during March for CEAREX'89. The average rms deviation for these data is 0.7 dB. The complex dielectric constant is related to the salinity. The greater the salinity, the greater the dielectric constant. In a less rigorous sense, the polarization ratio is shown to respond to salinity—the largest ratio being associated with the entity

TABLE 5-3. Measured surface-roughness statistics.

Summer/ Winter	Thickness, cm	Ice type	RMS roughness, cm	Correlation length, cm	Lab or field	Sample length, cm	Comments
W	0.3	C–FY	0.065	—	L	—	Estimated from backscatter response
W	0.8	F–FY	0.032±0.002	0.740±0.391	L	60	
W	1.0	F–FY	0.038±0.008	1.000±0.410	L	180	
W	1.5	C–FY	0.020±0.002	0.490±0.159	L	44	
W	2.6	F–FY	0.033±0.019	1.717±0.771	L	125	After 7-cm snowfall
W	3.0	C–FY	0.069±0.006	0.454±0.030	L	66	
W	5.0	C–FY	0.038±0.010	2.69±2.180	L	68	
W	5.2	F–FY	0.031±0.012	1.261±0.807	L	120	
W	6.5	C–FY	0.048±0.001	0.745±0.109	L	44	
W	7.1	F–FY	0.077±0.0245	1.527±0.692	L	100	
W	7.5	C–FY	0.096	1.1112	L	30	
W	9.0	C–FY	0.233±0.032	1.742±0.466	L	75	
W	10.0	C–FY	0.119±0.018	1.447±1.235	L	50	
W	12.0	C–FY	0.059±0.001	1.70±0.580	L	40	
W	15.0	C–FY	0.026±0.000	0.694±0.259	L	34	
W	20.0	C–FY	0.340±0.031	3.652±0.922	L	40	Desalinated first-year ice
W	150	C–FY	0.053	1.736	F	58	Formed under calm conditions
S	200+	C–FY–R	0.766	7.75	F	1000	Early summer (6/13): still cold
S	200+	C–FY-S	0.108	0.538	F	150	Early summer (6/13): still cold
S	200+	C–FY–R	0.493	2.778	F	400	Summer (6/20): showing superimposed ice
W	200+	Typical SY–HM	0.277±0.120	5.651±3.88	F	85	Formation site same as on 6/13: smooth
W	200+	SY–HM–WD	0.262±0.204	6.015±5.44	F	111	
W	200+	SY–MP–S	0.075±0.159	2.402±0.728	F	102	
W	200+	Cusped SY–MP	0.209±0.090	3.479±3.2	F	150	
W	200+	MY–HM–S	0.185±0.035	2.75±1.05	F	60	
W	200+	MY–HM–R	0.890±0.170	3.90±6.20	F	60	
W	200+	MY–MP–S	0.080±0.030	2.8±1.9	F	60	

C	=	congelation ice	HM	=	hummock	R = rough
F	=	frazil ice	MP	=	melt pool	S = smooth
						WD = well-developed

with the highest salinity (i.e., seawater), and the smallest ratio being associated with the entity with the lowest salinity (i.e., the pure uppermost portion of multiyear ice). The result shown for the multiyear ice has an additional twist that is equally important. Given that the responses for VV and HH are nearly identical, the discrete scatterers in the upper ice sheet must be dominated by spherical gas bubbles (i.e., there is no polarization preference). This result is expected and is based on examining the bubble shapes by studying numerous thin sections of ice.

5.3.5 Brine Surface Layer and Brine-Enriched Slush

A brine surface layer may form through the expulsion of brine from the upper ice sheet to the surface, through the melting of the ice surface (by warming or solar heating), or through the process of dehydration. A brine-enriched slush layer will form through the wicking action of snow on an ice surface. Quantifying the contributions of each of these processes is difficult and is the subject of laboratory study. The impact of the surface layer and slush is also discussed in Chapter 9. In the case of young ice, questions arise as to the importance and the relationship of the thickness of a brine surface layer and backscatter response. First-year ice is often observed to have a surface wet with brine. In the case of either a bare or snow-covered surface, the presence of brine does not necessarily dominate the backscatter process by presenting a semi-infinite half-space with a high dielectric. Measured bulk dielectric constants are biased toward values approaching those of pure ice, rather than those of seawater.

5.3.6 Importance of Low-Density Ice Layer for Multiyear Ice Backscatter Characterization

Microwave signatures and physical properties have been acquired at a large number of multiyear sea ice stations in order to examine the physical-property variation in the

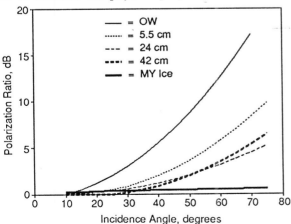

Fig. 5-16. The measured polarization ratio $P_r = \sigma^{\circ}_{VV}/\sigma^{\circ}_{HH}$ at 5.25 GHz is shown as a function of incidence angle for open water, 5.5-, 24-, and 42-cm-thick first-year ice, and multiyear ice. The average rms deviation for these data is 0.7 dB.

uppermost portion of the ice sheets [Onstott, 1991]. Variations in the thickness, density, bubble size, and roughness associated with ice of low density located in the uppermost portion of multiyear ice sheets were detected and documented. Selected examples are provided in Table 5-4 to illustrate the impact of these combinations of parameters on microwave signatures.

5.3.7 Physical and Microwave Properties of Multiyear Ice

Five multiyear ice sites have been selected for discussion. Four of these sites are from hummocks and one is from a freshwater melt pool. The low-density ice in the upper portion of the ice sheet is one of two sources that produce an enhanced backscatter. The pressure ridge, which is a topographical feature, is the second. To highlight the importance of the LDI layer, the backscatter response of three hummock areas (DS-7, DS-9, and DS-13) and a freshwater melt pool (DS-MP) that were within a 50-m radius and were resident on the same multiyear ice floe will initially be considered. Later, a fourth hummock case will be added to complete this discussion.

The critical difference between multiyear ice hummocks and melt pools is the number of discrete scatterers (i.e., gas bubbles) in the top 15 cm of the ice. In addition, the number of scatterers of a given bubble size may also be related to the density of the ice. The hummocks have densities of 457 to 517 kg/m³, while the melt pool has a density of 914 kg/m³ (i.e., the density of pure ice). As the LDI layer decreases, the number of discrete scatterers for the given bubble size increases. In the region from 30° to 60°, the difference in backscatter (>15 dB) between these two multiyear ice features is striking, and the dominant scattering mechanism becomes apparent.

The interface between snow and LDI may be smooth, moderately rough, or very rough. The rms heights range from 0.14 to 1.01 cm. Correlation lengths were very similar and range from 2.0 to 4.6 cm. It is necessary, typically, to characterize the LDI transitional layer, which is often composed of large globs and has a density that falls between the uppermost LDI layer and the pure ice below. Figure 5-5 illustrates the three ice layers and the range of the critical physical properties for the sites under discussion.

The angular responses of the backscatter for the three multiyear hummock sites discussed above are shown in Figure 5-17 for like- (VV and HH) and cross-polarizations (VH and HV). The width of the mean angular response interval for the drift station (DS) hummocks is about 5 dB for like-polarization; for cross-polarization, it is about one decibel wider. An examination of the ability to predict the ranking of these four sites according to absolute backscatter intensity has been performed. Basic rules of thumb have been derived for describing the impact on backscatter intensity for the range of physical-property parameters that were measured during the characterization of these sites. These rules have been supported through the use of a radiative transfer model of multiyear ice with a rough surface and embedded Rayleigh scatterers [Fung and Eom, 1982; Kim

TABLE 5-4. Characterization of the upper sheet of multiyear ice.

Parameter	Alpha-35	DS-7	DS-9	DS-13	DS-MP
σ, cm rms	0.33 ±0.2	0.14 ±0.02	0.7 ±0.12	1.01 ±0.31	0.08 ±0.03
ι, cm	4.7 ±0.21	2.0 ±1.3	4.6 ±1.2	3.2 ±1.2	2.8 ±1.9
LDI–1					
Thickness, cm	14	5.0 ±0.6	4.9 ±1.0	3.0 ±1.3	5.2 ±3
Salinity, ‰	0.0	0.0	0.0	0.0	0.0
Density, kg/m^3	815	457	513	513	914
Bubble diameter, mm	1.6	2.5	2.3	3.3	1.3
Void diameter, mm	0.5	8	7	5	0
LDI–2					
Thickness, cm	–	3.5 ±0.4	4.9 ±0.4	13.6 ±0.4	4.8 ±0.3
Salinity, ‰	–	0	0	0	0
Density, kg/m^3	–	728	728	929	919
Bubble diameter, mm	–	4	4.3	1	0
Void diameter, mm	–	2	0	0	0

Alpha-35, DS-7, DS-9, and DS-13 are hummock areas, DS-MP is a freshwater melt pool, and ι = correlation length.

et al., 1984a]. The results of a parametric study are summarized:

- Increasing the bubble diameter from 2 to 3 or 4 mm increases the like-polarization return by 5 and 8 dB and the cross-polarization return by 8 and 13.5 dB, respectively. Hence, the depolarization ratio ($\sigma°_{like}/\sigma°_{cross}$) decreases from 10 dB to 7 and 4.5 dB, respectively.

- Increasing the rms roughness from 0.125 to 0.5 causes a reduction in $\sigma°$ of 2 to 3 dB (θ_i = 0° to 55°), with the decrease increasing with incidence angle. Increasing the roughness to 1.0 cm causes an additional decrease of 3 to 6 dB.

- Increasing the density from 500 to 600 or 700 kg/m^3 causes a reduction in $\sigma°_{like}$ of 1 and 1.5 dB, respectively. For the same increases, $\sigma°_{cross}$ showed 2 and 3 dB reductions.

- If the thickness of the LDI layer is changed from 5 to 10 or 20 cm, then $\sigma°_{like}$ increases by 2.5 to 5.5 dB, respectively, and $\sigma°_{cross}$ increases by 5 and 10 dB, respectively.

- Varying the correlation length from 2 to 5 cm produces little effect.

The parameter set for Site DS-7 may be considered typical. Site DS-9 produced the weakest like-polarization response in the hummock set (DS-7, DS-9, and DS-13); this is largely attributed to its greater rms roughness. As is noted in the parametric study, an increase in surface roughness will cause a reduction in the like-polarization response, but little reduction in the cross-polarization response, although some reduction in the cross-polarization response occurs at the largest angles. The cross-polarization response for this site falls in the middle of the response range for these sites. Site DS-13 fits in the upper portion of both the like- and cross-polarization response ranges. Its LDI layer is unique in that the uppermost portion is relatively thin, but is of very low density (513 kg/m^3), and is followed by a layer very high in density (914 kg/m^3) that contains some extremely large globs (diameters of 1 cm). Its ranking is attributed to bubbles that are 30% larger than those found at Sites DS-7 and DS-9 and that contribute to large cross sections and reduced depolarization ratios.

Site Alpha-35 represents a case where the LDI thickness is large (15 cm), but of high density (815 kg/m^3). Bubble diameters are 50% smaller than the typical case. Both the high density and the smaller bubbles suggest a reduced backscatter intensity. This combination results in a backscatter that is 10 dB or more weaker than those of the multiyear hummocks described earlier.

In a simplistic sense, what these five responses show is that the backscatter intensity increases with decreasing density (i.e., the increase in the volume of discrete scatterers) in the upper portion of the ice (it is assumed that all other parameters are held constant). The development of the LDI layer is observed [R. G. Onstott, field observations, 1982] to be associated with either thick snow layers or snow drifts. During summer melt, these areas are the last to melt, or they melt more slowly. Areas of thin snow melt quickly and produce melt pools. Water is then free to drain from the drift areas into the surrounding melt pool strings, further slowing the melt process and allowing a transition to a layer with larger ice crystal diameters and lower density. It is reasonable to say that this layer is produced by a joint process that includes the metamorphosis of snow and the simultaneous erosion of the upper ice sheet.

5.4 ICE TYPE BACKSCATTER SUMMARY

In this section the active microwave behavior of sea ice is discussed further in terms of its response with angle and frequency time to evolution, season, and region.

5.4.1 Angle and Frequency Behavior

During winter, the microwave signatures of multiyear ice are clearly different from those of the more highly saline first-year ice. The situation in summer is more complex: Summer is the time of melting snow and ice, of melt pool formation, and desalination. In winter, the active microwave backscatter of first-year and multiyear ice increases linearly with increasing frequency (see Figure 5-6). This is true for ice surfaces that are smooth and rough, and for ice with air bubbles in its upper layers. Multiyear and first-year ice can be distinguished independent of frequency between 5 and 35 GHz. Discriminating between ice types becomes more difficult with the introduction of moisture into the snow and the warming of the ice sheets (see Figures 5-18, 5-19, and 5-20).

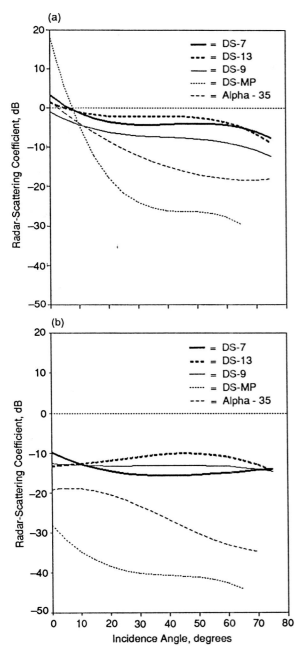

Fig. 5-17. The angular response of the scattering coefficients for four multiyear ice hummocks and a melt pool at 10 GHz and at (a) like- and (b) cross-polarizations. The responses at VV- and HH-polarizations are nearly identical, as are the VH and HV responses.

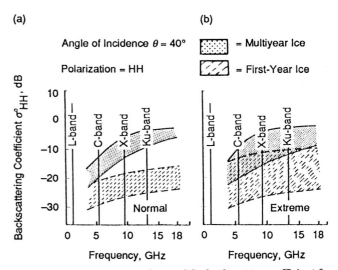

Fig. 5-18. Frequency dependence of the backscatter coefficient for first-year and multiyear ice at 40°. Theoretical $\sigma°$ for sea ice under (a) normal-winter and (b) extreme-summer conditions.

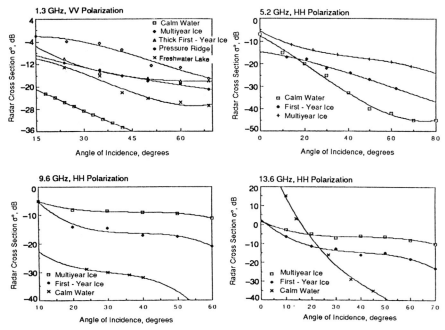

Fig. 5-19. Radar backscatter cross sections at L-, C-, X-, and Ku-bands during winter, illustrating the contrast in ice type and water signatures as a function of frequency and incidence angle.

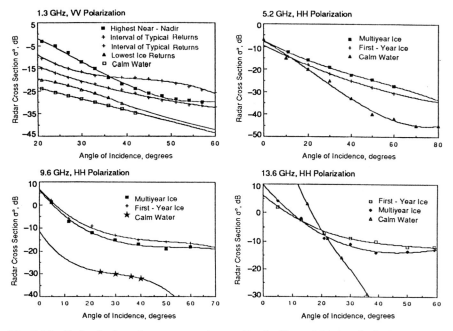

Fig. 5-20. Radar backscatter cross sections at L-, C-, X-, and Ku-bands during summer, illustrating the contrast in ice type and water signatures as a function of frequency and incidence angle.

5.4.2   Evolution of the Microwave Signature of First-Year Ice

The description of the evolution of the microwave signature of first-year ice begins with the freezing of open water and continues through the duration—up to nine months, at which time thicknesses of 200+ cm may be attained. Changes have been noted in the physical properties, the dielectric properties, and the surface-roughness statistics of an ice sheet, as well as the snow and frost flower layer. Figure 5-21 illustrates the change in the scattering coefficient as ice thickness increases from 0 to 200+ cm. Many facets of this function are of interest. The liquid ocean may take on a variety of cross-section levels due to wind and fetch conditions. New ice may produce a signature that is greater, the same, or less than that of the liquid ocean, depending on the environmental conditions at the time. In the calm conditions often encountered during lead formation, the new-ice signature may show an enhancement. As the dielectric constant of the ice sheet decreases (a function of ice aging), the backscatter intensity will reduce to a point where the lowest backscatter intensity for any ice condition (i.e., new ice that is smooth and cold) is reached. Within a few days, nilas transforms into young sea ice. When air temperatures are cold, ice crystals form as clumps (frost flowers) on the surface of the thin ice sheet. The colder the temperature, the thicker the ice becomes, and the more rapid the formation of frost flowers. As the ice sheet continues to thicken with time, the density (spatial) of frost flowers will increase, and the clumps of crystals may continue to increase in size until heights of about 3+ cm are attained. Brine from the thin ice sheet is wicked into the clumps, thereby producing a feature that has a dielectric constant similar to or higher than that of the ice sheet. These features represent roughness elements on a previously smooth dielectric plane. The temporal change in surface roughness (from smooth to very rough) appears to be linked to age and meteorological conditions (the nature of the temporal change is the topic of ongoing research) and provides the opportunity to use synthetic aperture radar (SAR) to obtain improved thickness

information for young sea ice. The greatest backscatter intensity for undisturbed first-year ice, almost that of multiyear ice, may be obtained when the frost flower layer is fully developed. However, with time and the eventual infiltration of snow and wind, the structure of the frost flowers becomes irreversibly altered. This has been observed to occur when a thickness of about 35 cm is reached [R. G. Onstott, field observations, 1987]. The snow and frost flower layer then undergoes a restructuring, which results in a steady reduction in the backscatter intensity for the ice sheet until thicknesses near 200 cm are reached.

5.4.3 Seasonal Evolution

The microwave signatures of sea ice may change dramatically with the season. With the continual monitoring of backscatter intensity levels, the determination of the physical state of ice sheets may be possible. This includes obtaining knowledge of the initiation of summer melt, the progression and intensity of summer melt, the duration of the melt cycle, the initiation of fall freeze-up, and the cooling of the ice sheets in late fall and winter (see Chapter 17). In Figures 5-22 and 5-23, a partial compilation of results from the many measurements presented in Table 5-1 is presented for both multiyear and first-year ice at 1.5, 5.25, and 9.5 GHz at an angle of 40° and at like-polarization (VV and HH are very similar in value). During late fall, winter, spring, and early summer, there is a significant separation between the scattering coefficients of multiyear and first-year ice at 5.25 and 9.5 GHz, whereas at 1.25 GHz there is little separation. For summer, however, the input of meltwater into the snowpack promotes a rapid merging of multiyear and first-year ice signatures at about June 15 [Onstott et al., 1984, 1987; Onstott and Gogineni, 1985; Livingstone et al., 1987b]. A period of indistinguishable ice signatures continues until midsummer. By midsummer, snow thickness is reduced to about half the year's high, allowing penetration of the radar signal to the snow–ice interface, which has experienced an important transformation. Percolation of free water onto an ice surface during late spring and early summer results in a superimposed ice layer, and processes associated with melt combine to enhance the small-scale surface roughness dramatically (Figure 5-24). The microwave response at midsummer marks a key backscatter trend reversal (see Figures 5-19, 5-20, and 5-25). The snow on multiyear ice is still thick and moist, and it dominates the backscatter process. In the case of first-year ice, the snow is thin and the surface is rough, producing a backscatter enhancement. Absorption losses decrease with increasing wavelength. Operation at 1.5 GHz was found to be optimal for discriminating between thin and thick ice types at this time. After midsummer, multiple contrast reversals may be observed. These are associated with melt and drain cycles. In a melt cycle, free water pools abundantly on an ice sheet. In a drain cycle, the ice sheet becomes less wet. By the end of summer, backscatter from multiyear ice once again becomes greater than that of first-year ice. However, this achievement is not associated with

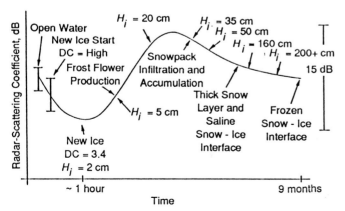

Fig. 5-21. The evolution of the microwave signature of first-year ice. (DC = dielectric constant and H_i = ice thickness.)

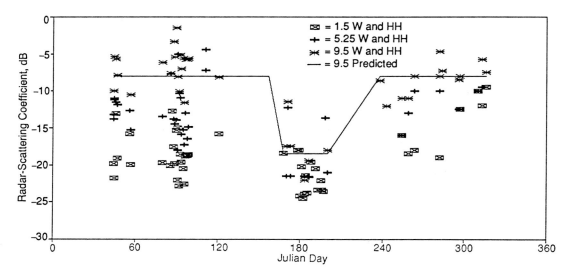

Fig. 5-22. Seasonal backscatter response for multiyear sea ice at 1.5, 5.25, and 9.5 GHz, an angle of 40°, and like-polarization (VV or HH). (W = water.)

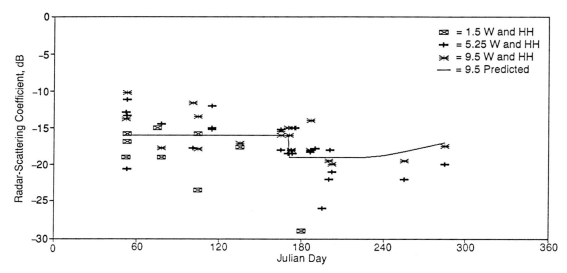

Fig. 5-23. Seasonal backscatter response for first-year sea ice at 1.5, 5.25, and 9.5 GHz, an angle of 40°, and like-polarization (VV or HH).

the return of volume scatter, as is the case in winter. On the small scale, the erosion effect of melt works to smooth exposed ice surfaces. First-year ice has large-scale surface undulations that are smaller in scale than the well-developed undulating topography of multiyear ice. A scene composed of many large-amplitude and rolling surfaces and edges has the potential to produce greater backscatter than a smoother scene. The transformations in the physical scene are illustrated in cross-sectional views provided in Figure 5-3.

Observations of the process of freeze-up during the fall show that by October, the signatures of first-year and multiyear ice are similar to those found during winter [Kim,

1984; Kim et al., 1984a; Onstott et al., 1984; Livingstone et al., 1987a; Onstott, 1990b, 1991]. However, the signature of young ice is greatly impacted. The lack of the severe cold temperatures characteristic of winter causes the congelation process to be disturbed and delayed and to become influenced by the accumulation of snow. This results in an ice sheet without a distinct snow–ice interface [Gow et al., 1990].

5.4.4 Regional Variation

Formation conditions impact the backscatter response of sea ice. For instance, first-year ice grown under calm conditions has a very smooth ice surface, whereas first-year

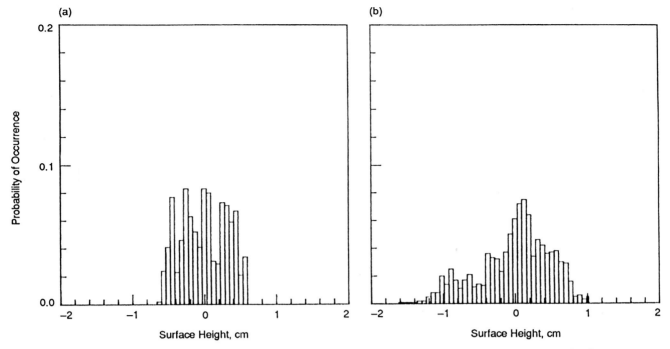

Fig. 5-24. First-year surface height histograms for a homogeneous ice sheet. (a) Data acquired June 14 are typical of early summer pre-melt, late fall, winter, and spring conditions. Height range, variance, and skewness were 0.932 cm, 0.044 cm², and 0.0003, respectively. (b) Data acquired June 20 are typical of early summer melt conditions. Superimposed ice increases the height range to 2.778, the variation to 0.243, and skewness to –0.588 [Onstott and Gogineni, 1985].

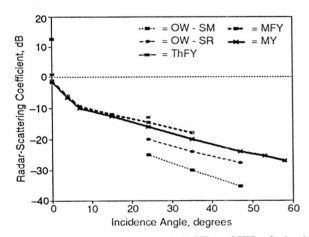

Fig. 5-25. The angular response at 5.25 GHz and HH-polarization for midsummer shown for open water, thin first-year (ThFY), medium first-year (MFY), and multiyear ice. The letters OW–SR indicate open water that is slightly rough, whereas OW–SM indicates open water that is smooth.

ice that began as pancake ice has a very rough surface. Their two backscatter responses are correspondingly different. In general, it is not well understood how or if the physical properties of a given ice type vary from region to region.

Passive microwave satellite data suggest that microwave property differences may be significant. Scatterometer observations have been carried out in five different geo-graphical regions during both cold (winter-like) and warm (summer-like) conditions. In general, it is found that a given type of ice is more similar than different in the various regions. Subtle differences are not always easily detected because of the limited number of observations. Observations over many years may be required to determine if region plays a critical role. It is anticipated that temporal weather fluctuations will play a dominant role in determining ice properties and signatures, except in regions (e.g., the Bering, Labrador, and Barents Seas) that are dynamic and characterized by deformation features. Moreover, major climatic differences may be important. For instance, the ice in the Antarctic Weddell Sea has so much snow that negative freeboards are common, and these affect radar backscatter.

5.5 SAR Observations

A large number of sea ice observations made by imaging radars were interpreted to determine the radars' ability to obtain geophysical information. It should be noted, precisely as was stressed by Moore [Ulaby et al., 1982], that these conclusions are based on a specific radar parameter set; one that is very often limited. In addition, the ability to obtain surface truth of these sensors is logistically difficult. This was especially true from 1956 to 1975. Much of the work to that point was qualitative and mainly supported by correlation with aerial photography.

The value of radar was noted, however. It was shown to be a significant tool and had an especially important prop-

(a)

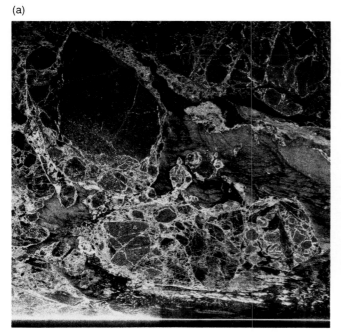

(b)

(c)

Fig. 5-26. Ice floe imagery obtained on March 21, 1989, during CEAREX with the ERIM/NADC P-3 SAR. Features include open water, nilas, gray ice, and multiyear ice. The polarization is VV and the incidence angle range goes from 20° to 70°. (a) An L-band (1.2-GHz) image, (b) a C-band (5.3-GHz) image, and (c) an X-band (9.4-GHz) image.

erty—that of being operable in all weathers and during the day or night, which is especially important in the dark and often cloud-covered regions in the Arctic—as well as the ability to provide geophysical information such as concentration, floe size, water openings, topographic features, fractures, and ice age [Ringwalt and MacDonald, 1956; Rouse, 1969; Johnson and Farmer, 1971; Loshchilov and Voyevodin, 1972; Ketchum and Tooma, 1973; Parashar, 1974; Gray et al., 1977, 1982; Campbell et al., 1978; Luther and Shuchman, 1980; Livingstone et al., 1981; Larson et al., 1981; Weeks, 1981; Lyden et al., 1984].

5.5.1 Examples of SAR Imagery

5.5.1.1. Sea ice during winter. A three-frequency (L-, C-, and X-band) image set is provided in Figure 5-26. These data were obtained in the Fram Strait during March when air temperatures were about –20°C. The instrument used was the Environmental Research Institute of Michigan/ Naval Air Development Center (ERIM/NADC) P-3 SAR, which has a resolution of about 1.8 m; incidence angles ranged from 20° to 70°, and the area coverage of an image was about 10 km.

5.5.1.2. L-band. Beginning with the L-band image, the features that stand out most strikingly are those associated with large-scale topographic features, which include pres-

sure ridges, ridge lines, fractures, and rubble. Other features include the open-water region (wind speed was about 5 m/s), the dark return from nilas, the bright return from some of the gray-ice areas, and the featureless return from homogeneous multiyear ice. The utility of an L-band radar [Onstott et al., 1991] to aid in the discrimination of open water, new ice, and nilas may be limited by the sensitivity of the radar (σ°_{min} = –35 dB), not the ice physics. Precise determination of the open water areas may require study or the utilization of HH-polarization data (water return areas are weaker than ice return areas by 5 to 7 dB). Determination of floe boundaries is often less precise at L-band than at other frequencies. The linear features seen in the 4-km multiyear floe are pressure ridges. Pressure ridges are composed of blocks of ice with tilted surfaces, and they may produce strong backscatter due to scattering from individual facets, multiple-facet scattering, and, in some cases, enhanced volume scattering due to a transition into a very low-density ice form [Gogineni, 1984; Livingstone et al., 1983].

5.5.1.3. C-band. Most striking in this image is the fact that floe boundaries are more defined and the old ice now has a much stronger backscatter than is the case at the lower frequency. This illustrates the important role volume scattering plays in the backscatter from sea ice. The areas that produce weak backscatter are either first-year ice or open water. The region of gray ice that runs diagonal in the image produces a moderate backscatter that is reasonably homogeneous and shows few ridge lines. It may be difficult to see from this image, but the open water provides the weakest backscatter and the nilas produces a slightly greater backscatter. Based on this image, it may be inferred that the calculation of the multiyear fraction may be obtained accurately.

5.5.1.4. X-band. Probably the most striking characteristic of this image is the similarity to C-band. It takes careful study to detect differences. It is not known what role the presentation (i.e., the normalization used in the preparation of the image) has played here, but scatterometer measurements have shown that backscatter at C- and X-band are more similar than different, with X-band showing a few decibels more dynamic range between ice types, but also a few decibels more variance.

5.5.1.5. Nilas and young ice. An image (Figure 5-27) obtained on March 20, 1989, during CEAREX at C-band (5.3 GHz) shows features that include open water, nilas, multiyear ice, and gray ice. The majority of the image is composed of gray ice of various ages and thicknesses that has the distinction of appearing like panes of broken glass. The brighter the image intensity, the thicker the gray ice in this case. The polarization is VV, and the incidence angle range goes from 20° to 70°. A vast floe located in the lower right-hand corner of the image moved at a drift rate different from that of the surrounding ice. Thin ice of various ages is found near the leeward side of this floe. The floes with linear boundaries and of similar sizes making up much of the center portion of this image are composed of young ice with thickness in the range of 5 to 50 cm. Their backscatter intensity is related directly to the development of frost flower formations and the infiltration of snow. The results of the CEAREX study will aid in determining if a continuum of ice thickness values may be retrieved. At present, only four major ice-type categories (open water, thin ice, first-year ice, and multiyear ice) are expected to be discriminated using satellite data [Kwok et al., 1991].

An image (Figure 5-28) obtained during a study of the fall freeze-up also shows that the contrast between the various ice types is very similar during winter and fall. As deter-

Fig. 5-27. Ice floe imagery obtained on March 20, 1989, during CEAREX with the ERIM/NADC P-3 C-band (5.3-GHz) SAR. Features include open water, nilas, gray ice, and multiyear ice. The majority of the image is composed of gray ice of various ages and thicknesses. The brighter the image intensity, the thicker the ice sheet. The polarization is VV, and the incidence angle range goes from 20° to 70°.

mined during CEAREX, the important forcing condition is a few days of cold weather. As noted above, during fall, nilas and young ice may present various backscatter intensities.

5.5.1.6. Sea ice during summer. With the input of free water into the snowpack during summer, microwave signatures may become progressively less variable. In Figure 5-29, three images (taken as part of MIZEX'84 on June 29) are presented to illustrate the character of sea ice at peak melt. The composition of the large multiyear floe includes an area where the snow was homogeneous (and thick) with no signs of surface pooling, areas with subsurface pooling of free water, and areas of first-year ice. The variability that is present is due to different degrees of surface wetness and different amounts of snow cover. By early summer to midsummer, the snowpack had experienced considerable melt, and depressions in the ice or snow with a low freeboard had collected varying quantities of meltwater. Fully open melt pools appear as "no return" areas, not dissimilar to the response of open water between floes, at both 1.2 and 9.4 GHz. In areas of melt pool formation and in areas where snow is being transformed into mixtures of snow, ice, and water, strong returns appear. For areas of thick and drained snow, weak returns appear at both frequencies, and the ice surface topography is well masked. As the frequency decreases, the contrast between areas with heavy and thin snow cover increases. In areas where melting has

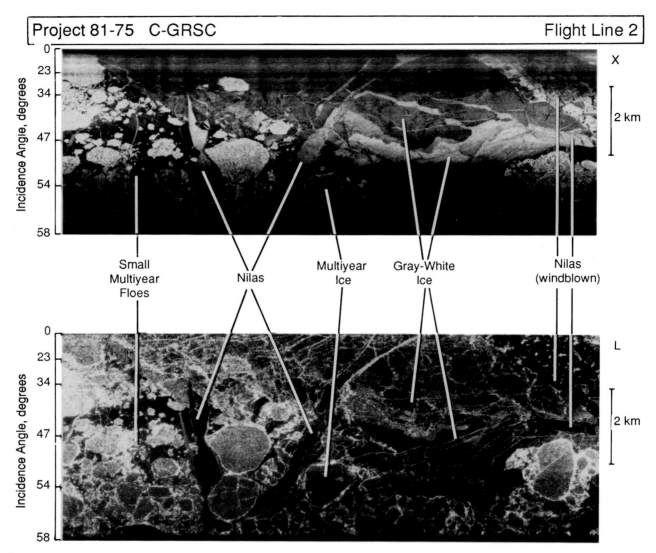

Fig. 5-28. Ice floe imagery obtained on November 1, 1981, during the Beaufort Sea Project at X-band (9.4 GHz) and L-band (1.2 GHz). Features include nilas, gray-white ice, and multiyear ice. The incidence angle range goes from 0° to 58°.

produced both a thin snow cover and enhanced small-scale surface roughness (on the order of 1 to 3 cm for first-year ice), strong returns are produced, especially at 1.2 GHz. Thus, later in summer, as the ice surface is exposed and surface melt ponds drain, the radar backscatter again reflects ice type differences [Onstott et al., 1987; Livingstone et al., 1987a, b; Cavalieri et al., 1990]. It was also determined that pressure ridges were very difficult to detect at midsummer, due to the heavy snow burden. At X-band, the brightest features detected at the multiyear floe were associated with thin first-year ice. Here, the presence of a thin snow cover and superimposed ice contribute to a relatively

strong backscatter. This is also apparent in the aerial photograph from the features that are gray in color. It should be noted that floes designated as multiyear ice are actually composites of multiyear and first-year ice of various ages. The first-year ice in effect welds the multiyear fragments together. This structure is apparent from further examination of the aerial photograph. The areas on a multiyear ice floe that experience the greatest melt are the first-year ice areas.

5.5.1.7. Post–peak melt. At the point after midsummer when 50% to 60% of the snow cover has melted, open-water

(a)

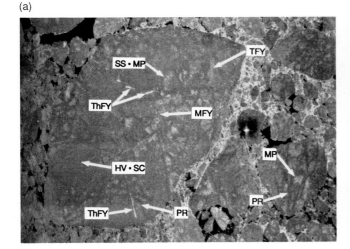

(b)

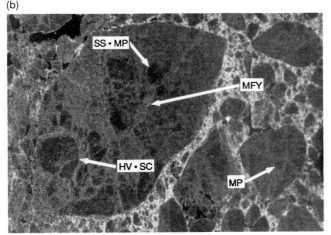

(c)

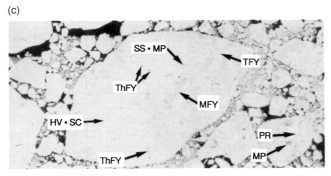

Fig. 5-29. Ice floe imagery obtained on June 29, 1984, during MIZEX'84 at (a) X-band (9.4 GHz), (b) L-band (1.2 GHz), and (c) with an aerial camera. Surface features include snow-covered ice (SC); heavily snow-covered ice (HV•SC); thick (TFY), medium (MFY), and thin first-year (ThFY) ice; pressure ridges (PR); sub-surface melt ponds (SS•MP); and melt pools (MP).

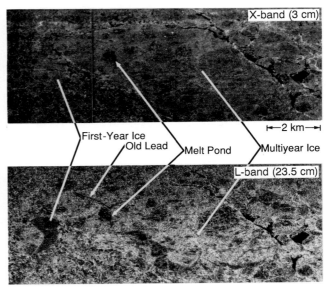

Fig. 5-30. Ice floe imagery obtained in July 1983, during MIZEX'83 at X-band (9.4 GHz) and L-band (1.2 GHz). Surface features include first-year ice, an old refrozen lead, a melt pond, and multiyear ice.

melt pools become common on thick ice. For the thin ice, the snowpack has now eroded into a 2-cm-thick granular snow–ice layer, and former melt pools consist of collections of candled ice tips that rise up to 1 cm above the freeboard of the thin, saturated ice sheet. Backscatter from multiyear ice is greater than or equal to that from first-year ice; a contrast reversal has taken place. After midsummer, first-year ice roughness elements have been eroded by melt to a point where they are small in relation to the radar wave-length; the surface appears smooth and produces weak backscatter (see Figure 5-30). Multiyear ice remains topo-graphically rough and has many tilted surfaces and a complex mixture of ice, snow, and water features that provide a strong surface scatter. At X-band, the image presented is almost featureless. First-year ice, an old lead, and multiyear ice produce similar backscatter levels. This indicates that the small-scale roughness for each of these features is very similar. Only the melt pool feature stands out. In the L-band image, first-year ice produces a weak backscatter (a trend reversal), but not as weak as that produced by open water between floes. It is spatially more homogeneous than the areas with probable subsurface pooling.

5.5.1.8. Tone and texture. To date, the primary exploita-tion of backscatter information is associated with intensity (tone) and standard deviation. The exploitation of spatial variation has been elusive to date, even though visual examination of imagery gives one the sense that it should be exploitable [Guindon et al., 1982; Burns and Lyzenga, 1984; Holmes et al., 1984; Barber, 1989; Shokr, 1990; Wackerman,

1991]. The spatial variability of physical and microwave properties for multiyear ice is considerable. The constituents that contribute to the signature of multiyear ice include pressure ridges, rubble areas, individual and strings of melt pools, and refrozen first-year leads. The spatial variability of first-year ice is primarily influenced by the distribution of pressure ridges, ridge lines, and fracture lines. Since there is no prescribed order to the physical characteristics of either first-year or multiyear ice, texture measures may be image dependent and not easily extrapolated. In addition, since many features are small in extent, the resolution impacts the results obtained. However, visual analysis uses textural clues, as well as floe shape, to establish ice-type information, with the requirement that the data are obtained at moderate resolution.

5.5.1.9. Pressure ridges. Multiyear ice ridges have cross sections only 3 to 4 dB greater than the background ice [Gray et al., 1982; Livingstone et al., 1983, 1987b]. First-year ridge returns are typically brighter than the background, with the probability that they scatter as strongly as the ridge returns of multiyear ice. Pressure ridges are characterized as long, linear features composed of blocks of broken ice. Strong backscatter is associated with tilted surfaces. It has been observed that old multiyear pressure ridges may be most easily detected if observed broadside and may not be detectable if observed along track. This may be attributed to the weathering of a ridge to a point where the backscatter is dominated by volume scattering, which is much less sensitive to the local incidence angle. This suggests that the reflection off the flat ice adjacent to an old ridge may be important in providing the needed backscatter enhancement that distinguishes the ridge from the background ice. The geometry is such that when the radar is aligned parallel to the ridge, this forward-scattering contribution is not present. In Figures 5-26 to 5-30 a number of ridges are identifiable. Note especially that during summer, the tilted blocks are coated with moist snow, which masks the backscatter from tilted blocks that is so important in winter. In addition, pressure ridges are more likely to be identified during this period as long, linear features of weak return surrounded by strings of melt pools.

5.6 SAR–Scatterometer Comparisons

A number of SAR and scatterometer intercomparisons have been made [Parashar, 1974; Livingstone et al., 1987a; Shuchman et al., 1989; Onstott and Shuchman, 1990]. These have included comparison of relative change between ice types, as well as of absolute change levels. One of the more important differences in these intercomparisons has to do with the sampling. Scatterometer observations are often conducted over homogeneous regions; for instance, for multiyear ice, a decision may be made to differentiate between pressure ridges and flat ice and to separate them into two populations. This is also true to a lesser extent in SAR analysis, and probably only for the prominent ridges. Melt pools and hummocks are often combined together. In general, the results from aircraft and in-situ observations are in agreement. Aircraft data often acquire more of a synoptic view of ice conditions, whereas in-situ observations obtain more detailed information, but at fewer locations.

5.7 Optimum Frequency, Polarization, and Incidence Angle

Many questions arise in the discussion of the optimum choices of frequency, polarization, and incidence angle. What if multiple frequencies and polarizations can be utilized? Furthermore, what if the coherent properties of clutter may be exploited? What then is optimum? Season and the set of ice types to be discriminated among may also have a bearing. At present, there is difficulty in specifying the best set of radar parameters. A general discussion of what appears reasonable follows.

5.7.1 Optimum Frequency

During winter, the critical mechanism in separating first-year and multiyear ice is discriminating backscatter dominated by volume scattering from that dominated by surface scattering. This works because the surface roughness for each of these ice types is similar. Observations suggest that surface scattering may increase with frequency squared. It has been shown [Kim, 1984; Kim et al., 1984a] that multiyear ice may be well modeled by spherical scatterers. In somewhat simplistic terms, this suggests that the optimum frequency will be one that exploits the fact that volume scattering dominates. Rayleigh-scattering cross sections increase with λ^6 (but losses reduce the overall effect to λ^4), so one should choose a wavelength λ such that surface scattering remains the principal backscatter mechanism for first-year ice; also, the wavelength should be short enough for strong multiyear ice volume scatter.

Frequency responses for frequencies from 1 to 100 GHz are shown in Figure 5-31 for multiyear and first-year ice. These data suggest that the contrast increases with increasing frequency and that a minimum occurs at frequencies below 2 GHz. The upper limit is not well-defined. A family of theoretical predictions (using radiative transfer theory) has been developed to examine the optimum frequency for the case of HH-polarization and an incidence angle of 40°. These results are presented in Figure 5-18. Measurement data are also included to provide a comparison of a typical backscatter response with the range of responses possible for observed physical-property values. This work suggests that the optimum frequency may occur in the Ku- or Ka-band portion of the microwave region. Observations and predictions argue that discrimination at C-band will be much better than at L-band, and that results at X-band will be similar to those at C-band, but with additional dynamic range.

During summer, it was shown [Onstott and Gogineni, 1985; Livingstone et al., 1987a, b; Onstott et al., 1987] that at high frequencies, the moist snowpack produced sea ice images that were essentially featureless, except that open water could easily be distinguished from sea ice. The ability to discriminate improved with decreasing frequency, with the greatest separation at L-band. The underlying process was attributed to the ability to penetrate and sense the rough ice–snow interface. For all-season capability, selection of a C-band frequency is a very reasonable choice when satellite operation is limited to one frequency.

5.7.2 Optimum Polarization

The choice of the optimum polarization is less obvious than the choice of the optimum frequency. For many ice forms, VV and HH are very similar. However, for open water and calm conditions, the cross section at VV-polarization is 5 to 7 dB greater than at HH-polarization. For very thin ice, VV-polarization may have a cross section 2 to 3 dB greater than that of HH-polarization. Cross-polarization has been shown to increase the range between multiyear and first-year returns by an additional 3 dB [Parashar, 1974; Onstott et al., 1979; Livingstone et al., 1983]. This is attributed to the very weak depolarization that occurs for smooth and slightly rough surfaces. Volumes composed of discrete scatterers or dielectric discontinuities give rise to a strong depolarization component. This is well illustrated by comparing a multiyear hummock (many discrete scatterers) and a melt pool (few discrete scatterers). However, the discrimination among young ice types is hampered if cross-polarization is used, because intensity differences due to surface scatter are suppressed. The choice of polarization is often driven more by system considerations, such as ease of antenna design or the amount of available transmitter power. An intercomparison of the contrast between multiyear ice, first-year ice, and thin ice is shown in Figures 5-32 to 5-34 for VV-, HH-, and VH-polarizations.

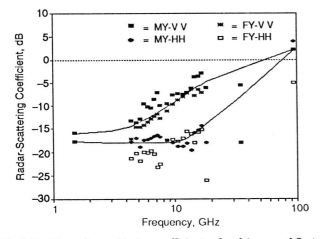

Fig. 5-31. The radar-scattering coefficients of multiyear and first-year ice are shown as a function of frequency and polarization, and at an angle of 40°. This response is for winter conditions.

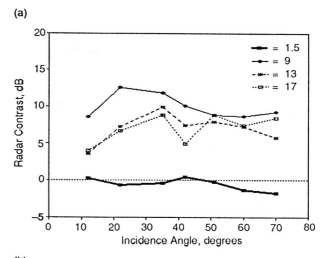

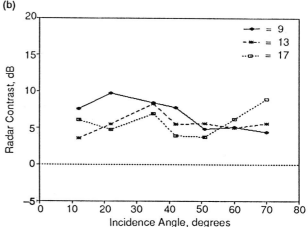

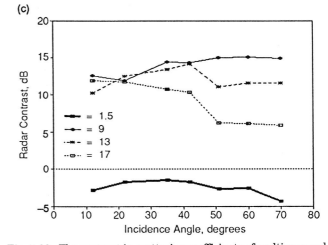

Fig. 5-32. The contrast in scattering coefficients of multiyear and first-year ice is shown as a function of polarization, angle, and frequency. In (a) the polarization is VV, in (b) it is HH, and in (c) it is VH.

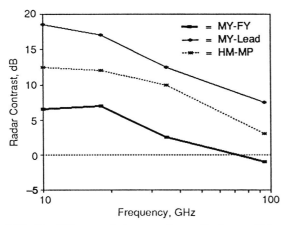

Fig. 5-33. The differences between radar-scattering coefficients of multiyear ice, first-year ice, lead ice (30-cm thick), and a multiyear hummock and melt pool are shown for fall conditions. The incidence angle is 40°, and the polarization is VV.

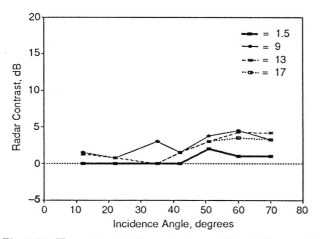

Fig. 5-34. The contrast in the scattering coefficient of gray-white ice and first-year ice is shown as a function of frequency. The polarization is VV.

TABLE 5-5.   Radar lookup tables developed for ASF GPS and ERS-1.

Ice season	Description	Period
Winter and early spring	Cold conditions	November–April
Late spring	Warming conditions	May and June
Early summer	Moist snow	June
Midsummer	Peak melt	Late June and July
Late summer	Post-peak melt	July and August
Fall	Freezing conditions	September and October

5.7.3 Optimum Incidence Angle

Based purely on backscattering phenomena, the optimum incidence angle range is from 20° to 70° (the middle angles), with the implication that the optimum angle may be polarization and ice-type sensitive. Angles near vertical are dominated by surface scattering and coherence effects. In the angle region from 10° to 15°, the various scattering responses intersect, causing confusion. At the small grazing angles, backscatter originates from topographical features, if it occurs at all. An example of the polarization response for the contrast between multiyear and first-year ice for winter conditions is presented in Figures 5-32 and 5-33. In Figure 5-34, gray-white ice is contrasted with thick first-year ice.

5.8 RADAR LOOKUP TABLES FOR ASF GPS

The results of the in-situ scatterometer observations have culminated in providing a set of tables that contain scattering coefficients for three to four ice-type combinations. There are tables for six seasons that have been identified according to changes produced in ice backscatter. They include winter to early spring, late spring, early summer, midsummer, late summer, and fall. The seasons, the ice state conditions, and the time periods are provided in Table 5-5. Tables have been produced for the European Space Agency's Remote Sensing Satellite (ERS-1 operates at C–VV–23°), the Japanese Earth Resources Satellite (JERS-1 operates at L–HH–35°), and the NASA Earth Observing System satellite (EOS is designed to operate at X–HH–25°, as well as other frequencies and polarizations). Entries in the tables include major ice-type categories, the thickness range associated with each category, and radar-scattering coefficients described by means, standard deviations, and slope (e.g., for 20° to 26° in the case of the ERS-1 radar lookup table [RLT] series). Tables 5-6, 5-7, and 5-8 are used in conjunction with the Alaska Synthetic Aperture Radar Facility (ASF) Geophysical Processor System (GPS) to discriminate among ice types. Clutter categories are segmented, statistics of each segment (mean $\sigma°$ and its standard deviation) are compared with values contained in the lookup tables, and a decision is made as to ice type (see Chapter 19).

The validation of ASF GPS ice-type products began in March 1992. Study of the accuracy of ice-type discrimination done with ERS-1 data and by using the ASF GPS is the subject of a major validation campaign. Questions such as the influence of region on accuracy, the need to develop RLT's with regional adaptability, and the approach in selecting RLT's for use in temporal and seasonal studies of ice sheet processes are topics under consideration. Efforts at improving ice-type discrimination will consider the use of ancillary data (such as air temperature, snowfall predictions, and time of year), the incorporation of electromagnetic models for absolute-level adjustment, and the exploitation of temporal SAR data.

TABLE 5-6. Radar lookup tables developed for ASF GPS and ERS-1.

Ice type	Thickness, cm	$\sigma°$, dB	Standard deviation, dB	Slope, dB/deg
Winter to early spring				
MY	>220	−8.6	2.2	−0.08
TFY	70 to 220	−11.5	2.1	−0.77
THFY	20 to 70	−13.3	1.2	−0.17
OW	0	< −19.7	4.0	−0.04
Late spring				
MY	>220	−10.7	2.1	−0.27
TFY	70 to 220	−13.2	1.1	−0.22
OW	0	< −19.7	4.0	−0.04
Early summer				
MY SY TFY MFY	>70	−16.3	1.1	−0.33
THFY	20 to 70	−13.1	1.4	−0.16
OW	0	< −19.7	4.0	−0.04
Midsummer				
MY and TFY	>120	−16.3	1.1	−0.33
MFY	70 to 120	−14.7	1.5	−0.27
THFY	30 to 70	−13.1	1.4	−0.16
OW	0	< −19.7	4.0	−0.04
Late summer				
MY and TFY	>120	−16.8	1.8	−0.60
THFY	30 to 70	−18.4	2.2	−0.45
OW	0	< −19.7	4.0	−0.04
Fall				
MY	>150	−10.5	1.7	−0.04
FY	120	−12.5	1.9	−0.21
THFY and OW	0 to 30	< −19.7	2.0 to 4.0	−0.04 to −0.63

MFY = medium first-year ice

TABLE 5-7. Radar lookup tables developed for ASF GPS and JERS-1.

Ice type	Thickness, cm	$\sigma°$, dB	Standard deviation, dB	Slope, dB/deg
Winter to early spring				
MY	>220	−7.0	TBD	−0.28
TFY MFY THFY	20 to 220	−23.4	TBD	−0.67
OW	0	< −30.7	TBD	−0.53
Late spring				
MY	>220	−15.5	TBD	−0.28
TFY	70 to 220	−23.4	TBD	−0.67
OW	0	< −30.7	TBD	−0.53
Early summer				
MY SY TFY MFY	>70	−15.1	TBD	−0.15
ThFY	20 to 70	TBD	TBD	TBD
OW	0	< −30.7	TBD	−0.53
Midsummer				
MY	>220	−22.7	2.0	−0.54
TFY MFY	70 to 220	−19.8	TBD	−0.55
OW	0	< −30.7	TBD	−0.53
Late summer				
MY and TFY	>120	−21.9	TBD	−0.54
ThFY	30 to 70	−28.1	TBD	−0.67
OW	0	< −30.7	TBD	−0.53
Fall				
MY	>150	−17.3	TBD	−0.43
FY	30 to 120	−18.4	TBD	−0.43
OW	0	< −30.7	TBD	−0.53

TABLE 5-8. Radar lookup tables developed for ASF GPS and EOS X–HH–23°.

Ice type	Thickness, cm	$\sigma°$, dB	Standard deviation, dB	Slope, dB/deg
Winter to early spring				
MY	>220	−3.6	1.9	−0.25
FY	20 to 220	−14.2	1.8	−0.10 to −0.31
OW	0	< −29.7	0.3	0
Late spring				
MY	>220	TBD	TBD	TBD
TFY	70 to 220	TBD	TBD	TBD
OW	0	< −29.7	0.3	0
Early summer				
MY SY TFY MFY	>70	−15.9	0.7	−0.42
ThFY	20 to 70	−15.1	0.4	−0.11
OW	0	< −29.7	0.3	0
Midsummer				
MY and TFY	>70	−15.7	0.7	−0.11 to −0.42
ThFY	20 to 70	−14.7	0.35	−0.11
OW	0	< −29.7	0.3	0
Late summer				
MY and TFY	>120	TBD	TBD	TBD
ThFY	30 to 70	TBD	TBD	TBD
OW	0	< −29.7	0.3	0
Fall				
MY	≥150	−6.9	2.4	−0.05
FY	>70	−13.5	0.9	−0.31
ThFY	30 to 70	−14.4	1.2	−0.10

5.9 FUTURE OPPORTUNITIES AND ISSUES

The ability to obtain temporal information about the polar regions is now becoming a reality. The polar regions in the United States have only been the subject of active microwave observation on an occasional basis, and then only for a few weeks at a time. This is beginning to change. Observations with moderate resolutions (i.e., about 10 m) will become commonplace rather than the exception. The European Space Agency launched the ERS-1 SAR on July 17, 1991. Japan's National Space Development Agency launched the JERS-1 SAR on February 11, 1992. The Canadian Space Agency has a 1994 launch scheduled for its Radarsat SAR (C–HH–30° to 50°). The United States is considering a multifrequency and multipolarization SAR for 2005. The former USSR launched a series of X-band RAR's (Kosmos/Ocean), starting in 1983, and S-band SAR's (ALMAZ) in 1987. With the potential of numerous sensors in space at once, it appears possible to have multifrequency, multiple polarizations, and multiple angle views. In addition, the frequency of coverage may allow for global mappings in enormous detail. The potential for retrieving very useful and timely information is upon us.

Future efforts will be directed toward the use of multiparameter and multitemporal data. As suggested above, to obtain global coverage, various sensors may be used, with probable differences in operating parameters. In the past, work has centered on the exploitation of a single image at a time. This will change. Both the study of the seasonal response (microwave and physical property) of sea ice and the availability of multitemporal SAR data allow for a very large next step, that is, exploitation of *time evolution*. With such an assessment, the ability to reduce signature ambiguities is possible, especially if multiple frequencies, polarizations, and angles are included. Given the proper information (including ancillary data) and continuous coverage, ice-thickness, mass-balance, and heat-transfer estimates will improve.

At present, discrimination between ice forms is by ice type, rather than by thickness. Ice thickness has been suggested to be an important indicator of global change and is known to be important in heat–mass balance predictions. Future research efforts will be directed to improving this ability to determine thickness. It is hypothesized that a continuum of thickness values will be retrieved by incorporating an understanding of the evolution of first-year ice (with links between microwave signatures, physical properties, and environmental conditions). This, coupled with temporal SAR data, certainly promises an exciting future and dramatic improvements in the retrieval of geophysical information about our polar regions.

REFERENCES

Anderson, V. H., High altitude, side looking radar images of sea ice in the Arctic, *Proceedings of the Fourth Symposium on Remote Sensing of the Environment*, pp. 845–857, Willow Run Laboratory, Ann Arbor, Michigan, 1966.

Barber, D. G., *Texture Measures for Sea-Ice Discrimination: An Evaluation of Univariate Statistical Distributions*, Report ISTS-EOL-TR89-005, Department of Geography, University of Waterloo, Waterloo, Ontario, Canada, 1989.

Burns, B. A. and D. R. Lyzenga, Textural analysis as a SAR classification tool, *Electromagnetics, 4*, pp. 309–322, 1984.

Campbell, W. J., J. Wayneberg, J. B. Ramseyer, R. O. Ramseier, M. R. Vant, R. Weaver, A. Redmond, L. Arsenault, P. Gloersen, H. J. Zwally, T. T. Wilheit, T. C. Chang, D. Hall, L. Gray, D. C. Meeks, M. L. Bryan, F. T. Barath, C. Elachi, F. Leberl, and T. Farr, Microwave remote sensing of sea ice in the AIDJEX main experiment, *Boundary-Layer Meteorology, 13*, pp. 309–337, 1978.

Cavalieri, D. J., B. B. Burns, and R. G. Onstott, Investigation of the effects of summer melt on the calculation of sea ice concentration using active and passive microwave data, *Journal of Geophysical Research, 95(C4)*, pp. 5359–5369, 1990.

Dunbar, M., *A Glossary of Ice Terms (WMO Terminology), Ice Seminar, Special Volume 10*, pp. 105–110, The Canadian Institute of Mining and Metallurgy, 1969.

Dunbar, M. and W. Weeks, *The Interpretation of Young Ice Forms in the Gulf of St. Lawrence Using Radar and IR Imagery*, DREO Report 711, Research and Development Branch, Department of National Defense, Ottawa, Canada, 1975.

Fung, A. K. and H. J. Eom, Application of a combined rough surface volume scattering theory of sea ice and snow, *IEEE Transactions on Geoscience and Remote Sensing, GE-20(4)*, pp. 528–536, 1982.

Glushkov, V. M. and V. Komarov, Side-looking radar system Toros and its application to the study of ice conditions and geological exploration, *Proceedings of the Seventh International Symposium on Remote Sensing of the Environment, 1*, p. 317, Willow Run Laboratory, Ann Arbor, Michigan, 1971.

Gogineni, S. P., *Radar Backscatter From Summer and Ridged Sea Ice, and the Design of Short-Range Radars*, Ph.D. dissertation, The University of Kansas, Lawrence, Kansas, 1984.

Gow, A. J., D. A. Meese, D. K. Perovich, and W. B. Tucker, The anatomy of a freezing lead, *Journal of Geophysical Research, 95(C10)*, pp. 18,221–18,232, 1990.

Gray, A. L., R. O. Ramsier, and W. J. Campbell, Scatterometer and SLAR results obtained over Arctic sea ice and their relevance to the problem of Arctic ice reconnaissance, *Proceedings of the Fourth Canadian Symposium on Remote Sensing*, pp. 424–443, 1977.

Gray, A. L., R. K. Hawkins, C. E. Livingstone, L. D. Arsenault, and W. M. Johnstone, Simultaneous scatterometer and radiometer measurements of sea ice microwave signatures, *IEEE Journal of Oceanic Engineering, OE-7(1)*, pp. 20–33, 1982.

Guindon, B., R. K. Hawkins, and D. G. Goodenough, Spectral-Spatial analysis of microwave sea ice data, *Proceedings of the IGARSS'82, 2*, pp. 4.1–4.8, Munich, 1982.

Holmes, Q. A., D. R. Nuesch, and R. A. Shuchman, Textural analysis and real-time classification of sea-ice types using digital SAR data, *IEEE Transactions on Geoscience and Remote Sensing, GE-22*, pp. 113–120, 1984.

Johnson, J. D. and L. Farmer, Use of side-looking radar for sea ice identification, *Journal of Geophysical Research, 76*, pp. 2138–2155, 1971.

Ketchum, R. D., Jr., and S. G. Tooma, Jr., Analysis and interpretation of air-borne multifrequency side-looking radar sea ice imagery, *Journal of Geophysical Research, 78*, pp. 520–538, 1973.

Kim, Y. S., *Theoretical and Experimental Study of Radar Backscatter From Sea Ice*, Ph.D. dissertation, University of Kansas, Lawrence, Kansas, 1984.

Kim, Y. S., R. K. Moore, and R. G. Onstott, *Theoretical and Experimental Study of Radar Backscatter From Sea Ice*, Remote Sensing Laboratory Technical Report 331-37, University of Kansas, Lawrence, Kansas, 1984a.

Kim, Y. S., R. G. Onstott, and R. K. Moore, Effect of snow cover on microwave backscatter from sea ice, *IEEE Journal of Oceanic Engineering, OE-9(5)*, pp. 383–388, 1984b.

Kwok, R., E. Rignot, B. Holt, and R. Onstott, An overview of the geophysical sea ice products generated at the Alaska SAR Facility, *Journal of Geophysical Research 97(C2)*, pp. 2391–2402, 1991.

Larson, R. W., J. D. Lyden, R. A. Shuchman, and R. T. Lowry, *Determination of Backscatter Characteristics of Sea Ice Using Synthetic Aperture Radar Data*, 110 pp., ERIM Final Report 142600-1-F, Environmental Research Institute of Michigan, Ann Arbor, Michigan, 1981.

Livingstone, C. E., R. K. Hawkins, A. L. Gray, K. Okamoto, T. L. Wilkinson, S. Young, L. D. Arsenault, and D. Pearson, Classification of Beaufort Sea ice using active and passive microwave sensors, *Oceanography From Space*, pp. 813–826, edited by J. F. R. Gower, Plenum, New York, 1981.

Livingstone, C. E., R. K. Hawkins, A. L. Gray, L. Drapier, L. D. Arsenault, K. Okamoto, T. L. Wilkinson, and D. Pearson, *The CCRS/Sursat Active-Passive Experiment 1978–1980: the Microwave Signatures of Sea Ice*, Canada Centre for Remote Sensing, Ottawa, 1983.

Livingstone, C. E., R. G. Onstott, L. D. Arsenault, A. L. Gray, and K. P. Singh, Microwave sea-ice signatures near the onset of melt, *IEEE Transactions on Geoscience and Remote Sensing, GE-25(2)*, pp. 174–187, 1987a.

Livingstone, C. E., K. P. Singh, L. D. Arsenault, and A. L. Gray, Seasonal and regional variations of active/passive microwave signatures of sea ice, *IEEE Transactions on Geoscience and Remote Sensing, GE-25(2)*, pp. 159–173, 1987b.

Loshchilov, V. S. and V. A. Voyevodin, Determining elements of drift of the ice cover and movement of the ice edge by the aid of the Toros aircraft lateral scan radar station, *Problemy Artiki i Antartiki, 40*, pp. 23–30, 1972.

Luther, C. A. and R. A. Shuchman, *The Proceedings of the Final Sursat Ice Workshop Summary Report*, 11 pp., edited by R. O. Ramseier and D. J. Lapp, Atmospheric Environment Service, Toronto, 1980.

Lyden, J. D., B. A. Burns, and A. L. Maffett, Characterization of sea ice types using synthetic aperture radar, *IEEE Transactions on Geoscience and Remote Sensing, GE-22(5)*, pp. 431–439, 1984.

Onstott, R. G., MIZEX'84 multifrequency helicopter-borne altimeter observations of summer marginal sea ice, *Proceedings of the IGARSS'90 Symposium,* pp. 2241–2244, College Park, Maryland, 1990a.

Onstott, R. G., Near surface microwave measurements of Arctic sea ice during the fall freeze-up, *Proceedings of the IGARSS'90 Symposium,* pp. 1529–1530, College Park, Maryland, 1990b.

Onstott, R. G., Active microwave observations of Arctic sea ice during the fall freeze-up, *Proceedings of the IGARSS'91 Symposium,* pp. 821–824, Espoo, Finland, 1991.

Onstott, R. G. and S. P. Gogineni, Active microwave measurements of Arctic sea ice under summer conditions, *Journal of Geophysical Research, 90(C3),* pp. 5035–5044, 1985.

Onstott, R. G. and R. A. Shuchman, Scatterometer measurements of wind, waves and ocean fronts during NORSEX, *Proceedings of the IGARSS'89 Symposium,* pp. 1084–1088, Vancouver, Canada, 1989.

Onstott, R. G. and R. A. Shuchman, Comparison of SAR and scatterometer data collected during CEAREX, *Proceedings of the IGARSS'90 Symposium,* pp. 1513–1516, College Park, Maryland, 1990.

Onstott, R. G., R. K. Moore, and W. F. Weeks, Surface-based scatterometer results of Arctic sea ice, *IEEE Transactions on Geoscience and Electronics, GE-17*, pp. 78–85, 1979.

Onstott, R. G., R. K. Moore, S. Gogineni, and C. V. Delker, Four years of low altitude sea ice broadband backscatter measurements, *IEEE Journal of Oceanic Engineering, OE-7(1)*, pp. 44–50, 1982.

Onstott, R. G., Y. S. Kim, and R. K. Moore, *Active Microwave Measurements of Sea Ice Under Fall Conditions: The Radarsat/FIREX Fall Experiment*, Technical report 331-30/578—Final, University of Kansas Remote Sensing Laboratory, Lawrence, Kansas, 1984.

Onstott, R. G., T. C. Grenfell, C. Mätzler, C. A. Luther, and E. A. Svendsen, Evolution of microwave sea ice signatures during early and midsummer in the Marginal Ice Zone, *Journal of Geophysical Research, 92(C7)*, pp. 6825–6835, 1987.

Onstott, R. G., R. A. Shuchman, and C. C. Wackerman, Polarimetric radar measurements of Arctic sea ice during the Coordinated Eastern Arctic Experiment, *Proceedings of the IGARSS'91 Symposium*, pp. 93–97, Espoo, Finland, 1991.

Parashar, S. K., *Investigation of Radar Discrimination of Sea Ice*, Ph.D. dissertation, CRES Technical Report 185-13, University of Kansas Center for Research, Inc., Lawrence, Kansas, 1974.

Parashar, S. K., R. M. Haralick, R. K. Moore, and A. W. Briggs, Radar scatterometer discrimination of sea ice types, *IEEE Transactions on Geoscience and Electronics, GE-15*, pp. 83–87, 1977.

Ramseier, R., P. Gloersen, W. Campbell, and T. Chang, Mesoscale descriptions for the principal Bering Sea Experiment, *Proceedings of the Final Symposium on the Results of the Joint Soviet–American Expedition*, pp. 231–236, edited by Y. Kondratyev et al., A. A. Balkema, Rotterdam, Netherlands, 1982.

Ringwalt, D. L. and F. C. MacDonald, *Terrain Clutter Measurements in the Far North*, Report of NRL Progress, Naval Research Laboratory, Washington, DC, 1956.

Rouse, J. W., Arctic ice type identification by radar, *Proceedings of the IEEE, 57(4)*, pp. 605–611, 1969.

Shokr, M. E., On sea-ice texture characterization from SAR images, *IEEE Transactions on Geoscience and Remote Sensing, 28(4)*, pp. 737–740, 1990.

Shuchman, R. A., C. C. Wackerman, A. L. Maffett, R. G. Onstott, and L. L. Sutherland, The discrimination of sea ice types using SAR backscatter statistics, *Proceedings of the IGARSS'89 Symposium*, pp. 381–385, Vancouver, Canada, 1989.

Ulaby, F. T., R. K. Moore, and A. K. Fung, *Microwave Remote Sensing—Active and Passive, Vol. II: Radar Remote Sensing and Surface Scattering and Emission Theory*, Addison–Wesley Publishing Company, Reading, Massachusetts, 1982.

Ulaby, F. T., R. K. Moore, and A. K. Fung, *Microwave Remote Sensing—Active and Passive, Vol. III: From Theory to Applications*, Artech House, Inc., Dedham, Massachusetts, 1986.

Wackerman, C. C., *Optimal Linear Combinations of Statistics and Texture Measures for SAR Sea Ice Classification*, presented at the IGARSS'91 Symposium, Espoo, Finland, 1991.

Weeks, W. F., Sea ice: The potential for remote sensing, *Oceanus, 4(3)*, pp. 39–47, 1981.

Wetzel, L. B., Sea Clutter, *Radar Handbook*, pp. 13.12–13.40, edited by M. Skolnik, McGraw-Hill, New York, 1990.

Chapter 6. Digital SAR Image Formation

CHRIS C. WACKERMAN

Environmental Research Institute of Michigan, P. O. Box 8618, Ann Arbor, Michigan 48107

6.1 INTRODUCTION

Synthetic Aperture Radar (SAR) systems differ from other sensors in that the data received and stored on the platform are not images, but an encoded version of the image that needs to be decoded before any analysis can be performed. This decoding, or image formation, process represents a significant computational burden and can cause artifacts within the final image data unconnected to the physical scattering processes being investigated. Since these artifacts can easily be misconstrued as part of the scene being imaged, it is very important to understand the process used to form a SAR image and the artifacts this process can introduce into the image before an accurate analysis can be performed. Section 6.2 will briefly describe this formation process and its image artifacts. After image formation, some postprocessing algorithms are applied to the SAR data and these are described in Section 6.3.

In the early days of SAR systems, image formation was performed optically, generating an image print that analysts could only inspect visually. Now almost all image formation is performed on computers, generating sampled images that allow a wider range of analyses. This has also prompted research into developing algorithms that automatically extract geophysical information from digital SAR imagery; a large amount of this effort is being performed in the area of sea ice properties. Such algorithms can be useful to an analyst in extracting the single geophysical quantity that is required (ice concentration, ridge locations, etc.) from the sometimes confusing information present in SAR images. Section 6.4 will briefly outline these algorithms for sea ice and how successful they have been.

6.2 DIGITAL IMAGE FORMATION

6.2.1 Image Formation Process

This section describes the general basis of strip-map SAR image formation algorithms. For more details, see Barber [1985], Brown et al. [1973], Elachi [1988], Jin and Wu [1984], Kirk [1975], Raney [1981], Rihaczek [1969], Skolnik [1990], Tomiyasu [1978], Van de Lindt [1977], Wu et al. [1982a, b], and Wu and Vant [1985].

In older radars, range resolution was determined by the length of the transmitted pulse. This required concentrating all the energy the pulse needed to propagate from the sensor to the ground and back again into a small time interval. However, it is not the pulse duration, but the pulse bandwidth that determines image resolution [Klauder et al., 1960]. Thus, a transmitted waveform with both a large bandwidth and a long duration can spread its energy over a relatively long amount of time and still generate fine range resolution. This solves the problem of having to push large amounts of energy through the transmitter in short amounts of time, but it creates the problem that the data received by the sensor have to be decoded before they can be analyzed. The extended pulses that are scattered back to the sensor from the ground have to be changed into narrow functions. This is done by correlating the received data with a copy of the transmitted waveform [Klauder et al., 1960].

The waveform transmitted by most SAR systems is referred to as a linear chirp waveform and is represented as

$$c(t) = w(t) \exp[i\pi b t^2 + i2\pi f_o t], \quad -T_p/2 \le t \le T_p/2 \qquad (1)$$

where $w(t)$ is the amplitude weighting of the waveform (usually a rectangular function), b is the chirp rate representing how quickly the waveform changes frequency, f_o is the carrier frequency determining the wavelength of the radar, and T_p is the pulse duration. Equation (1) represents both the transmitted waveform, as well as the function to correlate the received data to compress the extended pulses. Although Equation (1) represents the most common way to encode the pulse (i.e., spread the pulse energy over time), other methods are sometimes used [Hovanessian, 1984].

Similar to the range direction, azimuth resolution in older radars was generated by creating a very narrow azimuth beam. This required a very long antenna, which soon exceeded the practical limits for airborne systems. A SAR system uses a much shorter antenna, but generates fine azimuth resolution by creating a synthetic aperture in the azimuth direction (thus the term SAR), simulating the data that would have been collected by a much longer antenna. This is done by recording both the magnitude and phase of the signal backscattered by each scatterer during the time that the scatterer is within the beam of the radar. To form the image in the azimuth direction, these complex-valued responses need to be coherently summed with an appropriate phase shift to simulate what an antenna would generate if it were large enough in the azimuth direction to receive all of these responses simultaneously. The phase shifts are required because the sensor does not receive the data simultaneously, but rather over time as the sensor moves past the scatterer. This relative motion between the sensor and the scatterer causes a phase modulation that needs to be removed before the coherent sum is calculated. Just as in the range direction, this procedure is done by correlating the received data with the response in azimuth that the sensor would have received from a single scatterer on the ground. For most SAR systems, this azimuth response, $a(x)$, is approximated as

$$a(x) = \exp[i(k/R)x^2], \quad -VT_i/2 \leq x \leq VT_i/2 \quad (2)$$

where k is the wave number of the radar ($k = 2\pi/\lambda$ where λ is the wavelength of the transmitted waveform), R is the range from the sensor to the scatterer, V is the velocity of the sensor, T_i is the integration (i.e., the time that the scatterer is within the azimuth antenna beam), and x represents the azimuth distance. Equation (2) is derived by approximating the change in range, R, between the sensor and the scatterer as the sensor moves past the scatterer to be a quadratic function in x, then using this approximation to determine the appropriate phase corrections as a function of azimuth location [Elachi, 1988; Raney, 1981; Tomiyasu, 1978]. The azimuth response in Equation (2) is written as a one-dimensional function for clarity. Due to the change in range between the sensor and the scatterer, the azimuth response is actually defined along a curve that follows these range changes, often called the range migration problem. Note also that the azimuth response in Equation (2) depends on the range from the sensor to the scatterer, R, and thus the correlation function required to compress the data in the azimuth needs to change with the range location of the scatterer.

Putting the range and azimuth responses together, the data received by a SAR from a single scatterer in the scene are an extended two-dimensional function whose range extent is determined by the duration of the transmitted pulse and whose azimuth extent is determined by the azimuth beam width (i.e., how long the scatterer is within the azimuth beam). Figure 6-1 shows an example of this extended response from a single point scatterer for the (a) real and (b) imaginary parts of the response. The symmetry of the function results from the similarity of Equations (1) and (2). The data recorded by the SAR as it images a scene are a coherent sum of such responses overlapped from each scatterer in the scene.

The response in Figure 6-1 comes from a single point scatterer, and the image formation process turns this extended response into a point by correlating in the range direction with a copy of the transmitted pulse and correlating in the azimuth direction with a copy of the azimuth response. The output of this correlation process in either direction will be a sinc function ($\sin[\pi x]/[\pi x]$) whose magnitude is shown in Figure 6-2 plotted on a decibel scale. The mainlobe of the response contains the dominant energy and

its width, usually measured at the 3-dB locations, represents the resolution of the sensor. Unfortunately, the sinc function also contains sidelobes that leak the energy of the response away from the scatterer location. The peak sidelobe response or the integrated sidelobe energy are often used as measures of how significant the sidelobes will be in image distortions. In Figure 6-2, the first sidelobe is shown

(a)

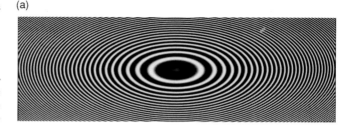

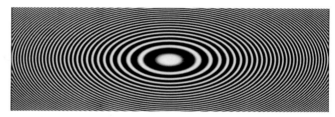

Fig. 6-1. Example of the extended response from a corner reflector recorded by a SAR sensor for the (a) real and (b) imaginary parts. Range is horizontal, azimuth is vertical.

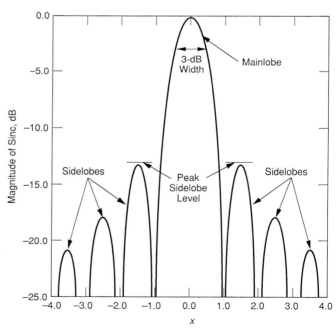

Fig. 6-2. The magnitude of a sinc function, the result of forming the image of a single scatterer. The 3-dB width of the mainlobe represents the sensor resolutions, and the sidelobes are sources of possible image artifacts.

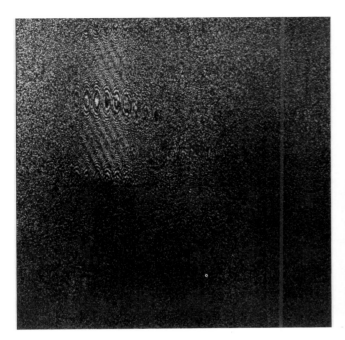

Fig. 6-3. Example of the real part of the data recorded by a SAR sensor over an actual scene. Range is vertical, azimuth is horizontal.

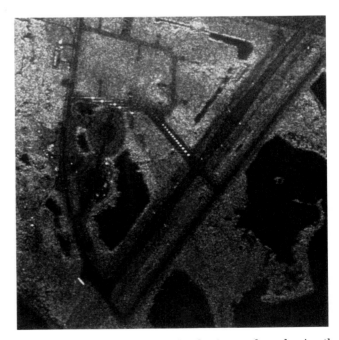

Fig. 6-4. Data in Figure 6-3 after having performed range compression.

13.5 dB down from the peak, but weighting techniques are routinely used to lower this considerably, although at the expense of broadening the mainlobe [Berkowitz, 1967].

This is a very general picture of SAR image processing and there are still many details that make the operation nontrivial, with range walk considerations and changes in the azimuth response for different ranges being the most important. Efficient ways of performing these operations provide the bulk of the work on SAR image formation algorithms [Jin and Wu, 1984; Wu et al., 1982a, b; Wu and Vant, 1985].

Figures 6-3 through 6-5 illustrate the SAR image formation procedure. Figure 6-3 shows the real part of a subset of SAR data recorded over a scene. Note that nothing is visually evident since these are the data received by SAR before image formation and thus each scatterer is represented by an extended two-dimensional function. There is a hint of a series of point responses in the upper left corner that resembles the model in Figure 6-1. Figure 6-4 shows the intensity of the same data after range compression. Each extended response has been compressed in the vertical, range, direction by correlating with the transmitted pulse, but is still elongated in the azimuth direction. Note that a significant amount of image information is now visible. Figure 6-5 shows the intensity of the data after azimuth compression, i.e., correlating in the azimuth direction with a copy of the azimuth response to generate the final SAR image. Note that much more detail is apparent, and the five bright points that caused the visible responses in Figure 6-3 can be seen.

Fig. 6-5. Data in Figure 6-4 after having performed azimuth compression, i.e., the final SAR image.

6.2.2 Image Artifacts

Due to the way SAR collects data and to the image formation process described above, the SAR image can contain artifacts. In either visually or digitally analyzing the image, these artifacts can be mistaken for properties of the radar cross section of the scatterers within the scene. Many of these artifacts can be corrected, but errors in the parameters required for these corrections can still cause residual artifacts within the imagery.

There are five dominant types of artifacts that the analyst can expect in SAR imagery: range and azimuth ambiguities, sidelobe effects, saturation, image modulations caused by the sensor, and image resolution degradation caused by misfocusing.

6.2.2.1. Range and azimuth ambiguities. SAR imagery can contain image ambiguities, dim copies of bright responses that are offset in range or azimuth from the original response. The two main causes of these are the inadequate sampling of the scatterer response and antenna sidelobes.

If the sensor does not adequately sample the response from a scatterer to support all of the frequencies within it, the higher frequencies will be aliased down to lower values. This causes the higher frequency portions of the original response to resemble the response of a scatterer shifted in either range or azimuth (depending on which direction is undersampled). When the data are processed into an image, these aliased portions cause dim, shifted versions of the original response to appear. Depending upon the strength of the original response, the degree of undersampling, and the brightness of the surrounding clutter, these ambiguities can significantly obscure the clutter responses. Often they can be recognized in the imagery since they are copies of brighter responses at the same range or azimuth location. The Fourier transform of the imagery can often give an indication as to whether it is adequately sampled. If the image has significant Fourier energy out to the highest sampled frequency in either range or azimuth, there is a high probability that the imagery is aliased.

If a scatterer is bright enough and the azimuth antenna sidelobes are high enough, an image of the scatterer can be made when it is in the sidelobe in addition to the mainlobe of the antenna. This will cause a dimmer version of the scatterer to appear in the imagery shifted in the azimuth direction. Techniques, such as bandpassing the data before they are digitized on the sensor, can minimize this effect. Range antenna sidelobes can also cause ghosts, especially if there are bright scatterers on the opposite side of the flight path from the direction the antenna is pointing. These artifacts are harder to remove since they are difficult to distinguish from legitimate returns. As mentioned below, saturation can also cause a form of ghosting.

6.2.2.2. Sidelobe effects. Because the response of a scatterer in the SAR image is a sinc function (shown in Figure 6-2), the sidelobes of the response spatially spread the energy of the scatterer. If a bright response (such as multiyear ice) is next to a darker response (such as open water), these sidelobes will cause energy from the brighter response to smear into the darker response. If the difference between the two responses is greater than the difference between the mainlobe peak and the maximum sidelobe peak, then the sidelobes of the brighter feature can completely obscure the darker feature. This artifact appears as a "leakage" of energy around the brighter response. The best way to test for this artifact is to determine the actual sidelobe structure of the scatterer response. This can be done if the image contains an isolated point scatterer from which the response can be extracted without contamination from surrounding clutter, but this is generally rare for Arctic images.

6.2.2.3. Saturation. If the response from a scatterer is strong enough to exceed the dynamic range of the sensor, the sensor will become saturated. This will cause the receiver to respond nonlinearly to the energy it receives and will introduce a ring of suppressed responses around the saturated target, often referred to as small signal suppression. In addition, saturation can cause the sidelobes of the scatterer response to be as bright as the mainlobe of unsaturated responses, resulting in ghosts of the the main response appearing in the imagery or in a continuous bright smear if the saturation is extreme. Saturated data can be detected from the raw phase history data before image formation. If the data are saturated, then the real and imaginary parts of the phase history will be more often set to their highest or lowest values. A histogram of the data will therefore indicate high peaks at either end, whereas in nonsaturated data, the histogram of either the real or imaginary part will peak at a value of zero and fall off at the extreme values.

6.2.2.4. Image modulations. Image modulations are caused by three factors: range falloff, range antenna pattern, and antenna motion during the collection. Range falloff refers to the decrease in energy received from scatterers at increasingly farther ranges from the sensor. This causes SAR images to go from very bright at the near range to very dark at the far range, even though the radar cross section of the scene may not change. The range elevation antenna pattern also imposes an image modulation since it illuminates different ranges of the scene with different amounts of energy. This will cause the image to be bright at the boresight of the antenna pattern and fade away on either side. Uncompensated motion in the antenna during the collection can cause an azimuth modulation in the strength of the scatterers as they move through the antenna beam. The product of all these functions will represent the SAR image modulation; note that it can be a function of both range and azimuth. This modulation can be removed if the collection geometry and antenna motions are known, but uncertainties can cause residual modulations to be left in the imagery. Usually these can be visually recognized because of their large spatial scales, extending throughout the azimuth or range extent of the image.

6.2.2.5. Image resolution degradation. To perform the image formation process, it is necessary to correlate the data with a model of the two-dimensional response from a scatterer. Generating this response requires knowing the collection geometry of the sensor, the range, velocity, and azimuth location as indicated in the azimuth response in Equation (2). Often there can be errors in the estimate of the collection parameters that cause a mismatch between the model and the data and result in an image that has decreased resolution, that is, an increase in the width of the mainlobe shown in Figure 6-2. This usually occurs only in the azimuth direction, since it is modeling the azimuth response that requires knowledge of the collection geometry. This artifact is usually of little concern to analysts unless they need the finest resolution possible to locate the signals of interest, in which case a number of autofocusing techniques are available [Brown and Ghiglia, 1988; Eichel et al., 1989; Li et al., 1985; Jin, 1986; Madsen, 1989] that require a sharp edge or isolated point target on which to focus. This mismatch between the data and the model can also cause an increase in the sidelobe levels that will effect the artifacts discussed in Section 6.2.2.2.

6.3 POSTIMAGE FORMATION CONSIDERATIONS

The ideal output product from SAR data for analysts is a spatial radar cross-section map. Even after properly forming the image and removing artifacts, two attributes cause the SAR image to differ from this ideal product—slant plane mapping of the data and noise in the data.

6.3.1 Slant-Range to Ground-Range Conversion

Since a SAR collects data by sampling the outputs of the receiver antenna evenly in time, the final SAR image is sampled evenly in range. The SAR image is therefore often referred to as being in the slant plane, whereas most analysts desire their data to be mapped to the ground plane to provide even sampling along the ground. The mathematics of this transformation are straightforward; however, it may be difficult to know precisely the parameters necessary to describe the flight collection, especially altitude. Uniform slant-range sampling maps into nonuniform ground-range sampling, so some manner of interpolation of the output values is required. In the near range, many ground grid points will be mapped to the same range sample value, while at the far range the mapping is closer to one to one. In this respect, the output image will look stretched in the near range, although this effect is significant only when the image contains samples that are near nadir (i.e., the location on the ground directly below the sensor), and thus will not be important for most satellite applications. This transformation maps all samples at a given range to the same ground location so that distortions due to changes in topography are still present.

6.3.2 SAR Image Noise Effects

SAR images will contain two types of noise, thermal and speckle. Thermal noise is the background energy that the radar receiver channel generates. It causes a noise floor throughout the image such that any received signal dropping below this noise floor will be indistinguishable from the thermal noise. For most bright return targets (such as multiyear ice) this will not be a problem, but for low return targets (such as new ice or open water) this can be significant. If both new ice and open water drop below the thermal noise level, then only thermal noise will be present in the SAR image and thus the two returns will appear identical, although they actually have different radar cross-section values.

Speckle occurs in any system [Goodman, 1976] that records the complex-valued nature of electromagnetic radiation received from a resolution cell containing many scatterers. Speckle causes the grainy appearance of SAR imagery and can be considered multiplicative in nature in that as the energy in the backscattered signal increases, the variance of the speckle increases [Goldfinger, 1982]. Speckle causes a randomness in any single sample value from an image, although the statistics of speckle are well understood [Goodman, 1976]. Many procedures have been developed to reduce the variance caused by speckle in image data (Crimmins [1985], Lee [1981], and Procello et al. [1976] are just some examples); most rely on some form of spatial averaging.

REFERENCES

Barber, B. C., Theory of digital imaging from orbital synthetic-aperture radar, International Journal of Remote Sensing, 6, pp. 1009–1057, 1985.

Berkowitz, R. S., *Modern Radar*, John Wiley and Sons, New York, 1967.

Brown, W. D. and D. C. Ghiglia, Some methods for reducing propagation-induced phase errors in coherent imaging systems, parts I and II, Journal of t*he Optical Society of America, 5*, pp. 924–957, 1988.

Brown, W. M., G. Houser, and R. Jenkins, Synthetic aperture processing with limited storage and presumming, *IEEE Transactions on Aerospace and Electronic Systems, AES-9*, pp. 166–175, 1973.

Crimmins, T. R., Geometric filter for speckle reduction, *Applied Optics, 24*, pp. 1438–1443, 1985.

Eichel, P. H., D. C. Ghiglia, and C. V. Jakowatz, Jr., Speckle processing method for synthetic-aperture-radar phase correction, *Optics Letters, 14*, pp. 1–3, 1989.

Elachi, C., *Spaceborne Radar Remote Sensing: Applications and Techniques*, IEEE Press, New York, 1988.

Goldfinger, A. D., Estimation of spectra from speckled images, *IEEE Transactions on Aerospace and Electronic Systems, AES-18*, pp. 675–681, 1982.

Goodman, J. W., Some fundamental properties of speckle, *Journal of the Optical Society of America, 66*, pp. 1145–1150, 1976.

Hovanessian, S. A., *Radar System Design and Analysis*, Artech House, Dedham, Massachusetts, 1984.

Jin, M. Y., Optimal Doppler centroid estimation for SAR data from a quasi-homogeneous source, IEEE Transactions on Geoscience and Remote Sensing, GE-24, pp. 1022–1027, 1986.

Jin, M. Y. and C. Wu, A SAR correlation algorithm which accommodates large-range migration, *IEEE Transactions on Geoscience and Remote Sensing, GE-22*, pp. 592–597, 1984.

Kirk, J. C., A discussion of digital processing for synthetic aperture radar, *IEEE Transactions on Aerospace and Electronic Systems, AES-11*, pp. 326–337, 1975.

Klauder, J. R., A. C. Price, S. Darlington, and W. J. Albersheim, The theory and design of chirp radars, *The Bell System Technical Journal, 34*, pp. 745–807, 1960.

Lee, J. S., Speckle analysis and smoothing of synthetic aperture radar images, *Computer Graphics and Image Processing, 17*, pp. 24–32, 1981.

Li, F. K., D. N. Held, J. C. Curlander, and C. Wu, Doppler parameter estimation for spaceborne synthetic-aperture radars, *IEEE Transactions on Geoscience and Remote Sensing, GE-23*, pp. 47–56, 1985.

Madsen, S. N., Estimating the Doppler centroid of SAR data, *IEEE Transactions Aerospace and Electronic Systems, AES-25*, pp. 134–140, 1989.

Procello, L. J., N. G. Massey, R. B. Innes, and J. M. Marks, Speckle reduction in synthetic-aperture radars, *Journal of the Optical Society of America, 66*, pp. 1305–1311, 1976.

Raney, R. K., Processing synthetic aperture radar data, *International Journal of Remote Sensing, 3*, pp. 243–257, 1981.

Rihaczek, A., *Principles of High Resolution Radar*, McGraw-Hill, Inc., New York, 1969.

Shokr, M. E., On sea-ice texture characterization from SAR images, IEEE Transac*tions on Geoscience and Remote Sensing, 28*, pp. 737–740, 1990.

Skolnik, M., ed., *Radar Handbook*, McGraw–Hill, Inc., New York, 1990.

Tomiyasu, K., Tutorial review of synthetic aperture radar (SAR) with applications to imaging of the ocean surface, Proceedings *of the IEEE, 66*, pp. 563–584, 1978.

Van de Lindt, W. J., Digital technique for generating synthetic aperture radar images, *IBM Journal of Research and Development, 21*, p. 415, 1977.

Wu, K. H. and M. R. Vant, Extensions to the step transform SAR processing technique, *IEEE Transactions on Aerospace and Electronic Systems, AES-21*, pp. 338–344, 1985.

Wu, C., K. Y. Liu, and M. Jin, Modeling and a correlation algorithm for spaceborne SAR signals, *IEEE Transactions on Aerospace and Electronic Systems, AES-18*, pp. 563–574, 1982a.

Wu, C., B. Barkan, W. J. Karplus, and D. Caswell, Seasat synthetic aperture radar data reduction using parallel programmable array processors, *IEEE Transactions on Geoscience and Remote Sensing, GE-20*, pp. 352–357, 1982b.

Chapter 7. Sea Ice Altimetry

FLORENCE M. FETTERER

Remote Sensing Branch, Naval Research Laboratories, Stennis Space Center, Mississippi 39529-5004

MARK R. DRINKWATER

Jet Propulsion Laboratory, California Institute of Technology, 4800 Oak Grove Drive, Pasadena, California 91109

KENNETH C. JEZEK

Byrd Polar Research Center, 1090 Carmack Road, The Ohio State University, Columbus, Ohio 43210

SEYMOUR W. C. LAXON

University College London, Mullard Space Science Laboratory, Holmbury St. Mary Dorking, Surrey, RH5 6NT, United Kingdom

ROBERT G. ONSTOTT

Environmental Research Institute of Michigan, P. O. Box 8618, Ann Arbor, Michigan 48107

LARS M. H. ULANDER

Department of Radio and Space Sciences, Chalmers University of Technology, S-41296 Göteborg, Sweden

7.1 INTRODUCTION

Using altimeter data for quantitative information about sea ice is an idea with relatively few proponents. Often the poor spatial sampling given by the altimeter's single-point measurements along widely spaced ground tracks and the difficulty of interpreting altimeter pulse echoes over ice are cited as reasons to avoid altimetry. It is becoming evident, however, that altimetry may be able to make unique measurements. For instance, the altimeter data record reveals the presence of small areas of open water within the pack at concentrations too low to be detected by a passive microwave sensor such as a Special Sensor Microwave/Imager (SSM/I) and too small to be resolved by satellite synthetic aperture radar (SAR).

At high latitudes the net laid down by a satellite altimeter's ground track becomes dense, and spatial sampling (over a period of days) can approach the resolution of SSM/I. A satellite altimeter's along-track resolution of less than a kilometer is better than that of SSM/I, and the altimeter's sensitivity in detecting the ice edge is greater. The low data rate makes it possible to process altimetry in near-real time. For these reasons, an ice edge product derived from Geosat altimeter data has been used operationally at the Navy/National Oceanic and Atmospheric Administration (NOAA) Joint Ice Center to supplement infrared imagery for estimating ice edge position [Hawkins and Lybanon, 1989].

Microwave Remote Sensing of Sea Ice
Geophysical Monograph 68
©1992 American Geophysical Union

Objections concerning the difficulty of interpreting altimetry over ice are more difficult to meet, in part because of the paucity of ground truth or imagery coincident with altimetry. In this chapter, we cover some of the research into understanding altimeter backscatter over sea ice and interpreting altimeter data in terms of parameters such as ice type, concentration, edge location, and wave penetration in the marginal ice zone.

7.2 INSTRUMENT DESCRIPTION

7.2.1 Instrument Design

Satellite altimeters have evolved steadily from Skylab, which in 1973 demonstrated the concept of altimetry from space. Development continued through GEOS-3 in 1975 until the 1978 launch of Seasat, which carried the first altimeter capable of measuring sea surface topography with an error of only a few centimeters. Precision in range or height measurement is obtained through a design that takes advantage of the ocean's near-Gaussian wave height distribution and uniform dielectric characteristics. Over ice, where surface height distributions and dielectric characteristics vary sharply from place to place, measurements of height and backscatter are often misleading. Therefore, it is necessary to understand how altimeters are designed to operate before one interprets altimeter backscatter from sea ice. Altimeters planned for the 1990's will refine rather than change the Seasat design, upon which the following description is based. Hereafter, similar altimeters such as Geosat and ERS-1 will be called ERS-1 class altimeters.

7.2.2 Pulse-Limited Altimetry

Satellite altimeters use pulse width-limited geometry. A wide-beam (with a 3-dB half-width of 0.8° for Seasat), relatively long (on the order of microseconds) pulse is emitted, and the pulse echo is processed in a way equivalent to measuring the travel time to the surface and back of a much shorter pulse only a few nanoseconds long. This technique of pulse compression gives high range resolution. (For a full explanation of pulse compression on satellite altimeters, see Chelton et al. [1989]). The area on the surface illuminated by the pulse grows with time until the trailing edge of the pulse leaves the lowest reflecting points at nadir (Figure 7-1). The maximum area simultaneously illuminated, then, depends on the surface height distribution and effective (compressed) pulse length. This area is usually referred to as the pulse-limited footprint (PLF). For Seasat, the diameter of the PLF was 1.6 km for a smooth sea and grew to 7.7 km for a significant wave height (SWH) of 10 m. As the pulse continues to move out over the surface, the footprint becomes an annulus of constant area and increasing circumference. It is important to note that due to irregularities in surface roughness and slope, the true illuminated area at any time may be irregular in shape or discontinuous. Over ridged sea ice, for instance, an off-nadir ridge sail may be high enough to be illuminated at the same time as the ice near nadir.

Altimeters that use beam width-limited geometry emit a narrow beam and do not rely on pulse-compression techniques for accurate range measurements. Footprint size depends only on beam width and satellite altitude. This geometry is not used for satellite altimeters because precise antenna pointing capability is a necessity and the antenna size and gain required to generate a narrow beam are large. However, use of a beam width-limited altimeter would simplify interpretation of signals over ice, because footprint area remains constant regardless of the surface height distribution [Rapley et al., 1983].

On board the satellite, the echo of the emitted pulse is differenced with a deramping signal. The frequency of the emitted pulse changes over the pulse length in a linear chirp or ramp. The total change in frequency over the ramp is the bandwidth of the instrument. The deramping signal (so-called because the effect of this processing step is to remove the ramp in frequency from the pulse echo) is generated at a delay time t_0 determined by the onboard tracking unit. This time corresponds to the expected arrival time at the receiver of the reflected pulse from mean sea level. The resulting signal is a mixture of returns from individual point scatterers. Echoes from wave crests arrive first and are therefore shifted earlier in time from tracking time t_0. For a pulse-limited altimeter, returns shifted later than t_0 come not only from points below mean sea level, but also, as the footprint expands, from scatterers distant from nadir. If the height distribution for the area over which the pulse expands remains Gaussian throughout, the shape of the signal $S(t)$ resulting from all scatterers is given by the convolution of functions describing the transmitted pulse shape, the antenna gain pattern, and the surface height distribution [Drinkwater, 1991]. This signal is the pulse echo waveform (Figure 7-2). The half-power point is at delay time t_0, which gives the range to mean sea level. Maximum power occurs at the delay time at which the maximum footprint area is achieved. At later delay times, power drops off due to the antenna gain pattern. Over the ocean, the slope of the leading edge (or ramp) of the waveform is inversely proportional to SWH, and the peak power is related to wind speed. The characteristic shape of this signal only holds for reflecting surfaces of meter-scale roughness for which specular point scattering can be assumed.

The digitally sampled waveform has a value at 60 (for Seasat) contiguous time intervals, centered, if the tracker is operating correctly, on the half-power point. The time intervals, or range gates, each correspond to a two-way travel time resolution of 3.125 nanoseconds, which is equivalent to a range resolution of about 1 meter. (Better precision in the range to nadir or height measurement is obtained through averaging). The power in the range gates

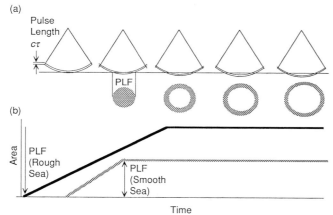

Fig. 7-1. (a) Area illuminated by a pulse hitting the smooth ocean's surface and (b) the growth of the footprint area with time for smooth and rough seas. The PLF is larger for meter-scale roughness, since the time required for the pulse to travel from crest to trough exceeds the effective pulse length. Graph is not to scale.

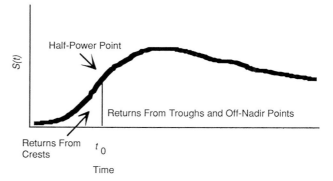

Fig. 7-2. The pulse echo for pulse-limited altimetry and a Gaussian surface height distribution. The half-power point at t_0 corresponds to the delay of returns from the mean sea level.

is the result of backscatter from range rings within the altimeter footprint (Figure 7-3). The area covered by each range ring is equal to that of the PLF

$$A_{\text{plf}} \approx \pi r_{\text{plf}}^2 \approx \pi c H \tau \qquad (1)$$

where r_{plf} is the radius of the PLF, H is satellite height above the surface, and τ is compressed pulse length (neglecting, for simplicity, the Earth's curvature). With this expression for area, it is possible to use the radar equation to arrive at an expression for the power received from any range ring

$$P_{\text{range ring}} = P_{\text{t}} \frac{G^2 \lambda^2 L_{\text{atm}} c \tau}{4(4\pi)^2 H^3} \sigma^\circ \qquad (2)$$

where L_{atm} is atmospheric attenuation loss. Using a form of this equation, backscatter can be retrieved from waveform range gate samples. For Geosat, the formula is

$$\sigma^\circ_{\text{i}} = 10\log_{10} P_{\text{i}} + AGC + K + 30\log_{10} \frac{H}{H_0} + L_{\text{atm}} \qquad (3)$$

where P_{i} is the power in range ring i expressed in counts, H_0 is a reference height, AGC is the attenuator setting (de-scribed in Section 7.2.4), and K is a constant dependent on system parameters [Ulander, 1987a]. The peak backscatter usually appears in one of the first few range rings for sea ice and is generally assumed to be the backscatter from nadir. When altimeter backscatter measurements are referred to in what follows, it is usually the peak backscatter that is meant.

While the PLF corresponds to that part of the waveform over which the maximum power is reached, the range window-limited footprint corresponds to the area mapped out by the pulse at the end of the time corresponding to the last range gate. For Seasat, the diameter of the range window-limited footprint was about 10 km, and the angle of incidence at the last range ring was about 0.3°.

7.2.3 Waveform Averaging

In the case of noncoherent scattering, the phase relation-ships of returns from all facets or specular points are random. The vector sum of amplitude and phase for reflections within a range ring is different for each range ring, which lends noise or fading characteristics to indi-vidual pulse echoes. Therefore, waveforms are averaged to reduce fluctuations and to provide a good estimate of height (given by the half-power point) and backscatter. Each waveform must be a statistically independent sample. The pulse repetition frequency (PRF) necessary to avoid corre-lation between pulse echoes depends on the surface roughness and orbital velocity. Seasat used a PRF of 1020 Hz. Fifty echoes were averaged, during which time the satellite nadir track moved about 330 m. For altimetry over the ocean, averaging 50 echoes is a good compromise that provides enough samples for an accurate power esti-mate while not degrading the along-track spatial resolution of the measurement beyond acceptable bounds, given the typical scale of changes in dynamic sea surface height, wind speed, and SWH. When surface characteristics are changing over the integration time (as is often the case over ice), or when pulse echoes are dominated by one or several strong scatterers, waveforms that result from averaging will not truly represent the surface.

7.2.4 The Adaptive Tracker and Processing Loops

Accurate estimates of height and other parameters de-rived from waveforms depend on accurate superposition of pulse echoes for averaging. This, in turn, depends on proper timing of the deramping pulse. The following description of the method by which this is achieved is based on the Seasat adaptive tracking unit, but Geosat and ERS-1 trackers are similar in concept. The adaptive tracker tracks waveforms through an automatic gain control (AGC) loop and tracking (or range and range rate) loops. The AGC component of the tracking system constantly adjusts the gain of the receiver to achieve the maximum signal-to-noise ratio. Power within the (averaged) waveform is computed according to

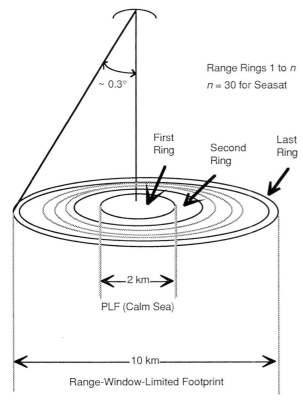

Fig. 7-3. Altimeter footprint range rings (not to scale). Each ring corresponds to a range gate in the digitally sampled waveform.

$$AGC_{\text{gate}} = Sc \sum_{i=1}^{60} P_i \qquad (4)$$

where i is range gate number, P is the power (expressed as a count value) in each gate, and Sc is a scale factor, Figure 7-4(a). An attenuation value is computed according to the recursion relation

$$AGC\,(n+1) = AGC\,(n) + \mathbf{a}\Delta AGC \qquad (5)$$

where ΔAGC is the difference between AGC_{gate} and some reference, and $\mathbf{a}$ is some constant (0.0265 for Seasat). The AGC loop works to keep the average power in the waveform constant. It is updated at a rate of 20 Hz (every 50 pulses), and the new value is applied to the next 50 pulse echoes. Because of the factor $\mathbf{a}$ in the recursion relation, an adjustment to the value of AGC necessitated by a change in backscatter from the surface takes place over 0.8 seconds, during which time the satellite moves 5.3 km. While the parameter AGC is a good estimator of backscatter over the ocean, this is not the case over ice, where peak power can change an order of magnitude in a much shorter distance. Therefore, to retrieve high-frequency variability in backscatter it is necessary to take into account the count values in each waveform, along with the AGC value that was the attenuator setting during the waveform integration time. In such cases where backscatter is changing quickly, AGC

is overestimating or underestimating power, and the signal-to-noise ratio will not be optimal.

In Figure 7-4 (typical of ocean signals), the value of AGC_{gate} is about equal to half the power in the waveform. The power in the center of the range window is measured in a middle gate M, which is sometimes called gate 30.5, Figure 7-4(b). The tracker seeks to minimize the range error Δr, where

$$\Delta r = AGC_{\text{gate}} - M_{\text{gate}} \qquad (6)$$

by shifting the range window (in time) to keep the half-power point in M. The width of gate M depends on the slope of the leading edge of the waveform, and therefore on SWH. The tracker adapts the width of M in range gates based on the difference between power in an early gate E at the bottom of the ramp and a later gate L at the top of the ramp. For a significant wave height of less than 2 m, gate M is a single range gate wide [Chelton et al., 1989].

The range error Δr is used to predict the range to the surface (and therefore the delay time t_0 at the center of the range window) for subsequent pulses according to

$$r\,(n+1) = r\,(n-1) + \mathbf{a}\Delta r\,(n) + r'(n)\,\Delta t \qquad (7)$$

where Δr is the range error for pulse n, $\mathbf{a}$ is the tracker time constant, Δt is the update interval of the loop, and

$$r'\,(n) = r'\,(n-1) + \mathbf{b}\,\frac{\Delta r(n)}{\Delta t} \qquad (8)$$

is the range rate. Time constants $\mathbf{a}$ and $\mathbf{b}$ can be adjusted to increase the agility of the tracker (at the price of increased noise). Different designs for tracking ocean and nonocean surfaces have been explored [McIntyre et al., 1986; Rapley et al., 1983; Ulander, 1987c]. The ERS-1 altimeter has both an ocean mode, which uses a Suboptimal Maximum Likelihood Estimator (SMLE) tracker, and an ice mode, which has a split-gate tracker unlike the Seasat/Geosat tracker [Griffiths et al., 1987]. The ice mode tracker has a wider range window for better tracking over undulating continental ice shelves. For sea ice, however, it is not height variations confounding the agility of the tracker, but rather backscatter variations within the footprint, which introduce errors exacerbated by interaction between the processing loops.

7.2.5 Errors Common Over Ice

Figure 7-5 illustrates how the rapid rise and fall of sea ice waveforms confounds the tracker selection of gates E, M, and L. Because of the steepness of the ramp and the rapid fall-off from the peak value in the return waveform in Figure 7-5(a), the power in M is not half the average power

(a)

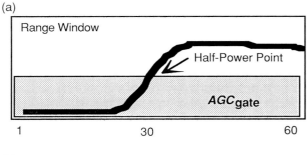

(b)

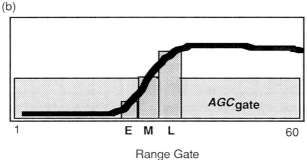

Range Gate

Fig. 7-4. The altimeter range window and gates. (a) Over the ocean, the power in AGC_{gate} should equal half the power in the waveform. The half-power point occurs at time t_0 if the tracker is operating correctly. (b) Early (E), middle (M), and late (L) gates in the center of the waveform determine the slope of the ramp (proportional to SWH). The half-power point should be in gate M.

in the waveform. Power within the AGC gate is much less than that within the middle gate. In extreme cases, this results in a large negative height error, causing the tracker to shift the range window so that subsequent pulses appear later in the range window, Figure 7-5(b). On the next update cycle, the middle gate measures the noise before the ramp, resulting in a large positive height error. The tracker corrects for this until the peak is in the middle gate again when the error repeats. This error, called tracker oscillation, is evident in data that exhibits height excursions. Tracker oscillation can sometimes be corrected by retracking the waveforms. Retracking involves (1) applying an algorithm to the waveform record that finds the point on the waveform that should corresponds to return from the surface at nadir, and (2) using the position of that point relative to the center of the range window to correct the height estimate based on the range window.

The range and range rate loops, like the AGC loop, were updated at 20 Hz (every 50 pulses) on Geosat and Seasat. Pulse echoes acquired with the same loop settings were averaged to make waveforms. However, waveforms were telemetered at 10 Hz, so the telemetered waveform is the average of two 50-pulse mean waveforms from two tracking loop updates. If the position of the range window is changing rapidly (due to changing topography or tracker oscillation), the resulting telemetered waveform will be a blurred or sometimes double-peaked composite, Figure 7-6(a). Errors resulting from the incorrect superposition of waveforms for averaging are termed telemetry summing errors [Rapley et al., 1987]. ERS-1 telemeters waveforms at the tracker loop update rate (20 Hz), thereby avoiding the problem that appeared in Seasat and Geosat data.

Tracker oscillation cannot be completely eliminated by retracking if telemetry summing errors are present, Figure 7-6(b). However, the probability that such errors exist can be estimated by examining the range rate parameter, which will peak when the range window is changing at the greatest rate and the probability of blurring is highest [Laxon, 1989]. The combined effect of tracker oscillation and telemetry summing, when it produces a sudden shift in range and blurred or double-peaked waveforms, is sometimes referred to as height glitch.

When the AGC loop sets the *AGC* value too low to accommodate power in the peaked portion of an ice waveform, saturation occurs. Saturation was common in Seasat data [Wingham and Rapley, 1987]. It is often associated with frequency splatter, Figure 7-6(c), where power from the saturated gate appears in earlier and later gates. Another artifact often present in ice data is snagging or "off-ranging." This occurs when a bright feature at nadir is followed by the tracker as it recedes, until it is sufficiently far from nadir for the true nadir return to once again dominate. Often this occurs at a transition from pack to fast ice, or from thin ice to multiyear (MY) ice [Laxon, 1989].

Antenna off-pointing was a serious problem with Geosat. Occasionally, the off-pointing was severe enough to cause loss of lock, a situation where the waveform disappears from the range window entirely for a few telemetry frames.

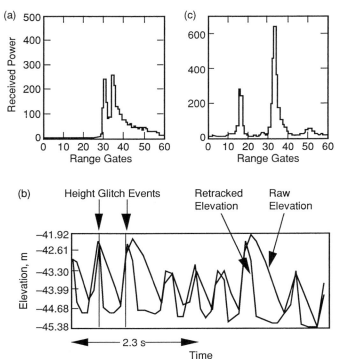

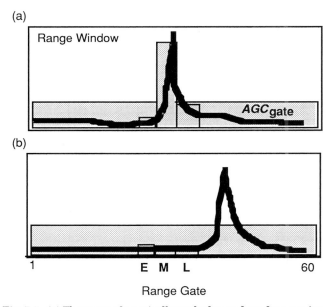

Fig. 7-5. (a) The ramp of a typically peaked waveform from sea ice can appear entirely in gate M. (b) This causes the tracker to shift the range window earlier in time for the next update cycle.

Fig. 7-6. Common errors in altimetry over sea ice. (a) An example of telemetry summing causing double peaking in a Seasat waveform over ice [Ulander, 1987b]. (b) Seasat data showing oscillation in the onboard height measurement. Retracking does not remove the oscillation due to telemetry summing errors [Laxon, 1989]. (c) A saturated Seasat waveform [Ulander, 1987b].

Over the ocean, off-pointing can be identified by the attendant rise in power in later gates, and a correction can be applied. Over sea ice, unpredictable waveform shapes make this impossible, resulting in altered waveform shapes and unidentifiable errors in height and backscatter measurements. The stabilizer used on ERS-1 should prevent off-pointing from being a serious problem.

Once the altimeter record over ice has been retracked and edited, the backscatter coefficient for any range gate can be found from the waveform using Equation (3) with the telemetered waveform range gate count values and the value of the attenuator setting (AGC). However, it is important to remember that the altimeter moves about 0.66 km between telemetered waveforms (for Geosat and Seasat). If the statistics of the surface are not stationary over that distance, the backscatter estimates may be misleading.

7.3 SEA ICE SCATTERING MODELS AT NORMAL INCIDENCE

7.3.1 Introduction to Altimeter Backscatter Models

The distinguishing characteristics of ice waveforms are that they are often narrow-peaked in shape, taper off quickly with incidence angle, have high backscatter coefficients at nadir, and are widely variable. Average annual values of Seasat backscatter over Arctic and Antarctic sea ice ranged from approximately 0 to 45 dB, with a mean greater than 20 dB [Laxon, 1989]. In contrast, backscatter over the ocean has a range of between 8 and 15 dB, depending on sea state [Chelton and McCabe, 1985]. On the basis of dielectric characteristics alone, one would expect ocean backscatter to be greater than that for ice, and indeed for surface measurements of backscatter at nadir this is usually the case. The dynamic range of altimeter measurements over ice cannot be explained by variability in the dielectric characteristics of ice alone. Since backscatter for airborne and satellite altimeters is dominated by surface rather than volume scattering, the nature of altimeter pulse echoes must be primarily determined by surface roughness. Specular or coherent reflection from smooth surfaces near nadir may produce the peaked component of waveforms, while diffuse reflection from rough surfaces may account for power in later, off-nadir gates. But while sea ice or ocean surfaces can in certain circumstances be extremely smooth, the question remains as to whether these surfaces are smooth over large enough areas to contribute coherent returns. Models that invoke only noncoherent (diffuse) reflection, on the other hand, may have difficulty describing the high backscatter from nadir and narrow peaked shape often seen over ice.

The physical characteristics of sea ice are quite variable, which compounds the modeling problem. Sea ice consists of both level and deformed ice of different ages, often with superimposed dry or wet snow layers. Water openings of different sizes and with variable wind conditions are often present. The different layers and interfaces of the snow, ice, and water may all contribute to the scattering of electromagnetic waves. Depending on the frequency used, however, it is often possible to neglect particular scattering terms and thus simplify the scattering computation. For example, the surface will appear rough and result in the dominance of diffuse scattering if the surface height variations are large compared with the radar wavelength. The dependence of surface roughness upon radar wavelength is expressed in parameters $k\sigma$ and kl, where $k = 2\pi/\lambda$, σ is the standard deviation of surface height, and l is the correlation length of surface roughness. These parameters determine the range of validity for commonly used models. In the following, we consider surface scattering close to normal incidence. The emphasis is on radar wavelengths in the Ku-band (1.7 to 2.4 cm), since most radar altimeters used for geophysical remote sensing operate in this band.

7.3.2 Backscatter From a Stationary Gaussian Random Surface

This section concerns backscatter from level and snow-free sea ice. The ice is modeled as a homogeneous medium with constant dielectric constant. Its surface has small-scale height variations represented by a stationary Gaussian process. The electromagnetic scattering from such a surface has been studied extensively in the past, but a general closed-form solution has not yet been found. A recently developed model for a perfectly conducting [Fung and Pan, 1987] and dielectric [Chen and Fung, 1991] surface has promise, but its range of validity has not yet been determined. High- and low-frequency solutions, however, are readily available and are sufficient for the present discussion.

7.3.2.1. Specular backscatter from a perfectly smooth surface. A surface is usually considered smooth if it meets the Rayleigh criterion. For Ku-band, σ must be less than about 0.3 cm. The scattering pattern for such a surface will show a large phase-coherent component of backscatter in the specular (nadir) direction and a much smaller diffuse component at other incidence angles. If a surface is perfectly smooth and planar, scattering is confined to the specular direction, with no diffuse component, and is equivalent to ray reflection. Such a surface will produce true specular reflection if it is perfectly smooth over at least two Fresnel zones. The edge of each circular Fresnel zone is defined where the phase of the reflected wave has changed by 0.5π compared with a reflection from the nadir point. The diameter of the first zone is given by

$$D_{\mathrm{F}} = (2\lambda H)^{1/2} \qquad (9)$$

or about 180 m for an ERS-1 class altimeter. Although possible in theory, it is unlikely that the rms height of the surface is perfectly smooth, or even smooth to within a few millimeters, over an area of several hundred square meters.

It is not convenient to describe true specular reflection in terms of a backscatter coefficient. However, it is possible to derive an effective backscatter coefficient that is dependent on the radar system. For example, the backscatter coefficient of an infinitely large dielectric mirror observed by a pulse-limited altimeter is given by [Ulander, 1987a]

$$\sigma^\circ_{coh} = |R(0)|^2 \frac{H}{c\tau} \qquad (10)$$

where $R(0)$ is the Fresnel reflection coefficient $R(\theta)$ evaluated at $0°$. The subscript indicates that reflection is, of course, coherent. The theoretical maximum return for an ERS-1 class altimeter, then, is about 48 dB, assuming an ice reflection coefficient of about –11 dB.

7.3.2.2. Coherent backscatter from a slightly rough planar surface. When a small rms surface roughness is superimposed on a flat, planar reflecting surface, an expression may be derived for the coherent backscattering coefficient, σ°_{coh}, which accounts for the sphericity of the wavefront and the radar system impulse response [Fung and Eom, 1983]:

$$\sigma^\circ_{coh} \approx \frac{|R(\theta)|^2}{\beta^2_{plf}} \exp\left[-4k^2\sigma^2 + \frac{\theta^2}{\beta^2_{plf}}\right] \qquad (11)$$

Equation (11) assumes a Gaussian beam described by the one-sided pulse-limited beam width β^2_{plf}, where $\beta_{plf} = c\tau/H$. At normal incidence this simplifies to [Ulander and Carlström, 1991]:

$$\sigma^\circ_{coh} = |R(0)|^2 \frac{H}{c\tau} \exp(-4k^2\sigma^2) \qquad (12)$$

Figure 7-7 illustrates the coherent backscatter function derived using Equation (11) with suitable parameters for the ERS-1 altimeter and the Rutherford Appleton Laboratory (RAL) airborne altimeter [Drinkwater, 1987] respectively. For the satellite case, the fall-off in σ°_{coh} is extremely rapid for all surface roughnesses. The maximum backscatter values of between 40 and 48 dB occur for slightly rough surfaces typifying new saline ice (i.e., $\varepsilon^*=4.0 + j0.4$). In contrast, the RAL altimeter, with a pulse-limited beam width of 0.63°, gives a much wider response function that is lower in peak power at all roughnesses. These results demonstrate the role that the compressed pulse length and altitude of the altimeter play in dictating the coherent response function of a slightly rough surface. In short, the system impulse response is the most important factor in determining the decay of the coherent scattering coefficient with incidence angle. Furthermore, the coherent power is sharply reduced as the surface roughness increases only slightly.

7.3.2.3. Noncoherent backscatter from a rough surface where kl is large, but kσ is small. Field measurements of sea ice surface roughness typically show that the rms height σ

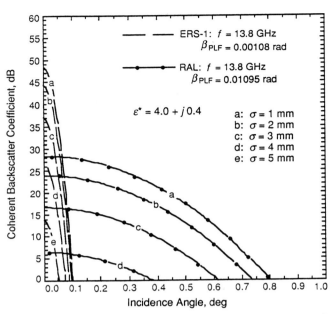

Fig. 7-7. Coherent backscatter versus incidence angle for space and airborne platforms.

is 0.1 to 1.0 cm and the height correlation function is exponential or Gaussian with a correlation length l of 2 to 10 cm [Ulander, 1991]. For radar wavelengths in the Ku-band, this ice surface is in the intermediate roughness range. Note that the correlation length is generally larger than the radar wavelength. The Kirchhoff approximation can therefore be used to evaluate the scattering integrals [Beckmann and Spizzichino, 1963]. This approach uses the diffraction integral for the scattered field and the tangent-plane approximation to compute the surface currents. The latter requires that the average radius of curvature of scattering elements be large compared to the wavelength. The rms surface slope must also be less than 0.25 radians in order to allow computation of an approximate (scalar) solution for dielectric surfaces. With these restrictions, the noncoherent backscatter coefficient σ°_{non} is given by [Ulaby et al., 1982]

$$\sigma^0(\theta) = |R(\theta)|^2 2k^2\cos^2\theta \exp(-g) \sum_{n=1}^{\infty} \frac{g^n}{n!} W^n(2k\sin\theta, 0) \quad (13)$$

where $g = (2k\sigma\cos\theta)^2$ and the first-order slope term that is important for larger incidence angles has been excluded. In Equation (13), the roughness spectrum $W^n(K_x, K_y)$ is related to the nth power of the correlation function defined by [Fung and Pan, 1987]

$$W^n(K_x, K_y) = \frac{1}{2\pi}\int_{-\infty}^{\infty}\int_{-\infty}^{\infty} \rho^n(\xi, \zeta) \exp(-jK_x\xi - jK_y\zeta) d\xi d\zeta \quad (14)$$

where $\rho(\xi,\zeta)$ is the two-dimensional normalized height autocorrelation function. Analytical expressions for the integral in Equation (14), when the surface is isotropic and has a Gaussian or exponential correlation function, can be found in Ulaby et al. [1982] and Kim et al. [1985], respectively.

7.3.2.4. Noncoherent backscatter from a rough surface where kl *and* kσ *are large.* In the Kirchhoff or physical optics theory, when rms height is much larger than the radar wavelength it is possible to evaluate the diffraction integral using a stationary phase approximation [Beckmann and Spizzichino, 1963]. This corresponds to the geometric optics limit, in which the surface is assumed to be made up of tangent planes or specular facets that are infinitely large with respect to a wavelength. Thus it is often known as specular point theory. For such rough surfaces, backscatter is completely noncoherent and the backscatter coefficient is given by [Ulaby et al., 1982]

$$\sigma(\theta) = \frac{|R(0)|^2}{s^2 \cos^4 \theta} \exp\left(-\frac{\tan^2\theta}{s^2}\right) \qquad (15)$$

where s^2 is rms slope in radians.

The scalar and tangent-plane approximations, Equations (13) and (15), respectively, both give the mean noncoherent backscatter from a stationary Gaussian random surface. Since backscatter is being summed noncoherently, if an altimeter were to scan such a rough surface the returned power would exhibit statistical fluctuations or signal fading (described in Section 7.2.3). The return power is then distributed according to a negative-exponential or Rayleigh distribution, provided that the following conditions are satisfied:
(1)   Many individual scattering elements contribute to the returned power, and no single one dominates.
(2)   The amplitude and phase of returns from the scatterers are independent, and the phase distribution is uniform.

7.3.3 Backscatter From a Heterogeneous Random Surface

Equations (11) and (12) show that the coherent return is a highly nonlinear function of the rms surface height. When the rms height changes only from 0.2 to 0.4 cm, the coherent component is reduced by 17 dB. This suggests that the normal-incidence response is dominated by the smoother parts of the surface for which the rms height is a tenth of a wavelength or less. The first model to recognize this fact was that proposed by Brown [1982]. The model assumes that the scattering originates from an ensemble of circular, horizontal, and flat dielectric patches near nadir. The patches are assumed to cover a small fraction of the surface, to be located at the same height, and to be randomly distributed in the horizontal plane. Brown shows that both a coherent and a noncoherent component result from this model. The former is caused by the patches being located at the same vertical level, to within a few mm, and is propor-

tional to the number of patches squared. The latter is caused by the random horizontal distribution of the patches and is directly proportional to the number of patches. Although the model recognizes the importance of the spatial variability of the surface characteristics, it inherently assumes that reflections from the patches add coherently. This assumption is probably unrealistic in practice, at least for spaceborne systems, due to the large first Fresnel zone.

Another model has been proposed which does not require the flat patches to add coherently [Ulander and Carlström, 1991]. The backscattered power is instead assumed to be dominated by reflections from horizontal and flat patches that add noncoherently. A narrow-peaked waveform may occur if the patches are sufficiently large. This assumption is more realistic, but field measurements are presently lacking to verify it. Using physical optics, the radar cross section from a single circular and flat dielectric patch is given by

$$\sigma(\theta) \approx A |R(0)|^2 \left(\frac{\pi D}{\lambda}\right)^2 \exp\left(-\frac{\theta^2}{\theta_D^2}\right) \qquad (16)$$

where A is the patch area, D is the patch diameter, and the angular response has been approximated by a Gaussian function with $\theta_D \approx 0.3\, \lambda/D$. For a collection of sparsely distributed and horizontal patches, we then obtain the backscattering coefficient according to

$$\sigma^\circ(\theta) = F\frac{\sigma(\theta)}{A} \approx A |R(0)|^2 \left(\frac{\pi D}{\lambda}\right)^2 \exp\left(-\frac{\theta^2}{\theta_D^2}\right) \qquad (17)$$

where F is the fractional coverage of flat patches.

Equation (17) does not take the altimeter system impulse response into account. When the patch diameter is sufficiently large to produce a narrow-peaked waveform, the effective backscattering coefficient is given by

$$\sigma^0 = \frac{1}{\tau} \int_0^\infty \sigma^0 \left(\frac{c\tau}{H}\right)^{1/2} d\tau \approx 0.9 FR |0|^2 \frac{H}{c\tau} \qquad (18)$$

Typical values for the narrow-peaked backscatter coefficient, as measured by spaceborne altimeters, are 20 to 40 dB, corresponding to $F \approx 0.2\%$ to 16% according to Equation (18). In this class of surface, the coherent component can exhibit signal fading as the number of patches or reflecting facets within the first Fresnel zone varies from pulse to pulse. Although the terms specular and coherent reflection are often used interchangeably, true specular reflection originates from a single infinitely large (when compared with the PLF) mirror-like facet and does not show fluctuations in amplitude and phase from pulse to pulse.

It is also possible to modify the Kirchhoff models according to Equations (13) and (15) to include the effects of a heterogeneous surface and to produce a narrow-peaked waveform. We here give the result for the geometric-optics model

$$\sigma^0 = \frac{1}{\tau}\int_0^\infty \sigma^0\left(\frac{c\tau}{H}\right)^{1/2} d\tau \approx FR\,|0|^2\frac{H}{c\tau} \qquad (19)$$

which only differs by a factor 0.9 from Equation (18). This shows that the geometrical optics and flat patch model predictions are indeed very similar, although the scattering mechanisms assumed are very different. The approach taken by Ulander and Carlström [1991] is more realistic in that it accounts for height distributions or surface tilting of patches and is not restricted by the approximation of scattering by individual facets to the scattering from smooth specular points. But while the assumptions of the other models are simplistic, the notion that scattering from many individual surface elements conforms reasonably closely to that from a specular surface appears not too far from the truth.

7.3.4 Discussion

In order to contrast normal incidence coherent and noncoherent components of a waveform, we consider here the relative difference in their power using an expression for the ratio of noncoherent to coherent backscatter. The expression is dependent upon the radar system parameters as well as the radar height above the surface and the rough surface statistics. At nadir, the expression is

$$\frac{\sigma^0_{non}}{\sigma^0_{coh}} = 2(kl)^2\,\beta^2_{plf}\sum_{m=1}^\infty \frac{k_0^m}{m!\,m} \qquad (20)$$

where $k = 2\pi/\lambda$, $k_0 = 4\,k^2\sigma^2$, and β_{plf} is the one-way half beam width (β^2_{plf} is equal to $c\tau/H$). Note that the ratio becomes independent of the surface reflection coefficient when this approximation is used.

Figure 7-8 puts into perspective the relative contributions of coherent and noncoherent backscatter to the maximum power in a peaked signal. The upper panel shows the ratio for a satellite-borne altimeter, and the lower for an airborne altimeter, calculated from Equation (20). In each case, the ratio is plotted against $k\sigma$, with isolines of kl. The figure indicates that either component of backscatter can dominate the signal peak, depending on the roughness of the surface and the altitude and system point target response of the radar. For the regime $k\sigma < 1.0$ and $kl \le 7.0$, the scattering should be predominantly coherent according to Kirchhoff theory. Such surfaces have been measured for artificial ice, labeled (a), and naturally occurring sea ice in the Gulf of Bothnia, labeled (b). Such surfaces, however, are rarely encountered in dynamic regions, where typical rms height and correlation lengths exceed 0.3 and 7.0 cm, respectively (i.e., $k\sigma > 1.0$ and $kl > 7.0$). Based on field measurements, the noncoherent component approaches and can exceed the coherent component in the roughness regimes of the Beaufort and Labrador Seas, labeled (c) and (d).

In Figure 7-8(a), the relevant system parameters are selected for the ERS-1 altimeter. Results indicate that it is likely that a dominant coherent component will be observed in the waveform peak over most consolidated ice surfaces except those such as in rough ice regime (d). In regimes (a), (b), and (c) the coherent component is more than 5 dB above the noncoherent component. To contrast with the satellite

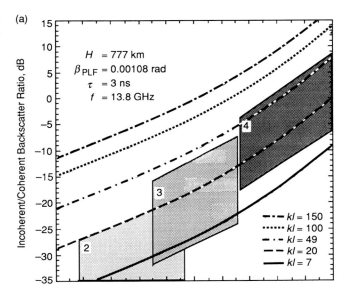

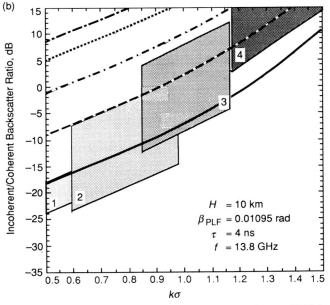

Fig. 7-8. Incoherent/coherent ratio for (a) spaceborne (ERS-1 altimeter) and (b) airborne (RAL altimeter) platforms. The labeled areas correspond to ice roughness regimes measured in several experiments: 1 = CRRELEX, $0.29 < k\sigma < 0.58$, $2.9 < kl < 8.7$; 2 = BEPERS, $0.58 < k\sigma < 0.97$, $1.2 < kl < 19.5$; 3 = Beaufort'90, $0.85 < k\sigma < 1.16$, $5.8 < kl < 34.7$; and 4 = LIMEX'87, $1.19 < k\sigma < 12.8$, $11.6 < kl < 66.5$.

situation, Figure 7-8(b) demonstrates that the simple increase in the antenna PLF beam width of the airborne altimeter results in increasing dominance of diffuse or noncoherent backscatter. Evidently, only in the smoother regimes of ice roughness will the coherent component approach or exceed the noncoherent returns. This may occur over calm ocean surfaces in marginal ice zones, over frazil slicks or smooth new ice sheets such as nilas, or even during summer melt, when maximum melt pond coverage is achieved. Under calm wind conditions melt ponds provide extremely effective specular reflectors.

Typical waveforms from the RAL altimeter show distinct examples where the peaked component in the waveform does dominate the diffuse component [Drinkwater, 1987; 1991], even in circumstances where the majority of ice is relatively rough. This cannot easily be explained by the simplified arguments presented above, because the ice surface in those cases is assumed to be uniform in its characteristics. Obviously there are situations where phase-coherent returns can occur from rough surfaces with mixed reflection coefficients and roughness scales. This idea is applied to a case of quasispecular RAL altimeter signals where the ratio of coherent to noncoherent powers is of the order of 25 dB [Drinkwater, 1987; 1991]. Such signals are shown to originate in areas of smooth open water between floes (such areas have a reflection coefficient about 10 dB greater than that of ice), and areas of new ice growth. For mixtures of specular ocean and diffuse ice scattering, the areal proportion of smooth ocean required to produce this type of signal is less than 1% [Drinkwater, 1991]. At least in the marginal ice zone, the dominant scattering mechanism therefore appears to be coherent reflection from smooth water between floes, or from new ice growth where centimeter-scale waves are damped out. Similar examples of extremely strong signals exceeding 30 dB (and often reaching 40 dB or more) have been observed in satellite altimeter measurements by both Ulander [1987b; 1991] and Laxon [1989]. Robin et al. [1983] calculated that only 0.01% of the surface within the PLF need be smooth to produce a characteristically peaked satellite waveform.

Given the importance of open water to waveform shape and power, it is necessary to understand under what conditions areas of open water might produce coherent returns. Long wavelength swell is almost completely damped out within a few tens of km of the ice edge. Shorter wind waves are damped in a short distance, on the order of km, but may be regenerated by wind in areas of open water [Wadhams et al., 1988]. Wadhams [1983] calculated the short wavelength roughness generated within the pack using

$$\sigma = 0.1265 U_* \frac{X}{g}^{1/2} \tag{21}$$

where U_* is the friction velocity determined from wind speed, X is fetch, and g is the acceleration due to gravity. The standard deviation of surface height versus fetch for various wind speeds, calculated from Equation (21), is shown in Figure 7-9. Figure 7-9 illustrates how wind speed can determine whether an area of open water of a given width will appear coherent to the altimeter. For a frequency of 13 GHz, this requires that σ be less than 2.88 mm to meet the Rayleigh criterion, or less than 0.72 mm to meet the more stringent Fraunhofer criterion. Although these relationships are only approximate, they at least give a first-order indication of the circumstances under which open water within the pack is likely to produce a near-specular return.

To conclude, the scattering mechanism for the narrow-peaked waveforms observed by altimeters has been a rather controversial issue in the past. There are not enough measurements of the surface roughness of ice at mm scales to provide a solid basis for theory. Investigators have arrived at different conclusions: that the waveforms originate from true specular reflection [Shapiro and Yaplee, 1975; Eom and Boerner, 1986] or from quasispecular noncoherent scattering [Robin et al., 1983; Rapley, 1984; Drinkwater, 1987]. The latter explanation assumes that the rms surface slope is small (less than the angle subtended by the PLF), such that the angular extent of the scattering is confined to within the altimeter angular response. It turns out that the effective backscattering coefficient for the two cases in fact agree, as noted by Robin et al. [1983]. This implies that it is impossible to unambiguously determine the scattering mechanism from waveform data alone [Ulander and Carlström, 1991].

One approach to resolving the question for sea ice is to compile probability density functions of echo energy over example areas. If the scattering is noncoherent, the probability density functions (pdf's) will result in a Rayleigh distribution. If indeed coherent returns are occurring, the pdf's will be more Gaussian in nature. The compilation of such pdf's should be undertaken for a variety of instruments of differing operating parameters (frequencies and beam widths) in order to establish the principal effects. Alternatively, it should be possible with an airborne system

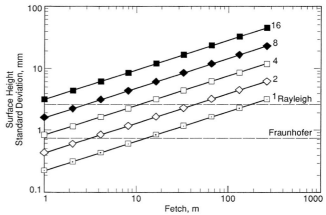

Fig. 7-9. Standard deviation of surface height versus fetch for wind speeds from 1 to 16 m/s. The cutoffs for the Rayleigh and the more stringent Fraunhofer criteria for a smooth surface are shown.

to overfly the same surface at different altitudes. This will enable the range dependency in the scattering signature to be evaluated and thus the scattering mechanism to be resolved completely. Comparing airborne and helicopter-borne results is an alternative method of analysis.

Note that the dependence on altitude in Equations (11) and (12) implies that it may be difficult to compare spaceborne and airborne measurements if a coherent component is present. Typically, the altitude of an airborne altimeter is two orders of magnitude less than that of a satellite, which reduces the coherent component from that measured by a satellite altimeter by about 20 dB. The noncoherent component, on the other hand, is unchanged.

Realistically, in considering coherent reflection one must acknowledge that many returns with different tracker biases or tracker jitters are added together to produce a mean satellite waveform. This inevitably results in a waveform which is the equivalent of the random superposition of mixed coherent and noncoherent components of the return signals. The fading of signals occurring between independent along-track views of the surface facets will inherently lead to some mis-superposition of pulses and thus noncoherent averaging of the signals during the tracking process.

7.4 THEORY VERSUS MEASUREMENT

7.4.1 Surface Measurements

To improve the retrieval of geophysical information from altimetry, it is necessary to better understand what an altimeter is sensing. Answers to questions such as—Where is the signal originating? and What are the effects of snow depth and ice type on backscatter? can be provided by surface measurements. Near-surface backscatter measurements from the Fram Strait marginal ice zone (MIZ) have been acquired at frequencies of 5.25, 9.6, 13.6, and 16.6 GHz using a frequency-modulated continuous-wave radar scatterometer with pencil-beam antenna patterns (two-way beam widths of 1.5° to 4.5°). The radar scattering coefficients were determined by comparing surface returns with returns from calibration targets of known radar cross section. These observations were made from a helicopter in conjunction with surface observations that allow backscatter to be related to a particular ice type or to ocean conditions [Onstott, 1991; Onstott et al., 1987]. Also given in this section is a synopsis of recent measurements in the Weddell Sea.

7.4.1.1. Vertical incidence scatterometer data from the Fram Strait marginal ice zone. The four major ice types present when the measurements were made in midsummer were multiyear (MY), thick first-year (TFY), medium first-year (MFY), and thin first-year (ThFY). Snow depths for MY and FY ice by midsummer were about half of what they were at the beginning of summer. This is important in that thick ice, such as TFY and MY ice, will have enough snow cover to mask much of the ice surface from microwave

radiation for at least half of the two-month-long summer. Data were acquired by flying scans from a position over the ocean about 20 km from the ice edge, across the ice edge, and into the MIZ until large floes were encountered (an additional 20 km). One such transect is shown in Figure 7-10. Illustrated in these data is the transition from the strong backscatter of the open ocean, to a region of calm water (5 km in extent) immediately adjacent to the ice edge, through a well-delineated and compact ice edge, into a region with bands of similarly sized ice floes and varying ice concentration, and then into a region of 90% ice concentration and large multiyear floes.

Using the open ocean signature as a reference, the area of dead water near the ice edge produced an enhancement of about 6 dB. Backscatter from ice is dramatically smaller (by 8 to 21 dB), in contrast to the usual observations by airborne and satellite altimeters moving from ocean to MIZ. This difference may be explained by the much larger Fresnel zone of airborne and satellite-borne instruments, which can encompass flat areas, with a resulting dominance of coherent reflection. As the ice concentration increases deeper in the MIZ, a lower average return is observed. The strongest returns are associated with the water between floes. These returns are specular in nature and are characterized by a cross section at vertical of greater than 25 dB and a reduction in intensity of almost five orders of magnitude by 25°. These results support the assertion that peaked returns in airborne and satellite data come from smooth water between floes. The range of scattering coefficients is also of interest. The range for the open ocean is about 6 dB, whereas in the MIZ it is about 40 dB. The high range in the MIZ is due to the presence of areas of open water, which produce a variety of microwave signatures influenced in large part by their position from the ice edge (Figure 7-11). Outside and within the ice edge, variations in ocean surface roughness are caused by complicated interactions between wind, waves, currents, and fresh water input. These interactions are driven by effects related to cold air from the pack

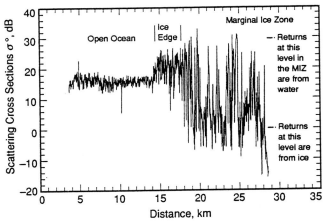

Fig. 7-10. Measurements from a single radar scatterometer transect across the ice edge on July 5, 1984 at 13.6 GHz and a vertical incidence angle.

crossing from the cold ocean (ice and sea surface temperature of 0° to 2.5° C) near the ice edge to the warmer open ocean (sea surface temperature of 4.4° C). One result of this movement of cold air is the area of calm water adjacent to the ice edge.

The transition from the open ocean to the calm water within the MIZ is very rapid in this case. It occurs within 600 m, and it produces large level shifts in backscatter. This sharp delineation was observed often during MIZEX'83 and '84. Ambiguities or errors in detecting the ice edge with altimetry may occur if the ocean near the ice edge damped by a cold air mass is mistaken as an ocean surface damped by a thin layer of ice. In the satellite case, the ice edge is marked by a similar increase in variability but higher backscatter values over the ice.

The MIZ is a dynamic region. Diffuse scattering from deformed ice is observed to produce returns 12 dB below that of flat ice. Returns from water in the shadows of floes are 2 to 5 dB above typical ocean returns from within the MIZ. Backscatter modulations are also found to be a function of ice concentration and, to a lesser extent, floe size. Based on other observations by Onstott, the backscatter response for winter is anticipated to be very similar. The distribution of features within the boundaries of a floe may also have an important impact on the backscatter response. Pressure ridges, melt ponds, and the blocks that pile around the floe edges require consideration. New pressure ridges may form at floe boundaries and are differentiated from older pressure ridges that may be covered with a layer of wet snow. Ridges result in weak backscatter, with the more recently formed ridges producing the more diffuse scatter response. Pools, often small and protected from wind by undulating ice, were found to produce backscatter cross sections of a magnitude similar to that of shadowed water signatures. On July 5, when these observations were made,

minimal floe area was covered by melt ponds. Melt pond formation was, however, under rapid acceleration, and by July 9, coverage on MY floes was estimated at about 30%.

7.4.1.2. Ice type and feature discrimination in Fram Strait measurements. During summer, the reflectivity at nadir of FY and MY ice is determined by the presence of moist snow, the areal extent of melt ponds, and topography. Melt ponds increase the reflectivity of a floe, and topography acts to reduce the intensity of the reflection. At 13.6 GHz, the reflectivities for the three primary ice types are within a very narrow 4.5 dB range. Table 7-1 shows that the backscatter response follows, in large part, surface topography. Thin first year ice is characterized by the thinnest snow cover (mean of 4 cm) and is tabular, with little if any deformation. It produces the largest reflectivity of the three ice types. Physically, MFY ice is similar to ThFY ice, except that it has a thicker snow cover (mean of 10 cm). MFY and MY share a similar signature, that of an ice sheet with a very thick snow layer. The data at 13.6 GHz for snow-covered MY appear high at first glance, but upon further examination they are consistent with the other observations shown in Table 7-1. A MFY floe produced a signature 3 dB lower than that of ThFY, where snow was about 10 cm. On the same MFY floe, however, there was a region where snow had accumulated to 14 cm. Here the reflectivity increased to 8.0 dB. This is identical to the reflectivity observed on the MY ice and within 1.2 dB of that of ThFY ice. Thick snow on the MY ice, rather than the snow–ice interface, dominates the MY signature.

While backscatter for winter and summer FY ice is very similar, significant differences in the seasonal response of MY ice are observed. These are related to the moisture contained in the snow and ice sheet [Onstott et al., 1987]. Moisture limits the penetration into the ice sheet, thereby reducing or negating the volume scattering that dominates the backscatter response of MY ice during winter. At vertical incidence, this causes a 9-dB signature enhancement (at 13.6 GHz) for MY ice during summer when compared to the winter response. Measurements of snow moisture show that the bulk wetness of the snow pack is about 5.5% by volume, and the majority of the snow pack is

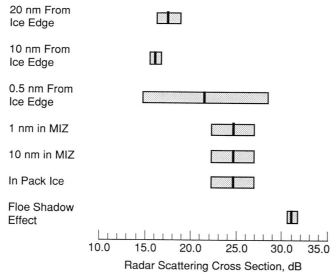

Fig. 7-11. Liquid ocean radar scattering cross sections shown as a function of position from the ice edge at 13.6 GHz and a vertical incidence angle.

TABLE 7-1. Radar scattering coefficients at vertical incidence angles expressed in decibels. Measurements were obtained July 5, 1984, in the marginal ice zone [Onstott, 1991].

Scene	Frequency, GHz			
	5.25	9.6	13.6	16.6
Ocean, open	13.3	14.5±1.6	15.9±0.7	
Ocean, edge		17.4±2.4	21.5±6.9	
Ocean, interior	31.3±10.1	22.8±22.6	24.5±2.4	
Thin FY ice	2.0±1.2	6.4±9.4	9.2±6.8	8.9±5.9
Medium FY ice	−0.8±2.4	2.9±3.0	4.7±3.9	7.0
MY thick snow	−1.3±3.0	2.6±4.0	7.9±3.9	6.1±5.0
MY with melt pools	10.4±8.3	8.3±9.8		

old and has a density of 400 to 500 kg/m^3 [Onstott et al., 1987]. The penetration depths calculated for these conditions are about 4.5 cm, 1.8 cm, 1.5 cm, and 1.3 cm for 5.25, 9.6, 13.6, and 16.6 GHz, respectively. The snow pack on MY ice up to midsummer may be 40 cm. Note that at 13.6 GHz, only 1 to 3 cm of snow is required before the snow layer completely dominates the backscatter response since for each penetration depth the incident signal experiences a 9-dB round-trip loss.

7.4.1.3. *Shipborne Ku-band radar observations in the Weddell Sea.*

Ku-band radar data were acquired during a transect of the R.V. Polarstern through the Weddell Sea during the austral winter of 1989 [Lytle et al., 1992]. A VV polarized network analyzer-based radar developed at the University of Kansas made observations at incidence angles spanning 0° to 30°. Most radar data were acquired on ice floes against which the ship was moored. Measurements of physical properties of the sea ice included surface roughness, snow cover thickness, snow and ice density, and grain size and salinity. Ice freeboard data were also collected by drilling numerous holes through the ice and measuring water level relative to the ice surface.

Flooding at the ice–snow interface seems to be a characteristic phenomenon of the western Weddell Sea [Lange and Eicken, 1991; Wadhams et al., 1987]. This process is apparently associated with enhanced snow accumulation depressing the ice surface below sea level. Flooding changes density and salinity in the lower parts of the snow layer, which in turn affects the radar response. Backscatter is plotted against freeboard and parameterized against slush layer presence in Figure 7-12. Backscatter is elevated when a slush layer is present, presumably because of the enhanced dielectric contrast. Flooding, then, may be another mechanism for situating a relatively bright target within the radar footprint. The possibility of flooding must be taken into account when attempting to interpret altimetry in terms of ice concentration or type. Laboratory measurement to further study the importance of flooding are described in Chapter 9.

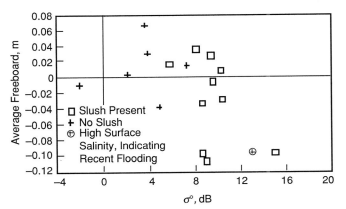

Fig. 7-12. Average ice freeboard versus $\sigma°$ parameterized against the presence of a slush layer.

7.4.2 *Aircraft Measurements*

Various work [McIntyre et al., 1986; Drinkwater, 1987, Fedor and Walsh, 1988] has documented the characteristics of airborne altimeter waveforms over mixtures of sea ice and water. Several of these studies [Powell et al., 1984; Cowan and Squire, 1984] document results from a pre-launch campaign for ERS-1, in which the RAL altimeter acquired data in experiments devised to indicate the characteristics of sea ice relevant to altimeter pulse analysis. Airborne altimeters generally have wider beams than satellite altimeters, in order to reduce the impact of antenna off-pointing caused by aircraft motion. A wide beam leads to different waveform characteristics than are observed with a satellite-borne instrument, as was discussed in Section 7.3.2. The surface area sensed by an airborne instrument is usually a factor of 100 smaller than for a satellite altimeter, while it subtends a wider range of incidence angles. This is a critical difference, because in areas consisting of mixtures of ice and water there is inherently a larger area of water available to the satellite altimeter for scattering from nadir than is available to the lower altitude airborne instrument. Another consequence of the airborne altimeter's wider beam is that backscatter fall-off with incidence angle can be measured.

7.4.2.1. *Waveform examples.*

Figure 7-13 is an example of the operation of the RAL altimeter over a variety of sea ice conditions in the marginal ice zone of the Fram Strait. Consecutive waveforms exhibit dramatic changes in proportion as the instrument passes over large ice floes and areas of open water in the pack. Note that some waveforms show saturation. One interesting finding illustrated by the waveform labeled *a* is that as the altimeter passes over a lead only a few meters wide, the altimeter responds extremely sharply if the lead is at nadir. Fedor and Walsh [1988] noted the same phenomenon in data from the Beaufort Sea. Since the power reflection coefficient of water is generally 11 dB higher than that for snow-covered sea ice, the water between ice floes dominates the backscattered signal near nadir. Other observations from this data set show that when the sea ice fills the footprint, waveform shape becomes suggestive of ocean returns as the signal level is modulated by the type and surface roughness of sea ice present. Fedor et al. [1989] note typical peak backscatter strengths of around 30 to 37 dB over areas of new sea ice and around 10 dB over deformed FY ice in the Beaufort Sea. These observations show close correspondence with the findings of Drinkwater [1991] over FY ice in the Fram Strait. In the latter study, accompanying surface measurements [Tucker et al., 1987] led to attempts to classify ice types using the altimeter waveforms. The study revealed four distinct surface classes based on surface roughness (or fall-off of backscatter with incidence angle) and normal incidence backscatter coefficient. These corresponded in near-coincident SAR images and aerial photographs with smooth, bare FY ice with minimal ridging; snow-covered FY

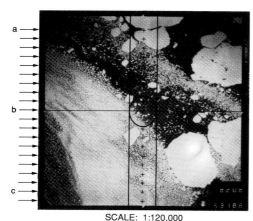

SCALE: 1:120,000

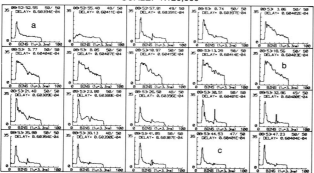

Fig. 7-13. MIZEX'84 RAL airborne altimetry over mixed ice concentrations in the Fram Strait. Waveforms show received power in range gate bins [Drinkwater, 1987].

ice; rotted, rough conglomerate ice floes; and MY ice with large-scale surface roughness.

In ice–water mixtures, the rapidity with which the power decays after the peak is largely a function of the wave slope (see Section 7.3.4). Some areas, such as for waveform *b*, provide non-fetch-limited regions of almost open water within the pack. Wind-generated gravity and capillary waves develop rapidly in the absence of floes (unlike at *c*). The waveform trailing edge responds dramatically to the presence of these waves and particularly to the higher wave slopes. The waveforms become more ocean-like, albeit with a relatively high normal incidence backscatter coefficient due to the lack of large gravity waves.

Volume scattering cannot be neglected when old ice has low-density, bubbly upper layers (Chapter 2). Volume scattering leads to a backscatter response function that tapers much more gradually with incidence angle. It is suggested as a mechanism for some Fram Strait observations where neither the rms surface roughness nor the residual aircraft motion can account for the spreading of the waveform leading edge. Together with the large-scale surface roughness observed on vast low-salinity MY floes, volume scattering appears to result in rms tracker delay variability or height fluctuations of between 0 and 4 m [Drinkwater, 1991]. The snow depth on these floes averaged 28 cm, and ridge sail heights varied between 3 and

5 m. In contrast, over large FY floes the measured rms surface height variability was less than 1 meter, and such tracking excursions were not noted. For the same effects to occur in satellite data, the radar would likely have to be directly over a vast MY ice floe or iceberg. This situation would be encountered in the Arctic during winter conditions and in the Antarctic, where vast tabular icebergs are often observed by altimetry [McIntyre and Cudlip, 1987; Laxon, 1989; and Hawkins et al., 1991].

7.4.2.2. Ice concentration derivation from airborne observations. Attempts have been made to use altimetry for ice concentration measurements. The models described in Section 7.3.3 yield fractional areas of smooth ice and open water facets, if reflection coefficient and size are specified. These do not, however, incorporate the slope distribution, which is portrayed for both ice floes and the water between in the heuristic model in Figure 7-14. This model illustrates the decline in significant wave height caused by sea ice wave attenuation [Rapley, 1984] and the associated decline in wave slope as short wavelength waves are preferentially damped. The resulting modulation of the waveform power directly from ice concentration will therefore vary as a function of both rms wave height and fractional coverage of sea ice (based on the premise that the returned power in the waveform decays as the ratio of low-reflectivity ice to high-reflectivity water increases). Figure 7-14 shows idealized waveforms at concentrations of 0, 20, 40, 60, 80, and 100% for a given wave regime. In each case the total height distribution is a combined effect of wave height, ice surface roughness, and penetration, while the form of the waveform trailing edges is a result of the wave slopes observed between ice floes.

Using altimeter data over ocean surfaces inside the MIZ edge, Drinkwater [1987] shows that the angular backscatter scattering function is predicted by noncoherent geometric-optics theory. The model fit to the data gives an rms

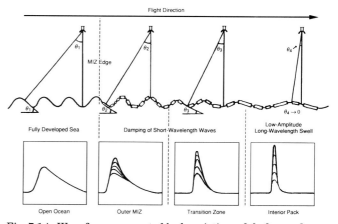

Fig. 7-14. Waveforms generated by heuristic model of wave decay with attenuation, and families of waveforms associated with varying ice concentration: 0, 20, 40, 60, 80, and 100% [Drinkwater, 1987].

slope 12.8°. This is consistent with waves measured in the ice edge for a fairly well-developed sea. With increasing distance into the MIZ, observations show a corresponding decrease of rms slope to around 3° at 40 km from the ice edge. The progressive damping of ocean waves as they pass into the MIZ results in more and more spiked echoes from the water between ice floes, as shown in Figure 7-14. Thus, waves and the wave spectrum in the MIZ mutually interact with the shape and strength of altimeter signals and must be considered when attempting to derive concentration from waveforms. In an analysis of the relationships between backscattered power and ice concentration in the Greenland Sea, Drinkwater [1991] investigates airborne Ku-band signals for which there are accompanying SAR and photography data. To obviate the requirement for compensating wave height information, altimeter waveforms are chosen beyond the point at which they are modulated by waves. These are subsequently compared with digitized, simultaneous aerial photographs over a track length of 130 km. Both the peak backscattered power and integrated power in the trailing edge of the waveform are shown to be related to the ice concentration. However, because of the saturation effects found in these data, the peak power could not be used for accurate ice concentration retrievals. The waveform integral, on the other hand, declines monotonically with increasing ice concentration and is reasonably robust. Regression analysis shows that the relationship between these two variables had a correlation coefficient of −0.81, and 65% of the signal energy variance is accounted for using this measure alone. The rms residual error is 13%, and the remainder of the variance is accounted for by varying surface conditions due to local wind roughening, varying ice types, ice roughness, and other factors.

7.4.3 Satellite Measurements

A common problem in the analysis of altimetry is to find high-quality surface truth data with which to determine ice conditions during the sensor overflight. For corroborating the interpretation of satellite altimetry the most successful data source has been SAR [Swift et al., 1985, Ulander 1987a, Fetterer et al., 1991, Ulander and Carlström, 1991]. Examples of investigations that rely on SAR are given in this section. Comparisons of altimetry with imagery from passive sensors is reserved for Section 7.4.4.

7.4.3.1. Echo waveform simulations compared with measurements. We noted earlier that it is impossible to determine unambiguously the scattering mechanism from a single narrow-peaked waveform. However, a series of such waveforms may provide this information as the altimeter moves from a smooth area into an adjacent area with diffusely scattering characteristics. For true specular reflection, the returned power essentially originates from the first Fresnel zone. This implies a horizontal resolution according to Equation (9), which is almost an order of magnitude finer than the pulse-limited resolution for an ERS-1 class altimeter. Hence, the echo waveform will, in

the specular case, respond more quickly to a change from a narrow-peaked to a diffuse type of scattering.

This observation has been used to determine the scattering mechanism of narrow-peaked waveforms over sea ice in the Baltic Sea [Ulander and Carlström, 1991]. A section of Geosat altimeter data was extracted over a newly frozen lead surrounded by very deformed ice. The former produced narrow-peaked waveforms and the latter essentially diffuse waveforms. A pulse echo series was then simulated by integrating over the illuminated surface at each platform position along-track, and an echo waveform was produced by averaging over one hundred pulse echoes. The surface composition is shown in Figure 7-15(a). It was derived from a high-resolution airborne SAR image that delineated the two surface types. Figure 7-15(b) shows the Geosat normal incidence backscatter coefficient and two simulations that assume a coherent and noncoherent mechanism, respectively. In the noncoherent simulation, the young ice was represented by a model according to Equation (17), whereas in the coherent simulation all the returned power originated from the first Fresnel zone with uniform weighing. Note the close agreement between the Geosat data and the noncoherent simulation. The model parameters in Equation (17) were chosen as $F = 3.3\%$ and $\theta_D = 0.068°$ (for $D \approx 5.5$ m). The coherent simulation, on the other hand, shows a much higher sensitivity to the changing surface characteristics than does the Geosat data. This particular example shows that the narrow-peaked waveforms over the frozen lead, corresponding to a high backscatter coefficient of 32 dB, were indeed noncoherent in nature. The backscatter value over the lead in Figure 7-15(b) is somewhat less due to distortion by the pulse echo simulation and waveform averaging.

The remaining issue to resolve is whether the noncoherent scattering is in accordance with the geometrical optics or flat patch model. Both models predict a backscatter coefficient of the same functional form and have essentially the same effective backscattering coefficient for narrow-peaked waveforms. Note, however, that the cause for the angular response is quite different. In the geometrical optics model it is determined by the rms surface slope, since the reflecting specular points are infinitely large compared to the wavelength; in the flat patch model it is determined by the flat patch diameter since the patches are assumed horizontal (i.e., with zero surface slope). There are, as yet, no field measurements which could help resolve this question. In general it would require that the joint height probability density function be determined for scale lengths between 1 cm and 100 m. Hence, our effort in the future must turn toward more sophisticated field measurements of the surface roughness.

7.4.3.2. Ice type signatures in satellite measurements. With an approach quite different from that outlined in Section 7.4.2 for airborne altimetry, Chase and Holyer [1990] extend the measurement of ice type and concentration to satellite altimetry. Their work uses a linear unmixing model, suggested by the fact that waveforms from the MIZ

(a)

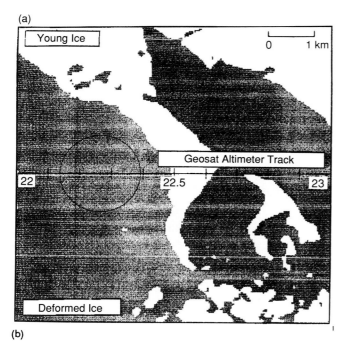

(b)

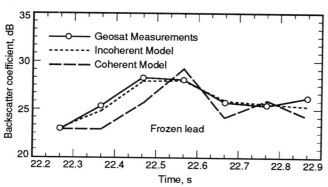

Fig. 7-15. (a) Simulated surface composition derived from an airborne SAR image of Baltic Sea ice, and (b) simulation result compared with Geosat measurements [Ulander and Carlström, 1991].

and pack are usually the result of backscatter from mixtures of ice and water in the footprint. Each waveform is represented as a multidimensional vector that can be expressed as a linear combination of end member vectors. Analysis consists of identifying the end members and associating each with a surface type. Although the model can be adapted for nonlinear combinations of waveforms, Chase and Holyer argue that it is not necessary for the data they examined. Using manually interpreted airborne passive microwave imagery collected in the Greenland Sea within hours of a Geosat overpass, Chase and Holyer found that the analysis of the concentration of three combinations of types (open water and brash ice, FY and old ice, and young and new ice) agreed with that from the passive microwave analysis to within 4.6% over the experiment area. Interestingly, the shape of the end member paradigm

waveforms suggests that concentration in this analysis is directly proportional to peak power and waveform integral. Possible explanations for this contrast with the airborne case are the differences in scale and range discussed in Section 7.3.2.

Most investigations of the ability of satellite altimeter measurements to distinguish between broad ice thickness classes and to detect water openings, starting with that of Dwyer and Godin [1980], have only used the altimeter data in a relative sense. This is because the data lack absolute radiometric calibration. It should be emphasized that it is only possible to obtain an effective backscatter coefficient when the angular waveform is narrow peaked. By assuming a particular form of the angular response, it may then be possible to obtain a better estimate of the backscatter coefficient at normal incidence. This inherently assumes that the system impulse response is sufficiently narrow to resolve the angular scattering response and that the scattering mechanism is known.

The correlation between radar backscatter at normal incidence and Arctic and Baltic winter ice types has been described by several investigators [Ulander, 1987a, 1988; Fedor et al., 1989; Ulander and Carlström, 1991]. Histograms of the normal incidence backscatter coefficient for some of these ice types are shown in Figure 7-16. The normal incidence backscatter coefficient decreases in general for increasing ice thickness, such that new ice has the highest values (30 to 40 dB) and MY ice the lowest (10 to 20 dB). Very deformed ice also has a low backscatter coefficient (10 dB). Note the high mean values for all classes when compared with the surface measurements of Section 7.4.1. These results suggest that broad ice classes can be inferred from the altimeter backscatter data, provided the surface conditions do not change rapidly along track. When the latter is not the case (for instance in the MIZ), the signature shows a large variability, which introduces difficulties in the interpretation.

Figure 7-17 shows curves of the backscatter coefficient and the echo waveform width (a measure of diffuse scattering) for a Seasat altimeter track in early October off the coast of Alaska [Ulander, 1987b]. The curves are overlaid on a Seasat SAR image that was acquired a few hours earlier. Narrow, peaked waveforms dominate over the pack, except where the nadir track crosses two large floes. Figure 7-18 shows a similar comparison of Geosat altimetry and airborne SAR in the Greenland Sea [Fetterer et al., 1991]. Here, however, the high-frequency variability in backscatter is much greater. Backscatter is from the 0.1-second waveforms. Note that *AGC* is a somewhat smoothed version of peak power. The voltage proportional to attitude, or VATT parameter, is a measure of power in later gates. Low values of VATT (plotted every second) indicate peaked returns. The parameters in Figure 7-18(c) delineate zones:

1) The MIZ, where backscatter rises from 9 to 30 dB as the ice edge is crossed. Waveforms are strong and peaked, suggesting reflection from water between the small floes, which give a bright return in the SAR image. Backscatter

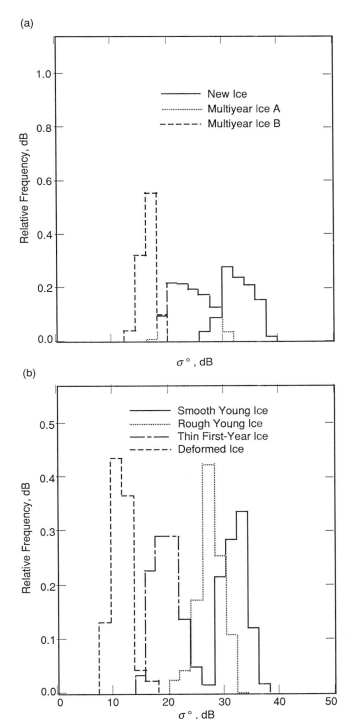

(a)

(b)

Fig. 7-16. Backscatter histograms for (a) Beaufort Sea ice from Seasat data [Ulander, 1988], and (b) Baltic Sea ice from Geosat data [Ulander and Carlström, © 1991 IEEE].

drops slightly over an expanse of open water several kilometers wide, where wind may roughen the surface.

2) The pack, where variability is high. Backscatter is somewhat lower in the western half, where there is a larger percentage of rough ice. Backscatter dips over two large old floes, which appear quite bright in the SAR image.

3) The fast ice, where backscatter reaches its lowest level and waveforms are not peaked but are more characteristic of ocean returns. This is evident in the waveforms plotted in Figure 7-18(a).

While changes in the nature and variability of waveforms are revealing, caution should be taken in interpreting the power and shape of waveforms from an inhomogeneous area such as this in terms of ice roughness or type. Instrumental effects and rapid changes in surface characteristics (described in Section 7.2.5) drastically alter individual averaged waveforms.

7.4.4 Satellite Altimetry Compared With Infrared, Visible, and Passive Microwave Imagery

In Sections 7.4.2 and 7.4.3 aerial photography or SAR imagery was used to gain insight into scattering mechanisms and to test the possibility of deriving concentration and type estimates from altimetry on a case-by-case basis. Further applications arise from the synergistic use of radar altimetry with imaging sensors. For example, if a certain ice type can be identified in imagery, this may aid greatly in the interpretation of the radar altimeter data, which may then be used to measure surface elevation and roughness (derived from waveform leading edge width). This section compares altimetry with wide-swath imagery on the regional and global scales required of operational algorithms. Such comparisons enhance our understanding of the operation of both sensors while aiding the development of altimeter data algorithms. Immediate applications lie in the area of algorithm validation. For example, the altimeter has shown great promise in validating ice extent mapping by passive microwave instruments. In addition, ERS-1 altimeter sea ice products can be validated using data from the ERS-1 Along Track Scanning Radiometer.

7.4.4.1. Waveform parameterization. Figure 7-19 shows the two parameters computed from altimeter waveforms to quantify the response over sea ice in the following analysis. The first, *SIGPK*, represents the peak backscatter measured in the return echo. The second, *SIGTD*, is computed as the difference between the average power in eight early and eight late range gates in the altimeter return. *SIGTD* is essentially a measure of the degree of peakiness in the return echo.

7.4.4.2. Observations of a compact ice edge in the Greenland Sea. The Advanced Very High-Resolution Radiometer (AVHRR) onboard the NOAA series of polar orbiting satellites provides visible and infrared imagery with a swath width of 2900 km and nadir resolution of 1.1 km. It is routinely used for operational ice analysis. Figure 7-20(a)

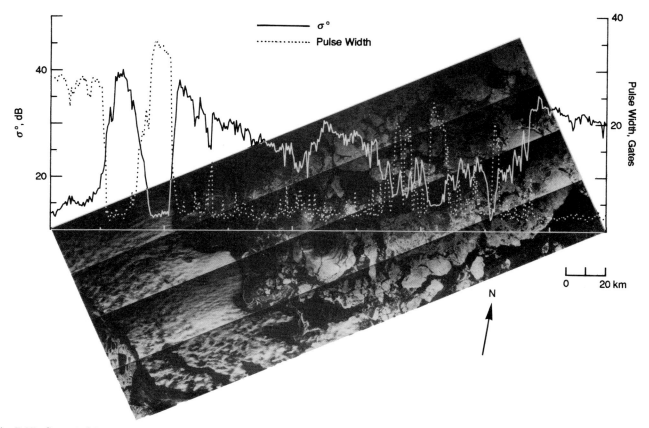

Fig. 7-17. Seasat altimeter measurements in the Beaufort Sea compared with Seasat SAR for October 3, 1978 [Ulander, 1987b].

shows four Geosat altimeter tracks overlaid on a channel 2 (visible band) image of the Greenland Sea. Both altimeter data and imagery were acquired on May 10, 1987. Parameters *SIGPK* and *SIGTD* are derived for every waveform. There is therefore a value every 0.1 seconds, or every 0.7 km along the ground track. A sharp rise in both parameters is observed close to the ice edge visible in the imagery. In some cases, notably tracks B and C, a rise in *SIGPK* is observed outside the ice edge and before any noticeable rise in *SIGTD*. The cause of this is unknown, but the examples demonstrate the sensitivity of the altimeter to the presence of sea ice. A similar rise in backscatter, related to the damping of waves due to the proximity of sea ice, is seen in the surface measurements of Section 7.4.1. The highest strength (with values up to 40 dB) and most peaked echoes are observed just inside the ice edge, which is normally populated by small (<100 m) ice floes. Further into the pack the waveform strength and peakiness decrease and become more variable. This is attributed to a less heterogeneous surface caused by the presence of larger ice floes. In some areas, such as where track A crosses track D, cloud cover can confuse identification of the ice edge. In such cases the altimeter may be used to better identify the location of the ice edge.

7.4.4.3. Observations of a diffuse ice edge in the Greenland Sea. Figure 7-20(b) shows three altimeter tracks overlaid on an AVHRR channel 4 (infrared) image acquired on March 18, 1987. In this instance an off-ice wind results in a more diffuse ice edge than is observed in the May 10 image. The two altimeter tracks labelled A and B show a gradual increase in echo strength over some distance with little increase in echo peakiness. This is attributed to a more gradual damping of ocean waves than is observed in the May 10 image. Again the strongest and most peaked returns are observed further into the ice pack.

Of particular interest in this image are the two large floes observed close to the coast and transected by altimeter track B. The floes are clearly delineated by low strength, ocean-like echoes. Unlike other areas of the pack, where smooth ice or open water can occur within the altimeter footprint, over these vast ice floes the altimeter is observing only a rough ice surface. Such a phenomenon may make it difficult to discriminate between the return from open water and that from a large floe. However, because such surfaces fill the footprint for several waveform integration times, they do provide the possibility of measuring meter-scale roughness and even ice freeboard. The AVHRR image was acquired some 13 hours after altimeter pass B. The southward offset of the two ice floes with respect to the

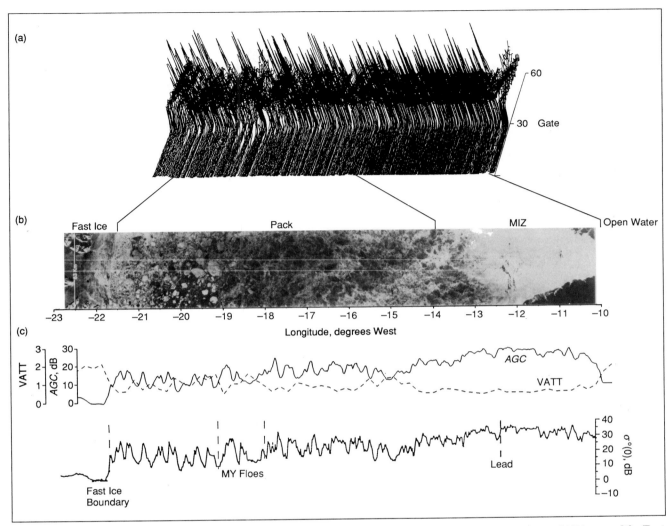

Fig. 7-18. (a) Geosat altimeter waveforms from April 13, 1987, with a (b) near-coincident (within 1 hour) airborne SAR image of the East Greenland Sea at 72° N. White lines on the image mark the approximate position of the 9-km-wide range-window-limited footprint. (c) Measures of peak power, $\sigma°(0)$, average power, AGC, and waveform shape, VATT, are plotted. Along-track registration of SAR image and altimeter data is estimated to be within 2 km [Fetterer et al., 1991].

signal in the altimetry can be attributed to motion in the East Greenland current.

A similar ocean-like signal is observed over the fast ice, with a sharp decrease in echo strength at the fast ice boundary followed by a slight step increase in echo strength further into the coast. Again this is due to the altimeter observing only a homogeneous rough ice surface. The delineation of two zones within the fast ice suggests that the altimeter is discriminating between an inner zone of fast ice that has survived the winter intact, and an outer, rougher zone that has been broken up at some time by a winter storm. (Similar zonation in the returns from fast ice is apparent in Figure 7-18).

7.4.4.4. Midwinter image of the Kara Sea. Figure 7-20(c) shows three altimeter tracks overlaid on an AVHRR channel 4 image of the Kara Sea acquired on February 21, 1986. Waveform echo data are not available for this period and therefore the AGC values are displayed. In this thermal image, darker areas indicate higher surface temperatures and therefore thinner ice. To the right of the island Novaya Zemlya an area of relatively new ice has formed. On the far right-hand side of the image a further large frozen lead, some 20 km across, is observed close to the continental land mass. For this image, echo strength shows a strong correlation with ice age. New, smooth ice produces stronger returns than older, rougher ice. This phenomenon is exactly what would be expected when considering normal incidence radar backscatter over surfaces with varying

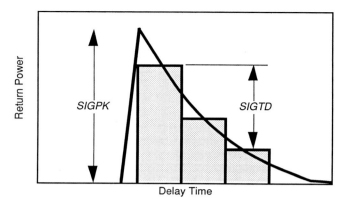

Fig. 7-19. Waveform parameters *SIGPK* and *SIGTD* are measures of the peak power in a waveform and of the peakiness, or rate of backscatter fall-off with incidence angle, in the waveform. *SIGTD* is calculated by taking the difference in average power between eight early and eight late range gates.

degrees of roughness. The highest echo strengths are observed over the narrow lead, which is slightly darker than the ice just east of Novaya Zemlya. Within the older ice some cracks are observed to correlate with individual peaks in the altimeter *AGC* value. In some cases peaks are observed in *AGC* that do not correlate with any cracks visible in the imagery. Smooth ice or open water not resolved in AVHRR imagery are believed responsible for these peaks. These observations demonstrate the sensitivity of altimeter data to areas of thin ice and leads. The mapping of leads has great importance to sea ice research, as leads modify ocean–atmosphere heat transfer in the high Arctic.

7.4.4.5. Midwinter image of the Chukchi Sea.
Figure 7-20(d) shows three altimeter tracks overlaid on an AVHRR channel 4 image of the Chukchi Sea acquired on March 14, 1988. All three tracks show numerous peaks in the echo strength, especially near coasts and islands. Here, leads are most likely to occur when winds blow older ice away from the coast, allowing new ice to form. Judging by high *SIGPK* and *SIGTD* values, a large area of new ice may be north of Wrangel Island, although this is difficult to observe in the imagery. This again shows the usefulness of altimetry as an adjunct to visible or infrared imagery.

7.4.4.6. Altimetry coincident with passive microwave imagery.
Satellite passive microwave data has been widely used for mapping global sea ice cover. Satellite altimetry provides the only practical alternative means for global synoptic monitoring of sea ice extent. The merging and comparison of data from these two sensors is therefore an obvious means to improve understanding of both [Laxon and Askne, 1991].

In the following analysis we examine data from a merged product produced using data from the Geosat altimeter and the SSM/I flown on board the Defense Meteorological Satellite Program (DMSP) satellite series. The five brightness

temperatures recorded by the SSM/I for the day and location of each altimeter data point are appended to the Geosat Geophysical Data Record (GDR). The spatial coincidence is therefore within 25 km (the gridded resolution of daily average SSM/I brightness temperatures) and the temporal coincidence within one day. The merged data set can then be used to directly compare the ice extent mapped by the two sensors. The SSM/I data are processed using the NASA Team algorithm. The algorithm used for detecting the ice edge in the Geosat data record uses the standard deviation of the 10 surface height values recorded in the GDR over 1 second. Over sea ice, the disruption of the tracker (described in Section 7.2) results in a significant increase in the standard deviation of the height measurement [Laxon, 1990]. Other algorithms have used the rise in AGC and the drop in SWH to detect the ice edge and the penetration of swell in the MIZ [Rapley, 1984] and the rise in the ratio of AGC to VATT that occurs as the edge is crossed [Hawkins and Lybanon, 1989].

Figure 7-21 shows a series of altimeter tracks for August 3, 1987 overlaid on a polarization difference image generated from the 85 GHz channel. Although the 85 GHz channel suffers from some weather contamination, the ice cover is reasonably visible. A data point is plotted in white where both sensors indicate that ice is present. Where only the altimeter indicates ice the data point is plotted in red. Where only the SSM/I indicates sea ice is present the data point is plotted in green. If neither sensor indicates that ice is present, a blue data point is plotted. Both sensors show an ocean signal in most areas away from the ice boundary, with the exception of some green points north of Scandinavia where the SSM/I incorrectly identifies some areas as being ice covered. It is well known that the SSM/I maps some areas of open water as ice due to weather effects. In the Beaufort and Chukchi Seas, various areas of both green and red points are observed. The red points appear to lie near the ice boundary, while the green points occur within the pack (in one case over what might be an area of open water). Small areas of red points are also observed in the Kara and Greenland Seas. These may delineate ice cover below the lowest concentration detectable using SSM/I.

In order to look at the Geosat altimeter and SSM/I differences on a larger scale, two 17-day merged data sets were compiled, one during midwinter and one during midsummer for both hemispheres. The data plots produced from these data sets do not show any points where both instruments indicate no ice to be present. This is to permit clearer delineation of areas of disagreements. Figure 7-22(a) shows data for August 1987 in the Northern Hemisphere. The amount of ice cover below 72° N (the northern limit of coverage) is limited, but significant clusters of red points are observed in the Greenland, Beaufort, Barents, and Kara Seas. At lower latitudes, such as around the British Isles, a large number of green data points illustrate the problem of weather effects in the SSM/I data. Figure 7-22(b) shows the corresponding image for February 1988. During maximum extent there is much better agreement overall between the altimeter and the SSM/I. The few

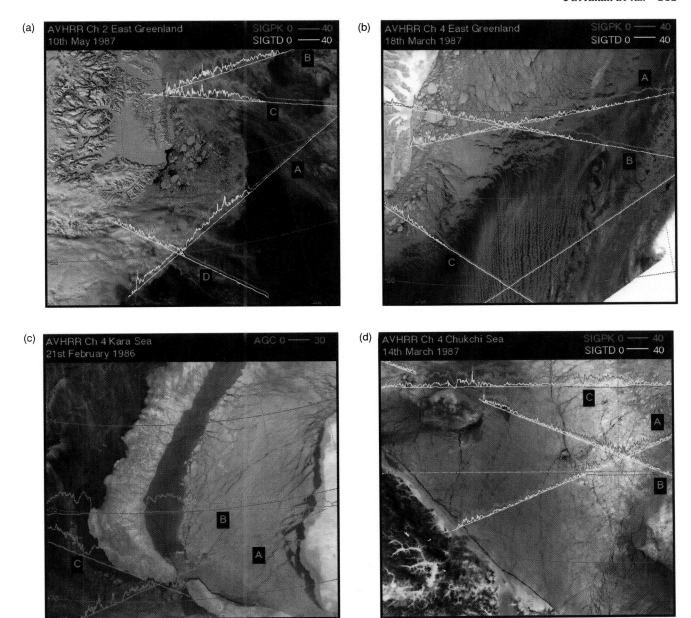

Fig. 7-20. (a) An AVHRR visible image of the Greenland Sea, with altimeter waveform parameters overlaid along the Geosat altimeter track (Scoresby Sound is near the center of the image); (b) as in (a), but with a less compact ice edge; (c) infrared band image of the Kara Sea, with Geosat *AGC* values plotted every 0.1 s along the nadir track; and (d) infrared band image of the Chukchi Sea, with waveform parameters.

exceptions are in the Chukchi Sea, around 165° E, and in the Greenland Sea, where some green points are observed. These points may be attributable to areas of fast ice and vast floes which, as seen in the comparisons with AVHRR data, result in an altimeter signal that is difficult to distinguish from open water.

Figure 7-22(c) shows the Southern Hemisphere during February 1988. There is a clear geographic zonation of green and red points. Since the altimeter samples the ice edge at random points during the 17-day period, the fact that the areas of disagreement are geographically correlated strongly suggests some geophysical cause. Red points in particular are observed mainly in the Weddell and Ross Seas, with green points occurring in the Bellingshausen Sea and along the eastern coast of the Antarctic continent. The cause of this discrepancy is not understood, but the

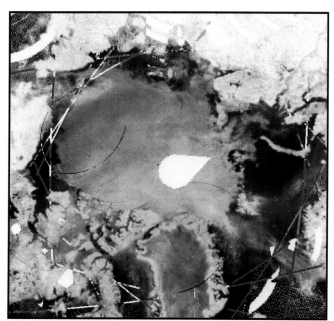

Fig. 7-21. An August 3, 1987, SSM/I 85-GHz polarization difference image.

implications for ice edge mapping are significant, since the altimeter is routinely identifying an ice edge nearly 100 km outside that identified by SSM/I in some areas. The green points on the eastern part of the continent might be attributable to fast ice, but those in the Bellingshausen Sea are more difficult to interpret. In contrast to the Northern Hemisphere weather effects do not seem to result in so many green points over the open ocean. Figure 7-22(d) shows the corresponding image for August 1987. Again there is better overall agreement during the winter period, with very few red points occurring anywhere. Some areas of green points are observed mainly in the Bellingshausen Sea. The reason for this is not fully understood. It may be attributable to areas of nearly 100% ice cover, resulting in an ocean-like altimeter signal.

7.5 SUMMARY

Potential users of radar altimetry will find sources for data from past, present, and future missions listed in Massom [1991], with a treatment of the applicability of the data to polar studies. Methods for extracting various ice parameters from the data record have been suggested in this chapter. Additional parameters and alternative ways of deriving parameters can be found in the literature (e.g., Rapley et al. [1987]).

Generally, ice parameters are arrived at by first retracking and editing the altimeter data record and then deriving waveform parameters that are linked to the desired ice parameters. The processing required to derive a waveform parameter and the strength of theory linking it to

an ice parameter vary. Ulander, for instance, calculates backscatter from waveforms and classes ice on the basis of backscatter [Ulander, 1991]. Drinkwater [1991] finds the theoretically inverse relationship between the integral of waveform power and ice concentration borne out in measurements. Chase and Holyer [1990] perform a sophisticated analysis of waveform shape, and empirically link shape to different ice types and concentrations. Other algorithms use parameters that are more distantly derived from waveform power and shape. For instance, SWH has been used to measure swell penetration of the MIZ, and AGC has been used to mark the ice edge (see Section 7.4.4). While AGC and SWH can be processed quickly and are more readily available than the waveforms themselves, they are the result of averaging waveforms and therefore have disadvantages. Yet another approach is to link variability in waveform parameters with ice conditions. Variability in waveform power aids in the delineation of zones in Figure 7-18, while the standard deviation of the height measurement is used in Section 7.4.4 to locate the ice edge.

With the exception of ice edge, it has not been demonstrated that any ice parameter can be reliably retrieved from altimetry on more than a case study basis. The research reported here strongly indicates that this will change. The best hope for progress in this direction may lie in seeking an empirical connection between waveform parameters and ice conditions using global satellite data sets.

Acknowledgments.

The work of F. M. Fetterer was supported by the Office of Naval Research under program element 0603704N, CDR P. Ranelli, Program Manager. S. W. C. Laxon's work was supported by the U.K. Science and Engineering Research Council. L. M. H. Ulander was supported by the Swedish Board of Space Activities for this study.

REFERENCES

Beckmann, P. and A. Spizzichino, *The Scattering of Electromagnetic Waves From a Rough Surface,* 503 pp., Pergamon Press, New York, 1963.

Brown, G. S., A theory for near-normal incidence microwave scattering from first-year sea ice, *Radio Science, 17(1),* pp. 233–243, 1982.

Chase, J. R. and R. J. Holyer, Estimation of sea ice type and concentration by linear unmixing of Geosat altimeter waveforms, *Journal of Geophysical Research, 95(C10),* pp. 18,015–18,025, 1990.

Chelton, D. B and P. J. McCabe, A review of satellite altimeter measurement of sea surface wind speed: with a proposed new algorithm, *Journal of Geophysical Research, 90,* pp. 4707–4720, 1985.

Chelton, D. B, E. J. Walsh, and J. L. MacArthur, Pulse compression and sea level tracking in satellite altimetry, *Journal of Atmospheric and Oceanic Technology, 6,* pp. 407–438, 1989.

(a)

(b)

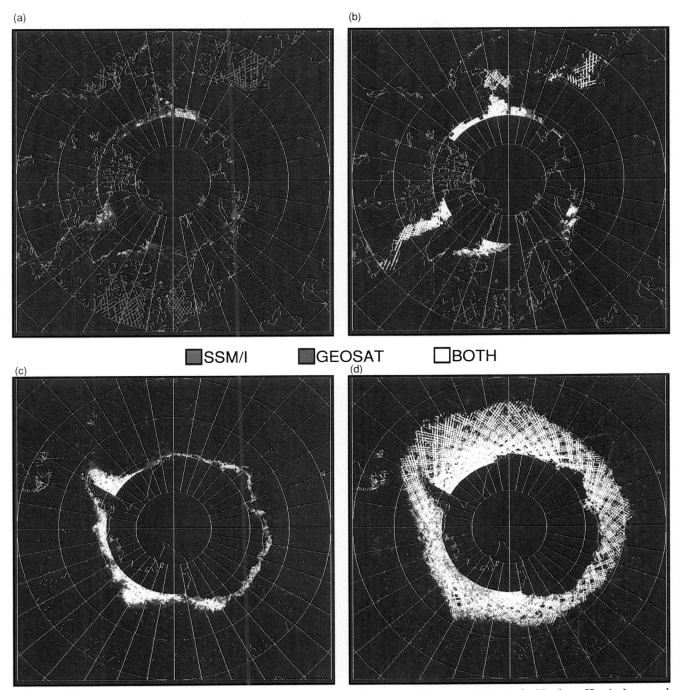

☐SSM/I ☐GEOSAT ☐BOTH

(c)

(d)

Fig. 7-22. A merged SSM/I and altimeter data product for (a) August 1987, and (b) February 1988, in the Northern Hemisphere, and (c) February 1988, and (d) August 1987, in the Southern Hemisphere. Colors indicate the detection of ice by either or both sensors.

Chen, K. S. and A. K. Fung, A scattering model for ocean surface, *Proceedings of the International Geoscience and Remote Sensing Symposium (IGARSS'91)*, pp. 1251–1254, European Space Agency, Helsinki, Finland, 1991.

Cowan, A. M. and V. A. Squire, *Sea Ice Characteristics Derived From Airborne Altimetry*, pp. 125–128, ESA SP-221, European Space Agency, Helsinki, Finland, 1984.

Drinkwater, M. R., *Radar Altimetric Studies of Polar Ice*, Ph.D. dissertation, Scott Polar Research Institute, University of Cambridge, Cambridge, England, 231 pp., 1987.

Drinkwater, M. R., Ku-band airborne radar altimeter observations of sea ice during the 1984 Marginal Ice Zone Experiment, *Journal of Geophysical Research Special MIZEX Issue, 96(C3)*, pp. 4555–4572, 1991.

Dwyer, R. E. and R. H. Godin, *Determining Sea-Ice Boundaries and Ice Roughness Using GEOS-3 Altimeter Data*, NASA Contractor Report 156862, 47 pp., National Aeronautics and Space Administration, Washington, DC, 1980.

Eom, H. J. and W. M. Boerner, A re-examination of radar terrain backscattering at nadir, *IEEE Transactions on Geoscience and Remote Sensing, 24(2)*, pp. 232–234, 1986.

Fedor, L. S. and E. J. Walsh, Interpretation of Seasat radar altimeter returns from an overflight of ice in the Beaufort Sea, *Conference Record: Oceans'88*, pp. 1697–1703 Marine Technology Society, Baltimore, Maryland, 1988.

Fedor, L. S., G. S. Hayne, and E. J. Walsh, Ice type classification from airborne pulse-limited radar altimeter return waveform characteristics, *Proceedings of the IGARSS'89*, pp. 1949–1952, European Space Agency, Vancouver, Canada, 1989.

Fetterer, F. M., S. Laxon, and D. R. Johnson, A comparison of Geosat altimeter and Synthetic Aperture Radar measurements over east Greenland pack ice, *International Journal of Remote Sensing, 12(3)*, pp. 569–583, 1991.

Fung, A. K. and H. J. Eom, Coherent scattering of a spherical wave from an irregular surface, *IEEE Transactions on Antennas and Propagation, AP-31(1)*, pp. 68–72, 1983.

Fung, A. K. and G. W. Pan, A scattering model for perfectly conducting random surfaces, I. Model development, *International Journal of Remote Sensing, 8(11)*, pp. 1579–1593, 1987.

Griffiths, H. D, D. J. Wingham, P. G. Challenor, T. H. Guymer, and M. A. Srokosz, *A Study of Mode Switching and Fast Delivery Product Algorithms for the ERS-1 Altimeter*, ESA Contract Report 6375/85/NL/BI, European Space Agency, Noordwijk, Netherlands, 1987.

Hawkins, J. D. and M. Lybanon, Geosat altimeter sea ice mapping, *IEEE Journal of Oceanic Engineering, 14 (2)*, pp. 139–148, 1989.

Hawkins, J. D., S. Laxon, and H. Phillips, Antarctic tabular iceberg multisensor mapping, *Proceedings of the IGARSS'91*, pp. 1605–1608, European Space Agency, Helsinki, Finland, 1991.

Kim, Y. S., R. K. Moore, R. G. Onstott, and S. Gogineni, Towards identification of optimum radar parameters for sea-ice monitoring, *Journal of Glaciology, 31(109)*, pp. 214–219, 1985.

Lange, M. A. and H. Eicken, The sea ice thickness distribution in the Northwestern Weddell Sea, *Journal of Geophysical Research, 96*, pp. 4821–4838, 1991.

Laxon, S. W. C., *Satellite Radar Altimetry Over Sea Ice*, Ph.D. thesis, 246 pp., University College London, Holmbury St. Mary Dorking, Surrey, UK, 1989.

Laxon, S., Seasonal and interannual variations in Antarctic sea ice extent as mapped by radar altimetry, *Geophysical Research Letters, 17(10)*, pp. 1553–1556, 1990.

Laxon, S. and J. Askne, Comparison of Geosat and SSM/I mapping of sea ice in the Arctic, *Proceedings of the IGARSS'91*, pp. 1593–1596, European Space Agency, Helsinki, Finland, 1991.

Lytle, V., K. Jezek, and S. Gogineni, Radar backscatter measurement during the winter Weddell Sea gyre study, in press, *South Antarctic Journal*, 1992.

Massom, R., *Satellite Remote Sensing of Polar Regions*, 307 pp., Belhaven Press, London, 1991.

McIntyre, N. F. and W. Cudlip, Observation of a giant Antarctic tabular iceberg by satellite radar altimetry, *Polar Record, 145*, pp. 458–462, 1987.

McIntyre, N. F., H. D. Griffiths, A. R. Birks, A. M. Cowan, M. R. Drinkwater, E. Novotny, R. J. Powell, V. A. Squire, L. M. H. Ulander, and C. L. Wrench, *Analysis of Altimetry Data From the Marginal Ice Zone Experiment*, ESA Contract Report 5948/84/NL/BI, 201 pp., European Space Agency, Noordwijk, Netherlands, 1986.

Onstott, R. G., *Multifrequency Helicopter-Borne Altimeter Observations of Summer Marginal Sea Ice During MIZEX'84*, Technical Report 239500-1-T, Environmental Research Institute of Michigan, Ann Arbor, Michigan, 1991.

Onstott, R. G., T. C. Grenfell, C. Mätzler, C. A. Luther, and E. A. Svendsen, Evolution of microwave sea ice signatures during early summer and midsummer in the marginal ice zone, *Journal of Geophysical Research, 92(C7)*, pp. 6825–6835, 1987.

Powell, R. J., A. R. Borks, C. L. Wrench, W. J. Bradford, and B. F. Maddison, *Radar Altimetry Over Sea Ice*, pp. 129–134, ESA SP-221, European Space Agency, Noordwijk, Netherlands, 1984.

Rapley, C. G., First observations of the interaction of ocean swell with sea ice using satellite radar altimeter data, *Nature, 307*, pp. 150–152, 1984.

Rapley, C. G., H. D. Griffiths, V. A. Squire, M. LeFebvre, A. R. Birks, A. C. Brenner, C. Brossier, L. D. Clifford, A. P. R. Cooper, A. M. Cowan, D. J. Drewry, M. R. Gorman, H. E. Huckle, P. A. Lamb, T. V. Martin, N. F. McIntyre, K. Milne, E. Novotny, G. E. Peckham, C. Schgounn, R. F. Scott, R. H. Thomas, and J. F. Vesecky, *A Study of Satellite Radar Altimeter Operation Over Ice Covered Surfaces*, 240 pp., ESA Contract Report 5182/82/F/CG(SC), European Space Agency, Noordwijk, Netherlands, 1983.

Rapley, C. G., M. A. J. Guzkowska, W. Kudlip, and I. M. Mason, *An Exploratory Study of Inland Water and Land Altimetry Using Seasat Data*, 377 pp. ESA Contract Report CR-6483/85/NL/BI, European Space Agency, Noordwijk, Netherlands, 1987.

Robin, G. de Q., D. J. Drewry, and V. A. Squire, Satellite observations of polar ice fields, *Philosophical Transactions of the Royal Society of London, A309*, pp. 447–461, 1983.

Shapiro, A. and B. S. Yaplee, Anomalous radar backscattering from terrain at high altitudes at nadir, *Proceedings of the IEEE, 63(4)*, pp. 717, 1975.

Swift, C. T., W. J. Campbell, D. J. Cavalieri, P. Gloersen, H. J. Zwally, L. S. Fedor, N. M. Mognard, and S. Peteherych, Observations of the polar regions from satellites using active and passive microwave techniques, *Advances in Geophysics, 27*, pp. 335–393, 1985.

Tucker, W. B., A. J. Gow, and W. F. Weeks, Physical properties of summer sea ice in the Fram Strait, *Journal of Geophysical Research, 92(C7)*, pp. 6787–6803, 1987.

Ulaby, F. T., R. K. Moore, and A. K. Fung, Microwave Remote Sensing—Active and Passive, Vol. II: Radar Remote Sensing and Surface Scattering and Emission Theory, Addison–Wesley Publishing Company, Reading, Massachusetts, 1982.

Ulander, L. M. H., Interpretation of Seasat radar-altimeter data over sea ice using near-simultaneous SAR imagery, *International Journal of Remote Sensing, 8(11)*, pp. 1679–1686, 1987a.

Ulander, L. M. H., *Seasat Radar Altimeter and Synthetic-Aperture Radar: Analysis of an Overlapping Sea-Ice Data Set*, Research Report 159, Department of Radio and Space Science, Chalmers University of Technology, Göteborg, Sweden, 1987b.

Ulander, L. M. H., Averaging radar altimeter pulse returns with the interpolation tracker, *International Journal of Remote Sensing, 8*, pp. 705–721, 1987c.

Ulander, L. M. H., Observations of ice types in satellite altimeter data, *Proceedings of the IGARSS'88*, pp. 655–658, European Space Agency, Edinburgh, UK, 1988.

Ulander, L. M. H., *Radar Remote Sensing of Sea Ice: Measurements and Theory*, Technical Report 212, Chalmers University of Technology, School of Electrical and Computer Engineering, Göteborg, Sweden, 1991.

Ulander, L. M. H. and A. Carlström, Radar backscatter signatures of Baltic sea ice, *Proceedings of the IGARSS'91*, pp. 1215–1218, European Space Agency, Helsinki, Finland, 1991.

Wadhams, P., A mechanism for the formation of ice bands, *Journal of Geophysical Research., 88(C5)*, pp. 2813–2818, 1983.

Wadhams, P., M. A. Lange, and S. F. Ackley, The ice thickness distribution across the Atlantic sector of the Antarctic Ocean in midwinter, *Journal of Geophysical Research, 92(C13)*, pp. 14,535–14,552, 1987.

Wadhams, P., V. A. Squire, D. J. Goodman, A. M. Cowan, and S. C. Moore, The attenuation rates of ocean waves in the marginal ice zone, *Journal of Geophysical Research, 93(C6)*, pp. 6799–6818, 1988.

Wingham, D. J. and C. G. Rapley, Saturation effects in the Seasat altimeter receiver, *International Journal of Remote Sensing, 8*, pp. 1163–1173, 1987.

Chapter 8. Microwave Sea Ice Signature Modeling

DALE P. WINEBRENNER

Polar Science Center, Applied Physics Laboratory, University of Washington, 1013 NE 40th Street, Seattle, Washington 98105

JONATHAN BREDOW AND ADRIAN K. FUNG

University of Texas, P. O. Box 19016, Arlington, Texas 76019

MARK R. DRINKWATER AND SON NGHIEM

Jet Propulsion Laboratory, California Institute of Technology, 4800 Oak Grove Drive, Pasadena, California 91109

ANTHONY J. GOW AND DONALD K. PEROVICH

Cold Regions Research and Engineering Laboratory, 72 Lyme Road, Hanover, New Hampshire 03755–1290

THOMAS C. GRENFELL

Department of Atmospheric Sciences, University of Washington, Seattle, Washington 98195

HSIU C. HAN AND JIN A. KONG

Massachusetts Institute of Technology, Cambridge, Massachusetts 02139

JAY K. LEE

Department of Electrical and Computer Engineering, Syracuse University, 121 Link Hall, Syracuse, New York 13244-1240

SABA MUDALIAR

Hanscom Air Force Base, Massachusetts 01731

ROBERT G. ONSTOTT

Environmental Research Institute of Michigan, P. O. Box 8618, Ann Arbor, Michigan 48107

LEUNG TSANG AND RICHARD D. WEST

Department of Electrical Engineering, University of Washington, Seattle, Washington 98195

8.1 INTRODUCTION

Remote sensing hinges on the interpretation of signatures and signature changes in terms of geophysical variables. To date, operational microwave remote sensing of sea ice has been based primarily on empirical relations between signatures and ice type or ice type concentrations. Empirical remote sensing has proven valuable in many geophysical studies, but a reliance on empiricism alone severely limits realization of the potential value of remote sensing methods.

Plots of ice signature data versus geophysical parameters often show considerable scatter. Some scatter is due to fundamental limitations in our instruments (e.g., Synthetic Aperture Radar [SAR] speckle), and some is almost surely due to random variations in ice or snow properties

with little or no relation to geophysically significant parameters. However, there is evidence that much of what we perceive as scatter is, in fact, caused by variations in interesting geophysical parameters. For example, signatures from ice typically labeled as first-year are variable partly because of an evolution in thin, first-year ice (i.e., nilas and gray ice) signatures as a function of thickness and perhaps other variables. This signature evolution can cause nilas and gray ice to appear brighter than both calm, open water and thicker ice in SAR images (Chapters 5, 14, and 25), and appear to passive microwave algorithms as mixtures of thick first-year and old ice (Chapter 14). Thus, these variations may contain thickness information. As a second example, variations in ice temperature, wind speed (over open water), and atmospheric water content cause variations in passive microwave signatures that look like noise in averaged data, but are at least partly invertible (Chapter 10). Thus, sea ice microwave signatures contain more geophysical information than is routinely utilized;

Microwave Remote Sensing of Sea Ice
Geophysical Monograph 68

much of the potential of sea ice microwave remote sensing remains untapped.

Realizing this potential requires an approach using both observations and physical insight. Ice morphology and properties are highly diverse, even within well-defined, traditional ice types. Signature variations of potential interest (passive or active, variations versus frequency, polarization, and so on) are legion. Thus, to find any but the most obvious links between signatures and geophysical properties, a purely empirical approach would require a huge (and hugely expensive) comprehensive data set, or risk missing valuable links. An exhaustive data set would surely contain much extraneous information, but there is no empirical way to predict which parts would later prove inessential. Moreover, recognizing valuable new links in such a thicket of data would be very difficult. More than a purely empirical approach is needed if we are to find and exploit valuable, but less than obvious, remote sensing opportunities.

Physical understanding and insight offer us the means to interpret data, uncover key regularities, and direct further efforts intelligently. Even a rough understanding of the physics underlying observations can lead to fruitful exploratory organizations of data and new insight. For example, Grenfell [1992] organized noisy passive microwave gradient ratios for old ice by plotting them against an approximate optical depth for the bubbly upper ice layer. The regularity he found suggests that, in addition to more precise work on this link, we investigate links between optical depth and more traditional geophysical parameters, such as ice freeboard.

Building a framework of physical understanding requires quantitative signature modeling and model testing. Several differing qualitative explanations for a given observation may be plausible; the difference in their implications for our understanding and for remote sensing may be considerable. Quantitative comparisons of differing explanations, i.e., models, against data provide a much sharper razor for separating the explanation closest to reality from the rest. The result is a stronger and broader framework of understanding that we can use to guide the next steps of investigation. This dialectic between theory and experiment is fundamental in all areas of physics, both basic and applied.

We derive several collateral benefits from physical understanding and signature models as well. Models allow us to simulate signature data for remote sensing system design, quantifying benefits and costs for various choices of system parameters. Recognizing the limits of our present understanding allows us to better identify situations in which present remote sensing techniques may fail. Thus progress in remote sensing depends on the development and judicious application of quantitative models for signatures.

Research over the past decade has produced a broad but imprecise understanding of the major physical effects determining sea ice signatures, as well as a number of sophisticated microwave signature models for application to sea ice. Microwave sea ice signatures are largely the product of scattering within the ice (volume scattering) and from its interfaces (rough surface scattering), and of interactions between these two types of scattering. A number of models focus on volume scattering to the exclusion of surface scattering, or vice versa; a few models treat both processes but make simplifying assumptions about interactions between the two. All these models are of course founded on Maxwell's equations of classical electrodynamics, and all assume linear, nonmagnetic dielectric behavior of the constituent materials of sea ice. The differences between models arise out of their significantly different sets of approximations, physical descriptions of the scattering material, and emphases on particular aspects of the problem at the expense of others. The highly developed crop of new models has brought us an array of choices in modeling the signature of any particular type of ice.

However, for any given combination of ice type and conditions, not all of the available choices can be appropriate. Many significant questions remain controversial. To give just one example, we have several models for volume scattering in first-year ice, but no generally accepted answer to the question: What is the role of volume scattering in cold first-year ice signatures, relative to surface scattering or other effects, as a function of ice thickness, snow cover, and radiation wavelength? Such questions remain because model development has outpaced model testing. The majority of sea ice model-data comparisons in the literature consist of reasonable-looking model fits to signature observations from ice for which there was little characterization. (Reasons for this include the dependence of scattering models on ice and snow parameters quite different from those measured in geophysical studies, as well as a lack of understanding of needed accuracies in ground truth data.) The problem is that most models contain several tunable parameters and the range of observations that can be fitted is large. Therefore, success with model fitting is, at best, limited evidence for validity of the model under consideration. On the other hand, a model can be excluded via this approach only when no plausible input parameter values produce results similar to observations. A finer criterion is needed to distinguish between competing explanations for the same observations.

That finer criterion is quantitative comparison of signature observations from a given scene with model results based on independent characterization of that same scene. Studies based on such comparisons, notably those by Ulander et al. [1992], Davis et al. [1987], Reber et al. [1987], Stogryn [1987], and Lin et al. [1988], are thus especially valuable (though the latter two works treat only microwave extinction rather than signatures per se, and those by Davis et al. and Reber et al. treat terrestrial snow pack rather than sea ice). However, the body of such work in the literature is far from sufficient. The net result of this situation is that, even as of 1992, we know of no operational program of microwave sea ice remote sensing in which quantitative signature models are used to link observations with geophysical properties (save ice type).

The purpose of this chapter is to clarify the state-of-the-art in microwave signature modeling for sea ice, and in the process elucidate signature-controlling properties of sea ice. Our approach is to compare signature computations from several models with observations in two relatively simple, particularly well-characterized cases. The data set in each case consists of nearly simultaneous, ground-based, active and passive signature observations and independent ice and snow characterization data. We use the data to constrain model inputs and/or compare with model inputs used to match the observations, and we provide information on model sensitivities to input parameter variations. Although we restrict ourselves to models previously documented in the literature, the results and comparisons we present here are new. Our collection of models is not exhaustive, but we believe we have examples of every major type of model presently used in connection with sea ice. Our focus on small, intensively characterized regions of ice makes possible relatively clean comparisons of models with data, and thus sharpens the inferences we can draw. However, this focus also complicates immediate application of our results in interpreting airborne and spaceborne sensor data. We think it worthwhile to accept this temporary complication for the sake of gaining physical insight.

The plan of this chapter is as follows. Section 8.2 presents an overview of the models we employ. The first part of the overview discusses generally some major physical issues that may be treated differently in different models and notes the resulting practical implications. The second part consists of summaries of the individual models, specifically noting the ways in which each treats the major issues (the discussions are primarily physical rather than mathematical, but provide ample references). Readers desiring only the most essential discussions may wish to read only Sections 8.2.1, 8.2.2.1, 8.2.2.5, and the first paragraph or so of each subsection in Section 8.2. Section 8.3 treats our first case study, that of a thin (8 cm) sheet of congelation ice, growing rapidly without snow cover and similar to ice in quiescent Arctic leads, that was studied as part of the Cold Regions Research and Engineering Laboratory Experiment (CRRELEX) in January 1988. Section 8.4 similarly addresses a second case study concerning cold, snow-covered old ice observed as part of the Coordinated Eastern Arctic Research Experiment (CEAREX) in October 1988. This second case study involves two different regions on a single old ice floe—first, a fresh, raised area having a very low density upper layer, and second, a refrozen melt pond with an upper layer density much closer to that of pure ice. Section 8.5 concludes the chapter with a discussion of our finings in the broader context of microwave signature modeling and remote sensing of sea ice. We summarize what we have learned about the state of signature modeling from this study and offer a few caveats about what our results do not imply. Finally, we consider directions for future work on signature physics and remote sensing algorithms.

8.2 Overview of Signature Models

Signature modeling for geophysical media is based on idealization and approximation. The geometric arrangement of constituent materials, that is, the morphology of geophysical media, is often complicated. Therefore, an initial step in modeling is to abstract from the actual morphology an idealized geometrical representation in the hope that scattering from the idealized medium mimics that from the actual medium while being more tractable to compute. The choice of idealizations for any particular combination of geophysical medium, electromagnetic wavelength, polarization, etc. is typically not straightforward, but rather a matter of judgment. Choices differ between modelers, even with respect to exactly the same geophysical medium. Moreover, solving the equations of scattering almost always necessitates a number of approximations. Appropriate approximations depend on wavelength, electromagnetic lossiness of the medium, and so on, and are often also matters of judgment. The choices of idealizations and approximations essentially define a scattering model.

Section 8.2.1 is meant as an observer's guide to spotting the key differences between models and interpreting the following model-data comparisons. We assume on the part of the reader a qualitative understanding of scattering, including rough surface versus volume scattering, single versus multiple scattering, and the propagation and scattering of coherent versus incoherent waves within sea ice. A sufficient background for our purposes is available in Chapter 3 of this book. Section 8.2.2.1 supplements this material. We also recommend as background the reviews of basic passive, active, and polarimetric signature characteristics given in Chapters 4, 5, and 25, respectively.

8.2.1 Fundamental Physical Mechanisms and Effects

The first issue is whether a model derives signatures based on scattering from dielectric inhomogeneities within the medium alone (volume scattering), scattering only from roughness at interfaces in the medium (rough surface scattering), or from some combination of the two. Research over the past decade has produced an array of volume scattering models that may plausibly apply to at least some types of sea ice, and a corresponding array of physical idealizations and approximation schemes. Rough surface scattering models are presently fewer in number; there are correspondingly fewer choices to discuss. We therefore begin with a discussion of the main points on which volume scattering models may differ.

A fundamental distinction between volume scattering models concerns layering. Ice properties typically show pronounced variations with depth. Snow may cover the ice. Thus, signature models typically treat the air–snow–sea ice–seawater system as a stack of horizontal layers with planar or rough interfaces (Figure 8-1). (Lateral variations in ice properties and morphology are important. However, because models incorporating such variations are much

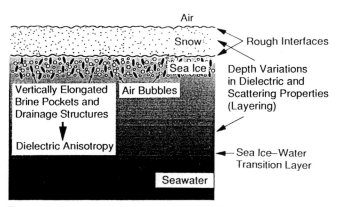

Fig. 8-1. The layered physical model of sea ice conceptually common to most sea ice signature models, including vertical variations in snow and sea ice properties and rough air–snow and snow–ice interfaces. The figure is schematic and not drawn to scale. Note that first-year ice contains many brine pockets at all depths, but few relatively small air bubbles. In old ice, the upper layers contain many relatively large air bubbles. Brine pockets are generally found in lower layers.

more difficult, and because we can partly account for lateral variation by using different models at different locations, current models do not generally address this issue. Neither has signature modeling for features such as ice ridges and thermal cracks received much attention; we therefore omit this topic from our discussion here. These matters may well require reconsideration, as we discuss in Section 8.5.) Models differ significantly in the number of scattering layers they treat. Perhaps even more significant is whether interactions of waves scattered or reflected from different layers are treated coherently or incoherently. When layer interfaces are nearly planar (measured in radiation wavelengths) and scattering in the layers is not too strong, waves reflected from layer interfaces remain coherent and thus interfere. Microwave signatures are strongly affected by this interference, and thus highly sensitive to layer thicknesses, radiation wavelength, and incidence angle (see, for example, Blinn et al. [1972]). Roughness of the layer interfaces, strong scattering within layers, or nonuniformity of layer thickness within the sensor footprint can destroy the coherence of contributions from different layers. Sensitivity to layer thickness is greatly decreased in this case. Most models treat field contributions as either completely coherent (e.g., strong fluctuation theory models) or completely incoherent (e.g., classical radiative transfer models). The appropriate choice of coherent or incoherent model in a given situation depends on the radiation wavelength, sensor bandwidth, uniformity of layer thickness and interface roughness, and so on, and may not be clear when first approaching a problem theoretically.

Another distinction between volume scattering models arises in the way they picture the spatial permittivity fluctuations inside the scattering medium. Discrete scatterer models envision a homogeneous dielectric background medium, in which are embedded discrete inclusions or

particles of materials having permittivities different from that of the background. The random nature of the scattering medium in this picture is due to randomness in particle positions, sizes, and perhaps composition. Computational tractability often constrains particle shapes to be relatively simple (spheres or ellipsoids) and, in some instances, also constrains particle size or the size distribution. In contrast are so-called continuous random medium models in which permittivity fluctuations may be modeled as an arbitrary random function of position, characterized by its mean, variance, and spatial correlation. The term "continuous random medium model" is actually a slight misnomer because there is no restriction that the function describing the random permittivity fluctuations be continuous; indeed, two-point permittivity fluctuation statistics in discrete random media have been computed and used [Stogryn, 1984a, 1985, 1986; Vallese and Kong, 1981; Jin and Kong, 1985; Reber et al., 1987]. In fact, both of the models based on the continuous random medium model that we employ in this chapter (Sections 8.2.2.6 and 8.2.2.7) actually assume discrete inclusions of scattering material in the ice. The point, however, is that continuous random medium models are not restricted to discrete scatterers and may treat geometrically complex media directly.

It might seem preferable to avoid the restrictions of discrete scatterer models entirely and employ only continuous random medium models. In practice, however, the choice between these types of models is linked to a third important distinction between models, namely the degree to which they treat multiple scattering of the incoherent field. Recall from Chapter 3 that volume scattering may be thought of in terms of a coherent field which propagates down into the medium, decaying due to both absorption and scattering into incoherent fields (Section 8.2.2.1). When absorption is small and scattering is strong, the incoherent fields may themselves be scattered repeatedly. Discrete scatterer models, especially those that can be cast in the form of classical radiative transfer theory, are generally able to treat multiple scattering of the incoherent field. However, practical solutions of the field equations in continuous medium models have, to date, been restricted to cases where the incoherent field is only singly scattered (specific examples and references are given in Sections 8.2.2.6 and 8.2.2.7—note that this is a restriction in practice, not in principle). This limitation is evidently significant in some strong scattering cases such as the one we consider in Section 8.4. Thus, the distinction between models that treat multiple scattering of the incoherent field and those that do not is linked in practice to the choice of description for the random scattering medium. There is, at least at present, a tradeoff between generality in describing the scattering medium on one hand and the strength of scattering that can be treated on the other.

Turning to rough surface scattering models, one major idealization stands out. The probability distribution of surface heights at any n points on the surface is assumed to be homogeneous and jointly Gaussian in virtually all models currently applied to sea ice. The roughness of some ice types (e.g., pancake ice) may fail to satisfy these assumptions; thus interpretation of model-data comparisons may be aided by awareness of this idealization. Most models also assume that surface roughness statistics (e.g., correlation length) are independent of orientation on the surface, i.e., that surface roughness statistics are directionally isotropic. Variation between models can occur because of differing assumptions about the form of the surface roughness correlation function, or equivalently the spectrum of surface height variations. The most common choice is an exponential form for the correlation function. This choice is not strictly compatible with models that are derived assuming that the surface height possesses at each point a well-defined tangent plane, but ignoring this conflict often leads to reasonable results. The main division between models is the choice of approximation scheme. Models relying on conventional perturbation theory are, roughly speaking, restricted to surfaces whose height variations are much smaller than the radiation wavelength. Models that use the tangent plane approximation (collectively called physical optics models) are restricted to surfaces whose roughness is smoothly undulating on horizontal length scales comparable to the wavelength. New rough surface scattering models generally attempt to supersede the restrictions of these classical models.

A few issues are common to volume and surface scattering models. The first of these arises because the brine pocket and drainage structures of congelation sea ice (Chapter 2) cause a directional anisotropy in dielectric properties for this ice [Sackinger and Byrd, 1972; Bogorodskii and Khokhlov, 1977; Golden and Ackley, 1981]. The permittivity and absorption measured with electric fields oriented vertically (along the preferred direction of the structure) are notably larger than those measured in orthogonal directions (Chapter 3). This may be important in understanding the signatures of some ice types. However, models differ in whether and how they take account of this phenomenon. Models for passive microwave signatures may or may not take into account the ice temperature profile within the effective range of emitting ice depths. Finally, models may be polarimetric, nonrigorously polarimetric, or simply nonpolarimetric. Volume scattering models derived systematically from Maxwell's equations are in principle polarimetric, whether or not the effort has been expended to make any particular implementation of the model polarimetric. The same holds true of classical radiative transfer models. However, models derived from radiative transfer after many simplifications typically are not polarimetric. Surface scattering models based on conventional perturbation theory are also polarimetric. However, physical optics-based surface scattering models are, at best, nonrigorously polarimetric, because of the nonrigorous tangent plane approximation they employ.

8.2.2 Volume Scattering Models

8.2.2.1. General comments on models in the form of classical radiative transfer. The classical theory of radiative transfer for scattering in volume was developed extensively to treat propagation, emission, and scattering problems in stellar and planetary atmospheres, as well as various other media [Ishimaru, 1978, and references therein]. It has since found application in signature modeling for vegetation and for sea ice [Fung and Eom, 1982; Tsang et al., 1985; Ulaby et al., 1982]. Though at least some snow and sea ice signatures computed using classical radiative transfer depart seriously from reality (see below), its intuitive form and the store of solution techniques from classical theory provide strong motivations to cast modern theories, including several that we apply in this chapter, in the classical form. We therefore begin this section with a discussion of features common to these models by reason of their form, as well as some of the fundamental physical insight about scattering that classical radiative transfer provides.

The fundamental quantity in radiative transfer is the specific intensity I defined at any point in space $\mathbf{r}$ as the power flowing in a given direction $\hat{\mathbf{s}}$ per unit solid angle per unit emitting area per unit bandwidth, assuming unpolarized radiation. (To construct a fully polarimetric theory, the scalar specific intensity I is generalized to a vector quantity $\mathbf{I}$ whose components are the Stokes parameters of the wave at $\mathbf{r}$ propagating in direction $\hat{\mathbf{s}}$ per unit solid angle per unit emitting area per unit bandwidth.) The point $\mathbf{r}$ need not lie in the volume containing the scatterers. In the classical theory, an integro-differential equation which governs I as a function of position and direction is derived heuristically [Ishimaru, 1978, Chapter 7]

$$\frac{d I(\mathbf{r},\hat{\mathbf{s}})}{d\hat{\mathbf{s}}} = -\kappa_e I(\mathbf{r},\hat{\mathbf{s}}) + \int_{4\pi} d\hat{\mathbf{s}} P(\hat{\mathbf{s}},\hat{\mathbf{s}}') I(\mathbf{r},\hat{\mathbf{s}}') + J(\mathbf{r},\hat{\mathbf{s}}) \quad (1)$$

where k_e is the extinction coefficient for coherent intensity, P is the so-called phase function relating scattering from direction $\hat{\mathbf{s}}'$ into direction $\hat{\mathbf{s}}$, and J is a thermal source term owing to emission within the scattering volume. In a polarimetric theory, the phase function and extinction coefficient generalize to 4×4 matrices, and the emission source term generalizes to a vector with four components (see, for example, Tsang et al. [1985]). This equation, known as the equation of radiative transfer or transport equation, describes the total change in specific intensity in the direction $\hat{\mathbf{s}}$ as a sum of effects, namely extinction, scattering from other directions into the direction $\hat{\mathbf{s}}$, and thermal emission into that direction. Note that the extinction coefficient k_e results not only from absorption but also from scattering out of direction $\hat{\mathbf{s}}$ into other directions. Because Equation (1) is a first order differential equation, it is necessary to specify boundary conditions, one for each region in which the equation is to be solved, to completely specify the problem.

An equation relating the phase function P, the single-scattering albedo, $\widetilde{\omega}$, and κ_e results from the statement of energy conservation:

$$\int_{4\pi} d\hat{\mathbf{s}}\, P\left(\hat{\mathbf{s}}, \hat{\mathbf{s}}'\right) = \kappa_e\, \widetilde{\omega} \le \kappa_e \qquad (2)$$

where the equality holds in the lossless case, i.e., when no absorption takes place in the background medium or in the scatterer. Physically, this equation states that the power scattered out of direction $\hat{\mathbf{s}}'$ must not exceed the total power loss for intensity traveling in the direction $\hat{\mathbf{s}}'$; in the case of no absorption, these powers must balance.

An approximate solution method for the transport equation can be used to gain some additional physical insight into the scattering process. Suppose for a moment that the albedo, $\widetilde{\omega}$, is small compared to one. Then it is reasonable to attempt a zeroth-order solution to the transport equation inside the scattering volume by simply ignoring the integral term on the right-hand side of Equation (1) (which is effectively proportional to $\widetilde{\omega}$). Neglecting also thermal emission for the moment, the specific intensity to zeroth-order in albedo satisfies the equation

$$\frac{d I_0(\mathbf{r}, \hat{\mathbf{s}})}{d\hat{\mathbf{s}}} = -\kappa_e\, I_0(\mathbf{r}, \hat{\mathbf{s}}) \qquad (3)$$

The extinction coefficient k_e is real and positive; thus the zeroth-order specific intensity is simply a real exponential function that decays on the length scale $1/k_e$ in the direction of propagation through the medium. (The boundary conditions merely determine position-dependent factors and additive constants.) $I_0(\mathbf{r}, \hat{\mathbf{s}})$ contains the effects of absorption and scattering out of the direction $\hat{\mathbf{s}}$, but no effect of scattering from other directions into $\hat{\mathbf{s}}$. The latter effect first enters in the first-order solution , the equation for which is obtained by substituting $I_1(\mathbf{r}, \hat{\mathbf{s}})$ for the unknown I in Equation (1).

$$\frac{d I_1(\mathbf{r}, \hat{\mathbf{s}})}{d\hat{\mathbf{s}}} = -\kappa_e\, I_1(\mathbf{r}, \hat{\mathbf{s}}) + \int_{4\pi} d\hat{\mathbf{s}}'\, P\left(\hat{\mathbf{s}}, \hat{\mathbf{s}}'\right) I_0(\mathbf{r}, \hat{\mathbf{s}}') \quad (4)$$

Thus, the specific intensity to first-order in albedo contains the effects of single scattering of the zeroth-order intensity which itself contains the effects only of absorption and scattering out of beam. This idea is similar in spirit (if not precisely in detail) to that in the distorted Born approximation discussed below (cf. Section 8.2.2.5): incoherent intensity is generated by single scattering of fields which themselves contain effects of only absorption and (multiple) coherent forward scattering. The idea extends naturally: in more strongly scattering cases, second-order intensities

result from scattering of the singly scattered first-order intensities, third-order intensities result from scattering of second-order intensities, and so on. This iterative approach is sometimes actually used for the numerical solution of radiative transfer and other multiple scattering theories. However, when the albedo is not small and scattering volumes are large (allowing more opportunities for multiple scattering), direct numerical solutions of equations such as Equation (1) are often more practical.

The chief assumptions in the derivation of Equation (1) are a lack of correlation between fields traveling in distinct directions and the independence of fields scattered from different locations or particles [Ishimaru, 1978; Tsang et al., 1985; Ulaby et al., 1982]. Thus coherent interaction, i.e., interference, between waves from different layers is not accounted for in classical radiative transfer. As noted at the beginning of this section, this neglect may in some cases be a source of error, while in others, especially those involving irregular layers, it may actually be an asset.

Solution of the transport equation requires the specification of boundary conditions at the layer interfaces. Boundary conditions for flat and rough interfaces have been developed and used in practice [Tsang et al., 1985; Ulaby et al., 1982]. Thus classical radiative transfer theory offers a way to treat the combined effects of volume and rough surface scattering. This is true as well for at least two of the models described below (dense medium radiative transfer and dense medium theory) that take the form of classical radiative transfer. However, this combined treatment is limited to incoherent interaction between waves scattered from the interfaces and in the volumes.

A classical radiative transfer model with rough interfaces has been used with some success to fit sea ice signatures [Fung and Eom, 1982]. However, a number of authors observed that radiative transfer models predicted erroneous signatures, especially for snow, when driven by input parameters derived from independent characterization measurements [Stogryn, 1986, and references therein]. Judging from passive signature observations, it appears that the classical assumption of independent scattering from distinct spatial regions or particles causes an overprediction for scattered intensities, thus lowering predicted brightness temperatures to unrealistic levels (however, see also Section 8.2.2.4). In retrospect, this may be understandable; the classical theory was developed to treat cases where scatterers occupied less than 1% of the total scattering volume. The scattering particles in snow and sea ice typically occupy 5% to 50% of the scattering volume; effects of one scatterer on the contribution of another, in addition to classical multiple scattering, can therefore become appreciable. Such effects are generically termed dense medium effects. While dense medium effects led Stogryn, Kong, and others to investigate strong fluctuation theory, others sought to develop alternative models in the form of classical radiative transfer that would account for nonindependent scattering. We employ three of the resulting models in our case studies and summarize the physical content of these models in subsections 8.2.2.3, 8.2.2.4, and 8.2.2.8. We first summarize a simpler model with its conceptual roots in the first-order physical picture of Equation (4).

8.2.2.2. Independent Rayleigh-scatterer layers. Most of the models used in this chapter are based on recent developments in the theory of wave scattering in random media; they are therefore sophisticated and, to nonspecialists, probably arcane. We do not yet understand which situations require the sophistication of the new models. It is therefore desirable to include in our study a physical model based on relatively simple, intuitive considerations, but which nonetheless has a reasonable basis in scattering physics. Drinkwater [1989, 1987], Drinkwater and Crocker [1988] and Livingstone and Drinkwater [1991] have recently applied such a model for backscattering from sea ice (the model is for backscattering only; passive signatures are not treated). The essence of the model was proposed by Attema and Ulaby [1978] for vegetation, and a version more appropriate for sea ice was later given by Kim et al. [1984a, b, 1985] (see also Section 11-5 of Ulaby et al. [1982]). In this chapter we employ a slight extension of the model to allow for two volume scattering layers within the ice. This feature permits us to address backscattering from low-density old ice with two layers in our second case study. The single layer version is similar to that used by Ulander et al. [1992] in connection with observations from the Bothnian Experiment in Preparation for ERS-1 (BEPERS).

This model pictures the snow-ice system as a snow layer overlying two ice layers (Figure 8-1). Each layer contains discrete volume scatterers; these are ice grains in snow, air bubbles in ice. (Generally, the snow layer may be wet and volume scattering from water inclusions is also modeled; however, this situation does not arise in our study.) The layers are assumed horizontally uniform with constant thickness; the scatterers are assumed uniformly distributed within each layer. The volume scatterers in each layer are assumed much smaller than the radiation wavelength and modeled as Rayleigh-scattering spheres. The layer densities and sphere radii (or more generally the distribution of sphere radii) are to be specified from ice and snow characterization data. The model does not presently treat any effect of dielectric anisotropy in congelation ice.

The scattering model is based on the idea illustrated in Equation (4), together with a partial accounting for rough surface scattering effects. The total backscattering cross section is modeled as an incoherent sum of component cross sections, each identified with a particular rough interface or volume scattering layer [Drinkwater and Crocker, 1988; Kim et al., 1985]:

$$\sigma_{total}^0 = \sigma_{as}^0(\theta) + T_{sa}^2(\theta)$$
$$\times \left\{ \exp\left(-2\kappa_e d / \cos\theta'\right)\left[\sigma_{si}^0(\theta') + \sigma_i^0(\theta')\right] + \sigma_s^0 \right\} \quad (5)$$

where q is the angle of incidence, θ' is the angle of refraction in the snow layer corresponding to incidence angle θ (assumed real because the imaginary part of the snow permittivity is assumed much smaller than the real part), $T_{sa}^2(\theta)$ is the Fresnel power transmission coefficient between air and snow (horizontally polarized for HH backscattering, vertically polarized for VV), d is the snow layer thickness, κ_e is the snow extinction coefficient, and σ_{as}^0, σ_s^0, σ_{si}^0, and σ_i^0 are the backscattering cross section contributions of the air–snow interface, the snow layer, the snow–ice interface, and the ice volume, respectively. The permittivity of the snow layer (which is dry in our case studies) is computed from an empirical formula and used to calculate κ_e for the snow and T_{sa}. The snow volume scattering term is given by

$$\sigma_s^0(\theta') = T_{sa}^2(\theta')\left[1 - \exp\left(-2\kappa_e d / \cos\theta'\right)\right] \quad (6)$$

where σ_v is the volume scattering cross section per unit volume of the snow crystals embedded in air. Equation (6) results by integrating the backscattered power contributions from each depth in the scattering layer, accounting at each depth for extinction of the illumination and backscattered contribution in the part of the layer above that depth [Ulaby et al., 1982, Section 11-5.3]; thus, the correspondence between this model and Equation (4). A further correspondence is the assumption that scattered intensity contributions from neighboring ice grains add incoherently. Thus $\sigma_v = \int dr\, N(r)\,\sigma_g(r)$ where σ_g is the Rayleigh backscattering cross section for a single spherical ice grain and $N(r)$ is the number of grains per unit volume of snow with radii between r and $r + dr$. The total volume of ice per unit volume of snow is $4\pi/3 \int dr\, r^3 N(r)$; this latter number can be determined independently by a measurement of bulk snow density.

The form of the equation for the ice volume scattering term, σ_i^0, in Equation (5) is similar to Equation (6) but contains two terms, one for each bubbly layer; the transmission coefficients in these terms refer to transmission at the snow–ice and ice layer interfaces, the extinction coefficients refer to extinction with the ice layers, refracted angles are computed within each layer, and so on. The permittivity of bubbly ice is estimated using a Polder–van Santen type formula given by Fung and Eom [1982]. The Rayleigh-scattering bubbles are then assumed to reside in an effective background having the effective permittivity (which is smaller than that of pure ice by an amount depending on density). This lowers the cross sections of individual bubbles in low density layers compared with identically sized but less numerous bubbles in higher density layers (Section 8.4.2.1). In the model results given below, extinction is computed on the basis of absorption alone. Scattering out of beam is neglected, but this should minimally affect computed backscattering levels for the scattering extinction coefficients and layer thicknesses in our cases.

Any reflection or scattering at the interface between ice layers is neglected. The rough surface scattering cross sections, σ^0_{as} and σ^0_{si}, are computed according to a physical optics model using interface roughness statistics from characterization data (Section 8.2.3.1). Note that the model treats no multiple scattering, scattering before or after reflections from layer interfaces, or coherent interaction between contributions from different layers. The model is not polarimetric, and in fact contains little polarization dependence. The backscattering cross sections for the air–snow and snow–ice interfaces contain some polarization dependence (Section 8.2.3.1), but volume scattering is polarization dependent only because transmission across layer interfaces differs for vertical and horizontal polarization. The backscattering within the volume is polarization independent. Note also that the treatment of surface scattering in Equation (5) accounts neither for the change in effective illumination of snow volume scatterers due to the sprays of energy from snow–ice and air–snow interfaces nor for any other interactions between surface and volume scattering.

To summarize, this model is based (like classical radiative transfer theory) on independent scattering. Moreover, it neglects a number of effects included in many radiative transfer models. Given the previous difficulties found in comparisons between observations and signatures computed using classical radiative transfer (Stogryn [1986] and Section 8.2.2.1), there is reason to question the applicability of this model to snow and sea ice. On the other hand, it does include significant nonclassical, if ad hoc, modifications to account for dense medium effects, particularly the reduction of scattering from bubbles in low-density ice layers. Thus it is uncertain, prior to a comparison such as ours, whether the model just described can accurately predict active signatures of sea ice. In view of this uncertainty, the relative simplicity of this model, and the ease with which we can isolate different physical effects in it, argue strongly for its inclusion in this study.

8.2.2.3. Dense medium radiative transfer. Dense medium radiative transfer (DMRT) [Tsang and Ishimaru, 1987; Tsang, 1987] is a discrete scatterer model, i.e., it proceeds from a physical model for the scattering medium (snow or sea ice) consisting of discrete, regularly shaped particles embedded in a homogeneous background medium. Particle positions are correlated for media in which the volume of particles exceeds a few percent of the total volume of scattering material. The essential physical effect in DMRT is interference between scattered field contributions from neighboring particles, even in the ensemble average over all particle arrangements. The interference is governed by the correlations between particle positions; for particles small compared with the radiation wavelength (i.e., for Rayleigh scattering particles), this interference is effectively destructive and results in less scattering than would be predicted based on an independent scattering assumption. Thus, for a given set of input parameters, higher brightness temperatures and lower scattering cross

sections are computed using DMRT than are computed using classical radiative transfer theory.

The derivation of DMRT proceeds from exact multiple scattering equations based on Maxwell's equations, using a series of consistent approximations for the coherent and incoherent scattered fields. Specifically, Dyson's general equation for the coherent field is first approximated using the quasicrystalline approximation with coherent-potential (QCA-CP). This approximation is sufficiently powerful to treat densely packed volumes of scatterers with permittivities differing strongly from their background; in this case, the real part of the effective propagation constant for the mean field may differ appreciably from the propagation constant in the background medium. Second, the ladder approximation for the intensity operator is applied in the general Bethe–Salpeter equation for the incoherent field. The ladder approximation accounts for that cascade of uncorrelated scattering events in the incoherent field that can each be described in terms of the two-point statistics of particle positions; DMRT thus presumes that such events dominate multiple scattering of the incoherent intensity. Within a densely packed medium, interscatterer separations range from near- to far-field values (i.e., some scatterers are within the near-field region of others). The wave interactions at all ranges are included in DMRT by using exact wave transformations from one scatterer center to the next. The result is a theory in the form of classical radiative transfer (the polarimetric version of Equation (1) where the extinction rate κ_e and albedo $\tilde{\omega}$ are given by expressions that agree with the classical expressions in the limit of small particle volume fractions but generally depend on correlations between particle positions). The phase matrix, extinction coefficient, and albedo satisfy energy conservation.

Tsang and Ishimaru [1987] and Tsang [1987] first developed DMRT for the case of a scattering layer containing Rayleigh-scattering spheres with a single radius. They assumed that the function describing the correlation between scatterer positions, that is, the pair-correlation function, could be approximated by the Percus–Yevick pair-correlation function derived in statistical mechanics. Under these assumptions, the complex effective wavenumber of the coherent field, including effects of both scattering and absorption, may be found by the procedure of Wen et al. [1990] (see, in particular, Section II). Tsang [1991] generalized the theory to treat Rayleigh-scattering spheres with a distribution of sizes. The DMRT phase matrix for these cases is identical to that in the classical theory, but albedos depend on correlations between scatterer positions. Expressions for the albedos are given with effective permittivity algorithms in each of the above references.

Because DMRT takes the form of radiative transfer, the array of solution methods developed for the classical theory may also be applied to this model. The present implementation of the theory employs the discrete eigenvalue-eigenvector method and treats two scattering layers over a nonscattering basement. The theory and implementation are fully polarimetric. Interference between waves reflected from the various layer interfaces is neglected just as in classical radiative transfer theory. Particle shapes in the present theory and implementation are restricted to spheres, and sizes are presently limited to the regime in which Rayleigh scattering is valid. The spherical particle restriction precludes modeling of any effects of dielectric anisotropy in the ice; thus DMRT is likely to be most appropriate in old ice or other ice that is, to a good approximation, dielectrically isotropic. The theory requires input information on the permittivities of the background material (ice in sea ice, air in snow), scatterers (air bubbles in old ice, ice grains in snow), and especially on the size and size distribution of scatterers in the various layers. Winebrenner et al. [1989] provide information on signature sensitivities due to variations in bubble size, ice salinity, and density in a DMRT model with a single bubble layer containing bubbles of a single size.

8.2.2.4. Dense medium theory. Dense medium theory [Fung and Eom, 1985] is also a discrete scatterer model for spheres much smaller than the radiation wavelength, the form of which is the same as that of classical radiative transfer, Equation (1). The essential physical difference between dense medium theory and the classical theory is a modification of the phase matrix, and therefore also the extinction coefficient. Classical radiative transfer uses the Rayleigh scattering phase matrix for small spheres. Fung and Eom [1985] rederived the phase matrix using Mie coefficients for terms in the spheres' fields that fall off in range faster than $1/r$. The total scattering cross section (per unit volume of the scattering material) is computed by integrating the modified phase function over all solid angles, then added to the absorption cross section from Mie theory to obtain the extinction coefficient. Thus this theory considers a key physical effect of densely packed scatterers to be a near-field interaction between neighboring scatterers. This theory does not rule out neighboring-scatterer interference but does not account for it. The theory does not at present specify conditions under which either its near-field terms or interference may predominate, or predict when both effects must be considered.

Based on its modifications to classical theory, dense medium theory predicts an increase in scattering over that computed from the classical theory. Fung and Eom [1985] show backscattering cross sections computed for snow at 7–10 GHz assuming a single particle size. The cross sections are 1–3 dB higher than those predicted by classical radiative transfer; computed snow brightness temperatures in the same frequency range are 1–8 K lower than classical values. Fung and Eom [1985] report a better match to snow backscattering data using dense medium theory than that achieved using classical theory, and extinction calculations based on this theory are consistent with optical data reported by Vedernikova and Kabanov [1974]. However, the dense medium passive signature predictions are opposite to what would narrow the discrepancy in snow noted by

Stogryn [1986] and others. The nonclassical effects predicted by dense medium theory decrease with increasing frequency and increasing particle radius, but increase monotonically for scatterer volume fractions between 0% and 30%, at least in the example given by Fung and Eom [1985].

In its present implementation, the theory treats a single scattering layer overlying a nonscattering basement. The layer need not be isothermal; this theory computes brightness temperatures directly, rather than emissivities, accounting fully for the temperature profile in the ice layer. The theory currently does not treat size distributions of scatterers, but rather fixes all scatterer radii at a single effective value to be determined on the basis of independent characterization information. The restriction to spherical scatterers also precludes modeling of the dielectric anisotropy in congelation ice. The classical form of the dense medium theory permits the use of the classical layer doubling method for solution [Ulaby et al., 1982]. The interaction of waves reflected from the layer interfaces is therefore incoherent, just as in the classical theory; results are therefore relatively insensitive to layer thickness. A rough surface boundary condition is employed at both the upper and lower layer interfaces. The results in this chapter were computed using the Integral Equation Method of Fung et al. [1991] for the elements of the boundary condition matrix (see Section 8.2.3.3).

8.2.2.5. General comments on strong fluctuation theory. The term strong fluctuation theory (SFT) refers to a class of volume scattering models that (1) employ the continuous random medium model to describe the scattering medium and (2) address the problem of strong contrasts in the permittivity of constituents of the scattering medium. For sea ice, the strong contrasts are those between the background pure ice and inclusions of brine and air. The correct treatment of such contrasts requires decomposition of integrals involving the dyadic Green's function for the electric field into sums of two terms. One term is a principal value integral with a volume around the source point excluded. The shape of this exclusion volume is determined by the shape and orientation of the scatterers and/or equicorrelation surfaces of the random permittivity [Stogryn, 1983a, b, 1984b; Tsang et al., 1985]. The second term consists of an integral over a product of terms, including a delta function centered on the source point times a dyad whose elements depend on the shape of the exclusion volume.

The fundamental random quantity in SFT is a second-rank tensor (which may be written in the form of a matrix), usually denoted $\bar{\bar{\xi}}$. Its elements depend on (1) the fluctuations of permittivity in the scattering medium and (2) the elements of the dyad associated with the delta function above, and thus the shape of the exclusion volume. The spatial cross-correlations between elements of $\bar{\bar{\xi}}$, as functions of lag, govern scattering of both the coherent and incoherent fields in SFT. For a scattering medium with only two components (e.g., pure ice and brine), a direct connection can be made between normalized (scalar) correlation functions of elements of $\bar{\bar{\xi}}$ and geometrical correlations of positions of inclusion material (e.g., brine) [Stogryn 1984a; Lin et al., 1988; Yueh et al., 1990]. Scattering media containing inclusions of more than one material (e.g., brine and air in ice, or ice and water in an air background) require the use of multiple geometrical correlations [Stogryn 1984a, 1985, 1987].

The coherent field in SFT is computed according to the bilocal approximation in Dyson's general equation for the coherent field [Stogryn, 1983a, 1984b; Tsang et al., 1985]. The bilocal approximation in Dyson's equation accounts for a cascade of uncorrelated single-scattering events, assuming that the coherent wave travels between events with a propagation constant equal to that in the background ice. The resulting effective permittivity includes an imaginary part due both to absorption and to extinction of the coherent field by scattering into the incoherent field. The bilocal approximation is believed to be accurate when the energy carried by the coherent field dominates the total energy flowing in the ice (due to both coherent and incoherent fields). The equation for the effective permittivity in the bilocal approximation can be solved analytically in the low frequency limit, i.e., when the correlation lengths of permittivity fluctuations are much smaller than the radiation wavelength [Tsang et al., 1985, Section 5.4; Stogryn, 1984b], and numerically at higher frequencies [Stogryn, 1986]. Both of the SFT models used in this chapter employ the low frequency, analytical solution for the bilocal approximation.

Both of the SFT models used in this chapter also compute the incoherent scattered field using the distorted Born approximation. This approximation accounts for contributions to the incoherent field that arise from scattering directly out of the coherent field, but not for any repeated, i.e., multiple, scattering of the incoherent field (cf. Section 8.2.2.1). Thus this approximation inherently limits the strength of scattering that can be treated. However, this limit is merely consistent with limitations inherent in the bilocal approximation for the coherent field [Stogryn, 1985]. Because of this sequence of approximations, the strong fluctuation theory models used in this chapter are sometimes termed multiple-forward-scatter, single-backscatter models.

The strong fluctuation theory models discussed below are purely volume scattering models, and interfaces between volume scattering layers are assumed planar. These models treat the interaction between waves reflected from layer interfaces coherently, and thus display layer thickness-dependent interference effects.

8.2.2.6. Polarimetric strong fluctuation theory. A fully polarimetric SFT model has recently been developed for a system of two scattering layers over a nonscattering half-space [Nghiem, 1991, and references therein]. We term this model polarimetric SFT because it is fully polarimetric in its present implementation, and to distinguish it conve-

niently from the other, quite distinct, SFT model that we also apply in this chapter. In polarimetric SFT, the strong permittivity fluctuations due to individual brine pockets or air bubbles are directly responsible for volume scattering within the sea ice. (Contrast this mechanism with that in the SFT model of Section 8.2.2.7.)

The individual scatterers are, in general, modeled as ellipsoidal particles of identical size. The spatial distribution of scatterer locations within the layers is uniform. In sea ice, the ellipsoids represent brine pockets and have their longest dimension aligned with the vertical to represent the preferred vertical direction observed in brine drainage structure. Individual brine pockets have an ellipsoidal cross section in the horizontal plane to represent the observed horizontal anisotropy of brine pockets sandwiched between ice platelets within a congelation sea ice crystal (Chapter 2). The azimuthal orientation of platelet structure, i.e., the horizontal direction of c-axis alignment, typically varies randomly between crystals (except in exceptional cases such as fast ice, where ocean currents may align the platelet structures). Thus, moving from scatterer to scatterer in a given realization of the scattering medium, the azimuthal orientation of the shortest axis of the ellipsoid varies randomly with a uniform distribution between 0 and 2π radians. When the lengths of the horizontal ellipsoid axes differ significantly, the local permittivity fluctuations have a pronounced azimuthal anisotropy, whereas the large-scale properties of the scattering layer are always azimuthally isotropic. In snow, the axes of the ellipsoids are chosen to have the same length to reflect the isotropy of typical snow [Vallese and Kong, 1981].

The general ellipsoidal scatterer shape, the consequent local anisotropy, and the fully polarimetric computations are key features of this most recent SFT model [Nghiem, 1991]. An important consequence of local azimuthal anisotropy in the sea ice layer is the prediction of substantial cross-polarized backscattering. Such a prediction differentiates this model from other models that use the first-order distorted Born approximation with a representation for the scattering medium that is azimuthally isotropic on the large scale. This model therefore predicts new and first-year ice signatures with much higher cross-polarized backscattering cross sections than those predicted by other SFT models.

Because the scatterers are nonspherical and randomly oriented, the exclusion volume (Section 8.2.2.5) in this model varies from scatterer to scatterer [Nghiem, 1991; Yueh et al., 1990]. Thus the normalized spatial correlation at lag $\mathbf{r}$ between any pair of elements of $\bar{\bar{\xi}}$ is expressed in terms of a correlation function, $R_\phi(\mathbf{r})$ which is conditioned on the azimuthal scatterer orientation angle ϕ. Expressed in local coordinates $\mathbf{r} = \widehat{\mathbf{x}}\, x + \widehat{\mathbf{y}}\, y + \widehat{\mathbf{z}}\, z$ appropriate to a given scatterer,

$$R_\phi(\mathbf{r}) = \exp\left[-\left(\frac{x^2}{l_x^2} + \frac{y^2}{l_y^2} + \frac{z^2}{l_z^2}\right)^{1/2}\right] \qquad (7)$$

Nghiem [1991] has coined the shorthand term "local correlation function" for this quantity. The correlation lengths l_x, l_y, and l_z are related to the axial dimensions of the scatterers (i.e., brine pockets or air bubbles in sea ice, and ice grains in snow). However, the correspondence is not precise for the following reasons. The actual scatterers in snow and sea ice have a distribution of sizes. The correlation lengths (and thus scatterers) are much smaller than the radiation wavelength, according to assumptions made in this model (Section 8.2.2.5). For scatterers much smaller than the radiation wavelength (i.e., Rayleigh scatterers), the strength of scattering increases rapidly and nonlinearly with increasing scatterer size. Thus larger, less numerous scatterers can contribute more to the total amount of scattering than the more abundant, smaller scatterers. Any single, effective correlation length used to characterize the scatterer size should therefore exceed the actual mean scatterer size by an amount depending on the shape of the size distribution [Jin and Kong, 1985]. At present, the effective correlation lengths in Equation (7) are chosen partly on the basis of model fitting and partly on the basis of independent thin section analysis such as that of Lin et al. [1988] and Perovich and Gow [1991]. (Thus this practice is a departure from that in work by Lin et al. [1988] using an earlier SFT model.) Typical correlation lengths in sea ice range from tenths of millimeters to millimeters, with the longer dimension in the vertical (l_z). The correlation functions for snow layers are spherically symmetric, i.e., $l_x = l_y = l_z$, and the correlation length is on the order of tenths of millimeters [Vallese and Kong, 1981; Reber et al., 1987].

As per the discussion above, the local azimuthal orientation of the correlation function in Equation (7) is assumed to be random with a uniform distribution over all possible horizontal directions. The effective permittivity, backscattering cross sections, and, for polarimetric signatures, elements of the Mueller and covariance matrices (Chapter 25) are computed by averages over the local azimuthal orientations. The resulting (tensor) effective permittivity is azimuthally isotropic but reflects a dielectric anisotropy between vertical and horizontal directions due to vertically elongated brine pockets [Nghiem, 1991]. The local azimuthal anisotropy locally couples electric fields of any given polarization into orthogonally polarized fields, and this is reflected in the backscattering signatures (for a somewhat simpler example of this phenomenon, see Yueh et al. [1990]).

Emissivity is computed in this model assuming that the ice is isothermal so Kirchhoff's law relates emissivities and reflectivities. However, because the model is polarimetric, polarimetric passive microwave signatures may also be computed.

8.2.2.7. Many layer strong fluctuation theory.
Stogryn has developed a model also based on strong fluctuation theory, but which is substantially different from that described in Section 8.2.2.6 [Stogryn, 1983a,b, 1984a,b, 1985, 1987]. He has applied this theory to study the effective (tensor) permittivity of sea ice [Stogryn, 1987], but to our knowledge has not yet published signature computations for sea ice in the literature. We apply here an implementation of Stogryn's theory by Grenfell which computes both active and passive signatures. We have termed this implementation the "many layer strong fluctuation theory" because (1) it treats problems with many layers (as many as 30 in the present implementation), (2) it is not fully polarimetric (although there is no fundamental barrier to making it so, and (3) this model is distinguished from that described in the previous subsection. Like polarimetric SFT, the many layer theory is based on the bilocal and distorted Born approximations. It also treats interference between waves reflected and transmitted through the various layer planar interfaces coherently. This theory is based fundamentally on the continuous random medium model; however, it assumes a very different picture for the sea ice scattering medium and for the ultimate cause of scattering than in the polarimetric SFT model.

Stogryn [1987] first applies strong fluctuation theory within individual sea ice crystals assuming an aligned array of ice platelets with vertically elongated brine pockets sandwiched between them. He computes a directionally anisotropic, polarization-dependent (i.e., tensor) effective permittivity for a single ice crystal of size on the order of 1 cm. Any air bubbles in the ice are assumed to lie around the edges of crystals. The effective permittivity varies from crystal to crystal because the orientation of platelet structures varies between crystals. These orientations are assumed uncorrelated and uniformly distributed through all azimuthal angles from 0 to 2π. It is the fluctuation of permittivity within the jumble of crystals that causes scattering, according to this model. The contrast in permittivities between crystals is smaller than that between brine and ice (the driving fluctuation in the previous model), but crystals are larger than brine pockets. This approach does produce a cross-polarized backscattering response, but the response is weaker than that produced by the polarimetric SFT model.

The many layer model takes as input profiles of ice temperature and salinity with depth. It uses the equations of Frankenstein and Garner [1967] to compute brine volume, and then computes effective and fluctuating permittivities using assumptions about brine pocket and crystal geometry and spacing. Thus the need for direct measurements of permittivity correlation functions is avoided at the expense of having to assume values for parameters specifying brine pocket geometries. Permittivity of the brine pockets is set using the equations of Stogryn and Desargent [1985]. The most significant tunable parameters are the mean tilt angle of the long axes of brine pockets with respect to vertical and the ratio of brine pocket length to width. Stogryn [1987] has suggested for these parameters values of 24° and 200, respectively, based on model fits to extinction data. However, these values differ from estimates taken directly from saline ice samples [Arcone et al., 1986; Lin et al., 1988]. Parameters relating to the geometry of liquid water in wet snow are also significant and poorly known, but do not affect the studies we present below.

Like polarimetric SFT, this model computes emissivities on the basis of Kirchhoff's law, which relates emissivity to reflectivity. Thus although the model accounts for effects of the true ice temperature profile on brine volume, and therefore on the effective permittivity, it does not fully account for any effects this profile may have on emission.

8.2.2.8. Modified radiative transfer. Modified radiative transfer (MRT) [Lee and Kong, 1988; Lee and Mudaliar, 1988; Mudaliar and Lee, 1990] is a model in the general form of classical radiative transfer based on the continuous random medium model. The model is derived on the basis of consistent multiple scattering approximations for both coherent and incoherent fields. The primary aim of MRT is to capture the general, partially coherent interaction between field contributions from different layers in the scattering medium (layer interfaces are assumed planar). To this end, the theory retains correlations between upgoing and downgoing waves at the same angles in each layer. The interactions appear in additional terms on the right-hand side of the MRT analog to Equation (1). This permits the study of coherent interaction effects as functions of volume scattering strength, layer optical depth, and other factors. Results from our first case study (Section 8.3) suggest the potential relevance of these effects.

The derivation of MRT begins with a continuous random medium in which permittivity fluctuations may be strong; the initial development is essentially the same as that in strong fluctuation theory. The dyadic Green's functions are decomposed into singular and nonsingular parts; the singular parts are treated carefully. However, MRT computes the coherent field using the nonlinear approximation in Dyson's equation. Recall from Section 8.2.2.5 that the bilocal approximation assumes that the coherent field travels between scattering events with the propagation constant appropriate for the background ice. By contrast, the nonlinear approximation assumes a propagation constant equal to the effective propagation constant of the sea ice, and thus accounts for additional multiple forward scattering events in the coherent field. The nonlinear approximation is roughly the continuous-medium analog to the QCA-CP approximation in DMRT (cf. Section 8.2.2.3). MRT then computes the incoherent scattered field, accounting for a (presumably predominant) class of multiple scattering events in the incoherent field as well. Specifically, MRT makes use of the general Bethe–Salpeter equation for the incoherent field and the ladder approximation for the intensity operator in this equation (also in a kind of continuous-medium analog to DMRT). The ladder approximation accounts for that cascade of uncorrelated scattering events in the incoherent field that can each be described in terms of two-point permittivity statistics. The combination of the nonlinear and ladder approximations produces a self-consistent theory in terms of energy conservation. Including some multiple scattering of the incoherent field would seem to permit MRT to treat stronger scattering than theories using the distorted Born approximation. However, solution of the nonlinear approximation equation for the coherent field remains restricted to the low-frequency regime. The net effect of this restriction is not presently clear.

MRT was first developed for electromagnetic wave scattering by Zuniga and Kong [1980] [Tsang et al., 1985, Section 5.5]. Lee and Kong [1985a, b, 1988] generalized the theory to treat dielectrically anisotropic scattering media such as saline ice. However, the solutions of the theoretical equations to date [Lee and Mudaliar, 1988; Mudaliar and Lee, 1990] are restricted to a single, infinitely thick scattering layer; the only layer interface is at the top of the layer. Thus there can be no interactions of waves from different layers. The present solutions are also limited to first-order scattering; this is analogous to Equation (4) and essentially equivalent to the distorted Born approximation. The present solution is also restricted to isothermal emitting media. Mudaliar and Lee [1990] have successfully matched some passive signature observations for old ice at microwave and millimeter-wave frequencies using this solution. However, because of the present restrictions we will not be able to demonstrate the full capability of MRT in this chapter. Rather, the initial results we present are intended as a spur to further research.

The continuous random medium model upon which MRT is built has yet to be specialized to model sea ice. Thus, the present version of MRT requires direct specifications for the mean, variance, and correlation length of permittivity fluctuations within the ice. In the case of dielectric anisotropy, two such sets of statistics are required, one for fluctuations in the preferred direction in the ice (which need not be vertical), one in an orthogonal direction. These statistics may in principle be supplied directly from independent ice characterization measurements such as those of Perovich and Gow [1991], but at present must often be set according to the modeler's judgment.

8.2.3 Rough Surface Scattering Models

Our concern with rough surface scattering in this chapter is confined almost entirely to its effect on backscattering signatures. Only one of the models we employ for passive signatures currently includes rough surface scattering effects (namely, the dense medium theory/integral equation method). We use this model to compute passive signatures in only one case, that of thin gray ice, and in this case the effects of surface roughness on emission are minor. Thus, the following summaries focus on backscattering characteristics of rough surface scattering models.

8.2.3.1. Physical optics under the scalar approximation. The common element in all models based on physical optics is the so-called tangent plane approximation. The fundamental unknowns in rough surface scattering problems are the source densities of Huygens' wavelets induced on the surface by the illumination. The tangent plane approximation replaces, at each point on the rough surface, the unknown source density by the density that would exist if, instead of the actual surface, there existed at that point a plane tangent to the actual surface, separating the same dielectrics actually separated by the rough surface. Thus physical optics models are valid only for surfaces that undulate smoothly on horizontal length scales comparable to the radiation wavelength; their validity at large incidence angles is also problematic [Thorsos, 1988]. A more quantitative statement of this restriction depends on the form of the surface roughness correlation function, but most authors [Ulaby et al., 1982; Thorsos, 1988] seem able to agree on a criterion $kL \gtrsim 2\pi$. When the standard deviation of surface heights is large compared with the wavelength, physical optics reduces to the familiar geometric optics approximation in which backscattering occurs from quasispecular surface points [Ishimaru, 1978; Ulaby et al., 1982].

The scalar approximation to physical optics uses a small slope assumption to further simplify the vector equations from the tangent plane approximation in the electromagnetic case to scalar equations [Ulaby et al., 1982; Eom, 1982]. The result is a cross section approximation with a relatively simple polarization dependence for application in cases where surface heights may not be large compared with the wavelength. The model predicts no cross-polarized backscattering. Kim et al. [1985] give a (misprinted) formula derived by Eom [1982] (where it is given correctly) for HH and VV cross sections in the scalar physical optics approximation, assuming an exponential form for the surface height correlation function:

$$\sigma_{pp}^0 = 2 |R_{pp}|^2 \cos^2\theta \exp\left(-k^2 h^2 \cos^2\theta\right)$$
$$\times \sum_{n=1}^{\infty} \frac{\left(4k^2 h^2 \cos^2\theta\right)^n}{n!} \frac{k^2(n/L)}{\left(4k^2 \sin^2\theta + n^2/L^2\right)^{3/2}} \quad (8)$$

where k is the free space wave number of the incident radiation, h is the standard deviation of rough surface height, L is the surface roughness correlation length, and R_{pp} is the Fresnel reflection coefficient (for the field, not the power) of the dielectric scattering material; the horizontal polarization reflection coefficient is to be chosen for R_{pp} when the pp = HH-polarized backscattering cross section is to be computed, and the vertical polarization reflection coefficient chosen for VV-polarized cross sections. Most dielectrics of interest in remote sensing display a Brewster angle in reflection; thus the reflection coefficient for horizontal polarization typically exceeds that for vertical polarization over a broad range of incidence angles. Consequently, the HH-polarized cross sections predicted by Equation (8) are often higher than their VV-polarized counterparts. This feature of this model is somewhat unusual and controversial.

8.2.3.2. Conventional Perturbation Theory. The treatment of rough surface scattering as scattering from a perturbed flat surface was first given by Rice [1951]. The theory has since been derived in alternate ways, but always with identical results [Ulaby et al., 1982; Tsang et al., 1985; Jackson et al., 1988]. The essential idea is to expand unknown scattered and transmitted fields in perturbation series with kh as the small parameter, where k is the radiation wave number and h is the standard deviation of surface heights. The dielectrics separated by the rough interface are assumed to extend to infinity both above and below. Expanding the boundary conditions in powers of kh allows an iterative solution of the perturbation equations, order by order. Ishimaru [1978] gives a convenient summary of results for the case of a dielectric with relative permittivity ε_r bounded above by free space. The first-order, or so-called Bragg scattering, backscattering cross sections for transmit and receive polarizations i and j are given by

$$\sigma_{ij}^0 = 16\pi k^4 \cos^4\theta \left|\alpha_{ij}(\theta)\right|^2 W(2k \sin\theta, 0) \quad (9)$$

where θ is the angle of incidence, W is the power spectrum of surface roughness defined by the correlation function of surface heights $\rho(x, y) = <f(x_0 + x, y_0 + y) f(x_0, y_0)>$,

$$W(K_x, K_y) = (2\pi)^{-2} \int_{-\infty}^{\infty} \int dx\, dy \exp\left(-iK_x x - iK_y y\right) \rho(x, y) \quad (10)$$

(we have for simplicity chosen the coordinate system such that the plane of incidence coincides with the x-z plane), and where

$$\alpha_{HH}(\theta) = \frac{\varepsilon_r - 1}{\left[\cos\theta + \left(\varepsilon_r - \sin^2\theta\right)^{1/2}\right]^2} = -R_h(\theta) \quad (11a)$$

$$\alpha_{VV}(\theta) = \frac{(\varepsilon_r - 1)\left[(\varepsilon_r - 1)\sin^2\theta + \varepsilon_r\right]}{\left[\varepsilon_r \cos\theta + \left(\varepsilon_r - \sin^2\theta\right)^{1/2}\right]^2} \quad (11b)$$

$$\alpha_{HV}(\theta) = \alpha_{VH}(\theta) = 0 \quad (11c)$$

There are several things to note here. First-order perturbation theory, like the physical optics model above, predicts no cross-polarized backscattering; cross-polarized backscattering appears in conventional perturbation theory as a second-order effect. Second, note that for any ε_r having a real part greater than one, VV-polarized backscattering will exceed that at HH-polarization, contrary to the situation in the model above. Finally, R_h in Equation (11a) is the Fresnel reflection coefficient for the field reflected from a flat interface between free space and the dielectric material beneath the rough interface.

Conventional, lowest order perturbation theory is generally accurate to within about 2 dB for surfaces with standard deviations of surface height less than 10% of the radiation wavelength, though very high surface slopes also degrade the accuracy of this approximation [Thorsos and Jackson, 1991]. The sensitivities of first-order cross sections to variations in ice permittivity and surface roughness statistics are discussed by Winebrenner et al. [1989].

8.2.3.3. Integral equation method. Fung and Pan [1987a, b] have developed an analytical model for electromagnetic scattering from perfectly conducting rough surfaces, named the integral equation method. This method has recently been extended by Fung et al. [1991] to treat rough interfaces between dielectrics. Thorsos [1988] has developed a numerical, Monte Carlo simulation method for the study of rough surface scattering, which he has also called the integral equation method. We employ in this chapter the method of Fung and coworkers, and thus in this context there is no opportunity for confusion. However, readers surveying the field of rough surface scattering more broadly should take care to avoid confusion of these two very different methods.

As we have noted, the fundamental problem in rough surface scattering is accurate approximation of the densities of Huygens' wavelet sources induced on the rough surface by the illuminating wave. The method of Fung and Pan [1987a, b] begins with an exact integral equation for the unknown source density (in this case, the surface current) on a given realization from an ensemble of rough surfaces. The zeroth-order solution to this equation is just the tangent-plane approximation for the source density. Fung and Pan iterate this equation once to produce an improved approximation (which is, roughly speaking, correct to higher order in surface curvature than the tangent-plane approximation [Dashen and Wurmser, 1991]). In the case where surface slopes are large, this improved approximation requires shadowing corrections to provide correct results [Ishimaru et al., 1991; Jin and Lax, 1990; Chen and Fung, 1990]. Thorsos and Jackson [1991] have argued further that this approach requires shadowing to limit long-range surface interactions and prevent divergent integrals in expressions for cross sections. However, in the case of moderate slopes and surface heights, Fung and Pan [1987a] derive like- and cross-polarized cross sections in terms of convergent integrals without explicit shadowing. They argue that convergence of the integrals results from the finite correlation length of surface roughness. Shadowing in this approach is applied as a correction to the final cross section results; the application is similar to that often made in classical physical optics theory [Ulaby et al., 1982]. The results agree with physical optics and perturbation theory in appropriate limits and compare favorably with numerical simulations for two-dimensional scattering problems by Chen et al. [1989] and data from controlled experiments [Fung and Pan, 1987a]. Thus, despite the disagreement

between authors over the role of shadowing, there is evidence for the validity of numerical results from this method.

The extension of the Integral Equation Method by Fung et al. [1991] to the case of a general dielectric interface requires a similar treatment of two coupled integral equations but yields analogous results. In both our case studies, the integral equation method has been used to derive rough surface boundary conditions for the dense medium (volume scattering) theory (Section 8.2.2.4), and is always used in conjunction with this theory. Characterization requirements for surface roughness in this model are identical to those for the classical models described above.

8.3   Case Study 1: A Thin Gray Ice Sheet

8.3.1 Ice History and Characterization

Our first case study concerns a thin, snow-free gray ice sheet grown as part of the 1988 CRRELEX experiment at the Cold Regions Research and Engineering Laboratory in Hanover, New Hampshire (Chapter 9). The sheet began on the night of January 11–12, 1988, as a snow-nucleated ice sheet growing in a pond of simulated seawater with salinity of 24‰. (Pond salinity is purposely set lower than that of natural seawater so that CRRELEX ice sheet salinities approximate natural values despite New Hampshire winter temperatures warmer than those in the Arctic.) The sheet at first grew slowly in temperatures near freezing, and even experienced a small amount of melting during its first two days. A change to clear, cold weather on January 13 caused the sheet to begin growing rapidly. A trace of snow blew onto the sheet on the night of the 13th, giving the ice surface a very fine scale gravelly visual texture, though the surface relief measured less than 1 mm. By the morning of the 14th, the sheet was approximately 6.5 cm thick and a band of crystals resembling frost flowers covered the center of the sheet (Figure 8-2). The night of the 14th was clear and cold with air temperatures below −28°C. The backscattering and emission measurements for this study were acquired between approximately 1630 and 2330 EST, on the night of the 14th. Grenfell and Winebrenner report ice thickness measurements of 8.0 and 8.3 cm at 1630 and 8.3 cm at 2130. Onstott reports a thickness measurement of 7.5 cm at 1910. Our experience indicates the variation in these measurements was likely due to spatial variability in the ice sheet thickness. The frost-flower-like crystals persisted on the ice sheet through the morning of the 15th; various investigators reported ice thicknesses ranging from 11.5 to 12.4 cm by 1030 that morning.

Perovich measured temperature and salinity profiles on the mornings of January 14 and 15. Figure 8-3 shows the measured temperature and salinity profiles; the salinities from the morning of January 14 are averages of two measurements made at opposite ends of the pond. Note that the cold night of the 14th evidently caused upward brine extrusion. The salinity in the top centimeter of ice increased

Fig. 8-2. Eight-centimeter thick gray ice in the CRRELEX pond used for the first case study. Detailed characterization is reported in Section 8.3.1.

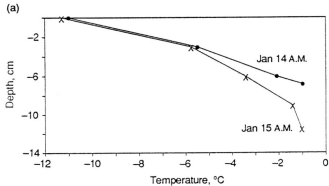

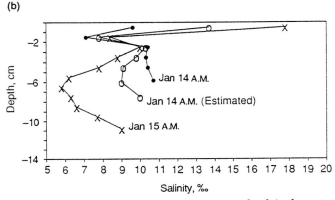

Fig. 8-3. (a) Temperature and (b) salinity versus depth in the gray ice sheet on the mornings of January 14 and 15 during CRRELEX in 1988. Grenfell and Winebrenner have heuristically estimated the salinity profile shown for late in the evening of January 14.

considerably, and salinity at depths 1 to 2 cm increased moderately; ice below 3 cm depth actually decreased in salinity. Ice temperature profiles were nearly linear on the 14th and 15th, except for the lowest point on the 15th which may be a temperature from a thermocouple in the water. We have no salinity profile for the night of January 14, i.e., at the time of the scattering and emission measurements. Grenfell and Winebrenner have therefore estimated the salinity profile at approximately 2000 on January 14 heuristically, with the result shown in Figure 8-3(b). We believe that more accurate salinity profiles than these would be difficult to acquire, given the horizontal variability in profiles at differing pond locations observed by Perovich and in nature by Tucker et al. [1984]. Reexamination of this issue may be necessary, however, if some signatures are conclusively shown to depend sensitively on details of this profile (Section 8.3.2.1). The ice surface temperature at the time of signature measurements was –16°C; we therefore estimate a linear temperature profile with this value at the

air–ice interface and a temperature of –1°C at the ice–water interface.

Perovich acquired a horizontal thin section photograph of the ice at a depth of approximately 6 cm on the morning of the 15th, which unfortunately lacked sufficient contrast to estimate the permittivity correlation length using the method of Perovich and Gow [1991]. However, a reasonable estimate of the permittivity correlation length is possible based on an examination of ice morphology and comparison with data from other CRRELEX ice sheets with similar histories, temperatures, and salinities. Perovich estimates the permittivity correlation length in our sample to have been 0.15 mm. Perovich and Gow [1991] show that estimates of this correlation length display a typical random variation of about 30% for different horizontal locations within a macroscopically homogeneous ice sheet. Variations in depth and temperature lead to larger but more predictable correlation length variations.

Estimates of the air–ice interface roughness for this sheet are available from photographs of an ice sample removed by Onstott during the day on the 14th. The sample was stored at low temperature (≈ –20°C) and later sectioned and photographed against a calibrated grid. The photographs were analyzed by two separate groups of investigators (by Bredow and Gogineni, and by Onstott) to derive independent estimates of surface roughness parameters. Both sets of estimates assume an exponential form for the correlation function of surface heights, $\rho(x) = h^2 \exp(-x/L)$, where x is spatial lag, and h is the standard deviation of surface height. Bredow and Gogineni [1990] estimate $h = 0.03 \pm 0.01$ cm and $L = 1.77$ cm with 90% confidence limits of 0.81 cm and 2.49 cm. Onstott estimates $h = 0.048$ cm and $L = 0.669$ cm. We have no information on under-ice surface roughness except a qualitative report by Onstott that the dendrite structure seemed uniform and displayed no obvious roughness on horizontal scales larger than the platelet size (which was less than 1 mm). We have no characterization of the frost-flower-like crystal aggregates appearing in the photograph in Figure 8-2.

Onstott acquired backscattering cross section measurements at 5.25 and 9.6 GHz for HH-, VV-, and cross-polarizations. He estimates the range of uncertainty to extend from 4 dB below to 2.6 dB above each data point. Overall, backscattering from this thin gray ice is weak. Co-polarized cross sections increase between 5 and 10 GHz whereas cross-polarized cross sections change relatively little.

Grenfell and Winebrenner acquired calibrated brightness temperatures for V- and H-polarizations over a range of incidence angles from 30° to 70° at frequencies of 6.7, 10, 18.7, 37, and 90 GHz. These brightness temperatures have been reduced to effective emissivities based on the measured ice surface temperature and sky brightness at each frequency. This facilitates the comparison of models that compute only emissivity with the data and with models that predict brightness temperature; outputs from the latter models are normalized exactly as are the observations. Grenfell estimates an accuracy of 0.02 in horizontally polarized emissivity and 0.01 for vertical polarization. The observed emissivities are generally high, especially at V-polarization. (Note that the H-polarization emissivity reported at 70° for 37 GHz is evidently contaminated by emission from the edge of the pond or some other structure.) A striking feature in these observations is the minimum in emissivity at 10 GHz, relative to 6.7 and 18.7 GHz, at both polarizations. The H-polarization emissivity at 50° incidence angle drops from 0.79 at 6.7 GHz to 0.70 at 10 GHz (a variation of 12%) before rising back to 0.77 at 18.7 GHz. While the corresponding variation at V-polarization is only 3%, this is still larger than the estimated measurement uncertainty. Data sets acquired before and after this data set show a similar anomaly, but data taken after alteration of the ice surface late on the 15th show no such feature. Data acquired earlier on the 14th by the University of Massachusetts Stepped Frequency Microwave Radiometer show a decreasing emissivity with increasing frequency between 4 and 7 GHz (K. St. Germain and C. Swift, personal communication). Grenfell has observed similar features in other CRRELEX ice sheets as well. Thus all indications are that the radiometers were functioning properly and the 10 GHz passive signature feature in these data is real.

8.3.2 Model Comparisons

8.3.2.1. Many layer strong fluctuation theory. The application of many layer SFT in this case is based on an eight-layer physical model for the ice. Temperatures and salinities are constant within each layer but vary with depth. The layers are of equal thickness (1 cm) and centered on the depth points in the estimated salinity profile for the night of January 14, except for the layer adjacent to the ice–water interface. Temperatures and salinities in each layer are set equal to the values of the estimated profiles at the centers of each layer, except for the layer adjacent to the air–ice interface. The thickness of the lowest layer and salinity of the uppermost layer were varied to examine model sensitivities to these poorly known parameters.

The model also requires assumptions about parameters such as brine pocket size and spacing, ice and brine permittivities (as functions of temperature), air bubble and ice crystal sizes, and mean brine pocket tilt and elongation. Although brine pockets are tilted from the vertical, their azimuthal orientations are random and uniformly distributed; thus the ice in this model possesses no macroscopic azimuthal anisotropy. Each of these parameters was set to the values used by Stogryn [1987] in his study of extinction, except for the values of mean brine pocket tilt and elongation. In the latter two cases, values suggested by previous characterization studies of artificial sea ice in CRRELEX [Arcone et al., 1986; Lin et al., 1988] were used. Specifically, these values are 4° for the mean brine pocket tilt (from vertical) and 10 for the ratio of brine pocket length to width (versus a tilt of 24° and ratio of 200 used by Stogryn). These parameters are assumed not to vary with depth in the ice sheet.

Figure 8-4 compares observations and signatures computed using two versions of the above physical model for the ice. Figure 8-4(a) shows computed and observed (effective) emissivities at 50° incidence angle, plotted versus frequency for V- and H-polarizations. One physical model is based on a total ice thickness of 8.3 cm, i.e., a lowest layer thickness of 1.3 cm, with layer salinities fixed at precisely those values specified by the estimated salinity profile. The uppermost layer salinity is 14‰ and results based on this model are labeled "8.3 cm, 14‰." An alternate model with total ice thickness 8.0 cm but an uppermost layer salinity of 20‰ is labeled "8.0 cm, 20‰." Both models show lower H-polarized emissivities at 10 GHz than at neighboring frequencies, though the feature is not quite as deep as that observed. The 8.3 cm model predicts an H-polarized 10 GHz emissivity of 0.74, whereas that observed is 0.70. The 8.0 cm model predicts 0.72, but agrees less well with the 6.7 GHz observation. Only the 8.0 cm model displays a feature at V-polarization, and this is also less pronounced than that in the data (0.96 predicted versus 0.93 observed). The predicted feature is an interference fringe caused by coherent interaction between field contributions from dif-

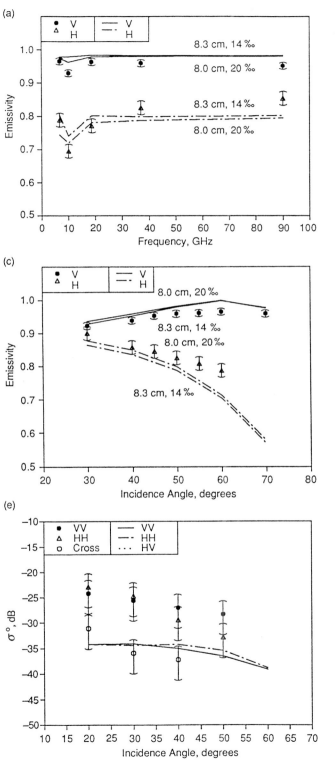

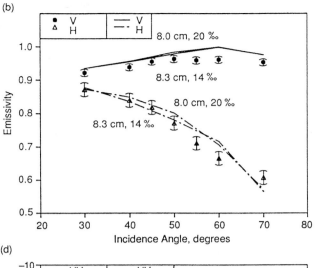

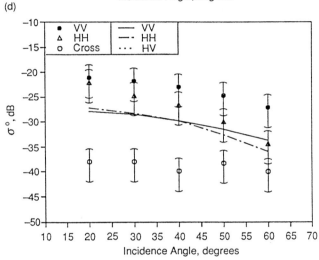

Fig. 8-4. Comparison of observations with emissivities and back-scattering cross sections computed using the many layer SFT model for the CRRELEX gray ice sheet. (a) Emissivity versus frequency at an incidence angle of 50°; (b) emissivity versus incidence angle at 18.7 GHz; (c) emissivity versus incidence angle at 37 GHz; (d) backscattering cross sections versus incidence angle at 10 GHz; and (e) backscattering cross section versus incidence angle at 5 GHz.

ferent layers within the ice. Model emissivities vary almost precisely with the computed reflection coefficients; thus the model predicts virtually no effect of volume scattering on any of the passive signatures, but rather emissivity variations due solely to variations in the reflectivity of the stack of layers comprising the ice. The reflectivity variations are due to variations in the salinity, and therefore effective permittivity, profile of the ice sheet; there is nothing in the frequency dependence of other dielectric properties (brine permittivity, etc.) sufficient to produce such a large variation.

The predicted emissivity minimum is robust in that it occurs between 5 and 12 GHz for much wider variations in lower layer thickness and surface salinity than we have shown here. However, the exact location (i.e., frequency) and depth of the minimum are sensitive to those parameters. For example, an 8.0 cm model with 14‰ surface salinity produces a minimum emissivity closer to 6.7 GHz. Adding a tenuous but uniform snow layer atop the ice, in an attempt to model the frost-flower-like layer, leads to alterations in the fringe that depend sensitively on snow depth and density. Neglecting the salinity profile entirely leads to unrealistic results; emissivities computed from any single-layer model oscillate, as functions of frequency and incidence angle, much more than is observed. Thus according to this model, coherent effects are essential in explaining the observed feature, but details of the feature depend sensitively on thickness, near-surface salinity, and perhaps other details of the salinity and temperature profiles. Different growth histories (thus, differing salinity profiles) and changes to the near-surface salinity profile by a frost-flower layer would therefore alter the fringe predicted by many layer SFT, but would not eliminate the feature. The model also displays a milder sensitivity to the values chosen for mean brine pocket tilt and elongation. The values we use for these parameters lead to less predicted absorption and thus to slightly stronger coherent interaction effects than do those values used by Stogryn [1987]. Figures 8-4(b) and (c) show computed and observed emissivities versus incidence angle 18.7 and 37 GHz, respectively, for each of the two physical models. Penetration depths are much reduced at these frequencies, making the near-surface salinity the most important parameter. Predicted signature sensitivities to upper-layer salinity are, however, relatively small for salinities in the range shown. Emissivity predictions at 50° differ from observations by less than 0.03 except for H-polarization at 37 GHz, where the predicted value is 0.04 too low. Agreement is better at smaller incidence angles but worse at larger angles, where Brewster-angle effects are more apparent in V-polarization predictions than in observations and H-polarization observations fall off less rapidly than predicted. Figure 8-4(a) shows this effect worsens at 90 GHz. Finally, Figures 8-4(d) and (e) show like- and cross-polarized backscattering cross sections at 5 and 10 GHz, respectively, as functions of incidence angle. Model results are shown only for the 8.3 cm case because sensitivities are modest, on the order of those for 18.7 and 37 GHz emissivi-

ties. The many layer SFT model predicts volume scattering levels too low to explain the observed level of backscattering, at least with the present set of assumptions for brine pocket and other parameters. Note that predicted like-polarized cross sections show a different polarization behavior than indicated by the observations. Predicted cross-polarized cross sections fall below −55 dB.

8.3.2.2. Polarimetric strong fluctuation theory. Polarimetric SFT model results for this case are based on a physical model consisting of a single ice layer overlying seawater. Physical properties do not vary within the ice layer; thus an effective layer temperature and salinity, resulting in a brine volume of 4.2%, characterize the layer. The layer thickness is fixed at 8.0 cm. The permittivities of the ice background, brine inclusions, and underlying seawater are frequency dependent and specified in Table 8-1. The only remaining physical model parameters are the permittivity correlation lengths, Equation (7). These are set at $l_x = 0.70$ mm, $l_y = 0.26$ mm, and $l_z = 1.2$ mm, independent of frequency, incidence angle, or any other parameters. The correlation lengths were chosen to match model results with backscattering observations at 5 GHz and then fixed for all subsequent model calculations. They are consistent with values used previously in comparisons of this model with 9 GHz backscattering data for first-year ice near Point Barrow, Alaska [Nghiem, 1991]. Though we have no independent estimate of l_z in this case study, the value of l_z is comparable to values found in another study of artificial sea ice in CRRELEX [Arcone et al., 1986; Lin et al., 1988]. The values of l_y and l_x are notably larger than Perovich's correlation length estimate (consistent with the reasoning in Section 8.2.2.6).

Figure 8-5 compares the computed signatures with backscattering and emission observations. Figure 8-5(a) is a plot of emissivities versus frequency at 50° incidence angle. Model emissivities agree with observations to within 0.025, except at 10 GHz (and at 90 GHz, where results are not available). V-polarized model results are 0.05 higher than experiment at 10 GHz, while at H-polarization the difference is 0.10. Angular emissivity responses at 18.7 and 37 GHz are shown in Figures 8-5(b) and (c). Although the differences between theory and experiment are similar to those in many layer SFT, the quantitative agreement here is better for 37 GHz, H-polarization. Passive signature results are insensitive to the choice of correlation length, indicating a minor role for scattering in this model for passive signatures. Ice sheet reflectivities, determined by the mean permittivity, govern the computed emissivities. The 6.7 GHz emissivities display a modest sensitivity to layer thickness. Because dielectric absorption is higher in the vertical direction (Table 8-1), this sensitivity is greater for horizontal than for vertical polarization.

Figures 8-5(d) and (e) compare observations with model predictions for like- and cross-polarized backscattering cross sections versus incidence angle, at 5 and 10 GHz, respectively. Theory for the 5 GHz, like-polarization cross

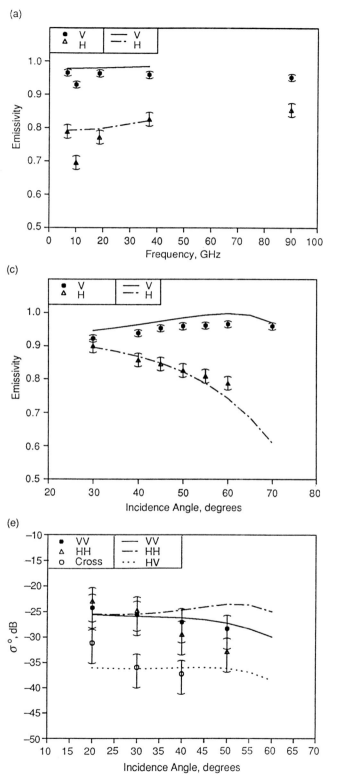

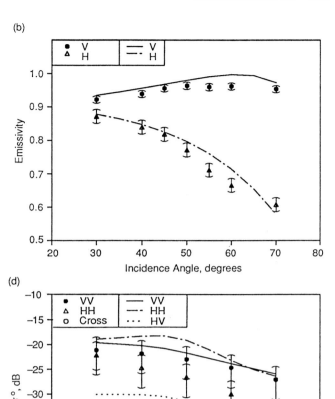

Fig. 8-5. Comparison of observations with emissivities and back-scattering cross sections computed using the polarimetric SFT model for the CRRELEX gray ice sheet. (a) Emissivity versus frequency at an incidence angle of 50°; (b) emissivity versus incidence angle at 18.7 GHz; (c) emissivity versus incidence angle at 37 GHz; (d) backscattering cross sections versus incidence angle at 10 GHz; and (e) backscattering cross section versus incidence angle at 5 GHz.

TABLE 8-1. Relative permittivities of materials in the Polarimetric SFT Model for the CRRELEX gray ice sheet.

Frequency, GHz	Seawater	Background ice	Brine inclusions	ε_{eff}	ε_{effz}
5.0	60 + i34	3.15 + i0.0015	37.5 + i43.1	3.53 + i0.072	4.10 + i0.40
6.7	60 + i34	3.15 + i0.0015	32.1 + i38.6	3.53 + i0.086	4.06 + i0.459
10.0	42 + i34	3.15 + i0.0015	24.0 + i32.0	3.51 + i0.116	3.92 + i0.562
18.7	30 + i34	3.15 + i0.0025	14.0 + i21.0	3.44 + i0.187	3.53 + i0.675
37.0	10 + i20	2.92 + i0.0030	9.8 + i12.0	3.07 + i0.182	2.92 + i0.433

sections agrees with observations to within better than 3 dB at 20° and 30°; the agreement is similar for cross-polarization at 30° and 40°. However, the model predicts HH-polarized backscattering higher than observations at larger incidence angles and a difference between HH- and VV-polarizations opposite to that observed. Model predictions at 10 GHz are 1.5 to 10 dB above the observations for incidence angles less than 50°, and the predicted like-polarization difference is again opposite to that observed. The predicted cross-polarized cross sections also exceed observed levels. The increased discrepancy at 10 GHz may result in part because permittivity correlation lengths were set on the basis of results at 5 GHz (see above). However, setting correlation lengths based on results at 10 GHz yielded less satisfactory results overall. In contrast to model predictions for emission, backscattering cross sections are sensitive to the scatterer size parameters, i.e., to the permittivity correlation lengths. The 5 and 10 GHz cross sections display a comparable or greater sensitivity to thickness of the ice layer as well; the backscattering level, frequency response, and VV-HH polarization contrast are all sensitive to this parameter.

8.3.2.3 Modified radiative transfer. The equations of modified radiative transfer have at present been solved only for a single, infinitely thick scattering layer. Thus the MRT results presented here are based on a physical model for the ice in which the ice–water interface plays no role and ice properties do not vary with depth. Dielectric properties are modeled as directionally anisotropic, however, with a specified tilt direction for the (single) preferred direction (i.e., the optic axis). Thus the tilt direction, as well as mean permittivities, normalized variances of permittivity (i.e., the variance divided by the mean squared) and permittivity correlation lengths, in directions parallel and perpendicular to the preferred direction, must be specified directly. Specification of these parameters was guided generally by knowledge of the ice structure and parameters, but detailed choices were based also on experience with this model. The parameters used to produce nearly all the results shown below are as follows. The preferred direction coincides with

the vertical. This reflects the macroscopic azimuthal isotropy of the ice sheet, while accounting for the vertical elongation of brine pockets. The mean permittivities are then 3.3 + i0.1 perpendicular to the preferred direction (horizontal) and 3.4 + i0.16 parallel (vertical). The normalized variances are 0.1 horizontal, 0.2 vertical. Finally, the permittivity correlation lengths are set at 1 mm in the vertical (reflecting the vertical extent of brine pockets) and 0.1 mm in the horizontal (reflecting the horizontal extent). Some results for backscattering sections were computed with different parameters as well for the purpose of studying sensitivities. We give the additional parameters below.

Figure 8-6(a) compares emissivities computed using MRT above with observations as a function of frequency at 50° incidence angle. Recall that the present solution of MRT is only a first-order scattering solution. Predictions at V-polarization, with the exception of 10 GHz, show good agreement even at 90 GHz. The predictions predict no feature at all at 10 GHz, however. Figures 8-6(b) and (c) show emissivity predictions and observations versus incidence angle for 18.7 and 37 GHz, with particularly good agreement at 37 GHz, V-polarization. Figures 8-6(d) and (e) show backscattering cross sections at 10 GHz computed using two different sets of input parameters. Like-polarized cross sections computed from the parameters given above fall approximately 20 dB below the observations, though the predicted polarization contrast is in approximate agreement. The most sensitive parameters in computing backscattering cross sections are the imaginary parts of the mean permittivities and permittivity correlation length. Reducing the imaginary parts and increasing correlation lengths increases predicted backscattering to a level compatible with the observations, as shown by the results in Figure 8-6(e). The parameters used to compute the latter result are mean permittivity of 3.1 + i0.001 and correlation length 0.15 mm in the horizontal direction, and 3.2 + i0.002 and 0.2 mm in the vertical. Although the computed VV-HH contrast is smaller than that observed, it has the correct sign. Note, however, that this model produces no cross-polarized response; this is a characteristic of most first-order scattering models when the scattering

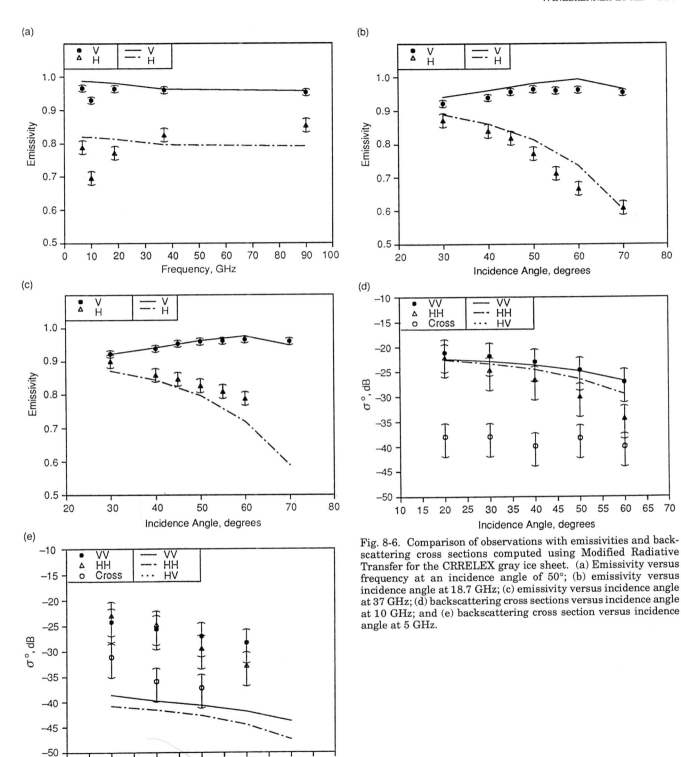

Fig. 8-6. Comparison of observations with emissivities and back-scattering cross sections computed using Modified Radiative Transfer for the CRRELEX gray ice sheet. (a) Emissivity versus frequency at an incidence angle of 50°; (b) emissivity versus incidence angle at 18.7 GHz; (c) emissivity versus incidence angle at 37 GHz; (d) backscattering cross sections versus incidence angle at 10 GHz; and (e) backscattering cross section versus incidence angle at 5 GHz.

medium is azimuthally isotropic.

8.3.2.4. Physical optics under the scalar approximation.
The model used by Drinkwater [1989, 1987] neglects volume scattering in high-density congelation ice. Thus the model reduces in this case to a purely surface scattering model based on physical optics under the scalar approximation. The model assumes that beneath the rough surface lies a directionally isotropic dielectric material with no depth variation. Thus the results below assume a physical model for the ice in which finite ice thickness and dielectric anisotropy play no role. The (scalar) permittivity of the ice must be computed in light of the penetration depth and temperatures and salinities over that depth. The permittivity is computed in this case using the model of Vant et al. [1978], assuming vertically oriented brine pockets 0.91 mm long and 0.15 mm wide. Ice temperature and salinity are set to the observed air–ice interface temperature of –16° and surface salinity of 14‰. Ice density is assumed to be 0.92 g/cm^3; the computed brine volume fraction is 5.4%. The resulting permittivities are $3.74 + i0.20$ at 5 GHz and $3.66 + i0.23$ at 10 GHz. The uncertainty in surface roughness statistics motivated computation of a range of model results based on the range of likely surface parameters.

Figure 8-7 shows observations and model predictions for like-polarized backscattering cross sections based on two sets of surface parameters. The first set of parameters, $h = 0.02$ cm, $L = 2.49$ cm (where L is the correlation length in the exponential correlation function of Section 8.3.1), correspond to the smoothest surface within the limits of uncertainty reported by Bredow and Gogineni [1990]. The second set of parameters, $h = 0.048$ cm, $L = 0.669$ cm, corresponds to the roughest surface consistent with the characterization data, namely that reported by Onstott. Figure 8-7(a) presents results at 10 GHz. (Recall that the physical optics model does not predict cross-polarized backscattering and that it treats backscattering only; it does not treat emission.) Using the roughest probable surface parameters, the predicted HH-polarized cross sections fall only 2–3 dB below the observations. This is encouraging given that the result relies only on independent characterization information. However, the polarization dependence is opposite to that observed; the predicted cross sections at VV-polarization fall below those for HH and well below the observations.

Figure 8-7(b) shows analogous results at 5 GHz. Here, the validity criteria for the model are rather severely violated (Section 8.2.3.1). The plot shows the best available fit to the observations using parameters within the range specified by independent characterization. Results at neither polarization compare well with observations.

8.3.2.5. Conventional perturbation theory.
The probable roughness parameters are within the range where conventional perturbation theory should apply at 5 and 10 GHz. Figure 8-8 compares observations with results from first-

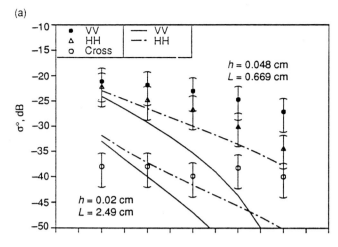

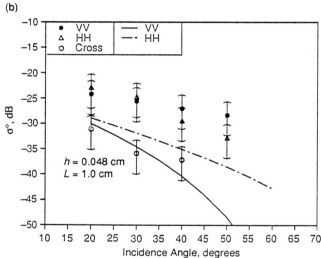

Fig. 8-7. Comparison of observations of the CRRELEX gray ice sheet with backscattering cross sections computed using the physical optics rough surface scattering model of Section 8.2.3.1 for backscattering cross sections versus incidence angle at (a) 10 and (b) 5 GHz.

order perturbation theory based on ice permittivities identical to those used in the physical optics model above.

Figure 8-8(a) shows the range of 10 GHz model predictions for the range of roughness parameters reported by Bredow and Gogineni [1990]. Using the upper limit of their height standard deviation and the shortest correlation length within their 90% confidence interval, model predictions are approximately 4 dB below the observations at 30° incidence angle for both polarizations; at 50° the figure is 6 dB. Model results for the roughness parameter values reported by Onstott, Figure 8-8(b), fall approximately 2.5 dB and 4 dB below the observations at 30° and 50°, respectively, for both polarizations. Thus the sign and magnitude of the VV-HH cross section difference predicted by this model agree well with the observations, but the overall level

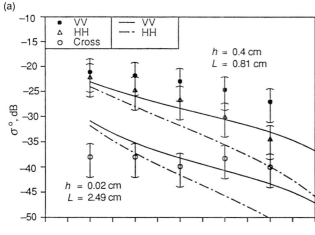

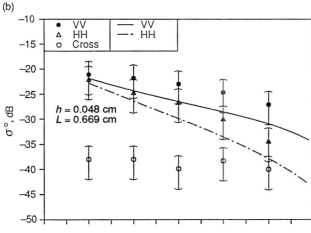

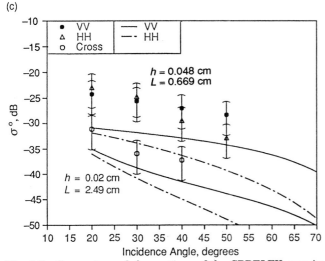

Fig. 8-8. Comparison of observations of the CRRELEX gray ice sheet with backscattering cross sections computed using conventional first-order perturbation theory. (a) and (b) Backscattering cross sections at 10 GHz for three sets of roughness parameters consistent with the reported measurements, and (c) backscattering cross sections at 5 GHz.

of backscattering is, at best, barely within the estimated uncertainties in the observations.

It is interesting to note that this comparison is not invalidated by the inference of finite-thickness effects from the many layer SFT model comparison. Recall that the first-order HH-polarized backscattering cross section is proportional to the power reflection coefficient, and that this result remains true even for layered media. The results above suggest that the power reflection coefficient of our CRRELEX '88 gray ice differs from that of infinitely thick ice by an amount on the order of 12 to 14%. This would alter the backscattering cross section of our gray ice sample by the same amount for HH-polarization. Yet here we find marginal agreement with results from an infinite-thickness model. This can be resolved by noting that backscattering cross sections are expressed in decibels, and the results for infinite thickness differ from those for finite thickness by less than 1 dB. Thus while 10 GHz passive signatures appear quite sensitive to finite thickness effects, 10 GHz backscattering is much less noticeably so.

At 5 GHz, the predicted VV-HH cross section differences again closely track the observations, but the predicted levels of backscattering are, at best, more than 6 dB too low. Figure 8-8(c) shows the range of possible results based on the independent roughness data. It appears that some physics other than, or in addition to, rough surface scattering from effectively infinitely thick ice must be involved in backscattering at 5 GHz.

8.3.2.6. Dense medium theory—integral equation method. The application of dense medium theory (DMT) in this case is based on a single-layer physical model for the ice, with constant ice properties within the layer and rough interfaces at both the top and bottom. The brine pockets are assumed spherical with radius 0.1 mm; brine volume is set at 5% based on the average temperature and salinity of the ice. The permittivities of the brine pockets and underlying seawater, as well as the computed mean ice permittivities, are given in Table 8-2. The depth of the layer is set at 7.5 cm. Roughness parameters for the air–ice interface are set at $h = 0.05$ cm and $L = 0.67$ cm. The corresponding parameters for the ice–water interface are $h = 0.03$ cm and $L = 0.96$ cm. The under-ice roughness parameters are plausible but constrained only loosely by Onstott's qualitative observation. Recall that DMT-IEM treats emission from nonisothermal layers; thus, in emission computations, a linear temperature profile is used assuming an ice–water interface temperature of −1°C and an air–ice interface temperature of −10°. The emitted brightness temperatures have been converted to effective emissivities here by dividing by the measured surface ice temperature, consistent with the reduction of the emission observations. Identical input parameters are used for all polarizations, frequencies, and incidence angles in both passive and active calculations.

Dense medium theory predicts only negligible volume scattering within the ice layer for these parameters. Thus

TABLE 8-2. Relative permittivities of materials in the Dense Medium–Integral Equation Model for the CRRELEX gray ice sheet.

Frequency, GHz	Seawater	Brine inclusions	ε_{eff}
6.7	60.4 + i39.4	50.4 + i40.3	3.40 + i0.19
10.0	35.0 + i38.0	33.5 + i38.7	3.35 + i0.17
18.7	18.4 + i30.2	18.4 + i28.2	3.30 + i0.16

predicted emissivities are dominated by emission from the lossy ice layer and backscattering results entirely from rough surface scattering at the upper and lower ice interfaces. Figure 8-9 compares observations with computed effective emissivities for 6.7 and 18.7 GHz (passive signatures at 37 GHz and other frequencies were not computed). Model results agree quite closely with H-polarization observations at 6.7 GHz, and do not show a strong sensitivity to thickness. Computed emissivities at 18.7 GHz are higher than the observations, but the angular trend for V-polarization shows no pronounced peak at large incidence angles, i.e., Brewster-angle effect. This trend, if not the actual predicted emissivities, is similar to that of the observations and notable among the models in this case study. V-polarized emissivities at 6.7 GHz, however, do show a strong Brewster-angle effect. The plot of computed emissivities versus frequency, Figure 8-9(c), does not show the 10 GHz feature present in the observations; though this plot contains model values at only 6.7 and 18.7 GHz, and thus could not show the feature, DMT does not predict such a feature in any case. Figures 8-9(d) and (e) show computed backscattering cross sections at 5 and 10 GHz, respectively. The 5 GHz results match the observations very closely. The predicted backscattering is due mostly to scattering at the ice–water interface. Although the assumed roughness at this interface is fairly small, the large dielectric contrast between ice and water causes scattering strong enough to dominate the overall response, even after accounting for absorption in the ice layer. Results at 10 GHz are computed using identical roughness parameters; these results agree with observations to within approximately 2.5 dB. In this case, scattering from the ice–water interfaces accounts for only about half the total; the rest is due to scattering at the air–ice interface.

8.3.3 Discussion

A few inferences seem reasonably firm, based on the model comparisons above. First, the thin, relatively saline gray ice in this study is not a strong scatterer. Passive signatures are dominantly influenced by the reflection properties of the ice; in none of the emission models does scattering play much of a role in determining the effective emissivity. Backscattering, which must be due entirely to scattering, is relatively weak; this gray ice sample clearly differs in some important way from the strongly backscattering gray ice mentioned in the introduction. Because backscattering is weak, it may be more likely that several weak processes combine to determine what we see.

Several of the present models compute effective emissivities for simple, gray congelation ice that agree with observations to within $\lesssim 0.05$, for both polarizations at 19 and 37 GHz, for incidence angles less than 55°. Both the strong fluctuation theory models, which include finite thickness effects, as well as the (effectively) infinite-thickness modified radiative transfer model achieve this accuracy, at least in our case. On the other hand, emissivities at the lower frequencies (6.7 and 10 GHz) evidently depend not only on ice thickness, but also on the profiles of ice temperature and salinity. We infer this because only many layer SFT treats coherent interactions between waves from several depths within the ice, and only this model reproduces even partly the emissivity feature we observe. However, the sensitivity of the present many layer SFT model may incorrectly imply a greater signature variability than we observe in nature. A partially coherent model, to account for the patchiness of frost flowers, irregularity of surface brine layers, and so on, may be necessary. We note also that none of our model results at 90 GHz show good agreement with observations, suggesting that essential physics remains unaccounted for at millimeter-wave frequencies.

The situation for backscattering depends strongly on frequency as well. At 10 GHz, rough surface scattering from the upper ice surface (assuming effectively infinite ice thickness) correctly predicts the observed difference in like-polarized cross sections, but the cross sections themselves are lower than the observations by amounts that increase with increasing incidence angle. Model predictions fall barely within the estimated range of uncertainty in the observations when we use the independently derived surface roughness parameters corresponding to the roughest surface. Thus surface scattering without finite thickness effects may explain the present observations, though the observations support such an explanation only marginally. As we have noted in Section 8.3.2.5, this result is consistent with the finite-thickness reflectivity variation at 10 GHz implied by the emissivity observations. It may also be interesting to note that using the same parameters in the Gaussian correlation function $\rho(x) = h^2 \exp[-x^2/L^2]$ produces results that agree with the observations to within 0.5 dB. This shift in the form of the correlation function is plausible given the present uncertainty in our roughness characterization, but the data cannot be said to motivate such a shift.

It seems much less likely that uncertainty in the surface roughness characterization can account for the differences between 5 GHz observations and predictions based on backscattering from the air–ice interface alone. While this mechanism again accurately predicts the difference between like-polarized backscattering cross sections, the pre-

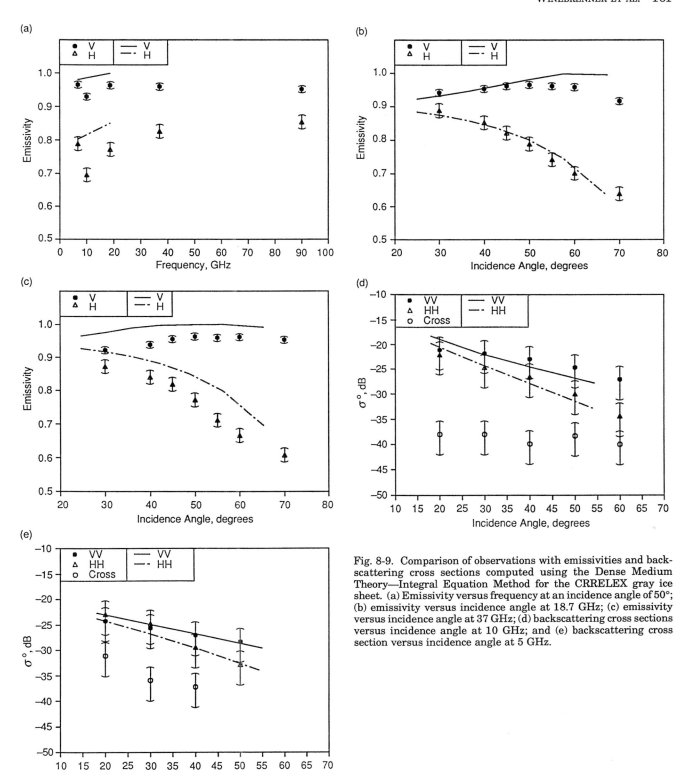

Fig. 8-9. Comparison of observations with emissivities and back-scattering cross sections computed using the Dense Medium Theory—Integral Equation Method for the CRRELEX gray ice sheet. (a) Emissivity versus frequency at an incidence angle of 50°; (b) emissivity versus incidence angle at 18.7 GHz; (c) emissivity versus incidence angle at 37 GHz; (d) backscattering cross sections versus incidence angle at 10 GHz; and (e) backscattering cross section versus incidence angle at 5 GHz.

dicted level of backscattering is well below the observations. Our findings here are similar to those of Bredow and Gogineni [1990] based on C-band observations of this same ice sheet earlier on January 14. All of our models estimate penetration depths on the order of the ice thickness (8 cm) at 5 GHz (wavelength 6 cm). The combined Dense Medium Theory—Integral Equation Method provides a close match to 5 GHz observations based on rough surface scattering at the ice–water interface with plausible parameters. However, we have no quantitative information on under-ice roughness with which to test this explanation, nor do we have observations that could rule out other potential explanations. Examples of the latter might include scattering due to the patchiness of surface brine layers, volume scattering from larger scale inhomogeneities in the ice, or reflectivity variations due to depth variations in ice properties.

8.4 CASE STUDY 2: COLD OLD ICE

8.4.1 Ice Characterization

The data for this case study were acquired from October 3 through 8, 1988, as part of the CEAREX experiment in the Arctic Ocean, north of the Barents Sea. Air temperatures were –16 ±2°C and had been well below freezing for more than two weeks. Snow–ice interface temperatures were –10 ±2°C or less. The ice was completely frozen to depths (from the snow–ice interface) greater than 50 cm [Wettlaufer, 1991]. The old ice floe on which our observations were made seemed in no way remarkable compared with other old ice floes in the area, or in the experience of the investigators present.

The microwave signatures of old ice and the physical properties of its upper layers are both highly variable (Chapters 2, 4, and 5). Raised old ice areas typically display strong backscattering and low brightness temperatures, whereas nearly refrozen melt ponds are often characterized by relatively low backscattering and brightness temperatures similar to those of first-year ice. Very low-density, bubbly upper layers are common to raised areas, whereas refrozen melt ponds are typically much denser. There may be systematic differences in the salinities, roughnesses, and other properties of these two old ice types as well. We have therefore structured this as a dual case study, selecting two old ice sites from the same floe that apparently bracket the extremes of upper layer density as well as signature behavior. For each site, Grenfell acquired vertically and horizontally polarized brightness temperatures at 6.7, 10, 18.7, and 90 GHz, for a range of nadir angles from 30 to 70°. Data at 37 GHz were also acquired at the melt pond site. The data have been reduced using measured sky brightness temperatures and we have again normalized these data by the snow–ice interface temperature to produce effective emissivities, for the same reasons as in the previous case study. Nearly simultaneously with the passive data, Onstott acquired HH-, VV-, and cross-polarized backscattering cross

sections at 10 GHz over a similar range of incidence angles.

Our first site is a raised area of ice known as drift station site 7 (abbreviated DS-7). The uppermost layers of ice at this site consisted of a fragile, geometrically complex matrix of air and ice containing many bubbles and some irregular, interconnected air spaces that we call voids. Beneath this matrix was a layer of ice that, while still bubbly, was more easily described in terms of discrete bubbles embedded in an ice background. Figure 8-10 shows a schematic drawing of ice structure versus depth, as well as profiles of ice salinity and density below 19 cm depth, based on analysis by Gow of a core sample taken from the site. Figure 8-10

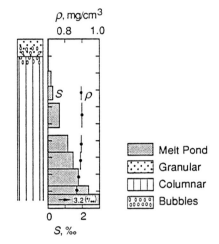

Snow: Layer Thickness = 10 cm
        Consists of Platelets and Rimed Needles
        Platelets 0.1 mm Thick and 0.5 mm Wide
        Needles 0.1 to 0.2 mm Thick and 0.3 to 0.6 mm Long
        Snow Density Approximately 0.1 gm/cm³

Ice: Consists of a 19-cm Decomposed Layer Over Columnar Ice

Density Profile		Salinity Profile	
cm	ρ, mg/m³	cm	‰
		0–19	0.0
19–66	0.895	19–28	0.1
66–77	0.89	28–36	0.3
77–87	0.88	36–38	Break in Core
87–92	0.87	38–53	0.7
		53–57	Break in Core
		57–67	1.2
		67–77	1.4
		77–87	1.9
		87–92	2.4
		92–98	3.0

Fig. 8-10. Structure, density and salinity in the upper ice layers at DS-7, numerical density and salinity profiles and snow cover observations.

also shows numerical salinity and density profiles, and semiquantitative observations by Grenfell of snow cover parameters at the site. Grenfell measured −19, −18, and −10°C for the air, air–snow interface, and snow–ice interface temperatures, respectively, with an estimated error in each case of ±2°.

Coring so badly disrupted the structure of the bubbly ice that quantitative parameters derived from the core cannot be considered reliable at depths less than 19 cm. To circumvent this problem, Onstott extracted larger samples of the upper ice layers using a chain saw. Based on these samples, he reports structure in the uppermost part of the ice consisting of two distinct layers, the first containing a great many bubbles as well as large voids, the second containing fewer, more discrete bubbles and fewer, smaller voids. Figures 8-11(a) and (b) each show two thick sections of each layer. Note that each photo is actually a juxtaposition of two separate photos, one of each layer against a calibrated background. The upper ice layer, in particular, is so riddled with inhomogeneities that it seems uncertain whether it can realistically be described in terms of discrete air bubbles, or even a size distribution of air bubbles, embedded in ice. Nonetheless, Onstott has visually derived

mean bubble sizes, characteristic void dimensions, and layer thickness for each layer, and has measured bulk layer density, salinity, and snow–ice interface roughness as well. He reports a standard deviation for snow–ice interface roughness of $h = 0.14 \pm 0.02$ cm with a correlation length $L = 2.0 \pm 1.3$ cm—assuming, as in the previous case study, a correlation function of the form $\rho(x) = h^2 \exp(-x/L)$. The upper, most porous ice layer is 5.0 ±0.6 cm thick, with salinity 0.0‰ and density 0.457 g/cm^3. Mean bubble diameter was estimated to be 2.5 mm with a characteristic air void dimension of 8 mm. The lower, less porous layer is reported to be 3.5 ±0.4 cm thick. Salinity in this layer was also 0‰, but the density was 0.728 g/cm^3. Mean bubble diameter was 4 mm with a characteristic void dimension of 2 mm. Based on Onstott's and Gow's measurements, the ice density below 8.5 cm depth increased to 0.895 g/cm^3, nearly that of pure ice (0.917 g/cm^3). We do not have measurements of total ice thickness at DS-7. However, we know that thickness was greater than 1.5 m and that ice salinity increased below 1 m to approximately 3‰; thus the lower portions of the ice were electromagnetically lossy, and it is virtually certain that neither the lower ice surface or underlying seawater affected the observed signatures.

HH- and VV-polarized backscattering cross sections for this site differ little; both are high (generally between 0 and −5 dB) and almost independent of incidence angle. Cross-polarized backscattering is also strong (above −20 dB). Effective emissivities decline sharply with increasing frequency and show little polarization dependence. Thus it appears that scattering at this site is indeed quite strong.

Contrast this with our second site, a refrozen melt pond some tens of meters from DS-7, known as the Del Norte melt pond. Backscattering cross sections for this site are much lower than those at DS-7, whereas effective emissivities are considerably higher. Figure 8-12 shows a thick section photograph of a sample acquired at this location by Onstott. (Note that this photo is not a juxtaposition; it shows only a

(a)

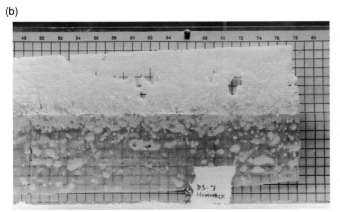

(b)

Fig. 8-11. The bubbly upper ice layers at DS-7. Each photograph is actually a juxtaposition of separate photographs of each of two distinct layers comprising the upper 8.5 cm of the ice. (a) and (b) show distinct samples taken at the site.

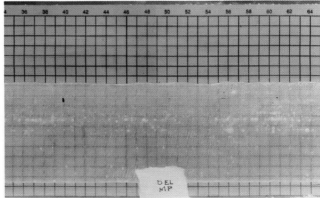

Fig. 8-12. A thick section of the upper layers of the Del Norte melt pond. Unlike Figure 8-11, this is not a composite but rather a single photo. Note the distinct layer of bubbles within the ice.

single layer.) The ice is clear and nearly bubble-free except for a thin, sparse layer of well-defined bubbles a few centimeters below the snow–ice interface. Onstott reports snow–ice interface roughness parameters of $h = 0.8 \pm 0.03$ cm and $L = 2.8 \pm 1.9$ cm, thicknesses of 3.2 ± 0.3 cm for the clear upper layer of ice and 2 cm for the bubbly layer, and a mean bubble diameter in the bubbly layer of 1.3 mm.

Figure 8-13 presents a schematic ice morphology and quantitative salinity and density profiles for this site, again based on Gow's analysis of a core sample. In this case, the ice density is high enough to obtain reliable parameter estimates from the core throughout its length. The

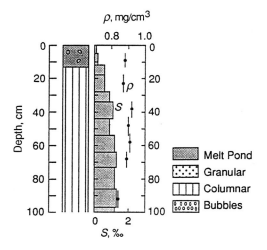

Fig. 8-13. Structure, density and salinity in the upper ice layers at the Del Norte melt pond, numerical density and salinity profiles and snow cover observations.

Snow: Layer Thickness = 8 cm
Consists of Platelets and Rimed Needles
Platelets 0.1 mm Thick and 0.5 mm Wide
Needles 0.1 to 0.2 mm Thick and 0.3 to 0.6 mm Long
Snow Density Approximately 0.1 gm/cm^3

Ice: Has a 12-cm Frozen Pond Layer that is Slightly Bubbly Over Columnar Ice Below

Density Profile		Salinity Profile	
cm	gm/cm^3	cm	‰
0–15	0.875	0–5	0.1
15–30	0.87	5–13	0.2
30–43	0.92	13–28	0.6
43–54	0.90	28–34	0.9
54–64	0.91	34–44	1.1
64–73	0.89	44–54	0.9
		54–64	1.2
		64–73	1.3
		73–86	1.2
		86–97	1.3
		97–100	1.2

upper 12 cm of ice is slightly bubbly with a bulk density of 0.875 g/cm^3, and ice below this layer is columnar. Total ice thickness was also greater than 1.5 m at this site; here again, we expect no effect of the lower ice boundary or seawater on signatures. Figure 8-13 also includes Grenfell's observations of parameters in a thin, light snow layer overlying the melt pond. Grenfell reports air, air–snow interface and snow–ice interface temperatures of –17, –16, and –11°C, respectively, with an estimated error in this case of ±1°C.

8.4.2 Model Comparisons

8.4.2.1. Independent Rayleigh scattering layers—physical optics. Application of this model at both sites is based on a physical model consisting of two bubbly ice layers overlain by dry snow. Volume scattering in the snow is negligible; the snow thus acts only as a layer of intermediate permittivity between the air above and ice below. This slightly increases transmission into the ice and reduces scattering at the snow–ice interface. Only the reduction in surface scattering has a noticeable effect in backscattering. Since this model does not compute emission, this is the sole effect of the snow layer in this section. Air bubbles in the two ice layers are modeled as Rayleigh-scattering spheres with a distribution of radii. Because we have no independent estimates of bubble size distributions, assumptions are made to fix the form of the bubble size distribution input to the scattering model. The form of the distribution is Gaussian with sharp truncations; the standard deviation is set arbitrarily at 50% of the mean radius and the resulting distribution is truncated on both sides at two standard deviations from the mean. The mean radius varies between sites and between layers at each site.

The characterization data indicate minimal salinity in the upper ice layers. The relative permittivity of ice between bubbles is therefore set at $3.14 + i0.01$, for all sites and all layers. As noted in Section 8.2.2.2, an effective permittivity for each bubbly ice layer is then computed according to a Polder–van Santen type formula and the measured layer density. The air bubbles are assumed to reside in an effective background medium with this effective permittivity for purposes of computing their scattering cross sections. Thus a given bubble scatters less, according to this model, when situated in a low-density ice layer than in a higher density layer because the effective dielectric contrast is lower in the latter case.

Layer thicknesses in the model are set equal to the measured mean layer thicknesses. Rough surface scattering is assumed negligible except at the air–snow and snow–ice interfaces; only scattering at the snow–ice interface significantly affects results at incidence angles greater than 20°. Roughness statistics for this latter interface are also drawn directly from the characterization measurements.

Figure 8-14(a) compares multipolarization observations with computed, HH-polarized backscattering cross sections at 10 GHz for the site DS-7 (recall that the model does not compute cross-polarized backscattering, and that its predictions for like-polarization differ chiefly when surface scattering is significant). According to the model, backscattering at this site is dominated by volume scattering from the two bubbly ice layers; surface scattering is insignificant for incidence angles between 20 and 60°. The lower, denser, bubbly layer at DS-7 contributes most of the volume scattering in this model. This results because the permittivity of bubbles in the lower layer contrasts more strongly with the density-dependent effective background permittivity than in the lower density layer above. The predominant model sensitivity is therefore to parameters in the lower ice layer. Figure 8-14(a) shows results computed using three mean

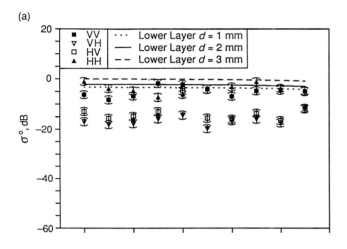

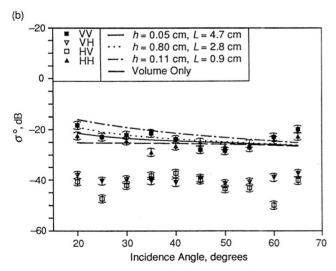

Fig. 8-14. Comparison of 10-GHz backscattering observations from DS-7 and the Del Norte melt pond with model predictions based on the independent Rayleigh scattering layer model of Section 8.2.2.2 for like-polarized backscattering cross sections at 10 GHz at (a) DS-7 for different assumed bubble sizes in the lower scatter layer, and (b) the Del Norte melt pond.

bubble diameters (but holding bulk density constant) in the lower layer. The bubble diameter of 4 mm in the lower layer is the mean diameter measured independently. Mean bubble diameter in the upper layer was set at 4.5 mm. The results span a range of approximately 3 dB for a 50% variation in mean bubble size (and, by way of the assumptions in the physical model, the same variation in the width of the size distribution). Results for the independently reported mean bubble size and for the smaller bubble size agree well with the HH-polarized observations.

Model results also compare well with observations at the melt pond site. Figure 8-14(b) shows four model curves for the 10 GHz, HH-polarized backscattering cross section as a function of incidence angle. The four curves at this site show the effects of different assumptions for snow–ice interface roughness. The lowest curve results from volume scattering alone (from the thin layer of bubbles within the melt pond), and falls approximately 2–5 dB below the observations. The upper three curves correspond to the smoothest, most probable, and roughest surfaces consistent with independent characterization. The middle curve, corresponding to the most probable roughness parameters, agrees notably well with the observations. Note that these curves are for HH-polarization; computed VV-polarization curves are similar but slightly lower at incidence angles less than 35°, where surface scattering plays some role (Sections 8.2.2.2 and 8.3.2.2).

8.4.2.2. Dense medium radiative transfer. The application of DMRT to this case is based on an ice model of two scattering layers overlying a homogeneous, nonscattering basement. The (scalar) permittivity of the basement is computed from the equations of Frankenstein and Garner [1967] and Vant et al. [1978] for moderately saline ice (3‰) at a temperature of –9°C, consistent with the characterization information. DMRT presently assumes planar layer interfaces; the snow–ice interface roughness is therefore neglected. The dry snow layer reported in the characterization data has at most a negligible effect on backscattering, though perhaps not on emission. A two-layer DMRT model for emission is presently under development but is not fully operational as of this writing. The DMRT results presented here are therefore limited to backscattering and this in turn permits neglect of the snow layer.

The densities in each layer, which determine the total volume fractions occupied by scatterers, are set equal to the independently measured layer densities. The scatterers in each layer are assumed spherical, Rayleigh-scattering air bubbles embedded in a background of ice. Because the characterization data specify only mean bubble size, it is desirable to employ a plausible single-parameter distribution of bubble radii in the ice model. A truncated Rayleigh distribution is employed in this case. (Note that use of a Rayleigh distribution is separate from the Rayleigh-scattering assumption.) The mode of the distribution is set equal to the reported mean bubble radius; this fixes the

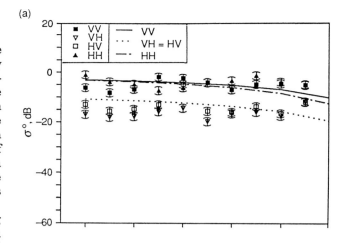

width of the distribution as well. The largest bubbles in the distribution are relatively few, but they scatter strongly because Rayleigh scattering increases rapidly with increasing particle size. If appreciable numbers of bubbles are present that because of their size act as Mie rather than Rayleigh scatterers, the total amount of scattering may be seriously overpredicted. The Rayleigh size distribution in this model is therefore truncated at an upper value of bubble diameter for which Rayleigh scattering remains a reasonable approximation, at least in view of the relative abundance and fraction of total scattering from bubbles near this size.

Figures 8-15(a) and (b) show computed backscattering cross sections at 10 GHz for site DS-7, based on densities in the upper and lower ice layers of 0.457 g/cm^3 and 0.728 g/cm^3, respectively, and upper and lower layer bubble mode diameters of 2.5 mm and 2 mm, respectively (the latter diameter was set at the reported mean void diameter for the lower layer—the assumed size distribution then includes substantial numbers of bubbles with the mean reported bubble size). The salinity of the background ice is taken to be 0‰. The curves in Figure 8-15(a) result from using a cutoff of 2 cm diameter in the bubble size distribution, while those in Figure 8-15(b) result from a cutoff of 0.96 cm. The former cutoff is a slightly liberal size limit for the Rayleigh size distribution (in our experience), whereas the latter cutoff is a slightly conservative limit. The like-polarization results in Figure 8-15(a) agree well with the observations, except perhaps at incidence angles greater than 55° where the observations seem to show a nonphysical upturn. The cross-polarized cross sections from DMRT, however, seem approximately 4 to 5 dB higher than those observed. Adopting the more conservative upper limit on bubble sizes, Figure 8-15(b), leads to like-polarized results approximately 1 to 3 dB lower than observed while the corresponding cross-polarized results agree with the data very closely.

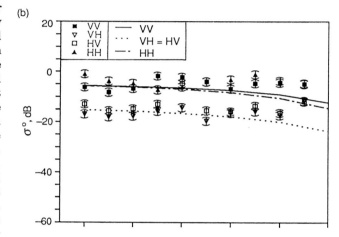

Figure 8-15(c) shows corresponding results for the Del Norte melt pond. The physical model in this case consists of a 2 cm thick layer containing bubbles with mode diameter 1.3 mm, overlain by a 3.2 cm thick layer containing very small bubbles (mode diameter 0.1 mm). The bubble sizes in this case are small enough to make cutoffs unnecessary. The density of both layers is 0.87 g/cm^3, and the relative permittivity of background ice is set in this case to 3.2 + i0.02 to reflect the slight ice salinity reported at this site. Model results agree with the like-polarized observations to within 3 dB or less for angles between 25° and 55°. The upturn in the observations at 20° is probably due to rough surface scattering, but this model indicates that the dominant mechanism in the observed return at larger angles remains volume scattering. Note, however, that the DMRT cross-polarized predictions at this site fall approximately 10 dB below the observations.

The strongest input parameter sensitivities in this model are to mean (or mode) bubble size. For size distributions other than Rayleigh, parameters controlling the relative

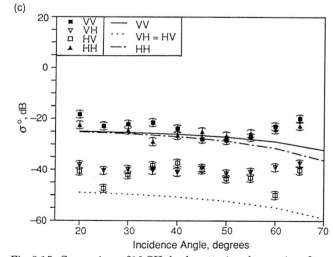

Fig. 8-15. Comparison of 10 GHz backscattering observations from DS-7 and the Del Norte melt pond with model predictions based on Dense Medium Radiative Transfer theory. Backscattering cross sections at 10 GHz at DS-7 for (a) 2-cm and (b) 0.90-cm truncations of the bubble size distribution in the lower bubbly layer, and (c) like-polarized backscattering cross sections at 10 GHz at the Del Norte melt pond.

abundance of larger radius particles are also sensitive model inputs. For example, varying the mode diameter of bubble sizes between approximately 1.1 and 1.5 mm in the bubbly melt pond layer above causes the VV-polarized cross section to vary by approximately 7 dB and the cross-polarized cross section to vary by 12 dB. A smaller sensitivity to salinity (i.e., to the imaginary part of the background ice permittivity), and much smaller sensitivities to layer thickness and density are also present in the model.

It is interesting that the independent scattering model in Section 8.4.2.1 predicts less volume scattering for the Del Norte melt pond site than does DMRT. This may seem contrary to our earlier statement that, for given input parameters, DMRT predicts less scattering than independent scattering. However, the physical models of this section and the last differ in their bubble size distributions, effective extinction coefficients, and other parameters. The large bubbles in the Rayleigh size distribution scatter disproportionately to their abundance, raising the level of scattering in the DMRT model. The independent scattering model includes a nonclassical decrease in scattering with decreasing ice density due to its decreased dielectric contrast between each bubble and the background ice. Thus the comparison between these two models is not simple. Comparison of DMRT and the model of Section 8.4.2.1 over a range of densities would give a clearer picture of the differences between them.

8.4.2.3. Dense medium theory—integral equation method. The physical model assumed in DMT-IEM presently includes only a single scattering layer. Thus application of DMT-IEM to the DS-7 and Del Norte sites requires the derivation of effective parameters for single scattering layers from the characterization data.

At DS-7, the physical model for the ice consists of an 8-cm thick layer of low-salinity ice (relative permittivity 3.3 + i0.003) containing single-sized, spherical, air bubbles of diameter 4 mm. The volume fraction of air bubbles is 40%. The relative permittivity of the underlying ice is the same as the effective permittivity of the bubbly ice, 2.2 + i0.01. Both the upper (snow–ice) and lower (ice–ice) interfaces are assumed rough with roughness parameters $h = 0.14$ cm, $L = 2.0$ cm. The snow layer is neglected. Model results for like-polarized, 10 GHz backscattering cross sections, shown in Figure 8-16(a), agree with observations to within approximately 3 dB for incidence angles less than 55°. Cross-polarized backscattering is not computed. The model predicts that volume scattering completely dominates rough surface scattering at DS-7 for incidence angles greater than 30°. Accordingly, the most sensitive model input parameter in this case is mean bubble size.

At the melt pond site, the physical model consists of a 6-cm thick layer containing single-sized, 1.5 mm diameter air bubbles; the air bubbles occupy 5% of the layer volume. The relative permittivities of the background ice and ice under the layer are both 3.1 + i0.01. The snow layer is again neglected, and identical roughnesses for the snow–ice and

ice–ice interfaces are again assumed, with parameters $h = 0.11$ cm, $L = 4.3$ cm. Figure 8-16(b) shows the resulting cross sections and comparison with observations; agreement is again good. At this site, DMT-IEM indicates a more important role for scattering from the snow–ice interface (because the permittivities of the ice layer background and underlying ice match, scattering from the lower layer interface is negligible).

It may seem puzzling that a model which predicts more scattering than independent scattering agrees approximately as well with like-polarized observations as do DMRT (which predicts less scattering than independent scattering) and the independent scattering model of Section 8.2.2.2. Note that the physical model used by DMT assumes a single bubble size set at the independently estimated mean size,

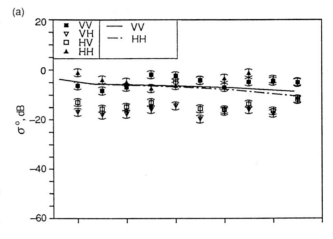

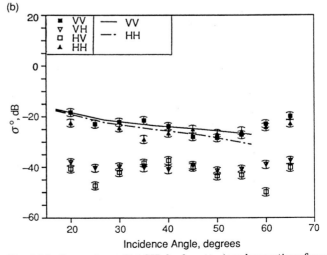

Fig. 8-16. Comparison of 10 GHz backscattering observations from DS-7 and the Del Norte melt pond with model predictions based on the Dense Medium Theory—Integral Equation Method. (a) Back-scattering cross sections at 10 GHz at DS-7, and (b) like-polarized backscattering cross sections at 10 GHz at the Del Norte melt pond.

whereas DMRT and the independent scattering model both assume size distributions with significant numbers of bubbles larger than the mean size. The larger bubbles scatter out of proportion to their relative abundance in the size wavelength (i.e., Rayleigh scattering) regime in this case study. Thus, DMT predicts roughly the same level of backscattering as the other two models, because it predicts greater scattering from smaller bubbles and does not assume the presence of larger bubbles.

8.4.2.4. Many layer strong fluctuation theory. The physical model for ice built into many layer SFT (bubbles only on the boundaries of 1 cm diameter ice crystals) makes treatment of ice like that in Figure 8-10 problematic; the assumptions connecting ice salinity, density, and so on to permittivity correlation functions seem unsuited to ice so different from congelation ice. The application of many layer SFT to site DS-7 therefore treats the upper, very low density ice layers as layers of relatively large, spherical ice particles embedded in air. Beneath the 10-cm, low-density snow layer, the model employs a snow-like layer of density 0.460 g/cm^3 consisting of 5-mm diameter ice particles. Beneath this is a second layer of 3-mm diameter ice particles with density 0.728 g/cm^3. Underlying these layers is high-density congelation ice with salinities given by those in the measured profile. A model snow cover with parameters set at the independently estimated values covers the ice. Figure 8-17(a) shows the computed effective emissivities for DS-7 as functions of frequency at incidence angle 50°. The model results roughly reproduce the drop in emissivities between 6.7 and 18.7 GHz, but then rise again with increasing frequency, contrary to observations. The relatively good agreement between the model and observations at 18.7 GHz carries over at all incidence angles, Figure 8-17(b). Computed like-polarized backscattering cross sections at 10 GHz, Figure 8-17(c), are also in good agreement with observation, but predicted cross polarized cross sections are well below the observations.

The situation is somewhat improved for the melt pond site. The physical model for the ice at this site is more nearly suited to the assumptions about ice morphology in many layer SFT. The melt pond site is modeled as very low salinity congelation ice with 1.2 mm diameter bubbles. The number density of bubbles is determined by bulk density, for each of five layers within the upper 54 cm of ice. Layer densities and salinities are set directly from the measured profiles. Figure 8-18 compares the results with observations. The 50° emissivity spectrum, Figure 8-18(a), compares well with observations at 18.7 and 37 GHz, though disagreements arise at 90 GHz and especially at the lower frequencies. The angular emissivity responses at 18.7 and 37 GHz, Figures 8-18(b) and (c), also agree reasonably well with observation, though the H-polarized emissivities oscillate as functions of angle due to coherent interactions from various ice layers to a greater degree than do the observations. The model is quite sensitive to details of the salinity

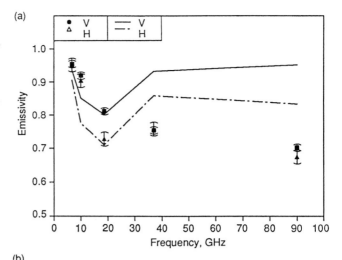

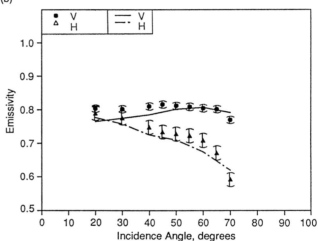

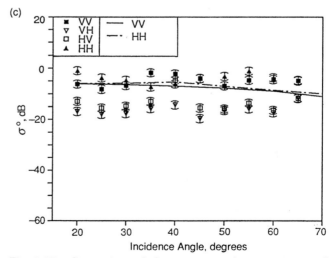

Fig. 8-17. Comparison of observations with emissivities and 10 GHz backscattering cross sections computed using the many layer SFT model for DS-7. (a) Emissivity versus frequency at an incidence angle of 50°; (b) emissivity versus incidence angle at 18.7 GHz; and (c) backscattering cross sections versus incidence angle at 10 GHz.

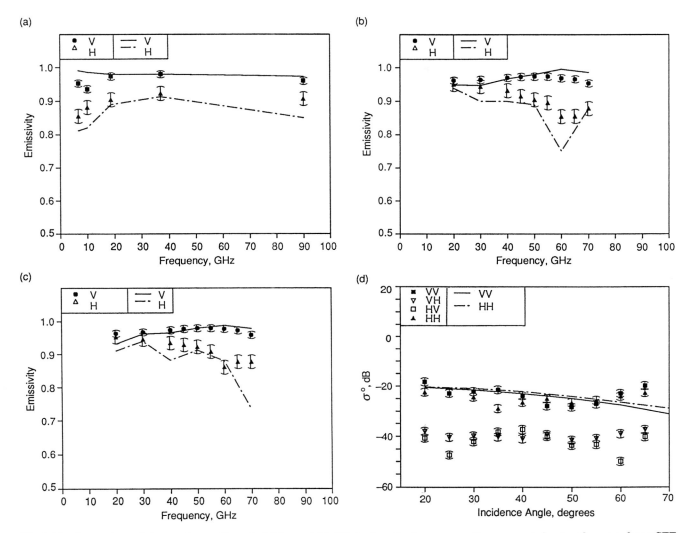

Fig. 8-18. Comparison of observations with emissivities and 10 GHz backscattering cross sections computed using the many layer SFT model for the Del Norte melt pond. (a) Emissivity versus frequency at an incidence angle of 50°; (b) emissivity versus frequency at 18.7 GHz; (c) emissivity versus frequency at 37 GHz; and (d) backscattering cross sections versus incidence angle at 10 GHz.

profile and the snow depth, suggesting that coherent effects may be overstated in the model. Note that the model has accurately tracked the change in like-polarized backscattering cross sections between DS-7 and the melt pond, Figure 8-18(d).

8.4.2.5. Modified radiative transfer. The physical model for MRT in this case consists of a uniform, infinitely thick scattering layer with a slight amount of azimuthal anisotropy, i.e., with a preferred direction tilted slightly off vertical and oriented in a particular way relative to the radar look direction. This physical model results in cross-polarized backscattering even with the current, first-order solution of MRT.

The dielectric parameters for site DS-7 are mean relative permittivities of 2.0 + i0.001 and 2.1 + i0.002 in directions

orthogonal to and parallel to the preferred direction, respectively. The preferred direction is tilted 4° from vertical and oriented 45° from the radar/radiometer look direction. The normalized variances of permittivity are 0.2 and 0.25 in the orthogonal and parallel directions, respectively; the corresponding permittivity correlation lengths are 2 mm and 2.5 mm, respectively. These parameters lead to the results in Figures 8-19(a) and (b) for backscattering and emission. Like- and cross-polarized backscattering fall 7 dB or more below observations for most angles, while computed emissivities show both levels and trends at variance with observation.

Dielectric parameters for the melt pond site are set at 2.8 + i 0.004 and 2.9 + i0.005 for orthogonal and parallel mean permittivities, respectively. Normalized variances are 0.17 orthogonal and 0.20 parallel; permittivity correlation

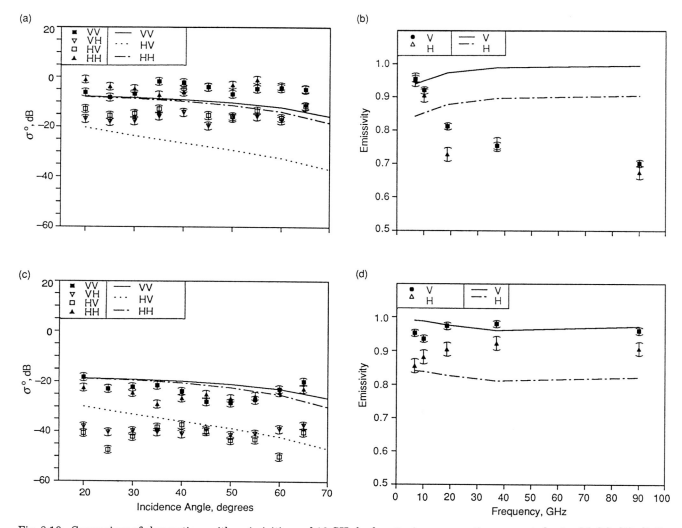

Fig. 8-19. Comparison of observations with emissivities and 10 GHz backscattering cross sections computed using Modified Radiative Transfer for DS-7 and the Del Norte melt pond. (a) 10-GHz backscattering cross sections versus incidence angle at DS-7; (b) emissivity versus frequency at an incidence angle of 50° at DS-7; (c) 10-GHz backscattering cross sections versus incidence angle at the Del Norte melt pond; and (d) emissivity versus frequency at an incidence angle of 50° at the Del Norte melt pond.

lengths are 0.3 and 0.4 mm orthogonal and parallel, respectively. The orientation of the preferred direction is the same as for site DS-7. Figures 8-19(c) and (d) show the results for backscattering and emission. Backscattering cross sections are generally above the data while computed emissivities show much more polarization difference than is observed.

Model sensitivity is again greatest for variations in permittivity correlation lengths. For small tilts of the preferred direction, such as that employed here, neither scattering cross sections nor emissivities show strong sensitivities to the azimuthal orientation. Additional model sensitivity information is provided by Lee and Mudaliar [1988] and Mudaliar and Lee [1990].

8.4.3 Discussion

DS-7 appears to be the opposite extreme to the gray ice in our first case study, not only in terms of ice age, but also of its strength as a scatterer. Though only two of the models we apply treat emission, both of these predict strong scattering effects in both backscattering and emission. Four of the models produce backscattering results that agree with like-polarized observations to within 3 dB at DS-7, and volume scattering is the dominant scattering mechanism according to all four. This is encouraging, and may be surprising given the difficulty of describing real ice at DS-7 in the physically idealized terms required by the models. However, the four models offer considerably different accounts of the physics behind the signatures. Clearly, differing assumptions about the physical model for the scattering medium and approximations in scattering physics can combine to produce results that are indistinguishable at a given single frequency, especially in the absence of bubble size distribution information. We expect that stringent multifrequency, multipolarization tests including characterization of bubble size distributions would distinguish which of the accounts is closer to reality. The strong scattering situation evidently makes accurate computation of emissivities difficult; neither of the models we applied at this site accurately explained passive observations.

Backscattering cross sections for the Del Norte melt pond are comparable to those of the gray ice in our first case study, though they fall off less rapidly with increasing incidence angle and show less polarization dependence. The melt pond surface has larger height variations but smaller surface slopes than the gray ice; of course the melt pond also has lower salinity and larger volume scatterers. Two of the models we apply at the Del Norte site indicate that rough surface scattering and volume scattering both contribute significantly to 10 GHz backscattering in the midrange of incidence angles. DMRT and many layer SFT account successfully for like-polarized backscattering in terms of volume scattering alone for incidence angles greater than 30°. The cross-polarization results from DMRT are lower than observations, but model sensitivities are such that a minor change in mean bubble size could bring them into near agreement. Again, the four models differ substantially in their accounts of the volume scattering physics behind the observations, but they agree that volume scattering is an essential part of backscattering from even a dense melt pond. Together with the DS-7 results, this indicates that volume scattering from bubbles is the predominant backscattering mechanism for nearly all old ice at 10 GHz.

The effect of scattering on emission from melt ponds remains unclear. Many layer SFT predicted accurate like-polarized cross sections based on volume scattering alone, but 19 and 37 GHz emissivities from this model showed an oversensitivity to depth variation in model parameters. This suggests the possibility that scattering at these two frequencies is weak enough for reflectivity to again control emissivity at this site. Emission calculations using DMRT and DMT-IEM, which were not available at the time of this writing but should soon be possible, should help in understanding this issue.

8.5 CONCLUSIONS

We set out in this chapter to clarify the state-of-the-art in ice signature modeling. To this end, our case studies have yielded several pertinent results.

We are able to compute emissivities for our smooth, bare gray ice sample accurate to within less than 0.05 for frequencies of 19 and 37 GHz and incidence angles less than 55°. This statement is based on results from several models, including polarimetric SFT, many layer SFT, and Modified Radiative Transfer. Ice 8 cm thick was, in our example, effectively infinitely thick at 19 and 37 GHz. The essential physics according to each of these models is the connection between the emission and the reflectivity properties of the ice, and thus the mean dielectric properties of the upper few centimeters of ice; emission from this ice type was modified only minimally by scattering. More accurate emissivity calculations for this ice will therefore depend on improved treatment of near-surface dielectric variations; note, however, that this conclusion may not hold for the type of strongly scattering gray ice reported in some radar observations (Chapters 5 and 14). At 6.7 and 10 GHz, we observe a distinct feature in the frequency dependence of emission. Emissivities computed using the many layer SFT model agree with observations to within 0.04. According to this model, the essential signature physics is coherent interaction between waves from various depths in the ice; the interaction is modulated by ice thickness and differing salinity profiles (i.e., differing growth histories). This frequency-dependent emissivity feature may prove valuable in sensing gray ice thickness from aircraft (or from spacecraft if the spatial resolution of radiometers can be improved sufficiently), provided we can learn to interpret the signatures correctly. The present model ignores thickness variations with the sensor footprint, however, and may overestimate signature sensitivity to frost flower and surface brine layers; this requires further investigation.

Backscattering from our gray ice sample at 10 GHz may be explicable in terms of scattering from the rough air–ice interface alone, but support for this explanation from our comparison is not strong. Conventional first-order perturbation theory correctly predicts the difference between like-polarized backscattering cross sections, but the cross sections themselves fall only barely within the estimated range of uncertainty below the observations. Uncertainty in surface roughness characterization may play a role. However, scattering from the air–ice interface alone is clearly insufficient to explain the observed backscattering level at 5 GHz. Neither did volume scattering models in our study reproduce the polarization and incidence angle dependencies observed for gray ice backscattering. The Dense Medium Model—Integral Equation Method provides a close fit to both 5 and 10 GHz like-polarized backscattering observations based on scattering from roughness at an abrupt ice–water interface in addition to the air–ice interface. All our models indicate considerable 5 GHz penetration to the lower part of an 8 cm gray ice sheet, making this explanation plausible. However, we unfortunately have no quantitative characterization data for the lower ice surface with which to test it. There are other presently plausible mechanisms which also cannot be ruled out on the basis of our data. Such mechanisms may figure in the anomalous brightness of some gray ice in 5 and 10 GHz SAR images (Chapter 5). On the other hand, the results of Ulander et al. [1992] indicate that scattering from the air–ice interface may indeed dominate 5 GHz backscattering from thicker first-year ice. The operational importance of 5 GHz SAR's for sea ice observation strongly motivates further investigation of these issues.

Four of our models computed 10 GHz, like-polarized cross sections for raised, bubbly old ice and for a melt pond that agree with observations to within 3 dB, using input parameters reasonably derived from independent ice characterization. One model (DMRT) also produced cross-polarized cross sections that agree with observations to within 5 dB at the bubbly ice site. The strongest input parameter sensitivities in each model are to bubble size parameters, but the natural variability in ice density is so large that it drives the cross section variation between sites in our study. A quantitative link between cross sections and density suggests that the variance of cross sections in an old floe is perhaps (or perhaps not, see below) related to the variance of ice density on the floe. The latter variance is evidently related to refrozen melt pond coverage, a parameter of geophysical interest. Thus, our results suggest a possible avenue of inquiry for remote sensing development. However, the four models otherwise offer considerably different accounts of the relevant scattering physics. Our data are presently not sufficient to determine which version is closer to reality. In contrast to backscattering, we find emission at the strongly scattering site difficult to model. This finding, together with our gray ice results, suggests that perhaps backscattering is more easily computed when scattering is very strong while accurate emission computations are more feasible when scattering is weak.

Several caveats are worth considering when applying our results or reasoning from them. First, our case studies examine only individual examples of three broad types of sea ice, and there exist few other such case studies in the literature. Moreover, we were compelled to focus on small areas of ice that could be characterized sufficiently to support our model comparison. We can say little about the ranges of natural variability for each type, and we cannot rule out the possibility that any given example constitutes an extreme within the range for its type. Pending replication, our results should therefore be considered provisional. Second, our case studies address only two relatively simple kinds of sea ice in a menagerie of ice types and conditions. Our work cannot support sweeping generalizations about scattering mechanisms in broad ice types or our present signature modeling abilities for those types. Finally, our comments on model sensitivities may or may not bear on signature variability for a given ice type when scattering dominates the signature. Present signature models compute a mean signature, i.e., a signature averaged over random medium realizations drawn from a single ensemble (with given means and variances of ice properties). Changing input parameters amounts to choosing a different ensemble, not just a different realization; thus, examining the mean signature variation for varying input parameters may not give an accurate picture of signature variance for different realizations drawn from the same ensemble.

Based on our study (and its limitations), we can draw a few conclusions about the state of microwave sea ice signature modeling and new work that would contribute to remote sensing. We now possess models to treat a variety of potentially significant effects and processes. However, the sophistication of our models exceeds our sophistication in knowing which model to apply when and why. Fundamental questions remain as to which effects must be treated to compute specified signatures for specified ice types and conditions. This problem can be addressed by further studies like those of Ulander et al. [1992] and ours covering a wider range of ice types and conditions. Good ice and snow characterization is central to this effort, and improved methods of estimating parameters, especially scatterer size distributions and ice surface roughness, are sorely needed. Further studies are more likely to edify than confuse if they begin with simpler conditions and move toward the more complex. To benefit remote sensing more immediately, the selected cases must also involve ice types and situations that play significant geophysical roles.

However, there is a gap between focussed studies at points on the ice and interpretation of data from airborne and spaceborne sensors. The spatial resolution of these sensors is typically larger than the areas we can characterize intensively enough for rigorous model tests. We presently know little about the magnitudes and length scales of inherent, natural variability in ice properties, either those that control signatures or, in some cases, those of geophysical significance. Lacking this understanding, it is problematic to link mean signatures for a small but sufficiently characterized ice region with signatures measured by sensors that average over some (perhaps much) larger area. This is, at the very least, a key validation problem for any signature model or remote sensing algorithm. It requires that we work out sampling strategies for characterization measurements both within single sensor resolution cells and for sufficient numbers of cells within a swath or scene. Sampling strategies must be based on better information about the horizontal length scales of both ice and signature variabilities as well as signature model sensitivities. The Canadian Sea Ice Monitoring Site (SIMS) experiment [Barber et al., 1991] is a valuable effort in this direction.

Better information alone might suffice if the problem were one of validation alone, but this may not always be the case. In virtually all signature models, we envision a slab of ice that is homogeneous, at least in a statistical sense, in the horizontal. If the sensor footprint covers an inhomogeneous area of ice, we assume, often only implicitly, that the actual ice can be replaced by some homogeneous slab, the signature of which we compute using model inputs derived from actual ice parameters. In remote sensing, we equate the signature we measure with that of the slab and, using a link between slab properties and slab signatures, infer some effective geophysical parameter for the actual, horizontally inhomogeneous ice. We assume the effective parameter has some simple relation (usually equality) to the actual geophysical parameter, appropriately averaged over the footprint. These are significant assumptions. Our guidelines as to when they may break down are mostly intuitive. Failure of these assumptions could fundamentally limit retrieval of some geophysical parameters; ways might also be found of exploiting such a failure. The issues here bear at least a superficial similarity to validation and parameterization problems in large- or mesoscale geophysical models; the relevant physical processes there often occur on scales much smaller than the model grid size (see, for example, Wettlaufer [1991]). In any case, a quantitative description of this process seems possible and should be undertaken. The problem is also amenable to ground-based experimental investigation; this work could begin modestly and might involve only analyzing existing data in a new way.

Finally, the benefits of signature modeling have so far scarcely reached operational sea ice remote sensing. We have made little use of the expanded information inherent in time series of observations, but this avenue looks very promising (Chapter 24). It is essential that we begin developing methods to estimate geophysical sea ice parameters based on what we understand now. Initial algorithms are likely to be limited in applicability and less accurate than may be desired, but a starting place is necessary before refinements can begin. Moreover, even imperfect results may show unexpected and valuable spatial or temporal patterns in geophysical variables; this is precisely the situation, for example, in the remote sensing of sea surface winds using microwave scatterometry [Freilich and Chelton, 1986]. The key thing is to begin. As we enter an era of routine floods of remote sensing data, the value of physical signature models in geophysical data interpretation must be made manifest.

REFERENCES

Arcone, S. A., A. J. Gow, and S. McGrew, Structure and dielectric properties at 4.8 and 9.5 GHz of saline ice, *Journal of Geophysical Research, 91(C12)*, pp. 14,281–14,303, 1986.

Attema, E. P. W. and F. T. Ulaby, Vegetation modelled as a water cloud, *Radio Science, 13*, pp. 357–364, 1978.

Barber, D. G., D. D. Johnson, and E. F. LeDrew, Measuring climatic state variables from SAR images of sea ice: The SIMS SAR validation site in Lancaster Sound, *Arctic, 44, Supplement 1*, pp. 108–121, 1991.

Blinn, J. C., III, J. E. Conel, and J. G. Quade, Microwave emission from geological materials: Observations of interference effects, *Journal of Geophysical Research, 77*, pp. 4366–4378, 1972.

Bogorodskii, V. V. and G. P. Khokhlov, Anisotropy of the microwave dielectric constant and absorption of Arctic drift ice, *Soviet Phys.—Tech. Phys., 22(6)*, pp. 747–749, 1977.

Bredow, J. W. and S. P. Gogineni, Comparison of measurements and theory for bare and snow-covered saline ice, *IEEE Transactions on Geoscience and Remote Sensing, GE-28(4)*, pp. 456–463, 1990.

Chen, K. S. and A. K. Fung, An iterative approach to surface scattering simulation, *Proceedings of the 1990 International Geoscience and Remote Sensing Symposium*, pp. 405–408, European Space Agency, University of Maryland, College Park, Maryland, 1990.

Chen, M. F., K. S. Chen, and A. K. Fung, A study of the validity of the integral equation model by moment method simulation—cylindrical case, *Remote Sensing of the Environment, 29(3)*, pp. 217–228, 1989.

Dashen, R. and D. Wurmser, Approximate representations of the scattering amplitude, *Journal of Mathematical Physics, 32(4)*, pp. 986–996, 1991.

Davis, R. E., J. Dozier, and A. T. C. Chang, Snow property measurements correlative to microwave emission at 35 GHz, *IEEE Transactions on Geoscience and Remote Sensing, GE-25(6)*, pp. 751–757, 1987.

Drinkwater, M. R., *Radar Altimetric Studies of Polar Ice*, Ph.D. dissertation, Scott Polar Research Institute, University of Cambridge, Cambridge, England, 231 pp., 1987.

Drinkwater, M. R., LIMEX'87 ice surface characteristics: Implications for C-band SAR backscatter signatures, *IEEE Transactions on Geoscience and Remote Sensing, GE-27(5)*, pp. 501–513, 1989.

Drinkwater, M. R. and G. B. Crocker, Modelling changes in the dielectric and scattering properties of young snow-covered sea ice at GHz frequencies, *Journal of Glaciology, 34(118)*, pp. 274–282, 1988.

Eom, H. J., *Theoretical Scatter and Emission Models for Microwave Remote Sensing*, Ph.D. dissertation, University of Kansas, Lawrence, Kansas, 1982.

Frankenstein, G. and R. Garner, Equations for determining the brine volume of sea ice from −0.5 to −22.9°C, *Journal of Glaciology, 6*, pp. 943–944, 1967.

Freilich, M. H. and D. B. Chelton, Wavenumber spectra of pacific winds measured by the Seasat scatterometer, *Journal of Physical Oceanography, 16*, pp. 741–757, 1986.

Fung, A. K. and H. J. Eom, Application of a combined rough surface and volume scattering theory to sea ice and snow backscatter, *IEEE Transactions on Geoscience and Remote Sensing, GE-20(4)*, pp. 528–536, 1982.

Fung, A. K. and H. J. Eom, A study of backscattering and emission from closely packed inhomogeneous media, *IEEE Transactions on Geoscience and Remote Sensing, GE-23(5)*, pp. 761–767, 1985.

Fung, A. K. and G. W. Pan, A scattering model for perfectly conducting random surfaces: I. Model development, *International Journal of Remote Sensing, 8(11)*, pp. 1579–1593, 1987a.

Fung, A. K. and G. W. Pan, A scattering model for perfectly conducting random surfaces: II. Range of validity, *International Journal of Remote Sensing, 8(11)*, pp. 1594–1605, 1987b.

Fung, A. K., Z. Li, and K. S. Chen, Backscattering from a randomly rough dielectric surface, *IEEE Transactions on Geoscience and Remote Sensing*, in press, 1991.

Golden, K. M. and S. F. Ackley, Modeling of anisotropic electromagnetic reflection from sea ice, *Journal of Geophysical Research, 86(C9)*, pp. 8107–8116, 1981.

Grenfell, T. C., Surface-based passive microwave studies of multiyear sea ice, *Journal of Geophysical Research, 97(C3), pp.* 3485–3501, 1992.

Ishimaru, A., *Wave Propagation and Scattering in Random Media*, Academic Press, New York, 1978.

Ishimaru, A., J. S. Chen, P. Phu, and K. Yoshitomi, Numerical, analytical, and experimental studies of scattering from very rough surfaces and backscattering enchancement, *Waves in Random Media, 1*, pp. S91–S107, 1991.

Jackson, D. R., D. P. Winebrenner, and A. Ishimaru, Comparison of perturbation theories for rough-surface scattering, Journal of the Acoustical *Society of America, 83(3)*, pp. 961–969, 1988.

Jin, Y.–Q. and J. A. Kong, Strong fluctuation theory for scattering, attenuation and transmission of microwaves through snowfall, *IEEE Transactions on Geoscience and Remote Sensing, GE-23(5)*, pp. 754–760, 1985.

Jin, Y.–Q. and M. Lax, Backscattering enhancement from a randomly rough surface, *Physical Review B, 42(16)*, pp. 9819–9829, 1990.

Kim, Y.–S., R. G. Onstott, and R. K. Moore, The effect of a snow cover on microwave backscatter from sea ice, *IEEE Journal of Oceanic Engineering, OE-9(5)*, pp. 383–388, 1984a.

Kim, Y.–S., R. K. Moore, and R. G. Onstott, *Theoretical and Experimental Study of Radar Backscatter From Sea Ice*, Remote Sensing Laboratory Technical Report 331-37, University of Kansas, Lawrence, Kansas, 1984b.

Kim, Y.–S., R. K. Moore, R. G. Onstott, and S. P. Gogineni, Towards identification of optimum radar parameters for sea-ice monitoring, *Journal of Glaciology, 31(109)*, pp. 214–219, 1985.

Lee, J. K. and J. A. Kong, Active microwave remote sensing of an anisotropic random medium layer, *IEEE Transactions on Geoscience and Remote Sensing, GE-23(6)*, pp. 910–923, 1985a.

Lee, J. K. and J. A. Kong, Electromagnetic wave scattering in a two-layer anisotropic random medium, *Journal of the Optical Society of America, A2(12), pp.* 2171–2186, 1985b.

Lee, J. K. and J. A. Kong, Modified radiative transfer theory for a two-layer anisotropic random medium, *Journal of Electromagnetic Waves and Applications, 2(3/4)*, pp. 391–424, 1988.

Lee, J. K. and S. Mudaliar, Backscattering coefficients of a half-space anisotropic random medium by the multiple scattering theory, *Radio Science, 23(3)*, pp. 429–442, 1988.

Lin, F. C., J. A. Kong, R. T. Shin, A. J. Gow, and S. A. Arcone, Correlation function study for sea ice, *Journal of Geophysical Research, 93(C11)*, pp. 14,055–14,063, 1988.

Livingstone, C. E. and M. R. Drinkwater, Springtime C-band SAR backscatter signatures of Labrador Sea marginal ice: Measurements versus modelling predictions, *IEEE Transactions on Geoscience and Remote Sensing, GE-29(1)*, pp. 29–41, 1991. (Correction, *IEEE Transactions on Geoscience and Remote Sensing, 29(3)*, p. 472, 1991.)

Mudaliar, S. and J. K. Lee, Microwave scattering and emission from a half-space anisotropic random medium, *Radio Science, 25(6)*, pp. 1199–1210, 1990.

Nghiem, S., *The Electromagnetic Wave Model for Polarimetric Remote Sensing of Geophysical Media*, Ph.D. dissertation, Massachusetts Institute of Technology, Cambridge, Massachusetts , 1991.

Perovich, D. K. and A. J. Gow, A statistical description of the microstructure of young sea ice, *Journal of Geophysical Research, 96(C9)*, pp. 16,943–16,953, 1991.

Reber, B., C. Mätzler, and E. Schanda, Microwave signatures of snow crusts: Modelling and measurements, *International Journal of Remote Sensing, 8(11)*, pp. 1649–1665, 1987.

Rice, S. O., Reflections of electromagnetic waves from slightly rough surfaces, *Commun. Pure Appl. Math. 4*, pp. 351–378, 1951.

Sackinger, W. M. and R. C. Byrd, *Reflection of Millimeter Waves From Snow and Sea Ice*, IAEE Report 7203, Institute of Arctic Environmental Engineering, University of Alaska, Fairbanks, Alaska, January 1972.

Stogryn, A., The bilocal approximation for the electric field in strong fluctuation theory, *IEEE Transactions on Antennas and Propagation, AP-31(6)*, pp. 985–986, 1983a.

Stogryn, A., A note on the singular part of the dyadic Green's function in strong fluctuation theory, *Radio Science, 18(6)*, pp. 1283–1286, 1983b.

Stogryn, A., Correlation functions for random granular media in strong fluctuation theory, *IEEE Transactions on Geoscience and Remote Sensing, GE-22(2)*, pp. 150–154, 1984a.

Stogryn, A., The bilocal approximation for the effective dielectric constant of an isotropic random medium, *IEEE Transactions on Antennas and Propagation, AP-32(5)*, pp. 517–520, 1984b.

Stogryn, A., Strong fluctuation theory for moist granular media, *IEEE Transactions on Geoscience and Remote Sensing, GE-23(2)*, pp. 78–83, 1985.

Stogryn, A., A study of the microwave brightness temperature of snow from the point of view of strong fluctuation theory, *IEEE Transactions on Geoscience and Remote Sensing, GE-24(2)*, pp. 220–231, 1986.

Stogryn, A., An analysis of the tensor dielectric constant of sea ice at microwave frequencies, *IEEE Transactions on Geoscience and Remote Sensing, GE-25(2)*, pp. 147–158, 1987.

Stogryn, A. and G. J. Desargent, The dielectric properties of brine in sea ice at microwave frequencies, *IEEE Transactions on Antennas and Propagation, AP-33(5)*, pp. 523–532, 1985.

Thorsos, E., The validity of the Kirchhoff approximation for rough surface scattering using a Gaussian roughness spectrum, *Journal of the Acoustical Society of America, 83*, pp. 78–92, 1988.

Thorsos, E. and D. R. Jackson, Studies of scattering theory using numerical methods, *Waves in Random Media, 3*, pp. S165–S190, 1991.

Tsang, L., Passive remote sensing of dense nontenuous media, *Journal of Electromagnetic Waves and Applications, 1(2)*, pp. 159–173, 1987.

Tsang, L., Dense medium radiative transfer theory for dense discrete random media with particles of multiple sizes and permittivities, Chapter 5, *Progress in Electromagnetic Research, 6,* edited by A. Priou, Elsevier, New York, 1991.

Tsang, L. and A. Ishimaru, Radiative wave equations for vector electromagnetic propagation in dense nontenuous media, *Journal of Electromagnetic Waves and Applications, 1(1)*, pp. 59–72, 1987.

Tsang, L. and J. A. Kong, Microwave remote sensing of a two-layer random medium, *IEEE Transactions on Antennas and Propagation, AP-24(3)*, pp. 283–288, 1976.

Tsang, L. and J. A. Kong, Scattering of electromagnetic waves from random media with strong permittivity fluctuations, *Radio Science, 16*, pp. 303–320, 1981.

Tsang, L., J. A. Kong, and R. T. Shin, *Theory of Microwave Remote Sensing*, John Wiley and Sons, New York, 1985.

Tucker, W. B., III, A. J. Gow, and J. A. Richter, On small-scale horizontal variations in salinity in first-year sea ice, *Journal of Geophysical Research, 89(C4)*, pp. 6505–6514, 1984.

Ulaby, F. T., R. K. Moore, and A. K. Fung, *Microwave Remote Sensing—Active and Passive, Vol. II: Radar Remote Sensing and Surface Scattering and Emission Theory*, Addison–Wesley Publishing Company, Reading, Massachusetts, 1982.

Ulander, L. M. H., R. Johansson, and J. Askne, C-band radar backscatter of Baltic Sea ice, *International Journal of Remote Sensing,* in press, 1992.

Vallese, F. and J. A. Kong, Correlation function studies for snow and ice, *Journal of Applied Physics, 52(8)*, pp. 4921–4925, 1981.

Vant, M. R., R. O. Ramseier, and V. Makios, The complex dielectric constant of sea ice at frequencies in the range 0.1–40 GHz, *Journal of Applied Physics, 49(3)*, pp. 1264–1280, 1978.

Vedernikova, E. A. and M. V. Kabanov, Scattering of optical radiation by a system of closely spaced scatterers, *Optical Spectroscopy, 37(1)*, 1974.

Wen, B., L. Tsang, D. P. Winebrenner, and A. Ishimaru, Dense medium radiative transfer theory: Comparison with experiment and application to microwave remote sensing and polarimetry, *IEEE Transactions on Geoscience and Remote Sensing, 28(1)*, pp. 46–59, 1990.

Wettlaufer, J., Heat flux at the ice–ocean interface, *Journal of Geophysical Research, 96(C4)*, pp. 7215–7236, 1991.

Winebrenner, D. P., L. Tsang, B. Wen, and R. West, Sea-ice characterization measurements needed for testing of microwave remote sensing models, *IEEE Journal of Oceanic Engineering, 14(2)*, pp. 149–158, 1989.

Yueh, S. H., J. A. Kong and R. T. Shin, Scattering from randomly oriented scatterers with strong permittivity fluctuations, *Journal of Electromagnetic Waves and Applications, 4(10)*, pp. 983–1004, 1990.

Zuniga, M. A. and J. A. Kong, Modified radiative transfer theory for a two-layer random medium, *Journal of Applied Physics, 51(10)*, pp. 5228–5244, 1980.

Chapter 9. Laboratory Investigations of the Electromagnetic Properties of Artificial Sea Ice

Calvin T. Swift and Karen St. Germain
Department of Electrical and Computer Engineering, University of Massachusetts, Amherst, Massachusetts 01003

Kenneth C. Jezek
Byrd Polar Research Center, 1090 Carmack Road, The Ohio State University, Columbus, Ohio 43210

S. Prasad Gogineni
Radar Systems and Remote Sensing Laboratory, University of Kansas Center for Research, Inc., 2291 Irving Hill Drive, Campus West, Lawrence, Kansas 66045

Anthony J. Gow and Donald K. Perovich
Cold Regions Research and Engineering Laboratory, 72 Lyme Road, Hanover, New Hampshire 03755

Thomas C. Grenfell
Department of Atmospheric Sciences, University of Washington, Seattle, Washington 98195

Robert G. Onstott
Environmental Research Institute of Michigan, P. O. Box 8618, Ann Arbor, Michigan 48107

9.1 Introduction

The electrical properties of sea ice growing on the open ocean are determined by the mechanical and thermodynamical influences of the ocean and atmosphere. Winds, currents, and air and water temperatures, among other variables, contribute to the eventual roughness, texture, chemical composition, and temperature gradient through the sea ice. These latter properties tend to be inhomogeneous over relatively short length scales (tens of meters horizontally and tens of centimeters vertically at best) and they tend to evolve with time as the boundary conditions and internal composition of the ice pack change. For these reasons, the analyses of electromagnetic data collected by spaceborne instruments over sea ice have tended to rely on empirical relationships between limited ranges of an electromagnetic variable (e.g., brightness temperature) and a geophysical property of the ice. Numerous papers document the success of this analysis approach for estimating sea ice concentration and sea ice motion (e.g., Swift and Cavalieri [1985]; Zwally et al. [1983]; Parkinson et al. [1987]). Yet for these same reasons, attempts to obtain a deeper understanding of the electromagnetic properties of naturally growing sea ice can be complicated. For example, it is logistically difficult to measure and sample young sea

ice. Some features such as the onset of flooding are transient phenomena, lateral inhomogeneity makes sampling difficult, and the opportunities to view a particular type of ice or even the same piece of ice throughout the seasonal cycle have been rare. Individually, almost all of these problems can and have been overcome by field investigators. But, taken together, there seems to be a strong argument for designing a new approach to answering fundamental questions about sea ice electromagnetic properties, namely, what are the important scattering and absorption mechanisms in sea ice and how are these mechanisms related to ice properties. It was felt that an approach could be found in the laboratory.

The Cold Regions Research and Engineering Laboratory (CRREL) Experiments were designed to confront some of the difficulties encountered in field work by growing sea ice in a carefully constrained, laboratory environment. Microwave experiments were conducted under an overarching set of principles that included: a uniform ice sheet of limited and well-documented physical properties could be grown and maintained; multiple sensors could collect data simultaneously from the same ice; and the interpretation of all electromagnetic observations of the ice with the measured ice physical properties has to be considered. As the experiments progressed, it also became apparent that two more facets needed to be added to the CRREL Experiment (CRRELEX). First, data relevant to a range of electromagnetic models (Chapter 10) needed to be collected. Second, time series data needed to be collected, partly because of initial observations that showed measurable changes in electrical properties at every stage of ice development. That CRRELEX could document this evolution is viewed as one of the more important contributions of this work.

9.2 Scope

As originally conceived, CRRELEX was designed to study the microwave response of young thin (10 to 20 cm) sea ice that forms on open leads. Facilities were constructed towards this objective and included a saline-water-filled pond fitted with a plastic liner (Figure 9-1). The pond was 12 m long, 5 m wide, and 1 m deep. These dimensions were chosen to accommodate the planned remote sensing instruments. For example, at a width of 5 m, negligible antenna beam spillover occurs at nadir for a C-band radiometer, which has the largest footprint of the remote sensing instruments used. The 12 m length allowed observations at incidence angles up to 50° and provided for acquiring several independent samples with the radars used. An enormous tent could be rolled on and off the pond to protect against changes in weather. Air-conditioning units were installed inside the tent to preserve the ice during inevitable warm temperature periods, but these units were only

rarely used and later were abandoned. Paving tiles placed around the perimeter of the pond allowed an instrumented gantry to be moved over the ice. Later, a second, concrete-lined pond 10 m long, 10 m wide, and 1 m deep was outfitted with an instrument shelter, a refurbished movable bridge, and a movable roof. Finally, an indoor test facility used to grow a sea ice simulant from an urea solution and a small refrigerated pit (8 × 5 × 1 m) were used at the later stages of CRRELEX. The two indoor facilities proved to be extremely valuable in achieving the controlled environmental conditions initially set forth as a CRREL objective. However, the enclosures themselves limited the observations to range-gated radar measurements.

Use of the outdoor facilities placed constraints on the range of simulated ice properties and the duration over which any one ice property could be studied. Thin ice could be grown in an essentially undisturbed fashion for periods of several days. When grown under quiescent conditions, CRRELEX ice sheets exhibited physical properties similar to those found for thin Arctic ice. Depending on whether the ice was seeded with a fine spray of supercooled water or was left undisturbed to develop spontaneously, the upper centimeter of the ice sheet would be composed of smaller or larger

Fig. 9-1. Typical experiment arrangement at the outdoor facility at the Cold Regions Research and Engineering Laboratory. Pictured from the left are five single-frequency radiometers, the Stepped Frequency Microwave Radiometer, and a cluster of radar systems.

crystals with vertical *c*-axes [Gow, 1986]. A transition in ice properties occurred at a 1 to 2 cm thickness. At that point the *c*-axes reorient toward the horizontal, and columnar organization of ice crystals, characteristic of Arctic congelation ice, becomes dominant (Figure 9-2).

Congelation ice sheets were routinely grown during CRRELEX with maximum thicknesses ranging from about 10 to 30 cm, depending on the experimental requirements. In each case, the ice was closely monitored for salinity, crystal structure, brine pocket distribution, and, as possible, surface roughness. An example of the structural and salinity characteristics of an ice sheet monitored from

initial growth to decay is presented in Figure 9-3. In this particular case, ice growth was initiated in January 1985, and for the first month this ice sheet exhibited typical C-shaped salinity profiles, Figure 9-4(a). Subsequent pooling of snow melt and rain on top of the ice sheet in late February to early March 1985 led to significant downward percolation of the water and concomitant desalination of the ice before the pooled water refroze to produce the salinity and structural characteristics of ice depicted in Figure 9-3. At this stage, the ice contained few brine inclusions and the outlines of crystals had become rounded, a situation similar to multiyear Arctic ice [Gow et al., 1987].

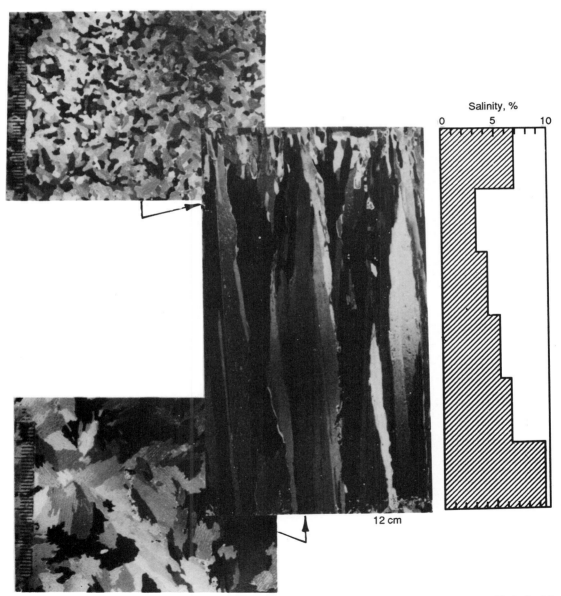

Fig. 9-2. The ice shown in these vertical and horizontal thin-section photographs and salinity profile is freshly grown saline ice from the CRREL test pond. Structurally, this ice closely simulates young Arctic sea ice. Scale subdivisions in horizontal sections measure 1 mm.

Salinity profiles documenting the progressive desalination of the January to March 1985 CRRELEX saline ice sheet are shown in Figure 9-4(b). Desalinated ice was also produced by cutting a pre-existing ice sheet into blocks and placing the blocks on wooden pallets. The blocks were allowed to weather and desalinate under ambient conditions. At the end of this process, the blocks were overlain on a second ice sheet growing in the pond; block salinities did not exceed 0.35‰.

Figure 9-5 documents salinity, temperature, brine volume, and crystalline structure more typical of 10 to 15 cm thick congelation ice sheets grown on the outdoor ponds at CRREL. Temperatures increase linearly with depth. Brine volumes are nearly constant (about 3%) in the upper 75% of the ice. Below that level, brine volumes increase rapidly in the region where the ice develops a more open dendritic structure. A characteristic C-shaped salinity profile is found in this ice with a maximum value at the surface of about 8‰ (averaged over three cm) and a minimum of 5.4‰ at a depth of 10 cm. Salinity increases below 10 cm again because of the structure of the dendritic layer. The horizontal distribution of brine pockets was quantified by digitizing horizontal thin sections taken at various levels through the ice sheet.

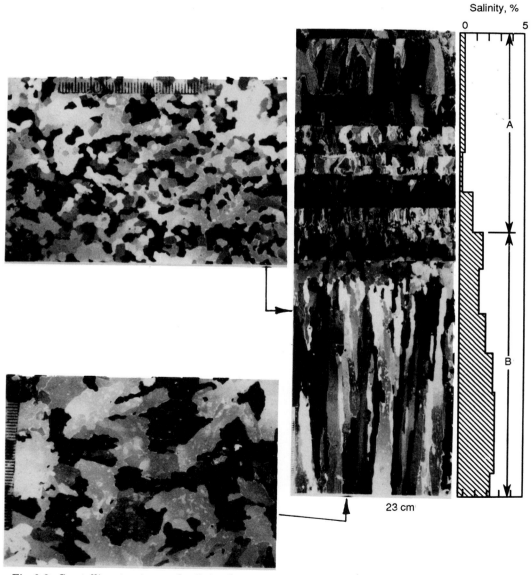

Fig. 9-3. Crystalline structure and salinity characteristics of ice desalinating in the CRREL test pond. In this example, several layers of melt pond ice (A) overlie desalinated columnar saline ice (B). Scale subdivisions in horizontal sections measure 1 mm.

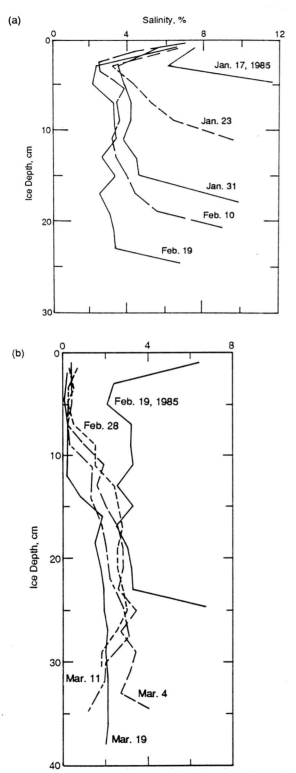

Fig. 9-4. (a) Salinity profiles at different stages of growth of sea ice grown in an outdoor pond at CRREL. Profiles closely simulate those found in Arctic first-year ice. (b) Transformation from a new ice to desalinated ice salinity profile in the outdoor pool at CRREL.

With additional experimentation, it was found that several other physical properties and ice-growth processes could be simulated in the laboratory. Varying surface roughness was introduced either by manually raking the ice surface or by spreading a layer of small chunks of ice on the growing ice sheet. In a few instances, larger blocks of ice were assembled into simulations of ridges. Other classes of complementary experiments could be included during each CRRELEX, and high-frequency acoustic [Stanton et al., 1986; Jezek et al., 1990] and optical frequency propagation measurements were often conducted simultaneously with the microwave observations.

Frazil ice, which forms in turbulent seas and has a granular texture, was simulated in three different ways. Snowfall onto the surface of the unfrozen pool produced a slush cover. Frozen slush results in an ice cover with salinity and structural characteristics very similar to frazil ice. Granular, frazil-like ice sheets up to 8 cm thick were produced in this manner. The second method for producing frazil ice involved a submersible pump to induce frazil crystal nucleation in the pool and the herding of crystals into a layer up to 4 cm thick. A third method simulated both frazil production and the agglomeration of frazil platelets into pancake ice. Here a wave generator consisting of a motor-driven panel was positioned at one end of the pond. Turbulence and wave action converted the semiconsolidated frazil ice sheet into a field of pancakes with pans measuring 20 to 30 cm in diameter and several centimeters thick. The vertical structure of a typical pancake including a salinity profile is presented in Figure 9-6.

Snow-covered ice was studied on the outdoor ponds and also under more controlled conditions in the indoor pit. Analyses of snow-covered ice is complicated by the subtle modification of the ice–snow surface when thin layers of snow are applied. A snow layer is absorbent and, over time, tends to wick surface brine up into the snow layer. This process seems to result in a dielectric roughening of the interface, although direct measurements of the amount or degree of roughening have yet to be made. A sufficient snow load will depress the ice surface below the unloaded freeboard, causing the surface to flood. Flooding was studied in the laboratory by loading a slab of ice with snow and then cutting the slab free from the restraining walls of the tank.

Finally, it is worth mentioning that frost flowers routinely developed in the indoor test basin and cold pit. Preliminary radar measurements were made on freshwater ice sheets completely covered by frost flowers and on urea and saline ice sheets with a low density of frost flowers.

As mentioned, many of the early CRRELEX's were exercises in learning about the ice. As each experiment was completed, it seemed possible to more clearly articulate several questions that drove the development of subsequent experiments:

- What are the microwave absorption and scattering coefficients of thin sea ice? How do they change with time?

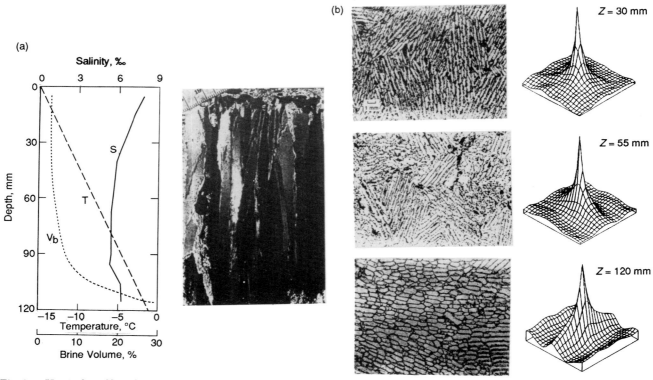

Fig. 9-5. Vertical profiles of temperature (T), salinity (S), and brine volume (V_b) for young sea ice. The vertical thin section shows that the ice structure is predominantly columnar with a thin surface layer of frazil. Also shown are horizontal thin sections from the top, middle, and bottom of the ice sheet with corresponding correlation functions [Perovich and Gow, 1991].

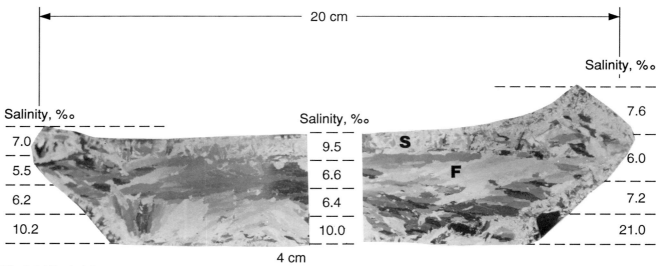

Fig. 9-6. Vertical thin-section structure and salinity profiles from a 4 cm thick pancake from CRRELEX'90. The sequence of ice textures from top to bottom of the pan is slush ice (s) underlain by platey frazil ice (f).

• To what thickness and in what fashion does the ice–water interface and the dendritic layer control electromagnetic response at a particular frequency?

• How important are the shapes, orientation, density, and distribution of brine pockets on the microwave response?

• Given that different classes of surface roughness are associated with different ice formation scenarios, which are the most important roughness classes; which classes dominate over, are equivalent to, or are less than volume scatter effects? Does the sequence of steps in the development and destruction of particular morphologies (e.g., frost flowers) result in a characteristic sequence of microwave responses?

• Does snow play a role in coupling electromagnetic energy between the air, snow, and ice? Is the modification of the ice surface due to the presence of thin snow an important effect? What happens to electromagnetic signatures when snow is sufficiently thick to cause flooding?

• Do thin ice, snow-covered ice, deformed ice, and ridges display azimuthal electromagnetic response? If so, why?

• How and why do signatures change when thin ice warms and desalinates and how is this related to the changing internal structure of the ice?

9.3 GROWTH PHASE OF THIN SALINE ICE

Microwave backscatter, transmission, and brightness temperature measurements were made by the CRRELEX team on saline ice during the winters from 1984 to 1989. Results from all measurements on undeformed congelation ice are similar. Results of representative experiments from this period are used to illustrate the range of measured responses.

Figure 9-7 shows the variation with time of the backscatter coefficient at 5.3 GHz as the surface evolves from open water to grey ice (14.5 cm thick) [Gogineni et al., 1990].

Vertically polarized data from 1984 are shown at 20° and 30° angles of incidence. For ice thicknesses greater than 3 cm, the scattering coefficients remained nearly constant, and possibly decreased when the ice thickness exceeded 12 cm. A more complicated response was observed for very thin ice during the transition from horizontal to vertical crystallographic orientation. The scattering coefficient increased immediately subsequent to ice formation, reached a maximum value at about 1 cm thickness, and then decreased, reaching a stable value after about 2 cm of growth. A similar behavior has been observed for measurements performed at 13.9 GHz on an ice sheet growing from an urea solution in the laboratory. Two mechanisms may contribute to the thin ice response: a high surface dielectric constant and surface roughening (e.g., frost flowers). Frost flowers or other macroscopic surface phenomena were not observed during the experiment. Although surface roughness changes cannot be ruled out on the basis of available observations, it seems more probable that a high dielectric constant explains the data. During initial ice growth, brine is expelled onto the ice surface and surface salinities as large as 70‰ have been reported [A. Kovacs, personal communication, CRREL, Hanover, New Hampshire, 1988].

Figures 9-8, 9-9, 9-10, and 9-11 show both the changing microwave signatures and the changing physical properties of the ice sheet for the 1985 experiments [Onstott, 1991]. Data were obtained at several frequencies and are shown at both like polarizations and at 0° and 40° angles of incidence. Two responses from open water were measured: at the beginning of the growth cycle with small ripples present and a rippleless case at the end of the growth cycle. Both open water points appear at the beginning of the time series.

In this experiment, at an oblique incidence angle (Figure 9-8), there is no evidence of a rise or fall in the thin ice backscatter response, though there is a weak suggestion of

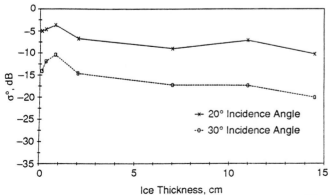

Fig. 9-7. Effects of ice growth on vertical polarization σ° at 5.3 GHz.

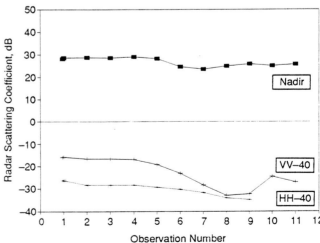

Fig. 9-8. The time series response of the backscatter coefficient at 5.0 GHz during the evolution of open water to new ice at nadir and 40° (VV and HH).

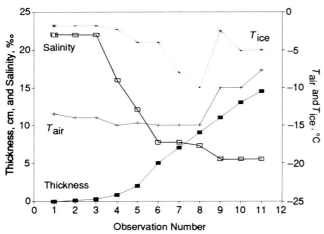

Fig. 9-9. Ice thickness, air temperature, ice temperature, and bulk salinity during the evolution of open water to new ice.

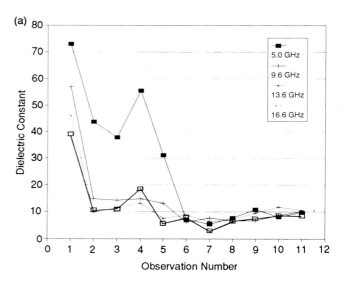

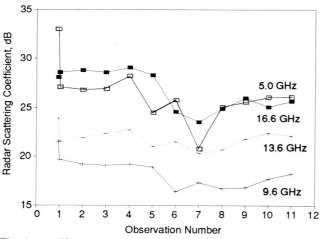

Fig. 9-10. The time series response of backscatter at vertical incidence during the evolution from open water to new ice shown as a function of frequency from 5.0 to 16.6 GHz.

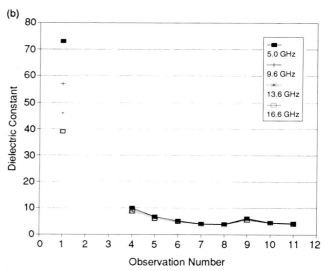

Fig. 9-11. The magnitude of the complex dielectric constant $|\epsilon_r^*|$ (a) as derived from the measurement of reflectivity and (b) by prediction.

that phenomenon at normal incidence angles (Figure 9-10). The backscatter is relatively constant (a slope of 0.05 with a mean square error of 0.4 dB) in the ice thickness range from 0 to 2 cm. Immediately subsequent to formation, the ice was very smooth visually, and measurements made later in the experiment indicated roughnesses of 0.05 cm root mean square (rms) with typical correlation lengths of 1 to 2 cm. The magnitude of the complex dielectric constant calculated from the scattering coefficient at nadir and at 5 GHz is (assuming a smooth surface) 42 ±9 (Figure 9-11). This high value for the dielectric constant is consistent with the inferences made from the data in Figure 9-7.

Near-surface ice temperatures and bulk salinities decreased as the ice thickened. A change in both ice physical

properties and microwave properties is evident at about OB5. The crystallographic orientation and growth process has undergone a transformation at this point. Evidently, the change in bulk properties and processes is driving a change in the electromagnetic response.

Even under relatively controlled conditions, the ice sheet surface will undergo a metamorphosis with time. Single snow particles blown onto the ice sheet (OB8–OB11) increase the roughness of new ice by creating an "orange peel" surface. Onstott [1991] observed that when snow particles are deposited onto new ice, brine is wicked from the ice layer and deposited about the snow particles. The particles melt, dilute the brine solution, and then refreeze. This process creates small roughness elements that modify the backscatter response. Bredow [1991] also investigated the relative contributions of volume scattering from brine pockets and scattering from the surface of thin saline ice. Modeled and predicted variations of X-band backscatter strength with incidence angle agreed well with a Kirchoff scattering model. Bredow also compared the measured radar pulse shape with simulations and found that the pulse shapes were primarily governed by surface scatter along with a weak volume scatter component that caused a slight broadening of the received pulse. At C-band, Bredow found that volume and surface scattering contribute about equally to the ice signature, presumably because of increasing penetration depth.

Linearly polarized radiometers operating at 10, 18, 37, and 90 GHz were used to observe the changing brightness temperature during the growth phase of thin ice. The variation in ice emissivity versus thickness during the growth phase is shown in Figure 9-12 for frequencies of 18, 37, and 90 GHz. The signature is characterized by a rapid increase in emissivity with increasing ice thickness up to a saturation level. Similar results at 10, 18, 37, and 90 GHz, vertical polarization and 50° incidence angle, are reported by Grenfell and Comiso [1986]. They show that emissivity increased from 0.525 to 0.95 at 10 GHz during the initial 50 mm of growth. They also found that emissivity at higher frequencies increased more rapidly with increasing ice thickness.

Results of concurrent radiometer measurements taken at C-band are shown in Figures 9-13 and 9-14. As expected, the emissivity at C-band saturates at substantially thicker ice than would be the case for higher frequencies. As discussed in the next section, the general shape of the curve

is as would be predicted by a model that only allows for the incoherent addition of energy propagating through the ice. However, the excursion observed during the initial growth phase is undoubtedly a coherent effect associated with the first half of an interference fringe. Figure 9-14 demonstrates that a change in wavelength influences the periodicity of the first minimum.

Grenfell and Comiso [1986] compare the degree of polarization P (defined as the ratio of the difference between the

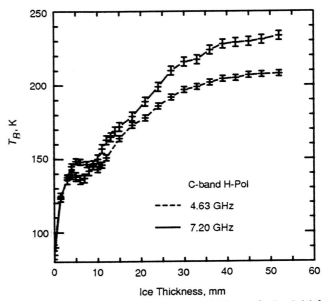

Fig. 9-13. Brightness temperature versus thickness during initial growth at two C-band frequencies and horizontal polarization. These observations cover the same ice sheet as in Figure 9-12 [Grenfell et al., 1988].

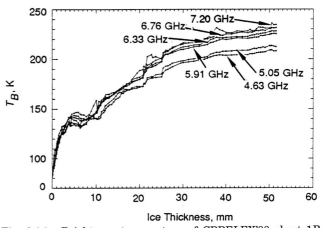

Fig. 9-14. Brightness temperature of CRRELEX'88 sheet 1B versus thickness at the six C-band frequencies of the Stepped Frequency Microwave Radiometer. The measurements were taken at horizontal polarization and a 35° incidence angle. Data from Figure 9-13 are included here without the associated uncertainty bars.

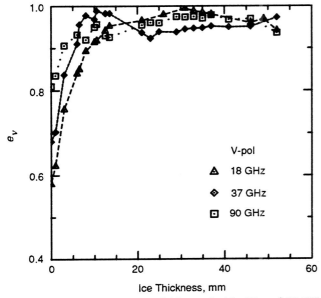

Fig. 9-12. Ice emissivity versus thickness for 18-, 37-, and 90-GHz vertical polarizations during initial growth of congelation ice [Grenfell et al., 1988].

two linear polarizations divided by their sum at a single frequency), during the early growth stages. Polarization decreased by about 60% at 18 GHz as the ice grew to a thickness of about 20 to 25 mm—somewhat less than was required for the emissivity to reach a comparable saturation level.

Arcone et al. [1986] discussed a series of transmission measurements on ice samples extracted from the CRREL pond, as well as in-situ observations acquired by placing antennas above the slab and in the water below the ice slab. Data were evaluated to estimate real and imaginary parts of the dielectric constant. Each experiment required that the ice maintain mechanical integrity and hence the minimum ice thickness sampled was about 7 cm, for which a well-ordered columnar structure was established. Slabs removed from the pond to a refrigerated laboratory were subjected to cyclic cooling and warming.

Loss was measured in situ at 4.5, 4.8, 8.9, 9.0, 9.15, and 10 GHz and was found to vary between 4.1 and 8.7 dB/cm. Higher loss rates were associated with higher frequencies. The highest loss rates were also associated with the thinnest ice. Such ice would be relatively warm and hence would have the highest brine volume fraction. Arcone et al. [1986] reported real permittivities for samples removed from the pond. Values ranged from a maximum of about 4.5 for a slab temperature of −4°C to about 3.2 for cold slabs. The authors modeled the permittivity of high salinity thin ice and concluded that the real part varied from about 10 to about 4 for the temperature range −2° to −10°C in the frequency range 4.8 to 9.5 GHz. In general, Arcone et al. found that both the real and imaginary parts varied linearly with brine volume.

9.3.1 Thin Ice Data Interpretation

Variations in radar backscatter and microwave emission data during thin ice growth seem to be associated with a transition in ice physical properties. The onset of freezing is characterized by the formation of a thin skim of ice that quickly evolves into a complex of stellar crystals. The optical axis of these ice crystals is horizontal, their crystalline texture looks granular in vertical cross section, and brine is expelled onto the ice surface. This upper, transitional layer of thin ice tends to be brine-free and bubble-free. After 1 or 2 cm of growth, a columnar structure develops wherein brine is expelled downward or is trapped within isolated pockets.

These evolving ice processes force progressive changes in the ice sheet's electrical properties. In particular, Figure 9-11 shows that very new sea ice has a dielectric constant much smaller (about 75%) than that of open water. Between thicknesses of 1 and 2 cm, the dielectric constant reaches a maximum value. Consequently, the backscatter response at normal incidence increases with the largest changes at higher frequencies (Figure 9-10). These radar observations are consistent with the facts that: brine is gradually being expelled to the surface as the freezing process progresses; brine accumulation peaks (in this study

at 1 cm ice thickness); and a large percentage change in the dielectric constant occurs.

At 2 cm thickness (OB5), Figure 9-8, backscatter response for VV-40 decreased significantly while the backscatter at nadir remained within about 1 dB. This indicates that multiple processes were at work. The nadir response at 5 GHz directly tracked the change in dielectric constant, while the responses at oblique incidence were sensitive to both the change in dielectric constant and changes in surface roughness statistics.

After the ice is about 2 cm thick, growth proceeds by the accretion of ice to the base of columnar crystals. Based on the normal incidence backscatter data, the magnitude of the dielectric constant after columnar growth is initiated was about 7 (OB6 in Figure 9-11) and reached a minimum about 6 hours later [Onstott, 1991]. The surface roughness at this point was 0.05 cm rms and the ice surface temperature had cooled to −4°C. By accounting for the change in the dielectric constant between the transitional ice growth phase and columnar growth, the roughness for the initial skim was determined to be about 0.065 cm rms. This estimate was based on predictions using the small perturbation model.

As ice growth continues, ice nearest the surface cools and the surface brine layer dissipates. The thickening layer of highly emissive ice shields the radiometrically colder water. This causes brightness temperatures to elevate at a rate proportional to their frequency. Dissipation of the thin, highly concentrated brine layer at the surface of the ice sheet will result in an additional emissivity increase. Also, as Arcone et al. [1986] pointed out, lower temperatures decrease the bulk dielectric constant, which in turn causes a decrease in the backscatter response. Figure 9-10 reveals a 7.5-dB reduction in backscatter (5.0 GHz and VV) between ice 0 to 1 cm thick and ice 4 to 7 cm thick.

To examine the frequency dependence during the transition to a columnar ice structure, the average value of the dielectric constant for OB1 through OB5 and the minimum value measured immediately at the beginning of columnar growth are used, Figure 9-11(a). Associated changes in backscatter are 5, 3, 2, and 2 dB at 5.0, 9.6, 13.6, and 16.6 GHz, respectively. Since cooling is a top-down process, changes at higher frequencies are expected to occur first. This is verified by correlating the physical observations in Figure 9-9 with the reflectivity response at ice thicknesses between 0 and 2 cm. The time at which the drop in dielectric constant occurs is frequency dependent. This level was reached at a thickness of 5 cm (OB6) for 5.0 and 9.6 GHz, and 2 cm (OB5) for 13.6 and 16.6 GHz. This trend further illustrates that the magnitude of the backscatter response is driven by the dielectric constant and that an effective bulk dielectric constant is determined by penetration depth, both of which are frequency dependent.

A modified version of a model developed by Kovacs et al. [1986] was used to predict dielectric profiles so that penetration depths and bulk dielectric constants could be calculated at each frequency. For comparison, these results are included in Figure 9-11(b). The C-shaped salinity profile further modulated the effective dielectric constant, especially at high frequencies, because of an increase in the brine volume in the uppermost portion of the ice sheet. Examination of the penetration depths provided in Table 9-1 shows that the backscatter response at 5.0 GHz was influenced by the entire dielectric layer for thicknesses from 0 to 2 cm. In contrast, at 16.6 GHz, the top 0.5 cm sets the absolute level of the backscatter response.

New ice grown under very calm conditions has a very smooth ice–air interface; however, the surface may become slightly roughened with the deposition of single snow crystals. This observation and Bredow's [1991] result at C-band that volume and surface scattering contribute about equally to the ice signature for very smooth ice argue that backscatter data at C-band and higher frequencies provide information about the upper layers of the ice. Radiometric results presented in Figures 9-12, 9-13, and 9-14 argue that at C-band frequencies and lower, brightness temperature data represent an integration of effects distributed throughout the thin ice sheet. In particular, Figure 9-14 shows an oscillatory dependence of brightness temperature of young sea ice with frequency and ice thickness. This dependence indicates that microwave emission from young sea ice also includes a coherent process.

Apinis and Peake [1976] derive an expression for coherent emission from a slab:

$$\varepsilon = 1 - r = \frac{(1 - r_i)(1 - Ar_w)}{(1 + Ar_i r_w + 2\sqrt{Ar_i r_w}\ \cos 2B\ell)} \quad (1)$$

where r is the composite power reflection coefficient
r_i is the power reflection coefficient at the air–ice interface
r_w is the power reflection coefficient at the ice–water interface
$A = \exp(-4\alpha\ell)$
α is the attenuation coefficient in ice
B is the phase constant in ice
ℓ is the radiation path length

This expression shows that the emissivity is a damped, periodic function of ice thickness. Of course, interface roughness will destroy coherence and, if the surface is sufficiently rough, the average emissivity is given by

TABLE 9-1. Predicted penetration depths for sea ice with thicknesses from 1 to 8 cm [Onstott, 1991].

Ice thickness,	Penetration depth, cm, for given frequencies			
cm	5.25 GHz	9.6 GHz	13.6 GHz	16.6 GHz
1	>1	0.8	0.5	0.4
2	2.0	1.0	0.7	0.5
4	3.3	1.6	1.0	0.8
8	3.7	1.6	1.0	0.8

$$\langle\varepsilon\rangle \cong \frac{(1 - r_i)(1 - Ar_w)}{(1 - Ar_i r_w)} \quad (2)$$

This very simple expression depends only on the reflection coefficients at the two interfaces and the product of the attenuation coefficient and path length. Therefore, if the bulk electrical properties are known, the total incoherent emissivity is a function only of the slab's mean thickness. The monotonic increase in emissivity predicted by Equation (2) is supported by data for ice with significant surface roughness [Swift et al., 1986]. (Equation (2) does not reduce to the proper value for zero slab thickness. An empirical equation has been introduced by Swift et al. [1986] to correct this limitation.)

Variations in emissivity shown in Figures 9-12 and 9-14 can be understood in terms of incoherent and coherent processes. The signatures in Figure 9-12 are characterized by rapid increases in emissivity with increasing ice thickness up to a saturation level—this is the functional form predicted by the incoherent propagation model. The frequency, ice thickness, and ice extinction coefficients will determine when the saturation level will be reached. To show this relationship more clearly, the attenuation coefficient, A, can be expressed as

$$A = e^{-4\alpha d_0} = \exp\left(\frac{-8\pi n_i d_0}{\lambda_0}\right) \quad (3)$$

where n_i is the imaginary part of the refraction index of sea ice. Because n_i is expected to be reasonably constant over the microwave band of frequencies, Equation (3) shows that the attenuation increases as the frequency of observation increases. This result agrees with the data trends in Figure 9-12, which indicate that for a given ice thickness, the emission from the underlying water decreases as the operating frequency increases.

The results of concurrent measurements taken at C-band are shown in Figure 9-14. Since the C-band wavelength is much larger, saturation is reached at substantially thicker ice. This signature is very similar to that predicted by the incoherent model, except for the excursion noted during the initial growth phase. This is undoubtedly a coherent effect associated with the first half-wave interference fringe. The interference patterns at 4.63 and 7.20 GHz are distinct. From Equation (1), we expect to observe a maximum when $l/\lambda_i = 0.25$ and a first minima at $l/\lambda_i = 0.5$.

Because the ice thickness was frequently measured, the data can be used to solve for the average dielectric constant of the ice layers. St. Germain and coworkers [St. Germain, K. M., C. T. Swift, and T. C. Grenfell, Determination of dielectric constant of young sea ice using microwave spectral radiometry, submitted to the *Journal of Geophysical Research*] derived a relationship between the real part of the dielectric permittivity and ice thickness given by

$$\varepsilon_r' = \frac{1}{2}\left\{\left(\frac{c\ell}{d_0 f \lambda_i}\right)^2 \pm \sqrt{\left(\frac{c\ell}{d_0 f \lambda_i}\right)^4 - 4\left(\frac{c\ell}{d_0 f \lambda_i}\right)^2 \sin^2\theta_i}\right\} \quad (4)$$

Table 9-2 presents a listing of thicknesses that correspond to the first and second quarter-wave interference fringes. Using data from Table 9-2 along with Equation (4) allows one to solve for the dielectric constant of ice. Results are presented in Figure 9-15, where a linear regression has been plotted through the data points. This figure shows that the dielectric constant of ice during the first centimeter of growth is approximately three times larger than results quoted for thick first-year ice, which is consistent with results reported above. St. Germain and coworkers evaluated the frequency dependence of the dielectric constant and found that n_i is equal to about 0.038 ($\varepsilon = 0.215$).

TABLE 9-2. Mean ice thicknesses measured at the first maximum $d/\lambda = 0.25$ and the first minimum $d/\lambda = 0.50$ in the time series brightness temperature data for each frequency.

Frequency, GHz	d_0 [$d = \lambda/4$], mm	d_0 [$d = \lambda/2$], mm
4.63	5.0	10.5
5.05	4.5	10.0
5.91	4.0	9.0
6.33	4.0	8.0
6.76	3.5	7.5
7.20	3.0	7.0

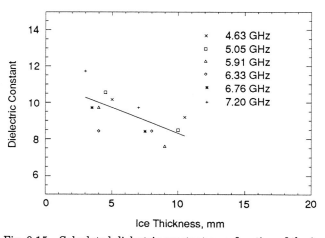

Fig. 9-15. Calculated dielectric constant as a function of the ice thickness.

9.4 UREA ICE

Urea ice is structurally identical to saline ice [Gow, 1984]. Under calm conditions it develops a columnar structure characterized by thin plates of pure ice that separate vertical planes of pockets filled with urea solution. Because urea is a covalent molecule, an urea solution is less conductive than brine. Nevertheless, at frequencies above 10 GHz and at temperatures near 0°C, the imaginary parts of the dielectric constant of saline water and even fresh water are identical [Ulaby et al., 1982, p. 2023]. The real part of the dielectric constant of saline water is a few percent less than that for fresh water. The complex permittivity of both fresh and saline water is strongly frequency dependent over 1 to 100 GHz.

Radar backscatter measurements of urea ice were acquired because urea is noncorrosive and, hence, large quantities can be easily managed in an indoor facility. Data were collected with 13.9 and 35 GHz polarimetric scatterometers during the growth and evolution of several urea ice sheets.

Normalized radar cross sections of smooth urea ice at different angles of incidence greater than 20° and at like and cross polarizations are shown in Figure 9-16 at 35 GHz. Figure 9-17 shows like-polarization cross sections at 13.9 GHz corrected for antenna beamwidth effects for both smooth urea ice and saline ice. Both figures show a strong, exponential decrease in cross section with increasing incidence angle although there is almost a 20 dB difference

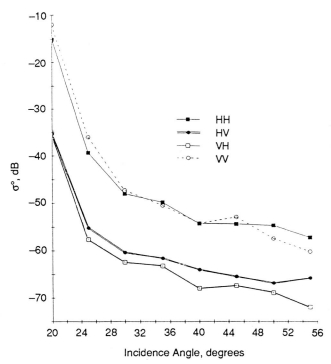

Fig. 9-16. Polarimetric normalized radar cross section ($\sigma°$) of smooth urea ice at 35 GHz [Colom, 1991].

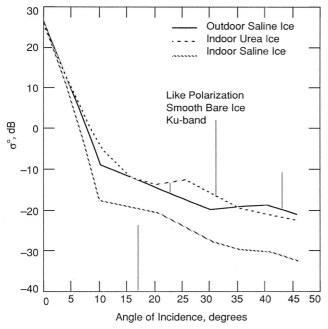

Figure 9-17. Beamwidth corrected backscatter coefficients for two saline ice sheets and an urea ice sheet. One of the saline sheets and the urea ice sheet were grown indoors. The second saline sheet was grown on an outdoor pond.

between the 13.9 and 35 GHz data at large angles at like polarization. Also at larger angles, the 13.9 GHz urea ice data are about 10 dB lower than those for saline ice, possibly because of the difference in dielectric constant between the two materials. Because the ice was smooth and there was no distinction between the HH- and VV-polarization components, any residual power at 13.9 GHz is attributable to volume scattering. The low power levels at large incidence angles indicate that the inclusions must be small compared to a wavelength.

An experiment was conducted to discriminate between surface and volume scattering. A smooth sheet of urea ice was grown indoors until it was 30.5 cm thick. The surface was smooth enough so that surface scattering was negligible. The temperature of the room was slowly raised over a period of 48 hours to observe any change in the resultant backscatter. It was anticipated that any changes could be primarily attributed to volume scattering. Figure 9-18(a) shows the results of this experiment at 35° and 45° of incidence where σ was plotted versus ice temperature. The trends shown are also consistent for each of the four polarizations plotted in Figure 9-18(b). As expected, σ is lower for a larger incidence angle. The value of σ is also low for the smooth ice at temperatures of approximately –9° and –8°C. The value of σ increases slowly until it reaches a maximum around –1°C and then drops off. During the experiment, air bubbles 2 to 5 mm wide were observed near the surface for a temperature range of –2° to 0°C. Bubbles formed because pockets of urea solution drained, leaving cavities partially or completely filled with air. The ice sheets free of

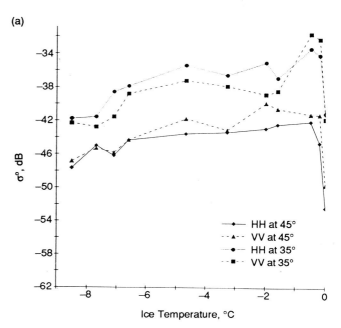

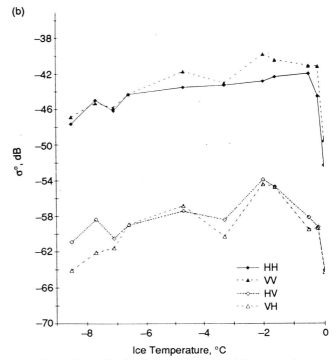

Fig. 9-18. Normalized radar cross section (σ°) versus ice temperature at (a) 35° and 45°, and (b) polarimetric cross sections at 45° [Colom, 1991].

an urea solution present low loss to the microwave signal, scattering more energy and explaining the maximum at about –1°C. Eventually, water filled the air pockets and covered the ice sheet, masking the scatterers located below the surface. Indeed, the scattered signal decreased considerably at 0°C.

9.5 Desalinated Ice

Ice sheets were left in place on the outdoor pond at CRREL for up to several months. One ice sheet, grown in January 1988, grew to a thickness of 16 cm, at which point ice cubes were distributed on the surface to simulate roughness. The ice sheet was left undisturbed until late January when unseasonably warm air temperatures caused the surface to melt. Fresh water so formed flushed much of the brine from the ice sheet. Periods of growth and melt followed until late February, when the ice was about 31 cm thick. The salinity of the ice in the top 5 cm of the ice sheet was less than 0.5‰. The density in the top 10 cm of the ice was about 0.89 mg/m³.

Measurements were made in 1989 on the desalinated ice blocks as mentioned in Section 9.2 [Gogineni, S. P., J. W. Bredow, A. J. Gow, T. C. Grenfell, and K. C. Jezek, Microwave measurements over desalinated ice under quasi-laboratory conditions, submitted to *IEEE Transactions on Geoscience and Remote Sensing,* 1992]. Salinities in the upper 5 cm were less than 1.0‰ with densities in the upper 10 cm varying between 0.84 and 0.87 mg/m³. Ice characteristics from both the 1988 and 1989 experiments appear in Table 9-3.

TABLE 9-3. Ice surface and internal characteristics.

Year	RMS height, cm	Standard deviation, cm	Correlation length, cm
1988	0.18	0.07	2.5
1989	0.08	0.15	3.3

Year	Bubble size, mm	Correlation length, mm	Bulk density
1988	0.5	—	—
1989	0.45	1.5–2.0	0.844

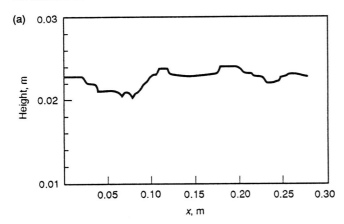

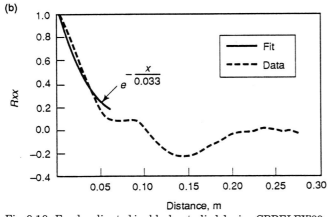

Fig. 9-19. For desalinated ice blocks studied during CRRELEX'89: (a) surface profiles, and (b) computed autocorrelations and minimum-mean-square-exponential fits.

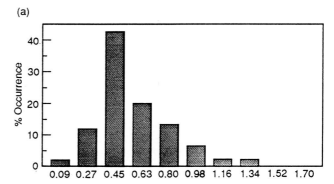

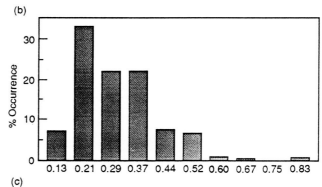

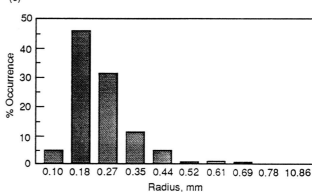

Fig. 9-20. Air-bubble size distribution from the (a) top, (b) middle, and (c) bottom of a block of desalinated ice studied during CRRELEX'89.

Figure 9-19(a) shows a typical surface roughness profile obtained using a contour gauge. The corresponding autocorrelation function is shown in Figure 9-19(b). Bubble size and distribution were determined from photographs of vertical and horizontal thin sections taken from the desalinated blocks. Bubbles were highlighted and the images digitized. Figure 9-20 shows a typical bubble distribution obtained from the desalinated ice.

Figure 9-21 shows the results of radar backscatter measurements made on the desalinated ice at C-band. The bars represent the total uncertainty in the scattering coefficient estimate resulting from random and systematic errors. The upper curve is the maximum value of σ obtained by increasing the rms height by one standard deviation and reducing the correlation length by one standard deviation. The minimum values of σ, represented by the lower dashed curve, were determined by decreasing the rms height by one standard deviation and increasing the correlation length by one standard deviation. The decay of σ with incidence angle is similar to that observed on saline ice. This decay indicates that surface scattering is the dominant source of scattering. To confirm this, data were collected over a 24-hour period to determine the temperature dependence of the scattering coefficient (Figure 9-22). At X-band, scattering from the ice sheet decreased by more than 10 dB when the ice temperature increased from about –5° to –3°C. There was negligible change at C-band. The decrease in scattering at X-band can be attributed to the presence of moisture on the surface, which reduced the volume contribution. However, because we did not observe a corresponding reduction at C-band, we concluded that scattering at C-band is dominated by the ice surface.

The dependence of emissivity on frequency and nadir angle are shown in Figure 9-23 for the desalinated ice measured during February 1988. The error bars indicate the standard deviation of the results of twelve sets of observations. The resulting frequency spectrum, Figure 9-23(a), shows a very weak frequency dependence. The emissivity spectra measured on the desalinated blocks created in 1989 show more structure (Figure 9-24). The maxima at 6 and 37 GHz are suggestive of interference fringes, however, the variation with incidence angle is quite weak (Figure 9-25). The differences between the 1988 and 1989 ice sheet simulations are pronounced. Removal of the blocks in 1989 no doubt allowed brine to drain from the entire slab. Low block salinities would have resulted in a decrease in the optical thickness of the ice. That the ice sheet was left in place in 1988, would suggest there was still substantial brine in the lower part of the ice slab. Higher salinities in the lower portion of the ice sheet could have caused sufficient attenuation to eliminate fringe effects.

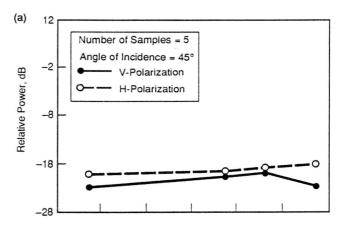

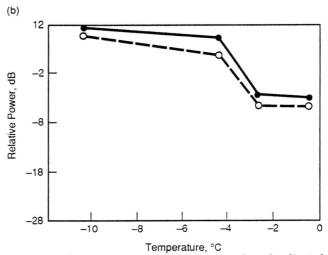

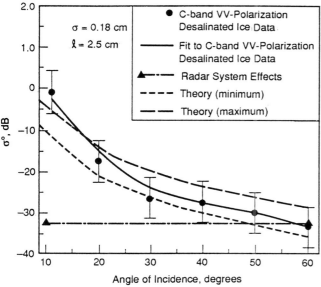

Fig. 9-21. Backscatter cross section of desalinated ice sheet at C-band, as determined from data obtained during CRRELEX'88, compared with predictions of the Small Perturbation Method.

Fig. 9-22. Effect of temperature on backscatter from desalinated ice at (a) C-band and (b) X-band as determined from data obtained during CRRELEX'88.

(a)

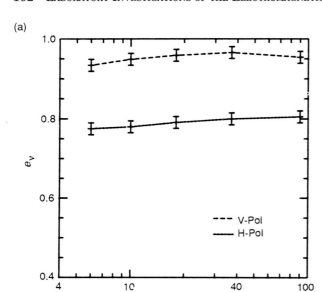

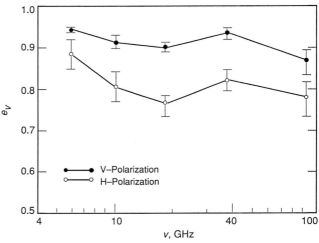

Fig. 9-24. Microwave emissivity versus frequency for the desalinated blocks in January 1989. The error bars denote the standard deviation for four independent samples.

(b)

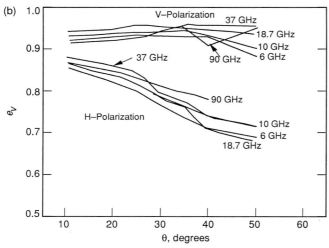

Fig. 9-23. (a) Emissivity spectra from 6 to 90 GHz of a desalinated ice sheet at V- and H-polarizations. (b) Microwave emissivity versus incidence angle for the desalinated pond ice. Curves give the results for the five observation frequencies.

9.6 PANCAKE ICE

Pancake ice was simulated by installing a wave-generating paddle at one end of the outdoor pond. Pancakes were formed with thicknesses of about 7 cm. The surface emissivity increased as the ice froze, but never achieved the values previously recorded for congelation ice. C-band observations indicated that the increase in brightness temperature was linear with thickness, contrary to the nonlinear relationships discussed earlier. This linear variation existed because the brightness temperature was being modulated more strongly by varying concentrations of ice and slush–water. As time and thickness increased so could the concentration of ice, however no quantitative method

Fig. 9-25. Microwave emissivity versus incidence angle (a) V and (b) H polarizations for the desalinated blocks.

for estimating the relative percentage of ice and slush was available.

9.7 SURFACE EFFECTS

The vertical distribution of brine changes rapidly as the ice sheet grows from the melt. As evidenced by the above results, this early redistribution of brine and the formation of brine pockets within the ice has a primary influence on microwave signatures. As the ice sheet continues to age, secondary metamorphic processes begin to rework the surface. Natural processes such as mechanical deformation, frost flowers, and snow cover will each produce different but characteristic scales of surface roughness and near-surface inhomogeneities that modify, and may eventually dominate, the electromagnetic response.

Several types of surface processes were duplicated in the indoor and outdoor laboratories. Results of those experiments are discussed in the following sections.

9.7.1 Roughened Surface

A roughened ice sheet was simulated by mechanically raking and gouging the ice surface. This produced an irregular texture with 2- to 3-cm height variations. Figure 9-26 shows the angular dependence of the backscattering coefficient for smooth, rough, and snow-covered ice at 9.6 GHz (HH-polarization). Roughening causes the scattering coefficient to increase by about 10 dB over the results for smooth ice at incidence angles greater than 20°. The shape of the backscattering curve is indicative of surface scattering. However, a volume contribution may be comparable to the surface term at large incidence angles because a volume term must be included to model the scattering from smooth saline ice. Of course, as the surface gets rougher, the volume term becomes less and less important.

Radiometer data were collected on the same ice sheet. Before roughening, the ice sheet was optically thick and there was no clear dependence of emissivity on frequency and incidence angle at vertical polarization. At horizontal polarization, the undisturbed ice showed the usual frequency and incidence angle dependence [Grenfell and Comiso, 1986]. After roughening, all of the horizontal observations at 18 and 37 GHz showed a large increase in emissivity (Figure 9-27). For vertical polarization, the 10- and 90-GHz emissivities decreased while the 18-GHz emissivity increased at 40° and 50° incidence. The 37-GHz emissivity increased at 60°. Grenfell and Comiso [1986] argued that the results at 18 and 37 GHz were explained by the fact that the scale of roughness was close to the wavelengths at these frequencies. All other results were within the range of expected values from conventional scattering and emission theory of rough surfaces [Ulaby et al., 1982]. When the rough layer was scraped off, the emissivities decreased further, suggesting that a thin liquid skim had formed on the smooth surface.

9.7.2 Rubble Surface

A rubbled ice surface was created by spreading fresh ice chunks on top of an 8-cm thick ice sheet. At this thickness, the physical properties of the ice sheet are only slowly changing, so sequential microwave observations essentially view the same material. The ice sheet grown for this experiment was very smooth (0.05 cm rms) and was snow-free. Ice surface temperatures prior to application of the ice chunks were about −16°C. Warmer air temperatures later in the experiment and after the chunks were applied caused the ice surface temperature to rise to about −8°C. The rubble layer was about 2.5 cm thick and photographs of vertical thick sections are shown in Figure 9-28. Environmental descriptions of bare and rubble-covered ice sheets are listed in Table 9-4. The scale of roughness simulated here is similar to that observed for pancake ice. Roughnesses greater than this are usually associated with large-scale deformation processes (i.e., strewn blocks and rubble).

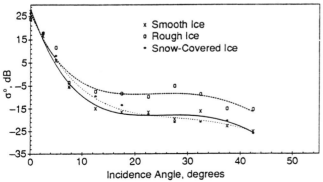

Fig. 9-26.   Effects of surface conditions on backscatter angle response at 9.6 GHz and HH polarization.

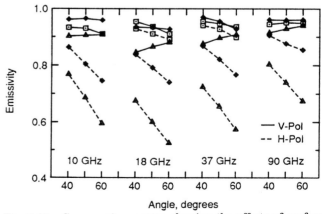

Fig. 9-27.   Comparative spectra showing the effects of surface roughness for the natural surface (diamonds), the same surface after roughening (squares), and after the rough upper surface was scraped off (triangles) [Grenfell and Comiso, 1986].

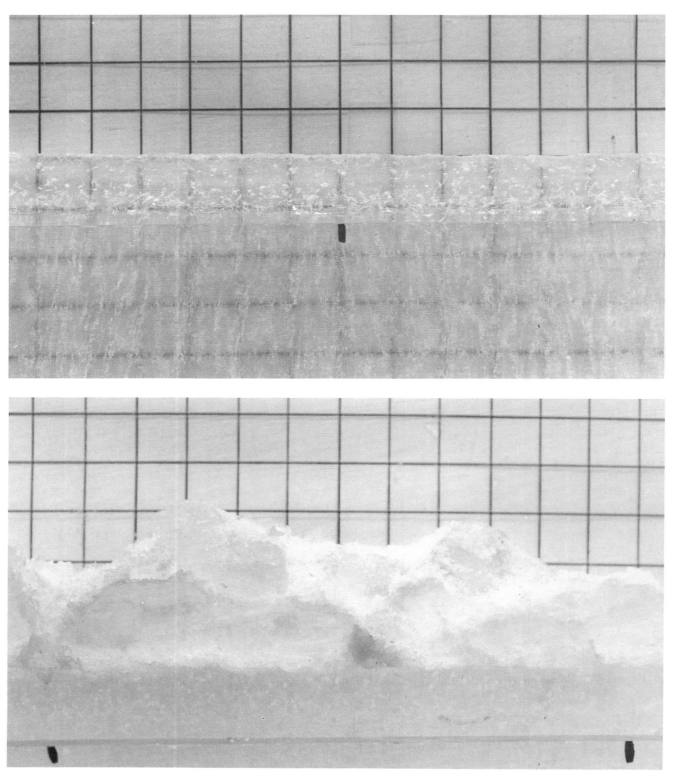

Fig. 9-28. Vertical thick sections showing the air–ice interface for (a) slightly rough and (b) very rough grey ice. The grid spacing is 1 cm.

TABLE 9-4. Summary of the physical property observations for a grey ice sheet with a smooth and rough surface [Onstott, 1990].

Description	Grey ice smooth	Grey ice rough
Ice age	70 hours	100 hours
Ice thickness	7.5 cm	8.3 cm
Snow cover	none	none
$\sigma_{roughness}$	0.048 cm rms	0.544±0.053 cm rms
$\ell_{correlation.length}$	0.669 cm	1.48 ±0.334 cm
T_{air}	−23.5 to −26.6°C	−14.6 to −18.1°C
T_{ice}	−16°C	−8°C
Bulk salinity	10.1‰	10.1‰ (ice sheet)
Top 5 cm		0 ‰ (roughness elements)

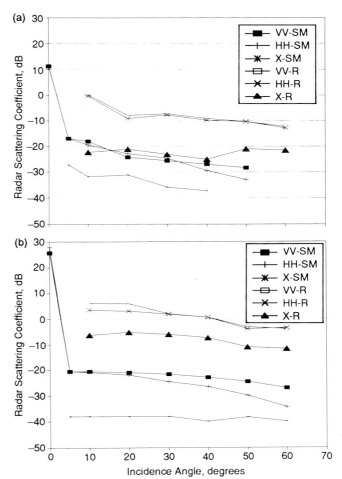

Fig. 9-29. Radar scattering coefficient response for ice with a rough (R) or smooth (SM) surface. The response is shown at (a) 5.0 and (b) 10.0 GHz.

A few comments are necessary before presenting and comparing data. First the magnitude of the dielectric constant of bare ice is about twice that of fresh ice. Consequently, the backscatter cross sections from rough ice require an increase of about 3.7 dB to account for this difference. Relative to the radar wavelength, there is almost an order of magnitude change in surface roughness between the bare and rubble-covered ice sheets (Figure 9-29). At 5 GHz and like polarization, backscatter increases by about 19 dB at 40°, while at 10.0 GHz it increases about 22 dB after the surface has been roughened.

Passive microwave observations of the same scene yielded results similar to those reported earlier for raked surfaces. At 6 GHz and vertical polarization, the emissivity was constant before and after addition of the rubble layer. As shown in Figure 9-30, the emissivity decreased at the higher frequencies after the rubble was added. This response is likely explained by the fact that the rubble layer is rough at higher frequencies but smooth at C-band. Horizontal polarization observations showed that the emissivity at low frequencies rose slightly, but decreased at higher frequencies. As presented in Figure 9-31, the C-band data between 4.9 and 7.2 GHz included a maximum at the low frequency and a minimum at the high frequency, consistent with a first-order fringe pair from a 2-cm layer with a dielectric constant of 1.15.

A rubble layer was deposited on the surface of an urea ice sheet. Chunks of urea ice 2 to 5 cm square were spaced over a new, smooth urea ice sheet. Figure 9-32 shows backscatter coefficients for rough ice versus the angle of incidence. As expected, backscatter is stronger for this case than for the smooth ice case (Figure 9-16), especially at large incidence angles.

9.7.3 Snow-Covered Surface

Snow cover is ubiquitous over sea ice and can significantly affect the microwave backscatter or emissivity [Kim et al., 1984; Comiso et al., 1989]. Snow cover overtly modifies the electromagnetic signature of sea ice simply by addition of another layer of scattering particles. But the influence of snow cover on sea ice is also more subtle. Snow acts as an insulator and a mechanical load. This combination serves to release brine from just below the ice surface. The brine is subsequently wicked up into the snow layer [Grenfell and Comiso, 1986]. If the snow is thick enough, the load can be sufficient to depress the ice sheet below its natural freeboard, and flooding will occur. Several laboratory experiments were carried out to assess the effects of a snow cover over sea ice and the subsequent flooding and refreezing of a basal slush layer on active and passive microwave signatures.

Naturally falling, dry snow accumulated on a saline ice sheet grown on the outdoor pond. The snow layer reached a thickness of 4.5 cm. Snow-covered and smooth ice returns at 9.6 GHz were similar in magnitude, indicating that thin snow cover has negligible effect on backscattering from

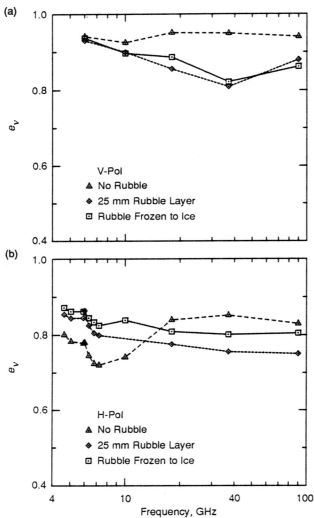

Fig. 9-30. Emissivity spectra of 150-mm-thick ice before and after the addition of a 25-mm layer of ice rubble at (a) vertical and (b) horizontal polarizations [Grenfell et al., 1988].

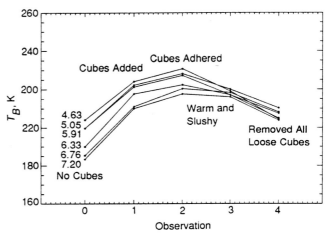

Fig. 9-31. C-band brightness temperature with varying surface roughness. Observation 1 was made immediately after adding approximately 2.5 cm of ice cubes to a 13.2-cm-thick ice sheet. Observation 2 was made the following morning, after most cubes had become frozen to the ice surface. Three hours later, the surface temperature had increased to –4°C for observation 3. Finally, for the last observation, the surface was raked to remove all loose cubes.

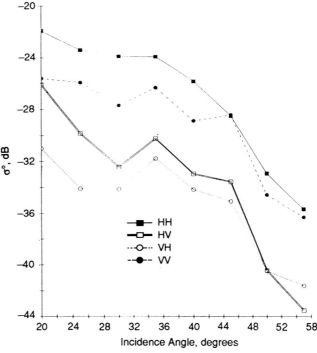

Fig. 9-32. Normalized radar cross section of rough ice at 35 GHz [Colom, 1991].

saline ice [Bredow, 1991]. However, backscatter coefficients of snow-covered ice at 13.6 GHz were about 3 dB higher than those of smooth ice at incidence angles greater than about 20° (Figure 9-33). Radiometer measurements of snow-covered saline ice recorded an increase in polarization and a decrease in emissivity, with a stronger decrease occurring at higher frequencies [Grenfell and Comiso, 1986]. Measurements at C-band (Figure 9-34) indicated the possibility of an interference effect, which, because of the snow depth, appeared to cause a spectral emissivity inversion to occur.

Lohanick [1992] conducted a series of detailed radiometric observations of snow-covered saline ice. Data were acquired at 10 and at 85 GHz before and for several weeks after a single snowfall event. A time series of 45° incidence-

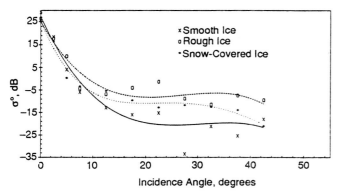

Fig. 9-33. Effects of surface conditions on backscatter angle response at 13.6 GHz and HH polarization.

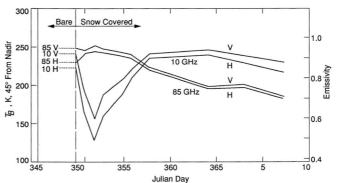

Fig. 9-35. Brightness temperatures and emissivities at 10 and 85 GHz and 45° incidence angle before and after the development of a snow cover on top of a saline ice sheet [Lohanick, 1992].

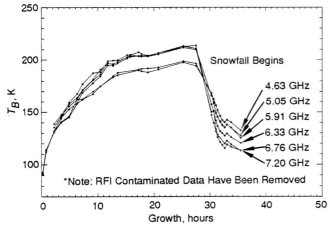

Fig. 9-34. C-band brightness temperature measurements illustrate the effect of light snowfall on a sheet of frazil ice. The snow thickness reached only 1 cm and became very moist over time.

mained constant. This suggested an explanation relying on increased volume scatter in the snow due to increased snow grain size.

A similar sequence of observations was made on urea and saline ice sheets growing in an indoor laboratory [Lytle, V. I., K. C. Jezek, R. Hosseinmostafa, and S. P. Gogineni, Laboratory backscatter measurements over urea ice with a snow cover at Ku-band, submitted to *IEEE Transactions on Geoscience and Remote Sensing*]. The 13.9-GHz radar backscatter observations were acquired during the growth phase and until the ice was approximately 9 cm thick (Figure 9-36). At that point, snow previously stored in a cold room was applied in three successive layers. Events 22, 23, and 24 represent the addition of three layers of snow that in total were 6.3, 11.2, and 14.8 cm thick, respectively. Event 25 represents the moment at which the snow-covered ice was freed from the restraining walls of the tank and began to flood. Event 35 represents the point at which the flooded snow layer was refrozen. Ice growth to 9 cm took about three days. The snow experiments, including flooding, were distributed over about 12 hours. The refrozen slush was measured about 15 hours after the intial flooding event.

As with the outdoor experiments, the snow caused the nadir return to decrease by about 4 dB, while oblique incidence returns increased by about 10 dB. Backscatter coefficients were essentially invariant with the thickness of the snow layer, which varied from 6 to 15 cm. Lytle and coworkers concluded that these observations could not be explained by volume scattering effects. Instead, and as concluded by Lohanick [1992], they attributed the observations to a metamorphosis of the snow–ice interface that caused additional surface scattering. The lower portion of the snow pack was observed to be wet because of the wicking phenomenon described earlier. Using a Kirchoff model, they found that an effective interface roughness of about 0.5 cm rms and a correlation length of 15 cm explained the observations.

After a 15-cm thick snow layer was applied to the ice surface, the section of the ice sheet covered with snow was freed from the restraining walls of the tank. The ice surface

angle radiometric temperature and emissivities for 10 and 85 GHz is shown in Figure 9-35. Snow fell late on Julian day 349. Dotted lines in Figure 9-35 show bare ice brightness temperatures. Emissivities at 10 GHz were calculated by dividing the brightness temperature by 268 K, the measured ice surface temperature; 85-GHz emissivities were calculated by dividing brightness temperature by snow surface physical temperature.

Figure 9-35 shows a significant drop (100 K) in 10-GHz brightness temperatures after snow deposition. Lohanick [1992] argued that the formation of a slush layer at the base of the snow was responsible for the behavior of the low frequency data. Brine wicked upwards into the snow caused a slush layer to form. This layer behaved radiometrically as a rough water interface. As the layer froze, it evolved to look more like water-saturated snow and finally a near perfect emitter at 10 GHz.

The 85-GHz data apparently were affected only by the snow layer. The 85-GHz brightness temperatures slowly dropped during the experiment and its polarization re-

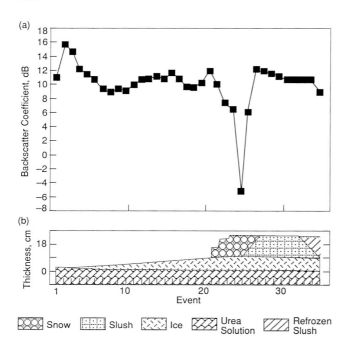

(a)

(b)

Snow   Slush   Ice   Urea Solution   Refrozen Slush

Fig. 9-36. (a) Evolution of backscatter coefficient with successive, controlled changes in ice sheet properties. Data were collected with a 13.9-GHz radar operating at normal incidence over an indoor tank filled with an urea solution. (b) The initial formation of ice from the melt (event 1) through the point at which the ice was about 9 cm thick (event 21) is shown. Events 22, 23 and 24 represent the addition of three snow layers. Event 25 represents the moment at which flooding begins.

depressed below the water level and the snow layer quickly flooded with urea solution. The entire column of snow eventually saturated with liquid. As soon as flooding occurred, normal incidence backscatter decreased by about 11 dB, but then rapidly increased to a level about 4 dB higher than backscatter measured prior to flooding. At 20° incidence angle, backscatter remained constant as the slush layer started to form; as the snow layer saturated the backscatter increased by about 5 dB. Increase in backscatter at all angles was attributed to the increased reflection coefficient at the flooded interface. Upon refreezing, backscatter at 20° incidence angle decreased 5 dB, similar to the level associated with an unflooded snow layer. This similarity was not unexpected. Although the overlying snow had been replaced by refrozen slush, the physical characteristics of this surface were expected to be similar to the ice–snow surface before flooding occurred.

Similar results have been derived from indoor laboratory measurements of snow-covered saline ice, and all observations are summarized in Figures 9-37(a), (b), and (c). These figures show that the backscatter coefficient consistently decreases at nadir with the application of a snow layer. They also show a consistent increase at oblique angles of incidence. Figures 9-38(a), (b), and (c) support the argument that interface roughness rather than volume scattering is the explanation for these data. These figures show the time domain response at 13.9 GHz over bare and snow-

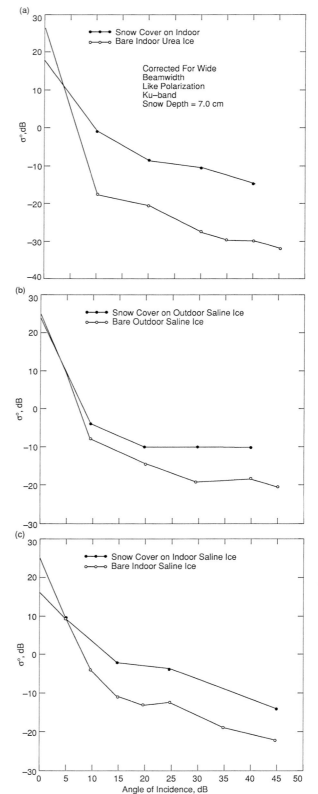

Fig. 9-37. Angular response of the backscatter coefficient at 13.9 GHz for three different bare and then snow-covered ice sheets: (a) urea grown indoors; (b) saline grown outdoors; and (c) saline grown indoors.

covered saline ice. The nadir signal from the ice surface is the large peak near the far left-hand side of Figure 9-38(a). Notice that application of a snow layer decreases the magnitude of the response, but application of additional snow does not change the pulse amplitude. Also there is no noticeable pulse broadening. At 45° incidence angle, the snow-covered return is about 10 dB higher than the equivalent bare ice result.

9.8 Significance of CRRELEX Results

Experiments using the outdoor and indoor test facilities at CRREL demonstrated that saline ice with a predefined range of physical properties can be grown. These saline ice types include thin ice with varying crystalline textures and salinity gradients, roughened ice that includes a range of roughness elements, desalinated ice, and snow-covered ice. That selected ice properties can be isolated, duplicated, and studied in the laboratory is an important achievement. This achievement has been exploited to better understand the relationship between particular saline ice properties and microwave propagation phenomena. For example, there is excellent agreement between the dielectric constant of thin ice as determined from radar backscatter, radar transmission, and microwave emission observations. Moreover, changes in the dielectric constant are clearly seen to be related to the complex distribution of brine and the migration of brine through the ice sheet as the ice ages. Similarly, snow cover was found to have a profound effect on both active and passive microwave observations. Again, multisensor data could be interpreted consistently in terms of the dielectrically rough, slushy layer that develops at the snow–ice interface.

The unique combination of attributes that were part of CRRELEX makes it seem unlikely that equivalent insight into the specific microwave properties of saline ice could be obtained from field observations alone. Many of these new insights are very exciting, still, interpretation must be tempered by the broader objective of studying sea ice growing on the polar oceans. Growth in the natural ocean environment occurs under the influence of many competing effects. In turn, these effects integrate to yield a complex physical structure and associated microwave response. This natural integration reveals the basic limitation of CRRELEX, namely, that it is difficult, perhaps impossible, to simulate simultaneously all the processes that drive the formation of natural sea ice. For example, while desalinated ice of a particular composition could be grown in the laboratory, experimental techniques are far from duplicating the complete properties of even first-year ice. The CRRELEX experiences have established a new methodology for combining laboratory research that tests specific hypotheses using field measurements to evaluate the significance of a particular phenomenon in a complex environment.

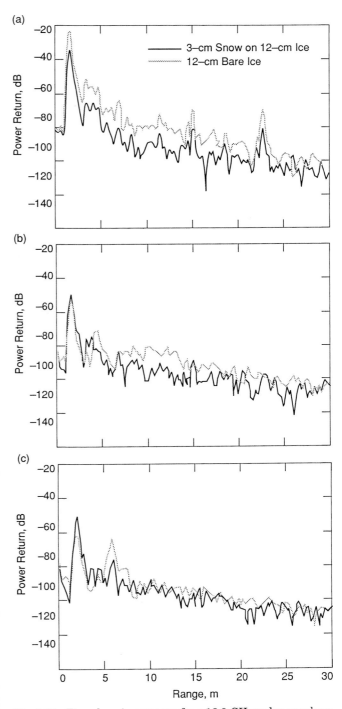

Fig. 9-38. Time domain response for a 13.9-GHz radar over bare and snow-covered saline ice. (a) The nadir signal from the ice surface corresponds to the large peak near the far left side. At (b) 25° and (c) 45° incidence angles, the backscattered signal with a slight time delay at oblique incidence also appears on the left side of each figure.

REFERENCES

Apinis, J. J. and W. H. Peake, *Passive Microwave Mapping of Ice Thickness*, Ohio State University Electrosciences Report 3892-2, Columbus, OH, 1976.

Arcone, S. A., A. J. Gow, and S. McGrew, Microwave dielectric structural and salinity properties of simulated sea ice, *IEEE Transactions on Geoscience and Remote Sensing, GE-24(6)*, pp. 832–839, 1986.

Bredow, J. W., *A Laboratory Investigation Into Microwave Backscattering From Sea Ice*, University of Kansas, Lawrence, Kansas, 173 pp., 1991.

Colom, J. G., *Development of a 35-GHz Network Analyzer Based Polarimetric Scatterometer*, University of Massachusetts, Amherst, Massachusetts, 107 pp., 1991.

Comiso, J. C., T. C. Grenfell, D. L. Bell, M. A. Lange and S. F. Ackley, Passive microwave in situ observations of winter Weddell sea ice, *Journal of Geophysical Research, 94(C8),* pp. 10,891–10,905, 1989.

Gogineni, S. P., R. K. Moore, P. Wang, A. J. Gow, and R. G. Onstott, Radar backscatter over saline ice, *International Journal of Remote Sensing, 11(4)*, pp. 603–615, 1990.

Gow, A. J., *Crystalline Structure of Urea Ice Sheet Used in Modelling Experiments in the CRREL Test Basin*, CRREL Report CR84-24, CRREL, Hanover, New Hampshire, 1984.

Gow, A. J., W. Tucker, and W. Weeks, *Physical properties of summer sea ice in the Fram Strait, June–July,* CRREL Report 87-16, U.S. Army Cold Regions Research and Engineering Laboratory, Hanover, New Hampshire, 1987.

Grenfell, T. C. and J. C. Comiso, Multifrequency passive microwave observations of first year sea ice grown in a tank, *IEEE Transactions on Geoscience and Remote Sensing, GE-24*, pp. 826–831, 1986.

Jezek, K. C., T. K. Stanton, A. J. Gow and M. A. Lange, Influence of environmental conditions on acoustical properties of sea ice, *Journal of the Acoustical Society of America, 88(4)*, 1903–1912, 1990.

Kim, Y.–S., R. G. Onstott, and R. K. Moore, The effect of a snow cover on microwave backscatter from sea ice, *IEEE Journal of Oceanic Engineering, OE-9(5)*, pp. 383–388, 1984.

Kovacs, A., R. M. Morey, G. F. N. Cox, and N. C. Valleau, *Modelling the Electromagnetic Property Trends in Sea Ice and Example Impulse Radar and Frequency Domain Electromagnetic Ice Thickness Sounding Results*, Technical Report, CRREL, Hanover, New Hampshire, 1986.

Lohanick, A. W., Microwave brightness temperatures of laboratory grown undeformed first-year ice with an evolving snow cover, *Journal of Geophysical Research*, in press, 1992.

Onstott, R. G., *Active Microwave Observations of the Formation Process of Simulated Sea Ice,* Technical Report 239500-2-T, Environmental Research Institute of Michigan, Ann Arbor, Michigan, 1991.

Parkinson, C. L., J. C. Comiso, H. J. Zwally, D. J. Cavalieri, P. Gloersen, and W. J. Campbell, *Arctic sea ice, 1973–1976: Satellite passive-microwave observations*, NASA SP-489, 296 pp., National Aeronautics and Space Administration, Washington, DC, 1987.

Perovich, D. and A. J. Gow, A statistical description of the microstructure of young sea ice, *Journal of Geophysical Research, 96*, 16,943–16,953, 1991.

Stanton, T. K., K. C. Jezek, and A. J. Gow, Acoustical reflection and scattering from the underside of laboratory grown sea ice: Measurements and predictions, *Journal of the Acoustical Society of America, 80*, 1486–1494, 1986.

Swift C. T. and D. J. Cavalieri, Passive microwave remote sensing for sea ice research, *Eos, Transactions of the American Geophysical Union, 66*, 1210–1212, 1985.

Swift, C. T., D. C. Dehority, and A. B. Tanner, Passive microwave spectral emission from saline ice at C-band during the growth phase, *IEEE Transactions on Geoscience and Remote Sensing, GE-24(6)*, pp. 840–848, 1986.

Ulaby, F. T., R. K. Moore, and A. K. Fung, *Microwave Remote Sensing—Active and Passive, Vol. II: Radar Remote Sensing and Surface Scattering and Emission Theory*, Addison–Wesley Publishing Company, Reading, Massachusetts, 1982.

Zwally, H. J., J. C. Comiso, C. L. Parkinson, W. J. Campbell, F. D. Carsey, and P. Gloersen, *Antarctic Sea Ice, 1973–1976: Satellite Passive-Microwave Observations,* 206 pp., NASA SP-459, National Aeronautics and Space Administration, Washington, DC, 1983.

Chapter 10. The Estimation of Geophysical Parameters Using Passive Microwave Algorithms

Konrad Steffen and Jeff Key

Cooperative Institute for Research in Environmental Sciences, University of Colorado, Boulder, Colorado 80309

Donald J. Cavalieri, Josefino Comiso, and Per Gloersen

Laboratory for Hydrospheric Processes, Goddard Space Flight Center, Greenbelt, Maryland 20771

Karen St. Germain

Department of Electrical and Computer Engineering, University of Massachusetts, Amherst, Massachusetts 01003

Irene Rubinstein

Ice Research and Development, York University, North York, Ontario, M3J 1P3, Canada

This chapter includes current algorithms based on passive microwave satellite data for use in ice-parameter retrieval in polar regions. Seven algorithms for ice concentration and two algorithms for ice temperatures are discussed in detail. With the exception of one algorithm, the derivation of ice concentration is based on a radiative transfer model for the surface and the atmosphere in combination with a mixing formalism that assumes that only sea ice and open water are within the footprint of the sensor. With the present state of the art, total ice concentration can be estimated with an accuracy of 5% to 10% during the dry period (fall, winter, and spring), and 10% to 20% during the wet period (summer). Existing passive microwave algorithms could be improved further by incorporating atmospheric data (e.g., opacity) in the radiative transfer equation. The major shortcoming of today's algorithms is that they do not unambiguously resolve ice classes; we hope this shortcoming will be overcome in the near future by the fusion of passive microwave data with synthetic aperture radar (SAR), thermal-infrared, and visible satellite data.

10.1 Introduction

10.1.1 Background of Passive Microwave Algorithm

Microwave radiation from the Earth is a complex function of the temperature, physical composition, and properties of the Earth's surface altered by the absorption, emission, and scattering of the atmosphere. The quantitative determination of environmental parameters is obtained from a limited set of microwave radiation measurements.

In general, these measurements are not sufficient for a unique determination of environmental parameters without using both a priori empirical knowledge and mathematical models of the relationship between parameters and the measured radiometric temperatures. The empirical information is used to impose limits within which the parameters can vary. The accuracy of the retrieved information is therefore affected by the noise in the data and the uncertainties in assumptions used in the model equations.

The potential of passive microwave remote sensing in polar regions is manifold. Monitoring the spatial and temporal variations of sea ice is one of the most important elements in studying the global heat budget, as sea ice is believed to be an extremely sensitive indicator of climate change. On the operational side, monitoring is also essential for navigating in polar seas. Therefore, the derivation of such parameters as ice extent, ice concentration, and ice type, first-year (FY) and multiyear (MY) ice, has been the main focus for the development of passive microwave algorithms over the past 10 years. This chapter presents techniques that make use of an atmospheric and ocean-surface radiative model, relating radiances at different frequencies from spaceborne microwave measurements to sea-ice parameters.

The NASA single-frequency algorithm was developed for the Electrically Scanning Microwave Radiometer (ESMR) data set, using a linear combination of water and ice brightness temperatures (Table 10-1). This derivation allows the mapping of ice concentration and ice extent for the Antarctic [Zwally et al., 1983], and the Arctic [Parkinson, et al., 1987]. The other six ice algorithms described in this chapter were developed for multifrequency, dual-polarized passive microwave data obtained from the Scanning Multichannel Microwave Radiometer (SMMR) and the Special Sensor Microwave/Imager (SSM/I). The Norwegian Re-

Microwave Remote Sensing of Sea Ice
Geophysical Monograph 68

TABLE 10-1. Passive microwave algorithm used for polar sea ice applications.

Algorithm	Method	Applications
NASA single frequency	Radiative transfer with mixing formalism, linear: 37 GHz	Gridded ESMR data, hemispheric
NASA Team	Radiative transfer with mixing formalism, nonlinear: 18 V/H, 37 V GHz; polarization and gradient	Gridded SMMR and SSM/I data, hemispheric with hemispheric tie points, regional with local tie points
AES–YORK and FNOC	Radiative transfer with mixing formalism, linear: 18, 37 GHz for ice concentrations >70%, 37 GHz for ice concentrations <70%, iterative technique for opacity	Gridded SMMR and SSM/I data, hemispheric and regional for navigation, hemispheric for operational ice forecasting
NORSEX	Radiative transfer with mixing formalism, linear: 10 V, 37 V GHz; iterative process for atmospheric opacity	Gridded SMMR and SSM/I data, regional with additional atmospheric input
U-Mass	Radiative transfer with mixing formalism, linear: 10 V/H, 19 V/H, 37 V/H GHz; iterative technique	SMMR and SSM/I orbital data, Northern Hemisphere only, marginal ice zone and high-density lead zone best performance
GSFC	Radiative transfer with mixing formalism, linear: 10 V/H, 18 V/H, 37 V/H GHz, Bootstrap technique	Gridded SMMR and SSM/I data, hemispheric
AI Systems	Neural network technique, expert systems 37 and 18 GHz	Gridded ESMR, SMMR, and SSM/I data, regional

mote Sensing Experiment (NORSEX) algorithm, basically a linear function, uses an iterative process to refine the atmospheric opacity for deriving final values of ice concentration. The NASA Team algorithm makes use of two radiance ratios as independent variables (polarization and spectral gradient ratio) and can be defined as a nonlinear algorithm. This algorithm is widely used by the research community, and a thorough validation program was carried out to estimate the algorithm accuracy (see Chapter 11). The Canadian Atmospheric Environmental Service at York (AES–York) and the Fleet Numerical Oceanographic Center (FNOC) algorithms use a linear combination of different brightness temperatures to derive the ice concentration. These algorithms are used operationally to provide navigational support in Arctic waters. The University of Massachusetts (U-Mass) algorithm retrieves ice temperature, atmospheric vapor, and open water and wind speeds, in addition to ice concentration using five brightness temperature channels. In an iterative mode, these additional parameters are used to improve the retrieval accuracy of the ice concentration. This algorithm can be applied only to orbital satellite data, whereas the other algorithms discussed in this chapter are usually applied to gridded brightness temperatures. The Goddard Space Flight Center (GSFC) Bootstrap algorithm derives ice concentration from a bootstrap approach, using the distribution of brightness temperature clusters in a scatter plot. This technique uses multidimensional cluster analysis as a valid alternative to the already mentioned algorithms, as more than two ice types can be identified using the scatter plots. Knowledge-based systems and neural networks are a promising avenue for future ice-concentration retrieval from satellite data. This new approach is described at the end of the ice-algorithm section.

The second part of this chapter is devoted to ice-temperature retrievals. The NASA Team algorithm uses 6.6-GHz brightness temperatures and derived ice concentration as input data. The U-Mass ice-temperature algorithm is similar to the NASA Team algorithm, with the exception that ice type distinctions are made. A second ice-temperature algorithm developed at U-Mass, called the decision-making algorithm, is part of their ice-concentration retrieval.

10.1.2 Radiative Transfer Equation

In the microwave wavelength region and for the physical temperatures encountered, the Rayleigh–Jeans approximation to the Planck radiation law pertains, and the radiated power (usually expressed as brightness temperature T_B) is therefore proportional to the physical temperature (T_S). However, most real objects emit only a fraction of the radiation that a perfect emitter would at the same physical temperature. This fraction defines the emissivity (e) of the object, so that brightness temperature is as shown in Equation (1). Spatial variations in the brightness temperature observed over the surface of the Earth are due primarily to variations in the emissivity of the surface material and secondarily to variations in physical temperature. For example, the emissivity of seawater at 19 GHz is about 0.44 compared to 0.92 for first-year sea ice and 0.84 for multiyear sea ice.

The microwave brightness temperature (T_B) of the Earth's surface depends on the electrical properties of the surface, embodied in its emissivity (e) and the physical temperature of the radiating portion of the surface (T_S). This may be expressed by the following relationship in terms of the wavelength (λ) and polarization (p); this relationship is true only for e and T_s independent of depths (a typical assumption for sea ice):

$$T_B[\lambda, p] = e[\lambda, p]\, T_S \qquad (1)$$

The radiative transfer equation is the basis for developing algorithms that convert the satellite radiance data into geophysical parameters. The microwave radiances received by the satellite are composites of various contributions from the Earth, atmosphere, and space and are illustrated schematically in Figure 10-1. The radiation received by the satellite is expressed as a brightness temperature $T_{B,ij}$, where the subscript i refers to wavelength and subscript j to polarization:

$$T_{B,ij} = e_{ij} T_S e^{-\tau_i} + T_{B,\mathrm{up},i} + (1 - e_{ij}) T_{B,\mathrm{down},i}\, e^{-\tau_i}$$
$$+ (1 - e_{ij}) T_{B,sp,i} e^{-2\tau_i} \qquad (2)$$

where e_{ij} is the surface emissivity; T_S, as before, is the physical temperature of the radiating portion of the surface; τ_i is the atmospheric opacity; $T_{B,\mathrm{up},i}$ is the atmospheric upwelling radiation; $T_{B,\mathrm{down},i}$ is the atmospheric downwelling component of radiation; and $T_{B,sp}$ is the cosmic background component. Both $T_{B,\mathrm{up}}$ and $T_{B,\mathrm{down}}$ may be expressed as a product of a constant weighted average atmospheric temperature $\langle T \rangle$ and the total atmospheric emission. In turn, $\langle T \rangle$ is assumed to be linearly related to T_S and so to $T_{B,ij}$, to the first order (neglecting the temperature dependence of e_{ij}).

10.2 ICE TYPE AND CONCENTRATION

10.2.1 NASA Single-Channel, Single-Polarization Algorithm

10.2.1.1. Equations and assumptions. If the sensor's field-of-view satellite includes a mixture of two materials, the observed brightness temperature is approximately a linear combination of the respective brightness temperatures according to the fractional areas of the two materials. For an area of seawater and sea ice with emissivities of e_W and e_I, with physical temperatures of T_W and T_I, and with atmospheric effects ignored,

$$T_B = (1 - C)\, e_W T_W + C e_I T_I \qquad (3)$$

where the sea-ice concentration is defined as C, and the fractional area covered by seawater is, therefore, $(1 - C)$. If the emissivities and physical temperatures are known or estimated, sea-ice concentration is determined by Equation (3) as a function of the observed brightness temperature at a particular frequency and polarization:

$$C = \frac{T_B - e_W T_W}{e_I T_I - e_W T_W} \qquad (4)$$

The full radiative-transfer equation applicable to passive microwave observations at a given wavelength within the sea-ice canopy includes terms that account for the attenuation of the radiation in the atmosphere, atmospheric emission, and reflection of background radiation from space (see Figure 10-1). The principal effect of the additional terms, which tend to be small in polar regions, is included in Equation (4) by interpreting $e_W T_W$ to be the brightness temperature of seawater as observed through a typical polar atmosphere. Variations in the atmosphere, as well as variations in the ocean emissivity due to foam and surface roughness, produce variations of the observed $e_W T_W$, with a standard deviation of about ± 5 K. Substantially larger variations, however, are caused by the presence of rainfall in the sensor's field of view. In the absence of atmospheric effects, variations of $e_W T_W$ are negligible because T_W tends to be nearly constant at 271 K in the presence of sea ice, and e_W at 19 GHz is inversely proportional to T_W, with a proportionality constant of approximately 130 K.

Over consolidated sea ice, the atmosphere is usually very dry, and cloudiness generally consists of underlying stratus that contains mostly ice crystals. These clouds have low opacity in the microwave region and could be neglected except for their influence on the sea-ice canopy surface temperature (T_I). Cloud effects on T_I have been observed to be as much as 7 K [Gloersen et al., 1973]. Annual variations in T_I are typically 30 K.

If the sea-ice pack consists only of FY ice and open water, the sea-ice concentration (C) can be determined by Equation (4) to about $\pm 15\%$, using estimated ice physical temperatures (T_I) [Comiso and Zwally, 1982; Zwally et al., 1983]. In estimating T_I, the approximate value is the temperature of the radiation layer, which for first-year sea ice is the saline ice just below the snow–ice interface. Consequently, T_I is usually somewhat warmer than the air temperature T_A, except near the melting point, and is approximately given by $T_I = T_A + 0.75\,(T_W - T_A)$.

If the sea-ice pack consists only of a mixture of FY and MY ice with emissivity e_{MY} and temperature T_{MY}, then the observed brightness temperature is

$$T_B = F e_{MY} T_{MY} + (1 - F) e_{FY} T_{MY} \qquad (5)$$

and solving for F gives

$$F = \frac{T_B - e_{FY} T_{MY}}{e_{MY} T_{MY} - e_{FY} T_{MY}} \qquad (6)$$

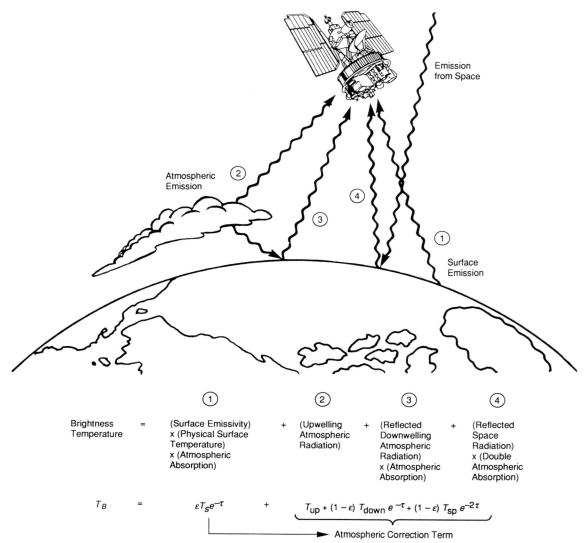

Fig. 10-1. Radiative transfer equation used to derive geophysical parameters from spacecraft microwave radiometer data [Swift and Cavalieri, 1985].

where F is defined as the fraction of ice cover that is MY ice. If open water is also present, the observed brightness temperature is

$$T_B = (1-C)e_W T_W + C(Fe_{MY} T_{MY} + (1-F)e_{FY} T_{MY}) \quad (7)$$

As can be seen in Equation (7), the values of both sea-ice concentration (C) and the MY fraction (F) cannot be obtained from only the single wavelength T_B obtained by a single-channel sensor.

10.2.1.2. Error analysis. Defining $T_{B,W} = e_W T_W$ as the brightness temperature over open water, including atmospheric effects, the partial differentials of C with respect to the four variables on the right-hand side of Equation (4) are:

$$\frac{\partial C}{\partial T_B} = \frac{1}{e_I T_I - T_{B,W}} \quad (8)$$

$$\frac{\partial C}{\partial T_{B,W}} = \frac{-(1-C)}{e_I T_I - T_{B,W}} \quad (9)$$

$$\frac{\partial C}{\partial e_I} = \frac{C T_I}{e_I T_I - T_{B,W}} \quad (10)$$

$$\frac{\partial C}{\partial T_I} = \frac{C e_I}{e_I T_I - T_{B,W}} \quad (11)$$

Because the four variables $T_B, T_{B,W}, e_I,$ and T_I are assumed independent, their total contribution to the error in calculating C may be obtained from the square root of the sum of their individual fluctuations multiplied by the above partials:

$$dC = \left\{ \left(\frac{\partial C}{\partial T_B} dT_B\right)^2 + \left(\frac{\partial C}{\partial T_{B,W}} dT_{B,W}\right)^2 + \left(\frac{\partial C}{\partial e_I} de_I\right)^2 + \left(\frac{\partial C}{\partial T_I} dT_I\right)^2 \right\}^{1/2}$$

(12)

where $dC, dT_B, dT_{B,W}, de_I,$ and dT_I are the errors in $C, T_B, T_{B,W}, e_I,$ and T_I, respectively.

Representative values for the coefficients represented by Equations (8) through (11) are given for the Nimbus-5 ESMR, which operated at a frequency of 19.35 GHz. All of these coefficients share a common denominator, which is approximately 100 K under winter conditions in the Central Arctic. Looking at the components of Equation (12) individually, the first term may be used to assess instrument errors. For the ESMR, the noise equivalent ΔT ($NE\Delta T$) was about 3 K, leading to a 3% contribution to the error budget for C, but newer instruments have lower $NE\Delta T$'s, typically less then 1 K. The second term in Equation (12) gives estimates of errors due to weather effects, which can give $\partial T_{B,W}$ values as high as 20 K. This represents dC's as high as 20% over open water ($C = 0$), which goes to zero for fully consolidated ice ($C = 1$). The third term represents errors that result from not knowing the ice type. For fully consolidated winter ice, the coefficient is approximately 2.3 at 19.35 GHz, the difference in emissivity is 0.08 [Parkinson et al., 1987], and the error in C that would be obtained by presuming all ice to be first year in a 100% multiyear region is 18%. The sensitivity to seasonal and spatial variations in ice temperatures is given by the fourth term in Equation (12). An uncertainty of 20 K in the ice temperature (T_I) leads to errors of 18% in C, which can be partially alleviated with the use of climatological temperature fields [Zwally et al., 1983].

Algorithm accuracy for the single-channel technique has been examined for ESMR Antarctic data by comparison with Landsat data [Comiso and Zwally, 1982]. The linear fit of Landsat versus ESMR data was given as an offset of +25% and a slope of 0.67, with a standard deviation of 12%.

10.2.2 NASA Team Algorithm

10.2.2.1. Equations and assumptions. Three types of surfaces are assumed to dominate the polar ocean environment; these are ice-free ocean (open ocean), first-year sea ice, and multiyear sea ice. The brightness temperature spectra for both the horizontal and vertical polarizations at Nimbus-7 SMMR frequencies for each of the three surface types are shown in Figure 4-18, as determined from three regions of the Arctic during February 3 to 7, 1979. The surface emission may be written as:

$$T_{B,ij} = C_W T_{B,W,ij} + C_{FY} T_{B,FY,ij} + C_{MY} T_{B,MY,ij} \quad (13)$$

where $T_{B,W,ij}, T_{B,FY,ij},$ and $T_{B,MY,ij}$ are the brightness temperatures of ice-free ocean, first-year sea ice, and multiyear sea ice, respectively, where the subscripts i refers to frequency and j to polarization. Here $C_W, C_{FY},$ and C_{MY} are the fractions of each of the three ocean-surface components within the field of view of the instrument; they add to unity. Equation (13) is fundamental to developing the algorithm described below.

The physical basis of the algorithm for distinguishing among the three ocean components is described by Figure 4-18. This figure illustrates two important characteristics. First, the difference between the vertically and horizontally polarized radiances is small for either ice type in comparison with that for the ocean. Second, the discrimination between ice types increases with decreasing wavelength. This discrimination is greatest at 37 GHz. The algorithm parameterizes these two characteristics through two radiance ratios that are used as independent variables [Cavalieri et al., 1984]. These are the polarization ratio (PR) and the spectral gradient ratio (GR) defined by

$$PR[18] = \frac{T_B[18V] - T_B[18H]}{T_B[18V] + T_B[18H]} \quad (14)$$

$$GR[37V/18V] = \frac{T_B[37V] - T_B[18H]}{T_B[37V] + T_B[18H]} \quad (15)$$

The advantage of using ratios of radiances is that they reduce the dependence of ice-temperature variations, which eliminates the problem of estimating ice temperatures both temporally (e.g., day-to-day and seasonal) and spatially (e.g., temperature gradients across the Arctic Basin). Equation (13) is used with the definitions of PR and GR, Equations (14) and (15), respectively, to solve for first-year and multiyear ice concentrations. The expressions for C_{FY} and C_{MY} are:

$$C_{FY} = \frac{F_0 + F_1 PR + F_2 GR + F_3 (PR)(GR)}{D} \quad (16)$$

$$C_{MY} = \frac{M_0 + M_1 PR + M_2 GR + M_3 (PR)(GR)}{D} \quad (17)$$

where

$$D = D_0 + D_1 PR + D_2 GR + D_3 (PR)(GR)$$

$$C_T = C_{FY} + C_{MY}$$

where C_T is the total sea-ice concentration. The numerical coefficients $F_i, M_i,$ and D_i for $i = 0,3$ are based on observed brightness temperatures and so depend on the microwave sensor [Gloersen and Cavalieri, 1986]. The polarization at 37 GHz has also been tested for obtaining concentration, using a different set of coefficients in Equations (16) and (17), at an improved spatial resolution, but at the expense of greater atmospheric interference.

Figure 10-2 shows a typical PR versus GR plot for the Northern Hemisphere. The three corner points indicate the regions in PR–GR space that are characteristic of open water ($C = 1$), first-year ice ($F = 1$, $C = 100$), and multiyear ice ($F = 0$, $C = 100$). The large spread of points in the vicinity of open water is due to an increase in received radiance by SMMR from atmospheric effects such as the roughening of the sea surface by wind, atmospheric water vapor, and cloud liquid water. Curved lines connecting the three surface reference points form a triangle; the physically meaningful solutions for C_T and C_{MY} lie within this triangle. Finally, a property of GR, that it is negative for areas of multiyear ice, approximately zero for areas of first-year ice, and positive for open ocean, allows the elimination of weather-related effects that would otherwise be interpreted by the algorithm as sea-ice concentration over open ocean. The horizontal line ($GR = 0.08$) in Figure 10-2 illustrates the effectiveness of this technique for eliminating most of the points that lie outside the algorithm triangle at low concentrations. This so-called weather filter is described in detail by Gloersen and Cavalieri [1986].

Examples of the March 1979, monthly averaged total ice-concentration and multiyear ice-concentration maps are shown in Figures 10-3(a) and (b). Total ice concentration in the central Arctic varies from about 88% to 100% with lower concentrations in the marginal seas, Figure 10-3(a). It is important to keep in mind the limitations of the algorithm when interpreting these images. For example, since the algorithm is limited to distinguishing between two ice types (generic names of FY and MY ice are assigned to these), other radiometrically different ice types (or ice surfaces) will result in either an overestimation or underestimation of total ice concentration. New ice types will result in a low concentration bias. Thus, the amount of open water cannot be assumed to be simply $1 - CT$, but instead is an indeterminate mixture of open water and new ice. This is the case at least during the cold months of the year. The degree to which variations in ice type result in ice-concentration errors depends on the degree to which PR and GR are affected. A detailed discussion of these and other errors is given in Section 10.2.2.3.

The multiyear ice-concentration map, Figure 10-3(b), covering the Central Arctic shows the dominant concentrations ranging from about 30% north of the Chukchi Sea to about 90% north of the Canadian Archipelago.

10.2.2.2. Hemispheric versus local tie points. Tie points are brightness temperature values of open water ($T_{B,W}$), first-year ice ($T_{B,MY}$), and multiyear ice ($T_{B,FY}$) as observed by the sensor and include atmospheric effects. They are critical for the accurate performance of the ice concentration retrieval algorithm and are empirically determined based on statistics of hemispheric brightness temperatures [Cavalieri et al., 1984]. If such hemispherically chosen tie points are used for the calculation of total and multiyear ice concentration, the variations of $T_{B,W}$, $T_{B,FY}$, and $T_{B,MY}$ over time and space are ignored. However, there are large variations in $T_{B,W}$ along the ice edge and in open pack ice areas caused by various factors, including surface roughness, foam, and atmospheric water vapor. A combination of these conditions can increase the brightness temperature of the ocean by as much as 40 K. Also $T_{B,FY}$ values are affected by spatial and temporal variations in physical temperature, characteristics of the emitting surface, and atmospheric conditions [Steffen and Maslanik, 1988]. To account for these variations, ice concentrations based on local tie points determined in the area of interest can also be derived using the procedure described by Steffen and Schweiger [1991]. Local ocean tie-point T_B values are usually higher than the hemispheric ones. This can be explained by the influence of wind since the brightness temperatures of the ocean surface increase with wind speed. Hemispheric ocean tie points are statistically derived from the entire SMMR or SSM/I data set for one year as the near-minimum brightness temperature, and therefore represent calm ocean conditions. The difference between hemispheric and local ocean tie points can be as large as 18 K [Steffen and Schweiger, 1991], which corresponds to a difference in wind speed of 18 m/s, based on an ocean-surface wind-speed algorithm developed by Hollinger et al. [1987].

10.2.2.3. Error analysis. The purpose of this section is to provide a quantitative estimate of the uncertainties in the ice-concentration values due to random and systematic

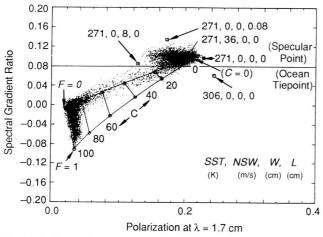

Fig. 10-2. Spectral gradient ratio versus polarization at 1.7 cm for the north polar area, February 3 to 7, 1979. The curved triangle is a representation of the algorithm used to calculate sea-ice concentration and age. Terms are listed as sea-surface temperature (K), near-surface wind (m/s), water vapor (cm), and cloud droplets (cm). The arrows indicate model calculation of GR and PR deviations from cold, specular, oceanic conditions [Gloersen and Cavalieri, 1986].

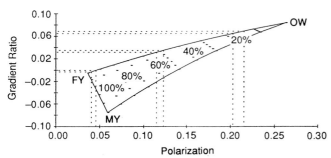

Fig. 10-4. Sensitivity of ice-concentration calculations to gradient ratio and polarization variations [Gloersen et al., 1992].

tually constant indicating that there is little variation in the sensitivity to GR with changes in total ice concentration.

10.2.3 AES–York and FNOC Algorithms

10.2.3.1. *Equations and assumptions.* Due to the noncoherent nature of the natural microwave emission from the surface, the surface radiation can be written as a sum of contributions from different ice types and open water. If C represents the fraction of the footprint covered with ice, the surface radiation term can be written as

$$T_{B,ij} = CT_{B,I,ij} + (1 - C)T_{B,W,ij} \qquad (18)$$

where subscript i refers to the frequency and subscript j to polarization, $T_{B,I,ij}$ is the ice brightness temperature of ice, and $T_{B,Wij}$ is the brightness temperature of open ocean. The surface emissivity can be written as a sum of emissivities from the ice fraction and open ocean:

$$e_{ij} = Ce_{I,ij} + (1 - C)e_{W,ij} \qquad (19)$$

where $e_{I,ij}$ and $e_{W,ij}$ are ice and ocean (ice-free) surface emissivities.

Replacing the surface emission term in Equation (2) by Equation (18) and e_{ij} by Equation (19), one can rewrite Equation (2) for vertical and horizontal polarization, at the same frequency brightness temperature:

$$T_{B,j} = C\big(T_{B,I,j} - T_{B,W,j} + \big(e_{W,j} - e_{I,j}\big)\big(T_{B,\text{down}} + T_{B,sp}e^{-\tau}\big)$$

$$+ \big(T_{B,W} + \big(1 - e_{W,j}\big)\big(T_{B,\text{down}} + T_{B,sp}\big)\big)\big)e^{-\tau} + T_{B,\text{up}} \qquad (20)$$

where $T_{B,\text{up}}$ and $T_{B,\text{down}}$ are proportional to the upward and downward microwave emission from the atmosphere, and $T_{B,sp}$ is the sky background radiation. The subscript j refers to vertical and horizontal components of the observed T_B; τ stands for the atmospheric opacity.

Taking the difference between vertical and horizontal polarization, Equation (20), and neglecting a contribution from space yields the following expression for C:

$$C = \frac{T_{B,\Delta j}e^{\tau} - T_{B,W,\Delta j}}{T_{B,I,\Delta j} - T_{B,W,\Delta j} + (e_{W,\Delta j} - e_{I,\Delta j})T_{B,\text{down}}} \qquad (21)$$

where the subscript Δj represents the difference between vertical and horizontal polarization.

The general form of Equation (21) can be written as:

$$C = AT_{B,\Delta j} + B \qquad (22)$$

Coefficients A and B are obtained from Equation (21) and are functions of the ice and ocean signature emissivities, signature brightness temperatures, atmospheric opacity, and atmospheric contribution. Note that the difference between vertical and horizontal signatures is used in these calculations. The dependence of total ice concentration included by the uncertainty in selecting atmospheric opacity is illustrated in Figure 10-5. For example, selecting atmospheric opacity equivalent to polar winter conditions (0.05) instead of midlatitude winter (0.1) would result in about a 5% difference in calculated total ice concentrations for pixels containing less than 25% ice.

For the SSM/I prelaunch algorithm (known as the Hughes algorithm), an assumption was made that $T_{B,\Delta j}$ is independent of the ice type, i.e., is the same for first-year and old ice samples. This type of algorithm can be tuned to be very sensitive to the sea-ice presence within the pixel, if the weather conditions are used as an input to calculations of A and B. In the absence of meteorological information, it can lead to false classification of an ice-free pixel for wind-roughened ocean or overcast sky. The errors in the estimation of the concentration can also occur if an inappropriate value of $T_{B,I,\Delta j}$ is selected as a threshold value for the ice presence within a pixel. For any two channels with different sensitivities to the atmospheric conditions and different sea-ice and open-ocean signatures, Equation (20) can be

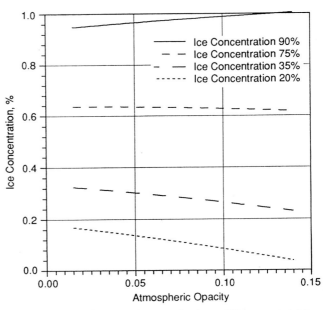

Fig. 10-5. The influence of atmospheric conditions on total ice-concentration retrieval (Hughes algorithm).

rewritten to be frequency dependent. Assuming that only two ice types are present within the pixel (where C_{F1} is the first ice type and C_{F2} is the concentration of the second ice type), and that C_W is the fraction of open water, these equations can be solved for the ice fractions under the following constraint:

$$C_{F1} + C_{F2} + C_W = 1 \tag{23}$$

Replacing $(1 - C)$ by $(1 - C_{F1} - C_{F2})$, one can show that solutions for C_{F1} and C_{F2} can be written as:

$$
\begin{aligned}
C_{F1} &= A_1 T_{B,i=1} + B_1 T_{B,i=2} + C_1 \\
C_{F2} &= A_2 T_{B,i=1} + B_2 T_{B,i=2} + C_2
\end{aligned} \tag{24}
$$

$$
\begin{aligned}
C &= C_{F1} + C_{F2} \\
&= A_C T_{B,i=1} + B_C T_{B,i=1} + C_C
\end{aligned} \tag{25}
$$

Coefficients A_1, B_1, C_1, A_2, B_2, C_2, A_C, B_C, and C_C, are obtained from Equation (20) with frequency dependence, and $T_{B,i=1}$ and $T_{B,i=2}$ are the brightness temperatures at 19 and 37 GHz, respectively.

The current knowledge of microwave sea-ice and open-ocean brightness temperatures allows retrieval of open-water, first-year, and multiyear ice fractions. The largest source of error in using brightness temperatures for calculations of the ice fractions is the uncertainty in the signatures. The spatial and seasonal variability of the ice signatures has to be included in the retrieval procedure. The analysis of several years of 37- and 18-GHz SMMR brightness temperatures, supplemented with SSM/I data, was used to generate an ice classification technique that includes information on the range of the variabilities. In addition, correction of spurious ice-concentration retrieval (labeled weather filter) caused by the wind, cloud, and precipitation effects on the brightness temperatures is included in the algorithm. The weather effects can be corrected imposing threshold values on $T_{B,i=1}-T_{B,i=2}$, but this results in eliminating ice concentrations of less than 15%. This cutoff is higher for newer ice-type areas. The use of the highest resolution channel for correcting the severity of the cutoff procedure allows reclassification of pixels identified as ice free. One of the shortcomings of a dual-frequency algorithm with current data sets is that the use of 18-GHz brightness temperatures (19 GHz for SSM/I) reduces the spatial resolution since the lower frequency channel has coarser resolution. This loss of resolution is not noticeable if the pixels within the field of view are located within the ice pack. For pixels from the marginal ice zone area (ice concentrations less than 70%) AES–York and FNOC algorithms use only the 37-GHz channel. An additional requirement on the FNOC algorithm was to provide flagging for the presence of 35% or higher old-ice fractions within pixels containing sea ice. The identification of the ice types present within the field of view is one of the most

difficult tasks imposed on the algorithm. All algorithms in use at the present time have problems in providing an accurate estimate of the amount of old ice or thin ice contained within the study area. The flagging procedure is very simple. The brightness temperature of the ice area within the sample is calculated by using the 19-GHz V-channel:

$$T_{B,I} = \frac{T_{B,j=19V} - (1 - C)T_{B,W,j=19V}}{C} \tag{26}$$

where $T_{B,j=19V}$ is the observed brightness temperature, $T_{B,W,j=19V}$ is the open-ocean 19-GHz vertical channel signature brightness temperature, and C is the calculated total ice concentration. If $T_{B,I}$ is below $(0.35\, T_{MY,j=19V} + 0.65\, T_{FY,j=19V})$, the pixel is flagged. The procedure for more exact ice-fraction calculations is more complicated. Due to the regional and seasonal variabilities of the ice signatures, the AES–York algorithm contains several ice brightness-temperature testing routines. Clustering observations containing data representative of different regions is used to define the range of pure signature classes (see also Figure 10-9). Pixels containing mixtures of pure classes can be identified as such if they are located within the limits of the following lines: points along line DA contain multiyear ice and first-year ice; open-ocean and first-year ice mixtures are assumed to be along lines HA. Pixels not containing old ice can be identified by using the procedure described previously. Since the calculated ice fractions should be real positive numbers, special tests are performed on the incoming brightness temperatures if any of the calculated fractions are negative. One of the reasons for negative fractions is an overestimation of the atmospheric opacity. New calculation of the ice fractions is then performed with the atmospheric opacity set to a minimum value. If one of the fractions is still negative, the multiyear ice signature is replaced by a different value. If that does not help, the pixel is classified as containing the ice type of a positive fraction. The operational usefulness of the AES–York algorithm is enhanced by additional procedures not used in any other algorithm. The brightness temperatures for the pixels identified as ice free are sent through an oceanic parameters algorithm. Using calculated wind speeds, additional testing of the measured brightness temperature is done.

In this testing, it is assumed that for the ice-free pixels the difference between the measured brightness temperatures at 19 and 37 GHz, corrected for wind-induced effects, and the calm ocean $AT_{B,\Delta j}[19]$ and $AT_{B,\Delta j}[37]$ at 19 GHz should be a fraction of the 37-GHz values, since 19 GHz is less affected by the atmosphere. For the pixels containing ice, the calculated difference for 19 GHz should be equal to or greater than the 37-GHz estimate of $AT_{B,\Delta j}[37]$. The AES–York algorithm is used to provide navigational support during sea-ice and ocean-research campaigns. The algorithm was coded at Canadian Meteorological Center (Dorval, Quebec), where SSM/I data are processed daily.

The retrieved sea-ice, ocean-surface wind speed, and precipitation information is available to the Canadian weather and ice-forecasting offices.

10.2.3.2. Error analysis. The modular structure of the AES–York algorithm demands error evaluation for different modules of the retrieving procedure. The errors in the first-guess total ice concentrations calculated using the dual-frequency equation are due to instrument noise and uncertainty in the signatures and atmosphere parameters. The instrument noise leads to 1.7% uncertainty in first-year ice retrieval and 3.7% uncertainty for old ice. The effect on total ice concentration is about 4.1%. The signature uncertainty and the uncertainty in atmospheric parameters, including the instrument noise effect, can lead to 9.5% uncertainty in total ice concentration and to about 6.5% in the ice fractions.

10.2.4 NORSEX Algorithm

10.2.4.1. Equations and assumptions. The radiation model used in the NORSEX algorithm is composed of two main parts, a model of the surface and a model of the atmosphere, Equation (2), Figure 10-1. Since the total atmospheric opacity (τ) is small, the approximation $e^{-\tau} = 1 - \tau$ is used, and the term $T_{B,sp}2\tau$ is neglected. Furthermore, downwelling and upwelling radiation have been assumed to be equal. The term $T_{B,\text{up}}$, Equation (2), has been replaced by δT_A, the weighted average atmospheric temperature in the lower troposphere [Gloersen et al., 1978; Shutko and Granov, 1982]. The first and dominant term in Equation (2) (eT_S) is replaced by T_B. In the remaining correction terms, e is written as T_B/T_S, and T_S is approximated as $\beta - 1T_A$, where β is a constant close to 1. From this procedure, the following equation is derived for the computation of brightness temperature at the surface [Svendsen et al., 1983]:

$$T_B = \frac{T_{B,H} - 2\delta T_A \tau + \delta T_B \tau^2 - T_{B,sp}}{1 - \tau - \beta\delta(\tau - \tau^2) - \beta(T_{B,sp}/T_A)} \quad (27)$$

where $T_{B,H}$ is the brightness temperature sensed at satellite height.

Values for τ in atmospheres with $T_{B,A} = 270$ K (subarctic winter) and $T_{B,A} = 250$ K (Arctic winter) are given by Reeves [1975]. The subarctic winter values compare well with the mean opacities measured during NORSEX (Table 10-3). However, large variations occurred in the measurements [NORSEX Group, 1983], indicating that the opacities vary considerably with conditions other than $T_{B,A}$, e.g., clouds.

The parameters β and δ are given the values of 0.95 and 0.9 in actual computation. These values are sufficiently good approximations, considering the uncertainties already present in the values for τ. The emitted brightness temperature T_B from a surface consisting of three components is now considered. The three components are (1) seawater, effective temperature 272 K, emissivity e_W, and concen-

TABLE 10-3. Modeled atmospheric zenith opacities (optical depth) compared with measured mean NORSEX values [after Svendsen et al., 1983].

Parameter	Frequency, GHz			
	4.9	10.4	21	36
Subarctic winter model atmosphere	0.011	0.015	0.050	0.082
Arctic winter model atmosphere	0.010	0.012	0.036	0.058
NORSEX ice edge Fall 1979 atmosphere	0.011	0.015	0.062	0.073

tration C_W; (2) multiyear ice, effective temperature T_I, emissivity e_{MY}, and concentration C_{MY}; and (3) first-year ice, effective temperature T_I, emissivity e_{FY}, and concentration C_{FY}. Concentrations are defined as area fractions:

$$1 = C_{MY} + C_{FY} + C_W \quad (28)$$

The emitted brightness temperature from the three-component surface is assumed to be the sum of individual emitted brightness temperatures, weighted by the concentration:

$$T_B = eT_S = C_{MY}e_{MY}T_I + C_{FY}e_{FY}T_I + C_We_W272 \quad (29)$$

The emissivities e_{MY}, e_{FY}, and e_W have all been measured by the NORSEX shipborne radiometer near the ice edge. However, comparison with aircraft measurements indicated that these emissivities are also representative of the ice farther into the pack.

As T_I is not readily measured, we need a relation between T_I and the surface temperature that is assumed to be equal to the surface atmospheric temperature T_A. On the basis of this reasoning that T_I is found somewhere between the surface and the underlying water, we have

$$T_I = \alpha T_A + (1 - \alpha)272 \quad (30)$$

To estimate α, the variation of T_B (calculated from the adjusted SMMR measurements) of an area north of Ellesmere Island was studied in relation to changes in surface air temperature measured from ice-drifting buoys [Svendsen et al., 1983]. Combining Equations (29) and (30) and taking the derivative $\delta/\delta T_A$, assuming the emissivities and α are weak functions of T_A, we have a rough estimate of $\alpha = 0.4$. Comparing this with NORSEX temperatures of multiyear ice covered with snow indicates that the temperature near the snow–ice interface represents the effective temperature T_I. With the correct choice of two SMMR channels and knowledge of the surface air temperature, C_{MY} and C_{FY} can be computed from Equations (27) through (30). The combination of 10-/37-GHz vertical polarization

(10 V/37 V) is chosen for estimates of C_{MY} and C_{FY}, but comparable estimates from the combination 18 V/37 V can also be made to give a better resolution.

The algorithm consists of the following computational steps:

(1) The available SMMR data are adjusted (see above) to give the brightness temperature $T_{B,H}$ at satellite height.
(2) An initial transformation of $T_{B,H}$ to emitted brightness T_B is performed using an atmosphere corresponding to T_p ($T_p = T_A$).
(3) By solving Equations (27) through (30) (using two channels gives two equations with two unknowns), we find the initial concentrations using $T_A = T_p$.
(4) A refined atmosphere (τ and T_A) is found by using the initial total ice concentration from step (3) to interpolate between surface temperature over water ($T_A = 272$ K) and over 100% ice ($T_A = T_p$). The refined opacities are then found from T_A.
(5) Steps (2) and (3) are repeated using the refined atmosphere, arriving at final values for the concentration.

10.2.4.2. Error analysis. During NORSEX, several flights were performed with the NASA C-130 aircraft carrying several microwave instruments. Data from the nadir-looking stepped frequency microwave radiometer (SFMR) and the multifrequency microwave radiometer (MFMR) were used to derive ice concentrations that were compared with the computed SMMR ice concentrations along the same line [Svendsen et al., 1983]. In the comparison, one has to remember that the aircraft measurements represent a line through the large SMMR footprints, and that the aircraft senses less atmosphere than the satellite. To account for some of this, a low pass (80-km running mean) was used on the aircraft data. The model used for the aircraft data is based on Equations (28) and (29).

A very good fit has been found between the SMMR using 10 and 37 GHz and the aircraft estimates of total ice concentration; there was a mean difference of 0.5 ±2.5%. For the ice concentration, the absolute accuracy of the aircraft total ice concentration estimates is ±3%. The difference between the SMMR 18- and 37-GHz channels and the aircraft total ice concentrations is −3.5 ±2.0%, indicating that with this channel combination the SMMR estimates are a few percent too low.

The multiyear ice estimates show a mean difference between those of the SMMR 10- and 37-GHz channels and those of the aircraft of −4 ±6%, and between those of the SMMR 18- and 37-GHz channels and those of the aircraft of −8 ±6%. The absolute accuracy of the aircraft multiyear ice concentration is estimated to be ±10%.

10.2.5 U-Mass Decision Algorithm

10.2.5.1. Equations and assumptions. A Decision algorithm for extraction of sea-ice fractions from SSM/I brightness temperatures was developed at the University of Massachusetts (U-Mass). The algorithm utilizes the vertical polarization channels at 19, 22, and 37 GHz, and the horizontal channels at 19 and 37 GHz. The uniqueness of the algorithm lies in a decision routine that chooses the variables of primary concern for each pixel, making it possible to account for more variables than were previously possible. The basis of the algorithm is the radiative transfer model presented in Equation (2). This includes an empirical model for atmospheric attenuation and emission in terms of integrated atmospheric vapor and integrated cloud liquid water. In addition, open-water emissivity is modeled as a function of wind speed. The resulting equation yields a frequency-dependent relationship between apparent brightness temperature (T_B) and a total of six unknowns: first-year ice (C_{FY}), multiyear ice (C_{MY}), water concentration (C_W), composite physical temperature (T_S), integrated atmospheric vapor (V), integrated cloud liquid water (L), and open-water surface wind speed (w). The radiative transfer equation then takes the form:

$$T_{B,i} = eT_S(-\tau[V,L]T_S) + T_{B,sp}(1-e)(-2\tau[V,L])$$

$$+ T_{B,\text{down}}[V,L](1-e)(-2\tau[V,L])$$

$$+ T_{B,\text{down}}[V,L](-\tau[V,L]) \tag{31}$$

where

$T_{B,i}$	=	brightness temperature with subscript i referring to frequency
e_W	=	emissivity of water
e_{FY}	=	emissivity of first-year ice
e_{MY}	=	emissivity of multiyear ice
$T_{B,sp}$	=	cosmic background radiation
$T_{B,\text{down}}$	=	atmospheric integrated brightness temperature
e	$\approx$	$F_{FY}e_{FY,i} + F_{MY}e_{MY,i} + F_w e_{w,i}$
e_W	=	$e_{cw} + \mu_w$
e_{cw}	=	"Klein and Swift" calm-water emissivity
w	=	wind speed, m/s
μ	=	empirically derived constant relating to excess emissivity
$\tau[V,L]$	=	$AV+BL+C$ = atmospheric attenuation (A, B, and C are empirically derived constants)
T_s	$\approx$	$(F_{FY}e_{FY} + F_{MY}e_{MY})T_I + F_W e_W(272)$

Initially, all variables are set to a "first-guess value," and the algorithm is set to "search mode": first-order parameters (C_{FY}, C_{MY}, V and) are allowed to vary, while all others are held constant. The Newton Raphson minimum square error (MSE) iterative technique is employed to arrive at a next guess for the varying parameters [Carnahali et al., 1964]. The algorithm then passes this information through a decision-making routine, which, based primarily on the surface-type fractions, decides which of the higher order parameters will have a significant effect. The decision tree is shown in Figure 10-6, where each level represents an iteration. For example, if the sum of the ice fractions is greater than 0.9, it would be assumed that open-water surface wind speed would have a negligible effect. Furthermore, under these conditions, integrated cloud liquid water is not expected to be detectable. This leaves four variables for which to solve. In successive iterations, under certain conditions, integrated atmospheric vapor may also be eliminated as a variable.

When a significant amount of open water is present in the footprint (>20%), the matrix inversion required for an MSE solution becomes unstable. This occurs because the emissivity of seawater is a function of temperature such that the derivative of the brightness temperature with respect to physical temperature approaches zero. For these cases, the physical temperature is set to a constant, and the iterations continue. For cases where less than 20% of the footprint consists of water, the composite surface temperature is treated as a variable. When the algorithm has completed its solution, a short routine converts the composite temperature to an ice temperature by eliminating the open water component as follows:

$$T_I = \frac{T_S - 272.0 F_W \frac{e_W}{e}}{(F_{FY} + F_{MY})\frac{e_I}{e}} \qquad (32)$$

$$e_I = \frac{F_{FY}e_{FY} + F_{MY}e_{MY}}{F_{FY} + F_{MY}} \qquad (33)$$

It is important to note that an algorithm of this sort, i.e., one that solves for parameters that may be rapidly varying, should not be run using daily average gridded brightness temperatures. The relationship between brightness temperature and the atmospheric parameters is nonlinear, that is, the parameter values retrieved from averaged data will not be equal to the average of the values retrieved from individual looks if the atmospheric vapor content is changing. The same is true of open-water wind speed. In fact, attempting to solve for these variables from time-averaged data can cause the algorithm to diverge or, at best, produce unbelievable results. Quantitatively speaking, this occurs because by linearly averaging nonlinear data, second-order information is lost and, in fact, this information looks like noise. This problem has been discussed several times in the open literature and can be demonstrated with data [Rothrock and Thomas, 1988].

10.2.5.2. Error analysis. The retrieved ice fractions compare favorably with those of previously validated algorithms. Data comparisons at several buoy locations indicate that the Decision algorithm retrieves 7% to 10% more total ice than the Modified NASA Team algorithm, which is well within the uncertainty of the algorithms. Scatter plots of ice-type retrievals shown in Figure 10-7 indicate that the Decision algorithm retrieves approximately 20% more multiyear ice under conditions of total ice coverage. This difference is probably due to the different emissivity tie points used and the methods of determining them. In addition, images of the integrated atmospheric vapor re-

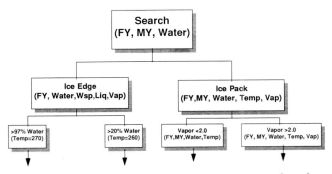

Fig. 10-6. Decision tree used in the U-Mass Decision algorithm. The algorithm variables are fractional coverage of first-year ice (FY), fractional coverage of multiyear ice (MY), fractional coverage of open water (Water), composite surface temperature (Temp) expressed in kelvin, integrated atmospheric water vapor (Vap) expressed in kg/m², integrated cloud liquid-water content (Liq) expressed in kg/m², and wind speed over open water (W_{sp}) expressed in m/s.

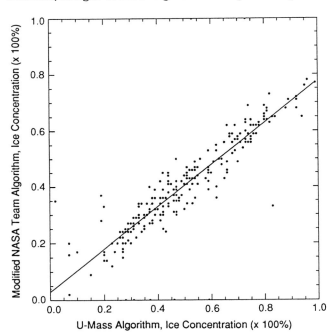

Fig. 10-7. Scatter plot of 1988 fractional multiyear ice concentration as retrieved by the Modified NASA Team algorithm versus the U-Mass Decision algorithm. Data points correspond to AOBP buoy locations in March and October of 1988.

trieval show that areas of increased vapor are coincident with areas of known storm activity or high lead density. Ice-temperature retrievals have been compared with buoy measurements from the Polar Science Center and the results are discussed in Section 10.3.2.2.

10.2.5.3. Sensitivity analysis. A sensitivity analysis was performed to investigate the stability and precision of the algorithm when noise is present in the brightness temperature signal. To isolate the noise effect, the analysis used simulated brightness temperature data generated for a typical range of surface and atmospheric conditions. White Gaussian noise was added to each set of temperatures, and the algorithm was run on the noisy data. Two different noise levels were tested. In the first case, the standard deviation was 1 K, allowing for easy comparison with algorithms created for both SSM/I and SMMR. The second method used a calculated standard deviation for each frequency and polarization combination. These standard deviations appear in Table 10-4, and represent the actual instrument noise present on each of the SSM/I channels.

The rms difference between the products retrieved from the noisy simulated brightness temperatures and the original values used to generate those brightness temperatures is presented in Table 10-5 for each noise level. This table indicates that the ice concentration and type are insensitive to instrument noise, which is on the order of that measured in SSM/I data. Even when noise with a standard deviation of 1 K is introduced, the error in total ice concentration is only 2%. The addition of noise introduces significant variability on the integrated atmospheric vapor product. The reason for this is most easily explained by noting that the

derivative of the radiative transfer equation with respect to g approaches zero if surface emissivity is close to one. This implies that scenes of nearly 100% ice cover (especially first-year ice) should be significantly affected by noise. The rms error for vapor in Table 10-5 includes the entire range of surface-type combinations, however, as expected, a much smaller difference was observed for low-emissivity scenes. The effect of instrument noise on the temperature measurement is relatively small (less than 1%). This difference arises because, to first order, the physical temperature is the observed brightness temperature divided by the emissivity. Since the emissivity of any surface type is less than 1, any noise in the brightness temperature will be magnified in the surface-temperature calculation. Clearly this effect is minimal over areas of 100% first-year ice (high emissivity), slightly greater over areas of 100% multiyear ice, and maximum over areas containing a large percentage of open water. Therefore, under normal winter conditions, the rms error associated with the surface temperature in Table 10-5 is expected to be a worst-case estimate.

To further estimate confidence levels in the output products, the sensitivity of these products to model errors was investigated. The two parameters expected to contribute the largest errors to the primary products (total ice concentration and multiyear ice concentration) are the emissivity of multiyear ice and the emissivity of first-year ice when thin ice is present. These effects were examined using simulated brightness-temperature data over a broad range of surface types.

The emissivity of multiyear ice was varied by ±5% and ±10% as the test brightness temperatures were generated. The rms differences between the output products and the original parameters are given in Table 10-6. This table suggests that the error in total ice concentration due to a variable multiyear ice signal is less than 1% for a +10% uncertainty in multiyear ice emissivity. The error in the multiyear ice concentration is larger, at just under 3% over the same range. Although the –10% case was tested, the results are not included because the data points fall so far outside the retrieval triangle that the algorithm cannot converge on a realistic result. This is not expected to introduce errors in practice because the emissivity tie points are empirically found to eliminate this problem. The difference in the temperature retrieval was significant as

TABLE 10-4. Typical standard deviation of instrument noise for each SSM/I channel in kelvin.

		Channel		
19V	19H	22V	37V	37H
0.45	0.42	0.75	0.37	0.39

TABLE 10-5. RMS difference between algorithm products and original values resulting from noise introduced in the brightness temperature data.

Noise	Total ice concentration, %	Multi-year ice concentration, %	Integrated atmospheric vapor, kg/m^2	Surface temp., K
SSM/I	0.533	2.51	0.817	0.906
1 K	1.820	3.35	1.185	1.980

TABLE 10-6. RMS difference between algorithm products and original values resulting from variable multiyear ice emissivity.

ΔT_B, %	Total ice concentration, %	Multi-year ice concentration, %	Integrated atmospheric vapor, kg/m^2	Surface temp., K
±5	0.17	2.52	0.165	5.73
+10	0.315	2.76	0.296	11.42

compared with the original input temperatures, again because of the direct relationship between brightness temperature, emissivity, and physical temperature.

Finally, the effect of thin ice within the footprint was examined. Thin ice has an emissivity between that of water and that of thick first-year ice (which depends on thickness and frequency). Since the algorithm does not allow for variable emissivity, such a signature would be interpreted as a combination of open water and first-year ice. The retrieved products behaved predictably when simulated data were processed. For example, when the pixel consisted of 50% water, 40% first-year ice, and 10% thin ice, the output of the algorithm varied between 60% water and 40% first-year ice, and 50% water and 50% first-year ice, depending on the chosen emissivity of the thin-ice area. This analysis assumed that linear increments in the emissivity occurred evenly across the spectrum. For example, when the emissivity at 19 GHz was half way between its open water value and its first-year ice value, it was also half way at 37 GHz. Although this is not strictly true in nature, it serves as a reasonable indicator of algorithm behavior. The degree to which this is not true would determine a tendency to overestimate first-year ice concentration. In the previous example, if the thin-ice emissivity at 37 GHz was nearly that of first-year ice, but the emissivity at 19 GHz has achieved 75% of the difference to its first-year value, the output first-year-ice product would fall somewhere in between, around 48% to 49%.

This study indicates that a conservative estimate of the overall uncertainty in total ice concentration would be approximately ±2%, while multiyear ice concentration uncertainty is approximately ±5%. The analysis assumed dry conditions (i.e., no melt ponds) and that the current model accounts for all significant physical parameters. The second-order products have larger uncertainties associated with them, consistent with expected behavior.

10.2.6 Bootstrap Algorithm

10.2.6.1. Background. The value of multichannel microwave data in the retrieval of geophysical parameters is illustrated in the three-dimensional scatter plots shown in Figure 10-8 of emissivities over the Central Arctic region. The emissivity data were derived from SMMR and temperature-humidity infrared radiometer (THIR) data, as discussed in Comiso [1983], and are useful for examining effects that are not associated with surface physical temperature. Data from three different frequencies at vertical polarization (37 V, 18 V, and 10 V) are plotted in Figure 10-8(a), while data from two different frequencies and another polarization (18 V, 37 V, and 37 H) are shown in Figure 10-8(b). The scatter plots provide a means whereby the representation of different ice surfaces can be examined in the multichannel microwave data. The clusters of points labeled A, B, C, and D are believed to be different types of radiometrically different surfaces of mainly consolidated sea-ice cover. The cluster of points labeled H

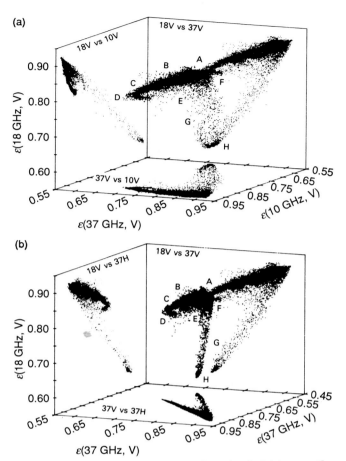

Fig. 10-8. Three-dimensional scatter plots of emissivities over the Central Arctic region in March 1979, for (a) 37 GHz versus 18 GHz versus 10 GHz, all at vertical polarization, and (b) 37-GHz horizontal versus 37-GHz vertical versus 18-GHz vertical polarization.

corresponds to ice-free ocean data. Data points between the consolidated ice clusters and H are either mixtures of ice and water or new ice. The plots also show two-dimensional projections of three-dimensional data that are the information content of various combinations of two-dimensional channel data. A coplanarity test and principal component analysis of the 10-channel data set actually revealed that the multispectral data are basically two-dimensional, since most of the channels are highly correlated with each other [Comiso, 1986; Rothrock and Thomas, 1988]. Thus, using conventional algorithms, three equations can be set up (two for the two independent channels and the mixing equation) to allow, at most, three different types of surfaces to be unambiguously identified. This creates a limitation in sea-ice applications since there are usually more than three types of radiometrically different surfaces, including open water, within the ice pack. Other surfaces observed to have different emissivities are new ice [Comiso et al., 1989] and second-year ice [Tooma et al., 1975]. Other effects such as spatial variability in the thickness and granularity of the snow cover further complicate retrieval because they modify the emissivity of a certain ice type in such a way that interpretation can be ambiguous.

10.2.6.2. Ice concentration—The bootstrap approach. A phenomenological technique for deriving ice concentration based on the distribution of radiometrically distinct clusters in the parameter space has been developed [Comiso, 1986; Comiso et al., 1984; Comiso and Sullivan, 1986]. An extension of this technique to enable surface classification will be presented later in this section. The approach used for deriving ice concentration has several advantages. First, it is not dependent on an absolute calibration of the instrument and an ability to exclude atmospheric effects. Good relative calibration that does not compromise the quality of the data is sufficient. Also, the technique assumes that atmospheric effects are part of the data set. Second, it utilizes general information inherent in the data. The general behavior of data points associated with the highly concentrated ice is analyzed and incorporated in the methodology. Also, the behavior of open-water data within and outside the ice pack is examined to obtain the best reference point possible and optimize the ability to detect the marginal ice zone. Third, the technique assumes only two types of surfaces, namely, open water and sea ice. The variability of both types of surfaces is considered and the behavior associated with this variability is utilized to optimize detection of a small fraction of open water within the ice pack. This assumption gives the flexibility of using only two channels, i.e., the best two channels for every situation. For example, the almost exclusive use of the 37-GHz channels (both polarizations) in the Central Arctic region optimizes resolution of derived data. An obvious weakness of the technique, which is also inherent with the other techniques, is its inability to characterize new-ice regions accurately. Depending on thickness and state of development, new ice is interpreted as having an ice concentration of less than 100%.

As with the other algorithms, this bootstrap technique uses the radiative transfer equation for the surface and the atmosphere in combination with a mixing formalism that assumes that only sea ice and open water are within the footprint of the sensor. Although consolidated ice data in the scatter plots are shown to have highly variable emissivities, they follow a predictable and highly correlated cluster of points. The two-dimensional scatter plot of 37 H versus 37 V is replotted in Figure 10-9(a), using brightness temperatures. In this plot, data from a highly concentrated ice region in the Central Arctic follow a locus of points along the line AD. The line is found to be well defined with a standard deviation of about 2 K. Furthermore, the slope of the line for this particular set of channels is also found to be consistent from one year to another and is approximately 1.0. In the same plot, areas of open water within the ice pack are represented by the points clustered at a position labeled H. In the formalism discussed by Comiso [1986], ice concentration can be derived using a functional relationship that utilizes as references the open water cluster H and the

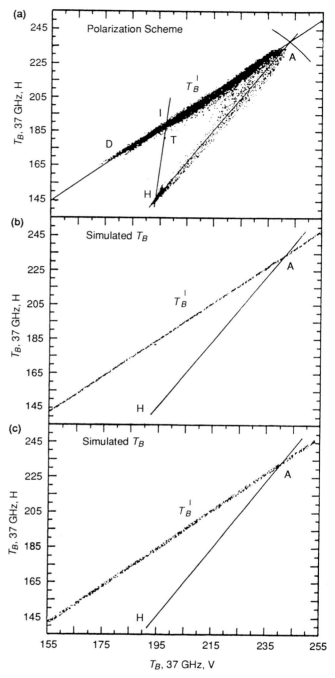

Fig. 10-9. Consolidated ice data for (a) two-dimensional scatter plot of 37-GHz horizontal 37-GHz vertical using brightness temperature data (the schematics for ice concentration determination are also shown). The sensitivity of data to variations in physical temperature of the ice from (b) 245–260 K and (c) 230–260 K generated at random.

consolidated ice line AD. The parameterization of the consolidated ice line is based on the assumption that within the ice pack there are large areas of 100% ice. Such an assumption has been verified by in-situ and aircraft observations [Comiso et al., 1989, 1991]. In this parameterization, the effect of varying surface temperatures is approximately taken into account as illustrated in Figures 10-9(b) and (c). These plots illustrate that along the consolidated ice line, effects of surface-ice temperature varying by as much as 30 K cause standard deviations of less than 1 K.

The formalism is similar to the single-channel algorithm but takes into account the spatial variability of the brightness temperatures of a 100% ice surface for the different channels utilized. Thus, for any set of brightness temperatures, $T_B I$ (37 H), $T_B I$ (37 V), at the point I along the consolidated ice line, any point T along line HI represents different mixtures of open water and consolidated ice. The ice concentration data represented by the point T is therefore given by

$$C = (T_B T - T_B H)/(T_B I - T_B H) \qquad (34)$$

where T_B's from either 37 H or 37 V can be utilized. Only two channels are required to derive ice concentration. The two channels at 37 GHz are utilized to obtain ice concentration in the Central Arctic region because the cluster distribution of consolidated ice data is very compact when these channels are used. Also, the slope of the ice line appears to be constant with the seasons over several years. Furthermore, in the algorithm, spatial variation in temperature is approximately but adequately taken into account with this set of channels as discussed in the following sections and in Comiso [1986].

For the seasonal sea-ice regions in both hemispheres, the use of two alternate channels was found necessary. For one reason, the distribution of data points in the ice-free ocean regions points overlap those of ice and water mixtures if only the 37-GHz channels are used. For another, the snow cover variability and flooding over first-year ice make discrimination from mixtures of ice and water [Comiso and Sullivan, 1986] difficult in this set of channels. These problems are better resolved by using a set of channels at two different frequencies, the most appropriate of which are the 18- or 19-GHz and 37-GHz channels because they effectively handle atmospheric and snow effects without compromising resolution too much. Although the horizontal polarization channels could also be utilized, the vertical polarization channels were chosen because they provided more consistent signatures for sea ice when SMMR data were used. Scatter plots for 18 V versus 37 V in the Antarctic region for areas that have (a) both open water and ice, (b) consolidated ice only, and (c) open water only are shown in Figures 10-10(a), (b), and (c), respectively. The corresponding plots for 37 H versus 37 V are shown in Figures 10-10(d), (e), and (f). The plots for consolidated ice show that better discrimination between open water and consolidated ice can be made when both the 18- and 37-GHz

channels are used. Also, a better discrimination of atmospheric effects in the open water from that of mixtures of first-year ice and open water can be made using the 18- and 37-GHz channels. The plot in Figure 10-10(c) is used as a basis for the open water cutoff in the algorithm. This procedure, which was first used by Comiso et al. [1984], minimized low ice concentration in the open ocean where ice is not expected and enabled a more accurate representation of the marginal ice zone where atmospheric effects are usually significant. The spatial variations in temperature are approximately accounted for, but not as well as those for the other set of channels (i.e., 37 H and 37 V). In the Weddell Sea, the spatial variations in the physical temperature of the snow–ice interface layer have been observed to average at about 5 K [Comiso et al., 1989], while surface air temperatures fluctuated from –1 to –30 K. This is believed to be the case with many of the seasonal sea-ice regions because of the effective insulation of the snow cover. Thus in these regions, errors caused by spatial changes in physical temperature are not as important as those caused by spatial changes in emissivity, as will be discussed later.

The bootstrap technique is employed because a more strict theoretical inversion technique that requires the use of a radiative transfer model of sea ice has not been successfully developed. The radiative transfer models have not advanced to a stage where the parameters affecting the satellite measurements (such as grain size, brine volume, and wetness) are adequately parameterized [Vant et al., 1974; Fung and Chen, 1981; Stogryn, 1985]. Furthermore, the lack of absolute calibration would make modeling very difficult. Nevertheless, the behavior of the cluster of points, especially for consolidated thick ice, is predictable enough to justify the parameterization. Monthly ice-concentration maps for an entire year's cycle of SSM/I data using the bootstrap technique are shown in Figures 7 and 8 of Comiso [1991] for the Arctic and the Antarctic, respectively. The set of images basically illustrates the growth and decay characteristics of the sea-ice cover on a global scale. Comparison of ice concentrations and ice edges derived from this technique have shown good general agreement with in-situ (about 15%) and other satellite data [Comiso et al., 1984; Comiso and Sullivan, 1986]. However, in areas of lead and polynya formations and ice edges, larger errors are incurred in the calculation of ice concentration because the emissivity of new ice varies continuously with thickness up to several centimeters [Comiso et al., 1989]. Thus, in the ice-concentration maps, new ice with thicknesses below a certain threshold is represented as a mixture of open water and thick ice. Other ambiguities are associated with surface melt during early spring, which enhances ice emissivities and causes abnormally high values in ice concentration [Grenfell and Lohanick, 1985]. Also, flooding or melt ponding lowers the emissivity of the surface and causes ice concentrations to be underestimated [Grenfell and Lohanick, 1985; Comiso, 1990]. Spatial variations in ice temperatures are approximately taken into account in the cluster technique, with slight errors in the Arctic as described in Comiso

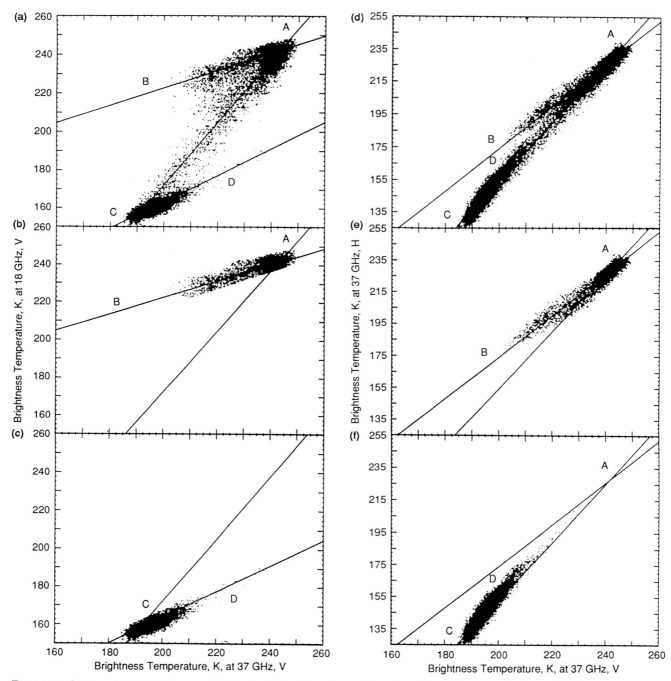

Fig. 10-10. Scatter plots for 18 V versus 37 V in the Antarctic region on November 7, 1983, for areas that have (a) both open water and ice, (b) consolidated ice only, and (c) open water only. Scatter plots for 37 H versus 37 V in the Antarctic region on November 7, 1983, for areas that have (d) both open water and ice, (e) consolidated ice only, and (f) open water only.

[1986] and significantly larger errors in the Antarctic. Spatial variations in snow–ice temperature are not as much of a factor in the Antarctic as in the Arctic because the Antarctic temperature averages about 7 K with a standard deviation of about 4 K [Comiso et al., 1989]. Other techniques that take into account temperature changes to the first order use the polarization and gradient ratios [Cavalieri et al., 1984]. The temperature of the emitting layer for the same wavelength (used for calculating polarization ratio) is likely the same. However, this is not necessarily the case for different wavelengths (used for calculating gradient ratio) because of different optical depths for different wavelengths; for some types of surfaces (e.g., multiyear ice), the temperature difference may be significant.

10.2.6.3. Ice type and surface classification—The cluster analysis. It is apparent that ice concentration maps do not provide complete information on the ice cover. Some techniques make use of the differences in the emissivity of first-year ice and multiyear ice to generate multiyear ice concentrations. This, of course, can be done, but the errors are large on account of large spatial variations in the emissivity of multiyear ice and also variations from one season to another. The signature of multiyear ice is a function of the scattering property of the material, especially the freeboard layer that includes the snow cover. Several aircraft and in-situ observations have documented such variability [Tooma et al., 1975; Comiso et al., 1991; Grenfell and Lohanick, 1985; Grenfell, 1992]. Furthermore, there is a problem with thin ice in the ice-concentration maps. Such an ice type is not adequately represented because, although the actual concentration might be very high, the derived value could be quite low.

An alternative (and perhaps complementary) way of characterizing the ice cover is to assess the radiometrically different surfaces and then to interpret each pixel of satellite data as belonging to one of these surface types. When the brightness temperature (or emissivity) data are plotted in a multidimensional space, the different surfaces (or ice types) are uniquely exhibited in the form of clusters of points as described earlier. These clusters were found to be well defined and persistent in autumn and winter. They may not represent pure ice types, but they could be used to identify different regimes in the ice pack as discussed below. The size and shape of the clusters depend on the type of surface, the channels used in the analysis, and the wavelengths of the channels used. For example, the contrast between water and ice tends to be exaggerated at low microwave frequencies while larger contrast is observed between first-year ice and multiyear ice at high microwave frequencies. Also, internal scattering in snow and ice occurs more at high microwave frequencies (where the wavelengths are comparable to the size of snow particles and air pockets within the ice) than at lower frequencies and varies from one region to another because of differences in thermal and formation histories. Furthermore, open water is highly polarized while sea ice is only slightly polarized.

Unambiguous ice classification is difficult to implement on the passive data because of the large footprint of the sensor, the small size of some surface features, and the overlap in the emissivities of some of the ice types. A supervised cluster analysis technically implemented by Comiso [1986] showed that there are a few large areas in the Central Arctic region with persistent signatures during the winter. The basic results have been reproduced using an automated cluster technique [Comiso, 1990]. The latter approach makes use of the Land Analysis System (LAS) routines originally developed at NASA's Goddard Space Flight Center for the analysis of Landsat data [Wharton and Lu, 1987]. The geographic regions represented by the clusters identified by this technique using various combinations of channels are shown in Figure 10-11. The images show a general consistency in the size and location of clusters using the different sets of channels. However, the results from the set of six channels are found to be most consistent. In the Central Arctic region, the geographic location of the clusters form several color-coded bands. While interpretation of these bands is still preliminary, the patterns are very similar to those predicted using buoy data for locations of ice with different ages [Colony and Thorndike, 1985]. This result is consistent with the emissivity of ice being modified every summer and with ice from each cluster having basically the same history. Some of the surfaces identified in the cluster analysis are actually new ice (like nilas, pancakes, and young ice) that has also been found in areas where such types of ice (e.g., marginal ice zones) are expected. In-situ and aircraft measurements also confirm the basic interpretation [Comiso et al., 1989, 1991]. While more detailed study is needed to better interpret the different clusters, this alternate way of characterizing the ice cover is promising. This approach would, for example, enable a comprehensive accounting of the various surfaces in the polar regions and help identify the different ice regimes. However, it will not provide complete information about multiyear ice, especially that with a first-year ice signature [Grenfell, 1992]. Such information is probably easier to obtain from the analysis of summer ice data, as was done in Comiso [1990].

10.2.6.4. Error analysis. The uncertainties in the derived products can be divided into two groups: those related to the instrument and peripheral hardware and those related to ambiguities in the interpretation and basic shortcomings of the data. Some examples of errors associated with the instrument are radiometer noise, calibration problems, and sidelobe and antenna-pattern effects. These sources of errors are common to all the algorithms and are expected to affect the latter in a similar fashion.

The largest source of error is believed to be that due to variations in the emissivity of the surface. Although many of these variabilities (such as variations in types of thick ice and snow cover) are taken into account by using a line as reference for highly concentrated ice, there are surfaces that cause lots of problems. Examples of these surfaces are

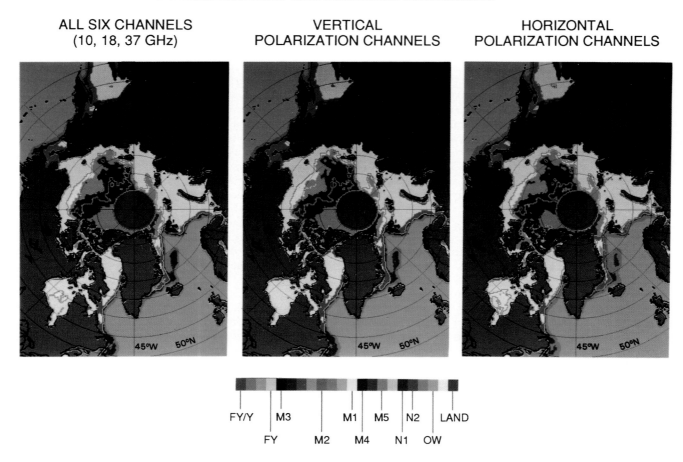

Fig. 10-11. The geographic regions represented by the March 1979 emissivity clusters identified by the unsupervised cluster analysis technique using various combinations of channels.

new ice, flooded ice, melt-ponded ice, and wet snow cover. These factors could sometimes cause very large errors (>50%) in the retrieval of geophysical parameters. In cases where these effects are predictable, as for example, the presence of only pancake ice and open water in the marginal ice zone, the errors can be minimized.

Another factor that can cause significant error is the spatial change in physical temperature of the emitting surface. Since the thickness of the layer of ice and snow from which the observed radiation originates varies with frequency, the error would depend on which of the channels are used in the algorithm. Because of high salinity, it is usually a thin layer about one wavelength thick for first-year and young ice. In the seasonal sea-ice region where a combination of the 19- and 37-GHz channels at vertical polarization is used, the concern would be the spatial variability in the ice-interface physical temperature. The relationship of surface air temperature to snow–ice temperature in the Weddell Sea is shown in Chapter 12. In the Central Arctic region, the standard deviation of the consolidated ice line in the plot of 37-GHz vertical versus the 37-GHz horizontal polarization data is only about 2 K,

corresponding to an uncertainty of less than 4%. This is partly because the effect of spatial variation in surface temperature is to move the data points along the consolidated ice line.

The Bootstrap algorithm is a multidimensional extension of the single-channel algorithm used for ESMR data [Zwally et al., 1983; Parkinson et al., 1987]. The use of a consolidated ice line as reference has been verified using aircraft and ship experiments. The results have uncertainties ranging from about 4% in the Central Arctic to 10% in the seasonal region when new ice, flooded or melt-ponded ice, and wet surfaces do not comprise a significant fraction of the field of view of the sensor. When a large fraction of these surfaces are in the field of view, the error can be as large as 50%. If only new ice and open water are in the field of view, the accuracy is also about 10%, but with slight modification of the reference temperatures. Time-series ice-concentration maps show good consistency overall, especially in the inner region during the fall/winter conditions.

10.2.7 Artificial Intelligence

Several methods exist that may improve the retrieval of ice concentrations, including the use of local tie points, corrections to reduce atmospheric attenuation and the effects of surface winds on open water, incorporation of such ancillary data as surface temperature, utilization of region-specific procedures, and the assimilation of observations with a physical model. The underlying theme of each of these approaches is that additional sources of information exist that can be used to reduce the uncertainty in ice classification. While methods such as the Kalman filter provide a means of combining data types, a more flexible unifying framework that allows for rapid algorithm prototyping is needed.

Knowledge-based systems (KBS) and neural networks may be useful for the retrieval of sea-ice parameters from satellite data because of their relatively simple yet flexible structures. Knowledge-based systems are an efficient and flexible way of encoding knowledge as "if–then" rules, without being bound to traditional sequential processing modes. Processing may occur numerically or symbolically and is controlled by the type of data available. The popularity of neural networks can be attributed to their ability to combine numeric and symbolic information (like knowledge-based systems), to accept degraded and even interrupted input streams, and to produce "fuzzy" output, alleviating some of the problems inherent in hard classifiers.

The use of artificial intelligence (AI) methods in remote sensing and the geosciences is increasing, and expert systems are being developed routinely for image classification. For sea-ice research, both KBS's and neural networks have been employed only in an exploratory sense. For example, Key et al. [1989] describe the development of a neural network that classifies collocated SMMR–AVHRR Arctic data into surface (sea-ice, water, and land) and cloud types. The network results were compared to those from the supervised maximum likelihood method, and were found to be more accurate when applied to other images. Additionally, the network was capable of dealing with fuzzy boundaries in both spatial and spectral dimensions. A KBS for detecting, mapping, and statistically analyzing sea-ice leads is described in Key et al. [1990]. While the system utilizes Landsat rather than passive microwave data, the basic framework is suitable to other sea-ice parameter retrievals. Individual lead fragments are identified and labeled with a region-growing procedure, then a set of rules is applied to generate hypotheses regarding the likelihood that each belongs to a lead system. As rules are fired, numeric procedures are invoked to determine various statistical and geometric properties of the lead fragments. Results indicate that the resulting leads and their derived width and orientation statistics are similar to those determined through manual interpretations. As a final example, a neural network has been developed that is capable of classifying SMMR data into ice seasons consisting of winter, premelt, and melt onset periods [Maslanik et al., 1990]. In that study, all five SMMR channels were utilized in a time series in the Beaufort Sea. The performance of the network was evaluated by comparing the resulting spatial and temporal patterns to visible-band imagery, surface albedo maps, and drifting buoy meteorological data.

For the retrieval of sea-ice concentration, two algorithms are currently under development, a KBS and a neural network, both for the estimation of sea-ice concentration. The neural network structure follows that of the Kalman filter, taking advantage of the temporal dependence of the ice signatures to reduce the ambiguity in the passive microwave signal, particularly during the summer [Thomas and Rothrock, 1989]. A physical model that describes the way in which ice concentration changes over time is an integral part of the network. Input includes brightness temperature observations, fractional coverages of each surface type estimated by the physical model (open water and first-year and multiyear ice), and time of year. The prototype network is trained with simulated observations based on assumed true-ice fractions: Monthly pure signatures for a single microwave channel and each of the surface types are combined with the fractional coverage of each surface and a random error based on the variability of the pure signatures (time and surface dependent). An inherent error that accounts for the unmodeled physics in the physical model is specified. An arbitrarily complex model can be used [Flato and Hibler, 1992] without change to the network, except for the specification of the model error. Preliminary results indicate that the network—like the Kalman filter—acts as a smoothing filter, reducing the influence of extreme brightness temperatures that correspond to, for example, summer melt.

The KBS design for sea-ice concentration retrieval consists of a set of modules to handle input, data checking and validation, classification, and certainty assessment. The NASA Team algorithm is the core of the classification step. The KBS is a controller and source of knowledge that supplies appropriate inputs to the algorithm and monitors the algorithm output in light of the time of year and geographic area. Inputs include brightness-temperature, geographical, and meteorological data. The rule base is used to impose upon the ice classification boundary conditions based on season and location, to determine when observed conditions indicate a potential problem (e.g., weather effects and melt), to develop confidence levels for the classified data, and to attempt to improve the classification by invoking correction schemes or by supplying such realistic information for classification as tie-points or probabilities of new-ice growth. The major processing steps are:

(1) The state of the system (e.g., valid brightness temperatures for each frequency and polarization), seasonally and regionally dependent gradient ratio thresholds for weather filtering, and tie points are defined by the initial facts.

(2) Each observation is then examined by the rule base to determine its validity. Depending on the season, buoy temperatures and other ancillary data may be used to determine the likelihood of special conditions such as surface melt.

(3) If the data are valid, the NASA Team algorithm is applied, possibly with the weather filter.

(4) The rules then examine the computed ice concentrations as well as the polarization and gradient ratios and determine a confidence level for that observation.

Results to date seem more realistic than ice-concentration estimates over the NASA Team algorithm alone, particularly during the summer months when surface melt and weather effects are at their maximum. However, weather effects in particular are still a problem, and the KBS will be extended to include additional atmospheric measurements of cloud and water vapor from, for example, AVHRR and TOVS.

The AI research described here has been largely exploratory, with the goal of examining the feasibility of these methods in the context of sea-ice parameter retrieval from satellite data. Overall, knowledge-based systems and neural networks show promise, although there are inherent problems. Neural networks, for example, are essentially "black boxes" that may perform remarkably well, but for unknown reasons. The weights on the connections between input, hidden, and output nodes are difficult if not impossible to decipher, especially when the number of nodes is large. Knowledge-based systems are more transparent, but the sequence of operations—being data driven—is not easy to predict. There are no claims that KBS's or neural networks provide the final solution to the types of problems described here. What they do provide, however, is a framework flexible enough to allow rapid prototyping of systems that include a diverse set of input variables for a complex and possibly poorly specified problem.

10.2.8 Comparison of NASA, AES–York, U-Mass, and Bootstrap Algorithms

Seven different ice-concentration algorithms have been discussed in this chapter. The regional ice-concentration differences in performance between the algorithms in the close pack ice and in the marginal ice zone, and the performances of these algorithms under severe weather over open water and ice are questions yet to be answered. In the following discussion, the NASA Team, AES–York, U-Mass, and Bootstrap algorithms will be compared for the Northern Hemisphere on March 13, 1988. Ice concentrations were derived for each of the four algorithms and are shown in polar stereographic projection maps (Figure 10-12). The following comparison is rather descriptive and speculative, as the true ice concentration and distribution is not known over the entire area, but only for relatively small areas in the Bering Sea and the eastern part of the Beaufort Sea. For these areas, high-resolution Landsat imagery (80-m pixels)

and Defense Meteorological Satellite Program (DMSP) imagery (800- to 1100-m pixels) were available.

10.2.8.1. Weather effects over the ocean. Weather effects on the microwave emissive properties of the open ocean are well known [Gloersen and Cavalieri, 1986]. Cloud liquid water contributes significantly to increased microwave emission over open water. The 37-GHz channels are more sensitive to clouds than are the 18-GHz channels. Also, wind roughening of the ocean surface increases the microwave emission at 19 GHz, whereas the horizontal polarization is almost twice as sensitive as the vertical polarization to near-surface winds [Webster et al., 1976; Gloersen and Barath, 1977]. To reduce these erroneous values of computed sea-ice concentration over open water, "weather filters" are used to set to zero ice concentrations that are less than or equal to a certain percentage. This method is applied to the NASA Team algorithm with a threshold value of 8% to 12% for the SMMR data and approximately 15% for the SSM/I data. The Bootstrap algorithm uses a threshold of 8% to 10%. The AES–York algorithm applies one weather filter to potentially high ice concentrations and a second weather filter to lower ice concentration, where ice–ocean discrimination is made using 37-V and 37-H channels. The U-Mass algorithm uses a model for atmospheric attenuation and emission to reduce the weather effects.

Comparing the performance of the four algorithms to reduce weather related effects over the open water, the AES–York and the Bootstrap algorithms show the best results, Figures 10-12(c) and (d). The NASA Team algorithm depicts some erroneous ice concentrations at the south tip of Greenland and in the Pacific sector, top left corner, Figure 10-12(a). The U-Mass algorithm successfully reduced the weather effect in the Pacific sector; however, a large area of 30% to 34% ice concentration remained at the southern tip of Greenland. From this comparison, it can be concluded that a threshold value on the order of 10% to 15% worked best for the given atmospheric conditions on March 13, 1988. However, it has to be mentioned that a large threshold value reduces the ice-concentration resolution along the marginal ice zone and in areas of low ice concentrations such as polynyas.

10.2.8.2. Regional differences in ice concentration. The total ice concentration for the Arctic Ocean is expected to be close to 100% during spring (March). Comparison of the four algorithms shows good agreement within a few percent (Figure 10-12). The expected accuracy for the NASA Team algorithm for clear sky conditions is 2% to 4%, based on an intercomparison with Landsat data [Steffen and Schweiger, 1991]. Most of the ice-concentration differences are within that percentage range for the four algorithms discussed in this comparison. In the following, we will concentrate on two regions in the southern Beaufort Sea. The derived ice concentrations from passive microwave were compared against DMSP visible and thermal infrared satellite imag-

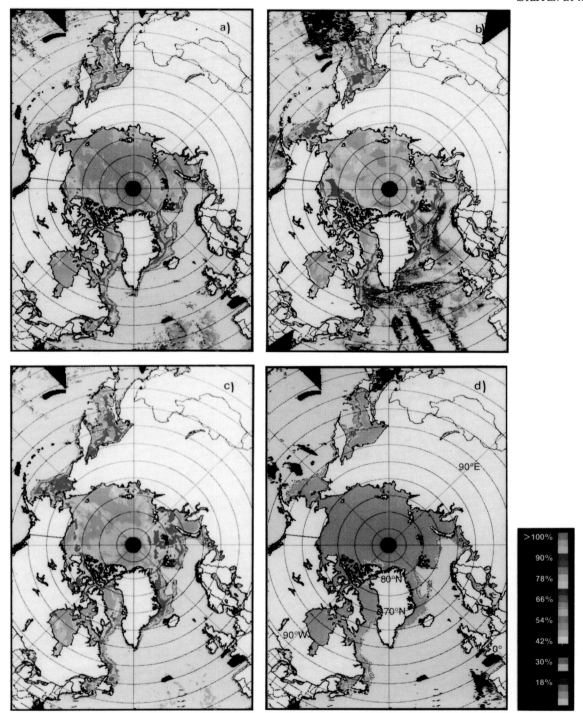

Fig. 10-12. Total ice concentrations derived from (a) NASA Team, (b) U-Mass, (c) Bootstrap, and (d) AES–York algorithms for the Northern Hemisphere on March 13, 1988.

ery. A total of 18 different satellite scenes for the Beaufort Sea was available for March 12, 13, and 14, 1988. With this sequence of DMSP imagery, the dynamics and short-time changes in the ice cover as well as cloud conditions can be discussed. A low-pressure system (cyclone) with its center in the western part of Alaska extending into the Beaufort Sea and a high-pressure system (anticyclone) with its center in the Canadian Archipelago resulted in strong surface winds in the NNW direction. This surface wind was responsible for the opening of a shore lead running several kilometers in width and approximately parallel to the coast. The DMSP satellite imagery revealed also numerous small leads close to the Canadian sector of the Beaufort Sea and to the west of the Canadian islands. This reduces ice concentration close to the coast in the eastern Beaufort Sea and is well reproduced by the NASA Team algorithm, Figure 10-12(a); however, the size of the reduced ice concentration along the Canadian islands is overestimated compared with the DMSP imagery. The Bootstrap algorithm resolves only part of the reduced ice concentration along the coast, whereas the AES–York algorithm seems insensitive to small changes in ice concentration. The U-Mass algorithm resolves the area of reduced ice concentration along the Canadian mainland very well, but overestimates the area of reduced ice concentration to the west of the Canadian islands and in the channels east of the Beaufort Sea.

A second area of reduced ice concentrations in the eastern Beaufort Sea can be seen in Figures 10-12(a), (b), and (c). This band of reduced ice concentration runs from Banks Island (Canadian Archipelago) to the northwestern tip of Alaska, best seen in Figure 10-12(c). The DMSP satellite imagery showed a homogeneous ice cover on March 12 and 14, 1988. On March 13, a low-level cloud could be seen in both the DMSP visible and thermal infrared imagery. The physical temperature of the cloud top as measured by the thermal DMSP channel was close to 0°C, whereas the ice surface temperature of the cloud-free scene was on the order of −20°C. This low surface stratus must have been at the maximum air-temperature inversion height and therefore was much warmer than the ice surface. The occurrence of such low-level stratus clouds at inversion height is rare as revealed by a case study of numerous DMSP thermal infrared images for the Arctic pack ice region. Three out of the four algorithms showed a reduced ice concentration at the same location where the low-level stratus occurred. This ice concentration for that region is reduced by 10% for the NASA Team algorithm and by 10% to 14% for the Bootstrap and U-Mass algorithms, with the later showing the largest reduction. The effect in the U-Mass algorithm could be explained by the fact that the atmospheric model is based on a standard polar atmosphere that does not include surface inversion. The AES–York algorithm showed no sensitivity to this weather effect as the whole Arctic Basin showed ice concentrations at the 100% level.

Another area of reduced ice concentration is located in northern Baffin Bay, between Greenland on the east and the Canadian islands Ellesmere and Devon on the west. This area is well known as the North Water, a recurring polynya with a well-defined southern boundary. This boundary is well resolved in the NASA Team, Bootstrap, and U-Mass ice-concentration maps, and missing in the AES–York ice-concentration map. It seems that the AES–York algorithm is insensitive both to weather effects and very low ice-concentration changes.

In the following, the ice concentration in the Bering Sea derived from passive microwave data is compared with high-resolution Landsat data. A total of five consecutive Landsat images covering the eastern part of the Bering Sea from Cape Nome to the ice edge (185 × 925 km) was collected on March 13, 1988. From the Landsat satellite data, ice concentrations were derived using a tie-point method [Steffen and Schweiger, 1991]. This comparison is of particular interest as large areas of young ice such as nilas and gray ice were present. The NASA Team algorithm clearly underestimates the ice concentration due to the presence of young ice in the concentration range of 85% to 100%, Figure 10-13(a). The underestimate was in the range of 5% to 15%. The U-Mass algorithm also underestimates the ice concentration, and the error due to the young ice is in the same range as that for the NASA Team algorithm, Figure 10-13(b). The Bootstrap algorithm shows an underestimate of 5% to 15% for the high ice concentration range between 95% and 100%, but a much smaller ice-concentration difference between 85% and 95% as compared with the Landsat results, Figure 10-13(c). It seems that the Bootstrap algorithm did actually resolve ice concentrations to within a few percent as compared with the Landsat high-spatial-resolution analysis for these young ice regions. However, for the ice concentration range between 65% and 85%, the Bootstrap algorithm overestimated by 5% to 15%. The AES–York algorithm performed reasonably well for the 95% to 100% ice concentrations, but overestimated at lower ice concentrations. It seems that the AES–York algorithm is not affected by young ice, but it is not sensitive to small changes in ice concentration.

10.3 ICE TEMPERATURE

Knowledge of surface temperature in the polar regions is extremely desirable for those studying global climatology, hydrology, and biology. This information has been inferred for the Arctic from data gathered by buoys drifting on ice floes [Thorndike and Colony, 1980]. Considerable difficulty with this approach provides incentive for the remote sensing community to develop an alternate method. Satellite-based infrared sensors such as AVHRR offer greater Arctic coverage and high resolution, but information retrieval is seriously constrained by cloud cover contamination. Satellite-based passive microwave instruments provide a solution to this problem at the cost of spatial resolution.

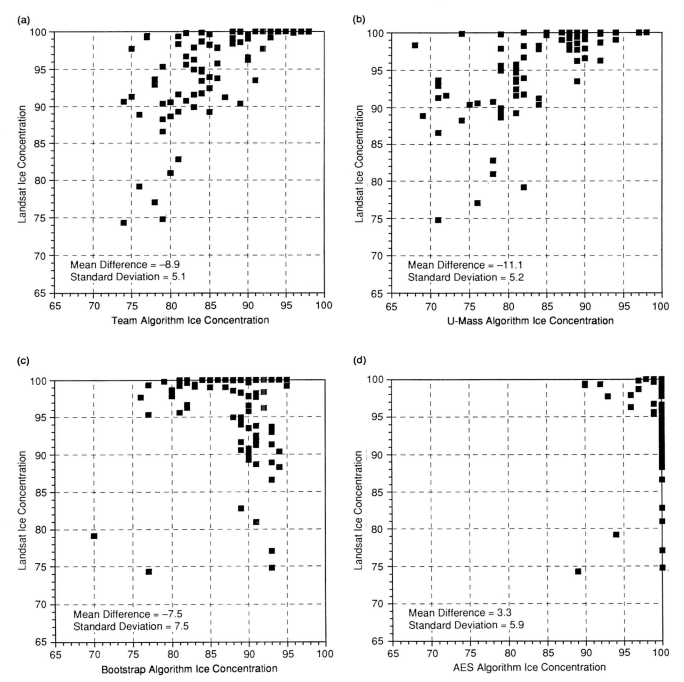

Fig. 10-13. Ice concentrations derived from (a) NASA Team, (b) U-Mass, (c) Bootstrap, and (d) AES–York algorithms plotted versus ice concentrations from high-resolution Landsat imagery of March 13, 1988.

10.3.1 NASA Team Algorithm

10.3.1.1. Equations and assumptions. The calculation of sea-ice temperatures makes use of the microwave spectral property of sea ice that at 6.6 GHz (4.6-cm wavelength) the microwave emissions of first-year and multiyear ice types are indistinguishable (Figure 10-3). Based on a study of Arctic sea-ice emissivities obtained from a combination of SMMR and infrared radiances of the ice [Comiso, 1983], a value of 0.96 is used with the 6.5-GHz vertically polarized radiances. The equation used to calculate the sea-ice temperature is

$$T_I = (T_B[6.5V] - (1 - C)T_{B,W}[6.5V])/e[6.5V]C \qquad (35)$$

where $T_B[6.5V]$ and $T_{B,W}[6.5V]$ are the observed and open-water brightness temperatures at 6.5 V and C is the total ice concentration in percent.

In addition, because of the sensitivity of this calculation to the total ice concentration when C is 80% or less, T_I is set to 271 K on the basis that ice temperatures in the vicinity of open water are likely to be near the melt point. It is important to note that the derived ice temperature is not a surface temperature, but the physical temperature of the radiating portion of the ice. This will tend to be close to the snow–ice interface temperature for first-year ice, but for multiyear ice it will be a weighted temperature of the freeboard portion of the ice.

The nine-year SMMR average sea-ice temperature map for February is illustrated in Figure 10-3(c). The ice temperatures range from about 240 K just off the coast of Ellesmere Island to 270 K in the seasonal sea-ice zones of the Bering, Okhotsk, and Labrador Seas. The 30-K difference in the observed ice temperature results in a large ice-temperature gradient across portions of the Arctic Basin and from Fox Basin to the southern portion of Hudson Bay. Note that because of the restriction to more than an 80% ice concentration in the calculation of ice temperature, the ice temperature field generally ends before the ice edge is reached.

Since weighted ice-temperature measurements that are spatially and temporally coincident with SMMR observations over a 150-km (6.5-GHz) footprint are nonexistent, Figure 10-3(c) is compared with climatological surface air temperature for the month of February, Figure 10-3(d). A comparison between the two figures shows that in general there are large-scale similarities between the temperature distributions. Climatology shows the coldest temperatures occur north of the Canadian Archipelago and the warmest temperatures are found in the peripheral seas. A tongue of warmer air in the Chukchi Sea is also apparent in the SMMR ice temperature, but the sharp gradient across Hudson Bay observed in the SMMR data is not apparent in the climatological map. The SMMR temperatures are generally warmer from a few degrees north of the Canadian Archipelago to about 10 degrees in the Chukchi Sea. Since the SMMR temperatures represent values close to the

snow–ice interface or to some weighted mean portion of the ice, it is not surprising that these temperatures are up to 10° warmer than the climatological surface air temperature. This difference, if taken to represent the difference between the ice-surface and snow-surface temperatures, results from the insulating properties of the snow cover and is in fair agreement with observed and modeled temperature profiles during this time of year [Maykut and Untersteiner, 1971].

10.3.1.2. Error analysis. The accuracy of the calculated sea-ice temperatures depends on the accuracy of the 6.5-GHz vertically polarized brightness temperature, the assumed ice emissivity of $e = 0.96$, and the calculated ice concentration. The uncertainty in T_I is given by

$$\Delta T_I = \left(\frac{\Delta T_B}{e_I C}\right)^2 + \left(\frac{(T_B - T_{B,W})\Delta C}{e_I 2C}\right)^2$$
$$+ \sqrt{\left(\left(\frac{T_B - T_{B,W}}{e_I 2C} + \frac{T_{B,W}}{2e_I}\right)\Delta e_I\right)^2} \qquad (36)$$

where the coefficients of ΔT_B, Δe_I, and ΔC are measures of the sensitivity of T_I to each parameter. Instrument-noise equivalent ΔT at 6.5 GHz is 0.4 K [Gloersen and Barath, 1977], the uncertainty in e_I is +0.02 [Comiso, 1983], and the observed standard deviation in computed ice concentration over the Central Arctic in winter varies from ±2% to ±4% [Cavalieri et al., 1984]. Although the concentration values contain both noise and real variability in ice-concentration, we use the ±4% as a conservative estimate of ice concentration uncertainty for this purpose. Using Equation (36), we find the uncertainty in T_I varies strongly with ice concentration, from 9.7 K at 50% concentration to 6.6 K at 100% concentration.

Winter ice temperatures from SMMR were compared with Arctic Ocean Buoy Program (AOPB) data for seven winter months, and a linear relationship was found for the two temperatures, as shown in Figure 10-14. A linear least-squares fit yields a slope of 0.5 and an offset of −3.5 K. The standard error of the estimate is ±4.7 K, with a correlation

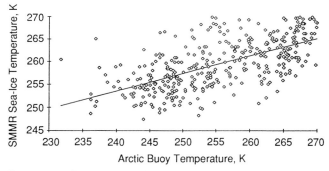

Fig. 10-14. Comparison of SMMR-derived sea-ice temperatures versus AOBP temperatures from seven winter months in the Central Arctic.

of 0.6. Conclusions drawn from this comparison are limited by inherent differences between the measurements. For example, the AOBP temperatures represent point measurements at a fixed depth, whereas the SMMR-derived temperatures are interpreted as a spatial average at a depth that is a function of ice type and snow cover. In addition, the buoy data may be contaminated by drifting snow, solar heating, or faulty operation. In general, the warmer SMMR temperatures are consistent with the subsurface nature of the microwave radiation.

10.3.2 U-Mass Algorithm

10.3.2.1. Modified NASA Team algorithm. A modification to the NASA Team SSM/I algorithm was developed to investigate the possibility of a temperature product from the SSM/I passive microwave data. This addition to the NASA Team algorithm involves estimates of ice emissivity as a function of frequency and ice type, and a model of the Arctic atmosphere as a function of surface temperature (assuming near saturation conditions [Orvig, 1970]).

The atmospheric model was developed after a first comparison with buoy data indicated that the temperature swings retrieved by the algorithm were significantly damped as compared to the buoy in-situ temperature swings (i.e., the slope of a regression line was significantly less than unity). The model gives atmospheric attenuation based on standard atmospheric profiles for temperature, pressure, and density at the latitudes, but does not allow for atmospheric temperature inversions.

The basic principle of operation is similar to the NASA Team SMMR temperature algorithm, except that ice-type distinctions are made. The inclusion of the atmosphere model produces the following transcendental equation:

$$T_B = e_{TOT} T_S e^{-\tau T_S} + T_{\mathrm{AIR}}(T_S)(1 - e^{-\tau T_S}) \qquad (37)$$

where:

T_B = measured brightness temperature

e_{TOT} = total surface emissivity based on retrieved types and emissivity tie points

$T_{\mathrm{AIR}}(T_S)$ = weighted atmospheric temperature as a function of T_S

$\tau(T_S)$ = atmospheric attenuation coefficient as a function of T_S

Except for T_S, all variables are either estimated or measured via the original NASA Team algorithm, and an iterative routine solves for surface temperature. As described in the Decision algorithm (Section 10.2.5.1), the ice temperature is derived from T_S and surface ice-type retrievals are solved by Equations (31) and (32). Figure 10-15(a) shows a comparison with AOPB data for March 1988; this comparison suggests that the two data sets bear a linear relationship. The linear least-square fit

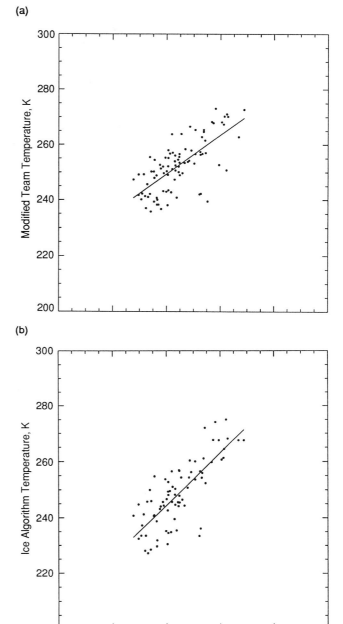

Fig. 10-15. Scatter plot of (a) ice temperature as retrieved by the NASA Modified Team algorithm versus air temperature measured by AOBP buoys throughout the high Arctic in March 1988, and (b) ice temperature as retrieved by the Decision algorithm versus air temperature measured by AOBP buoys in March 1988. These buoys are the same as those used in the comparison in Figure 10-14.

gives a slope of 0.7 with an intercept of 134.3 K. The correlation coefficient is 0.71 and the standard deviation is 6.7 K. Since air temperatures are expected to be extremely low in Arctic regions during March, problems associated with melting and storms are expected to be at a minimum. When those data are combined with comparisons for October 1988, the least-square-fit slope decreases to 0.8 and the intercept is 135.3 K. The standard deviation decreases to 6.2 K, but the correlation coefficient also decreases to 0.57. There are several reasons for the slope to be less than unity. First, the insulation effect of a snow layer would cause the snow–ice interface temperature swings to be damped relative to the air-temperature swings. In addition, the snow, like the atmosphere, has attenuating properties that depend heavily on water content. When the physical temperature increases, the attenuation, generally speaking, also increases, thereby damping out the expected rise in brightness temperature. An attempt has been made to correct for the atmospheric contribution under normal conditions in the algorithm, but the snow effect has not been resolved. The emissivities used in this routine are empirically derived, and thus the retrieved ice temperatures will always fall into a reasonable range. If the slope is less than one in the comparison, there will be an according offset to the regression line, which should not be interpreted as a bias in the retrieved ice temperature.

10.3.2.2. Decision making algorithm. The good quality calibration and extremely small ΔT of the SSM/I instrument give rise to the possibility of treating atmospheric parameters as variables, reducing the need for assumptions regarding the weather. This approach offers great advantage in the marginal ice zones and regions of increased storm activity. The operation of the algorithm is described in Section 10.2.5 and the results of a comparison with the buoy data are presented here.

The derived surface temperatures are compared with the AOBP temperatures for March 1988, in Figure 10-15(b). The regression line plotted through the data has a slope of 0.95 and an intercept of 16.6 K. The standard deviation is 7.4 K and the correlation coefficient of the two data sets is 0.78. As in Section 10.3.2.1, when data from October are included, the slope of the regression line decreases to 0.63 and the intercept becomes 93.2 K. The standard deviation increases to 8.7 K and the correlation coefficient changes to 0.57. As expected, algorithm performance is enhanced during the winter months.

The comparisons presented here and in the previous section indicate that the quality of ice-temperature retrievals (and perhaps the ice-concentration and -type products) improves if the varying atmospheric attenuation is accounted for. If the Decision algorithm is used on orbital data, the attenuation may be treated as a variable, thus improving the surface retrievals a step further. In either case, the time series comparison in Figure 10-16 shows the ice physical temperature trends generally agree with those recorded by in-situ buoys. The ice temperatures are warmer

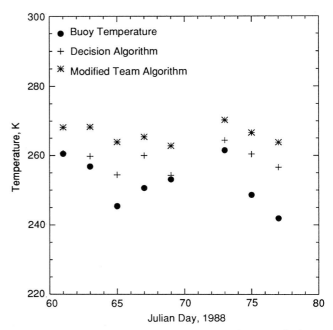

Fig. 10-16. Time-series comparison for Buoy 727, which was located at approximately 70.05° N, 226.36° E. The data points correspond to every odd day in March of 1988. The algorithm-retrieved temperatures are significantly warmer than the air temperatures, as expected.

than the buoy temperatures, which is expected because the buoy data used in this study represent local air temperatures whereas the brightness temperatures used in the algorithms originate at the warmer snow–ice interface. Although further comparison with buoys is needed to estimate the seasonal limits to the accuracy of the temperature products, the fundamental problems of comparing a 25×25-km² area to a single point measurement (which may not be representative of the area) and sorting out corrupted buoy data should be recognized.

10.4 Summary and Outlook

Today, passive microwave algorithms provide valuable information for navigation and large-scale sea-ice studies in polar regions. Important sea-ice parameters such as total ice concentration, ice-edge location, multiyear ice concentration, and ice temperatures can be retrieved with different algorithms. The scope of this chapter is to present the most common algorithms used for ice-parameter retrieval in polar regions based on passive microwave satellite data. The achievements reached today are remarkable, considering the fact that seasonal and interannual variations of the sea-ice cover on a global scale were hardly known some 15 years ago. Seven algorithms were presented in this chapter, all developed at different institutions; most likely there are additional algorithms used today by the polar-science community, either modified versions of the ones discussed in this chapter or new ones. We have limited our selection to the algorithms published in the open literature.

Most of a satellite-observed brightness temperature is due to the emitted radiation at the surface. Owing to the size of the satellite microwave footprint, each observation will see radiation from a mixture of surface types (assumed to be open water, first-year ice, and multiyear ice). It is for resolving those mixtures that an algorithm is needed. There is evidence suggesting that the pure-type signatures vary spatially and temporally (e.g., [Comiso, 1983]), implying that one needs to know local pure-type signatures to accurately resolve concentrations of the surface types. For large-scale studies (e.g., the whole Arctic for many years) it is impractical, if not impossible, to identify enough local signatures. The best we can do is use hemispheric pure-type signatures. Because of spatial variability, hemispheric signatures will always give rise to erroneous concentration estimates in many locations. That is unavoidable; we should still aim for zero-mean estimation errors; that is, the estimation errors should be unbiased. The invariance of the brightness-temperature ratios with respect to surface temperature is true only if brightness temperature is proportional to surface temperature. During summer, the emissivities of first-year and multiyear ice are known to vary as functions of surface wetness, eventually becoming indistinguishable and thus precluding ice-type resolution. An alternate to the choice of unbiased, pure-type signatures as tie points is proposed by D. Thomas (personal communication, 1991). Because of the spatial and temporal variability, the data chosen to represent each pure type should come from a large geographical region throughout the year. Because of the large tie-point regions, one should not assume that all the data from a region represents a single surface type. Instead, define the pure-type signatures as the mean of the main cluster of data from the region. Whether this approach will improve the accuracy of the NASA Team algorithm will be seen in future.

Validation studies have been carried out for some of the algorithms (see Chapter 11), but others need to be validated before an overall judgment can be made. However, the details given for each algorithm are sufficient for the choice of an algorithm that best serves the reader's application. Table 10-1 is a summary of the seven algorithms and their prime application. The accuracy range expected for the retrieved total ice concentration is on the order of 3% to 10% for the dry seasons (fall, winter, and spring), and on the order of 10% to 20% for the melt season (summer). A limiting factor to achieve higher accuracies is the coarse resolution of the satellite radiometer of 20 to 40 km, which depends on the frequencies used. An algorithm has been developed by Svendsen et al. [1987] that makes use of the high-frequency 85-GHz microwave channel from the SSM/I. It is based on radiation physics and has a spatial resolution of 12 km. Due to the large effects on the signals caused by time-varying atmospheric conditions and radiation properties of the ice, the algorithm is made self-adjusting. This new approach has been applied to aircraft passive microwave measurements with good success, but a validation for a hemispheric SSM/I data set remains to be done. Of even greater utility will be spaceborne radiometers with higher spatial resolution, such as the upcoming ESA Multifrequency Imaging Microwave Radiometer planned for launch on both ESA and NASA platforms.

First-year ice typing of different thickness classes with a passive microwave algorithm is still an unresolved problem. For sea-ice–atmosphere interaction, the knowledge of ice-thickness class distribution is essential, and one of the major goals for the development of new algorithms. There is hope to resolve this problem in the near future with an algorithm that combines passive microwave and SAR data (ERS-1, JERS-1, and Radarsat) and would also reduce part of the ambiguities of the present ice-concentration algorithms. Existing passive microwave algorithms could be improved further by incorporating atmospheric data sets to reduce some of the unknown variables in the radiative transfer models. Therefore, next-generation ice algorithms will most likely be based on multisensor satellite data sets, including those from passive microwave, SAR, atmospheric sounders, and possibly thermal infrared. This may sound a bit futuristic, but by the time this book is in press, we will have all the above-mentioned satellite data sets at our disposal for additional remote sensing studies of the polar regions.

References

Carnahali, J. B., H. A. Luther, and J. O. Wilkes, *Applied Numerical Methods*, 421 pp, John Wiley and Sons, Inc., New York, 1964.

Cavalieri, D. J., P. Gloersen, and W. J. Campbell, Determination of sea ice parameters with the Nimbus 7 SMMR, *Journal of Geophysical Research, 89(D4)*, pp. 5355–5369, 1984.

Colony, R. and A. Thorndike, Sea ice motion as a drunkard's walk, *Journal of Geophysical Research, 90(C1)*, pp. 965–974, 1985.

Comiso, J. C., Sea ice effective microwave emissivities from satellite passive microwave and infrared observations, *Journal of Geophysical Research, 88(C12)*, pp. 7686–7704, 1983.

Comiso, J. C., Characteristics of Arctic winter sea ice from satellite multispectral microwave observations, *Journal of Geophysical Research, 91(C1)*, pp. 975–994, 1986.

Comiso, J. C., Multiyear ice classification and summer ice cover using Arctic passive microwave data, *Journal of Geophysical Research, 95*, pp. 13,411–13,422 and 13,593–13,597, 1990.

Comiso, J. C., Satellite remote sensing of the Polar Oceans, *Journal of Maritime Systems*, pp. 295–434, 1991.

Comiso, J. C. and C. W. Sullivan, Satellite microwave and in situ observations of the Weddell Sea ice cover and its marginal ice zone, *Journal of Geophysical Research, 91*, pp. 9663–9681, 1986.

Comiso, J. C. and H. J. Zwally, Antarctic sea ice concentration inferred from Nimbus 5 ESMR and Landsat imagery, *Journal of Geophysical Research, 87,* pp. 5836–5844, 1982.

Comiso, J. C., S. F. Ackley, and A. L. Gordon, Antarctic sea ice microwave signatures and their correlation with in-situ ice observations, *Journal of Geophysical Research, 89(C1),* pp. 662–672, 1984.

Comiso, J. C., T. C. Grenfell, D. Bell, M. Lange, and S. Ackley, Passive microwave in situ observations of Weddell sea ice, *Journal of Geophysical Research, 88,* pp. 7686–7704, 1989.

Comiso, J. C., P. Wadhams, W. Krabill, R. Swift, J. Crawford, and W. Tucker, Top/Bottom multisensor remote sensing of Arctic sea ice, *Journal of Geophysical Research, 96(C2),* pp. 2693–2711, 1991.

Flato, G. M. and W. D. Hibler III, On a simple sea ice dynamics model for climate studies, *Annals of Glaciology,* in press, 1992.

Fung, A. K. and M. F. Chen, Emission from an inhomogeneous layer with irregular interfaces, *Radio Science, 16,* pp. 289–298, 1981.

Gloersen, P. and F. T. Barath, A scanning multichannel microwave radiometer for Nimbus-G and Seasat-A, *IEEE Journal of Oceanic Engineering, OE-2(2),* pp. 172–178, 1977.

Gloersen, P. and D. J. Cavalieri, Reduction of weather effects in the calculation of sea ice parameters with Nimbus 7 SMMR, *Journal of Geophysical Research, 93,* pp. 3913–3919, 1986.

Gloersen, P., W. Nordberg, T. J. Schmugge, T. T. Wilheit, and W. J. Campbell, Microwave signatures of first-year and multiyear sea ice, *Journal of Geophysical Research, 78,* pp. 3564–3572, 1973.

Gloersen, P., H. J. Zwally, T. C. Chang, D. K. Hall, W. J. Campbell, and R. O. Ramseier, Time-dependence of sea ice concentration and multiyear fraction in the arctic basin, *Boundary Layer Meteorology, 13,* pp. 339–359, 1978.

Gloersen, P., W. J. Campbell, D. J. Cavalieri, J. C. Comiso, C. L. Parkinson, and H. J. Zwally, *Arctic and Antarctic Sea Ice 1978–1987: Satellite Passive Microwave Observations and Analysis,* NASA SP-511, 200 pp., National Aeronautics and Space Administration, Washington, DC, 1992.

Grenfell, T. C., Surface-based passive microwave studies of multiyear sea ice, *Journal of Geophysical Research, 97(C3),* pp. 3485–3501, 1992.

Grenfell, T. C. and A. W. Lohanick, Temporal variations of the microwave signatures of sea ice during the late spring and early summer near Mould Bay Northwest Territories, *Journal of Geophysical Research, 90(C3),* pp. 5063–5074, May 1985.

Hollinger, J. R., R. Lo, G. Poe, R. Savage, and J. Pierce, *Special Sensor Microwave / Imager User's Guide,* 120 pp., Naval Research Laboratory, Washington, DC, 1987.

Key, J. R., J. A. Maslanik, and A. J. Schweiger, Classification of merged AVHRR and SMMR arctic data with neural networks, *Photogrammetric Engineering and Remote Sensing, 55,* pp. 1331–1338, 1989.

Key, J., A. J. Schweiger, and J. A. Maslanik, Mapping sea ice leads with a coupled numeric/symbolic system, *ACSM / ASPRS Proceedings, 4,* pp. 228–237, Denver, Colorado, March 18, 1990.

Maslanik, J., J. Key, and A. J. Schweiger, Neural network identification of sea-ice seasons in passive microwave data, *IGARSS'90 Proceedings, 2,* pp. 1281–1284, Washington DC, May 1990.

Maykut, G. A. and N. Untersteiner, Some results from a time-dependent thermodynamic model of sea ice, *Journal of Geophysical Research, 76,* pp. 1550–1575, 1971.

NORSEX Group, Norwegian remote sensing experiment in the marginal ice zone north of Svalbard, Norway, *Science, 220(4599),* pp. 781–787, 1983.

Orvig, S., World survey of climatology, Chapter 14, *Climates of the Polar Regions,* 260 pp., Elsevier Publishing Co., New York, 1970.

Parkinson, C. L., J. C. Comiso, H. J. Zwally, D. J. Cavalieri, P. Gloersen, and W. J. Campbell, *Arctic Sea Ice, 1973–1976: Satellite Passive-Microwave Observations,* NASA SP-489, 296 pp., National Aeronautics and Space Administration, Washington, DC, 1987.

Reeves, R. G. (editor), *Manual of Remote Sensing,* American Society of Photogrammetry, Falls Church, Virginia, 1975.

Rothrock, D. A., D. R. Thomas, and A. S. Thorndike, Principal component analysis of satellite passive microwave data over sea ice, *Journal of Geophysical Research, 93(C3),* pp. 2321–2332, 1988.

Shutko, A. M. and G. Granov, Some peculiarities of formulation and solution of inverse problems in microwave radiometry of the ocean surface and atmosphere, *IEEE Journal of Oceanic Engineering, OE-7(1),* pp. 40–43, 1982.

Steffen, K. and J. A. Maslanik, Comparison of Nimbus 7 Scanning Multichannel Microwave Radiometer radiance and derived sea ice concentrations with Landsat imagery for the North Water area of Baffin Bay, *Journal of Geophysical Research, 93(C9),* pp. 10,769–10,781, 1988.

Steffen, K. and A. Schweiger, DMSP-SSM/I NASA algorithm validation using Landsat satellite imagery, *Journal of Geophysical Research, 96(C12),* pp. 21971–21987, 1991.

Stogryn, A., *A Study of Some Microwave Properties of Sea Ice and Snow,* Report 7788, Aerojet Electrosystems Co., Azusa, California, 1985.

Svendsen, E., K. Kloster, B. Farrelly, O. M. Johannessen, J. A. Johannessen, W. J. Campbell, P. Gloersen, D. J. Cavalieri, and C. Mätzler, Norwegian Remote Sensing Experiment: Evaluation of the Nimbus 7 Scanning Multichannel Microwave Radiometer for sea ice research, *Journal of Geophysical Research, 88(C5),* pp. 2781–2792, 1983.

Svendsen, E., C. Mätzler, and T. C. Grenfell, A model for retrieving total sea ice concentration from a spaceborne dual-polarized passive microwave instrument operating near 90 GHz, *International Journal of Remote Sensing, 8(10)*, pp. 1479–1487, 1987.

Swift, C. T. and D. J. Cavalieri, Passive microwave remote sensing for sea ice research, *Eos, Transactions of the American Geophysical Union, 66*, pp. 1210–1212, 1985.

Thomas, D. R. and D. A. Rothrock, Blending sequential scanning multi-channel microwave radiometer and buoy data into a sea ice model, *Journal of Geophysical Research, 94(C8)*, pp. 10,907–10,920, 1989.

Thorndike, A. S. and R. Colony, *Arctic Ocean Buoy Program Data Report, 1 January 1979–31 December 1979*, Contract Report, University of Washington, Polar Science Center, Seattle, Washington, 1980.

Tooma, S. G., R. A. Mannella, J. P. Hollinger, and R. D. Ketchum, Jr., Comparison of sea-ice type identification between airborne dual-frequency passive microwave radiometry and standard laser/infrared techniques, *Journal of Glaciology, 15*, 1975.

Vant, M. R., R. B. Gray, R. O. Ramseier, and V. Makios, Dielectric properties of fresh and sea ice at 10 and 35 GHz, *Journal of Applied Physics, 45*, pp. 4712–4717, 1974.

Webster, W. J., T. T. Wilheit, D. B. Ross, and P. Gloersen, Spectral characteristics of the microwave emission from a wind-driven foam-covered sea, *Journal of Geophysical Research, 81*, pp. 3095–3099, 1976.

Wharton, S. W. and Y.–C. Lu, The Land Analysis System (LAS): A general purpose system for multispectral image processing, *Proceedings of the IGARSS'87 Symposium*, pp. 18–21, Ann Arbor, Michigan, 1987.

Zwally, H. J., J. C. Comiso, C. L. Parkinson, W. J. Campbell, F. D. Carsey, and P. Gloersen, *Antarctic Sea Ice, 1973–1976: Satellite Passive-Microwave Observations*, 206 pp., NASA SP-459, National Aeronautics and Space Administration, Washington, DC, 1983.

Chapter 11. The Validation of Geophysical Products Using Multisensor Data

DONALD J. CAVALIERI

Laboratory for Hydrospheric Processes, Goddard Space Flight Center, Greenbelt, Maryland 20771

11.1 INTRODUCTION

Spaceborne passive microwave radiometers have provided global radiance measurements with which to map, monitor, and study the polar sea ice canopies. Almost two decades of satellite radiometer data have been collected since the launch of the Electrically Scanning Microwave Radiometer (ESMR) on the Nimbus 5 spacecraft in 1972. The advent of multichannel microwave sensors, starting with the Scanning Multichannel Microwave Radiometer (SMMR) on Seasat-A and Nimbus 7 in 1978 and continuing with the Defense Meteorological Satellite Program (DMSP) Special Sensor Microwave/Imager (SSM/I) in 1987, 1990, and 1991, has enabled not only more accurate determination of sea ice concentration, but has also extended the list of parameters to include ice type and temperature [Cavalieri et al., 1984; Gloersen et al., 1992].

Sea ice products derived from the Nimbus ESMR and SMMR sensors have been published as atlases [Zwally et al., 1983; Parkinson et al., 1987; Gloersen et al., 1992] and have been used widely in various geophysical applications. While this work has clearly demonstrated the utility of these data for geophysical research, there remains a need to quantify the accuracy of the microwave-derived sea ice parameters. The lack of confidence in the accuracy of these parameters stems from the relatively small number of validation case studies, the inability of the algorithms to perform consistently in different regions and for different seasons, and the dissimilar algorithms used. All of these factors preclude a clear, simple statement of the precision and accuracy of sea ice products derived from microwave satellite sensors.

Two recently completed validation programs associated with the DMSP SSM/I have provided the most comprehensive validation of geophysical products derived from a single passive microwave imager [Hollinger, 1989, 1991; Cavalieri, 1991]. Both programs aim to provide the operational and research communities with a quantitative measure of the accuracy of the SSM/I geophysical products. The launches of the first three SSM/I's are to be followed by five additional SSM/I instruments scheduled for launch over the next two decades as part of the Department of Defense program to obtain global coverage of atmospheric, oceanic, and terrestrial geophysical parameters. To validate

Microwave Remote Sensing of Sea Ice
Geophysical Monograph 68
©1992 American Geophysical Union

these geophysical products, scientists compare remote observations with known measurements. These comparisons are complicated by issues of spatial and temporal averaging, regional differences, and acquisition of comparison data sets. As a minimum requirement, studies should address regional differences by including regions of perennial ice cover, seasonally ice-covered seas, and marginal ice zones in both the Arctic and Antarctic under winter and summer conditions.

This chapter summarizes the results from the principal sea-ice validation studies carried out since the launch of ESMR, with emphasis on the more recent Naval Research Laboratory (NRL) and NASA validation programs for the DMSP SSM/I. Many of the algorithms used in these validation studies are described in Chapter 10. This summary will provide a measure of the overall accuracy of the principal sea ice products in different regions and seasons, thus establishing a level of confidence with which these data can be used.

11.2 VALIDATION DATA SETS

Since the launch of the Nimbus 5 ESMR, data sets have been compiled to validate sea ice products derived from satellite microwave radiances. The principal data sets used as validation tools include visible and infrared satellite imagery, visual observations from aircraft, and airborne measurements with high-resolution active and passive microwave sensors. These data sets range in size and complexity from a pair of coincident images to coordinated surface, aircraft, and satellite measurements acquired during large multinational field campaigns. Airborne microwave sensors have proved to be an important source of validation data, because they provide spatially and temporally coincident coverage with spacecraft measurements. Furthermore, they serve as a link between satellite and surface measurements [Svendsen et al., 1983; NORSEX Group, 1983; Cavalieri et al., 1986; Burns et al., 1987; Gloersen and Campbell, 1988; Ramseier et al., 1991; Cavalieri et al., 1991; Carsey et al., 1989].

Although many of the early Nimbus 5 ESMR studies were qualitative in nature, they established the credibility of satellite microwave observations and provided a basis for further study and in-depth quantitative analyses. The first quantitative comparison of sea ice concentrations derived from the Nimbus 5 ESMR utilized Landsat Multispectral Scanner (MSS) imagery [Comiso and Zwally, 1982]. Nimbus 7 SMMR and DMSP SSM/I comparative studies that took advantage of observations from coincident or nearly coincident passes by other satellites and spacecraft included studies using the National Oceanic and Atmospheric Administration (NOAA) Advanced Very High Resolution Radiometer (AVHRR) imagery [Cavalieri et al., 1983; Emery et al., 1991], Landsat MSS imagery [Steffen and Maslanik, 1988; Steffen and Schweiger, 1991], and Synthetic Aperture Radar (SAR) imagery from space shuttle flights [Martin et al., 1987].

The combined use of aircraft and spacecraft sensors during many of the field campaigns noted earlier provides, in principle, the basis for cross-checking different combinations of sensors, thereby establishing a relative measure of the accuracy of the ancillary data sets. This multisensor approach served as the basis for the NASA validation program [Cavalieri, 1991]. The NASA and Navy SSM/I underflights helped demonstrate consistency among the various aircraft sensors flown [Cavalieri et al., 1991] and helped explain discrepancies between SSM/I and AVHRR sea ice concentrations through a comparison between Landsat MSS and NOAA AVHRR sea ice concentrations [Emery et al., 1991].

Published sea ice validation studies have focused on spaceborne passive systems, including the Nimbus 5 ESMR, the Nimbus 7 SMMR, and the DMSP SSM/I; there are no corresponding studies for active systems. Except for data from the Seasat-A SAR which operated for only 105 days in 1978, from the Shuttle Imaging Radar-B (SIR-B) in 1984, and from the Geosat altimeter launched in 1985, no other spaceborne active sensor data have been made available to the science community. Nonetheless, these data sets permitted sea ice comparisons utilizing SAR, Seasat-A Scatterometer System, and SMMR data from Seasat-A [Swift et al., 1985; Carsey and Pihos, 1989], coincident SIR-B and Nimbus 7 SMMR data [Martin et al., 1987], and coincident Geosat altimeter and Nimbus 7 SMMR data [Laxon, unpublished]. Active microwave systems on aircraft have proven extremely valuable as a validation tool for satellite microwave radiometers. Many of the validation data sets acquired include wide-swath, high-resolution imagery obtained with aircraft side-looking airborne radar (SLAR) and SAR.

11.3 SEA ICE PRODUCT ACCURACY

Only since the advent of the dual-polarized multifrequency radiometer has it been possible to substantially reduce the ambiguity resulting from a mixture of radiometrically different ice types and open water within a satellite's field of view (FOV). This capability has not only resulted in improved sea-ice concentration accuracy, but has provided important information on ice type and temperature distributions. The results summarized in this section are based on comparative studies of sea ice extent, sea ice concentration, and ice type. While sea ice temperatures (subsurface temperatures) derived from the Nimbus 7 SMMR are not discussed here, preliminary comparisons

with climatological surface air temperatures and with Arctic buoy temperatures show that the SMMR temperatures are consistent with model predictions [Gloersen et al., 1992].

11.3.1 Ice Extent

In principle, the sharp contrast in microwave emission between ocean and sea ice should make it easy to locate the edge of the sea ice with microwave radiometers. In practice though, errors in locating the ice edge may be substantial. A principal source of error is the radiometer's low resolution, which results in a smearing of the ice edge, thereby limiting the location of the ice edge to about half the resolution of the sensor or to the grid resolution of the mapped data. Another important source of error in determining the ice-edge position arises from the enhanced microwave emission due to wind roughening of the ocean surface and to the presence of atmospheric water vapor and cloud liquid water in the intervening atmosphere. These effects result in spurious sea ice concentrations over ice-free ocean areas. Techniques developed to minimize these effects are discussed in Chapter 10. Finally, some forms of frazil are classified visually as ice, but are seen by the passive microwave sensors as indistinguishable from open water (Chapter 14).

Comparisons of brightness temperature gradients and contours of sea ice concentration derived from SMMR algorithms with ice-edge positions as determined from AVHRR, aircraft observations, and surface ship reports have been used to check the accuracy to which the SMMR could be used to map the location of the ice edge. Results from these studies are summarized in Table 11-1. Shortly after the launch of Nimbus 7 SMMR, NORSEX investigators determined the position of the ice edge to an accuracy of 10 km by using the 37-GHz horizontal polarization (H-pol) SMMR radiances [Svendsen et al., 1983]. A Navy P-3 ice reconnaissance flight over the Bering Sea on March 9, 1979, provided another opportunity to test the ability of the SMMR to provide accurate ice-edge locations [Cavalieri et al., 1983]. The SMMR ice edge was defined by a sharp concentration gradient ranging from 25% to 55% over a distance of 50 km. This 50-km width results partly from the SMMR FOV, and partly from the width of the physical transition from open water to consolidated pack ice. The P-3 ice-edge position was within 25 to 30 km of the SMMR edge, or within one SMMR map grid spacing. Using aircraft ESMR imagery obtained over the Greenland Sea during MIZEX-East, Campbell et al. [1987] showed that, for a diffuse ice edge, the 30% SMMR ice concentration contour is closest to the ice edge, whereas for a compact ice edge, the SMMR's 40% to 50% isopleths are closest.

Other studies have been conducted using Canadian operational sea ice algorithms to evaluate the accuracy of SMMR-derived ice edge positions. Moreau et al. [1985], using aerial reconnaissance observations and SLAR data to determine the position of the ice edge, showed that the

TABLE 11-1. Summary of results for satellite-derived ice-edge positions.

Region	Month	Sensor	Ice concentration, isopleth, %	Accuracy, km	Reference
Beaufort Sea	June–September	SSM/I	16	≤20	Ramseier et al. [1991]
		SMMR	10	31	Moreau et al. [1985]
		SMMR	35	21	Moreau et al. [1985]
Bering Sea	March	SMMR	25–55	≤30	Cavalieri et al. [1983]
	March	SSM/I	15 (ice band)	≤25	Cavalieri et al. [1991]
	March	SSM/I	38 (main pack)	≤25	Cavalieri et al. [1991]
Greenland Sea	September	SMMR	T_B[a]	10	Svendsen et al. [1983]
	June	SMMR	40–50 (compact)	≤25	Campbell et al. [1987]
	June	SMMR	30 (diffuse)	≤25	Campbell et al. [1987]

[a] This study used the 37-GHz H-pol brightness temperatures to map the ice edge.

mean difference between the SMMR ice edge, defined as the 10% SMMR ice-concentration contour, and the position of the ice edge, as determined from the aircraft data, was 4.4 km with a root-mean-square (rms) distance of 31.2 km. They also found that the 35% SMMR ice-concentration contour correlated better with the aircraft observations and reduced the rms to 21.2 km.

More recent ice-edge comparisons were made as part of the NRL and NASA SSM/I validation programs. During the NASA and Navy SSM/I aircraft underflights in March 1988 [Cavalieri et al., 1991], the NASA DC-8 overflew the Bering Sea ice edge and, with the aid of a global positioning navigation system, provided the precise location of ice-edge features for determining the appropriate SSM/I ice-concentration contour to use for locating the ice edge on a 25-km SSM/I grid. Results from these comparisons showed that, on average, the 15% concentration contour corresponds to the position of the initial ice bands as determined from the aircraft, while the location of the edge of the main ice pack corresponds to an SSM/I concentration of about 38%, consistent with the results of Campbell et al. [1987] and Moreau et al. [1985].

Bjerkelund et al. [1990] and Ramseier et al. [1991] assessed the accuracy of two sea ice algorithms in support of the NRL validation program. The first algorithm was developed at Hughes Aircraft Company for the Navy (Navy–Hughes); the second algorithm was developed by the Canadian Atmospheric Environmental Service and York University (AES–York). These studies showed that the AES–York algorithm performed better than the Navy–Hughes algorithm in locating the position of the ice edge. In particular, using the Navy–Hughes algorithm, the 0% ice-concentration contour as determined from aircraft radar data taken in the Beaufort Sea corresponded to SSM/I concentrations between 25% and 50%, with an average of

35% depending on ice type. In contrast, the AES–York algorithm gave ice concentrations ranging from 0% to 25%, with an average ice concentration of 16% at the radar-determined ice-edge position. The latter result is consistent with that obtained during the NASA SSM/I aircraft underflights, where the location of the initial seaward ice bands encountered by the NASA DC-8 corresponded to an average SSM/I ice concentration of 15% [Cavalieri et al., 1991].

A summary of results from the validation of the sea-ice edge location is presented in Table 11-1. Most of the comparisons involve algorithms that use either the 18-GHz channels in the case of SMMR or the 19.4-GHz channels in the case of SSM/I. The accuracies listed in the table correspond to approximately half the sensor FOV at these frequencies, or about 20 to 30 km. While the particular ice concentration contour closest to the actual ice edge depends largely on the nature of the ice edge itself, recent NRL and NASA SSM/I studies show that the 15 or 16% contour corresponds best to the 0% ice-edge location defined by aircraft observations. The SSM/I 85-GHz data with its higher spatial resolution held promise for a superior ice-edge definition, but, because of the greater atmospheric microwave emission at this frequency, techniques had to be employed to reduce the atmospheric contamination at the expense of using lower frequencies with their lower spatial resolutions.

11.3.2 Ice Concentration

Sea ice concentration is a measure of the fraction of ocean surface covered by sea ice within a specified area. In remote sensing, this area is usually a satellite footprint, image pixel, or map grid. Thus, a necessary requirement for the validation of sea ice concentration from satellite radiometers is that coincident spatial coverage be attained on scales of hundreds of square kilometers corresponding to at least one satellite radiometer footprint.

The earliest evaluation of a SMMR sea ice algorithm involved a comparison of Greenland Sea ice concentrations measured from aircraft radiometers along selected flight tracks during NORSEX with SMMR concentrations obtained with the NORSEX algorithm using different combinations of SMMR channels [Svendsen et al., 1983]. The total ice concentration results reported by Svendsen et al. [1983] were $0.5\pm2.5\%$ using the 10- and 37-GHz combination of channels and $-3.5\pm2.0\%$ using the 18- and 37-GHz combination. These results are surprisingly good given that the satellite and aircraft data sets were not spatially and temporally coincident.

In an effort to obtain spatially coincident sea-ice concentration estimates for comparison with SMMR sea-ice concentrations in the Weddell Sea, Comiso and Sullivan [1986] utilized helicopter-borne visual observations covering several SMMR footprints. Their results (Figure 11-1) show an offset of about 16% and a standard deviation of 13%.

Only two studies using the NASA SMMR Team algorithm have provided spatially coincident quantitative comparisons. The first study compared SMMR concentrations with those derived from imagery obtained with SIR-B in the Weddell Sea during October 1984 [Martin et al., 1987]; the second compared SMMR and Landsat MSS ice concentrations in Baffin Bay for selected days in March, April, May, and June [Steffen and Maslanik, 1988]. Results of these studies are presented in Figures 11-2 and 11-3, respectively.

The most extensive validation studies to date have resulted from the recent NRL [Hollinger, 1991] and NASA

[Cavalieri, 1991] validation programs for the DMSP SSM/I. From the NRL program, Figure 11-4 illustrates the results of a comparison between sea ice concentrations obtained with both the Navy–Hughes and the AES–York SSM/I sea ice algorithms and concentrations obtained from aircraft radar [Ramseier et al., 1991]. Examination of the figure shows a mean difference in total ice concentration of 23% with the Navy–Hughes algorithm over the range of

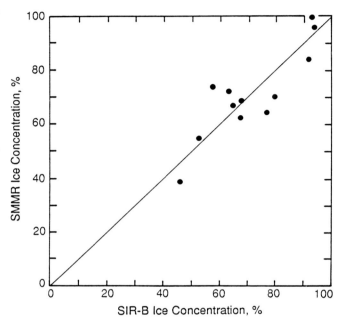

Fig. 11-2. SMMR versus SIR-B sea ice concentrations obtained over the Weddell Sea in October 1984 [Martin et al., 1987].

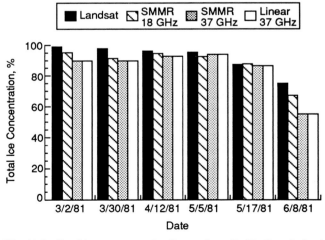

Fig. 11-3. Total ice concentration for northern Baffin Bay derived from SMMR data using the SMMR NASA Team algorithm with 18 and 37 GHz, and a linear interpolation algorithm with 37 GHz. Landsat-derived concentrations are also shown [Steffen and Maslanik, 1988].

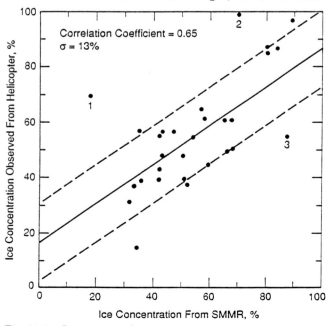

Fig. 11-1. Comparison of ice concentration derived from SMMR and helicopter visual observations [Comiso and Sullivan, 1986].

concentrations, while the mean difference drops to 14% with the AES–York algorithm.

As part of the NASA sea ice validation program, a comparison of sea ice concentrations derived from SSM/I radiances with those from Landsat imagery was undertaken by Steffen and Schweiger [1991]. Over 100 comparisons were made in regions of the Arctic and Antarctic for different seasons. Mean differences (SSM/I–Landsat) in the central Arctic, Baffin Bay, and the Greenland Sea ranged from –3.7±1.4% to +0.5±2.5% during fall, winter, and spring. In the Bering Sea, which has large areas of new and young ice, the mean difference was –9±6.1%. In the Antarctic, the mean differences were smaller, but the standard deviations were larger. Except during summer months and in areas where large amounts of new ice were present, the total ice concentration could be determined with little bias (generally slightly negative) and with standard deviations ranging from 2% to 3%. Comparison statistics for all fall, winter, and spring cases were computed using both

locally and globally defined (actually hemispherically defined) NASA algorithm tie-points (Figure 11-5). Use of locally defined tie-points reduces the bias from 3.6% to 1.5% and the standard deviation of the differences from 6.7% to 4.5%. Direct comparison of SSM/I and aircraft microwave concentrations from data obtained during the NASA- and Navy-coordinated SSM/I underflights for regions having at least 80% aircraft coverage indicates that the SSM/I total ice concentration is lower on average by 2.4±2.4%, which is consistent with the winter Landsat comparisons for the Beaufort Sea [Cavalieri et al., 1991].

A summary of results from those studies that provide quantitative estimates of total sea-ice concentration accuracy is presented in Table 11-2. Except for the large mean differences (Figure 11-4) reported by Ramseier et al. [1991], the mean differences for the central Arctic, Baffin Bay, and the Greenland Sea range from –3.7±1.4% to 0.5±2.5% during fall, winter, and spring. In the Bering Sea, which has large areas of new and young ice, the mean difference is –9±6.1% [Steffen and Schweiger, 1991]. Under melt conditions, a large negative bias of –10% is reported [Steffen and Maslanik, 1988]. In the Antarctic, with the exception of the results reported by Comiso and Sullivan [1986], the mean differences are smaller, but the standard deviations are larger.

11.3.3 Ice Type

Current passive microwave algorithms can distinguish between two ice types, namely first-year and multiyear ice. Other ice types such as new, young, and second-year ice cannot be resolved unambiguously. Thus, this section centers on the validation of distinguishing between first-year and multiyear ice types.

Since sensors operating at visible and infrared wavelengths cannot provide an accurate measure of multiyear ice concentration even under the most suitable light levels and atmospheric conditions, the only source of data with sufficient accuracy and spatial resolution for use as a validation tool for distinguishing between first-year and multiyear ice is high-resolution active and passive imagers flown on aircraft. These airborne sensors have a demonstrated capability to discriminate between these two ice types [e.g., Johnson and Farmer, 1971; Wilheit et al., 1972; Gloersen et al., 1973; Campbell et al., 1976], and they provide the requisite high-resolution coverage needed to ensure a sufficiently accurate measure of multiyear ice concentration.

During NORSEX, SMMR underflights provided an early test of the capability of a multichannel algorithm to measure multiyear concentration [Svendsen et al., 1983], but only along individual flight lines. Comparisons between multiyear concentrations from a multichannel aircraft radiometer flown along a flight track in the Greenland Sea and from SMMR radiances obtained with the NORSEX algorithm revealed differences of –4±6% with the 10- and 37-GHz channels and –8±6% with the 18- and 37-GHz channels, indicating in both cases that the SMMR under-

(a)

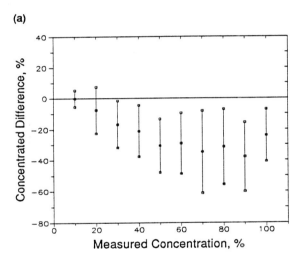

(b)

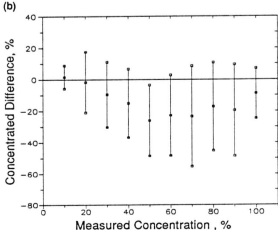

Fig. 11-4. Mean difference (SSM/I – radar) and standard deviation for total ice concentration derived from airborne radar and from (a) the Navy–Hughes and (b) AES–York algorithms [Ramseier et al., 1991].

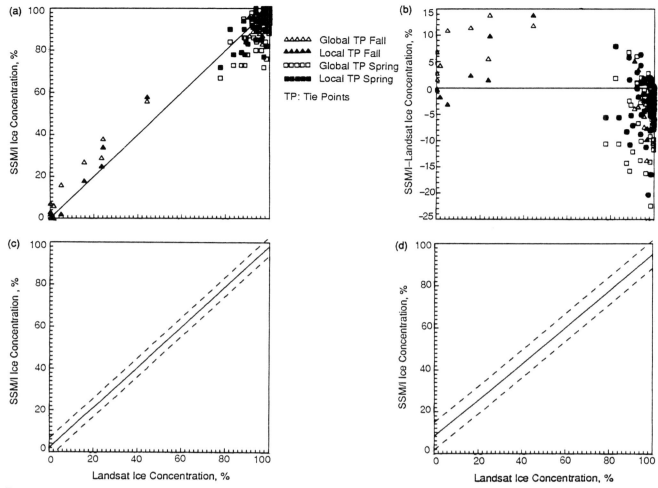

Fig. 11-5. Comparison of sea ice concentration derived from Landsat MSS imagery and from SSM/I radiances using the NASA Team algorithm with both hemispheric and local tie-points: (a) SSM/I versus Landsat MSS concentrations; (b) concentration differences versus Landsat MSS concentrations; (c) least squares best fit using local tie-points; and (d) least squares best fit using hemispheric tie-points [Steffen and Schweiger, 1991].

estimated the concentration relative to the aircraft measured values. The estimated absolute accuracy of the aircraft-determined concentrations was ±10% [Svendsen et al., 1983].

More recently, the NASA and Navy SSM/I underflights across the Beaufort and Chukchi Seas during the NASA validation program [Cavalieri et al., 1991] provided data with which to validate multiyear ice concentrations obtained with the NASA SSM/I algorithm. The airborne fixed-beam and imaging radiometers established that the algorithm correctly maps the large-scale distribution of multiyear ice: the zone of first-year ice off the Alaskan coast, the large areas of mixed first-year and multiyear ice, and the region of predominantly multiyear ice north of the Canadian Archipelago. The comparisons also revealed spurious multiyear concentrations of up to 25% in areas of first-year ice in the Chukchi and Beaufort Seas and a zone of anomalously low concentrations in the coastal region

north of Ellesmere Island. An imaging radiometer and a C-band SAR were used to generate mosaics covering 35 50-km SSM/I map grids. Comparisons between the analyzed aircraft mosaics and the corresponding SSM/I multiyear ice concentration grids having at least 80% aircraft coverage reveal that the SSM/I algorithm overestimated multiyear ice concentration by 4±5% on average in the Chukchi Sea and by 12±11% on average in the Beaufort Sea. Results from these comparisons for all SSM/I grids including those grids with less than 80% aircraft coverage are illustrated in Figure 11-6.

A summary of the multiyear ice comparisons is presented in Table 11-3. In contrast to the total sea ice concentration comparisons in Table 11-2, these results reveal a considerably greater variability. This variability is attributed to a number of factors, including multiyear ice signature variability on individual floes [Grenfell, 1992], variability due to physical temperature differences at

TABLE 11-2. Summary of results for satellite-derived total sea ice concentrations.

Region	Month	Sensor	Mean difference ± one standard deviation, %	Validation data set	Reference
Arctic					
Baffin Bay	March–May	SMMR	−3.5	Landsat MSS	Steffen and Maslanik [1988]
Baffin Bay	June	SMMR	−10	Landsat MSS	Steffen and Maslanik [1988]
Bering Sea	February	SMMR	±5[a]	Ship reports	Cavalieri et al. [1986]
Beaufort and Chukchi Seas	September–November	SSM/I	0.6±7.4	Landsat MSS	Steffen and Schweiger [1991]
Beaufort Sea	March	SSM/I	−2.1±3.1	Landsat MSS	Steffen and Schweiger [1991]
Bering Sea	March	SSM/I	−9.4±6.1[b]	Landsat MSS	Steffen and Schweiger [1991]
Greenland Sea	September	SSM/I	−3.7±1.4	Landsat MSS	Steffen and Schweiger [1991]
	October	SMMR	0.5±2.5[c] 3.5±2.0[d]	Aircraft radiometers	Svendsen et al. [1983]
Beaufort Sea	March	SSM/I	−2.7±2.4	NADC/ERIM SAR[f]	Cavalieri et al. [1991]
Chukchi Sea	March	SSM/I	−1.8±2.0	NOARL/ KRMS[g]	Cavalieri et al. [1991]
Beaufort Sea	June 1987–	SSM/I	−23[e]	Aircraft radar	Ramseier et al. [1991]
	September 1988	SSM/I	−14[e]	Aircraft radar	Ramseier et al. [1991]
Antarctic					
Weddell Sea	October	SMMR	0.3±7.6	SIR-B	Martin et al. [1987]
Weddell Sea	November	SSM/I	−1.1±3.1	Landsat MSS	Steffen and Schweiger [1991]
Amundsen Sea	December	SSM/I	1.3±3.6	Landsat MSS	Steffen and Schweiger [1991]
Weddell Sea	November	SMMR	16±13	Helicopter visual observations	Comiso and Sullivan [1986]

[a] In regions of thick first-year ice.

[b] Includes new ice.

[c] NORSEX algorithm using 10- and 37-GHz channels.

[d] NORSEX algorithm using 18- and 37-GHz channels.

[e] Estimated from Figure 11-4.

[f] NADC/ERIM SAR = Naval Air Development Center/Environment Research Institute of Michigan SAR.

[g] NOARL/KRMS = Naval Oceanographic and Atmospheric Research Laboratory/Ka-band Scanning Imaging Radiometer.

penetration depths corresponding to 19- and 37-GHz (the frequencies most commonly used in deriving multiyear ice concentrations) [Gloersen et al., 1992], and the influence of a variable snow cover (particularly in areas of heavily ridged first-year ice) [Cavalieri et al., 1991].

11.4 FUTURE REQUIREMENTS

Validation studies of sea ice parameters obtained with spaceborne multichannel passive-microwave radiometers have shown that the multichannel algorithms provide the location of the ice edge to an accuracy of approximately 25 to 30 km, while the total ice concentration can be determined with little bias (generally slightly negative) and with standard deviations ranging from 2 to 3% during winter in the central Arctic. Larger negative biases and standard deviations occur in regions of new ice production (e.g., Bering Sea) and during summer when surface melt effects are appreciable. Three of the four Antarctic comparisons show very small biases (±1%) with standard deviations ranging from 3% to 8%. More Antarctic studies are needed, especially in the Ross, Bellingshausen, and Amundsen Seas, which are second only to the Weddell Sea in terms of areal ice cover; better seasonal coverage is also needed in all regions.

There is still a great deal of uncertainty in the accuracy of Arctic multiyear sea ice concentrations. This is very likely the result of the inherent variability of the emissive character of the ice itself, but there are too few studies from which to draw firm conclusions. If we cannot significantly

TABLE 11-3. Summary of results for satellite-derived multiyear ice concentration.

Region	Month	Sensor	Mean difference ± one standard deviation, %	Validation data set	Reference
Greenland Sea	October	SMMR	−4±6[a] 8 ±6[b]	Aircraft radiometers	Svendsen et al. [1983]
Beaufort Sea	March	SSM/I	5.3±3.9[c] 14.2±11.3[d]	NADC/ERIM SAR	Cavalieri et al. [1991]
Chukchi Sea	March	SSM/I	3.8±4.7	NOARL/KRMS	Cavalieri et al. [1991]
Beaufort Sea	March	AMMR	−6.0±14	JPL C-band SAR	Cavalieri et al. [1991]

[a] NORSEX algorithm using 10- and 37-GHz channels.

[b] NORSEX algorithm using 18- and 37-GHz channels.

[c] Excluding data from one of four flights that gave anomalously large biases.

[d] Including data from all four flights.

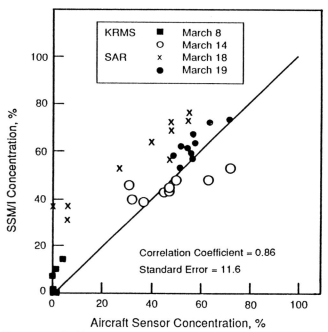

Fig. 11-6. DMSP SSM/I–aircraft multiyear sea ice comparison [Cavalieri et al., 1991].

improve upon the accuracy of multiyear ice concentration retrievals from multichannel radiometers alone, other techniques and sensors or combinations of sensors need to be employed.

The recent launch of the ERS-1 complement of sensors, both active and passive, will lead to the development of new single-sensor and multisensor algorithms. Clearly, the geophysical products derived from SAR algorithms currently in place at the Alaskan SAR Facility in Fairbanks and at the European Space Agency (ESA) centers will need to be

validated along with those algorithms currently being planned for use with the SAR's on JERS-1 and Radarsat. Spatially and temporally coincident data sets from visible, infrared, and microwave sensors are becoming more accessible. New approaches, including models that utilize temporal information from satellite radiometers along with relevant atmospheric and oceanic parameters, are on the horizon. Other techniques, including artificial intelligence, are now being applied to these multisensor data sets. Whether or not these models, new data sets, and techniques will provide more accurate sea ice information must await future validation studies.

REFERENCES

Bjerkelund, C. A., R. O. Ramseier, and I. G. Rubinstein, Validation of the SSM/I and AES/York algorithms for sea ice parameters, in *Sea Ice Properties and Processes*, edited by S. F. Ackley and W. F. Weeks, Monograph 90-1, Cold Regions Research and Engineering Laboratory, Hanover, New Hampshire, pp. 206–208, February 1990.

Burns, B. A., D. J. Cavalieri, M. R. Keller, W. J. Campbell, T. C. Grenfell, G. A. Maykut, and P. Gloersen, Multisensor comparison of ice concentration estimates in the Marginal Ice Zone, *Journal of Geophysical Research, 92*, pp. 6843–6856, 1987.

Campbell, W. J., P. Gloersen, W. J. Webster, T. T. Wilheit, and R. O. Ramseier, Beaufort sea ice zones as delineated by microwave imagery, *Journal of Geophysical Research, 81*, pp. 1103–1110, 1976.

Campbell, W. J., P. Gloersen, E. G. Josberger, O. M. Johannessen, P. S. Guest, N. Mognard, R. Shuchman, B. A. Burns, N. Lannelongue, and K. L. Davidson, Variations of mesoscale and large-scale sea ice morphology in the 1984 Marginal Ice Zone Experiment as observed by microwave remote sensing, *Journal of Geophysical Research, 92*, pp. 6805–6824, 1987.

Carsey, F. D. and G. Pihos, Beaufort–Chukchi Seas summer and fall ice margin data from Seasat: Conditions with similarities to the Labrador Sea, *IEEE Transactions on Geoscience and Remote Sensing, 27*, pp. 541–551, 1989.

Carsey, F. D., S. Digby–Argus, M. Collins, B. Holt, C. Livingstone, and C. Tang, Overview of LIMEX'87 ice observations, *IEEE Transactions on Geoscience and Remote Sensing, 27*, pp. 468–482, 1989.

Cavalieri, D. J., NASA sea ice validation program for the DMSP SSM/I, *Journal of Geophysical Research, 96(C12)*, pp. 21,969–21,970, 1991.

Cavalieri, D. J., S. Martin, and P. Gloersen, Nimbus 7 SMMR observations of the Bering sea ice cover during March 1979, *Journal of Geophysical Research, 88, pp. 2743–2754, 1983.*

Cavalieri, D. J., P. Gloersen, and W. J. Campbell, "Determination of Sea Ice Parameters With the NIMBUS 7 SMMR," *Journal of Geophysical Research, 89,* pp. 5355–5369, 1984.

Cavalieri, D. J., P. Gloersen, and T. T. Wilheit, Aircraft and satellite passive microwave observations of the Bering Sea ice cover during MIZEX West, *IEEE Transactions on Geoscience and Remote Sensing, GE-24(3)*, pp. 368–377, 1986.

Cavalieri, D. J., J. P. Crawford, M. R. Drinkwater, D. T. Eppler, L. D. Farmer, R. R. Jentz, and C. C. Wackerman, Aircraft active and passive microwave validation of sea ice concentration from the DMSP SSM/I, *Journal of Geophysical Research, 96*, pp. 21,989–22,008, 1991.

Comiso, J. C. and C. W. Sullivan, Satellite microwave and in-situ observations of the Weddell sea ice cover and its marginal ice zone, *Journal of Geophysical Research, 91*, pp. 9663–9681, 1986.

Comiso, J. C. and H. J. Zwally, Antarctic sea ice concentration inferred from Nimbus 5 ESMR and Landsat imagery, *Journal of Geophysical Research, 87,* pp. 5836–5844, 1982.

Emery, W. J., M. Radebaugh, C. W. Fowler, D. J. Cavalieri, and K. Steffen, An intercomparison of sea ice parameters computed from AVHRR and Landsat satellite imagery and from airborne passive microwave radiometry, *Journal of Geophysical Research, 96*, pp. 22,075–22,085, 1991.

Gloersen, P. and W. J. Campbell, Satellite and aircraft passive microwave observations during the Marginal Ice Zone Experiment in 1984, *Journal of Geophysical Research, 93*, pp. 6837–6846, 1988.

Gloersen, P., W. Nordberg, T. J. Schmugge, T. T. Wilheit, and W. J. Campbell, Microwave signatures of first-year and multiyear sea ice, *Journal of Geophysical Research, 78*, pp. 3564–3572, 1973.

Gloersen, P., W. J. Campbell, D. J. Cavalieri, J. C. Comiso, C. L. Parkinson, and H. J. Zwally, *Arctic and Antarctic Sea Ice 1978–1987: Satellite Passive Microwave Observations and Analysis*, NASA SP-511, 200 pp., National Aeronautics and Space Administration, Washington, DC, 1992.

Grenfell, T. C., Surface-based passive microwave studies of multiyear sea ice, *Journal of Geophysical Research, 97(C3)*, pp. 3485–3501, 1992.

Hollinger, J., *DMSP Special Sensor Microwave/Imager Calibration/Validation, Final Report Volume I*, Naval Research Laboratory, Washington, DC, 1989.

Hollinger, J., *DMSP Special Sensor Microwave/Imager Calibration/Validation, Final Report Volume II*, Naval Research Laboratory, Washington, DC, 1991.

Johnson, J. D. and L. D. Farmer, Use of side-looking airborne radar for sea ice identification, *Journal of Geophysical Research, 76*, pp. 2138–2155, 1971.

Martin, S., B. Holt, D. J. Cavalieri, and V. Squire, Shuttle Imaging Radar B (SIR-B) Weddell sea ice observations: A comparison of SIR-B and Scanning Multichannel Microwave Radiometer ice concentrations, *Journal of Geophysical Research, 92(C7)*, pp. 7173–7179, 1987.

Moreau, T. A., E. L. Sudeikis, I. G. Rubinstein, and F. W. Thirkettle, *A Study and Report on Operational Ice Mapping, Nimbus-7 SMMR*, prepared by PhD Associates, Inc., for Atmospheric Environment Service, Ice Branch, DSS Contract No. 01SE.KM147-4-0501, Downsview, Ontario, 1985.

NORSEX Group, Norwegian remote sensing experiment in the marginal ice zone north of Svalbard, Norway, *Science, 220(4599)*, pp. 781–787, 1983.

Parkinson, C. L., J. C. Comiso, H. J. Zwally, D. J. Cavalieri, P. Gloersen, and W. J. Campbell, *Arctic sea ice, 1973–1976: Satellite passive-microwave observations*, NASA SP-489, 296 pp., National Aeronautics and Space Administration, Washington, DC, 1987.

Ramseier, R. O., L. D. Arsenault, K. Asmus, C. Bjerkelund, T. Hirose, and I. G. Rubinstein, Sea ice validation (Chapter 10), in *DMSP Special Sensor Microwave/Imager Calibration/Validation, Final Report Volume II*, edited by J. Hollinger, Naval Research Laboratory, Washington, DC, 1991.

Steffen, K. and J. A. Maslanik, Comparison of Nimbus 7 Scanning Multichannel Microwave Radiometer radiance and derived sea ice concentrations with Landsat imagery for the north water area of Baffin Bay, *Journal of Geophysical Research, 93*, pp. 10,769–10,781, 1988.

Steffen, K. and A. Schweiger, NASA Team algorithm for sea ice concentration retrieval from DMSP SSM/I: Comparison with Landsat satellite imagery, *Journal of Geophysical Research, 96(C12)*, pp. 21,971–21,987, 1991.

Svendsen, E., K. Kloster, B. Farrelly, O. M. Johannessen, J. A. Johannessen, W. J. Campbell, P. Gloersen, D. J. Cavalieri, and C. Mätzler, Norwegian Remote Sensing Experiment: Evaluation of the Nimbus 7 Scanning Multichannel Microwave Radiometer for sea ice research, *Journal of Geophysical Research, 88(C5)*, pp. 2781–2792, 1983.

Swift, C. T., W. J. Campbell, D. J. Cavalieri, P. Gloersen, H. J. Zwally, L. S. Fedor, N. M. Mognard, and S. Peteherych, Observations of the polar regions from satellite using active and passive microwave techniques, Chapter 9, in *Satellite Oceanic Remote Sensing*, edited by B. Saltzman, Academic Press, New York, 1985.

Wilheit, T. T., W. Nordberg, J. Blinn, W. Campbell, and A. Edgerton, Aircraft measurements of microwave emission from Arctic sea ice, *Remote Sensing of the Environment, 2*, pp. 129–139, 1972.

Zwally, H. J., J. C. Comiso, C. L. Parkinson, W. J. Campbell, F. D. Carsey, and P. Gloersen, *Antarctic Sea Ice, 1973–1976: Satellite Passive-Microwave Observations,* 206 pp., NASA SP-459, National Aeronautics and Space Administration, Washington, DC, 1983.

Chapter 12. Microwave Remote Sensing of the Southern Ocean Ice Cover

JOSEFINO C. COMISO

Laboratory for Hydrospheric Processes, Goddard Space Flight Center, Greenbelt, Maryland 20771

THOMAS C. GRENFELL

Department of Atmospheric Sciences, University of Washington, Seattle, Washington 98195

MANFRED LANGE

Director, The Arctic Centre, University of Lapland, P. O. Box 122, SF-96101 Rovaniemi, Finland

ALAN W. LOHANICK

Cold Regions Research and Engineering Laboratory, 72 Lyme Road, Hanover, New Hampshire 03755-1290

RICHARD K. MOORE

Radar Systems and Remote Sensing Laboratory, University of Kansas Center for Research, Inc.,
2291 Irving Hill Drive, Campus West, Lawrence, Kansas 66045

PETER WADHAMS

Director, Scott Polar Research Institute, University of Cambridge, Lensfield Road, Cambridge, CB2 1ER, United Kingdom

12.1 INTRODUCTION

The Southern Ocean sea ice cover grows dramatically to about 20×10^6 km^2 in the spring and breaks up abruptly to about 4×10^6 km^2 in the summer (e.g., Zwally et al. [1983b]). This makes it one of the most seasonally variable climate parameters on the surface of the globe. Compared to the Northern Hemisphere, the ice cover in the Southern Ocean is about 20% greater at its maximum extent [Comiso and Zwally, 1984; Gloersen and Campbell, 1988]. By virtue of its size alone, the impact of the Southern Ocean ice cover on the regional and global climate can be considerable, since sea ice drastically changes surface albedo and roughness and insulates the ocean from the atmosphere. Seasonal and interannual variations in the spatial distribution of sea ice also cause the redistribution of salts, which, in turn, cause changes in the vertical stratification of the ocean. Compared to the Arctic, the environmental geographical background for sea ice in the Antarctic is also very different. Land surrounds most of the southern limits of the Arctic ice cover, whereas in the Antarctic, there is no corresponding land boundary in the north. The Southern Ocean ice cover is more divergent since it is more vulnerable to dynamic forcing than its Arctic counterpart. Accurate estimation of the percentage of open water is important because heat and salinity fluxes increase considerably even with just small

Microwave Remote Sensing of Sea Ice
Geophysical Monograph 68

increases in the fraction of open water or new ice [Maykut, 1978; Allison, 1981]. These fluxes are, in turn, closely linked with bottom water formation, ocean circulation, and momentum exchange between the ocean and the atmosphere. Oceanic heat flux is also believed to be a major determinant of ice growth rate [Bagriantsev et al., 1989].

Since the Southern Ocean sea ice cover is primarily seasonal, the percentage of ice with a first-year ice signature is expected to be very high. Even the ice that survives the summer has been observed to have physical characteristics (e.g., salinity) similar to those of first-year sea ice [Gow et al., 1982]. This is partly because of flooding or the intrusion of seawater into the snow–ice interface, and partly because, unlike the Arctic, there are almost no melt ponds in the Antarctic [Andreas and Ackley, 1982]. Much of the ice that survives the summer is second-year ice, since a large percentage of the ice in the perennial ice region gets advected out of the area during the fall and winter. However, because the ice is divergent, the percentage of new and flooded ice, the emissivity and backscatter of which differ from those of dry and cold first-year ice, is also expected to be significant within the ice pack. Variability in the snow cover characteristics can also cause spatial variations in the signature.

The physical and radiative characteristics of sea ice in the Antarctic region have not been as extensively studied as in the Arctic, because the former is generally more inaccessible. However, there have been some Antarctic programs, mostly in the Weddell Sea, with good in-situ measurements, that have been used to advance our knowledge of the microwave characteristics of Antarctic sea ice. In this chapter, the basic physical, radiative, and backscatter properties of sea ice in the Southern Ocean will be presented. Also, techniques used to derive geophysical parameters, including ice extent and concentration from space-based systems, will be evaluated.

12.2 THE ICE REGIMES—OUTER AND INNER ZONES

To understand the microwave satellite signatures of Antarctic sea ice, it is useful to divide the ice regimes into two zones: the outer and inner zones. The outer zone includes the marginal ice zone covered by new ice and ice floes up to several meters in diameter surrounded either by open water, grease ice, or slush. The inner zone is the region beyond the outer zone in which the ice cover is consolidated over several tens (or hundreds) of kilometers with breaks only in leads and polynyas. The outer zone can thus be characterized as a zone of small floes and new ice, while the inner zone is a region of very large floes. The outer zone can be further associated with big waves, while the inner zone has small or no waves. Waves often make the signature of the same type of surface different (either directly or indirectly) in the two zones. The coastal region, which is usually the site of coastal polynyas, is a special case, because although it is part of the inner zone, its radiometric signature is sometimes very much like that of the outer zone.

A large fraction of the outer zone is usually covered by new ice, such as grease ice, pancakes, gray ice, and white nilas, the emissivity or backscatter of which are usually different from those of the thicker ice types. Also, the floe size, thickness, and roughness change from the ice edge to the inner pack [Wadhams et al., 1987], and at the same time floes become thicker and rougher. The outer zone is also the region where ice banding is prevalent, especially during late winter and summer. These bands, which consist mainly of broken up and flooded thick ice, are typically a few hundred meters wide and several kilometers across. The regions between these bands are usually covered by open water, but, during subfreezing temperatures, these are covered by small pancakes and grease ice. Throughout the year, satellite data indicate that along the periphery of the ice cover, while growth is occurring in some places, decay (or retreat) is going on in other places, partly due to changing wind, temperature, and wave patterns.

The inner zone, on the other hand, is expected to have much more stable radiative and backscatter characteristics than the outer zone. The reason for this is that the ice cover in the inner zone is dominated by snow-covered saline thick ice that generally has a first-year ice signature. In addition, the inner zone is not as dynamic as the outer zone, so the snow–ice interface that contributes to much of the observed brightness temperatures (or backscatter) is less likely to be disturbed by atmospheric and oceanic forcing. Thus, except

in areas where leads and polynyas are abundant or large, and where snow cover thickness varies considerably [Ackley et al., 1990; Lange and Eicken, 1991], the emissivity or backscatter of the ice cover in the inner zone is not expected to change very much spatially. However, during storms, flooding and rafting could change the signature considerably.

An early stage of ice growth in the Weddell Sea (April 9, 1988), as observed almost simultaneously by the Special Sensor Microwave Imager (SSM/I) and the Advanced Very High-Resolution Radiometer (AVHRR) sensors, is shown in Figure 12-1. Figure 12-1(a) is an ice concentration map derived from the SSM/I data using an algorithm described by Comiso et al. [1984] and Comiso and Sullivan [1986]. The inner zone is seen in the passive microwave data as a region of very high concentration, while the outer zone is shown to have the largest gradient in ice concentration. The corresponding AVHRR visible channel image is shown in Figure 12-1(b). In the AVHRR image, cloud cover is obviously a problem, but in cloud-free areas, the inner zone is seen as

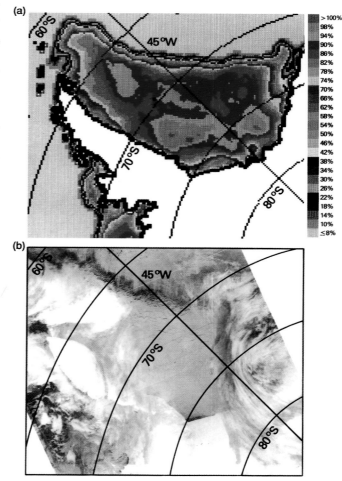

Figure 12-1. Color-coded SSM/I ice concentration map and AVHRR visible channel image of Southern Ocean sea ice during ice growth on April 9, 1988.

an area of almost continuous ice cover interrupted by only a few leads. The outer zone as shown by the same image is, on the other hand, a much more active ice area and consists of a wide variety of ice types from new ice (grayish) to thicker ice in ice bands (white).

12.3 Physical and Electrical Characteristics of Antarctic Sea Ice

There are some basic differences between the physical characteristics of Antarctic sea ice and Arctic sea ice. Core studies indicate that Antarctic sea ice consists mainly of frazil ice (>50%), while Arctic sea ice is primarily congelation (ca. 90%) [Gow et al., 1982; Lange et al., 1989; Tucker et al., 1987]. The total thickness of Antarctic sea ice is, to a large degree, due to rafting, while that for the Arctic is mainly due to thermodynamic freezing at the bottom. Also, ridging is much less intense in the Antarctic than in the Arctic [Weeks et al., 1989; Lytle and Ackley, 1991]. Furthermore, there is less snow in the Central Arctic than in the Antarctic [Barry, 1989]. The ablation seasons for the two regions are also

distinctly different with melt ponds prevalent in the inner zone of the Arctic, but not observed in the Antarctic [Andreas and Ackley, 1982]. These differences provide some of the reasons why the radiative and backscatter characteristics of ice in the two hemispheres are different.

The physical properties of sea ice most relevant to remote sensing are ice thickness, snow cover, salinity, snow–ice interface temperatures, porosity, roughness, and wetness. A general discussion of these properties and their effects on emissivity and backscatter is presented in Chapters 2, 4, 5, 15, and 16. Sketches of surface core properties of Antarctic sea ice taken during the Winter Weddell Sea Project in 1986 [Comiso et al., 1989] are shown in Figure 12-2.

Examples of ice and snow thickness distributions along short transects for both first-year and multiyear ice floes are shown in Figure 12-3 [Wadhams, 1991]. In the seasonal ice region, the average thickness of first-year ice has been observed to be about 0.7 m for undeformed ice [Wadhams et al., 1987] and about 1 m for both deformed and undeformed ice [Lange and Eicken, 1991], excluding the snow cover that

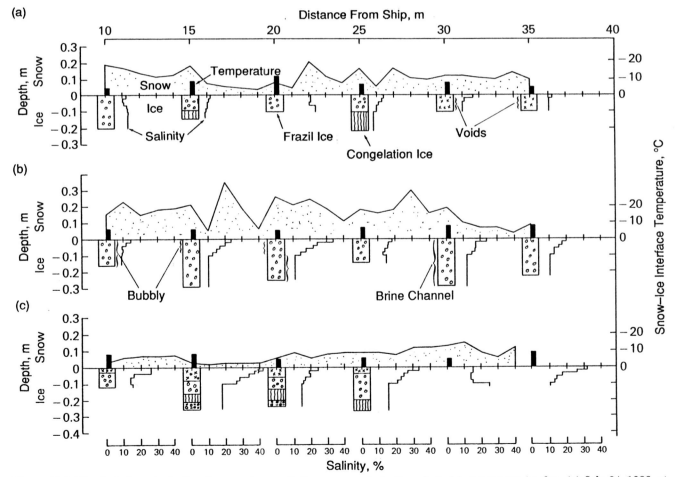

Figure 12-2. Physical characterization (temperature, salinity, snow, and stratigraphy) of Antarctic sea ice for: (a) July 24, 1986, at 64.75° S, 2.32° W; (b) August 6, 1986, at 66.66° S, 5.05° W; and (c) August 22, 1986, at 66.38° S, 1.63° E [Comiso et al., 1989].

(a)

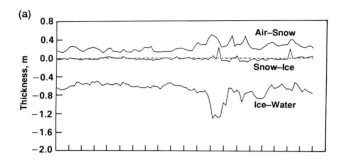

(b)

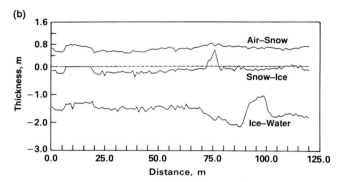

Figure 12-3. Ice thickness and snow profiles for (a) first-year ice and (b) multiyear ice [Wadhams, 1991].

varied in thickness from 5 to 30 cm. Antarctic multiyear ice was observed to be significantly thicker (>2 m) [Gow et al., 1982; Garrison, et al., 1986; H. Eicken, private communication, 1989], but now enough measurements have been made to obtain a statistically good average. The density of sea ice in the Weddell Sea is typically 0.90 to 0.92 mg/m^3, while snow densities vary from 0.2 to 0.4 mg/m^3.

Snow cover can affect the brightness of sea ice in at least two ways: it can increase emissivity because it emits radiation, or it can decrease the emissivity because it serves as a scatterer to radiation. In addition, it acts as an insulator that keeps the ice relatively warm compared to the surface. For the usually dry Weddell Sea snow cover of about a 15 to 30 cm thickness, the increase in emissivity due to snow emission is small because the absorption coefficient of dry snow is relatively low. The microwave consequences of snow cover are discussed in Chapter 16.

Variations in snow–ice interface temperature are due partly to differences in the thickness of the snow cover and partly due to variations in air temperature. The air temperature over sea ice in the Weddell Sea was observed to vary typically from 0° to –30°C from July through September 1986, while the snow–ice interface temperature varied from –2° to –12°C [Comiso et al., 1989]. The two temperatures are approximately linearly related as shown in Figure 12-4, with the correlation coefficient of about 0.8 and standard deviation of about 2.5°C. The mean snow–ice

temperature is about –6°C with a standard deviation of 2°C. Thus, the spatial variability in the brightness temperature due to variations in the physical temperature may be relatively small.

The salinity profiles in Figure 12-2 show a large range in salinity at the snow–ice interface (10‰ to 32‰), indicating that sometimes the snow–ice interface is extremely saline. These values are typical for Antarctic sea ice and probably result from the infiltration of seawater (flooding) into the snow–ice interface and the subsequent formation of snow ice. This is observed to be a common process in the Weddell Sea [Lange and Eicken, 1991; Ackley et al., 1990]; in some regions, like the Western Weddell Sea, the percentage of flooded floes was found to be as high as 30% [D. Meese, private communication, CRREL, Hanover, New Hampshire, 1989]. A fraction of the floes sampled were second-year or older ice with a very thick snow cover. The snow load causes the floes to have negative freeboard and can be a main factor causing the infiltration of seawater. The profiles in Figure 12-3 show more snow for multiyear ice than first-year ice, and hence a more negative freeboard for the former.

Snow ice is also characterized by a relatively high porosity, which could make the brightness temperature and backscatter of infiltrated sea ice significantly different from those of normal first-year ice. The profiles in Figure 12-3 provide information about surface roughness at meter resolution. It is apparent that the snow surface roughness may not reflect the roughness of the snow–ice interface.

In the outer zone, the surface of ice floes is generally wet, sometimes because of flooding and sometimes because of melting since in this region air temperature is occasionally above freezing temperature [Comiso et al., 1984]. In many cases, especially near the ice edge, water is trapped on the surface by raised edges of the floes. In the inner zone, the

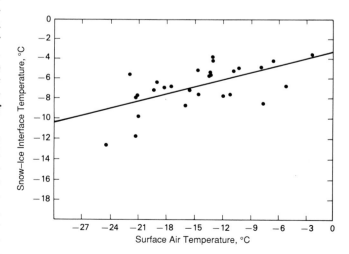

Figure 12-4. Snow–ice versus surface–air temperature from the Weddell Sea for surface measurements from July 20 to September 2, σ = 2.5 [Comiso, et al., 1989].

surface of the ice cover in spring changes progressively from that of a slightly wet surface to one saturated by water. During this time period, the radiative characteristics of the ice become less predictable and errors in the determination of geophysical parameters are greater.

12.4 RADIATIVE AND BACKSCATTER PROPERTIES

12.4.1 Passive Microwave Emissivity

The brightness temperatures of the Antarctic sea ice cover have been observed to have large spatial variations from the marginal ice zone to the coastal region [Zwally et al., 1983a]. This is true despite the predominance of relatively thick first-year ice, the radiative characteristics of which are supposed to be stable and unique. The variation in the outer zone is due largely to the transitory state of the components of the ice cover. In a matter of hours, the emissivity of pancake ice or nilas, which are predominant in the region, can change considerably because of increases in thickness, changes in surface properties, or atmospheric effects. In the inner zone, some of these variations are due mainly to dynamical atmospheric and oceanographic forcing, causing the formation of leads and polynyas. However, variations in emissivity over consolidated ice areas have also been observed, especially at 37 GHz and higher frequencies.

Brightness temperature maps of the Antarctic region using six channels of SSM/I data are shown in Figure 12-5. The color-coded images correspond to those for the vertical and horizontal polarizations at 19, 37, and 85 GHz. Data from both polarizations at 19 and 37 GHz have been most often utilized for deriving sea-ice parameters because they are less affected by atmospheric effects. Since the 85-GHz data offer finer resolution (12.5 km) and good sensitivity to the snow cover, useful information about snow thickness, granularity, and wetness, may be possible to obtain as suggested by Comiso et al. [1989].

The various images show a larger contrast in brightness temperature between ice and water at horizontal than at vertical polarization. They also show that the brightness temperature of water goes down with frequency while that of ice is similar at all frequencies. The 19 and 37 GHz data clearly show the location of the ice edge around the continent while the 85 GHz data show the location of the ice edge only in a few regions, because in other regions the ice edge is obstructed by atmospheric effects. In the inner zone, the brightness temperature of sea ice is observed to be more uniform at lower than at higher frequencies. This is mainly due to a larger sensitivity caused by scattering in the snow (or the atmosphere) at shorter than at longer wavelengths (i.e., 0.35 cm at 85 GHz compared to 1.55 cm at 19 GHz) as discussed earlier.

All the images in Figure 12-5 show consistently lower brightness temperatures near the coast of the Bellingshausen and Amundsen Seas region than in other parts of the Antarctic. Ice concentration maps derived from

the technique described by Comiso et al. [1984] and Comiso and Sullivan [1986] indicate highly consolidated ice in the area. It is a region of perennial ice cover and, although the signature is not identical to multiyear ice signature in the Arctic, the pattern is similar. A possible cause of the difference is that in the Arctic, the ice is thicker and less saline, whereas in the Antarctic, the ice is thinner but with thicker snow cover [W. Weeks and M. Jeffrey, personal communication, Fairbanks, Alaska, 1990]. Low brightness temperatures in the inner zone may also be due to open water as is apparent at 45° E and 65° S, the same area where the Cosmonaut polynya has been recurring [Comiso and Gordon, 1987]. In this particular case, the reduced values probably correspond to a reduction in ice concentration.

To better illustrate multichannel characteristics of Antarctic sea ice, scatter plots of 19 versus 37 GHz data at vertical polarization in the Weddell Sea, Bellingshausen Sea, Amundsen Sea, and Ross Sea regions, using SSM/I data, are shown in Figures 12-6(a) and (b). In these figures, data points along line AB are consistent with those observed for consolidated ice cover from ships, helicopters, and airplanes [Comiso and Sullivan, 1986; Comiso et al., 1989, 1991]. The standard deviation of highly concentrated ice along line AB is about 2.5 K. Significant variations in the brightness temperatures of sea ice are observed as shown in the scatter plots. The variation is due partly to snow cover which either suppresses the emissivity (through internal scattering) when the surface is dry, or enhances the emissivity when the surface is wet. The data points along line CD correspond to open ocean areas outside the ice pack. Much of the open water area within the ice pack would have brightness temperatures corresponding to location C in the scatter plot. As the open water surface gets rougher (with foam), the emissivity goes up along line CD [Swift, 1980]. Mixtures of ice and water are therefore mainly confined to the region between AB and C. When the horizontal polarization at 37 GHz is plotted against the vertical polarization, Figures 12-5(c) and (d), similar effects are observed. However, the data points along line AC are now closer to those along line AB, which means that areas covered by consolidated ice are not separable from those with mixtures of open water and ice. Furthermore, open water regions that are well separated from ice-covered areas in Figures 12-5(a) and (b) are shown to have similar signatures to those of mixtures of ice and water in Figures 12-5(c) and (d). Thus, the sole use of the 37-GHz channels (at two polarizations) would cause ambiguities in the determination of sea ice concentration. Similar ambiguities also applied when data from the 18-GHz channels were used. The differences in the distribution of data in the scatter plots for various regions also illustrate the spatial variability of the emissivity. The variations are likely due to differences in the history of the formation of the ice cover.

The effect of varying snow cover characteristics on brightness temperatures during spring, is illustrated in Figure 12-7, which shows a time series of scatter plots of brightness temperatures from November 7 through Decem-

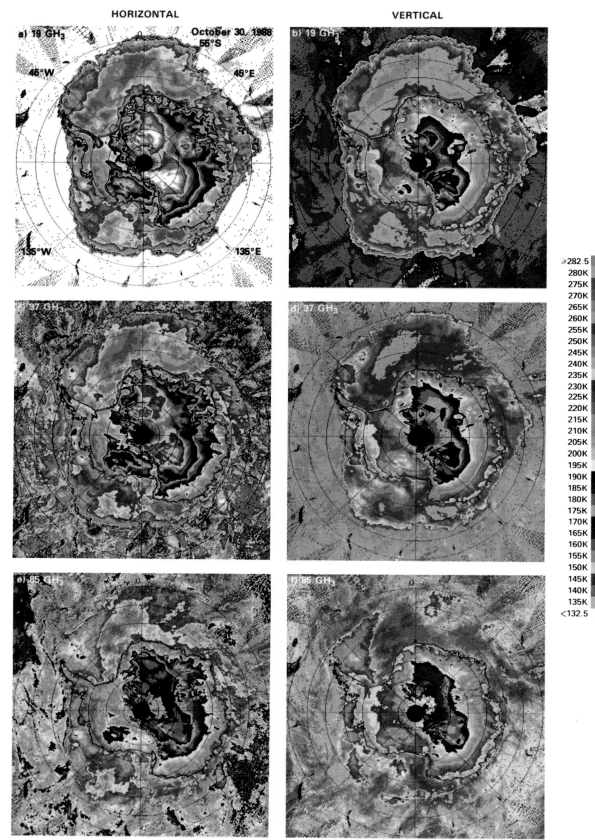

Figure 12-5. SSM/I brightness temperatures of Antarctic sea ice.

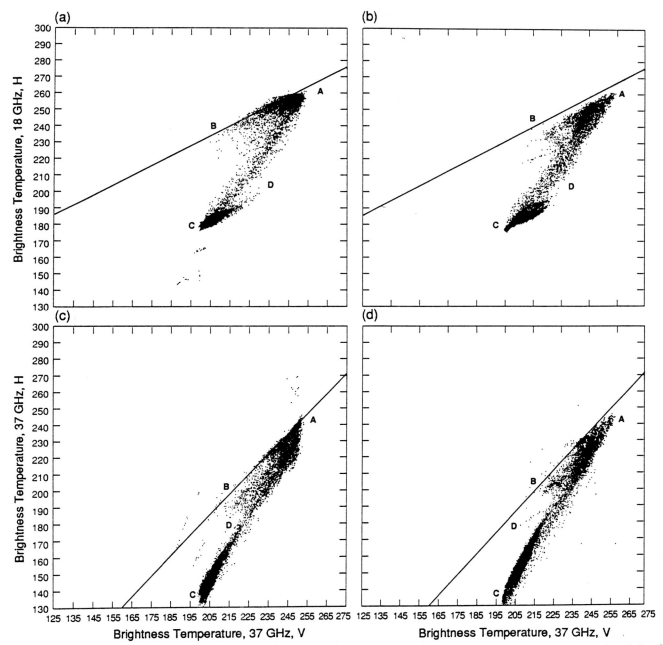

Figure 12-6. Scatter plots of 19 V versus 37 V at (a) 0° to 90° W and (b) 90° to 180° W, and of 37H versus 37V at (c) 0° to 90° W and (d) 90° to 180° W using SSM/I Antarctic data on September 29, 1989.

ber 3, 1983, taken in the vicinity of the AMERIEZ study area (near 45° W). It is apparent from Figure 12-7 that the consolidated ice data points move along line AB from one day to another. This migration of data points was observed by Comiso and Sullivan [1986] to be highly correlated with variations in the observed surface temperature. Most of the consolidated ice data points are shown to be on the lower left corner of the plots (near B) on November 13, 25, and 29. The surface temperature data show drops of 5°C in values during these same days. The decrease in physical temperature during these periods was only about 5 K, while the corresponding drop in brightness temperature at 37 GHz was as much as 40 K. The effect was due, therefore, not just to changes in surface physical temperature. It was postulated that the main cause of the oscillations was due to refreezing of the snow cover. When the snow cover is wet, as when the surface temperature is above freezing, the emission comes from the surface of the snow, as discussed earlier. When the temperature is below freezing, the snow cover becomes dry and the emission comes from the snow–ice interface. However, the refreezing of the snow causes it to become granular and a more effective scatterer of radiation, thereby causing the observed reduction in the emissivity. Similar occurrences, sometimes referred to as freeze–thaw cycles, have been observed during field studies [Grenfell and Lohanick, 1985; Comiso et al., 1989].

12.4.2 Active Microwave Backscatter

Only a few studies based on radar measurements of Antarctic sea ice appear in the literature. The only reported observations of Antarctic sea ice from imaging radars are from SIR-B [Carsey et al., 1986; Martin et al., 1987], which imaged data only to 59° S, and from the Kosmos-1500/Okean series of real-aperture radars of the USSR [Kalmykov et al., 1986; Krasyuk and Andreeve, 1988]. Other studies include the use of the radar altimeter in the study of ice extent [Laxon, 1990]. Radar altimetric studies of sea ice are covered in detail in Chapter 7. Generally, however, the radar backscatter behavior of the Southern Ocean sea ice should be similar to that of Arctic sea ice, especially under similar ice conditions.

Some features of sea ice in the Western Weddell Sea that require special consideration for backscatter are a thicker snow cover and thinner ice than on most Arctic sea ice, and the resulting negative or low freeboard and flooding for much of the Antarctic ice.

Various operating active systems include the X-band side-looking airborne radar (SLAR) on board the Kosmos-1500/Okean, the C-band synthetic aperture radar (SAR) on board the first European Remote Sensing Satellite (ERS-1) SAR (and for the future Radarsat), and the L-band SAR on board the Japanese Earth Resources Satellite (JERS-1) (and Shuttle Imaging Radar, SIR-B). The L-band is known to give poorer sea-ice contrast (at least in the Arctic during winter) than the higher frequencies. One would expect similar results in the Antarctic for L-band

measurements of similar types of surface. The same may not be true for the higher frequencies because of the effects of the thicker snow cover.

The SIR-B imaging L-band radar was in an orbit that allowed imaging of the MIZ in the Weddell Sea at about 50° S in early October 1984. Some of these images, as analyzed by Carsey et al. [1986] and Martin et al. [1987], indicated an ability to detect long surface gravity waves, icebergs, and the transition from small to large floes in the ice margin. Significant differences between the wet-ice/stormy-sea contrast at the ice edge and cold-ice/calmer-sea conditions associated with leads farther into the pack were observed. In the former, the backscatter from the sea was stronger than that from the ice (ice appeared black on a white sea). In the latter, the water was black (weak echo) and the ice was white (strong echo). This is not unexpected because similar results are obtained in Arctic experiments. Presumably the backscatter from the ice-edge region was from very wet, saline ice, perhaps flooded or with ponding, which is known to give weak echoes. These were accompanied by strong backscatters from the open ocean, which is known to be rough at around 50° S The backscatter from the inner zone (but not very far in) was probably from colder, drier snow-covered ice and smooth open water.

The only in-situ scatterometer measurements in the Antarctic ice pack were conducted during the 1989 Winter Weddell Gyre Experiment. Broadband scatterometers operating at C-band and Ku-band were mounted at deck level on the ship. Most measurements analyzed were at stations where the ship stopped and surface physical measurements were made. Preliminary results, reported in Lytle et al. [1990], indicated a significant difference in the backscatter at Ku-band for sea ice with slush at the snow–ice interface compared with dry snow at the interface. Also, at C-band and at angles of incidence near 40°, a difference of 7 dB in the scattering coefficients between first-year and second-year ice were found, consistent with measurements in the Arctic for similar types of ice by Kim [1984]. The scattering coefficients over second-year ice in the Weddell Sea were also found to be very similar to those made in 1983 at Mould Bay, Northwest Territories, Canada, over similar ice types. Also, at 65°, the scattering coefficients in the Weddell Sea for the same ice type were about 6 dB weaker than those at Mould Bay. Over first-year ice, some difference was observed at angles of incidence of only up to 30°. First-year ice in the Weddell Sea was found to have scattering coefficients larger by about 5 dB at 20° and 30° than comparable first-year ice at Mould Bay. Further analysis and a check on the calibration of the instruments are necessary before meaningful interpretation of these results can be made.

12.5. DERIVED GEOPHYSICAL PARAMETERS

12.5.1 Ice Extent and Ice Concentration

The large contrast in the emissivity of sea ice and that of open water at certain frequencies makes it possible to

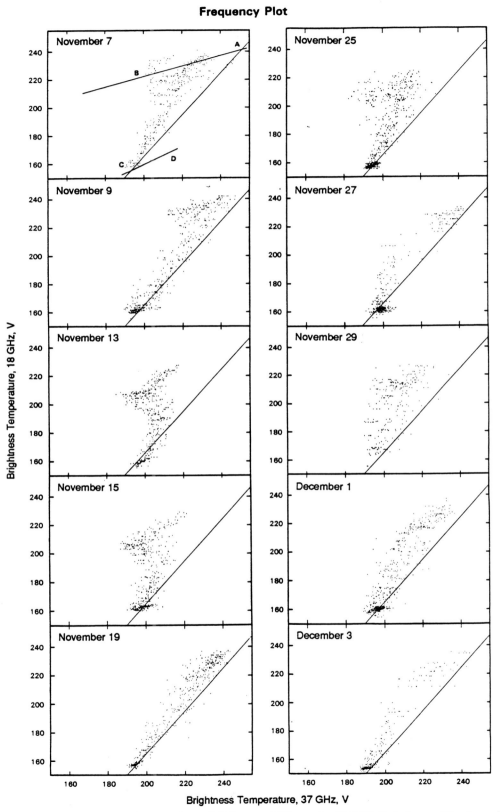

Figure 12-7. Time series of scatter plots of 18 V versus 37 V over the AMERIEZ study area [Comiso and Sullivan, 1986].

obtain reasonably accurate ice extent information from one-channel data. Ice extent is usually defined as the area of ice-covered ocean with ice concentrations greater than a certain value. For one-channel Electrically Scanning Microwave Radiometer (ESMR) data, a threshold of concentration of >15% ice extent has been utilized to define ice extent [Zwally et al., 1983a]. However, since the launch of the Scanning Multichannel Microwave Radiometer (SMMR), the accuracy in retrievals of ice concentration was improved because of the availability of more frequencies and polarizations, which enables removal of some of the ambiguities. Also, the ability to filter atmospheric effects near the marginal ice zone [Comiso et al., 1984; Comiso and Sullivan, 1986; Gloersen and Cavalieri, 1986] has made it possible to use >10% ice extent. The accuracy of the extent depends on the accuracy of the ice concentration data, the percentage coverage of the marginal ice zone, and the time of the year. Errors in the determination of ice concentration are usually largest in the marginal ice zone and polynya regions because of the abundance of new ice. However, the determination of the total extent of Antarctic sea ice is highly accurate (<2%) because the size of the marginal ice zone is small compared to the total ice area. The accuracy in the determination of regional extents, especially small regions, is not as good because of the aforementioned uncertainties.

Ice concentrations have been derived with the use of a mixing formulation that assumes that the ice cover consists of two types of sea ice (first year and multiyear) and open water (see Chapter 10 for more details). The basic technique is to solve for three equations in three unknowns to obtain percentage concentrations of these three surface types. However, in the Antarctic, the percentage of multiyear ice cover is small compared to the Arctic. Moreover, most of the sea ice in the perennial ice region has signatures similar to that of first-year ice. This has made it possible to generate ice concentration maps from a one-channel system, as was done in Zwally et al. [1983a], in which variations in physical temperature were accounted for by the use of climatological data. Multichannel systems provide the flexibility of using additional channels to account for spatial variations in physical temperature. This has also been done using ratios of brightness temperatures to eliminate the effect of physical temperature in the polarization ratio and reduce this effect to a second-order one in the case of the spectral gradient ratio [Gloersen and Cavalieri, 1986].

In the Arctic, the distinctive microwave signature of multiyear ice is due primarily to desalination that transforms the freeboard layer of the ice into a low-loss material. The relatively low emissivity for multiyear ice in the Arctic is thus due to the presence of ice lenses near the surface and brine pockets within the ice that cause scattering of the emitted radiation. In the Antarctic, however, this signature is not as distinctive, with the possible exception of data in the Bellingshausen–Amundsen Seas area. There is no longer a distinct cluster of points at very low brightness temperatures, instead there is a more continuous spread of data points from the first-year cluster in the direction of

lower brightness temperatures. It is probable that thicker (and/or more granular) snow cover and frequent flooding of the snow cover noted by many Antarctic field observers are responsible for masking the multiyear ice signature and the range of lower brightness temperatures.

Algorithms aimed at deriving Antarctic ice concentration have been forced to adapt to these peculiarities of the Antarctic sea ice pack. One class of algorithms (e.g., the bootstrap technique by Comiso and Sullivan [1986]) assumes that the physical effects mentioned above lower the brightness temperatures of consolidated first-year ice in a linear fashion when data of vertically polarized brightness temperatures from individual locations are plotted [see Figure 12-8(a) for Antarctic data from 0° to 90° W]. Thus, 100% ice is approximated by line AB in this figure. Similarly, open water is approximated by line CD, where C represents undisturbed water (commonly seen within the ice pack), and the data trending toward point D are caused by wind-roughened surfaces (seen outside the ice pack). Ice concentration of any data point is then calculated as the ratio of the distance of that point from C to the distance between C and line AB along the line passing through C at the data point. The data points in the consolidated ice region (above line AB) are studied by binning all data points along an axis perpendicular to line AB. The resulting plot is shown in Figure 12-9(a), the distribution of consolidated ice data being labeled A with a standard deviation of ±4.6% (ice concentration) the distance between A and C. This technique can be sensitive to large fluctuations in the temperature of the emitting ice layer. However, the effects of fluctuations of 5 and 10 K (see Figure 12-4 for typical values) in physical temperatures are shown in Figures 12-9(b) and (c), respectively, and may not be that critical since the effect is already included in the distribution of data near A in Figure 12-9(a).

The SMMR Team algorithm based on polarization and gradient ratios [Cavalieri et al., 1984] accounts for the spread of different microwave signatures along line AB by defining two tie-points at each end of the distribution of points between A and B, when these data are plotted as polarization versus gradient ratio, Figure 12-8(b). This scheme also assumes 100% concentration along the line between these two tie-points and calculates concentrations for other points according to their position between this line and the open water tie-point C and using a standard three-component mixing formula, see Figure 12-8(b). Comparative analysis of the results from this technique with Landsat data is reported in Steffen and Schweiger [1991].

While ice concentration maps derived using the two techniques are basically similar, in the Bellingshausen and Amundsen Seas differences as large as 20% in calculated ice concentrations have been noted. There is also a difference in the derived ice concentrations along coastal regions with the technique by Cavalieri et al. [1984], showing less open water in polynya areas. This may indicate differences in the detection of new ice in these regions. Validation data in regions where differences between algorithms are large is

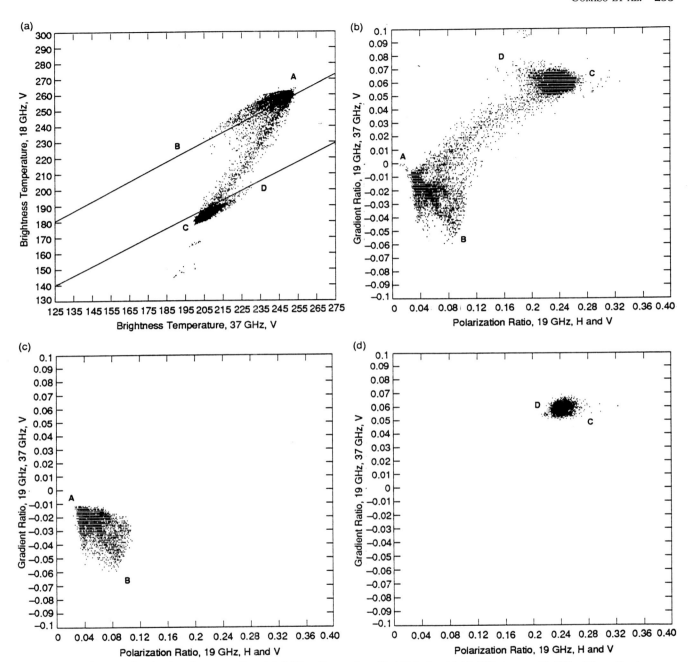

Figure 12-8. Scatter plot for data between 0° and 90° W on September 29, 1989, for: (a) 19 GHz versus 37 GHz; (b) gradient versus polarization ratios; (c) gradient versus polarization ratios for data points above the line AB in (a); and (d) gradient versus polarization ratios for data points covering a 2 by 2 K area at point C in (a).

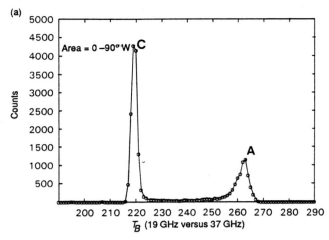

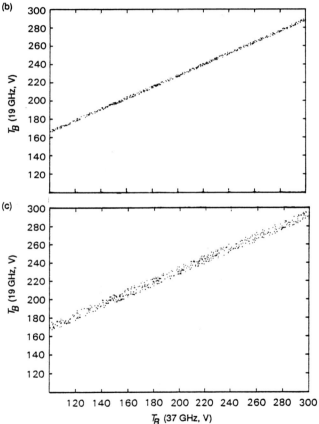

Figure 12-9. (a) Data plotted in Figure 12-8 along a line perpendicular to AB; (b) scatter plot of 19 GHz versus 37 GHz for uncertainties in a surface temperature of 5 K, randomly distributed; and (c) scatter plot of 19 GHz versus 37 GHz for uncertainties in a surface temperature of 10 K, randomly distributed.

inadequate to resolve the discrepancies at this time. The two types of algorithms discussed above differ in their quantifications of the microwave data and probably differ in their sensitivities. Using the polarization and gradient ratio greatly reduces the temperature sensitivity. The bootstrap technique is independent of horizontal polarization data, which could be beneficial since Mätzler et al. [1984] have observed a greater sensitivity of horizontally, rather than vertically, polarized brightness temperatures to roughness, ice layering, flooding, and surface wetness. This sensitivity could be responsible for the greater spread of data in the ice region [Figure 12-8(c)] and near the open-water point C—i.e., data points inside a 2 by 2 K area at location C in Figure 12-8(a) when plotted as polarization versus gradient ratio become redistributed as shown in Figure 12-8(d). Further investigation is necessary, with the aid of ancillary data, to find out the real characteristics of the ice cover and how these results relate to actual ice conditions.

Ice concentrations derived from SAR might provide needed data to improve interpretation of the passive microwave data. However, many of the same factors that make it difficult to interpret the passive microwave data also affect the SAR backscatter values. Thresholding techniques as well as mixing algorithms have been utilized for deriving ice concentration from SAR data, but the problem remains that in many cases open water is difficult to discriminate from new ice, young ice, and some forms of first-year ice. Martin et al. [1987] compared ice concentrations obtained from the SIR-B images with those derived from Nimbus-7 SMMR using a technique that utilized the polarization ratios as discussed above (but used the 37-GHz channel data). Ice concentration was derived from the SIR-B data using both an automated technique developed at the Jet Propulsion Laboratory and a manual technique implemented by the Scott Polar Research Institute. An original image and the corresponding automated classification of the ice cover are shown in Figure 12-10. The passive and active data were found to agree within 2% in the mean values with 7% standard deviation for ice concentrations higher than 40%. Since a large part of the study area was near the marginal ice zone where both passive and active microwave data are difficult to interpret, due to abundance of new and flooded ice and unpredictable ocean surface conditions, it is not clear how the derived values compare with actual ice concentrations. The use of C-band in ERS-1 should improve ice measurement compatibility compared with the L-band Seasat and SIR-B. However, it is too early to tell how successful this will be.

12.5.2 Surface Type and Snow Cover

Because of poor resolution, it is difficult to conduct ice-type classifications with passive microwave data. The different ice types are normally located in leads and polynyas, and these features are typically represented only as reductions in ice concentration. In most cases, the widths

(a)

(b)

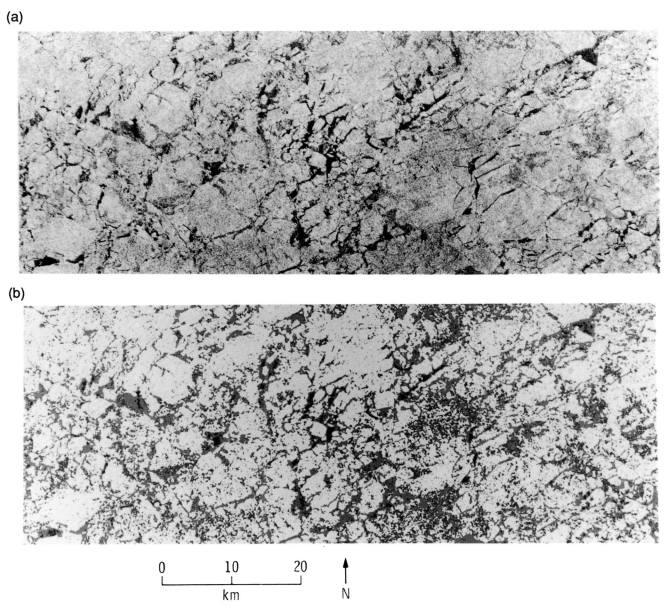

Figure 12-10. SIR-B observation over the Weddell Sea showing (a) original SAR image and (b) derived ice classes [Martin et al., 1987].

of leads and the diameter of polynyas are smaller than the typical 900 km² size of the satellite grid. The new and young ice coverages in these leads and polynyas are even smaller in extent. However, the information content of the satellite multispectral data is very rich and potentially allows retrieval of parameters beyond that of simply getting ice extent and ice concentration. If resolution is not a problem, the emissivities of some of the ice types as observed from the field are actually distinct. Averages of emissivities for various ice types measured from the Weddell Sea in 1986 over a period of six months [T. C. Grenfell, personal commu-

nication, University of Washington, Seattle, Washington, 1991] are shown in Table 12-1. Represented in the sampling are gray and white nilas, pancake ice, and brash ice, as well as cold and melting first-year ice. Because the radiometers used are not dual-polarized, the vertical and horizontal polarization measurements cannot be made simultaneously and may represent different regions of the ice cover.

Another application of passive microwave observation would be the delineation of different regimes of snow cover within the ice pack. The passive microwave data over Antarctica and Greenland show a large variability, which

TABLE 12-1. Emissivities of different Weddell Sea ice types.

Ice type	Pol	6.7 GHz	10 GHz	18 GHz	37 GHz	90 GHz
Open water	V	—	0.517 (0.020)	0.538 (0.020)	0.645 (0.030)	0.785 (0.050)
	H	—	0.317 (0.020)	0.321 (0.020)	0.397 (0.030)	0.530 (0.060)
Light nilas	V	—	0.907 (0.019)	0.947 (0.022)	0.965 (0.027)	0.961 (0.024)
	H	—	0.859 (0.038)	0.893 (0.053)	0.940 (0.031)	0.927 (0.018)
Gray nilas	V	—	0.741 (0.065)	0.842 (0.052)	0.906 (0.039)	0.922 (0.020)
	H	—	0.603 (0.154)	0.699 (0.121)	0.802 (0.115)	0.841 (0.051)
Pancake ice	V	—	0.822 (0.081)	0.887 (0.067)	0.953 (0.024)	0.941 (0.058)
	H	—	0.795 (0.062)	0.854 (0.053)	0.943 (0.026)	0.892 (0.063)
Cold first-year ice	V	0.919 (0.060)	0.937 (0.060)	0.951 (0.023)	0.962 (0.024)	0.954 (0.036)
	H	0.903 (0.070)	0.905 (0.057)	0.937 (0.060)	0.938 (0.045)	0.919 (0.026)
Melting first-year ice	V	—	0.932 (0.018)	0.949 (0.029)	0.965 (0.018)	0.921 (0.039)
	H	—	0.898 (0.016)	0.899 (0.040)	0.930 (0.010)	0.881 (0.027)
Brash	V	0.851 (0.040)	0.861 (0.031)	0.926 (0.025)	0.952 (0.013)	0.953 (0.025)
	H	—	0.795 (0.062)	0.854 (0.053)	0.943 (0.026)	0.892 (0.063)
V/2 average first year	V	0.830 (0.050)	0.921 (0.019)	0.960 (0.019)	0.958 (0.022)	0.875 (0.058)
	H	0.815 (0.050)	0.878 (0.031)	0.912 (0.032)	0.931 (0.028)	0.860 (0.063)
V/3 average first year	V	0.890 (0.056)	0.911 (0.034)	0.939 (0.038)	0.948 (0.038)	0.908 (0.091)
	H	0.848 (0.050)	0.859 (0.050)	0.886 (0.040)	0.915 (0.034)	0.848 (0.095)

has been postulated to correspond to variations in the granularity of the snow cover [Zwally, 1977; Chang et al., 1976]. Snow parameters over land have been determined in continental Europe, the United States, and Canada from the 37- and 18-GHz channels [Künzi et al., 1982; Chang et al., 1990]. Over certain types of sea ice (i.e., first-year ice), the ability to derive snow parameters should be even more promising than over land, because the material underneath the snow is more spatially uniform in the former. The use of the 85- or 90-GHz channel in combination with other channels has indicated even more potential for snow. In a pilot study, Comiso et al. [1989] showed that the emissivity at 90 GHz decreases with thickness of fresh snow. If atmospheric effects can be accounted for in the 85-GHz data, the snow over sea ice can be characterized on a global scale much more accurately than is presently possible. Weekly averages of SSM/I 85-GHz data (with atmospheric filter) in fact indicate persistence of low brightness temperatures in regions believed to be covered by thick snow. Having a determination of the depth of the snow cover could also indirectly provide information about multiyear ice. The field experiments in the Antarctic (mentioned earlier) indicated that the multiyear ice in the Weddell Sea has a very thick snow cover. Potential candidates for these types of ice cover are the low brightness temperatures in the Western Weddell Sea and in the Bellingshausen and Amundsen Seas, as shown in Figure 12-5.

The ability to do ice classification with SAR in the Antarctic is not well determined because of lack of data, but the potential is very strong because of its high resolution. While a new SAR system has been launched on board the ERS-1 satellite, validation experiments are needed to ensure that the interpretation of the data is correct. An example of ice classification done from the SIR-B image is shown in Figure 12-10. It was assumed that different ice types could be identified by visual inspection and would correspond to unique ranges of backscatter values, as shown in Figure 12-10. This technique is a sensible approach, especially if done by a person familiar with the region. However, there is large spatial variability in the backscatter for a certain ice type from the ice edge to the inner pack and it is not known yet how well the ambiguities can be resolved.

Other systems including SLAR have also been launched and the data are found to be valuable for Antarctic studies. Antarctic sea ice radar data have been collected almost continuously since 1983 by the Kosmos-1500 series of real-aperture radars [Kalmykov et al., 1986], now called Okean. Unfortunately, a six-month gap occurred during the 1989 Weddell Gyre Project cruise, which obtained scatterometer data. The resolution for this X-band SLAR is on the order of 1 to 2 km, and the swath width is 450 km. The images were used to navigate ice breakers in the Antarctic [Krasyuk

Figure 12-11. Kosmos-1500 image of Southern Ocean sea ice from the ice edge (between New Zealand and Antarctica) to the Ruppert Coast. The image illustrates the ability of the sensor to identify different ice regimes and the effects of rafting and ridging. (Courtesy of A. Kalmykov and S. Velichko.)

and Andreev, 1988], including the rescue of the Mikhail Somov that was stuck in the ice pack. Although the authors do not know of any systematic study of the entire Antarctic sea-ice pack using these data, many maps have been made of parts of the Ross Sea, an example of which is shown in Figure 12-11 [Krasyuk et al., 1987]. Potential applications for the Kosmos-1500/Ocean series of radars were summarized in Nazirov et al. [1990] and include identification of leads and polynyas, ice edge, multiyear ice edge, and homogeneous zones. The data would also be useful for dynamics studies and comparative studies with passive microwave radiometer data.

12.5.3 Actual Open Water and New Ice

Accurate determination of actual open water and new ice within the ice pack is necessary to be able to quantify heat and salinity fluxes. The actual area of ice-free open water in the inner zone is difficult to derive from passive microwave data because of the relatively small size of typical open water areas compared to the satellite footprint and the ambiguities introduced by the presence of new ice. The emissivity of new ice is also variable (see Table 12-1 and Chapter 14) and depends on the stage of development and the region in which it was formed. For an area the size of a satellite footprint, this emissivity is difficult to distinguish from that of a mixture of open water and ice. To obtain a better idea about the effect of the presence of new ice, a sensitivity study was undertaken. Assuming that there is only one type of new ice in the field of view of the sensor, which means either pancake ice, as is normally the case in the outer zone, or nilas, which is more dominant in the inner zone, ice concentration values could thus be interpreted as shown in Table 12-2, where 80% ice concentration of nilas, pancake, or light nilas (or young ice) would be represented in derived ice concentration maps as 49%, 62%, or 79%, respectively. Several examples are shown to provide a guide for the interpretation of the ice concentration maps.

The aforementioned problem with new ice is possible to overcome with the use of SAR data. The fine resolution from SAR enables the detection of leads and polynyas that are as small as a few tens of meters wide. However, it is still not clear that new (or flooded) ice can be unambiguously discriminated from open water, especially if it is just one-channel SAR data. Furthermore, near simultaneous global

TABLE 12-2. Actual versus derived ice concentration for different ice types.

Ice concentration (actual)	100%	90%	80%	50%	30%
Nilas	62%	55%	49%	31%	19%
Pancake	78%	70%	62%	39%	23%
Light nilas/young ice	99%	89%	79%	49%	30%

coverage is difficult to obtain because of the SAR narrow swath width. A short-term solution might be to use SAR in combination with passive microwave data to better interpret and use the data to extrapolate limited SAR coverages.

12.6 DISCUSSION AND CONCLUSIONS

The ability to monitor the ice surface globally and consistently from space provides a means of improving our understanding of geophysical processes in the polar regions. Global extent and actual ice area can be derived at high precision using the microwave data. The Southern Ocean has more first-year ice, thicker snow cover, and more flooding than the Arctic. Ice concentrations can be derived with acceptable accuracies using unsupervised techniques that can account for many surface and atmospheric effects over thick ice. However, more accurate estimates of the fraction of open water and new ice within the ice pack are required if good values for heat and salinity fluxes are to be obtained. The main problem is the presence of new ice, which has an emissivity different from either open water or thick ice and can occur in different stages of development and in various forms during the year. The problem is further confounded by the large footprint of the passive microwave sensor. The ability to resolve some of these new ice features when using SAR data is intriguing and might provide a solution to some of these problems. However, it is not yet known how well and how consistently SAR data can be used for ice classification in the Southern Ocean.

Even if the ice concentration data do not show the absolute amount of open water, the data provide good information about spatial distribution of ice and a means of identifying the ice edge as well as regions of divergence and polynya activity within the ice pack. With the advent of new SAR systems, drift characteristics of floes can also be examined in greater detail. The use of combined active and passive remote sensing systems also offers a strong potential for more accurate monitoring of sea ice. Techniques using satellite data alone, however, may not be enough. More in-situ data are needed to obtain basic characteristics of the ice cover that are needed to interpret spatial and temporal changes in the signature. For example, accurately knowing rates of ice growth and the divergent characteristics of the ice pack could provide better estimates of the amount of open water and new ice, as well as fluxes of brine at the ocean surface and of heat and moisture into the atmosphere.

REFERENCES

Ackley, S. F., M. Lange, and P. Wadhams, Snow cover effects on Antarctic sea ice thickness, *Sea Ice Properties and Processes*, edited by S. F. Ackley and W. F. Weeks, CRREL Monograph 90-1, 300 pp., Cold Regions Research and Engineering Laboratory, Hanover, New Hampshire, 1990.

Allison, I., Antarctic sea ice growth and oceanic heat flux, *Sea Level, Ice, and Climatic Change*, International Association of Hydrological Sciences, Guildfold, UK, pp. 161–170, 1981.

Andreas, E. L. and S. F. Ackley, On the differences in ablation seasons of the Arctic and Antarctic sea ice, *Journal of Atmospheric Science, 39(3)*, pp. 440–447, 1982.

Bagriantsev, N. V., A. L. Gordon, and B. A. Huber, Weddell Gyre: temperature maximum stratum, *Journal of Geophysical Research, 94(C6)*, pp. 8331–8334, 1989.

Barry, R. G., The present climate of the Arctic Ocean and possible past and future states, *The Arctic Seas*, edited by Y. Herman, Van Nostrand, New York, pp. 1–46, 1989.

Carsey, F. D., B. Holt, S. Martin, L. McNutt, D. Rothrock, V. A. Squire, and W. F. Weeks, Weddell–Scotia Sea marginal ice zone observations from space: October 1984, *Journal of Geophysical Research, 91*, pp. 3920–3924, 1986.

Cavalieri, D. J., P. Gloersen, and W. J. Campbell, Determination of sea ice parameters with the Nimbus 7 SMMR, *Journal of Geophysical Research, 89(D4)*, pp. 5355–5369, 1984.

Chang, A. T. C., P. Gloersen, T. Schmugge, T. T. Wilheit, and H. J. Zwally, Microwave emission from snow and glacier ice, *Journal of Glaciology, 16(74)*, pp. 23–30, 1976.

Chang, A. T. C., J. L. Foster, and D. K. Hall, Satellite sensor estimates of Northern Hemisphere snow volume, *International Journal of Remote Sensing, 11(1)*, pp. 167–171, 1990.

Comiso, J. C. and A. L. Gordon, Recurring polynyas over the Cosmonaut Sea and the Maud Rise, *Journal of Geophysical Research, 92(C3)*, pp. 2819–2834, 1987.

Comiso, J. C. and C. W. Sullivan, Satellite microwave and in-situ observations of the Weddell Sea ice cover and its marginal ice zone, *Journal of Geophysical Research, 91*, pp. 9663–9681, 1986.

Comiso, J. C. and H. J. Zwally, Concentration gradients and growth/decay characteristics of the seasonal sea ice cover, *Journal of Geophysical Research, 89(C5)*, pp. 8081–8103, 1984.

Comiso, J. C., S. F. Ackley, and A. L. Gordon, Antarctic sea ice microwave signatures and their correlation with in-situ ice observations, *Journal of Geophysical Research, 89(C1)*, pp. 662–672, 1984.

Comiso, J. C., T. C. Grenfell, D. Bell, M. Lange, and S. Ackley, Passive microwave in-situ observations of winter Weddell sea ice, *Journal of Geophysical Research, 94(C8)*, pp. 10,891–10,905, 1989.

Comiso, J. C., P. Wadhams, W. Krabill, R. Swift, J. Crawford, and W. Tucker, Top/Bottom multisensor remote sensing of Arctic sea ice, *Journal of Geophysical Research, 96(C2)*, pp. 2693–2711, 1991.

Garrison, D. L., C. W. Sullivan, and S. F. Ackley, Sea ice microbial community studies in the Antarctic, *Bioscience, 36*, pp. 243–250, 1986.

Gloersen, P. and W. J. Campbell, Variations in the Arctic, Antarctic, and global sea ice covers during 1978–1987 as observed with the Nimbus 7 SMMR, *Journal of Geophysical Research, 93*, pp. 10,666–10,674, 1988.

Gloersen, P. and D. J. Cavalieri, Reduction of weather effects in the calculation of sea ice parameters with Nimbus 7 SMMR, *Journal of Geophysical Research, 93,* pp. 3913–3919, 1986.

Gow, A. J., S. F. Ackley, W. F. Weeks, and J. W. Govoni, Physical and structural characteristics of Antarctic sea ice, *Annals of Glaciology, 3,* pp. 113–117, 1982.

Grenfell, T. C. and A. W. Lohanick, Temporal variations of the microwave signatures of sea ice during the late spring and early summer near Mould Bay, Northwest Territories, *Journal of Geophysical Research, 90(C3),* pp. 5063–5074, 1985.

Kalmykov, A. I., V. B. Efimov, A. S. Kurekin, B. A. Nelepo, A. P. Piguschin, A. B. Fetisov, B. E. Khmyrov, V. N. Tsymbal, and V. P. Shestopalov, The radar system of the Cosmos-1500 Satellite, *Soviet Journal of Remote Sensing, 4,* pp. 827–840, 1986.

Kim, Y.-S., R. G. Onstott, and R. K. Moore, The effect of a snow cover on microwave backscatter from sea ice, *IEEE Journal of Oceanic Engineering, OE-9(5),* pp. 383–388, 1984.

Krasyuk, V. S. and M. D. Andreev, O gidrometeorologicheskom obespechenii zimneyi navigatsii b Antarktidye, (On hydrometeorological security of winter navigation in Antarctica), *Meteorologiya i Gydrologiya, 11,* pp. 124–127, 1988.

Krasyuk, V. S, M. Nazirov, P. A. Nikitin, and E. V. Bukhman, Ob ekspress-analizye radiolokatsnionnikh izobranzhenni morskoikh lybda (On express analysis of radar observations of sea ice), *Meteorologiya i Gydrologiya, 2,* pp. 70–75, 1987.

Künzi, K. F., S. Patil, and H. Rott, Snow cover parameters retrieved from Nimbus 7 Scanning Multichannel Microwave Radiometer (SMMR) data, *IEEE Transactions on Geoscience and Remote Sensing, GE-20,* pp. 452–467, 1982.

Lange, M. A. and H. Eicken, The sea ice thickness distributions in the Northwestern Weddell Sea, Antarctica, *Journal of Geophysical Research, 96,* pp. 4821–4837, 1991.

Lange, M. A., S. F. Ackley, P. Wadhams, G. S. Dieckmann, and H. Eicken, Development of sea ice in the Weddell Sea, *Annals of Glaciology, 12,* pp. 92–96, 1989.

Laxon, S. W. C., Seasonal and interannual variations in Antarctic ice extent as mapped by radar altimetry, *Geophysical Research Letters, 17(10),* pp. 1553–1556, 1990.

Lytle, V. I. and S. F. Ackley, Sea ice ridging in the Eastern Weddell Sea, *Journal of Geophysical Research, 96(10),* pp. 18,411–18,416, 1991.

Lytle, V. I., K. C. Jesek, S. Gogineni, R. K. Moore, and S. F. Ackley, Radar backscatter measurements during the winter Weddell Gyre study, *Ant. Journal of U.S., XXV(5),* National Science Foundation, Washington, DC, pp. 123–125, 1990.

Martin, S., B. Holt, D. J. Cavalieri, and V. Squire, Shuttle Imaging Radar B (SIR-B) Weddell sea ice observations: A comparison of SIR-B and Scanning Multichannel Radiometer ice concentrations, *Journal of Geophysical Research, 92(C7),* pp. 7173–7180, 1987.

Mätzler, C., R. O. Ramseier, and E. Svendsen, Polarization effects in sea-ice signatures, *IEEE Journal of Oceanic Engineering, OE-9(5),* pp. 333–338, 1984.

Maykut, G. A., Energy exchange over young sea ice in the Central Arctic, *Journal of Geophysical Research, 83,* pp. 3646–3658, 1978.

Nazirov, M., A. P. Pichugin, and Yu. G. Spiridonov, *Radiolokatsiya Poverchnosti Zemli iz Kosmosa (Radar Observations of the Earth's Surface From Space),* edited by L. M. Mitnika and S. V. Viktorova, 200 pp., Gidrometeoizdat (Hydrometeorological Publishing Office), St. Petersburg, 1990.

Steffen, K. and A. Schweiger, NASA Team algorithm for sea ice concentration retrieval from defense meteorological satellite program Special Sensor Microwave Imager: Comparison with Landsat satellite imagery, *Journal of Geophysical Research, 96,* pp. 21,971–21,988, 1991.

Swift, C. T., Passive microwave remote sensing of the ocean—A review, *Boundary Layer Meteorology,* pp. 25–54, 1980.

Swift, C. T., L. S. Fedor, and R. O. Ramseier, An algorithm to measure sea ice concentration with microwave radiometers, *Journal of Geophysical Research, 90,* pp. 1087–1099, 1985.

Tucker, W. B., III, A. J. Gow, and W. F. Weeks, Physical properties of summer sea ice in the Fram Strait, *Journal of Geophysical Research, 92(C7),* pp. 6787–6804, 1987.

Wadhams, P, Atmospheric–ice–ocean interactions in the Antarctic, *Antarctica and Global Change,* edited by C. Marris and B. Stonehouse, Belhaven Press, London, pp. 65–81, 1991.

Wadhams, P., M. A. Lange, and S. F. Ackley, The ice thickness distribution across the Atlantic sector of the Antarctic Ocean in midwinter, *Journal of Geophysical Research, 92(C13),* pp. 14,535–14,552, 1987.

Weeks, W. F., S. F. Ackley, and J. W. Govoni, Sea ice ridging in the Ross Sea, Antarctica, as compared with sites in the Arctic, *Journal of Geophysical Research, 94,* pp. 4984–4988, 1988.

Zwally, H. J., Microwave emissivity and accumulation rate of polar firn, *Journal of Glaciology, 18,* pp. 195–215, 1977.

Zwally, H. J., J. C. Comiso, C. L. Parkinson, W. J. Campbell, F. D. Carsey, and P. Gloersen, *Antarctic Sea Ice, 1973–1976: Satellite Passive-Microwave Observations,* 206 pp., NASA SP-459, National Aeronautics and Space Administration, Washington, DC, 1983a.

Zwally, H. J., C. L. Parkinson, and J. C. Comiso, Variability of Antarctic sea ice and changes in carbon dioxide, *Science, 220,* pp. 1005–1012, 1983b.

Chapter 13. Microwave Study Programs of Air–Ice–Ocean Interactive Processes in the Seasonal Ice Zone of the Greenland and Barents Seas

OLA M. JOHANNESSEN

Nansen Environmental and Remote Sensing Center, Solheimsvik, Bergen N-5037, Norway

WILLIAM J. CAMPBELL

U.S. Geological Survey, University of Puget Sound, Tacoma, Washington 98416

ROBERT SHUCHMAN

Environmental Research Institute of Michigan, P. O. Box 8618, Ann Arbor, Michigan 48107

STEIN SANDVEN

Nansen Environmental and Remote Sensing Center, Solheimsvik, Bergen N-5037, Norway

PER GLOERSEN

Goddard Space Flight Center, Greenbelt, Maryland 20771

JOHNNY A. JOHANNESSEN

Nansen Environmental and Remote Sensing Center, Solheimsvik, Bergen N-5037, Norway

EDWARD G. JOSBERGER

U.S. Geological Survey, University of Puget Sound, Tacoma, Washington 98416

PETER M. HAUGAN

Nansen Environmental and Remote Sensing Center, Solheimsvik, Bergen N-5037, Norway

13.1 INTRODUCTION

Following the International Geophysical Year (1957 through 1958), a series of large field experiments was performed, culminating in the Arctic Ice Dynamics Joint Experiment, AIDJEX (1972 through 1976). These experiments considerably aided our understanding of the growth, motion, and decay of sea ice in the interior of the Arctic Ocean [Untersteiner, 1986; Pritchard, 1980; Campbell et al., 1978; Gloersen et al., 1978]. With these experiments concluded and coupled nonlinear sea ice dynamic–thermodynamic models in hand [Hibler, 1979; Coon, 1980], attention shifted to the problem of understanding the processes that occur near the open ocean boundaries of polar sea ice covers. The seasonal ice zone (SIZ) is the crucial region in which the polar atmosphere, sea ice covers, and oceans interact with the bordering atmosphere and oceans. The air–ice–ocean processes and exchanges that take place there determine the advance and retreat of the sea ice and profoundly influence the global climate. These processes also exert a significant effect on marine productivity, commercial fisheries, petroleum exploration and production, and naval operations [Johannessen et al., 1984a; Wadhams, 1986; Sandven and Johannessen, 1990].

Remote sensing is an essential tool in the study of the SIZ. Some examples of different remote sensing data used in the study of SIZ processes are shown in Figure 13-1. Figure 13-1(a) shows the total ice concentration of the Northern Hemisphere for February 1989, as observed by the Special Sensor Microwave Imager (SSM/I). These observations provide daily information on ice edge position and concentration of first-year and multiyear ice with a resolution of 30 km. They show many regional features that repeatedly occur in the SIZ. For example, an historically important feature, the Odden (a region of rapid ice growth and advance that generally occurs every winter), can be seen projecting northeastward into the central Greenland Sea. The long time series of similar images acquired from the Nimbus 5 Electrically Scanning Microwave Radiometer (ESMR) from 1973 to 1976 [Zwally et al., 1983; Parkinson et al., 1987], the Nimbus 7 Scanning Multichannel Microwave Radiometer (SMMR) from 1978 to 1987, and the SSM/I from 1987 to the present provides a

Microwave Remote Sensing of Sea Ice
Geophysical Monograph 68
©1992 American Geophysical Union

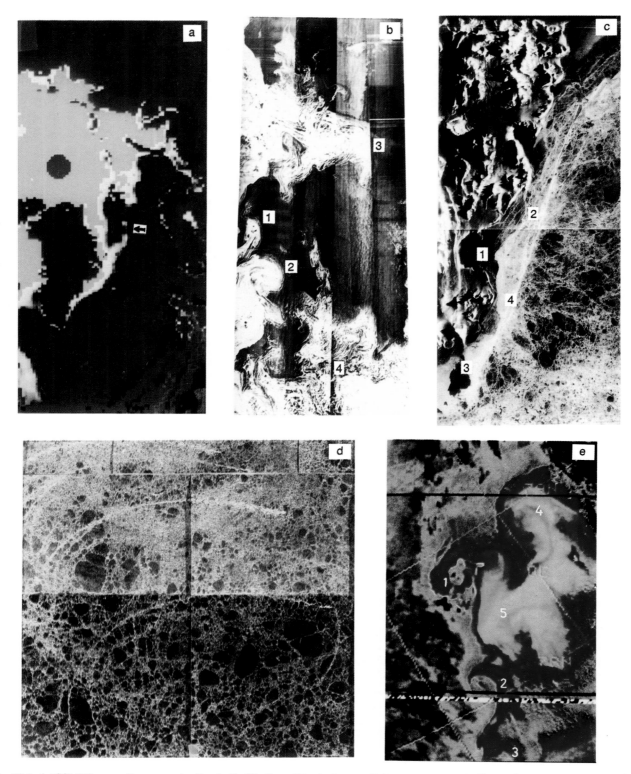

Fig. 13-1. (a) SSM/I image of ice concentration in the Northern Hemisphere on February 18, 1989. Yellow indicates concentration above 90% and dark blue is open water. The arrow shows the location of the ice tongue Odden. (b) Airborne SAR image from the Greenland Sea obtained on March 28 and 29, 1987, showing a (1) vortex-pair, (2) anticyclonic ice-ocean eddy, (3) family of eddies over the Molloy Deep, and (4) newly formed ice. (c) Airborne SAR image from February 23, 1989, of the ice off the east coast of Svalbard, showing (1) land-fast ice, (2) shear zones, (3) polynyas, and (4) ridges. (d) SAR image of the Barents Sea ice cover showing the curved tracks of grounded icebergs. The image was obtained on March 23, 1988. (e) NOAA AVHRR image of ice edge eddies (1 through 5) in the Greenland Sea on July 4, 1984.

unique 16-year record that is extremely important for SIZ and global change studies [Gloersen and Campbell 1991a, b; Gloersen et al., 1992].

Of all the passive and active microwave sensors, synthetic aperture radars (SAR's) provide the highest resolution ice information. Sequential SAR images, with a resolution as fine as 5 m, provide quantitative information about the growth, motion, and decay of ice edge eddies and other mesoscale features [Johannessen et al., 1983a, 1987; Shuchman et al., 1987; Campbell et al., 1987; Johannessen, 1987] in addition to ice concentration, type, and kinematics [Burns et al., 1987]. The SAR image from the Greenland Sea acquired during MIZEX'87 on March 28 and 29, Figure 13-1(b), reveals how extremely complex the morphology of the marginal ice zone can be and how powerful a tool SAR can be for the study of ice edge structure. On the day this image was acquired, the surface wind speed was low and the air was cold (−10°C); therefore, the ice mirrored the mesoscale ocean circulation and freezing processes. In this image, bright signatures derive from various types of ice, while the dark signatures are from ice-free water. In the western central Greenland Sea, Figure 13-1(b) shows a narrow jet shooting out from the ice edge and ending in a vortex-pair with a scale of 10 to 20 km [MIZEX Group, 1989]. Slightly south of this vortex-pair is a clearly defined anticyclonic ice-ocean eddy with a diameter of about 30 km. In the northern Greenland Sea in this image, a family of eddies are centered over the 5500-m depression known as the Molloy Deep, shown in the location map, Figure 13-2 [MIZEX Group, 1986]. In the southernmost Greenland Sea in this image, newly formed ice is present at the northern edge of a large tongue of ice projecting eastward from the main ice edge approximately at the Greenland Fracture Zone.

The SAR image from the Barents Sea, Figure 13-1(c), acquired during the Seasonal Ice Zone Experiment, SIZEX'89 [SIZEX Group, 1989], shows characteristic ice features along the east coast of Svalbard, such as land-fast ice, shear zones, and polynyas. In the shear zone between the land-fast and drifting ice, the ice pack converges and forms large ridges in a belt along the coast. Adjacent to the shear zone, the transition from very large floes to small to medium floes is clearly seen.

SAR images can also be used to detect icebergs and evaluate their effects on the surrounding sea ice pack. For example, a SAR image, Figure 13-1(d), of the Barents Sea shows the interaction between grounded icebergs and the ice pack driven by the strong tidal current and mean ice advection [Johannessen et al., 1991b]. This interaction can be seen as tracks cut into the ice cover by icebergs. This is an important mechanism for fracturing large ice floes. Tracks are also generated by drifting icebergs due to the differential motion between the ice pack and the icebergs. The combined effects of this mechanism and the wave penetration causes the small size of ice floes in the western Barents Sea. The tracks left in the ice pack makes the detection of the icebergs easy.

On cloud-free days, infrared and visible-channel data from the Advanced Very High Resolution Radiometer (AVHRR) onboard the National Oceanic and Atmospheric Administration (NOAA) satellites provide images with 1-km resolution showing sea surface temperature and ice distribution. These images have revealed ice edge eddies, jets, and vortex-pairs along the ice edge of the Greenland Sea, Figure 13-1(e) [Johannessen et al., 1987]. The abundance of eddies (five overall) in this region show that eddies 1 through 4 strongly interact with the ice cover, while eddy 5 is seen to be an ocean eddy in the infrared image. However, the frequent occurrence of clouds in the SIZ makes it difficult to obtain a time series of images that can be used to study the dynamics of the ice edge region.

The images in Figure 13-1 are examples of an extensive data set collected in the Greenland and Barents Seas since 1978. In this chapter, we will review the main results accomplished during a decade of experiments that emphasized the use of microwave remote sensing techniques. In Section 13.2, the campaigns are summarized and the most important physical processes in the SIZ are presented. Recent modeling results of SIZ processes are summarized in Section 13.3. An example of regional ice forecasting using microwave observations is presented in Section 13.4, and in Section 13.5 the role of the SIZ in the climate system is discussed.

13.2 CAMPAIGNS

From 1979 to 1989, five major international field experiments were carried out in the Greenland Sea, Fram Strait,

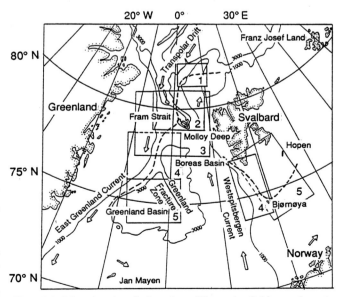

Fig. 13-2. Map showing the location of the major field experiments from 1979 to 1989: (1) NORSEX'79, (2) MIZEX'83, (3) MIZEX'84, (4) MIZEX'87, and (5) SIZEX'89. The arrows show the main currents and ice motion. The dashed lines indicate the mean ice edge location in the Greenland Sea during the summer experiments, and south of Svalbard during the winter experiments.

and Barents Sea: the Norwegian Remote Sensing Experiment, NORSEX'79 [Johannessen et al., 1983a]; the Marginal Ice Zone Experiments, MIZEX'83, MIZEX'84, and MIZEX'87 [Johannessen et al., 1983b; Johannessen, 1987]; and the Seasonal Ice Zone Experiment, SIZEX'89 [Johannessen et al., 1991a]. The locations of the experiments are shown in Figure 13-2, and a summary of the experiments' objectives and available sensors is presented in Table 13-1. A brief review of the oceanography of this region has been published by Johannessen [1986a].

The overall objectives of the experiments have been to improve our understanding of the air–ice–ocean processes

TABLE 13-1. Experiment objectives and available sensors.

Campaign	Objectives	Remote Sensing Observations
NORSEX'79 September–October 1979	1) The first study of mesoscale processes in the MIZ by combined remote sensing and in-situ observations. 2) Development of a passive microwave algorithm for ice concentration, ice types, and ice edge location.	Satellite: Nimbus-7 SMMR (every two days) Aircraft: CV-990: L-band SAR (two flights) C-130: passive radiometer and scatterometer Ship: Radiometer
MIZEX'83 June–August 1983	1) Pilot study of mesoscale physical and biological processes in the summer MIZ in the Fram Strait region. 2) Validation of active and passive microwave observation of ice and ocean processes in the summer MIZ.	Satellite: Nimbus-7 SMMR (every two days) Aircraft: CV-990: passive microwave imager and radiometer NRL-P3: passive microwave imager CCRS/ERIM CV-580: SAR Ship: Radiometer and scatterometer
MIZEX'84 June–August 1984	Extensive use of remote sensing techniques combined with in-situ measurements to study: 1) Ocean and ice circulation in the Greenland Sea during summer. 2) Mesoscale ice edge eddies. 3) Characteristics of summer ice. 4) Boundary layer meteorology. 5) Biological activity.	Satellite: Nimbus-7 SMMR (every two days) Aircraft: CV-990: passive microwave and imager radiometer NRL-P3: passive microwave imager and radiometers CCRS/ERIM CV-580: SAR CNES B-17: SLAR Ship: Radiometers and scatterometers
MIZEX'87 March–April 1987	1) Demonstrate the importance of daily SAR imagery in the study of ice–ocean eddies, ice motion, and other MIZ processes. 2) Study ice distribution, physical ice properties, water masses, deep water formation, boundary layer meteorology, biological activity, and ocean acoustics in the Greenland Sea in winter. 3) Carry out the first SAR study of the sea ice in the Barents Sea.	Satellite: Nimbus-7 SMMR (every two days) Aircraft: Intera X-band SAR (24 flights, three in the Barents Sea) NRL-P3: passive microwave imager Ship: Radiometers and scatterometers
SIZEX'89 Part 1 (The Barents Sea February 1989) SIZEX'89 Part 2 (The Greenland Sea March 1989)	1) Pre-ERS-1 validation of SAR ice parameters: ice concentration, ice kinematics, and ice types. 2) Demonstrate use of remote sensing data in ice forecasting. 3) Process studies: eddies, jets, vortex pairs, fronts, chimneys, deep water formation, and their influence on the ambient noise.	Satellite: DMSP SSM/I (daily) Aircraft: CCRS CV-580: SAR X- and C-bands ERIM/NADC P-3: SAR X-, C-, and L-bands Ship: Radiometers and scatterometers

in the SIZ and to develop and validate remote sensing techniques. During this decade, there has also been a development from basic research towards application and operational use of remote sensing techniques in sea ice monitoring and forecasting in the seasonal ice zone. One of the aims of this chapter is to present this development.

The concept and the strategy of all these experiments were to collect data from a three-level observational system—satellites, aircraft, and in-situ observations. During these experiments, data were collected from ice-strengthened vessels with helicopters, from operating inside the ice pack, from oceanographic research vessels operating in the open ocean adjacent to the ice edge, from drifting buoys tracked by the Argos system and bottom-moored buoys, and from aircraft and satellites. Instruments on these varied platforms acquired a diverse suite of ice, ocean, and atmospheric data.

13.2.1 NORSEX'79

The first remote sensing experiment in this series took place north and west of Svalbard in September and October 1979. Active and passive microwave instruments were used to study the location and structure of the ice edge, distribution of ice types, ice concentration, eddies, and ice edge upwelling in the seasonal ice zone region [NORSEX Group, 1983]. SAR imagery of the area was obtained for the first time using the NASA CV-990 L-band SAR with 25-m resolution. Two flights were performed: one on September 19 during easterly winds of 10 to 15 m/s parallel to the ice edge, and the next on October 1 when calm conditions had prevailed for the two preceding days (Figure 13-3).

On September 19, the SAR image showed a straight ice edge with several trailing ice plumes indicating the presence of a jet along the ice edge [Johannessen et al., 1983a]. Buoys drifting in the jet indicated a speed of up to 0.30 m/s along the ice edge. Sections of conductivity and temperature versus depth (CTD) obtained from the ship indicated upwelling outside the ice edge. During the calm wind situation of October 1, the SAR image showed the ice had a wavelike edge with approximate length scale of 20 to 40 km and several eddies with horizontal dimensions of 5 to 15 km. Johannessen et al. [1983a] showed that during these wind conditions the SAR image mirrored the ocean circulation.

Another aim of the experiment was to use the integrated data sets to develop and test an algorithm for estimating total and multiyear ice concentration from passive microwave measurements [NORSEX Group, 1983]. This set of emissivity observations from various ice types and water was used in the development of an algorithm [Svendsen et al., 1983] designed for use in the Greenland–Barents Seas SIZ. Comparison of simultaneous Nimbus 7 SMMR and aircraft data indicated that the total ice concentration in the experimental area could be estimated with an accuracy of ±3%, and multiyear ice concentration with ±10%. The geographical positioning of the ice concentration estimates, including the ice edge location, are accurate to ±10 km. The

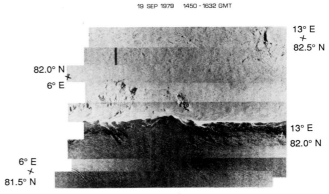

NORSEX JPL L-BAND SAR
19 SEP 1979   1450 - 1632 GMT

Fig. 13-3. Two airborne SAR mosaics obtained during NORSEX'79. The upper image is from October 1, 1979, when calm conditions had prevailed for the two preceding days. The lower image is from September 19, 1979, when easterly winds of 10 to 15 m/s blew parallel to the ice edge.

ice in this region varied from newly frozen, thin pancakes to snow-covered multiyear floes. During the experiment the snow cover remained dry, making the microwave signature of first- and multiyear ice clear [NORSEX Group, 1983].

13.2.2 MIZEX'83 and 84

NORSEX led to the creation of a more comprehensive and international program MIZEX, which consisted of a series of summer [MIZEX Group, 1989] and winter experiments [Johannessen, 1986b]. While NORSEX'79 focused on remote sensing and selected ice-ocean processes, the objective of MIZEX was a much broader attempt to improve our understanding of mesoscale air–ice–ocean interactive processes and the role they play in the climate system. In addition to coordinated sea ice, oceanographic, meteorological, and remote sensing programs as carried out in NORSEX, MIZEX included acoustical and biological studies [MIZEX Group, 1986]. A comprehensive discussion of all aspects of these experiments is given in the thematic Marginal Ice Zone issues of the *Journal of Geophysical Research* [*92(C7)*, 1987, and *96(C3)*, 1991], and in *Polar Oceanography* by Muench [1990].

The remote sensing program obtained sequential synoptic images of the ice morphology and evolution of the ice edge region utilizing a variety of remote sensing aircraft and satellites. Of great importance in the execution of MIZEX was the ability to acquire real-time SMMR data from Nimbus 7, AVHRR data from the NOAA satellites, and airborne SAR data of the experiment areas. These observations were used to position the ships for mesoscale mapping. Remote sensing was also a key element in the oceanographic program that focused on obtaining a synoptic history of ocean currents and structure and to assess the role of fronts, eddies, and large-scale currents in the movement and decay of the marginal ice zone (MIZ). These field experiments were carried out in the Fram Strait–Greenland Sea region and the Bering Sea, however, emphasis was placed on the Fram Strait region because of its primary importance for the exchange of water, heat, and other quantities between the Arctic Ocean and the world's ocean.

The first MIZEX in the Fram Strait, which was a pilot for the main experiment in the summer of 1984, was carried out during a two-month period in the summer of 1983. This pilot experiment consisted of three major parts. The first was a study of the ice deformation of the interior of the MIZ [Leppäranta and Hibler, 1987] and internal waves [Sandven and Johannessen, 1987]. The second was a detailed study of the system of topographically controlled eddies in the Molloy Deep area [Johannessen et al., 1984b; MIZEX Group, 1986]. In the Fram Strait (Figure 13-4), warm water is carried northwards along the coast of Svalbard while sea ice and cold water are transported southwards in the western part of the Strait. In this dynamic region, eddies frequently occur and several were observed in the open ocean and along the ice edge (Figure 13-4). Over the Molloy Deep depression, a system of eddies were present. Deep CTD observations showed that these eddy signatures extended all the way to the bottom and thus were generated by the topography in the region. The third part of the experiment was an ice edge dynamics study south of the Molloy Deep based on transponder data transmitted from drifting ice buoys to the ship.

MIZEX'84 was the largest research program ever conducted in a marginal ice zone [Johannessen, 1987]. A multidisciplinary team of more than 200 scientists from 11 countries participated in the two-month experiment, which utilized seven ships, four helicopters, eight aircraft, and four satellite systems. A new experimental design, employed in MIZEX'84, utilized near-real-time remote sensing data acquired by aircraft and satellites to direct the ships to interesting ice-ocean features in which to deploy

drifting Argos buoys, some with current meters below, and to obtain oceanographic and in-situ microwave observations. Figure 13-5(a) shows the aircraft SAR mosaic acquired on July 5, 1984. This image was used to direct ships to the eddy for in-situ observations and CTD profiles that were used to construct the three-dimensional eddy structure shown in Figure 13-5(c). An oblique aerial photograph, Figure 13-5(d), acquired at the same time as the SAR image, shows the surface manifestation of this eddy. Subsequent interpretation of the SAR image, Figure 13-5(b), shows the eddy along with other ice edge and ocean features.

The results from the MIZEX's [Campbell et al., 1987; Johannessen et al., 1987; Johannessen, J. A., et al., 1987] show that a number of eddies of 20 to 40 km dominate the mesoscale circulation in the Fram Strait. The rotational direction is mainly cyclonic and the maximum orbital speed is up to 0.40 m/s. Life times of these eddies have been observed to exceed 20 to 30 days. Some of the eddies remained stationary due to topographical trapping, while others propagated 10 to 15 km per day driven by the mean current. When these eddies propagate along the ice edge, ice is transported out into open water and warm water is advected into the ice pack, as accomplished by eddy 1 in Figure 13-1(e). Thus, the melting of the ice is enhanced and the ice edge can retreat 1 to 2 km per day due to the action of the eddies [Johannessen, J. A., et al., 1987].

While the aircraft SAR observations of the experimental area provided detailed information on the ice morphology at the mesoscale 30 to 100 km, the Nimbus 7 SMMR observations provided sequential information of the ice edge and ice concentration variations on a regional scale. A comparison of the aircraft active and passive microwave and visual observations with the SMMR ice concentration distributions indicates that for the summer MIZ the 30% isopleth correlates best with the actual ice edge position under diffuse conditions, while the 50% isopleth correlates best with a compact ice edge [Campbell et al., 1987]. Using these results, a composite of the derived SMMR ice edges for the experimental period (Figure 13-6) showed that the envelope of the edge variations is quite narrow, on the order of 50 km. This envelope correlates well with the shelf break south of 79° N, implying a strong regional topographical steering. Northeast of 79° N, the ice edge location is primarily determined by the inflow of warm Atlantic water [Johannessen et al., 1983a].

These results from the summer MIZEX'83 and '84 show the ability of active and passive microwave techniques to observe ice-ocean eddies and ice morphology during melting conditions. Physical processes are significantly different during the freezing season and one must understand both seasons to model the interannual variability of the MIZ. Complementary winter programs therefore took place in the Greenland and Barents Seas during March and April 1987 (MIZEX'87) and in February and March 1989 (SIZEX'89). SIZEX'89 was also an approved European Space Agency (ESA) prelaunch experiment for ERS-1 [Johannessen, 1986b].

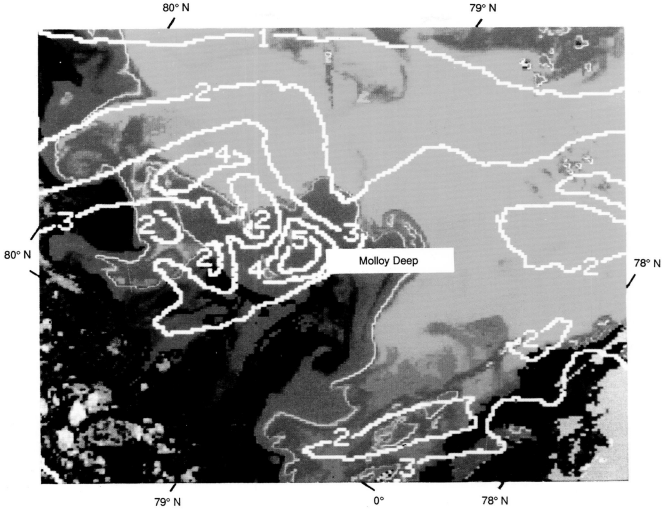

Fig. 13-4. NOAA AVHRR infrared image obtained in the Fram Strait during MIZEX'83. Yellow and red indicate warm water at 5° to 6°C, while blue represents a mixture of cold water and ice. The white lines show the bottom contours (for example, 1 indicates 1000 m).

13.2.3 MIZEX'87

The winter MIZEX'87 investigations were based on the need to understand the atmosphere–ice–ocean processes responsible for the advance of the winter ice edge and to validate the microwave remote sensing accuracies under conditions different from those in the summer.

MIZEX'87 was the first field experiment wherein daily regional SAR images were downlinked to the ship in real-time, which was essential to the execution of the experiment and efficient ship navigation in the ice. During the experiment, a total of 24 SAR flights were completed, 13 of which were on consecutive days in the Greenland Sea and three in the Barents Sea [MIZEX Group, 1989]. The previous experiments had shown the importance of having SAR coverage of the same area every day in order to map the

rapid changes of the ice edge characteristics [Campbell et al., 1987].

In Figure 13-7, a unique sequence of 12 images from March 28 to April 8 is shown for the first time. The salient feature of this sequence is the extreme structural variability of the ice edge at the daily time scale. The cause for this is the evolution of eddies, jets, and vortex-pairs that are modified by wind and surface wave variations.

What follows is a description of each mosaic of the location and evolution of key ocean and ice events. On March 28, slightly north of the center of the mosaic, an ice tongue extended eastwards from the ice edge opposing a northeasterly wind of 6 m/s. South of this tongue, an anticyclonic eddy was indicated. Twelve hours later, when the wind had decreased to 3 m/s from the north, this anticyclonic eddy was clearly shown with a diameter of

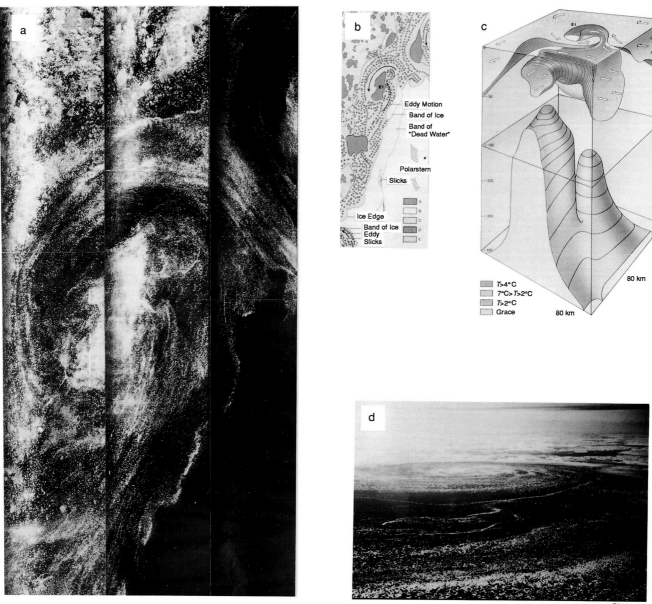

Fig. 13-5. (a) Aircraft SAR image acquired on July 5, 1984, of a cyclonic ice edge eddy in the Greenland Sea. (b) Interpretation of the whole SAR mosaic. (c) Three-dimensional structure of the eddy obtained from CTD sections. (d) Oblique aerial photograph of the eddy.

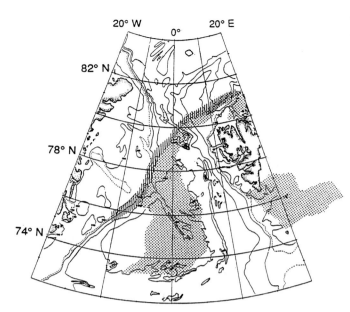

Fig. 13-6. Envelope of ice edge positions observed by SMMR during the summer experiment MIZEZ'84, ice edge variations are typically within a 60-km range (hatched area) and during the winter experiment MIZEX'87 (dotted area).

approximately 40 km. At this time the tongue to the north of the anticyclonic eddy had evolved into a jet with a vortex-pair less than 20 km. In the northern part of this image, the system of eddies over the Molloy Deep is clearly seen, as well as new ice freezing in the southern part of the image. This image serves as an excellent example of how the ice floes mirror the complex mesoscale ocean circulation under moderate to light wind conditions. Such phenomena can only be observed by sequential SAR images.

During the next several days from March 30 to April 2, the wind varied from 6 to 8 m/s between the north and the northeast. In these images, the ice edge was meandering and was compact due to the on-ice wind forcing and surface wave radiation pressure. In spite of the on-ice wind forcing on March 31, a new tongue evolved below the center of the image and projected eastwards. During the next two days, this tongue evolved and was advected to the south. Note that at the extreme end of this tongue a small vortex-pair was observed on March 31 and April 1. On April 2, in the center of the mosaic, a cyclonic eddy was present, extending into the ice pack. In the southern part, a new jet with the indication of a vortex-pair was present. During the next four days the wind decreased to 4 m/s, and this caused the ice pack to relax and once again mirror the mesoscale ocean circulation. The vortex-pair that was present in the southern part on April 2 evolved into a large anticyclonic eddy with a diameter of approximately 60 km, which persisted until April 6.

During this experiment, the smaller eddies and vortex-pairs clearly seen in the SAR mosaics were also documented

by CTD observations, which were composed of polar surface water plumes extending from the ice edge. On April 5, a smaller tongue of ice projected southwestwards from the center of the 60-km anticyclonic eddy to form a new vortex-pair, which was clearly seen on April 6. North of this eddy several jets and vortex-pairs projected eastwards from the ice edge on April 6, revealing exceptionally energetic mesoscale ocean circulation. This condition persisted throughout the following three days.

Downlinking SAR information enabled us to acquire a CTD section of this anticyclonic eddy, Figure 13-8. The fields of salinity and density show this eddy extended to the bottom. The geostrophic velocity field computed from these data, Figure 13-8(d), and velocities obtained by drifting buoys verified the anticyclonic nature of this eddy. This section across the eddy center indicated a core of warm Atlantic intermediate water that circulated in the eddy between 100 and 200 m depth. The generation of this anticyclonic eddy can be explained by conservation of potential vorticity as the East Greenland current passed over a major ridge, the Greenland fracture zone.

The April 6 image shows the greatest number of eddies, jets, and vortex-pairs. The entire ice edge, approximately 300-km long, is composed of these energetic mesoscale phenomena. On April 7 and 8, the ice morphology is similarly complex but less structured than on April 6. On April 7 and 8, in the southwest of each image large fields of ice plumes composed of grease ice and small pancakes appeared, indicating active freezing. These are potential areas of convection driven by brine exclusion. In our long experience of acquiring SAR images during NORSEX, MIZEX, and SIZEX, the April 6 image is the most complex we have ever obtained and presents a major challenge to the modeling community.

In general, the ice edge region is characterized by a 20- to 30-km wide zone of a bright SAR signature. In-situ observations established that this zone consists of broken-up floes of mainly multiyear ice with rough surfaces and edges that cause high backscatter from the SAR. In this zone, individual floes cannot be identified. The band pattern is indicative of strong horizontal velocity shear. Inside the ice edge zone, the most predominant feature is the occurrence of large multiyear floes of high backscatter that are broken into pieces. Between the multiyear floes, thin new-frozen ice with low backscatter is found. By comparing consecutive mosaics, the interior of the pack ice is observed to drift southwards at a speed of about 20 km per day, while the velocity of the ice edge is 30 to 40 km per day. This ice edge jet is caused by the baroclinicity associated with the East Greenland polar front, the frontal boundary between the cold, fresh waters of the East Greenland current and the warmer, more saline Atlantic intermediate water seaward of the ice edge. This ice edge jet is enhanced during northerly winds because of the higher atmospheric drag coefficient over the rough ice edge zone [Guest and Davidson, 1987]. Similar jets were observed during NORSEX'79 [Johannessen et al., 1983a].

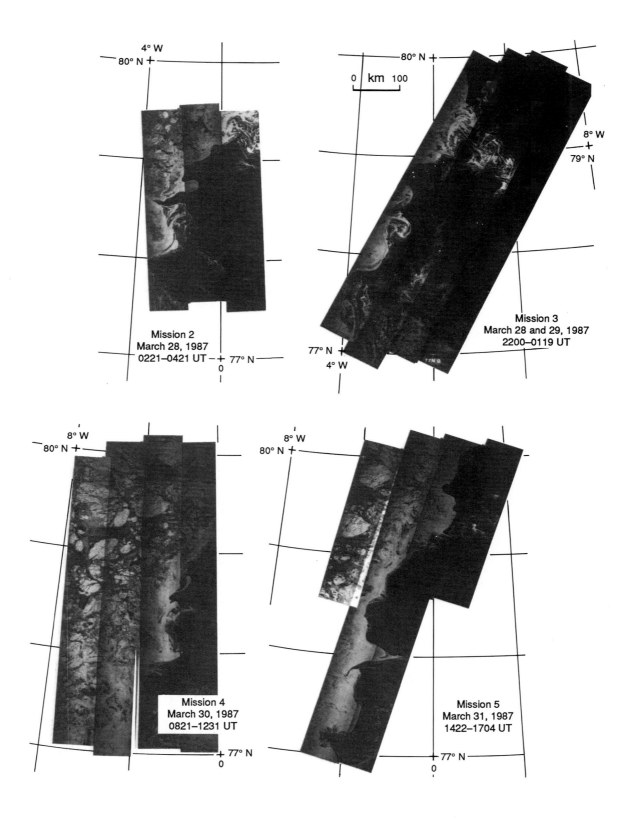

Fig. 13-7. Sequence of 12 SAR images in the Greenland Sea obtained from March 28 to April 8, 1987.

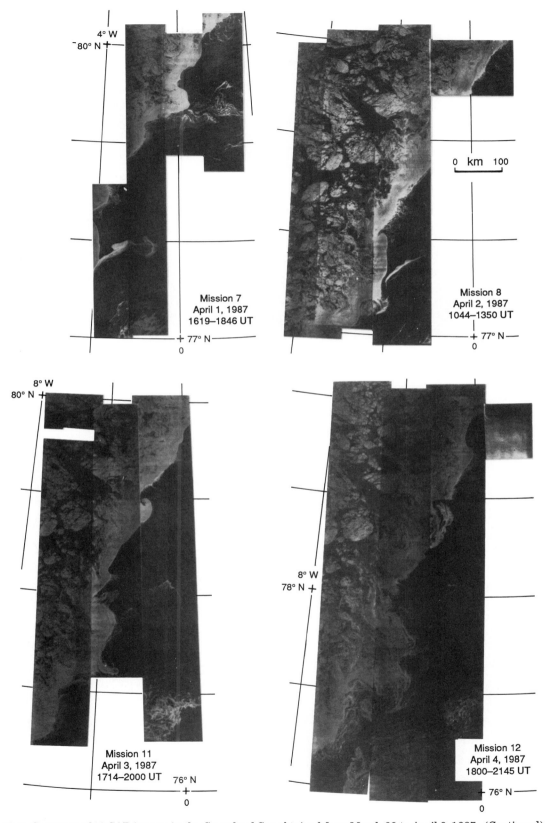

Fig. 13-7. Sequence of 12 SAR images in the Greenland Sea obtained from March 28 to April 8, 1987. (Continued)

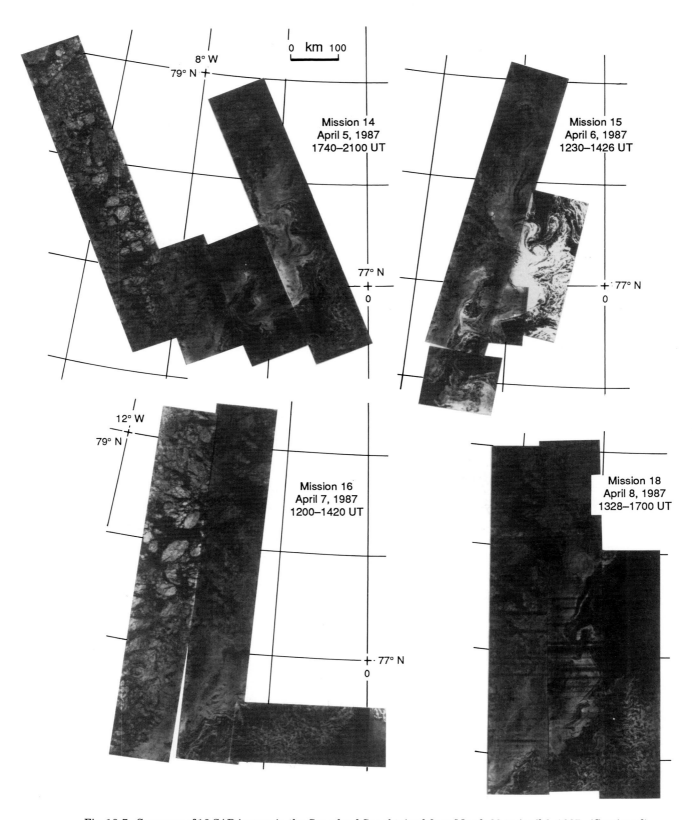

Fig. 13-7. Sequence of 12 SAR images in the Greenland Sea obtained from March 28 to April 8, 1987. (Continued)

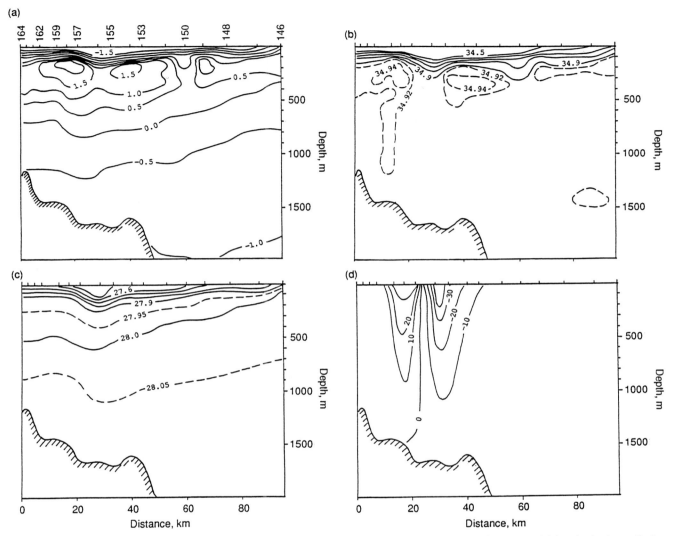

Fig. 13-8. (a) The vertical section of temperature in degrees Celsius, (b) salinity in parts per thousand, (c) potential density in sigma-theta, and (d) geostrophic velocity in cm/s across the anticyclonic eddy shown in the SAR images of April 4 and 5. Positive values of the geostrophic velocity indicate flow into the page, negative values out of the page.

Regional observations of the SIZ were accomplished with SMMR data. A sequence of nine SMMR images from the Greenland and Barents Seas from March 1 to April 10 is shown in Figure 13-9. Even at this coarse resolution (25 km), the extremely dynamic nature of the SIZ can be observed on a regional scale much larger than the mesoscale of the SAR images. It is interesting to compare these images with the SAR images presented above in Figure 13-7. For example, on March 29 the large area of ice plumes in the southern part of the SAR image (Figure 13-7) appears in the SMMR image as a tongue of low-concentration ice projecting from the ice pack eastwards at 77° N. This ice tongue is most clearly seen in the SMMR data on March 31.

This sequence of SMMR images also serves to illustrate how different the ice edge morphologies in winter are from those in summer. For example, from March 1 through 31 a large area of ice projected eastward towards Svalbard with the ice edges far east of those found during the summer, as can be seen in the comparison of the winter and summer envelopes of ice edge positions shown in Figure 13-6. By comparing these envelopes with the bathymetry, it can be seen that the winter ice-edge positions are not forced by the bottom topography as are the summer positions. Rather, the eastern extent of the ice edge during winter is determined by the location of the polar ocean front in the Greenland Sea [Johannessen, 1986a].

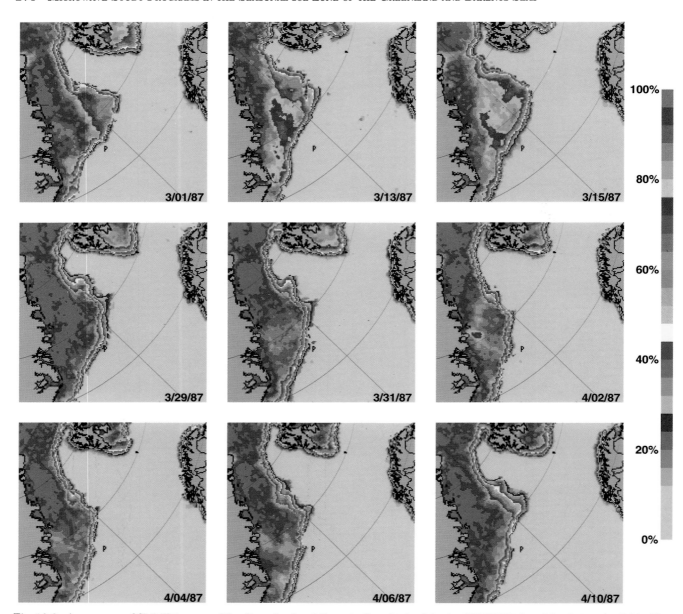

Fig. 13-9. A sequence of SMMR images of the Greenland and Barents Sea obtained during MIZEX'87 from March 1 to April 10. The vertical bar gives the color code for the derived ice concentrations.

During the winter, the western Barents Sea is primarily divided into a warm and cold part, separated by the Bear Island polar front located at the shelf break near Bjørnøya (Figure 13-2). This division is caused by the inflow of warm Atlantic water and the outflow of cold polar water along the shelf between Hopen and Bjørnøya [Johannessen and Foster, 1978]. When cold polar air flows southward, rapid freezing of new ice occurs over large areas. When warm air is advected in from the south by rapidly moving low pressure systems, this new ice is destroyed by melting and wave action. Some of these rapid advances in ice growth can be seen in the sequence of SMMR images shown in Figure 13-9,

especially during March. The southern limit of the ice edge in the Bjørnøya region coincides with the warm boundary of the Bear Island polar front. When the ice is forced across this front, it will quickly melt in the warm Atlantic water.

The first SAR mosaics of the Barents Sea were obtained during MIZEX'87 (Figure 13-10) starting on April 9, 1987 [Sandven and Johannessen, 1992]. They provided detailed information on the ice edge morphology and processes, as well as ice floe distribution, ice concentration, and ice motion. A pronounced feature in the mosaics is the rapid growth of an ice jet that developed into a vortex-pair near the southern tip of Svalbard. This information, coupled

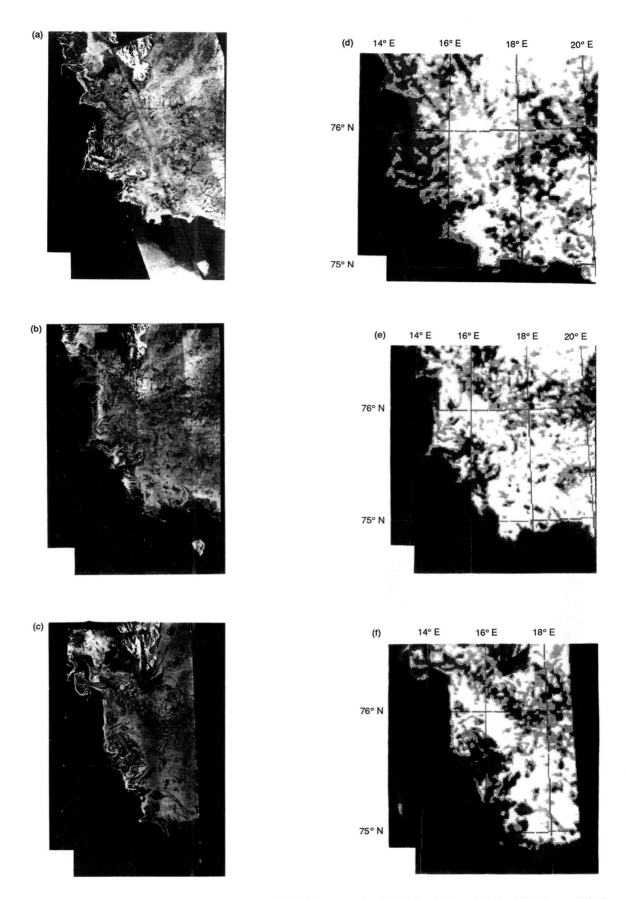

Fig. 13-10. Three SAR images obtained in the Barents Sea during MIZEX'87 on (a) April 9, (b) April 10, and (c) April 11. For each SAR image, the ice concentration is estimated using the algorithm proposed by Sandven et al. [1991b] is shown to the right of the image. White indicates ice concentration above 90%, orange from 70% to 90%, and light blue from 50% to 70%.

(a)

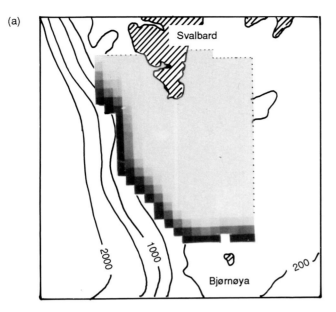

(b)

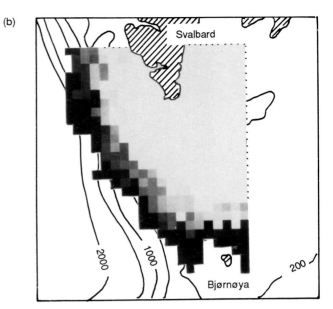

(c)

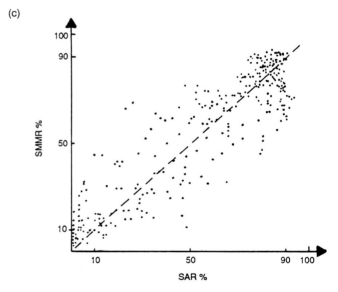

with that from the Greenland Sea, suggests that similar jets and vortex-pairs are common features on shelf breaks of the world's oceans [Fedorov and Ginsburg, 1986; Johannessen, J. A., et al., 1989; MIZEX Group, 1989]. Thus the SAR images of the seasonal ice zones and infrared images from other parts of the world's oceans have been very instrumental in focusing our attention on this phenomenon.

Ice concentration is a fundamental parameter for both heat budget calculations and ice navigation. An algorithm for determining ice concentration from SAR mosaics based on the thresholding technique has been developed [Sandven et al., 1991b]. This algorithm has been applied to the SAR images and the results are shown adjacent to each image in Figure 13-10. The resolution of the ice concentration images has been averaged to 1 km. In order to validate this technique, a comparison was made between ice concentrations derived by the SAR algorithm and by SMMR data using the NORSEX algorithm [Svendsen et al., 1983], which has an accuracy of less than +3% for winter conditions in the MIZ. Since the resolutions of SAR and SMMR are very different, a common pixel size of 12 km was used. A comparison between the two algorithms, shown in Figure 13-11, indicates that the SAR ice concentration estimates have an accuracy on the order of 20%. These results are promising for the application of SAR ice concentrations in operational ice monitoring and forecasting, but need to be improved for heat flux estimates in climate studies.

13.2.4 SIZEX'89

The SIZEX program was accepted by ESA as a key ice validation program for the first European Remote Sensing Satellite (ERS-1). A prelaunch experiment, SIZEX'89, was carried out in the Barents and Greenland Seas during

Fig. 13-11. Two images of ice concentration from April 10, 1987, in the Barents Sea, are superimposed on a bathymetric map, and displayed by a gray-scale ranging from black (10% to 20% concentration) to white (>90% concentration). (a) Ice concentration derived from the SAR image, averaged to a resolution of 12 km; (b) ice concentration derived from SMMR data using the same resolution as above; and (c) a scatter diagram of the two estimates.

February and March 1989. It was the first major seasonal ice zone experiment in a shallow water region with strong tidal currents.

The ERS-1 SAR C-band was simulated by flying the Canadian Centre for Remote Sensing X- and C-band SAR onboard the CV-580 every third day, which was the ERS-1 repetition cycle for the ice phase from January 1 to March

31, 1992 [SIZEX Group, 1989]. Figure 13-12 shows the experiment area, the ice edge positions derived from SSM/I and AVHRR, selected buoy tracks, current vectors near the ice edge, and ship tracks where oceanographic data were obtained.

The four mosaics from February 17, 20, 23, and 26, which covered the ice edge region from about 18° E to 23° E, provide detailed information about ice features, bands, tongues, and eddies (Figure 13-13). In general, there are characteristic differences in ice edge conditions between the western Barents Sea, from Bjørnøya to Spitzbergen, and the area east of Bjørnøya. East of Bjørnøya the ice edge region usually consists of fairly compact ice, except for a few nautical miles near open ocean. Outside the main ice edge, relatively cold water, which is favorable for freezing in winter, is present. This cold water is, however, usually confined to the shelf region shallower than 100 to 200 m and limited by the Bear Island polar front, which inhibits freezing south of the slope of the Bjørnøya Channel and the shelf break between Bjørnøya and Spitzbergen, as discussed in Section 13.2.3. Between Bjørnøya and Spitzbergen, the ice that drifts towards the open ocean, as shown by the Argos buoys drift (Figure 13-12) and SAR-derived ice kinematics, rapidly melts when it comes in contact with the warmer Westspitzbergen current. Some of the ice is transported northwards along the west coast of Spitzbergen where it eventually melts.

In this section we describe and quantify some of the characteristic ice features observed in these SAR images. The results have been presented by Sandven et al. [1991a] and are summarized in Table 13-2. The most pronounced

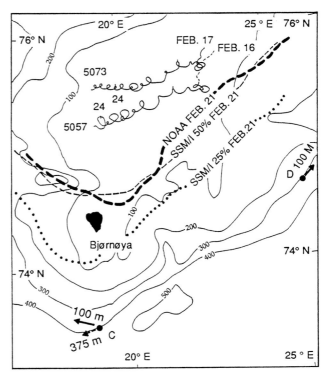

Fig. 13-12. Map of the SIZEX'89 experiment area in the Barents Sea during February. The ice edge positions from NOAA AVHRR and SSM/I are shown in bold dashed and dotted lines, respectively, the trajectories of two drifting ice buoys (5057 and 5073) are shown for the period from February 16 to 25. Two bottom-moored current meters are indicated with vectors illustrating the mean current at 100 m (mooring D) and 100 and 375 m (mooring C).

TABLE 13-2. Quantification of ice features.

Feature	Reference in Figure 13-13	Length, km	Width, km	Orientation/ Location	Time scale, days	Motion
Bands outside the ice edge	1	5–30	0.2–5	Mostly N–S meanders	<3	Propagating
Zones of reduced ice concentration inside the ice edge (leads)	2	5–30	0.5–5	Mostly N–S meanders	1–5?	Westward 10 km/day
Bights	3	20–50	20–50	E of Bjørnøya NW of Bjørnøya	Permanent? Permanent?	Stationary stationary
Tongues	4	20–50	10–20	NS at 23° E NS at 19° E	? 2–5?	Stationary propagating
Eddies	5	5–15	5–15	Around Bjørnøya 23° E	1–2? ?	Propagating ?
Polynyas	6	20–30	20–30	Hopen area	10	Westward propagation
Tracks produced by grounded icebergs	7	10–50	0.100–0.200	Ellipses from Bjørnøya to Hopen	1–3	Propagating
Fronts	8	>20	≈0.100	NE along the Bjørnøya Channel, S and W of Bjørnøya	Permanent	Stationary
Internal waves	9	Wavelength 1–3		E of Bjørnøya	<3	Propagating

(a)　　　　　　　　　　　　　　　　　　　(b)

February 17　　　　　　　　　　　　　　February 20

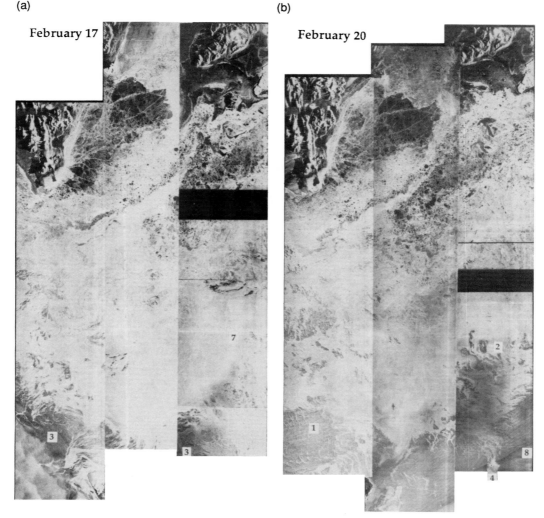

Fig. 13-13. SAR mosaics obtained during SIZEX'89 on (a) February 17, (b) February 20, (c) February 23, and (d) February 26.

features are bands of ice outside the main edge, indicated by the number 1 in the mosaics of Figure 13-13. East of Bjørnøya these bands extended out from the ice edge and were oriented normal to the wind direction. West of Bjørnøya the bands were also normal to the wind direction, but along the ice edge. None of the ice bands observed on February 23 could be recognized three days later. In spite of moderate winds during this period, the ice bands changed rapidly, implying that their life times were less than three days. The typical length of the bands was from 5 to 30 km and the width varied from 100 m to 2 km.

While ice bands outside of the edge of the pack are a prominent feature in these mosaics, elongated zones of reduced ice concentration were observed inside the main ice edge, denoted by the number 2 in the mosaics of Figure 13-13. These zones are identified by darker signatures, indicative of areas with lower ice concentration or thin ice, and had a prevailing north–south orientation with consid-

erable meanders. Their length scale was similar when compared to the bands outside the ice edge, while their widths were larger. Two of the larger zones observed on February 23 can be recognized three days later. They had moved about 30 km westwards in three days. Such ice zones can also reflect some of the ocean circulation in the area, such as tidal currents and eddies.

A characteristic feature in all the mosaics is a bight of open water with scattered ice bands to the north of Bjørnøya, located by the number 3 in the mosaics of Figure 13-13. This bight has a horizontal scale varying from 20 to 50 km. A second bight is found east of Bjørnøya on February 17 and 20, with a diameter of 50 to 60 km. In the two last mosaics, this bight becomes less pronounced. The bights are associated with submarine troughs, which are filled with warmer water. The bights are therefore stationary and probably a permanent feature when the mean ice edge is located in the same position as in February 1989. These bights, which are

(c)

February 23

(d)

February 26

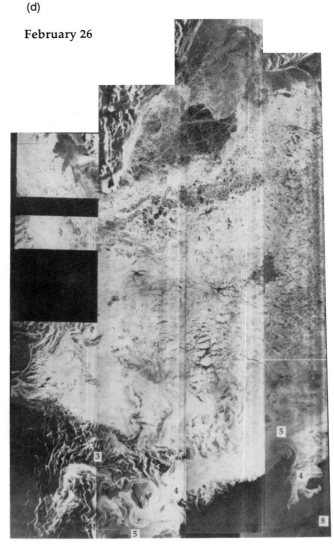

Fig. 13-13. SAR mosaics obtained during SIZEX'89 on (a) February 17, (b) February 20, (c) February 23, and (d) February 26. (Continued)

essentially like polynyas, are important elements in determining the heat flux from the ocean to the atmosphere in this region, and they demonstrate the usefulness of SAR observations for climate studies on regional and global scales.

Another characteristic feature of all the mosaics is a tongue of ice projecting out of the ice edge, with a typical scale in the range of 20 to 40 km, which is located in the mosaics by the number 4. A tongue of ice is advected southwards on the eastern side of Bjørnøya, and is present in all four mosaics, especially in Figures 13-13(c) and (d); a second tongue is located approximately 100 km east of Bjørnøya. On February 17, observations from the ship showed that this tongue was made up of pancake ice. Three days later the tongue was less pronounced, but in the images it becomes clearly defined. The ice tongues are

associated with a southerly extent of cold water. In many cases they are associated with the occurrence of eddies, located on the mosaics by the number 5. For example, on February 26, two eddies are clearly seen within a tongue. In the second tongue, several eddy features and meanders develop. The time scale of the tongues and eddies is estimated to be a few days. It is possible that the features can propagate, but the data set does not give an indication of a propagation speed.

Polynyas are also conspicuous in these mosaics and are located by the number 6. The 50-km polynya, which was observed in the mosaics of February 23 and 26 west of Hopen, changed its shape as it moved westwards. Greater ice speed on its northern side compared to its southern side caused a stretching and rotation of the polynya as it propagated westward. For the three-day period between the SAR images, the polynya propagated westward at 10 to 20 km per day. The thin ice within the polynya, which can clearly be seen from the fracture patterns in the images, appears to grow thicker during this three-day interval, as indicated by an increase in the radar backscatter. These images show how useful sequential SAR observations can be in observing and quantifying the growth of thin ice.

The Barents Sea ice cover is exposed to frequent and large swell penetration generated by storms in the North Atlantic and the Norwegian Sea. These swells propagate deep into the ice pack and are a major mechanism for fracturing floes, resulting in the small floe size west of Hopen. An example of this swell penetration is seen in the SAR image acquired on February 23 (Figure 13-14). The swell patterns in the ice cover are clearly visible throughout the entire image. These observations and two-dimensional fast Fourier transform (FFT) analysis indicate that the typical wavelength is 200 m. The swell propagates in a northeasterly direction and the waves refract around Bjørnøya. Another important mechanism for the production of small ice floes in this region is the fracturing of large floes by icebergs, as shown in Figure 13-1(d) [Johannessen et al., 1991b]. These combined mechanisms produce the small floe-size distribution in the Barents Sea west of Hopen. East of Hopen, where the ice floes consistently have large sizes, on the order of tens of kilometers, this combined effect is absent.

13.3 MODELING SEASONAL ICE ZONE PROCESSES

The series of experiments discussed above has demonstrated that the processes in the seasonal ice zone are complicated and not fully understood. The results of these experiments have provided quantitative information on which to base new models. The NORSEX results that showed upwelling along the ice edge prompted Røed and O'Brien [1983] to simulate this process with a coupled nonlinear model. This model included internal ice stress, had a variable ice concentration, and was forced by wind stresses on the ice and ocean. The model demonstrated that winds blowing along the ice edge and to the left when facing the ice cause upwelling along the ice edge. The same model was used by Smedstad and Røed [1985] to study ice edge breakup and banding. These studies emphasized the sensitivity of the response of the coupled system to wind direction and drag coefficients. Häkkinen [1986a, b] extended Røed and O'Brien's work by including nonlinear ocean dynamics. The width of the upwelling zone is determined by the baroclinic Rossby radius of deformation. The jet associated with upwelling is more stable than the one generated in a downwelling situation.

The field programs have also shown that mesoscale eddies are an important feature and dynamical component of the SIZ, e.g., Figures 13-1, 13-3, 13-4, and 13-7. This prompted many investigators to model eddy generation, evolution, and decay. Häkkinen [1986b] showed that a spatial variation in the ice cover together with uniform wind forcing can lead to pycnoclinic changes similar to those generated by a small travelling storm system. Shedding of eddies from the ice edge, with the dominance of cyclonic eddies towards the open ocean, and dissipation of anticyclones underneath the ice cover was also demonstrated in this study.

Smith et al. [1984] studied topographic eddy generation near the Molloy Deep in the Fram Strait using a two-layer primitive equation model for the ocean coupled to a free drift ice model. Häkkinen [1987] used a similar, but linearized, ice–ocean model in a study with a smooth shelf break ridge or canyon. For low wind situations, the bottom topography generates both trapped and travelling ice edge features, which do not always correspond directly to oceanic mesoscale features below. Winds that generate downwelling enhance ice edge meanders, while the ice edge stays compact during upwelling conditions, even in the case when a strong baroclinic vortex is developed underneath the ice. The vortices generated in this modeling study always had strong vertical shear. The combined presence of an ice cover and topographical features were thus shown to be a possible cause of baroclinic instability with subsequent mesoscale variability in the ocean. Smith et al. [1988] performed a systematic parameter study to determine the relative importance of various terms in the ice momentum equation for ice adjustment to evolving eddy–jet interaction. An interesting result from this model is that when the wind blows along the ice edge with the ice on the right, anticyclonic eddies rapidly dissipate. This agrees with the field observations from MIZEX'84 when northerly winds dominated and 12 out of 14 observed eddies were cyclonic [Johannessen, J. A., et al., 1987]. Smith and Bird [1991] continued this study by simulating the interaction of an ocean eddy with an ice edge jet. There are many other modeling studies of SIZ processes, including Ikeda [1986, 1989] who developed models for the Labrador Shelf SIZ, and Kantha and Mellor [1989] who included tides and thermodynamics in a model of the Bering Shelf SIZ.

The field investigations showed that ice bands are a common feature in the SIZ, e.g., Figures 13-7, 13-10, and 13-13. As a result, they have been modeled by many investigators. Wadhams [1983] used wave radiation pressure to generate bands parallel to the ice edge, under off-ice wind conditions. Muench et al. [1983] proposed a model where internal waves generated surface divergence and convergence zones that resulted in ice band formation. In the ice breakup and ice band model by Smedstad and Røed [1985], the formation of ice bands takes approximately 12 days because the ice breakup depends on the development of strong, along ice-edge velocities in the ocean. Chu [1987a] used thermally generated surface winds blowing from ice to water (ice breeze) to force the drift of ice floes. Ice bands resulted from changes in the surface temperature gradients. Another model takes into account the effects of the ocean, so the three media are linked through the surface temperature gradient and the interfacial stresses [Chu, 1987b]. All of these band models use different physics but

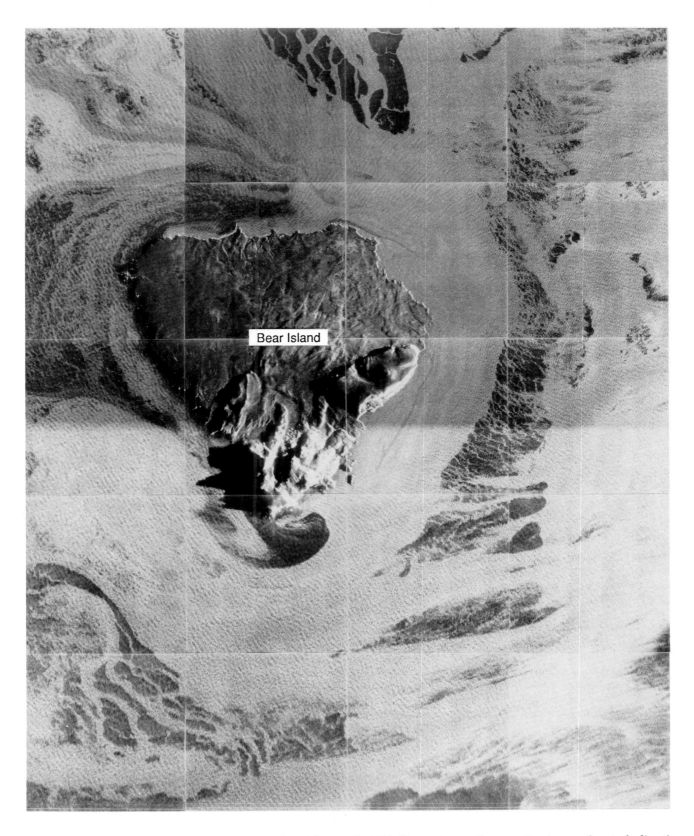

Fig. 13-14. SAR image of the Bjørnøya area obtained on February 23, 1989, illustrating swell propagation in a northeasterly direction around Bjørnøya.

generate similar results, which indicates the need for more detailed observations and model development.

Leads and polynyas have previously been treated in large-scale studies of the interior Arctic [Campbell et al., 1984] and in the Antarctic [Zwally et al., 1983]. The SIZ field experiments have shown that leads and polynyas are also a common mesoscale phenomena, e.g., Figures 13-7 and 13-13. Lead and polynya formation results from the complex interaction of many of the processes that occur in the SIZ. Smedstad and Røed [1985] have proposed a model that simulates ice divergence in the ice-edge region. Ikeda [1985, 1991] has developed wind-driven models for ice motion, including the generation of shore leads and eddy-like features. However, no existing model adequately explains the formation of leads and polynyas within the seasonal ice zone.

The results from the SIZ experiments show that the structure of the seasonal ice zone is fundamentally different from that of the interior ice packs. Compared to the interior ice pack, the SIZ ice pack contains relatively more new ice types and smaller ice floes, is subject to stronger divergence and convergence, undergoes wave penetration, and receives greater snowfall. All presently known SIZ models rely on the continuum hypothesis, which states that all state variables (ice velocity, thickness, etc.) should be continuous functions of position on scales down to the model resolution, e.g., Hibler [1979]. This criteria is not satisfied for SIZ models with resolutions on the order 1 km because of the large variations in floe sizes found in the East Greenland and Barents Seas, as shown in Figures 13-7 and 13-10. None of the models that have been described take specific account of floe size distribution nor try to predict it. However, in an idealized collisional rheology [Shen et al., 1987], the internal ice stresses depend quadratically on the floe size.

Based on the SAR images obtained during the campaigns since 1979, a conceptual SAR model for the seasonal ice zone has been formulated to identify the different ice edge processes (Table 13-3). In this scheme, the ice edge configurations from SAR images have been classified and related to atmospheric and ocean conditions [Johannessen, J. A., et al., 1991]. This conceptual SAR model provides useful guidelines for further numerical modeling of the seasonal ice zone. A realistic ice model should be coupled to the ocean both dynamically and thermodynamically. Thereby the ice can be influenced by ocean currents and waves and melting/freezing can take place. Internal ice stress should be parameterized in a way that is relevant for the SIZ. The wind forcing should act on both ice and ocean, using drag coefficients that depend on ice roughness and concentration. It is not practical to include all these effects in a model at the same time. A useful approach is to divide the overall modeling problem into subproblems that are studied separately and can be verified by SAR observations. Such subproblems could include: developing an ice rheology based on variations in floe size, ice concentration, ice type, ice motion, ridging and rafting, and wave penetration;

determining the temporal and spatial aspects of the ice edge variability; observing melting and freezing events and using them to develop a realistic thermodynamic model; and improving modeling of ice bands.

13.4 Regional Forecasting Using Microwave Observations

The Barents Sea is an important region for fisheries, as well as oil exploration. Ice forecasting is therefore a needed service for these industries. The Hibler–Preller ice model for the Barents Sea has been implemented at the Nansen Center for operational use in the Norwegian Ocean Monitoring and Forecasting System with a spatial resolution of 20 km [Hibler, 1979; Preller et al., 1989; Skagseth et al., 1991]. The model is initialized using ice concentration estimates from SSM/I data transferred from the Civilian Navy Ocean Data Distribution System in Monterey, California, to the Nansen Center in near-real time. The model predicts ice motion and concentration for 2 to 5 days with wind forecasts as the driving force. The 24- and 48-hour forecast of ice concentration is compared with updated SSM/I observations. Every 48 hours, the model is restarted with updated ice concentration data from SSM/I. An example of the ice forecast is shown in Figure 13-15. This forecasting scheme represents a considerable improvement compared to running the model without SSM/I data. With a 20-km resolution, the model is not capable of resolving

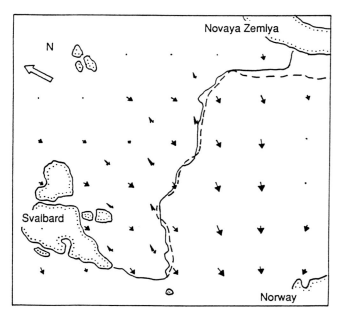

Fig. 13-15. Ice forecasting model for the Barents Sea using real-time ice concentration from SSM/I. The fully drawn line is the observed ice edge, represented by the 30% isoline for ice concentration. The dashed line is the 24-hour forecast for this line. The 24-hour wind forecast is indicated by arrows. The predicted ice motion is 10° to 20° to the right of the wind vectors (indicated by some V-shaped twin vectors inside the ice edge).

TABLE 13-3. Conceptual SAR model for the seasonal ice zone.

SAR Image Example	SAR Ice Edge Configuration	Ice Edge Process	Atmospheric Conditions	Upper Ocean Conditions
	Straight	Upwelling	Parallel Strong to Moderate Wind	Along Ice Jet and Divergent Ekman Flow and Convection
	Meandering and Eddies	Ice-Ocean Eddies	Moderate to Calm Wind	Ocean Eddies, Precondition, Convection
	Ice Jet Perpendicular to the Edge	Momentum Pulse, Ice Jet, and Vortex-Pair	Moderate to Calm Wind	Shallow Upper-Layer Ocean Jet
	Low Backscatter off the Ice Edge and in Leads	Ice Freezing (Winter) — Ice Melting (Summer)	Calm or Off-Ice Wind — Sun Radiation	Salinity and Density Increase, Convection — Fresh Water, Increased Stratification
	Wave Pattern and Compact Small Floes	Wave Propagation, Refraction, and Attenuation	On–Ice Wind	Wind Waves and Swell Propagating Towards the Ice
	Ice Banding, Streamers, and Internal Waves	Ice Bands and Internal Waves	Off-Ice Wind or Varying Wind	Convergence, Divergence, Internal Waves

mesoscale processes, however rapid changes of ice edge position due to wind forcing or freezing/melting is simulated reasonably well by the model. The microwave observations have proven to be indispensable in operational monitoring and forecasting on a regional scale. At present, SAR data from ERS-1 is implemented in the ice monitoring scheme at the Nansen Center and will also be assimilated into mesoscale models.

13.5 The SIZ in the Climate System

Sea ice is an important part of the global climate system. The two key aspects of the complex and dynamic sea ice covers are ice extents and the area of open water within the ice. The sea ice extent affects the amount of solar radiation absorbed in its hemisphere and alters the atmosphere–ocean exchanges of heat, moisture, and momentum. Ice-free oceans generally have an albedo of 10 to 15% [Lamb, 1982], whereas sea ice albedos average about 80% [Washington and Parkinson, 1986]. Because of this great contrast between ice and ocean albedos, the presence of sea ice considerably reduces the amount of solar radiation observed at the Earth's surface. Since sea ice covers such a large area of the Earth's surface, a number of investigators have studied the variations of the ice extent of the polar regions for evidence of climate change. They have examined the existing satellite record of ice extents of both the Antarctic sea ice cover [Zwally et al., 1983] and the Arctic sea ice cover [Campbell et al., 1984; Parkinson et al., 1987; Gloersen and Campbell, 1988; Parkinson and Cavalieri, 1989]. The sea ice extent extrema were first studied on a global scale by Gloersen and Campbell [1988] for both ESMR and SMMR. Recently, Gloersen and Campbell [1991a, b] have extended the SMMR analysis using the corrected data set.

While the ice extent is determined by SIZ processes, open water occurs in varying amounts throughout the whole ice cover [Campbell et al., 1984; Parkinson et al., 1987]. The area of open water within the ice pack, small even compared with the ice extent, is critically important because the fluxes of heat and moisture through a given area of open water are 2 to 3 orders of magnitude greater than through the same area of sea ice [Maykut, 1978]. The multispectral and dual-polarized observations with the SMMR allow determinations of ice extent and open water with greater accuracies than those obtainable with earlier satellite sensors [Gloersen and Campbell, 1991a].

In this section, we present the SMMR results for ice extent and open water amounts and variations for the Arctic and Antarctic ice covers and for the Greenland and Barents–Kara Seas. Gloersen and Campbell [1991b] used a band-limited regression technique [Lindberg, 1991] to obtain ice extent trends of the Arctic and Antarctic ice covers for the almost nine-year-long SMMR record. The time series of the SMMR records for each hemisphere and their trends are shown in Figure 13-16. These data show that the trends of the polar ice extents were not in phase:

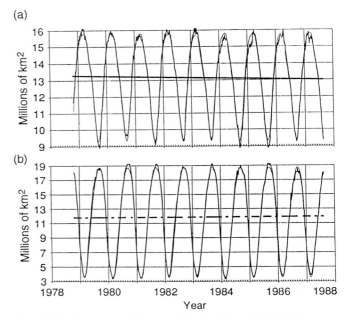

Fig. 13-16. Time series of ice extent derived from SMMR data from the (a) Arctic and (b) Antarctic [Gloersen and Campbell, 1991b]. The solid line for the Arctic ice extent shows the statistically significant decline of approximately 2%. The dashed line for the Antarctic ice extent indicates no statistically significant trend.

The Arctic had a statistically significant negative trend of ~2%, whereas the Antarctic had no significant trends.

The Greenland Sea SMMR records of ice extent, annually averaged ice extent and trends, and open water amount are shown in Figure 13-17 [Gloersen et al., 1992]. There are a number of noteworthy features in the oscillatory record of the sea ice in these figures. The amplitude of the annual cycle varies by an average of 50% during the nine-year period, with the smallest peak in ice extent occurring in 1984, the only year in which the Odden was absent. The area of open water is 2 to 4×10^5 km^2, about one-third of the ice-covered area. There is generally more open water during the winter than in the summer because the winter ice extent is greater and the pack undergoes divergence due to winter storms. Significantly large amounts of open water occur during all seasons. The SMMR record of the annually averaged ice extent in the Greenland Sea shows a negative trend of ~4%.

The SMMR records for the Barents and Kara Seas are shown in Figure 13-18 [Gloersen et al., 1992]. In this combined region, the ice extent undergoes even larger annual variations than in the Greenland Sea, expanding from about 0.7×10^6 km^2 in September to about 1.9×10^6 km^2 in April. Similar to the Greenland Sea, the Kara and Barents Seas region exhibits strong interannual difference in both the summer and the winter months [Gloersen et al., 1992]. The SMMR record of the annually averaged ice extent for the Barents and Kara Seas reveals a large negative trend of ~14%. The SMMR record of open water

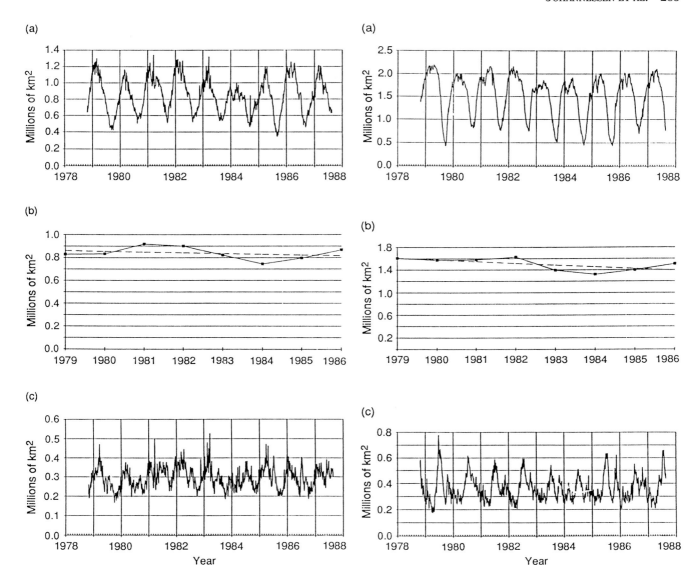

Fig. 13-17. (a) Time series of SMMR data from the Greenland Sea ice extent, (b) annually averaged sea ice extent, and (c) open water within the ice pack [Gloersen and Campbell, 1991a].

Fig. 13-18. (a) Time series of SMMR data from the Barents and Kara Sea ice extent, (b) annually averaged sea ice extent, and (c) open water within the ice pack [Gloersen and Campbell, 1991a].

amounts in the Barents and Kara Seas shows that generally more open water occurs in the summer than in the winter.

During the SMMR years, the negative trend of the annually average ice extents that occurred in the Greenland, Barents, and Kara Seas occurred in only one other Arctic Sea, the Sea of Okhotsk. Positive trends occurred in the Arctic Ocean, Bering Sea, Hudson Bay, Baffin Bay, Daris Strait, and Labrador Sea. But the combined effect of these positive trends was not sufficient to cancel out the negative trends of the other seas [Gloersen et al., 1992]. Therefore, the negative trends of the Greenland, Barents, and Kara Seas played a dominant role in determining the overall

negative trend in ice extent of 2% for the entire Arctic during the SMMR period.

The dominant retreat of sea ice in the passive microwave data record occurs in the region of the Atlantic approach to the Arctic Ocean. Many different mechanisms may contribute to such variations. It may be part of a self-sustained interdecadal cycle involving changes in the atmospheric circulation, surface freshwater flux, and deep water formation [Dickson et al., 1988; Mysak et al., 1990; Ikeda, 1990; Johannessen et al., 1991a]. Subtle changes in the complicated mesoscale processes discussed in the previous section may contribute to this kind of variability.

Recent results from global ocean–atmosphere climate models [Stouffer et al., 1989; Manabe et al., 1990, 1991a, b; Cubasch et al., 1991] show a large sensitivity of the present climate in this region to increased greenhouse gas forcing, as previously indicated in the ice–ocean model by Semtner [1987] and discussed by a large number of authors. Numerical models that both resolve the in- and outflows to the North Atlantic [Oberhuber, 1990] and include proper sea ice dynamics are needed to investigate whether these sensitivities are realistic.

The present time series of passive microwave observations from satellites is obviously too short to distinguish between natural variability and anthropogenic change. It is also too short to make any definitive statement of a climate trend. However, the quality of the data offers unprecedented possibilities for model testing. Figure 13-19 shows ice concentrations in the Northern Hemisphere from an ice–ocean model [Oberhuber, 1990] and SSM/I observations for the month of February. More detailed model–data comparisons should at least allow a proper assessment of the quality of present large-scale models.

Continuous, accurate monitoring of sea ice variations from passive microwave instruments will be able to determine future fingerprints of an anticipated global warming. Use of these data in large-scale numerical models, which include parameterizations of the smaller scale processes discussed previously, will be crucial to achieving a better understanding of the role of regional climate processes in the global system. Both active and passive microwave data can therefore be expected to play a major role in climate research in the future.

Acknowledgments. This extensive work reported in this chapter has received major funding from the Office of Naval Research, the two Norwegian research councils, the University of Bergen, the Norwegian Space Center, the European Space Agency, NASA, and the Operators Committee North (a consortium of 12 oil companies operating on the Norwegian Continental Shelf). The preparation of this manuscript has been supported by the International Space Year Office, European Space Agency Technology Center.

REFERENCES

Burns, B. A., D. J. Cavalieri, M. R. Keller, W. J. Campbell, T. C. Grenfell, G. A. Maycut, and P. Gloersen, Multisensor comparison of ice concentration estimates in the marginal ice zone, *Journal of Geophysical Research, 92(C7)*, pp. 6843–6856, 1987.

Campbell, W. J., J. Wayenberg, R. O. Ramseier, M. R. Vant, R. Weaver, A. Redmond, L. Arsenault, P. Gloersen, H. J. Zwally, T. T. Wilheit, T. C. Chang, D. Hall, L. Gray, D. C. Meeks, M. L. Bryan, F. T. Barath, C. Elachi, F. Leberl, and T. Fan, Microwave remote sensing of sea ice in the AIDJEX main experiment, *Boundary Layer Meteorology, 13*, pp. 309–337, 1978.

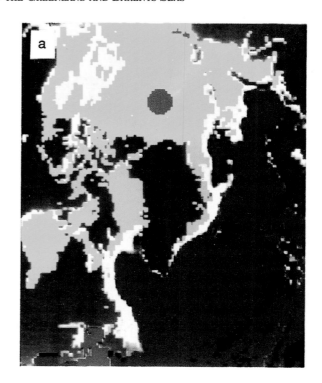

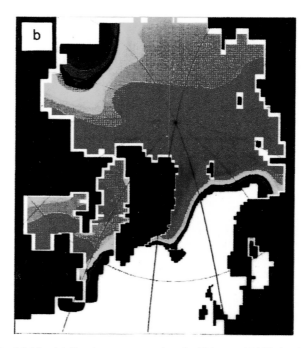

Fig. 13-19. (a) Sea ice concentration for February 1989, in the Northern Hemisphere as observed by SSM/I (yellow: >90% and blue: <10%); and (b) simulated ice concentrations for a typical February from the model by Oberhuber [1990] (red: >90% and dark blue: <10%).

Campbell, W. J., P. Gloersen, and H. J. Zwally, Aspects of Arctic sea ice observable by sequential passive microwave observations from the Nimbus-5 satellite, *Arctic Technology and Policy*, edited by I. Dyer and C. Chryssostomidis, pp. 197–222, Hemisphere Publishing Corporation, New York, 1984.

Campbell, W. J., P. Gloersen, E. G. Josberger, O. M. Johannessen, P. S. Guest, N. Mognard, R. Shuchman, B. A. Burns, N. Lannelongue, and K. Davidson, Variation of mesoscale and large-scale ice morphology in the 1984 Marginal Ice Zone Experiment as observed by microwave remote sensing, *Journal of Geophysical Research, 92(C7)*, pp. 6805–6824, 1987.

Chu, P. C., An ice breeze mechanism for an ice divergence-convergence criterion in the Marginal Ice Zone, *Journal of Physical Oceanography, 17(10)*, pp. 1627–1632, 1987a.

Chu, P. C., An instability theory of ice–air interaction for the formation of ice edge bands, *Journal of Geophysical Research, 92*, pp. 6966–6970, 1987b.

Coon, M. D., A review of AIDJEX modeling, *Sea Ice Processes and Models, Proceedings of the Arctic Ice Dynamics Joint Experiment International Commission on Snow and Ice Symposium*, University of Washington Press, 474 pp., 1980.

Cubasch, U., K. Hasselmann, H. Höck, E. Maier–Reimer, U. Mikolajewicz, B. D. Santer, and R. Sausen, *Time-Dependent Greenhouse Warming Computations With a Coupled Ocean–Atmosphere Model*, Report 67, 18 pp., Max-Planck-Institute für Meteorologie, Hamburg, Germany, 1991.

Dickson, R. R., J. Meincke, S.–A. Malmberg, and A. J. Lee, The great salinity anomaly in the Northern North Atlantic 1968–1982, *Programmetric Oceanography, 20*, pp. 103–151, 1988.

Fedorov, K. N. and A. I. Ginsburg, Mushroom-like currents (vortex dipoles) in the ocean and in a laboratory tank, *Annals of Geophysics, 4*, p. 507, 1986.

Gloersen, P. and W. J. Campbell, Variations in the Arctic, Antarctic, and global sea ice covers during 1978–1987 as observed with the Nimbus 7 Scanning Multichannel Microwave Radiometer, *Journal of Geophysical Research, 93*, pp. 10,666–10,674, 1988.

Gloersen, P. and W. J. Campbell, Variations of extent, area, and open water of the polar sea ice covers: 1978–1987, *International Conference on the Role of the Polar Oceans in Global Change*, Fairbanks, Alaska, 18 pp., 1991a.

Gloersen, P. and W. J. Campbell, Recent variations in Arctic and Antarctic sea-ice covers, *Nature, 352*, pp. 33–36, 1991b.

Gloersen, P., H. J. Zwally, A. T. C. Chang, D. K. Hall, W. J. Campbell, and R. O. Ramseier, Time dependence and distribution of sea ice concentration and multiyear ice fraction in the Arctic Basin, *Boundary Layer Meteorology, 13(1–4)*, pp. 339–360, 1978.

Gloersen, P., W. J. Campbell, D. J. Cavalieri, J. C. Comiso, C. L. Parkinson, and H. J. Zwally, *Arctic and Antarctic Sea Ice 1978–1987: Satellite Passive Microwave Observations and Analysis*, NASA SP-511, 200 pp., National Aeronautics and Space Administration, Washington, DC, 1992.

Guest, P. S. and K. L. Davidson, The effect of observed ice conditions on the drag coefficient in the summer East Greenland Sea marginal ice zone, *Journal of Geophysical Research, 92*, pp. 6943–6954, 1987.

Häkkinen, S., Coupled ice–ocean dynamics in the marginal ice zones: Upwelling/downwelling and eddy generation, *Journal of Geophysical Resarch, 91(C1)*, pp. 819–832, 1986a.

Häkkinen, S., Ice banding as a response of coupled ice–ocean systems to temporally varying wind, *Journal of Geophysical Research, 91*, pp. 5047–5053, 1986b.

Häkkinen, S., Feedback between ice flows, barotropic flow, and baroclinic flow in the presence of bottom topography, *Journal of Geophysical Research, 92*, pp. 3807–3820, 1987.

Hibler, W. D, III, A dynamic thermodynamic sea ice model, *Journal of Physical Oceanography, 9*, pp. 815–846, 1979.

Ikeda, M., A coupled ice–ocean model of a wind-driven coastal flow, *Journal of Geophysical Research, 90*, pp. 9119–9128, 1985.

Ikeda, M., A mixed layer beneath melting sea ice in the marginal ice zone using a one-dimensional turbulent closure model, *Journal of Geophysical Research, 91*, pp. 5054–5060, 1986.

Ikeda, M., A coupled ice–ocean mixed-layer model of the Marginal Ice Zone responding to wind forcing, *Journal of Geophysical Research, 94(C7)*, pp. 9699–9709, 1989.

Ikeda, M., Decadal oscillations of the air–ice–ocean system in the Northern Hemisphere, *Atmosphere–Ocean, 28(1)*, pp. 106–139, 1990.

Ikeda, M., Wind-induced mesoscale features in a coupled ice–ocean system, *Journal of Geophysical Research, 96*, pp. 4623–4629, 1991.

Johannessen, J. A., O. M. Johannessen, E. Svendsen, R. Shuchman, T. Manley, W. J. Campbell, E. G. Josberger, S. Sandven, J. C. Gascard, T. Olaussen, K. Davidson, and J. Van Leer, Mesoscale eddies in the Fram Strait Marginal Ice Zone during the 1983 and 1984 Marginal Ice Zone Experiments, *Journal of Geophysical Research, 92*, pp. 6754–6772, 1987.

Johannessen, J. A., E. Svendsen, S. Sandven, O. M. Johannessen, and K. Lygre, Three-dimensional structure of mesoscale eddies in the Norwegian Coastal Current, *Journal of Physical Oceanography, 19*, pp. 3–19, 1989.

Johannessen, J. A., O. M. Johannessen, S. Sandven, and P. M. Haugen, A characterization of air–ice–ocean interactive processes in the MIZ based on SAR-derived ice edge configurations, *IGARSS'91 Proceedings*, pp. 429–432, European Space Agency, Helsinki, Finland, 1991.

Johannessen, O. M., Brief overview of the physical oceanography, *The Nordic Seas*, edited by B. G. Hurdle, pp. 103–127, Springer Verlag, 1986a.

Johannessen, O. M., *Seasonal Ice Zone Experiment, ERS-1 Pre- and Postlaunch Experiments in the Barents Sea, Fram Strait, and Greenland Sea*, NERSC Special Report 18, Nansen Environmental and Remote Sensing Center, Bergen, Norway, 1986b.

Johannessen, O. M., Introduction: Summer Marginal Ice Zone Experiments during 1983 and 1984 in the Fram Strait and the Greenland Sea, *Journal of Geophysical Research, 92*, pp. 6716–6717, 1987.

Johannessen, O. M. and L. A. Foster, A note on the topographically controlled oceanic polar front in the Barents Sea, *Journal of Geophysical Research, 83*, pp. 45–67, 1978.

Johannessen, O. M., J. A. Johannessen, J. H. Morison, B. A. Farrelly, and E. A. S. Svendsen, Oceanographic conditions in the marginal ice zone north of Svalbard in early fall 1979 with emphasis on mesoscale processes, *Journal of Geophysical Research, 88*, pp. 2755–2769, 1983a.

Johannessen, O. M., W. D. Hibler III, P. Wadhams, W. J. Campbell, K. Hasselmann, I. Dyer, and M. Dunbar, *A Science Plan for a Summer Marginal Ice Zone Experiment in the Fram Strait/Greenland Sea*, CRREL Report 83-12, 47 pp., Cold Regions Research and Engineering Laboratory, Hanover, New Hampshire, 1983b.

Johannessen, O. M., W. D. Hibler III, P. Wadhams, W. J. Campbell, K. Hasselmann, and I. Dyer, Marginal ice zones: A description of air–ice–ocean interactive processes, models, and planned experiments, *Arctic Technology and Policy*, edited by I. Dyer and C. Chryssostomidis, pp. 133–146, Hemisphere Publishing Corporation, New York, 1984a.

Johannessen, O. M., J. A. Johannessen, B. Farrelly, K. Kloster, and R. A. Shuchman, Eddy studies during MIZEX'83 by ship and remote sensing observations, *Proceedings of the 1984 IGARSS*, pp. 365–368, ESA SP-215, European Space Agency, Helsinki, Finland, 1984b.

Johannessen, O. M., J. A. Johannessen, E. Svendsen, R. A. Shuchman, W. J. Campbell, and E. G. Josberger, Ice edge eddies in the Fram Strait marginal ice zone, *Science, 236*, p. 427, 1987.

Johannessen, O. M., S. Sandven, and J. A. Johannessen, Eddy-related winter convection in the Boreas Basin, *Deep Convection and Deep Water Formation*, edited by P. C. Chu and J. C. Gascard, pp. 87–105, Elsevier Science Publishers, New York, 1991a.

Johannessen, O. M., S. Sandven, and K. Kloster, Icebergs in the Barents Sea during SIZEX'89, *Proceedings of the Eleventh International Conference on Port and Ocean Engineering Under Arctic Conditions (POAC'91)*, Memorial University of Newfoundland, St. John's, Newfoundland, Canada, 1991b.

Kantha, L. H. and G. L. Mellor, Application of a two-dimensional coupled ocean–ice model to the Bering Sea marginal ice zone, *Journal of Geophysical Research, 94*, pp. 10,921–10,935, 1989.

Lamb, H. H., The climatic environment of the Arctic Ocean, *The Arctic Ocean. The Hydrographic Environment and the Fate of Pollutants*, edited by L. Rey, pp. 135–162, Comite Arctique International, London, 1982.

Leppäranta, M. and W. D. Hibler III, Mesoscale sea ice deformation in the East Greenland Marginal Ice Zone, *Journal of Geophysical Research, 92(C7)*, pp. 7060–7070, 1987.

Lindberg, C. R., Band-limited regression, Part I: Simple linear models. *Journal of the Royal Statistical Society*, Part B, in press, 1991.

Manabe, S., K. Bryan, and M. J. Spelman, Transient response of a global ocean–atmosphere model to a doubling of atmospheric carbon dioxide, *Journal of Physical Oceanography, 20*, pp. 722–749, 1990.

Manabe, S., R. J. Stouffer, M. J. Spelman, and K. Bryan, Transient responses of a coupled ocean–atmosphere model to gradual changes of atmospheric CO_2, Part I: Annual mean response, *Journal of the Climate, 4*, pp. 785–818, 1991a.

Manabe, S., R. J. Stouffer, K. Bryan, and M. J. Spelman, The transient response of the Arctic climate to gradual changes of atmospheric carbon dioxide, AGU Fall Meeting December 9–13, *Supplement to EOS Transactions of AGU*, 1991b.

Maykut, G. A., Energy exchange over young sea ice in the central Arctic, *Journal of Geophysical Research, 83*, pp. 3646–3658, 1978.

MIZEX Group, MIZEX East 83/84: The summer Marginal Ice Zone Program in the Fram Strait/Greenland Sea, *EOS, 67(23)*, pp. 513–517, 1986.

MIZEX Group, MIZEX East 1987: The winter Marginal Ice Zone Program in the Fram Strait/Greenland Sea, *EOS, 70(17)*, pp. 545–555, 1989.

Muench, R., Mesoscale phenomena in the Polar Oceans, *Polar Oceanography*, edited by W. O. Smith, Jr., pp. 223–286, Academic Press, Inc., 1990.

Muench, R. D., P. H. LeBlond, and L. E. Hachmeister, On some possible interactions between internal waves and sea ice in the marginal ice zone, *Journal of Geophysical Research, 88(C5)*, pp. 2819–2826, 1983.

Mysak, L. A., D. K. Manak, and R. F. Marsden, Sea-ice anomalies observed in the Greenland and Labrador Seas during 1901–1984 and their relation to an interdecadal Arctic climate cycle, *Climate Dynamics, 5*, pp. 111–133, 1990.

NORSEX Group, The Norwegian remote sensing experiment in a marginal ice zone, *Science, 220(4599)*, pp. 781–787, 1983.

Oberhuber, J. M., *Simulation of the Atlantic Circulation With a Coupled Sea-Ice–Mixed-Layer Isopycnal General Circulation Model*, Report No. 59, Max-Planck-Institut für Meteorologie, Hamburg, Germany, 1990.

Parkinson, C. L. and D. J. Cavalieri, Arctic sea ice 1973–1987: Seasonal, regional, and interannual variability, *Journal of Geophysical Research, 94*, pp. 14,499–14,523, 1989.

Parkinson, C. L., J. C. Comiso, H. J. Zwally, D. J. Cavalieri, P. Gloersen, and W. J. Campbell, *Arctic Sea Ice, 1973–1976: Satellite Passive-Microwave Observations*, NASA SP-489, 296 pp., National Aeronautics and Space Administration, Washington, DC, 1987.

Preller, R. H., S. Riedlinger, and P. G. Posey, *The Regional Polar Ice Prediction Systems Barents Sea (RPIPS-B): A technical description*, NORDA Report 182, Naval Oceanographic and Atmospheric Research Laboratory, Stennis Space Center, Mississippi, 1989.

Pritchard, R. S., *Sea Ice and Models, Proceedings of the Arctic Ice Dynamics Joint Experiment, International Commission on Snow and Ice Symposium*, University of Washington Press, Seattle, Washington, 474 pp., 1980.

Røed, L P. and J. J. O'Brien, A coupled ice–ocean model of upwelling in the marginal ice zone, *Journal of Geophysical Research, 88*, pp. 2863–2872, 1983.

Sandven, S. and O. M. Johannessen, High-frequency internal wave observations in the marginal ice zone, *Journal of Geophysical Research, 92*, pp. 6911–6920, 1987.

Sandven, S. and O. M. Johannessen. Seasonal ice zone studies, *The Sea. Ocean Engineering Science. Volume 9*, John Wiley and Sons, Inc., pp. 567–592, 1990.

Sandven, S. and O. M. Johannessen, Sea ice studies in the Barents Sea, *Journal of Photogrammetry and Remote Sensing*, in press, 1992.

Sandven, S., O. M. Johannessen, and P. M. Haugan., *A Study of the Marginal Ice Zone, Summary Report*, NERSC Technical Report 61, Nansen Environmental and Remote Sensing Center, Bergen, Norway, 1991a.

Sandven, S., K. Kloster, and O. M. Johannessen, *SAR Ice Algorithms for Ice Edge, Ice Concentration, and Ice Kinematics,* Nansen Environmental and Remote Sensing Center Technical Report 38, Bergen, Norway, 1991b.

Semtner, A. J., Jr., A numerical study of sea ice and ocean circulation in the Arctic, *Journal of Physical Oceanography, 17*, PP. 1077–1099, 1987.

Shen, H. H., W. D. Hibler III, and M. Leppäranta, The role of floe collisions in sea ice rheology, *Journal of Geophysical Research, 92*, pp. 7085–7096, 1987.

Shuchman, R. A., B. A. Burns, O. M. Johannessen, E. G. Josberger, W. J. Campbell, T. O. Manley, and N. Lannelongue, Remote sensing of the Fram Strait Marginal Ice Zone, *Science, 236*, pp. 429–431, 1987.

SIZEX Group, *SIZEX'89, A Prelaunch ERS-1 Experiment*, Nansen Environmental and Remote Sensing Center Technical Report 23, Bergen, Norway, 1989.

Skagseth, Ø., P. M. Haugan, S. Sandven, O. M. Johannessen, K. Kloster, and Å. R. Nilsen, Demonstration of Operational Ice Forecasting in the Barents Sea, Nansen Environmental and Remote Sensing Center Special Report 10, Bergen, Norway, 1991.

Smedstad, O. M. and L. P. Røed, A coupled ice–ocean model of ice break-up and banding in the marginal ice zone, *Journal of Geophysical Research, 90(C1)*, pp. 876–882, 1985.

Smith, D. C., IV, and A. A. Bird, The interaction of an ocean eddy with an ice edge ocean jet in a marginal ice zone, *Journal of Geophysical Research, 96(C3)*, pp. 4675–4690, 1991.

Smith, D. C., IV, J. Morison, J. A. Johannessen, and N. Untersteiner, Topographic generation of an eddy at the edge of the East Greenland Current, *Journal of Geophysical Research, 89*, pp. 8205–8208, 1984.

Smith, D. C., IV, A. A. Bird, and W. P. Budgell, A numerical study of mesoscale ocean eddy interaction with a marginal ice zone, *Journal of Geophysical Research, 93(C10)*, pp. 12,461–12,473, 1988.

Stouffer, R. J., S. Manabe, and K. Bryan, Interhemispheric asymmetry in climate response to a gradual increase of atmospheric CO_2, *Nature, 342*, pp. 660–662, 1989.

Svendsen, E., K. Kloster, B. Farrelly, O. M. Johannessen, J. A. Johannessen, W. J. Campbell, P. Gloersen, D. Cavalieri, and C. Mätzler, Norwegian Remote Sensing Experiment: Evaluation of the Nimbus 7 Scanning Multichannel Microwave Radiometer for sea ice research, *Journal of Geophysical Research, 88(C5)*, pp. 2781–2792, 1983.

Untersteiner, N., *The Geophysics of Sea Ice*, pp. 1–8, Plenum Press, New York, 1986.

Wadhams, P., A mechanism for the formation of ice edge bands, *Journal of Geophysical Research, 88(C5)*, pp. 2813–2818, 1983.

Wadhams, P., The seasonal sea ice zone, *The Geophysics of Sea Ice*, edited by N. Untersteiner, pp. 825–992, Plenum Press, New York, 1986.

Washington, W. M. and C. L. Parkinson, *An Introduction to Three-Dimensional Climate Modeling*, University Science Books, Mill Valley, California, 1986.

Zwally, H. J., J. C. Comiso, C. L. Parkinson, W. J. Campbell, F. D. Carsey, and P. Gloersen, *Antarctic Sea Ice, 1973–1976: Satellite Passive-Microwave Observations*, 206 pp., NASA SP-459, National Aeronautics and Space Administration, Washington, DC, 1983.

Chapter 14. Considerations for Microwave Remote Sensing of Thin Sea Ice

THOMAS C. GRENFELL

Department of Atmospheric Sciences, University of Washington, Seattle, Washington 98195

DONALD J. CAVALIERI AND JOSEFINO C. COMISO

Laboratory for Hydrospheric Processes, Goddard Space Flight Center, Greenbelt, Maryland 20771

MARK R. DRINKWATER

Jet Propulsion Laboratory, California Institute of Technology, 4800 Oak Grove Drive, Pasadena, California 91109

ROBERT G. ONSTOTT

Environmental Research Institute of Michigan, P. O. Box 8618, Ann Arbor, Michigan 48107

IRENE RUBINSTEIN

Ice Research and Development, York University, 4700 Keele Street, North York, Ontario, M3J 1P3, Canada

KONRAD STEFFEN

Cooperative Institute for Research in Environmental Sciences, University of Colorado, Boulder, Colorado 80309–0449

DALE P. WINEBRENNER

Polar Science Center, Applied Physics Laboratory, University of Washington, 1013 NE 40th Street, Seattle, Washington 98105

14.1 INTRODUCTION

Thin ice, consisting of the World Meteorological Organization's (WMO's) group of types for new and young sea ice less than about 0.3 meters in thickness, is of considerable importance both for the energy exchange between the atmosphere and ocean in the polar regions and for the dynamics of the sea ice cover. Integrated over the annual cycle, thin ice can be found over an area of at least 8×10^6 km^2 and 16×10^6 km^2 in the Arctic and Antarctic, respectively, and the maximum instantaneous coverage is estimated at about 1×10^6 km^2. The brightness temperatures, emissivities, and radar scattering cross sections of thin ice evolve during growth. Their behavior for thin ice is determined by its bulk dielectric properties, but can be strongly modified by the presence of snow and by the properties of the near surface layers, where the salinity can be very large.

In this chapter, we will show that surface-based radiometric results and principal component analysis indicate that some thin ice types can be resolved under favorable circumstances. Mixtures of thick ice and open water, however, can still give rise to ambiguities in ice type

identification in available satellite data. Initial comparisons of concurrent radiometric and radar data show the potential to improve discrimination of thin ice on the basis of emitted and backscattered intensities. We expect the ability to distinguish thin ice using satellite imagery will improve considerably with the combination of Special Sensor Microwave/Imager (SSM/I) data, high-resolution results from the First European Remote Sensing Satellite (ERS-1), and microwave models of the ice.

The spatial and temporal distribution of thin ice plays a central role in both the thermodynamics and dynamics of the world's sea ice. Thin ice refers to all sea ice types younger than first-year ice and less than 0.3 meters thick. The polar packs consist of complex mixtures of many different thicknesses. A typical mesoscale region can be expected to contain open water, new ice, and young ice in varying proportions, depending on season and geographic location. Because of thermodynamic mass changes and dynamic motions that rearrange existing ice, the area covered by any particular thickness category undergoes continual change. Given sufficient data about the velocity field and mechanical properties of the ice and the energy fluxes at the upper and lower boundaries one can calculate the magnitude of these changes using thickness-distribution-dependent dynamical models [Rothrock, 1975; Thorndike et al., 1975]. Reviews by Rothrock [1986] and Thorndike [1986] show

Microwave Remote Sensing of Sea Ice
Geophysical Monograph 68

that the ice thickness distribution plays a central role in controlling the large-scale dynamics of the ice and that thin ice is particularly important. Observational verification of such modeling, however, is not yet available in sufficient detail to know how well the amount of thin ice is predicted.

Many properties of sea ice are strongly dependent on its thickness, including surface temperature, strength, growth rate, salinity, and dielectric constant. Thus, some knowledge of the way in which ice thickness is distributed in a particular part of the ice pack in a particular season is needed before its large-scale behavior can be predicted accurately. The extent and distribution of the areas covered by open water and thin ice are of special concern in modeling heat exchange with the atmosphere and the salinity balance of the upper ocean. In the central Arctic, open water appears to make up no more than 1% of the ice pack during the winter, but estimates from submarines of the amount of ice less than 1 m in thickness give a range from 8% to 12% [Wittman and Schule, 1966], about an order of magnitude greater. These latter estimates are consistent with independent theoretical calculations [Thorndike et al, 1975]. For the southern ocean [Wadhams et al., 1987; Lange and Eicken, 1991], up to 7% of thin ice is associated with floes, but, because of logistical limitations, this is an underestimate that does not include new ice.

Modeling of heat transport through young sea ice [Maykut, 1978, 1982] predicts that in the central Arctic during the cold months, the net heat input to the atmosphere from ice in the 0 to 0.4 m range is between one to two orders of magnitude larger than that from perennial ice. Maykut concludes that with the present estimates of ice thickness distribution in the central Arctic, total heat input to the atmospheric boundary layer from regions of young ice is equal to or greater than that from regions of open water or thick ice.

Similar considerations hold for the salt and buoyancy fluxes in the ocean [Smith et al., 1990; Morison et al., 1991]. The rejection of salt due to the growth of thin ice significantly enhances convection in the mixed layer and affects the oceanic heat flux to the bottom of the ice. In certain areas of the marginal ice zones (MIZ) and areas of persistent polynyas, the salt flux appears to be influential in generating oceanic bottom water. Over the Arctic Ice Dynamics Joint Experiment (AIDJEX) array, thin ice rather than open leads had the greatest influence on ice production, heat input to the atmosphere, and salt input to the ocean.

In the Antarctic and in the northern MIZ areas, there is often greater divergence and even more open water and thin ice than in the central Arctic. Cruise results from the Antarctic [Wadhams et al., 1987; Eicken and Lange, 1989; Worby and Allison, 1991] indicate that much of the winter Antarctic ice pack is highly dynamic, drifting at speeds of tens of kilometers per day, and that this motion is often divergent, leading to a large percentage of new ice within the pack. Under these conditions, the larger areas (even as low as 80%) of thin ice result in greater turbulent heat loss than from open water.

Microwave imagery from aircraft and satellites provides monitoring on intermediate to large geographic scales and has already demonstrated its usefulness for understanding the behavior of the global sea ice cover. Because of limitations in spatial resolution and the small relative area covered by freezing leads and polynyas, however, thin ice determination has proved difficult. Techniques to refine our knowledge of the thickness distribution of new and young ice are still being developed. This chapter aims to review the present state of the art.

14.2 TERMINOLOGY AND CLASSIFICATION

Because of the diversity of conditions that can accompany the formation and growth of thin ice, its physical properties can cover a range that is wide even for ice in a particular thickness interval. The WMO division into the subcategories of new and young ice based on visual appearance is useful. The visually distinct surface forms often appear to be accompanied by differences in their dielectric properties.

The WMO system defines the categories usually used to distinguish various stages of evolution for both calm and wavy conditions. This sequence is shown diagrammatically in Figure 14-1, adapted from Weeks [1976], along with the approximate range of thicknesses associated with each category. A certain amount of crossover is possible if calm conditions become wavy or vice versa. For example, grease ice can quickly develop into nilas if the waves subside, and nilas can be broken up into small pieces that quickly form

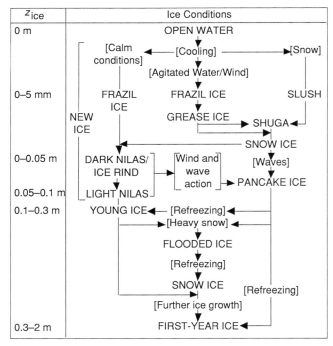

Fig. 14-1. Evolutionary sequence for thin sea ice. Ice types are shown in capital letters and related environmental processes are enclosed in square brackets.

pancakes. Young ice develops both from the consolidation and thermal growth of pancake ice and from light nilas. The snow cover is quite variable and is often thick enough to make visual identification of the underlying ice difficult.

Since the WMO system is based on the visible albedo and morphology of the ice, in general it will not have a one-to-one relationship with the microwave characteristics. For example, variations in snow cover can easily confuse a particular visible categorization while they may have a much smaller effect on microwave signatures. This sort of situation could be exploited to allow greater discrimination of ice types using a combination of sensors. The quantities we wish to determine by remote sensing are the physical properties of the ice (e.g., thickness, salinity, temperature, density, and snow cover characteristics) and the attendant environmental conditions, rather than WMO types per se. The ultimate evaluation of any classification system is based on how well it succeeds in doing this.

14.3 Geographical and Temporal Distribution of Thin Ice

Thin ice must occur during the freeze-up phase in seasonal sea ice zones—those areas covered by ice in the winter, but free of ice in summer. It is also formed in polynyas and leads throughout the winter in the perennial ice. Consequently, thin ice will occur over large areas of the polar regions, but because the growth of new and young ice tends to be rapid, a smaller area would in general be covered with thin ice at any one time. An overview of the subject by Maykut [1986] indicates that, as a result of freezing alone, ice will form and grow to 0.3 m in thickness in approximately 5 to 10 days if $T_{air} = -30°C$ and in about 7 to 15 days if $T_{air} = -20°C$. Then, for a satellite sampling rate of once every three days, a given area of thin ice would be observable between one and five times before developing into another type of ice. Consequently, we would expect that in many places the area covered by thin ice would not appear as a persistent or static feature.

Microwave satellite imagery has greatly refined our knowledge of the extent and cycling of the seasonal sea ice zones (SSIZ) of both the Northern and Southern Hemispheres since 1973 [Zwally et al., 1983; Parkinson et al., 1987; Parkinson and Cavalieri, 1989; Parkinson, 1991]. This has in fact provided the most precise identification and discovery of many areas of thin ice. In the SSIZ regions, thin ice occurs in abundance and can be readily delineated from the seasonal information provided by passive satellite imagery. These regions are shown in Figures 14-2 and 14-3 for the Arctic and Antarctic, respectively, from four-year averages of Electrically Scanning Microwave Radiometer (ESMR) data. The greatest area is around Antarctica, where the SSIZ covers 16×10^6 km^2, but large areas are also found around the margins of the Arctic Basin, in the Bering Sea, and in the Sea of Okhotsk covering 8×10^6 km^2.

Cracking and divergence of the pack continually provide new areas of ice formation and growth. Large areas where this occurs, polynyas and major lead systems, have been detected by microwave satellite imagery in many places throughout the Arctic and Antarctic Sea ice zones [Martin and Cavalieri, 1989].

A striking and recurrent winter feature in the Greenland Sea is the Odden, a large tongue of sea ice jutting out from

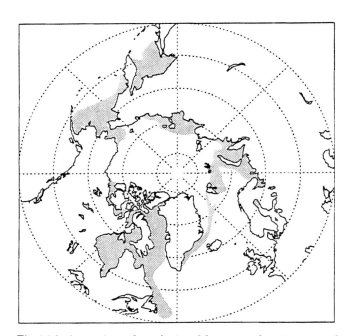

Fig. 14-2. Approximate boundaries of the seasonal sea ice zones of the Northern Hemisphere derived from ESMR satellite imagery [Parkinson et al., 1987].

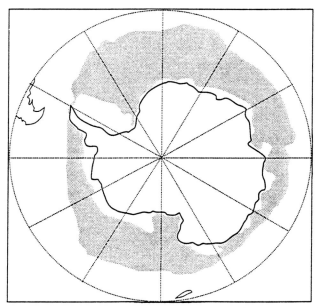

Fig. 14-3. Approximate boundaries of the seasonal sea ice zones of the Southern Hemisphere derived from ESMR satellite imagery [Zwally et al., 1983].

the thick ice of the east Greenland drift stream, which has routinely been detected by the Scanning Multichannel Microwave Radiometer (SMMR) and SSM/I. This feature can cover an area of more than 10^5 km^2 and can appear and dissipate within a few days to a week. Ship-based observations during recent field experiments have shown that the Odden is made up in large part of thin ice. The rapid appearance and disappearance of the Odden reflects fluctuations in the delicate balance between the atmospheric and oceanic heat fluxes in the region; indeed this may be a region of considerable importance in the ventilation of the deep ocean [Gascard, 1990].

Estimates of the amount of thin ice in the central Arctic have been made on the basis of empirical growth rate formulae from Anderson [1961] and from the results of dynamic and thermodynamic model results by Maykut [1982]. They give regional percentages of thin ice of 20% during fall freeze-up, 1.4% to 4% in midwinter, and 10% in midsummer. Corresponding estimates for the Antarctic [Weller, 1980] give wintertime concentrations of 20% for the inner zone and 60% for the outer zone.

For the entire Arctic and Antarctic, including the peripheral seas, an independent estimate can be made directly using satellite imagery to determine rates of change in total ice area [Zwally et al, 1983; Comiso and Zwally, 1984; Parkinson and Cavalieri, 1989]. These are then integrated over the lifetime of thin ice to give areal coverage at a particular time. Figure 14-4 shows the resulting estimates for each month, assuming that it takes 7 days on average for the ice to become 0.3 m thick during growth and that no ice grows during summer breakup. Estimated areas are as high as 0.5×10^6 to 1×10^6 km^2 (about 5 and 25% of minimum extent) in the Arctic and Antarctic, respectively. The current estimates are only approximate, but thin ice is clearly present in substantial quantities in both hemispheres.

14.4 MICROWAVE SIGNATURES OF THIN ICE

The concept of microwave signatures for thin ice and its subcategories implies a well-defined relationship between some aspect of the physical properties of the ice and an appropriate combination of brightness temperatures, emissivities, and/or scattering coefficients that will identify that ice type. The first case to consider is that of an instantaneous image where the signature makes it possible to extract the distribution of a particular ice type in the image.

14.4.1 Radiometry

Because of the coarse resolution of satellite passive microwave sensors to date, thin ice has been difficult to identify in the imagery and can be easily confused with mixtures of thick ice and open water. However, even though thin ice may be difficult to identify, the ice types in this category do have significantly different microwave signatures than open water or thicker ice. Thus, the effects of thin ice must be present in the satellite record and the above distribution estimates suggest that the effects should be significant under appropriate circumstances.

Representative emissivity spectra for thin ice types, from surface and aircraft-based observations, are presented in Chapter 4. These show that thin ice has emissivity spectra that lie between the values for open water and those of first-year ice and have different slopes. Figure 14-5 shows laboratory measurements illustrating the progression of emissivity with ice thickness for calm growth conditions [Grenfell and Comiso, 1986]. A complication associated with near-surface observations is the contribution of interference fringes. These are due to multiple reflections from layering in spatially homogeneous areas of ice and snow, and they generate T_B fluctuations in small-scale observational results (Chapter 9). In aircraft or satellite studies, the footprints are much larger, usually enough to include many different ice conditions and to average out the fringes.

For ice growth under both calm and wavy conditions, the rate at which the emissivities evolve with ice thickness depends strongly on the development of the dielectric properties of the ice. This is determined by the distribution of brine in the ice. In calm water, this is determined by the

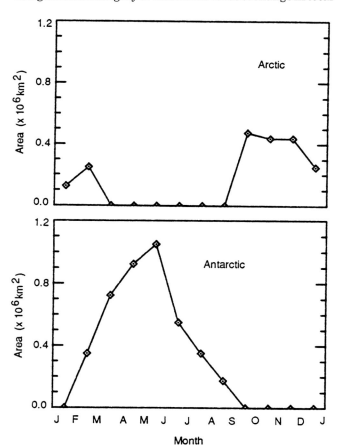

Fig. 14-4. Estimates of areal coverage by thin sea ice in both the Arctic and Antarctic based on ESMR data.

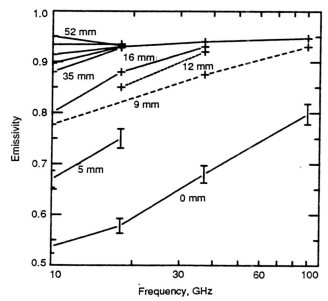

Fig. 14-5. An example of the progression of emissivity versus frequency at vertical polarization of thin ice from open water to young ice. Each curve gives the ice thickness. Maximum thickness for this set of observations is about 100 mm. V-pol observations are at a nadir angle of 50°.

growth rate of the ice and the salinity of the water [Weeks and Ackley, 1986]. Laboratory observations [Grenfell, 1986; Grenfell and Comiso, 1986; Wensman, 1991] show that the ice becomes optically thick at physical thicknesses between 50 and 80 mm for 10 GHz, decreasing with increasing frequency to about 10 mm for 90 GHz. Beyond this point, the high reflectivity of the underlying water no longer makes a significant contribution to the emitted radiation. Subsequent emissivity changes then arise primarily from changes in the distribution of brine in the uppermost layers.

For grease ice and pancake ice, the situation is quite different. Salt entrapment is not due to growth at a planar interface, but involves the consolidation of a crystal slurry that results in entrapment of brine throughout the slurry's volume. After consolidation, flushing of the surface continues through the development of pancake ice and further modifies the salinity distribution. Since the mechanisms of brine entrapment, drainage, and expulsion are different than those associated with congelation ice growth, we would in general expect differences in the microwave signatures.

A cluster plot of observations from the Marginal Ice Zone Experiment (MIZEX) 1987 (Figure 14-6) shows results for thin ice types grown under wavy conditions. The brightness temperatures (T_B's) span the range from open water to thick first-year (FY) ice. Since the footprint size for each of these observations was about 5 m across, small but decreasing amounts of open water between the pancakes were included in the radiometer's field of view. Observed signature development thus combines changes due to ice growth with changes due to increasing ice concentration.

Corresponding observations were carried out during an experiment in the Weddell Sea [Comiso et al., 1989] over areas where bare, rafted, and snow-covered nilas covered most of the area around the ship. The emissivities for nilas were within 0.05 to 0.2 of the values for FY ice, consistent with laboratory results. The results at 18 and 90 GHz, for example, showed several clusters associated with thin ice (Figure 14-7, labeled new ice) that were due to undisturbed and rafted areas with different amounts of snow cover. In this case, the brightness temperatures were tightly grouped rather than spread out evenly between open water and first-year ice as in Figure 14-6.

The polarization ratio, $PR(\nu) = [T_B(\nu, \text{V-pol}) - T_B(\nu, \text{H-pol})]/[T_B(\nu, \text{V-pol}) + T_B(\nu, \text{H-pol})]$, where H- and V-pol denote horizontal and vertical polarization, respectively, is also used to indicate thin ice in SMMR and SSM/I imagery. It was selected on the basis of observed differences between open water and thick ice. During new or young ice growth PR_V decreases from open water values near 0.3 to young ice values of about 0.1 [Grenfell and Comiso, 1986]. This is shown in detail in Chapter 9, and is determined by the relative contributions to the total reflectivity of the upper and lower surfaces of the ice. Both surface and aircraft observations in the Bering Sea (Figure 14-8) show a similar behavior [Cavalieri et al., 1986] as do satellite results from the Chukchi Sea and northern Baffin Bay [Steffen and Maslanick, 1988; Carsey and Pihos, 1989; Steffen, 1991].

SSM/I data from the Bering Sea [Cavalieri, 1988] shown in Figure 14-9 indicate clustering in the circled region,

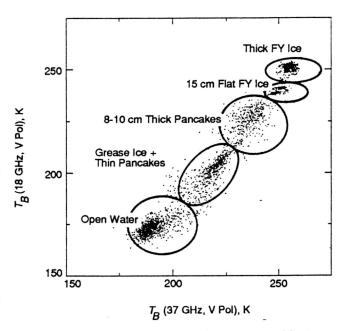

Fig. 14-6. Cluster observations of some thin ice types and first-year ice from MIZEX'87 in the northern Greenland Sea [Grenfell, T. C. and D. L. Bell, unpublished, University of Washington, Seattle, Washington, 1988]. V-pol observations are at a nadir angle of 50°.

where $GR(\nu1, \nu2) = [T_B(\nu2, \text{V-pol}) - T_B(\nu1, \text{V-pol})]/[T_B(\nu2, \text{V-pol}) + T_B(\nu1, \text{V-pol})]$. The density of points in this cluster is higher during the presence of polynyas and appears to indicate the presence of young ice. A recent analysis of thin ice [Wensman, 1991] gives growth trajectories of undisturbed young ice, for which representative results are shown in Figure 14-10. Growing ice passes quickly through the early phase, but moves much more slowly toward the end of the sequence just when the signatures lie within the area defined by the oval in Figure 14-9. Although young ice

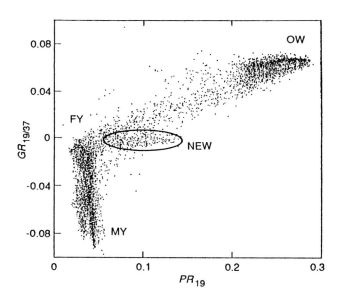

Fig. 14-9. Cluster plot of preliminary data from SSM/I during January. The cluster of points in the oval are from the Bering Sea in January 1988.

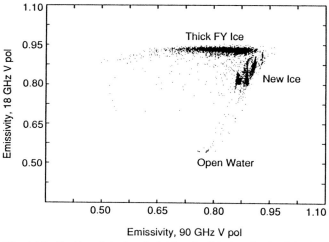

Fig. 14-7. Scatter plot of emissivities for thin ice obtained during the winter of 1986 in the Weddell Sea. New ice indicates nilas in different stages of development—undisturbed and overthrusted with and without snow cover.

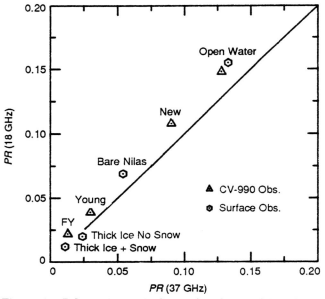

Fig. 14-8. Polarization ratio for combined aircraft-based and surface observations in the Bering Sea [Cavalieri et al., 1986]. The nadir angle is 45°.

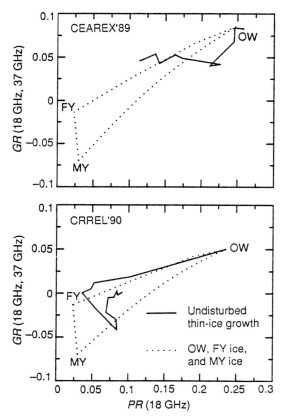

Fig. 14-10. Results for new ice from the northern Greenland Sea (CEAREX'89) and for young ice at the CRREL pond (CRREL'90).

may not account for all the points within the oval, it apparently clusters there. A recent aircraft study also found a similar result over the Beaufort Sea for 100% ice concentration [Drinkwater et al., 1991a].

The work by Wensman [1991] also presents a supervised principal component analysis that strongly suggests that under favorable conditions both young and new ice can be distinguished using passive microwave data. The separation of new and young ice in principal component space is illustrated in Figure 14-11 based on data from the Cold Regions Research and Engineering Laboratory (CRREL) pond experiment and Coordinated Eastern Arctic Research Experiment (CEAREX). A thin ice algorithm based on this analysis and applied to SSM/I data appears to explain quite well the ice distribution for a test case from the Bering Sea. Further development is needed, however, before this technique can be used operationally.

14.4.2 Radar

Studies of radar signatures of thin ice have also become available recently. Figure 14-12(a) shows combined results from a progression of surface-based backscatter cross sections at C- and X-bands from open water through thick ice covering a dynamic range of about 20 dB [R. G. Onstott,

personal communication, 1991]. Minima are found for grease and frazil ice about 0.02-m thick and for medium first-year ice, and there is a strong maximum for 0.05- to 0.3-m thick ice, which spans the range from light nilas to young ice. Corresponding results from Soviet satellite observations using the X-band side-looking real aperture radar on Cosmos 1500 [Nazirov et al., 1990] are shown in Figure 14-12(b) for the eastern Arctic. These results depict the optical density of the imagery that is approximately proportional to scattering in decibels. These results show the same strong maximum for gray nilas as Onstott's results. Another broad minimum is evident for FY ice, after which the scattering undergoes a gradual increase, and finally a strong jump for MY ice. Nazirov and coworkers

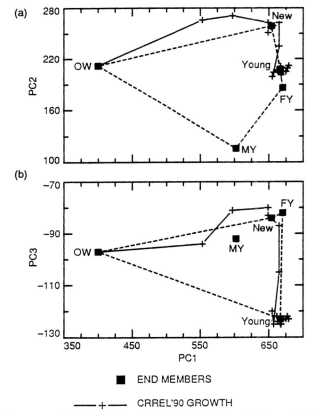

Fig. 14-11. Principal components derived from a 10-channel data set consisting of dual-polarization brightness temperatures from 6 to 90 GHz.

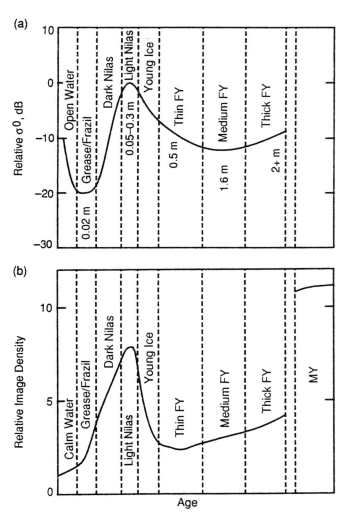

Fig. 14-12. (a) Radar backscatter sequence for FY ice. Combined surface backscatter cross sections at C- and X-bands' VV and HH polarizations showing the progression from open water through thick ice. (b) Satellite real-aperture backscatter observations at 9.7-GHz VV pol from Cosmos 1500 showing the same progression from open water through thick ice. The dynamic range is not reported, but can be inferred relative to the jump from thick FY to MY ice.

quote results for calm open water with lower scattering than for new ice, while Onstott shows results for roughened water.

The maximum for nilas in Figure 14-12(a) is related to the presence of frost flowers whose density can become quite high and blanket the entire surface. The individual flowers (Figure 14-13) are large enough to present a significant cross section to radar radiation, but their density and the size of the dendrites indicates that they should be transparent at frequencies up to at least X-band. However, frost flowers can also affect the near-surface brine distribution [Grenfell and Comiso, 1986; Drinkwater and Crocker, 1988] and are frequently characterized by a lump at the base due to wicking of surrounding brine from the ice surface and subsequent metamorphism (as can be seen directly in Figure 14-13). These structures are sufficiently large and have a high enough dielectric constant to present a significant enhancement in the radar cross section. The second minimum for thicker ice is attributed to further densifying of the frost layer due to snow accumulation combined with an increase in brine volume due to the thermal insulation. This brine could be partly incorporated into the snow by capillary action and would introduce attenuation to mask the rough interface.

Aircraft SAR observations at P-, L-, and C-bands [Drinkwater et al., 1991b; Winebrenner, D. P. and L. D. Farmer, On the L-band polarimetric SAR response to ice thickness in new and thin sea ice types, *Journal of Geophysical Research*, in preparation] were carried out over the Beaufort and Bering Seas during March 1988. These studies confirm the strong influence of frost flowers and indicate that their absence results in the near disappearance of the maximum of σ for nilas. They also show that at

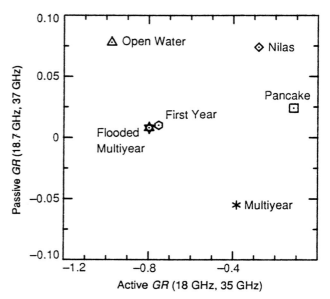

Fig. 14-14. Comparison of microwave observations for selected ice types measured during MIZEX'87 in the northern Greenland Sea in early spring.

L-band, both the ratio of backscatter cross sections, σ_{VV}/σ_{HH}, and the copolar phase are sensitive to certain thin ice types. To first order, the ratio σ_{VV}/σ_{HH} depends on the dielectric constant of the scattering material and not on surface roughness, and values for thin ice are between those of open water and FY ice.

Some surface and aircraft-based comparisons of radiometric and radar data for thin ice from MIZEX'87 [Tucker et al., 1991] show that selected combinations of active and passive gradient ratios appear to separate nilas and pancake ice from open water and FY ice (Figure 14-14). Although these data were for a few selected sites for which the ice properties had been extensively measured, the distinct ice types were clearly identified. More data are needed to test the robustness of this result.

Passive–active comparisons have been reported for the Bering Sea in 1988 from NASA DC-8 observations. Cross-polarization scattering cross section (σ_{HV}) versus T_B (37 GHz, V-pol) reported by Drinkwater et al. [1991a] are shown in Figure 14-15(a). The young ice is separated as indicated from the locus of points for varying concentrations of small first-year floes and open water (squares) by more than 20 K in T_B and 4 dB in σ_{HV}. This locus is curved because the contribution to scattering by floe edges is not directly proportional to concentration and hence to T_B. Dual-frequency radar data at like polarizations, including one very low frequency, also separate out the young ice, Figure 14-15(b). The decrease in scattering cross section by the young ice suggests that frost flowers were not present in this case.

Fig. 14-13. A single frost flower on growing young ice observed during LEADEX'91. (Courtesy of R. G. Onstott.)

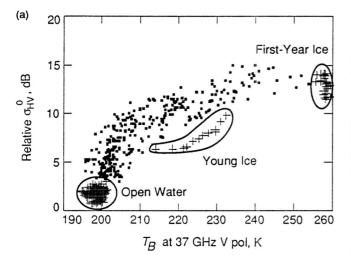

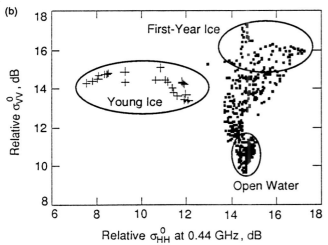

Fig. 14-15. Comparison of relative radar backscatter at 5.3 GHz and (a) HV and (b) VV pol measured by aircraft over the Bering Sea MIZ on March 21, 1988.

14.4.3 The Influence of Ice Structure on Microwave Signatures

In addition to total ice thickness and bulk salinity, several phenomena related to the small-scale structure of the surface layers are important in modulating the microwave signatures of thin ice. The first class of effects includes those that affect the brine volume of the ice that in turn can modify the permittivity quite strongly. These effects combine considerations of both temperature and salinity. One of the most important of these is the existence of a highly saline surface layer where the salinities can exceed 80 ppm. This gives rise to a very high brine content that has produced detectable changes in T_B [Grenfell and Comiso, 1986; Wensnahan et al., 1992]. This surface layer is known to exhibit horizontally spatial inhomogeneity. Other considerations that influence the salinity are the ice growth

rate and whether the growth occurs under quiescent or wavy conditions [Weeks and Ackley, 1986]. A second class of significant effects consists of those that affect the surface roughness, including the growth of frost flowers and the occurrence of ridging or overthrusting.

The importance of snow cover for ice signatures is complex (Chapter 16). Snow cover on thin ice affects both the salinity distribution and the surface roughness. It insulates the ice, increasing the temperature, but also acts to wick up the extra brine, and as metamorphism proceeds can introduce significant surface roughness [Lohanick, personal communication, 1990]. If the snow cover is thick enough, it can submerge the ice and cause flooding [Weeks, 1976; Wadhams et al., 1987; Eicken and Lange, 1989; Tucker et al., 1991]. Recent laboratory measurements have demonstrated that each of these effects can modify the microwave signatures of thin ice (for more detail see Chapter 9), but it remains to show how to characterize the effects in general.

14.5 IMPLICATIONS FOR SATELLITE ALGORITHMS

Operational satellite algorithms that do not account for thin ice can produce significant errors, particularly in MIZ. For example, thin ice has a high enough PR to look like almost 30% open water, and this offers a potential explanation of certain ambiguities in the interpretation of both SMMR and SSM/I data. There are frequent occurrences in the satellite records of low ice concentration events during midwinter persisting over very long time periods compared to known ice growth rates. Present evidence suggests that these low concentration areas may include extensive amounts of thin ice. Thin ice might also account for part of the jump in ice extent during the transition from ESMR to SMMR in the course of the 15-year record of Arctic Sea ice from 1973 to 1987 analyzed by Parkinson and Cavalieri [1989]. This jump is most noticeable in the Kara, Barents, and Greenland Seas and the Sea of Okhotsk where higher abundances of thin ice are expected.

The trajectories of brightness temperature, emissivity, and backscatter cross section for thin ice during growth show behavior that suggests that T_B, e, and σ° vary independently to some degree. The maximum scattering coefficients due to frost flower formation, for example, do not appear to have a corresponding expression in the emissivity data.

Even though the number of different processes that can modify the ice structure may appear to be large, satellite imagery suggests that the microwave signatures develop in relatively reproducible orderly sequences, and that some measure of ice thickness resolution is practical. Thus, although the relative importance of the various physical processes is not yet well understood, these processes probably occur in regular progression and give relative statistical homogeneity over large areas.

From the point of view of ice dynamics and mass balance, even a single additional thin ice category would greatly improve our present understanding of the ice. More resolvable categories would correspondingly improve the definition of the thickness distribution.

14.6 POTENTIAL FOR MULTISENSOR DATA

These studies of thin ice indicate that additional information is available from radiometer imagery. ESMR and SMMR have identified areas where thin ice should be found, but techniques capable of resolving thin ice types have not yet succeeded in producing unambiguous retrievals. The improved spatial resolution and radiometric precision of SSM/I have produced some promising results, and analysis based on surface observations has also made progress in that regard. Progress in thin ice identification should further benefit, though, by comparison with data from other types of sensors in combination with theoretical studies. On the basis of results presented here, it appears that it is precisely for thin ice that the combination of active and passive imagery has the greatest potential.

Additional sensors exist that should assist in improving the resolution of thin ice types. In the visible and near infrared, ice signatures respond to different physics than at microwave frequencies because the scattering inhomogeneities are much larger than the wavelength. Upwelling solar radiance above the ice is determined primarily by multiple scattering and the entire volume of the ice can contribute. Snow cover is very important and masks the underlying ice if it is thicker than a few centimeters. In the thermal infrared, sea ice and snow act as blackbodies and observations at these wavelengths would be useful for determining the changes in surface temperature associated with ice growth. Visible, infrared, and microwave signatures will not be completely independent, however, because the physical properties that determine them are linked. For example, brine pocket size and density, snow cover, surface roughness, and ice thickness change together as the ice ages.

14.7 FUTURE DIRECTIONS

Thin sea ice can occur in substantial quantities, and it appears that the technology to detect thin ice is becoming available. Ambiguities in interpretation, however, are still present. Current limitations stem from spatial resolution, instrument accuracy, and analysis techniques. Developments in these areas include regional cluster analysis, Kalman filtering analysis, principal component analysis, and image texture analysis (see Chapters 4, 12, and 23).

Understanding the small-scale structure and development of the surface layers with and without snow cover is important in interpreting the microwave signatures of thin ice, but many of the details remain to be investigated. The transition from the signature of young ice to that of FY ice is also poorly understood. These topics should have high priority in future investigations.

Development of new satellite microwave sensors with higher spatial resolution is needed. In addition to SAR, lightweight large-aperture interferometric radiometers are under development, and they can achieve surface resolutions near one kilometer. Lower-frequency microwave observations should also be explored because of their improved penetration depth, reduced sensitivity to volume scattering by inhomogeneities, and greater sensitivity to the brine content of the ice. The required technology is practical now, and we expect considerable progress in the near future.

REFERENCES

Anderson, D. L., Growth rate of sea ice, *Journal of Glaciology, 3*, pp. 1170–1172, 1961.

Cavalieri, D. J., Preliminary observations of polar sea ice with the special sensor microwave imager, *Proceedings of the IGARSS '88 Symposium*, Edinburgh, Scotland, ESA SP-284, European Space Agency, Noordwijk, Netherlands, August 1988.

Cavalieri, D. J., P. Gloersen, and T. T. Wilheit, Aircraft and satellite passive microwave observations of the Bering Sea ice cover during MIZEX West, *IEEE Transactions on Geoscience and Remote Sensing, GE-24(3)*, pp. 368–377, 1986.

Comiso, J. C. and H. J. Zwally, Concentration gradients and growth/decay characteristics of the seasonal sea ice cover, *Journal of Geophysical Research, 84(C5)*, pp. 8081–8103, 1984.

Comiso, J. C., T. C. Grenfell, D. L. Bell, M. A. Lange, and S. F. Ackley, Passive microwave in situ observations of winter Weddell sea ice, *Journal of Geophysical Research, 94(C8)*, pp. 10,891–10,905, 1989.

Drinkwater, M. R. and G. B. Crocker, Modelling changes in the dielectric and scattering properties of young snow-covered sea ice at GHz frequencies, *Journal of Glaciology, 34(118)*, pp. 274–282, 1988.

Drinkwater, M. R., J. P. Crawford, and D. J. Cavalieri, Multi-frequency, multi-polarization SAR and radiometer sea ice classification, *Proceedings of the IGARSS 91, vol. 1*, catalog number 91CH2971-0, pp. 107–111, IEEE, New York, 1991a.

Drinkwater, M. R., R. Kwok, D. P. Winebrenner, and E. Rignot, Multifrequency polarimetric Synthetic Aperture Radar observations of sea ice, *Journal of Geophysical Research, 96(C11)*, pp. 20,679–20,698, 1991b.

Eicken, H. and M. A. Lange, Development and properties of sea ice in the coastal regime of the southeastern Weddell Sea, *Journal of Geophysical Research, 94*, pp. 8193–8206, 1989.

Gascard, J. C., Deep convection and deep-water formation, *EOS, 71(49)*, pp. 1837–1839, December 1990.

Grenfell, T. C., Surface-based passive microwave observations of sea ice in the Bering and Greenland Seas, *IEEE Transactions on Geoscience and Remote Sensing, GE-24(3)*, pp. 378–382, 1986.

Grenfell, T. C. and J. C. Comiso, Multifrequency passive microwave observations of first-year sea ice grown in a tank, *IEEE Transactions on Geoscience and Remote Sensing, GE-24(6)*, pp. 826–831, 1986.

Lange, M. A. and H. Eicken, The sea ice thickness distribution in the Northwestern Weddell Sea, *Journal of Geophysical Research*, *96*, pp. 4821–4838, 1991.

Martin, S. and D. J. Cavalieri, Contributions of the Siberian shelf polynyas to the Arctic Ocean intermediate and deep water, *Journal of Geophysical Research, 94(C9)*, pp. 12,725–12,738, 1989.

Maykut, G. A., Energy exchange over young sea ice in the Central Arctic, *Journal of Geophysical Research*, *83*, pp. 3646–3658, 1978.

Maykut, G. A., Large-scale heat exchange and ice production in the Central Arctic, *Journal of Geophysical Research*, *87*, pp. 7971–7984, 1982.

Maykut, G. A., The surface heat and mass balance, *The Geophysics of Sea Ice*, edited by N. Untersteiner, pp. 395–464, NATO ASI Series B: Physics vol. 146, Plenum Press, New York, 1986.

Morison, J. H., M. G. McPhee, T. Curtin, and C. A. Paulson, The oceanography of leads, *Journal of Geophysical Research*, in press, 1991.

Nazirov, M, A. P. Pichugin, and Yu. G. Spiridonov, *Radar Observations of the Earth's Surface From Space (Radiolokatsiya Poverchnosti Zemli iz Kosmosa)*, edited by L. M. Mitnika and S. V. Viktorova, 200 pp., Gidrometeoizdat (Hydrometeorological Publishing Office), St. Petersburg, 1990.

Parkinson, C. L. and D. J. Cavalieri, Arctic sea ice 1973–1987: seasonal, regional, and interannual variability, *Journal of Geophysical Research*, *94*, pp. 14,499–14,523, 1989.

Parkinson, C. L., J. C. Comiso, H. J. Zwally, D. J. Cavalieri, P. Gloersen, and W. J. Campbell, *Arctic sea ice, 1973–1976: Satellite passive-microwave observations*, NASA SP-489, 296 pp., National Aeronautics and Space Administration, Washington, DC, 1987.

Rothrock, D. A., The mechanical behavior of pack ice, *Annual Review of Earth and Planetary Science, 3*, pp. 317–342, 1975.

Rothrock, D. A., Ice thickness distribution—measurement and theory, *The Geophysics of Sea Ice*, edited by N. Untersteiner, pp. 551–576, NATO ASI Series B: Physics vol. 146, Plenum Press, New York, 1986.

Smith, S. D., R. D. Munch, and C. H. Pease, Polynyas and leads: An overview of physical processes and environment, *Journal of Geophysical Research, 95(C6)*, pp. 9461–9479, 1990.

Thorndike, A. S., Kinematics of sea ice, *The Geophysics of Sea Ice*, edited by N. Untersteiner, pp. 489–550, NATO ASI Series B: Physics vol. 146, Plenum Press, New York, 1986.

Thorndike, A. S., D. A. Rothrock, G. A. Maykut, and R. Colony, The thickness distribution of sea ice, *Journal of Geophysical Research*, *80(33)*, pp. 4501–4513, 1975.

Tucker, W. B. III, T. C. Grenfell, R. G. Onstott, D. K. Perovich, A. J. Gow, R. A. Shuchman, and L. L. Sutherland, Microwave and physical properties of sea ice in the winter marginal ice zone, *Journal of Geophysical Research*, *96*, pp. 4573–4587, 1991.

Wadhams, P., M. A. Lange, and S. F. Ackley, The ice thickness distribution across the Atlantic sector of the Antarctic Ocean in midwinter, *Journal of Geophysical Research*, *92(C13)*, pp. 14,535–14,552, 1987.

Weeks, W. F., Sea ice conditions in the Arctic, *AIDJEX Bulletin No. 34*, pp. 173–205, December 1976.

Weeks, W. F. and S. F. Ackley, The growth, structure, and properties of sea ice, *The Geophysics of Sea Ice*, edited by N. Untersteiner, pp. 9–164, NATO ASI Series B: Physics vol. 146, Plenum Press, New York, 1986.

Weller, G., Spatial and temporal variations in the south polar surface energy balance, *Monthly Weather Review*, *108*, pp. 2007–2014, 1980.

Wensman, M. R., *Microwave emission from thin saline ice: field observations and implications for remote sensing*, Master's thesis, University of Washington, Seattle, 1991.

Wensnahan, M. R., T. C. Grenfell, D. P. Winebrenner, and G. A. Maykut, Observations and theoretical studies of microwave emission from thin saline ice, *Journal of Geophysical Research*, in press, 1992.

Wittmann, W. I. and J. Schule, Jr., Comments on the mass budget of Arctic pack ice, *Proceedings of the Symposium on the Arctic Heat Budget and Atmospheric Circulation*, edited by J. O. Fletcher, pp. 215–246, Memorandum RM-5233-NSF, Rand Corporation, Santa Monica, California, 1966.

Worby, A. P. and I. Allison, Ocean–atmosphere energy exchange over thin Antarctic pack-ice of variable concentration, *Annals of Glaciology*, *15*, in press, 1991.

Zwally, H. J., J. C. Comiso, C. L. Parkinson, W. J. Campbell, F. D. Carsey, and P. Gloersen, *Antarctic sea ice, 1973–1976: Satellite passive-microwave observations*, NASA SP-459, 206 pp., National Aeronautics and Space Administration, Washington, DC, 1983.

Chapter 15. Microwave Remote Sensing of Polynyas

SEELYE MARTIN

School of Oceanography, University of Washington, Seattle, Washington 98195

KONRAD STEFFEN

Cooperative Institute for Research in Environmental Sciences, University of Colorado, Boulder, Colorado 80309-0449

JOSEFINO COMISO AND DONALD CAVALIERI

Laboratory for Hydrospheric Processes, Goddard Space Flight Center, Greenbelt, Maryland 20771

MARK R. DRINKWATER AND BENJAMIN HOLT

Jet Propulsion Laboratory, California Institute of Technology, 4800 Oak Grove Drive, Pasadena, California 91109

15.1 INTRODUCTION

A polynya is a large region of open water and thin ice that occurs within much thicker pack ice. The World Meteorological Organization [1970] states that a polynya consists of open water and an associated area of thin ice with thicknesses up to 0.3 m. Whereas a lead is a long linear feature, a polynya has a rectangular or oval aspect ratio and is surrounded by large floes and thick ice. Although polynyas occur in both winter and summer, this chapter is restricted to the winter case. The region of open water or reduced ice concentration that makes up the polynya persists for periods of several days, so that the polynya is the site of large exchanges of heat between the atmosphere and ocean. The systematic study of polynyas began with the advent of passive microwave satellites; in particular, the observation from the Electrically Scanning Microwave Radiometer (ESMR) of the large Weddell Sea polynya was one of the ESMR's most intriguing discoveries. As this chapter shows, a major advantage of the passive microwave is that the observations provide frequent data on ice concentrations for polynya regions.

15.1.1 Types of Polynyas

Smith et al. [1990] summarized polynya processes and described two basic types of polynyas. In the first type, called a coastal polynya by Gordon and Comiso [1988], the ice is advected away from an adjacent coast by winds or currents, so that new ice forms continuously within the polynya. In the second type, which generally occurs away from the coasts and is called an open-ocean polynya by Gordon and Comiso, the upwelling of warm ocean water

prevents ice formation within the polynya, so that the atmospheric cooling is supplied by the oceanic cooling of the water column. Figures 15-1 and 15-2, which show the approximate size and location of the Northern and Southern Hemisphere winter polynyas, show both of these types. Examples of coastal polynyas include the St. Lawrence polynya in the Bering Sea, the polynyas around Novaya Zemlya in the Barents Sea, and the coastal polynyas around Antarctica. Because the open water areas of these polynyas are highly dependent on local wind speed, their ice-free areas vary greatly with time. Similarly, examples of open ocean polynyas include the Kashevarova Bank polynya in the Sea of Okhotsk, and the Weddell, Cosmonaut, and Maud Rise polynyas in the Antarctic Ocean. These polynyas are more persistent than the coastal polynyas. Finally, some polynyas are of a mixed type, the best example of which is the North Water in Baffin Bay.

15.1.2 Ice Formation in Polynyas

The strong winds that create and maintain a coastal polynya's open water also generate wind waves, which lead to the formation of frazil and pancake ice (Chapter 14). Frazil ice forms when water at its freezing point is agitated by wind waves and is subjected to a heat loss. Under these conditions, sea ice forms throughout the water column as small crystals (frazil ice), which rise to the surface to form a slurry [Martin, 1981]. The slurry, which grows to 10 to 30 cm thick, is an efficient damper of short waves, those with a wave number on the order of 1/frazil thickness. As the slurry increases in thickness, ice floes 0.3 to 1 m in diameter and 5 to 10 cm thick form on the surface. As time proceeds, these cakes of pancake ice oscillate in the long wave field, raft over one another, and grow rims 10 to 50 mm high. Finally, the wind blowing over the polynya generates a Langmuir circulation, which herds the frazil and pancake ice into streaks parallel to the wind. These streaks yield a

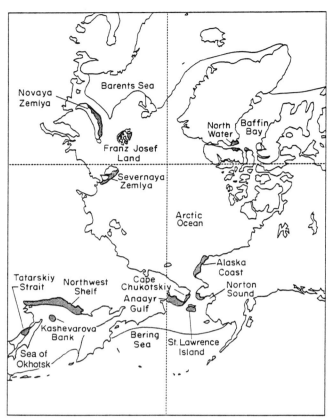

Fig. 15-1. The location of the Arctic polynyas. The approximate polynya locations are shaded; the fine line shows the winter ice edge.

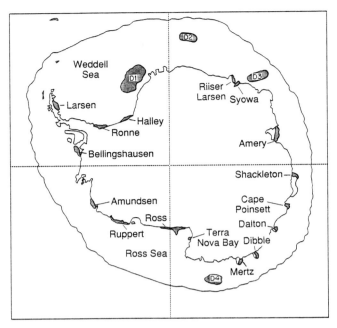

Fig. 15-2. The location of the Antarctic coastal and open-ocean polynyas. The approximate polynya locations are shaded; the fine line shows the winter ice edge. The deep ocean polynyas are: D1, Weddell Sea; D2, Maud Rise; D3, Cosmonaut Sea; and D4, Balleny Islands.

unique signature on aircraft active and passive microwave images (for example, see Figure 4-8).

A case study of the St. Lawrence polynya demonstrates how its signature changes as a function of instrument, frequency, and resolution. On March 22, 1988, at 0340 UT, the NASA DC-8 equipped with SAR overflew the west side of St. Lawrence Island. For the same day, Figure 15-3 shows the surface pressure, a visible satellite image, and a passive microwave image. First, Figure 15-3(a) shows the surface pressure chart at 00 UT, where St. Lawrence Island is shaded, and the line marked with the arrowheads shows the aircraft flight path. The geostrophic winds are parallel to the isobars, and the surface winds are to their right. The air temperature was −16°C, the wind velocity was 14 m/s out of 030°, and the open water heat loss was about 500 W/m^2. Second, Figure 15-3(b) shows an Operational Line-Scan System (OLS) image, where ice and clouds are white and regions of open water or thin ice are dark. The image shows that polynyas occurred in Norton Sound, south of St. Lawrence Island, in the Gulf of Anadyr, and south of Cape Navarin. Third, Figure 15-3(c) shows the ice concentrations derived from the NASA team algorithm and Special Sensor Microwave/Imager (SSM/I) data. The SSM/I ice concentrations generally agree with the visible image.

For the St. Lawrence polynya, aircraft SAR imagery provides a more detailed look at ice behavior. Figure 15-4 shows the aircraft polarimetric SAR data and a handheld visible photograph of the polynya region. In the visible image, open water is blue and ice and clouds are white. The photograph shows that the Langmuir circulation organizes the ice into streaks that lie approximately parallel to the surface wind. The other images show the SAR data at three frequencies, corresponding to 0.44 GHz (P-band), 1.25 GHz (L-band), and 5.3 GHz (C-band), with respective wavelengths of 0.68, 0.24, and 0.056 m [Drinkwater et al., 1991]. Examination of the radar images shows that as the radar wavelength increases, the backscatter from the ice bands changes. Comparison with the handheld photograph shows that at C-band, the streaks of ice are bright and the surrounding open water is dark. At P-band, the situation is reversed; the water is bright and the ice is dark. At L-band, the ice and water have intermediate returns. The reason for the frequency-dependence of the ice and water signatures is that the radar return from the surface is due primarily to Bragg scattering. For the ice streaks containing both frazil and pancake ice, since frazil ice damps out the short water waves, the radar return is due to scattering from the raised pancake edges. The ice streaks that are bright at C-band become darker at P-band because the pancake edges are small compared to the longer radar wavelength, which reduces the rough surface backscatter. The ice streaks that remain dark at all three frequencies are probably composed entirely of frazil ice, which damps all of the Bragg frequen-

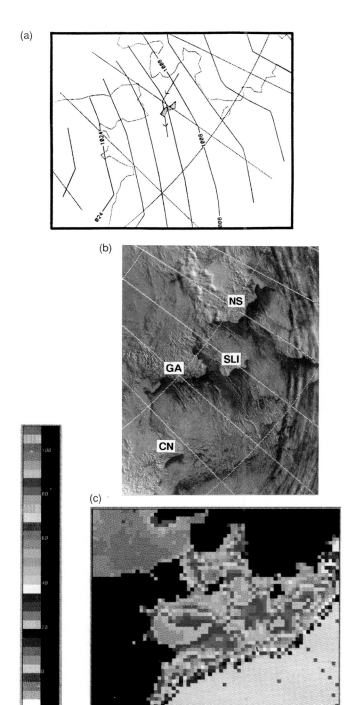

Fig. 15-3. The Bering Sea for March 22, 1988: (a) the surface pressure map, where the numbers on the isobars give pressure in millibars (the map is derived from gridded data provided by the National Meteorological Center); (b) an OLS visible image (0.4 to 1.1 μm) with a 0.56-km resolution from the DMSP satellite, where NS = Norton Sound, SLI = St. Lawrence Island, GA = Gulf of Anadyr, and CN = Cape Navarin (courtesy of K. Steffen); and (c) the ice concentration derived with the NASA team algorithm from SSM/I data, where the color bar to the left shows the ice concentrations (courtesy of D. Cavalieri).

cies. Finally, the large floes in the lower left corner are smooth and therefore reflect most of the radar energy, appearing dark at all frequencies. (For further background on thin sea ice and polarimetric SAR and thin sea ice, see Chapters 14 and 25.)

15.2 MICROWAVE DETECTION AND MAPPING OF POLYNYAS: LIMITATIONS AND POTENTIAL

Satellites used in the study of polynyas include Landsat, the Advanced Very High Resolution Radiometer on the National Oceanic and Atmospheric Administration satellites, the OLS on the Defense Meteorological Satellite Program (DMSP) satellite, and the various scanning microwave instruments (ESMR; the Scanning Multichannel Microwave Radiometer, or SMMR; and SSM/I). The advantage of visible/infrared sensors is their finer resolution (40 m to 1 km) and that their infrared channels also have the possibility of yielding open water surface temperature. The disadvantage of these sensors is that clouds frequently obscure the view of the polynya, so that it is difficult to obtain time series of the polynyas opening and closing. The advantage of the passive microwave instruments is that, in spite of their low resolution, they provide daily or every-other-day coverage of the polynyas.

Chapter 14 discusses the general accuracy of microwave thin ice retrievals. For the specific case of coastal polynyas, the low resolution and sidelobes of the passive microwave instruments mean that some of the open water or thin ice pixels contain a fraction of land, fast ice or glacier ice, which must be accounted for in the applicable algorithm. For the North Water, Steffen and Maslanik [1988] derive the magnitude of the errors in the open water estimates introduced by the surrounding land and ice.

15.3 THE OCEANOGRAPHIC AND CLIMATE ROLE OF POLYNYAS

For a winter coastal polynya located in shallow water where the underlying water column is at the freezing point, ice growth within the polynya generates a large salt flux, which increases the water column salinity. This saline water mass then moves offshore to enrich the adjacent oceanic water masses. At the same time, the desalinated ice generated within the polynya is exported to the surrounding pack. Thus, the combination of ice growth within the polynya and ice export divides the seawater into a salty layer of rejected brine and a freshened mass of sea ice.

Given a polynya and its meteorological forcing, there are at least two approaches to calculating heat loss. First, Martin and Cavalieri [1989] assumed that all of the heat loss within the polynya takes place through the open water. They then used either local weather station or National Meteorological Center (NMC) gridded data to calculate the heat flux and the associated ice production. Their results showed that the ice production strongly depends on the sensible heat flux, which is proportional to the product of

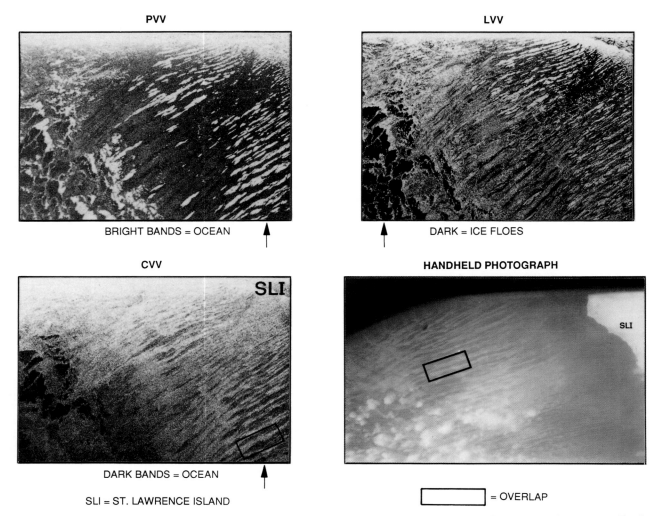

Fig. 15-4. Images of the ice in the St. Lawrence Island polynya taken from the NASA CV-580 overflight on March 22, 1988. The first three images are SAR images, where P, L, and C refer to the radar bands, and VV refers to the vertical polarization. Each SAR image measures about 6 by 8 km. The long axes of these images are parallel to the flight path; the image white-out at the top of each image is due to nadir saturation of the radar. The last image is a handheld photograph of the polynya. The box outlined in black on the handheld photograph shows the same region of ice as the box in the CVV radar image. (Courtesy B. Holt and M. Drinkwater.)

the wind speed and temperature depression below the freezing point of seawater. Second, Steffen [1991], in his investigation of the North Water, used the 18- or 37-GHz polarization ratio to determine the ice type of each pixel. He assumed that the heat flux depended only on the ambient air temperature and the ice thickness, then used the air temperature and ice type information to calculate a regional heat budget. The storms and strong wind events entered only indirectly in his model, to mix and alter the nature of the ice cover.

15.3.1 Northern Hemisphere

The Northern Hemisphere polynyas occur in the Arctic Basin, the Bering and Japan Seas, the Sea of Okhotsk, and

in Baffin Bay. Except for the Kashevarova polynya in the Sea of Okhotsk, these are all coastal polynyas and, with the possible exception of the North Water, they each contribute to the adjacent oceanic water masses.

15.3.1.1. Arctic Basin. Figure 15-1 shows the polynyas that contribute to the Arctic Basin. These polynyas are located in the Barents, Kara, and Laptev Seas in the Eastern Arctic, and the Bering, Chukchi, and Beaufort Seas in the Western Arctic [Martin and Cavalieri, 1989; Cavalieri, D. and S. Martin, Contributions to Arctic intermediate water production by Alaskan, Siberian, and Canadian coastal polynyas, submitted to the *Journal of Geophysical Research*, 1992]. Because these polynyas must export ice to exist, they tend to occur in regions adjacent to open water.

Even though the Bering Sea is in the North Pacific, the dense water generated within the Bering Sea polynyas flows north through the Bering Strait into the Arctic Basin. These coastal polynyas contribute a substantial portion of a cold saline water mass within the Arctic Ocean called the cold halocline layer, which occurs in the top 200 m. Martin and Cavalieri [1989] and Cavalieri and Martin [ibid.] summarized the chemical and physical evidence that this water is generated on the shelves, and showed that these polynyas provide much of this water.

15.3.1.2. North Water. The North Water is located in northern Baffin Bay between Greenland to the east and Ellesmere and Devon Island to the west (Figure 15-1). The North Water, which covers an area of approximately 10^5 km^2, is bounded to the north by a fast-ice arch at the narrowest point between Greenland and Canada, and consists of an area of loose pack ice, bounded by the three polynyas of Smith Sound, Lady Ann Strait, and Lancaster Sound/Barrow Strait [Steffen, 1985].

Observations indicate that the North Water polynya is maintained by a combination of wind-driven ice divergence and an upwelling of warm water. As evidence of upwelling, Steffen [1985] carried out a series of overflights during the winter and spring of 1979 and 1981 that observed regions of open water with sea surface temperatures above the freezing point of sea water. These warm water areas are caused by coastal upwelling induced by the strong northerly winds in Smith Sound. The other effect of the strong winds is to mix and alter the ice types. Figure 15-5 shows a time series of the ice type variability, derived from SMMR data, for selected winters from 1978 to 1985 within North Water.

15.3.1.3. Sea of Okhotsk. In the Sea of Okhotsk, the dominant northerly winds during winter generate polynyas along the northern coast. The brine generated within these polynyas contributes to the regional oceanography. Studies of the Sea of Okhotsk show that, between depths of 150 to 800 m, there is a layer of cold, low-salinity intermediate water, which is colder, less saline, and more oxygenated than water of the same density outside the Okhotsk in the Pacific. Alfultis and Martin [1987] showed that this water is generated within the northern polynyas at a rate of about 1 to 2 Sverdrups (10^6 m^3/s), which yields a renewal time for

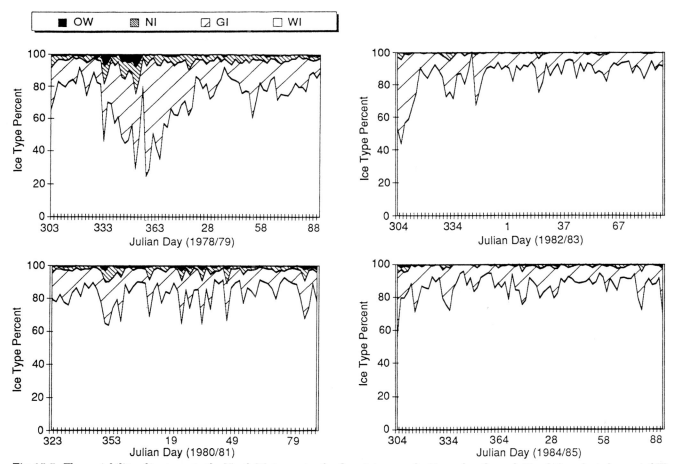

Fig. 15-5. The variability of ice types in the North Water region for the winter months November through March for selected years. OW is open water, NI is nilas, GI is gray ice, and WI is white ice (adapted from Steffen [1991]).

this water of 10 to 40 years. The Sea of Okhotsk is also the location of the centrally located Kashevarova Bank polynya, which occurs over the shallow (less than 200 m depth) Kashevarova Bank [Alfultis and Martin, 1987].

15.3.1.4. Japan Sea.
The Tatarskiy Strait region of the Japan Sea is a small region of vigorous ice growth. The Strait is a seasonally ice-covered embayment of the Japan Sea, with an area of 4×10^5 km^2, and a minimum sill depth of about 4 m at the northern constriction. Because the Japan Sea is a 3500-m deep basin with passes shallower than 200 m connecting it to the world ocean, all of the water at depths greater than 200 m must form within the basin, either from evaporation along the Siberian coast or by freezing in the northern part of the sea. For the Tatarskiy Strait during winter, about 0.8 to 1×10^5 km^2 of ice form each year. Within this ice cover, Martin et al. [Martin, S., E. A. Munoz, and R. Drucker, The Effect of Severe Storms on the Ice Cover of the Northern Tatarskiy Strait, 1992] showed that an episodic polynya forms that is driven by the infrequent, very strong winds associated with low pressure systems moving through the region.

15.3.2 Southern Hemisphere

The Antarctic Ocean is the site of at least four open-ocean polynyas and of many coastal polynyas (Figure 15-2). Martinson [1991] and Gordon [1991], who reviewed the contributions from polynyas to the regional oceanography, show that coastal and open-ocean polynyas contribute in different ways to the properties of the Antarctic bottom water. For coastal polynyas, Gordon described how the ice production in these polynyas contributes to the generation of the cold salty shelf water that flows down the continental slope and mixes with the deeper water to form the bottom water. For open-ocean polynyas, both Gordon and Martinson cited evidence that shows that the heat loss from the open-ocean polynyas leads to a deep convection that cools the Antarctic bottom water.

15.3.2.1. Coastal polynyas.
Zwally et al. [1985] and Cavalieri and Martin [1985] surveyed the location of coastal polynyas, and estimated the ice production within these polynyas and the impact of the salt flux on the underlying ocean. Because coastal polynyas are located over continental shelves, the salt rejection from the polynya ice growth yields a salinity increase within the shelf water. This cold saline shelf water then mixes with the adjacent water masses, or flows down the continental slope to contribute to the bottom water.

For 1974 observations, Zwally et al. [1985] used ESMR data to estimate the variability and the oceanic impacts of 16 coastal polynyas at different locations around Antarctica (Figure 15-2). They estimated the total ice production for 1974 at 4.6 to 7.8 m of sea ice over the entire Antarctic continental shelf—these numbers were several times the observed ice thicknesses. However, they found that the predicted salination of the shelf region by this ice growth

was much greater than expected from oxygen isotope measurements. This suggests that more accurate measurements of polynya sizes and their heat loss will be necessary. Cavalieri and Martin [1985] used the linear 37V algorithm to derive the areas of the polynyas that form along the coast from the Mertz Ice Tongue to the Shackleton Ice Shelf. They calculated that the polynyas grew about 10 m of ice per season per unit area, as compared to a 1-m growth for shorefast ice. The authors found a good geographic correlation between elevated water salinities on the shelf and the salt production in the polynyas, but a weak correlation between polynya variability and wind speed.

15.3.2.2. Open-ocean polynyas.
The Weddell polynya, the largest polynya discovered, occurred during the austral winters 1974, 1975, and 1976 [Zwally and Gloersen, 1977; Carsey, 1980; Martinson et al., 1981; Gordon and Comiso, 1988]. Figure 15-6, an every-other-month sequence of one month averages of the growth and decay of the Antarctic sea ice for February to December 1974, shows the formation and decay of this polynya. The figure shows that as the ice cover spread over the Weddell Sea, a large polynya formed inside of the ice cover during June and July. This polynya persisted until November, after which the entire ice cover melted. The oceanographic evidence suggested that the open-water area of the polynya was maintained by the upwelling of warm water. Gordon [1991] described the results of a 1979 cruise to the polynya region following the last appearance of the polynya. At the polynya location, he observed a large region of cold low salinity water that extended 5 km to the bottom, and the remnants of an oceanic convective chimney. This observation inspired several theoretical investigations summarized in Lemke et al. [1990].

Although many ship-based observations have been conducted in the Weddell Sea during the past few years, and the oceanography of the region is much better understood now than during the occurrence of the Weddell polynya, its formation and persistence from 1974 to 1976 is still largely a mystery. Part of the reason for this mystery is that since 1976, no polynya with the size and persistence of the Weddell polynya has occurred. What has been observed in the Weddell Sea from SMMR and SSM/I are reductions in ice concentrations over the same general area during spring [Comiso and Gordon, 1987]. These reductions are shown in Figure 15-2 as the Maud Rise polynya. Recent observations show that this polynya is associated with the upwelling that occurs over the Maud Rise, where the mixed layer is shallow and the contrast in temperature and salinity is slight [Gordon and Huber, 1990].

15.3.2.3. Other Antarctic polynyas.
A smaller open-ocean polynya that occurs annually is the Cosmonaut polynya, located about 45° E and 65° S [Comiso and Gordon, 1987]. Although this polynya, which occurs over the Cosmonaut rise, has an area that can be as large as 10^5 km^2, it does not persist for more than a month. The sequence of six images in Figure 15-7 shows several images that illustrate the

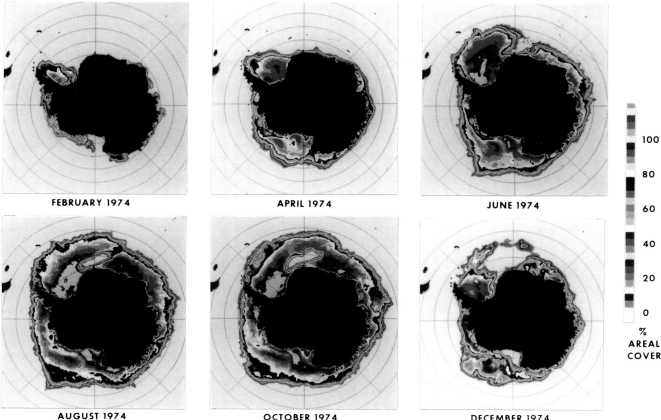

FEBRUARY 1974   APRIL 1974   JUNE 1974

AUGUST 1974   OCTOBER 1974   DECEMBER 1974

Fig. 15-6. Monthly averages of the Antarctic ice concentration for every other month during February through December 1974 derived from the Nimbus-5 ESMR. The color bar to the right gives the percent concentration (courtesy of F. Carsey).

opening of the Cosmonaut polynya during 1988, and demonstrates the episodic nature of its opening and closing; Comiso and Gordon [1987] show a similar time series from 1980. In Figure 15-7, the July 26 image shows the formation of a low concentration embayment over the Cosmonaut rise. This low concentration region closed, then reopened on August 6 and remained open through August 10. For this opening, the August 8 image shows that the polynya was large and well defined, with ice concentrations as low as 30%. Following August 10, the polynya remained closed until September 26, when, as the September 28 image shows, a well-defined low concentration region again occurred, which remained open until October 4. For the rest of the season, the polynya remained closed, with the formation of less well-defined low ice concentration regions on October 26, and again during the spring breakup on November 12. Comiso and Gordon [1987] showed from a simple model that the inability of this polynya to survive may have to do with its size, where the polynya and its accompanying convection are terminated by an invasion of sea ice from the sides of the polynya.

Other polynyas are not so easy to classify and identify. For example, Jacobs and Comiso [1989] observed two polynyas near the continental shelf-break in the Ross Sea. Both polynyas were more obvious in autumn and spring, but occurred as areas of reduced concentration in winter as well. Because they were adjacent to the continental slope region where warm water upwelling may have occurred, they also appeared to be topographically controlled; however, the influence of wind forcing from the continent could not be discounted.

15.4 FUTURE DIRECTIONS

The above discussion shows in both polar oceans that polynyas contribute to the regional oceanography, and in the Southern Hemisphere to the globally important Antarctic bottom water. The discussion also demonstrates the episodic nature of the opening and closing of the coastal polynyas. Observations from the Arctic Ocean, the North Water, Tatarskiy Strait, and the Ross Sea show that these polynyas are highly variable and episodic in extent, and only open in response to strong winds, which are frequently accompanied by low temperatures. The dependence of the ice production within the polynyas on wind speed means that accurate weather data will be as important to future polynya studies as the improvement in ice concentration estimates associated with the forthcoming SAR's. With the

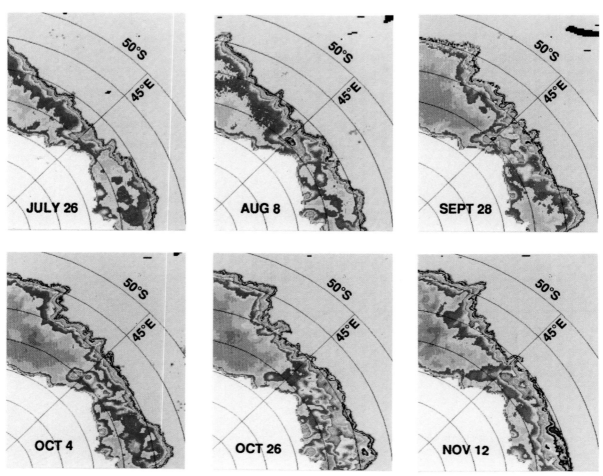

Fig. 15-7. Images of the opening and closing of the Cosmonaut Sea polynya derived from SSM/I data from July 26 through November 12, 1988. (Courtesy of J. Comiso.)

possible exception of the Tatarskiy Strait and the Sea of Okhotsk coast, each of the coastal polynyas is located in a region of sparse weather stations. Any future study would require a combination of active and passive microwave to delineate the surface ice types, and local meteorological observations or surface sounder data to determine the surface pressure, temperature, and winds.

Acknowledgments. The authors gratefully acknowledge the support of the Oceanic Processes Branch at NASA Headquarters and of the Office of Naval Research under contract NOOO14-90-J-1061. We thank Ms. Esther Munoz for her help with the NMC data and for producing the surface pressure chart shown in Figure 15-3(a). This is contribution 1934 of the School of Oceanography, University of Washington.

REFERENCES

Alfultis, M. A. and S. Martin, Satellite passive microwave studies of the Sea of Okhotsk ice cover and its relation to oceanic processes, 1978–1982, *Journal of Geophysical Research, 92,* pp. 13,013–13,028, 1987.

Carsey, F. D., Microwave observations of the Weddell polynya, *Monthly Weather Review, 108,* pp. 2032–2044, 1980.

Cavalieri, D. and S. Martin, A passive microwave study of polynyas along the Antarctic Wilkes Land coast, *Antarctic Research Series 43,* pp. 227–252, American Geophysical Union, Washington, DC, 1985.

Comiso, J. C. and A. L. Gordon, Recurring polynyas over the Cosmonaut Sea and the Maud Rise, *Journal of Geophysical Research, 92(C3),* pp. 2819–2833, 1987.

Drinkwater, M. R., R. Kwok, D. Winebrenner, and E. Rignot, Multifrequency polarimetric Synthetic Aperture Radar observations of sea ice, *Journal of Geophysical Research, 96(C11),* pp. 20,679–20,698, 1991.

Gordon, A. L., Two stable modes of southern ocean winter stratification, *Deep Convection and Deep Water Formation in the Oceans*, edited by P. C. Chu and J. C. Gascard, pp. 17–35, Elsevier Science Publishers, New York, 1991.

Gordon, A. L. and J. C. Comiso, Polynyas in the southern ocean, *Scientific American, 256(6)*, pp. 90–97, 1988.

Gordon, A. L. and B. A. Huber, Southern ocean winter mixed layer, *Journal of Geophysical Research, 95*, pp. 11,655–11,672, 1990.

Jacobs, S. S. and J. C. Comiso, Sea ice and oceanic processes on the Ross Sea continental shelf, *Journal of Geophysical Research, 94*, pp. 18,195–18,211, 1989.

Lemke, P., W. B. Owens, and W. D. Hibler III, A coupled sea ice-mixed layer pycnocline model for the Weddell Sea, *Journal of Geophysical Research, 95*, pp. 9513–9525, 1990.

Martin, S., Frazil ice in rivers and oceans, *Annual Review of Fluid Mechanics, 13*, pp. 379–381, 1981.

Martin, S. and D. Cavalieri, Contributions of the Siberian shelf polynyas to the Arctic Ocean intermediate and deep water, *Journal of Geophysical Research, 94*, pp. 12,725–12,738, 1989.

Martin, S., E. Munoz, and R. Drucker, The Effect of Severe Storms on the Ice Cover of the Northern Tatarskiy Strait, *Journal of Geophysical Research*, in press, 1992.

Martinson, D. G., Open ocean convection in the southern ocean, in *Deep Convection and Deep Water Formation in the Oceans*, edited by P. C. Chu and J. C. Gascard, Elsevier Science Publishers, New York, pp. 37–52, 1991.

Martinson, D. G., P. D. Killworth, and A. L. Gordon, A convective model for the Weddell polynya, *Journal of Physical Oceanography, 11*, pp. 466–488, 1981.

Smith, S. D., R. D. Muench, and C. H. Pease, Polynyas and leads: An overview of physical processes and environment, *Journal of Geophysical Research, 95(C6)*, pp. 9461–9479, 1990.

Steffen, K., Warm water cells in the North Water, northern Baffin Bay during winter, *Journal of Geophysical Research, 90(5)*, pp. 9129–9136, 1985.

Steffen, K., Energy flux density estimation over sea ice based on satellite passive microwave measurements, *Annals of Glaciology, 15*, pp. 178–183, 1991.

Steffen, K. and J. A. Maslanik, Comparison of Nimbus 7 Scanning Multichannel Microwave Radiometer radiance and derived sea ice concentrations with Landsat imagery for the north water area of Baffin Bay, *Journal of Geophysical Research, 93(C9)*, pp. 10,769–10,781, 1988.

World Meteorological Organization, *WMO Sea-Ice Nomenclature*, WMO Report 259, 155 pp., Secretariat of the World Meteorological Organization, Geneva, Switzerland, 1970.

Zwally, H. J. and P. Gloersen, Passive microwave images of the polar regions and research applications, *Polar Record, 18*, pp. 431–450, 1977.

Zwally, H. J., J. C. Comiso, and A. L. Gordon, Antarctic offshore leads and polynyas and oceanographic effects, *Antarctic Research Series 43*, pp. 203–226, American Geophysical Union, Washington, DC, 1985.

Chapter 16. Characterization of Snow on Floating Ice and Case Studies of Brightness Temperature Changes During the Onset of Melt

CAREN GARRITY

Alfred-Wegener-Institut für Polar- und Meeresforschung, Postfach 120161, Columbusstrasse, D-2850 Bremerhaven, Germany
(On secondment for one year by the Microwave Group-Ottawa River, Canada.)

Significant changes in the physical properties of snow on sea ice occur during the onset of melt in the Arctic and Antarctic. These changes have been quantified for the Arctic (Greenland and Barents Seas) and Antarctic (Weddell Sea) based on 318 snow pit measurements. Using a surface-based dual-polarized 37-GHz radiometer and the Special Sensor Microwave/Imager (SSM/I), changes in brightness temperature of sea ice have been observed for increases in snow depth, wetness, stratification, and slush at the snow–ice interface. A representative model for the passive microwave response to snow metamorphism shows changes in brightness temperature due to increases in snow wetness.

16.1 BACKGROUND

Ground-based passive microwave measurements, made in conjunction with electrical and physical property measurements of snow, have shown that microwave effects of snow properties must be understood before algorithms are fully developed for determining ice concentration by type in the microwave region from satellites. Many studies have been made of the snow layer over land where a correlation of snow measurements to passive microwave measurements was completed [Künzi et al., 1982; NORSEX Group, 1983; Mätzler et al., 1984; Chang et al., 1985]. However, it was not until the 1980's that correlations of brightness temperature (T_B) with snow properties were made for snow over sea ice.

Mätzler et al. [1982] found that for a snow wetness greater than 1% there was an effect on the emissivity from sea ice. The T_B's from different ice types with a wet snow cover are needed before algorithms can be generalized in space and time. Dry snow structure, such as ice lenses, could cause a decrease in the horizontal T_B for the 5- to 35-GHz range. Radiometric measurements made at 94 GHz were unaffected by layers throughout the snow cover due to skin depth.

During the late spring and early summer of 1982, near Mould Bay, Beaufort Sea, ground-based passive microwave measurements at 10, 18.7, and 37 GHz were made to further the understanding of changes in T_B due to a snow cover, as well as of melt puddle formation [Grenfell and Lohanick, 1985]. It was found that the T_B for multiyear (MY) ice was higher than that for first-year (FY) ice, which had also been observed in the Greenland Sea in 1987, during the evening hours [Garrity, 1991]. This is the opposite of winter, when the snow is dry. They also observed that by late summer, MY and FY ice had similar T_B's, which also occurred in the Greenland Sea (1990) and Arctic Ocean (1991) [Garrity, 1991; Garrity et al., 1992]. When the snow was dry, the FY T_B was independent of frequency and approximately equal to the physical temperature of the ice surface. The MY T_B decreased with frequency. During the summer, T_B increased with frequency for both FY and MY ice. Daily changes in T_B were observed due to snow wetness and structure when cold nights followed warm days [Garrity, 1991].

Comiso [1985] stated that more reliable snow cover interpretation was achieved by using two polarizations of the same frequency, rather than two frequencies with the same polarization. From the polarization data, he found that the T_B for dry snow decreased due to volume scattering in the snow cover. He also found that for a snow cover with less than 5% free water content by volume, but greater than 0% wetness, the emissivity was higher and the MY T_B was similar to the FY T_B. During freeze–thaw cycles, the decrease in T_B was attributed to volume scattering in the dry snow cover.

Onstott et al. [1987] presented representative snow models for different seasons from winter to late summer. These models showed changes in free water content in snow covers for different seasons, but were not quantitative. Descriptive snow models for early spring to summer have been quantified by Garrity [1991]. Onstott et al. [1987] stated that a decrease in emissivity in the frequency range from 5 to 94 GHz was due to the upper snow layer that had metamorphosed, causing an increase in grain size. The larger grain size enhanced volume scattering. By the second half of the summer, the freeze–thaw cycles and melt puddles caused a further decrease in emissivity, with large standard deviations. There were large interday fluctuations in emissivity, which led to the realization that the snow–ice interface and snow cover are extremely important to the measured emissivity from sea ice in the spring, summer, and fall.

16.2 INTRODUCTION

To understand the T_B of sea ice, the overlying snow layer must be studied. This chapter discusses snow as a distinct

Microwave Remote Sensing of Sea Ice
Geophysical Monograph 68

material and changes in passive microwave signatures due to floating ice. Ice types can be identified based on their T_B when the snow cover is dry. However, from the spring to fall, the snow cover undergoes extensive metamorphism, causing changes in T_B, such as a fresh snowfall and a snow cover that has free water present and is stratified. A sharp increase in T_B will occur for a small increase in free water in the snow.

Snow can be classified by its physical properties. One classification scheme developed by a research group at the University of Bern [Schanda et al., 1983] is applicable to physical properties of snow observed on sea ice in the Arctic and Antarctic. Spring snow in this scheme has two subgroups: wet spring snow and dry spring snow. Wet spring snow is defined as a "snow pack surface of thick (at least several centimetres deep), firm layers of wet, quasispherical ice crystals (1 to 3 mm in diameter) formed during the day at temperatures above the freezing point and usually associated with either the passage of warm fronts or sunny, clear-sky conditions." Dry or refrozen spring snow has a surface snow pack "of a layer of refrozen, firm snow that forms during clear, cold nights and is several centimetres thick." This chapter goes a step further by presenting descriptive snow models for early, mid-, and late spring in the Arctic and Weddell Sea based on different ice regimes. Such descriptive snow models are needed to understand the interaction of microwave radiation with the surface of sea ice.

The results presented here cover a wide spectrum of the effects of snow on T_B measurements. Table 16-1 lists the study areas along with their major contributions to this chapter. Since the results from these experiments are dependent on uncontrollable environmental conditions and are subject to variations due to geographical location, many data sets were required to establish the general behaviour of snow metamorphism on floating ice.

The onset of melt is defined here as the spring period during which the snow cover experiences changes in free water content throughout its depth due to redistribution of the water. The average free water content of the snow cover is between 0% and 3% for this period. This average wetness does not include the slush sometimes present at the snow–ice interface.

Simple snow conditions are illustrated by case studies using ship-based and satellite (SSM/I) T_B's. A representative passive microwave snow metamorphism model summarizes and explains the T_B behaviour for a change in the snow cover from the onset of melt to advanced (or later) melt periods.

16.3 METHOD OF MEASUREMENTS

16.3.1 Snow Measurements

A snow pit was dug into the snow until the ice surface was reached. The snow pit was kept as small as possible to reduce heating from solar radiation on the exposed walls of the pit. Layers in the snow were identified, and changes in the hardness of the snow were observed qualitatively. Once all layers had been noted, quantitative snow property measurements were made for each layer, as described below.

The amount of free water in a snow cover will influence the dielectric properties and thus the apparent microwave emission from an underlying surface such as sea ice. The instrument used to measure snow wetness was a resometer [Denoth et al., 1984], which measured the dielectric characteristics of a material in a volume determined by the sensor element. Nondestructive measurements can be made of the snow permittivity and hence of its liquid water content. The sample size is limited to a layer of about 5 mm in thickness.

TABLE 16-1. Experiments discussed.

Experiment	Year start–end	Main contribution
St. Lawrence River	January–March 1986	Snow wetness (<1%) and T_B
Gulf of St. Lawrence	February 1987–February 1988	Dry snow and T_B
Greenland Sea	May–June 1987	Snow characteristic and T_B
St. Lawrence River	January–March 1988	Slush and T_B
Gulf of Bothnia	March 1988–March 1989	Snow wetness and structure
St. Lawrence River	December 1989–April 1990	Snow wetness (>1%) and T_B
Barents Sea	April 1989–April 1990	Snow characteristic, snow wetness, and satellite T_B
Greenland Sea	May 1989–May 1990	Snow characteristic, snow wetness, and T_B
Weddell Sea	September–October 1989	Snow characteristic, snow wetness, and satellite T_B
Barents Sea	April 1990–April 1991	Snow characteristic, snow wetness, and satellite T_B
Resolute Passage	May–June 1990	Diurnal changes in snow

The free water content in the snow can be calculated as a percent by volume using an empirical relationship developed by Ambach and Denoth [1980]. Care was taken to obtain reasonable averages of snow wetness, since variability in wetness depended on surface topography and structure of the snow cover. The average of all snow wetness measurements from the surface and throughout the vertical snow profile was taken as a representative percent wetness of the snow cover. Density was determined from the mass of a carefully trapped 100 cm^3 of snow.

To measure the snow grain size, individual grains were gently removed from the different snow layers and placed onto millimetre graph paper, where their sizes and shapes were recorded. The temperature of each layer was measured using a termistor probe inserted 5 cm into each snow layer. The air temperature was recorded in the shade about 10 cm above the snow surface. At least one snow pit per site was completed. The underlying ice type was also recorded based on surface characteristics and freeboard.

The error of a snow wetness measurement was mainly due to error in the density measurements. The error for density measurements was ±0.05 g/cm^3 based on 80 density measurements compared using two different snow samplers and weighing devices [Garrity and Burns, 1988]. This error was the upper limit, since the snow samples were weighed on site using a balance and the wind often made the balance unstable. All density measurements made for the current analysis used an electronic balance protected from the wind. Using this upper error for a snow permittivity of 2, the error in snow wetness was ±0.4%. The reported accuracy in obtaining percent wetness, if the density was correct, was about 0.01% for dry snow (approximately less than 1%) and 0.05% for wet snow (approximately greater than 3%) An upper estimated error for snow wetness was 0.5% [Mätzler, 1985]. The snow wetness measurements were repeatable to ±0.05%, except for the surface snow layer, for which the deviation could be as high as ±1%.

16.3.2 Radiometric Measurements

A dual-polarized 37-GHz radiometer was mounted at a 53° incidence angle on the sides of a ship and tower, 17 and 5 m above the ice surface, respectively. A similar procedure was used for both radiometric measurements. Once the ship was alongside an ice floe, 1000 T_B measurements were made with a 6.5° corrugated horn antenna, giving a footprint of approximately 5 m. The T_B measurements were averaged with a standard deviation between ±1.2 and ±1.4 K. After these measurements were made, the field team was lowered onto the ice in a cage off the side of the ship. Measurements of the snow properties were then carried out within the footprint of the radiometer as described above.

16.4 EVOLUTION OF A SNOW COVER

Significant differences in snow properties on sea ice have been observed between the Arctic and Antarctic (Weddell Sea). In the Arctic, there is a progressive change in the snow cover properties as spring advances from early to mid- to late spring, Figures 16–1(a) to (c), respectively. In the Antarctic, different ice type regimes in the Weddell Sea have characteristic snow covers. The short time (two months) over which snow measurements were made in the Weddell Sea did not allow a representative time series of changes in snow properties to be derived; however, we could establish descriptive snow models for a characteristic set of snow depths.

During spring and into summer, the physical properties of a snow cover change significantly. Tables 16-2(a) to (i) illustrate the variability in snow covers for the Arctic and Antarctic. Table 16-2(i) is based on the snow measurements contained in Tables 16-2(b), (c), and (f). Measurements on second-year (SY) ice were made in the Antarctic and no measurements were made on MY ice. The measurements were made at random on an ice floe, and thus some ridged areas on FY ice are included in the averages. Snow depth in the Arctic in the summer, Table 16-2(h), is much less than that in the spring, Figures 16-2(a) to (e). Holt and Digby [1985] observed snow depth decreasing by half in six days on FY fast ice in Mould Bay, Canadian Archipelago. Twelve days later, all of the original 30 cm of snow had disappeared. The variability in snow wetness, ranging from ±0.6% to 1.5% for the Arctic and ±1.4% for the Antarctic, will cause a significant change in T_B throughout a season.

In the Weddell Sea, there is a large variation in snow depth due to different ice regimes. The snow depth on SY ice is about three times as large as on FY ice. The extensive areas of SY ice define a regime of thicker snow cover often associated with slush at the snow–ice interface. Table 16-2(i) shows a significant difference between the percentage of slush observed on FY and SY ice. The opposite is true for the Arctic, where slush was not as prominent on MY as on FY ice. Ice lenses that occurred in the Antarctic during the spring were mainly close to the snow surface and were not well defined, whereas, in the Artic, lenses could be found throughout a snow cover.

The descriptive snow models shown in the cases below for the Arctic and Antarctic are based on a total of 318 snow pits and are representative of a snow cover for each of the Greenland and Barents Seas and the Weddell Sea during the onset of the melt period.

16.4.1 Arctic

Figure 16-1 illustrates the metamorphism for a snow cover during the spring, which is better defined as the onset of melt for the Arctic. The descriptive snow models are based on 127 distinct vertical snow profiles (snow pits) (350 individual measurements) obtained during 1987 and 1989 in the Greenland Sea and 1989 and 1990 in the Barents Sea [Garrity, 1991]. The cases are for a snow cover on both FY and MY ice; however, it was observed that a snow cover on FY ice can be further advanced in the metamorphosis for a given period. For example, in 1987, the snow cover on MY

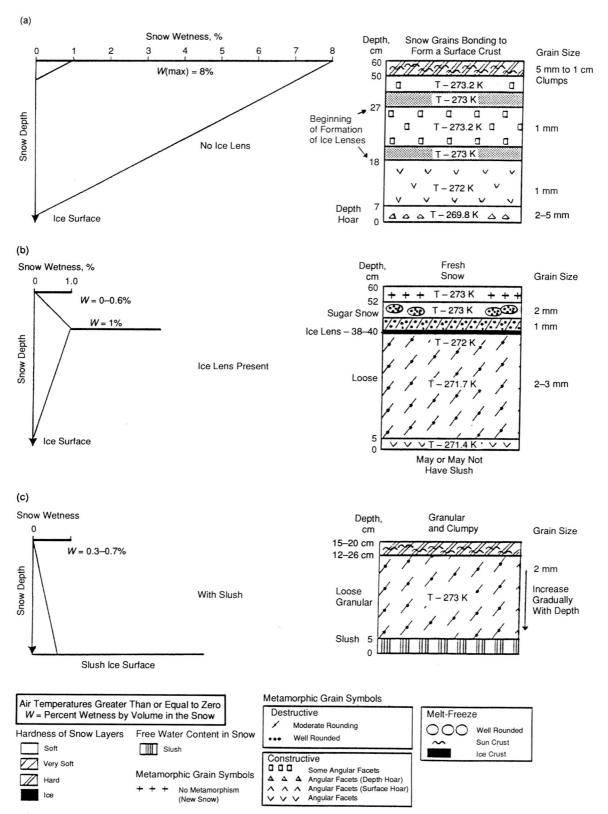

Fig. 16-1. Snow models for spring in the Arctic for: (a) early spring, when a snow cover with no ice lenses is present; (b) mid-spring, when ice lenses are present in the snow cover; and (c) late spring, when slush is present at the snow–ice interface.

TABLE 16-2a. Average snow properties measured in the Greenland Sea during the 1987 spring.

Parameter	Sample size	Average	Standard deviation	Minimum	Maximum
	FY, MY	FY, MY	FY, MY	FY, MY	FY, MY
Wetness, %	6, 7	0.8, 0.6	0.9, 0.9	0, 0	2, 2.3
Depth, cm	6, 9	24, 29	22, 11	10, 7	67, 45

TABLE 16-2b. Average snow properties measured in the Barents and Greenland Seas during the 1989 spring.

Parameter	Sample size	Average	Standard deviation	Minimum	Maximum
	FY, MY	FY, MY	FY, MY	FY, MY	FY, MY
Air, °C	14, 29	0.1, −0.8	1.1, 2.3	−1.8, −6.7	1.9, 2.4
Snow, °C	17, 34	−1.0, −2.2	0.7, 1.7	−2.2, −5.9	0.0, −0.1
Wetness, %	16, 34	1.2, 1.0	1.1, 1.5	0.0, 0.0	4.0, 7.0
Depth, cm	16, 34	22, 31	15, 15	4.5, 7.5	58, 70
Density, g/cm^3	16	0.351	0.085	0.224	0.546

TABLE 16-2c. Average snow properties measured in the Barents Sea during the 1990 spring.

Parameter	Sample size	Average	Standard deviation	Minimum	Maximum
	FY, MY	FY, MY	FY, MY	FY, MY	FY, MY
Air, °C	13, 3	−1.7, −1.7	1.7, 3.8	−5.1, −4.1	1.6, 2.7
Snow, °C	19, 3	−2.0, −2.4	1.3, 1.6	−4.3, 3.4	−0.1, −0.5
Wetness, %	18, 3	0.2, 0.6	0.6, 0.3	0.0, 0.4	2.5, 0.9
Depth, cm	25, 5	22, 20	9, 8	6, 12	45, 31
Density, g/cm^3	18, 3	0.371, 0.365	0.070, 0.030	0.188, 0.343	0.461, 0.400

TABLE 16-2d. Average snow properties measured in the Greenland and Barents Seas during 1987, 1989, and 1990 springs.

Parameter	Sample size	Average	Standard deviation	Minimum	Maximum
	FY, MY	FY, MY	FY, MY	FY, MY	FY, MY
Air, °C	27, 32	−0.8, −0.8	1.7, 2.4	−5.1, −6.7	1.9, 2.7
Snow, °C	36, 37	−1.5, −2.2	1.2, 1.6	−4.3, −5.9	0.0, −0.1
Wetness, %	40, 44	0.7, 0.9	1.0, 1.4	0.0, 0.0	4.0, 70
Depth, cm	47, 48	22, 29	13, 14	4.5, 7.0	67, 7.0
Density, g/cm^3	34, 37	0.362, 0.355	0.077, 0.058	0.188, 0.241	0.546, 0.497

TABLE 16-2e. Average snow properties measured from both FY and MY ice in the Greenland and Barents Seas during 1987, 1989, and 1990 springs.

Parameter	Sample size	Average	Standard deviation	Minimum	Maximum
Air, °C	59	−0.8	2.1	−6.7	2.7
Snow, °C	73	−1.9	1.5	−5.9	0.0
Wetness, %	84	0.8	1.2	0.0	7.0
Depth, cm	95	26	14	4.5	70
Density, g/cm^3	71	0.358	0.068	0.188	0.546

TABLE 16-2f. Average snow properties measured in the Weddell Sea during the 1989 spring.

Parameter	Sample size	Average	Standard deviation	Minimum	Maximum
	FY, SY	FY, SY	FY, SY	FY, SY	FY, SY
Air, °C	90, 21	−8.1, −2.8	4.7, 3.9	−18.3, −13	−0.3, 1.5
Snow, °C	94, 21	−6.5, −3.0	3.0, 1.8	−13.0, −6.8	−1.5, −1.0
Wetness, %	89, 22	1.1, 0.5	1.5, 0.6	0.0, 0.0	7.2, 2.3
Depth, cm	146, 22	15, 44	12, 21	0.5, 14	66, 110
Density, g/cm^3	90, 22	0.275, 0.368	0.052, 0.044	0.177, 0.283	0.466, 0.420

TABLE 16-2g. Average snow properties measured from both FY and SY ice in the Weddell Sea during the 1989 spring.

Parameter	Sample size	Average	Standard deviation	Minimum	Maximum
Air, °C	111	−7.1	5.0	−18.3	1.5
Snow, °C	115	−5.8	3.1	−13.0	−1.0
Wetness, %	111	1.0	1.4	0.0	7.2
Depth, cm	168	19	17	0.5	110
Density, g/cm^3	112	0.293	0.062	0.177	0.466

TABLE 16-2h. Average snow properties measured in the Greenland Sea and Arctic Ocean during 1990 and 1991 summers.

Parameter	Sample size	Average	Standard deviation	Minimum	Maximum
Air, °C	25	1.3	1.6	−4.7	4.2
Snow, °C	25	−0.1	0.2	−1.0	−0.1
Wetness, %	26	1.2	1.4	0.0	4.7
Depth, cm	42	3.1	3.2	0.5	14.5
Density, g/cm^3	27	0.300	0.100	0.163	0.490

TABLE 16-2i. Percentage of ice lenses and slush measured on sea ice in the Greenland and Barents Seas during 1989 and 1990 springs plus the Weddell Sea 1989 spring.

Year	Ice type	Lenses, %	Slush, %
Arctic			
1989	FY	29	18
	MY	34	3
1990	FY	33	22
	MY	20	0
1989 and	FY	28	18
1990	MY	26	2
	FY and MY	26	10
Antarctic			
1989	FY	3	19
	SY	54	59
	FY and SY	10	26

ice was similar to that shown in Figure 16-1(a), whereas the snow cover on FY ice had advanced to Figure 16-1(b) for the same period. The different snow metamorphic stages for FY and MY ice constitute a major cause. The snow cover on FY ice is generally thinner than on MY ice [Tucker et al., 1987; Perovich et al., 1988; Garrity, 1991]. Since snow is a good insulator, it takes longer for air temperature and solar radiation to influence the thicker snow cover on MY ice.

There is no descriptive snow model included in this chapter for summer. However, a few words will describe the snow cover during the summer (July and August 1990 in the Greenland Sea and August and September 1991 in the Arctic Ocean). It is surprising, but understandable, that the snow wetness is similar to that of spring and is not high (1.2±1.4%), Table 16-2(h). The snow cover is thin, averaging 3.1±3.2 cm, with a loose snow layer over a hard, almost ice layer. The snow–ice interface is not distinguishable, and thus an accurate snow depth is difficult to measure. Since the snow grains are large, about 0.5 to 2 cm, water is able to drain freely from the snow cover. This results in a snow cover that has largely disappeared, leaving a grainy texture and a rough ice surface. This has also been observed by Carsey [1985].

16.4.1.1. Descriptive Snow Models. In general, during early spring, the upper portion of the snow cover is warmer than the lower by about 4 K, whereas for the mid-spring, the temperature difference has been reduced to 2 K. There is no difference for the late spring, when the snow cover is near the melting point.

The descriptive early spring model, Figure 16-1(a), shows a decrease in snow wetness with snow depth. This is characteristic of a snow cover on sea ice during the onset of melt. After the winter months, this is the start of free water in the upper part of the snow cover, which can be as high as

8% due to the passage of a warm weather front. However, 1% is more representative for the snow surface wetness. Some properties of a snow cover, such as a decreasing snow temperature with depth, are shown beside the snow wetness profile in Figure 16-1(a). This type of snow cover generally has no ice lenses; however, there are indications of structural changes that may lead to the formation of ice lenses. A surface crust exists, and it can be described as being composed of a granular layer of many snow crystals bonded together forming crystal aggregates [Colbeck, 1979; Colbeck et al., 1988] or clumps. The average snow grain diameter is 5 mm, the size of the clumps is about 1 cm. The larger grains in the surface layer are a result of the increased amount of incoming radiation during the spring, which enhances melt at the grain boundaries and impurities. Refreezing of water on the grains in the surface layer during the night increases their size. The remainder of the snow cover has a grain size of 1 mm except for the formation of 2- to 5-mm depth hoar crystals at the snow–ice interface due to an upward vapour flux [Colbeck et al., 1988].

Often thin layers of very hard snow or ice with a thickness of only a few millimetres to centimetres are in a snow cover. Refreezing of water in a snow cover can produce flat horizontal ice lenses in less than 24 hours.

By mid-spring, Figure 16-1(b), very often there is at least one well-defined ice lens present within the top 10 to 20 cm of the snow cover. An ice lens will form when downward motion of water in the snow is prevented from percolating further by a less permeable snow layer, such as a previously wind-blown surface covered by a new snowfall, followed by freezing snow temperatures. A decrease in the surface snow wetness during early spring is due to redistribution of free water in the snow cover followed by either refreezing to form an ice lens or accumulation as a slush layer at the snow–ice interface. If a slush layer is present, the snow depth generally is reduced compared to early spring. The ice lens is an obstacle causing a decrease in the amount of free water below its location in the snow cover. Snow wetness just above the ice lens is often significant, for example 1%. The soft snow layer above the ice lens could be described as sugar snow that is advanced in its metamorphism compared to snow grains in early spring. The snow grains are rounded, about 2 mm, and very loose, forming a medium that has the texture of large raw sugar grains, an excellent medium for free water to move through. Below the ice lens, there is commonly a layer containing elongated grains, which are moderately rounded and 2 to 3 mm. This layer facilitates the free water percolation through the snow to form slush at the snow–ice interface.

The number and density of ice lenses can vary spatially on an ice floe, but when present they are generally on nearly all ice floes within an area comparable to a few SSM/I satellite footprints (a footprint at 37 GHz is 37×29 km^2). Large variations in the occurrence of ice lenses have been observed from year to year. Sometimes in mid-spring less well-developed, discontinuous ice lenses are present in a snow cover, which is more consistent with the descriptive early spring snow model, Figure 16-1(a). The formation of

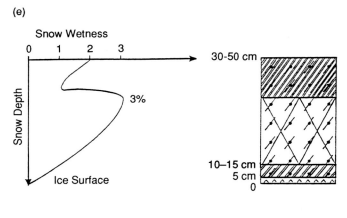

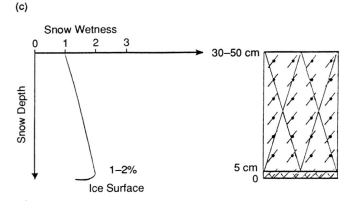

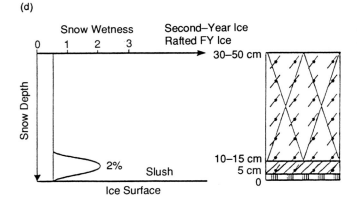

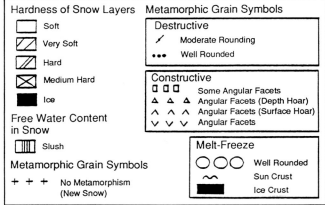

Fig. 16-2. Snow models for spring in the Weddell Sea, Antarctic, are shown for air temperatures less than 268 K (a) without and (b) with slush at the snow–ice interface. The same features are shown for an air temperature between (c) 268 and (d) 273 K. The snow wetness distribution changes significantly when (e) the air temperature is greater than 273 K. (a), (b), (c), and (d) are representative snow covers found at the western Weddell Sea ice edge; (d) and (e) are found at the eastern ice edge. (e) is also characteristic of the interior of the Weddell Sea. There is no descriptive snow model close to Antarctica presented.

ice lenses depends on the stage of metamorphism of a snow cover, which depends in turn on the environmental history and snow thickness, which change from year to year.

The late spring is characterized by a snow cover that increases in snow wetness with depth. The snow wetness is not necessarily large, ranging from 0.3% near the surface to 1% above a slush layer. The snow depth has decreased from the previous stage, and the snow cover has become more homogeneous. A surface crust exists, composed of a layer only a few millimetres thick, with an average grain size of 2 mm. The increasing intensity of solar radiation, approaching a maximum in the afternoon, can cause the amount of free water to increase at the snow surface. The free water around a snow grain can cause another grain to connect to it due to the increased surface tension of the water-covered grain. Grain size increases gradually with depth, reaching a maximum size of approximately 4 mm just above the slush layer. The bulk of the snow can be described as loose and granular, with a temperature near 273 K. The water is able to move through the bulk of the snow with ease because the looseness of the snow and its large grains produce channels for water movement. This often results in a slush layer at the snow–ice interface and has been observed by Grenfell and Lohanick [1985] in the Beaufort Sea in 1982, during the late spring. The slush layer is usually 1 to 5 cm thick and consists of a mixture of snow, melted snow, and sometimes rotting ice. If the latter is present, the slush on FY ice is saline.

16.4.2 Weddell Sea, Antarctica

During the Weddell Sea experiment in September and October 1989, 191 distinct snow profiles (384 individual measurements) were obtained from FY and old ice. The old ice present in the western area of the Weddell Sea was mainly SY ice, which was also observed near the ice edge. The average air temperature decreased from 270 K near the ice edge to less than 268 K very close to Antarctica. Table 16-3 summarizes the different ice regimes, which have characteristic snow covers, near the ice edge. Thicker snow covers were found near the ice edge, which is consistent with measurements made in 1986 [Wadhams et al., 1987]. The

snow depths generally ranged from 50 cm to 1 m on SY ice and up to a maximum of 50 cm (level ice) on FY floes, which were sometimes rafted and/or deformed near the ice edge. The large snow depths caused loading of the ice, and the negative freeboard allowed for slush to infiltrate the snow–ice interface. These substantial snow depths were characteristic of SY and FY ice for the western and eastern Weddell Sea areas, respectively. These two regimes correspond to what is defined as the outer zone in Section 12.2. FY ice near the ice edge also exhibited slush at the snow–ice interface. An observed slush thickness for both the eastern and western Weddell Sea was generally 6 cm. The slush extended over the entire ice floe for both SY and FY ice. Undeformed FY ice often did not have slush at the snow–ice interface.

The presence of slush was visible from the air as a gray area on floes from July–September 1986 [Wadhams et al., 1987]. During September and October 1989, the slush layer was hidden by a snow cover [Garrity, 1991]. There was in fact no visible melt on the ice or snow surface during the spring of 1989 in the Weddell Sea, in contrast to spring in the Arctic. Most of the melt in the Weddell Sea was from the bottom of the ice [Andreas and Ackley, 1982].

Profiles in the snow are also considerably different from those observed in the Arctic. For example, a surface crust is common in the Arctic; however, any surface crust in the Antarctic is only millimetres thick. This wind crust has small snow grains ranging from 0.5 to 1 mm. In general, the vertical structure of the snow cover is more homogeneous in the Antarctic than in the Arctic.

For air temperatures ranging from 261 to 268 K, a descriptive model for a snow cover without the presence of slush is shown in Figure 16-2(a). The snow wetness ranges from 0.1% to 0.4% and is constant throughout the snow cover. The snow wetness distribution with the presence of slush is shown in Figure 16-2(b). The absorption of water from the slush layer at the snow–ice interface is analogous to a sponge soaking up water. This soaking up, or wicking, of water causes the snow wetness to increase to 2% just above the slush layer. When slush is present at the snow–ice interface, T_B decreases due to the high free water content (greater than 40%) of the slush. These descriptive

TABLE 16-3. Predominant ice regimes near the ice edge in the Weddell Sea.

Ice type	Location	Snow depth, cm	Comments
	Western Weddell Sea		
Second year	64° 19′S 46° 44′W 65° 25′S 40° 17′W	50–100	Slush present/ negative freeboard
Second year	64° 19′S 46° 44′W 65° 25′S 40° 17′W	50	Dry snow–ice interface
	Eastern/Western Weddell Sea		
First year	63° 47′S 02° 23′W	50	Slush present/ deformed ice
First year	63° 47′S 02° 23′W	50	Dry snow–ice interface

models, Figures 16-2(a) and (b), are for snow on SY ice in the western Weddell Sea.

For air temperatures between 268 and 273 K, there is often significant free water in the snow cover. The resulting snow physical properties are shown in Figures 16-2(c) and (d) (slush present). Figure 16-2(c) shows snow characteristics observed in the western and eastern ice edges where there is a gradual increase in snow wetness with depth, reaching a maximum of 2% near the snow–ice interface. Wicking is shown again above the slush layer, Figure 16-2(d), which is generally present on rafted FY floes in the eastern ice edge. The snow surface layer is wetter in Figure 16-2(d) than in Figure 16-2(b) due to warmer air temperatures. Only a few snow measurements were made for air temperatures above freezing in the eastern ice edge and interior of the Weddell Sea, Figure 16-2(e). The surface was wetter than the immediate underlying layer, and snow wetness increased near the middle of the snow cover. The wetness decreased gradually to 0% at the snow–ice interface. There was also more structure in the snow cover compared with Figures 16-2(a) to (d), mainly due to higher air temperatures. Only Figure 16-2(c) is in the inner zone; the remainder of the illustrations in Figure 16-2 are in the outer zone.

16.5 Case Studies of Brightness Temperature From Snow on Floating Ice

For surface-based radiometric measurements, T_B is equal to the product of emissivity with physical temperature. Since the changes in T_B are an order of magnitude larger than changes in the snow temperature, T_B can be used instead of emissivity.

16.5.1 New Snowfall

An example of the change in T_B due to a new snowfall over MY ice in the Greenland Sea is shown in Figure 16-3. These measurements were made by a 37-GHz radiometer mounted on the RV Polarstern. As only one snow depth measurement was made, approximately 3 hours after the start of snowfall, it is assumed the rate of snowfall was constant during this time. In this particular example, T_B increased from 165 to 205 K for the horizontal polarization, and from 195 to 225 K for the vertical polarization, with an increase in snow depth from 0 to 2 cm. The T_B response to the snowfall was rather quick and large, followed by smaller variations in T_B. The increase in T_B is due to a small amount of free water (0.5%) within the new snow, which increases the absorption by, and therefore the emission from, the snow. High absorption and small optical depth mask the emission from the underlying surface.

16.5.2 Dry to Wet Homogeneous Snow

An increase in emission from a homogeneous snow cover due to increased snow wetness was first observed by Edgerton et al. [1971] for snow over sediments and lake ice and by

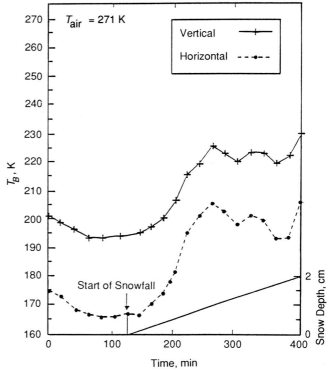

Fig. 16-3. Brightness temperature (T_B) at 37 GHz and a 53° incidence angle increases due to a snowfall on a bare MY ice floe.

Hofer and Mätzler [1980] for snow-covered land at a high-altitude alpine test site. On sea ice, as on most natural substrates, a homogeneous snow cover is rare. Because of the many external forces, namely further precipitation, storms, diurnal insulation causing partial melting, and wind, only the initial snowfall on bare ice is uniform, and then only for a short time. Such a snowfall was observed by radiometers in the laboratory by Grenfell and Comiso [1986]. The total snowfall was 45 mm and air temperature was near –12°C. They found that emissivity decreased when snow was added, with a larger decrease at higher frequencies up to 37 GHz. The new snow also tended to increase the polarization ratio, $(EV - EH)/(EV + EH)$. Grenfell and Lohanick [1985] observed wet established snow of similar thickness at the same frequencies on MY ice in the Canadian Archipelago. The angular dependences of brightness temperatures and polarization ratios were similar to the laboratory cases of bare ice [Grenfell and Comiso, 1986], but the vertical structure of the snow and its moisture content were not measured. Similar trends were seen in the emission from snow-covered FY and old ice in the Arctic and Antarctic, as well as from snow-covered freshwater ice on the St. Lawrence River. The measured emissivities presented in Figure 16-4 show an increase as a dry snow cover evolves to a wetness of 1%. The large variation in emissivities for 0% snow wetness is due to different snow depths, causing varying amounts of volume scattering, and

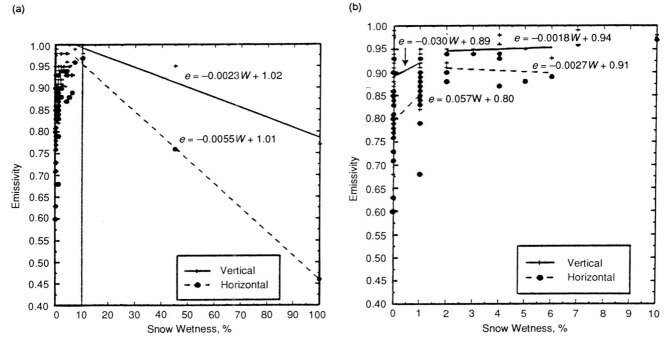

Fig. 16-4. (a) Composite of snow wetness measurements, expressed as a percent of free water by volume, combined with satellite, ground-based, and ship-based radiometric measurements at 37 GHz and a 53° incidence angle. The measurements are from snow-covered floating ice. The scatter in emissivity near 0% wetness is mainly due to the different ice types present. (b) An enlargement of the right half of (a) shows more detail.

to various ice types underlying the snow. The general trend indicates an increase in emissivity for a snow wetness from 0% to 1%.

Comparisons of snow wetness measurements in areas of 100% ice cover were made to the SSM/I T_B's in the Barents [Ramseier et al., 1991; Garrity, 1991] and Weddell [Garrity, 1991] Seas. The snow cover wetness caused a spatially homogeneous T_B for an area greater than the size of the SSM/I footprint. The measured amount of free water within the snow cover in the Weddell Sea is compared with the emissivity from SSM/I data. As in Figure 16-4, these measurements indicate a sharp increase in the emissivity with increasing free water content from 0% to 1% for both 19 and 37 GHz.

16.5.3 Stratified Snow

As mentioned, unstratified snow is rare on sea ice. According to Mätzler [1985], development of stratification in a snow cover is one cause for an increase in the polarization ratio, $P = (EV - EH)/(EV + EH)$, at microwave frequencies. Grenfell and Lohanick [1986] found an average P somewhat less than 0.1 at a 45° nadir angle on dry snow-covered FY ice. The snow cover varied in thickness from near zero to about 25 cm and had an ice layer within it from a mid-winter melt. Snow density was about 300 kg/m³ both above and below the ice layer. For wet (and presumably stratified) snow on MY ice, Grenfell and Lohanick [1985] found that P ranged from about 0.03 to 0.1 for snow thicknesses from 1

to 10 cm and was smaller at higher frequencies up to 37 GHz.

In other studies, observations from dry snow show that stratification can decrease the horizontal T_B more than the vertical T_B by about 10 K [Garrity, 1991], causing an increase in the V–H difference of at least 20 K; a 65 K difference has been observed [Garrity, 1991]. The V–H data [Garrity, 1991] also show that a surface crust can cause as great a V–H difference as a stratified snow cover, depending on how well developed the varying layers are.

On April 29, 1989, ship-based measurements of T_B were made at different incidence angles at 0645 UT, Figure 16-5(a), and again at 1200 UT, Figure 16-5(b). Two different snow pits, each with a 50-cm snow depth, were studied in close proximity to each other, with the radiometer resolving each snow pit individually at a 53° incidence angle. The snow wetness was near 0% for both snow pits. The average snow temperature was 270 K, and grain sizes were in the submillimetre range, except at the base of the snow cover, where depth hoar grains of 5 mm were found. The presence of an ice lens in the first snow pit resulted in a highly polarized microwave emission, Figure 16-5(a). A 0.2% wetness was also measured for the snow surface, which caused an increase in T_B as compared to a dry snow surface, Figure 16-5(b). The snow cover with little structure did not polarize the T_B, Figure 16-5(b). The depolarization and low T_B's in Figure 16-5(b) are due to volume scattering from the dry snow cover.

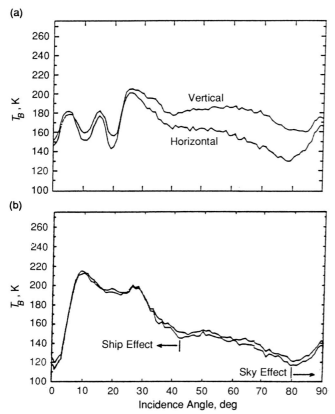

Fig. 16-5. The brightness temperature incidence angle profile from a MY ice floe (a) with and (b) without an ice lens in the snow cover. The ice lens caused polarization in the microwave signature at 37 GHz. Incidence angles less than 40° are affected by the ship's side. The increase in brightness temperature for incidence angles greater than 80° is due to the contribution from the sky.

If there are no multiple reflections between the ice lens and ice surface and no interference effects from reflections at different boundaries, the reflectivity from the ice lens can be calculated using the Fresnel equations [Ulaby et al., 1981]. The reflectivity is calculated for a situation characterized by an infinite snow depth, a permittivity of 1.45 (dry snow), a dielectric loss of 0.0013, and the presence of an ice lens. Assuming the ice lens is pure ice, with a permittivity of 2.92 and dielectric loss of 0.01 [Vant et al., 1974] at 37 GHz, the vertical T_B is calculated to be 269 K and the horizontal T_B to be 201 K for a 53° incidence angle. Similarly, if there were no ice lenses in the snow cover, the vertical T_B would be 270 K and the horizontal 259 K. This results in a V–H difference of 68 K for the snow cover with an ice lens and 11 K for a homogeneous snow cover. Comparing these calculations to measured V–H differences in Figure 16-5(a), the calculated V–H difference is 49 K larger than the measured one. The measured T_B is also lower than the calculated T_B. Volume scattering would both reduce the T_B and violate the application of the Fresnel equations.

However, this calculation and Figures 16-5(a) and (b) illustrate that the V–H difference will increase if an ice lens is present in a dry snow cover.

16.5.4 Slush at the Snow–Ice Interface

Drinkwater and Crocker [1988] and Crocker and Wadhams [1989] have reported salinity and brine layers at the base of snow on sea ice, which can significantly affect the microwave properties of the snow and even cause the snow to mask the signature of the underlying ice at frequencies at which the snow is optically thin. Grenfell and Lohanick [1986] showed T_B's of snow-covered FY ice that suggested the presence of a thin snow layer roughening the ice surface, which increased the transmission at the snow–ice interface. Lohanick [1990] showed that an established snow cover on sea ice, in interacting with the ice through such mechanisms as upward wicking of brine, caused significant contributions to the T_B. Lohanick [1992] has shown that brine wicked into a snow layer from FY ice causes the formation of a slush layer that freezes into an added frazil ice layer, with accompanying large changes in T_B at frequencies below about 40 GHz.

Effects of a slush layer on satellite-measured T_B's is illustrated with data from the Weddell Sea in 1989 [Garrity, 1991]. The western Weddell Sea was covered predominantly with SY ice up to 2 m thick [Casarini, 1989]. The snow depth on the SY ice generally ranged from 40 cm to 1 m, with slush at the snow–ice interface. The Atmospheric Environment Service of Canada (AES) algorithm described in Chapter 10 indicated old ice concentrations of up to 98% due to a combination of lower radiances, caused by volume scattering in the deep and relatively dry snow, Figure 16-2(a), and the slush, Figure 16-2(b), at the snow–ice interface.

The eastern ice edge of the Weddell Sea had a significant amount of FY ice, which was sometimes rafted. The snow depth was often as high as 50 cm, with slush at the snow–ice interface, Figure 16-2(d). This ice also appears as old ice on the AES SSM/I-derived ice charts. The attenuation of microwave energy was small enough for microwave emissions to come from the total snow depth and slush layer, giving results similar to those for the western Weddell Sea. In this case, old ice classified by the AES/ISTS algorithm was actually FY ice with a low T_B due to volume scattering from the snow and absorption loss from the slush layer.

16.5.5 Variation in Brightness Temperature Between Ice Types

In the Greenland Sea, from late May to mid-June (mid-spring) 1987, the FY ice T_B decreased, whereas the MY ice T_B increased [Garrity, 1991]. The MY ice T_B increased due to a homogeneous snow cover, at least for the upper 10 to 20 cm, with a snow wetness greater than about 0.2% and less than 1%, which caused a reduction in penetration depth (Chapter 2), and thus a reduction in the volume scattering from MY ice.

By late spring, the vertical snow structure on FY ice became more homogeneous, with larger snow grains and a free water content greater than 0.2% but less than 1%, causing an increase in T_B. The snow cover on MY ice had become stratified in the upper portion of the snow cover, much like snow on FY ice in the mid-spring, causing a decrease in T_B for MY ice.

The insulating property of the thicker snow cover on MY ice caused a delay in metamorphism of the snow as compared to the thinner snow on FY ice. In addition, the saline ice surface of FY ice caused brine to be present in the snow near the snow–ice interface, therefore decreasing the melting point of the snow and enhancing the metamorphic process.

16.6.6 Summary of Changes in Brightness Temperature Due to a Snow Cover

Observed ranges in T_B for FY and MY ice are shown in Table 16-4 and Figure 16-6. Only cases where the calculated skin depth indicated that the snow cover was the main medium contributing to the measured T_B are included in Table 16-4. The vertical T_B is always greater for FY and MY ice during the onset of melt than it is when the snow is dry. The range in T_B is largest for MY ice, where a snow cover can increase the horizontal T_B by as much as 83 K and the vertical T_B by as much as 76 K. This is significant considering that the difference in T_B between FY and MY ice for a dry snow cover is about 55 K for the horizontal and 52 K for the vertical. Depending on the time of day, there are overlaps in T_B from FY and MY ice during the spring [Garrity, 1991], and FY and MY T_B's are equal during the summer at all times of the day [Garrity et al., 1992].

16.7 A DESCRIPTIVE METAMORPHISM MODEL OF HOMOGENEOUS SNOW FOR PASSIVE MICROWAVE SENSING

Several regimes of snow metamorphism, with accompanying qualitative behavior of T_B, are shown in Figure 16-7. The snow cover is assumed homogeneous in wetness. This is a useful assumption for microwave properties since for a snow wetness of 1% or greater the emission from the snow is governed primarily by how much free water is present. A representative T_B for sea ice during the winter when the snow is dry is labelled on the sketch, providing a reference for the influence of snow wetness on T_B. Volume scattering from dry snow dominates [Ulaby and Stiles, 1980], causing depolarization and a decrease in T_B. There is a point (around 0.2%) at which snow wetness will cause T_B to increase. A small amount of free water in the snow, for example in Figures 16-1(b) and (c) and 16-2(a) and (d), increases the T_B and reduces the emission from the ice surface as compared to a dry snow cover [Grenfell and Lohanick, 1985].

The case for emissivity values of unity (T_B equals the physical snow temperature), which is the plateau extending from about 1% to 10% snow wetness in Figure 16-7, is representative of Figure 16-1(a) when the snow surface is wet (8%). Volume scattering is negligible during this stage because the skin depth of the wet snow is millimetres.

A reduction in T_B of greater than 6% to 10% free water in snow, Figure 16-7, coincides with Colbeck et al.'s [1988] transition to the funicular regime for snow on land. The pendular regime is for snow with a low free water content (3% to 8%), in which case air can move in a continuous path throughout the snow. The funicular regime (8% to 15%) is for a higher free water content where isolated air bubbles in the ice–water–air mixture are present. It is not until the pore volume is completely filled with water that the snow is saturated.

TABLE 16-4. Influence of snow on brightness temperature at 37 GHz and 53° incidence angle.

Snow type	T_B, K horizontal	T_B, K vertical	T_{BD}, K	Comments
Dry snow	238	250	12	First year
Dry snow	171	186	15	Multiyear, frozen melt puddles, and hummocks
Dry snow	195	210	15	Multiyear
Homogeneous	254	262	9	
Surface crust	202	224	22	
Slight layering	226	242	16	
Stratification	223	241	21	
Slush	179	259	20	
New snowfall	240	260	20	Increase in T_B for both ice types
Wet snow (1–6% by volume)	243	259	16	
Snow depth (10 cm)	225	250	25	
Snow depth (50 cm)	191	227	36	

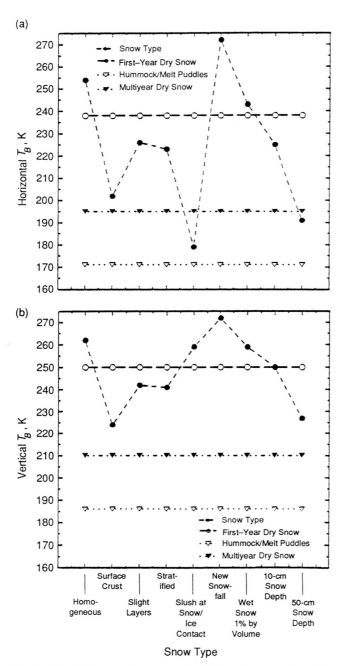

Fig. 16-6. Influence of snow on brightness temperature at a frequency of 37 GHz and an incidence angle of 53° for (a) horizontal and (b) vertical polarizations.

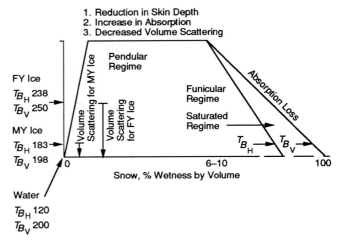

Fig. 16-7. The different regimes of snow metamorphism and the effect on brightness temperature (T_B horizontal and T_B vertical) for 37 GHz. The calm water, FY ice, and MY ice brightness temperatures are shown in degrees kelvin for reference.

Drainage of water by gravity depends significantly on the snow texture, grain size, and grain shape. The range in snow wetness where drainage occurs has been reported to vary from 3% [Colbeck et al., 1988] to 7% [Denoth, 1980]. The free water will migrate to lower levels in the snow cover until eventually a slush layer starts to form at the snow–ice interface, Figure 16-1(c). The high free water content of the slush layer causes a radiometrically homogeneous layer. Eventually, for slush, as in Figures 16-1(c) and 16-2(b) and (d), the low T_B of bulk water dominates the signal.

16.8 CONCLUSION

Snow models have been developed for the Greenland and Barents Seas, as well as the Weddell Sea during the onset of the melt period. Metamorphic processes in a snow cover described in these snow models allow changes in snow conditions to be related to changes in observed sea ice T_B's.

Free water content in the snow is the main parameter causing a change in the apparent T_B of floating ice. Brightness temperature increases for a snow wetness between 0% and 1%, then decreases for high free water content associated with slush layers. The increase in snow wetness can cause old ice to be classified as FY ice using a present-day ice-type algorithm. The decrease in T_B due to slush can cause the ice-type algorithm to classify the ice as old, even if it is FY ice. Such response of the microwave T_B to snow cover changes allows information to be extracted from satellite radiometric data on the onset of melt period, extent of slush in a region, freeze-up period, melt–freeze cycles, and possibly snow depth.

Acknowledgements. The following are gratefully acknowledged for their financial support: Alfred-Wegener-Institut für Polar-und Meeresforschung of Bremerhaven, Germany, the Atmospheric Environment Service of Canada, and the National Science and Engineering Counsel. I thank Dr. H. Eicken, Dr. A. W. Lohanick, and Dr. B. A. Burns for their suggestions in reorganizing and editing the text.

REFERENCES

Ambach, W. and A. Denoth, *The Dielectric Behaviour of Snow: A Study Versus Liquid Water Content,* NASA CP 2153, National Aeronautics and Space Administration, Washington, DC, 1980.

Andreas, E. L. and S. F. Ackley, On the differences in ablation seasons of Arctic and Antarctic sea ice, *Journal of Atmospheric Sciences, 39(3),* pp. 440–447, March 1982.

Carsey, F. D., Summer Arctic sea ice character from satellite microwave data, *Journal of Geophysical Research, 90(C3),* pp. 5015–5034, 1985.

Casarini, M. P., editor, *Ice Observation Log, Winter Weddell Gyre Study (ANT VIII 2),* Scott Polar Research Institute, University of Cambridge, Cambridge, UK, 1989.

Chang, A. T. C., J. L. Foster, M. Owe, D. K. Hall, and A. Rango, Passive and active microwave studies of snowpacks, *Nordic Hydrology, 16,* pp. 57–66, 1985.

Colbeck, S. C., Grain cluster in wet snow, *Journal of Colloid and Interface Science, 72(3),* pp. 371–384, December 1979.

Colbeck, S. (chair), E. Akitaya, R. Armstrong, H. Gubler, J. Lafeuille, K. Lied, D. McClung, and E. Morris, *The International Classification for Seasonal Snow on the Ground,* 23 pp., International Committee on Snow and Ice, World Data Center, University of Colorado, Boulder, Colorado, 1988.

Comiso, J. C., Remote sensing of sea ice using multispectral microwave satellite data, *Advances in Remote Sensing Retrieval Methods,* edited by A. Deepak, H. E. Fleming, and M. T. Chahine, Deepak Publishing, Hampton, Virginia, pp. 349–359, 1985.

Crocker, G. B. and P. Wadhams, Modelling Antarctic fast-ice growth, *Journal of Glaciology, 35(119),* pp. 3–8, 1989.

Denoth, A., The pendular-funicular transition in snow, *Journal of Glaciology, 25(91),* pp. 93–97, 1980.

Denoth, A., A. Fogar, P. Weiland, C. Mätzler, H. Aebischer, M. Turi, and A. Sihvola, A comparative study of instruments for measuring the liquid water content of snow, *Journal of Applied Physics, 56,* pp. 2154–2160, 1984.

Drinkwater, M. R. and G. B. Crocker, Modelling changes in the dielectric and scattering properties of young snow-covered sea ice at GHz frequencies, *Journal of Glaciology, 34(118),* pp. 274–282, 1988.

Edgerton, A. T., F. Ruskey, D. Williams, A. Stogryn, G. Poe, D. Meeks, and O. Russell, *Microwave emission characteristics of natural materials and the environment: A summary of six years research,* final technical report 9016R-8, contract N00014-70C-0351, Aerojet General Corporation, El Monte, California, 1971.

Garrity, C., *Passive Microwave Remote Sensing of Snow Covered Floating Ice During Spring Conditions in the Arctic and Antarctic,* Ph.D. dissertation, York University, CRESS department, North York, Ontario, September 1991.

Garrity, C. and B. A. Burns, *Electrical and Physical Properties of Snow in Support of BEPERS-88,* Technical Report MWG 88-11, York University, CRESS Department, North York, Ontario, 1988.

Garrity, C., K. W. Asmus, and R. O. Ramseier, *International Arctic Ocean Experiment: WAGB-10 Polar Star Platform Remote Sensing Group,* Norland Science and Engineering, Contract 159H, Ottawa, Canada, 1992.

Grenfell, T. C. and J. Comiso, Multifrequency passive microwave observations of first-year ice grown in a tank, *IEEE Transactions on Geoscience and Remote Sensing, GE-24(6),* pp. 826–831, 1986.

Grenfell, T. C. and A. W. Lohanick, Temporal variations of the microwave signatures of sea ice during the late spring and early summer near Mould Bay, Northwest Territories, *Journal of Geophysical Research, 90(C3),* pp. 5063–5074, 1985.

Grenfell, T. C. and A. W. Lohanick, Variations in brightness temperature over cold first-year sea ice near Tuktoyaktuk, Northwest Territories, *Journal of Geophysical Research, 91(C4),* pp. 5133–5144, 1986.

Hofer, R. and C. Mätzler, Investigations of snow parameters by radiometry in the 3-mm to 60-mm wavelength region, *Journal of Geophysical Research, 85,* pp. 453–460, 1980.

Holt, B. and S. A. Digby, Processes and imagery on fast first-year sea ice during the melt season, *Journal of Geophysical Research, 90(C3),* pp. 5035–5044, 1985.

Künzi, K. F., S. Patil, and H. Rott, Snow cover parameters retrieved from Nimbus 7 Scanning Multichannel Microwave Radiometer (SMMR) Data, *IEEE Transactions on Geoscience and Remote Sensing, GE-20,* pp. 452–467, 1982.

Lohanick, A. W., Some observations of established snow cover on saline ice and their relevance to microwave remote sensing, *Sea Ice Properties and Processes,* edited by S. F. Ackley and W. F. Weeks, CRREL Monograph 90-1, 300 pp., Cold Regions Research and Engineering Laboratory, Hanover, New Hampshire, 1990.

Lohanick, A. W., Microwave brightness temperatures of laboratory-grown undeformed first-year ice with an evolving snow cover, *Journal of Geophysical Research,* in press, 1992.

Mätzler, C., *Resometer Manual,* University of Bern, Switzerland, 1985.

Mätzler, C., E. Schanda, and W. Good, Towards the definition of optimum sensor specifications for microwave remote sensing of snow, *IEEE Transactions on Geoscience and Remote Sensing, GE-20,* pp. 57–66, 1982.

Mätzler, C., R. O. Ramseier, and E. Svendsen, Polarization effects of sea ice signatures, *IEEE Journal of Oceanic Engineering, OE-9(5),* pp. 333–338, 1984.

NORSEX Group, Norwegian remote sensing experiment in the marginal ice zone north of Svalbard, Norway, *Science, 220(4599)*, pp. 781–787, 1983.

Onstott, R. G., T. C. Grenfell, C. Mätzler, C. A. Luther, and E. A. Svendsen, Evolution of microwave sea ice signatures during early summer and midsummer in the marginal ice zone, *Journal of Geophysical Research, 92(C7)*, pp. 6825–6835, June 30, 1987.

Perovich, D. K., A. J. Gow, and W. B. Tucker III, Physical properties of snow and ice in the winter marginal ice zone of Fram Strait, *Proceedings of the IGARSS '88 Symposium*, Edinburgh, Scotland, ESA SP-284, European Space Agency, Noordwijk, Netherlands, pp. 1119–1123, August 1988.

Ramseier, R. O., K. W. Asmus, M. Collins, and C. Garrity, *Scientific Cruise Reports of Arctic Expeditions ARK VI/1-4 of RV "Polarstern" in 1989*, edited by G. Krause, J. Meincke, and H. J. Schwarz, Alfred Wegener Institute, Bremerhaven, Germany, 1991.

Schanda, E., C. Mätzler, and K. Künzi, Microwave remote sensing of snow cover, *International Journal of Remote Sensing, 4*, pp. 149–158, 1983.

Tucker, W. B., III, A. J. Gow, and W. F. Weeks, Physical properties of summer sea ice in the Fram Strait, *Journal of Geophysical Research, 92(C7)*, pp. 6787–6803, June 30, 1987.

Ulaby, F. T. and W. H. Stiles, The active and passive microwave response to snow parameters, 2. Water equivalent of dry snow, *Journal of Geophysical Research, 85(C2)*, pp. 1045–1049, February 20, 1980.

Ulaby, F. T., R. K. Moore, and A. K. Fung, *Microwave Remote Sensing: Active and Passive, Vol. I: Microwave Remote Sensing Fundamentals and Radiometry*, Addison–Wesley Publishing Company, Reading, Massachusetts, 1981.

Vant, M. R., R. B. Gray, R. O. Ramseier, and V. Makios, Dielectric properties of fresh and sea ice at 10 and 35 GHz, *Journal of Applied Physics, 45*, pp. 4712–4717, 1974.

Wadhams, P., M. A. Lange, and S. F. Ackley, The ice thickness distribution across the Atlantic sector of the Antarctic Ocean in midwinter, *Journal of Geophysical Research, 92(C13)*, pp. 14,535–14,552, 1987.

Chapter 17. The Effects of Freeze-Up and Melt Processes on Microwave Signatures

S. Prasad Gogineni and Richard K. Moore

Radar Systems and Remote Sensing Laboratory, University of Kansas Center for Research, Inc., 2291 Irving Hill Road, Lawrence, Kansas 66045

Thomas C. Grenfell

Department of Atmospheric Sciences, University of Washington, Seattle, Washington 98195

D. G. Barber

Earth-Observations Laboratory, University of Waterloo, Ontario, N2L 3G1, Canada

Susan Digby and Mark Drinkwater

Jet Propulsion Laboratory, California Institute of Technology, 4800 Oak Grove Drive, Pasadena, California 91109

17.1 Introduction

The physical characteristics of sea ice and snow change rapidly during the warm season in the Arctic. This, in turn, causes both active and passive microwave signatures to fluctuate rapidly. The processes involved are snow metamorphosis, snow melt, pond formation and drainage, and both horizontal and vertical ice ablation. These processes significantly affect the entire annual cycle of fluxes between atmosphere and ocean. Moreover, ice ablation during the summer contributes to the supply of fresh water in the ocean. As the ice deteriorates, floes break up (and new boundaries are formed), thereby changing the ice concentration in the ocean both during summer and the following winter.

Summer processes also determine the characteristics of the ice during the subsequent winter. The amount of open water present at the end of the summer determines the extent of new ice that forms in the fall. This new-ice extent, in turn, influences the ice–ocean brine fluxes of early winter and the amount of new ice that can deform and produce thick first-year (FY) ice during the winter.

17.2 Physical and Electrical Properties of the Ice During the Melt Season

17.2.1 Physical Properties

During summer, ablation and breakup of sea ice occur in several stages. First-year ice and multiyear (MY) ice behave differently, although the bases for the behavior are the same. Livingstone et al. [1987] and Barber et al. [1992b] characterize the stages as early melt, melt onset, and advanced melt.

17.2.1.1. First-year ice. The first-year situation is depicted, in part, in Figure 17-1. Here, Figure 17-1(a) shows the premelt conditions, with the ice covered by a thin layer of snow. During the early-melt period, solar insolation moistens the upper snow and the crystal structure of that snow changes—the snow consolidates, see Figure 17-1(b). As insolation increases in early melt, hoar ice and meltwater begin to form under the snow when water from the moist snow percolates down to the impermeable surface of the ice. This meltwater freezes by contact with the cold ice, often forming a rough layer of superimposed ice, Figure 17-1(c).

As the air temperature rises, the melt-onset stage, during which the snow melts completely, begins. This melting results in a surface, made of superimposed ice, see Figure 17-1(d), that is much rougher than the surface of first-year ice in winter. The small puddles of water have albedos between 40% and 60%, whereas snow-covered regions and drained white ice (which is found in areas above the local water table) have albedos between 60% and 80%.

Advanced melt stage begins when air temperatures exceed 0°C, see Figure 17-1(e). The surface of the ice melts preferentially below the puddles [Untersteiner, 1961]. This pattern increases the surface relief, with the result that the ridge-to-trough height may exceed 0.5 m.

In late summer, water drains off the surface through leads, seal breathing holes, and thaw holes, see Figure 17-1(f). There is also some evidence for through-ice drainage via enlarged brine-drainage channels [Holt and Digby, 1985]. During this period, the salinity of the ice decreases, and meltwater draining from the ice results in a thin layer of "fresher" water directly beneath the ice.

17.2.1.2. Multiyear ice. The multiyear-ice situation in winter to late summer is illustrated in Figure 17-2. The multiyear (MY) ice is rough and hummocky during winter as a result of the processes of the previous summer. Its

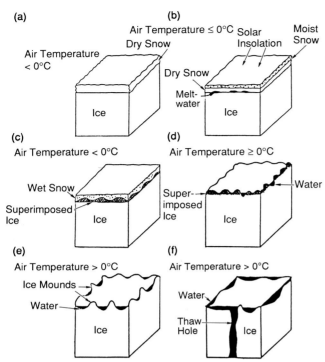

Fig. 17-1. The changes that take place in FY ice during the summer melt season, which has the stages (a) premelt, (b) very early summer, (c) early summer, (d) early midsummer, (e) midsummer, and (f) late summer [Gogineni, 1984].

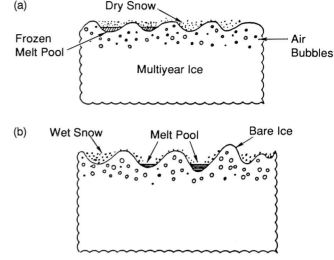

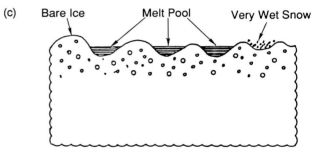

Fig. 17-2. The changes in MY ice during (a) winter, (b) early summer, and (c) late summer [Gogineni, 1984].

surface contains hummocks surrounded by flat surfaces; the flat surfaces are caused by refreezing of the previous summer's melt pools. In winter, the entire MY ice surface is covered with snow of varying depths—thinner on the hummocks and thicker over the melt pools. In early summer, the snow begins to melt, and in late summer, the upper ice layers melt, together re-forming the pools. Because of the surface relief of the MY ice, melt pools cover less area in midsummer than they do in midsummer for first-year (FY) ice.

17.2.2 Electrical Characteristics

The electrical characteristics of both sea ice and of snow are complex, as is discussed in Chapter 3. Both the real and imaginary parts of the dielectric constant increase rapidly as the temperature rises toward the melting point. In considering the melt season, one needs to consider air temperatures below 0°C. This is because of the effects of radiation loading and because of the presence of high brine contents in the lower portion of the snow cover.

For low temperatures (less than about −5°C), the relative permittivity is on the order of 2.5 for MY ice and 3 for FY ice. As the temperature gets within about 5° of freezing, the

permittivity tends to the value for water (60 to 80, depending on the frequency). Similarly, the loss factor of saline ice increases very rapidly, with differences between frazil ice and columnar ice (shown in Figure 17-3). Snow dielectric properties behave similarly. While the morphology of cold snow is important, melting snow has losses that are very high, with the result that the microwave signal depends almost exclusively on a very thin layer of melting snow near the surface [Grenfell and Lohanick, 1985]. With radar the exact impact of the melting snow cover is unknown. It does appear that the high-salinity hoar layer affects the scattering return once water is available in liquid form (i.e., temperatures greater than approximately −10°C) [Barber et al., 1992a].

17.3 SUMMARY OF RESEARCH RESULTS

17.3.1 Radar

Because of the dominating effect of water in wet snow, wet ice, and ponds, the measured contrast in summer between the σ^0's of different types of ice is much lower than the measured contrasts in winter. Radar scatter from summer ice was measured several times in the early 1980's

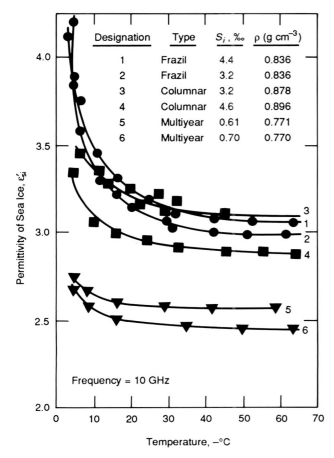

Fig. 17-3. Temperature variations in the loss factor for three types of sea ice at 10 GHz.

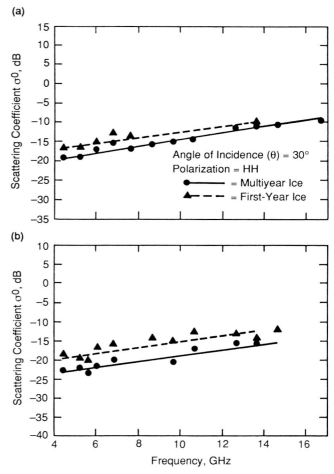

Fig. 17-4. Change in the contrast between σ^0 of FY and MY ice over a six-day period in June, 1982. Most of the snow had melted from the FY ice by the end of the time. The measurements were taken at Mould Bay, Northwest Territories, Canada, beginning (a) June 19 and ending (b) June 24 [Gogineni, 1984].

at Mould Bay, and results from the Marginal Ice Zone Experiment (MIZEX) in the Fram Strait contributed considerable additional information. Further data for the early part of the melt season came from the Seasonal Sea Ice Monitoring Site (SIMS) program in Lancaster Sound/Barrow Strait, Northwest Territories, Canada. Data for low-latitude pack ice came from two Labrador Ice Margin Experiments, LIMEX'87 and LIMEX'89. In this chapter, we summarize the essence of the results, but one must consult the literature for more details [Onstott and Gogineni, 1985; Livingstone et al., 1987; Holt and Digby, 1985; Carsey, 1985; Carsey et al., 1989; Drinkwater, 1989; Barber et al., 1992a, b].

The following discussion relates only to the high microwave frequencies (C-band and above). For L-band, the greater depth of penetration gives quite different results that show less contrast between ice types.

To illustrate the variability of the summer-ice signatures, we consider four specific examples: (1) measurements made in 1982 of fast ice at Mould Bay, Northwest Territories, Canada [Gogineni, 1984; Onstott and Gogineni, 1985; Holt and Digby, 1985; Livingstone et al., 1987]; (2) pack ice

measured in Fram Strait with MIZEX; (3) Arctic Archipelago fast ice measured during the SIMS experiment; and (4) low-latitude LIMEX'87 and LIMEX'89 experiments.

17.3.1.1. Mould Bay'82.
The experiments at Mould Bay demonstrate the extremely rapid changes that take place during different stages of the melt period. Figure 17-4 shows the changes that occurred in less than a week, from June 19 to 24, early summer at Mould Bay's location.

On June 19, when the ice was covered with wet snow, the backscatter contrast between FY and MY ice was low because the volume scatter from the MY ice was masked. The data at 30° show almost no difference between FY and MY ice returns on June 19. However, by June 24, the winter contrast had reversed (in winter, MY ice gives more return than FY ice), and the FY ice gave about 4 dB more return than the MY ice. This corresponds to the disappearance of snow from the FY ice and the emergence of rough superimposed ice. By June 30, only six days later, enough superimposed ice had melted that the σ^0 values were once again nearly the same for the FY and MY ice (for both like polarizations), as shown in Figure 17-5. Two days later (July 2), more superimposed ice had melted, causing the σ^0 of the MY ice to exceed that of the FY ice again (see Figure 17-6).

17.3.1.2. MIZEX.

Onstott et al. [1987] and Williams et al. [1987] analyzed backscatter data collected with a helicopter-borne spectrometer at 5.2, 9.6, 13.6, and 16.6 GHz during the MIZEX in 1984. Measurements took place during June 22 to July 9, 1984. During the measurement period, many MY ice floes were covered with snowpacks in excess of 60 cm in depth. Ambient temperatures transformed ice and snow from early summer's melt conditions at the start of the experiment to peak melt conditions by the end of the experiment.

Figure 17-7 shows, for FY and MY ice during early summer's melt conditions, the scattering coefficient's angular response at different frequencies. At 5.2 GHz, there is very little difference in the σ^0 of FY ice and that of MY ice. At 13.6 and 16.6 GHz, MY ice returns are only slightly higher than those from FY ice. However, this difference is

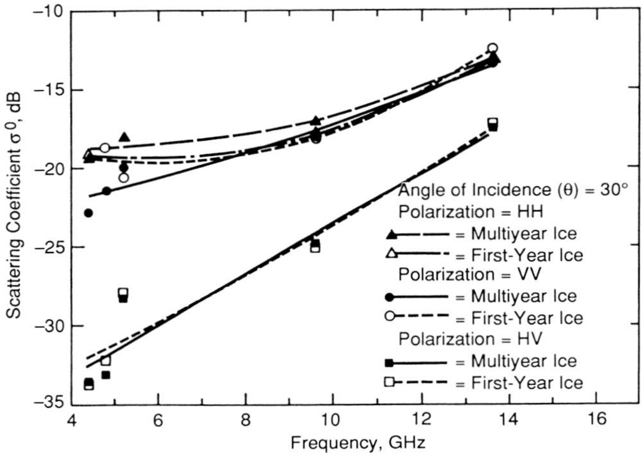

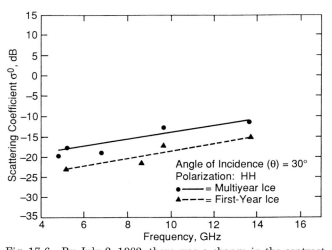

Fig. 17-5. Reduced contrast between σ^0 of FY and MY ice five days after the measurements pictured in Figure 17-4. Much of the superimposed ice had melted from the FY ice by this time. The measurements were taken at Mould Bay, Northwest Territories, Canada, on June 30, 1982 [Gogineni, 1984].

Fig. 17-6. By July 2, 1982, there was a change in the contrast between the σ^0 for FY ice and that of MY ice; there was again a larger value for the MY ice σ^0 (as compared with the similarity between FY and MY ice σ^0's in Figure 17-5). The measurements were taken at Mould Bay, Northwest Territories, Canada [Gogineni, 1984].

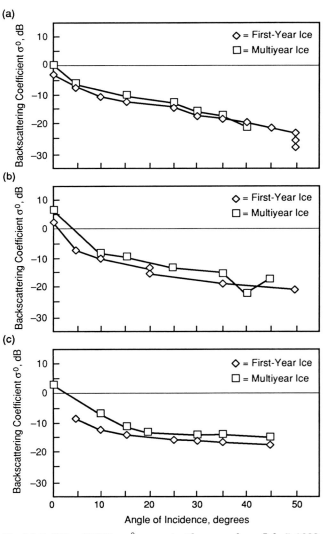

Fig. 17-7. FY and MY ice σ^0 versus incidence angle on July 5, 1982; the data were collected at (a) 5.2 GHz, (b) 13.6 GHz, and (c) 16.6 GHz.

not large enough for distinguishing MY from FY ice by using algorithms based on image intensity. These results are consistent with earlier summer-season observations reported by Onstott and Gogineni [1985].

Figure 17-8 shows the temporal backscatter response at 30° angle of incidence for data gathered at four frequencies. The results show an overall 5-dB variation in σ^0 for each ice type on any given day, with very little change in contrast between ice types from one day to another.

17.3.1.3. SIMS'90 experiment. SIMS'90 involved field data collection from May 15 to June 8, 1990, on fast ice near Resolute Bay, Northwest Territories. Synthetic Aperture Radar (SAR) data were collected using STAR-2, the Sea Ice and Terrain Assessment Radar (9.25 GHz, HH-polarization), coincident in space and time with the SIMS'90 sample sites. These data are used here to illustrate the diurnal and seasonal changes in microwave scattering from FY and MY ice types.

Diurnally, a dramatic change in X-band SAR signatures is shown by the two data sets collected on June 13, 1990. In Figure 17-9(a), data collected at 0818 hours (local time) clearly show the location of two MY formations (a conglom-

erate floe and the SIMS'90 MY ice site). These floes are not distinguishable in imagery acquired three hours later, Figure 17-9(b). We postulate that the volume scattering of the snow and ice dominated the 0818 hour image and that surface scattering from the snow–air interface dominated the 1100-hour image [Barber et al., 1992a]. Since the snow surface roughness was similar for both the FY and MY ice types, there is no way to distinguish the types once the snow surface begins to dominate the returns.

Seasonally, the evolution of the scattering cross section would be a useful parameter both from an ice classification perspective and for the detection and monitoring of the evolution of the energy-balance parameters of the snow-covered sea ice volume. To assess the potential of SAR for monitoring this seasonal evolution, we have used a multivariate analysis of variance (MANOVA) to separate the effects of SAR scattering season (winter, early melt, melt onset) from those of ice type (our three sites were MY ice, FY ice, and home ice); see [Barber et al., 1992b] for details. To illustrate the seasonal evolution of image tone and texture, the MY ice site is presented in Figures 17-10(a) and (b). The metric used in this analysis was the contrast statistic from the gray level co-occurrence matrix (GLCM); see [Barber

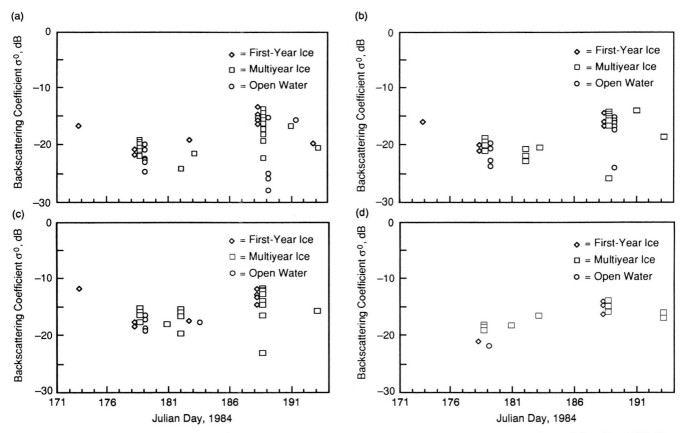

Fig. 17-8. FY ice, MY ice, and open water σ^0 plotted against observation day. (Julian Day 178 is June 26; July 5 is Julian Day 187.) The backscatter response is 25°. The data were gathered at (a) 5.2 GHz, (b) 9.6 GHz, (c) 13.6 GHz, and (d) 16.6 GHz. The observations for each type of ice and the water are displaced slightly on the "Day" axis to improve legibility.

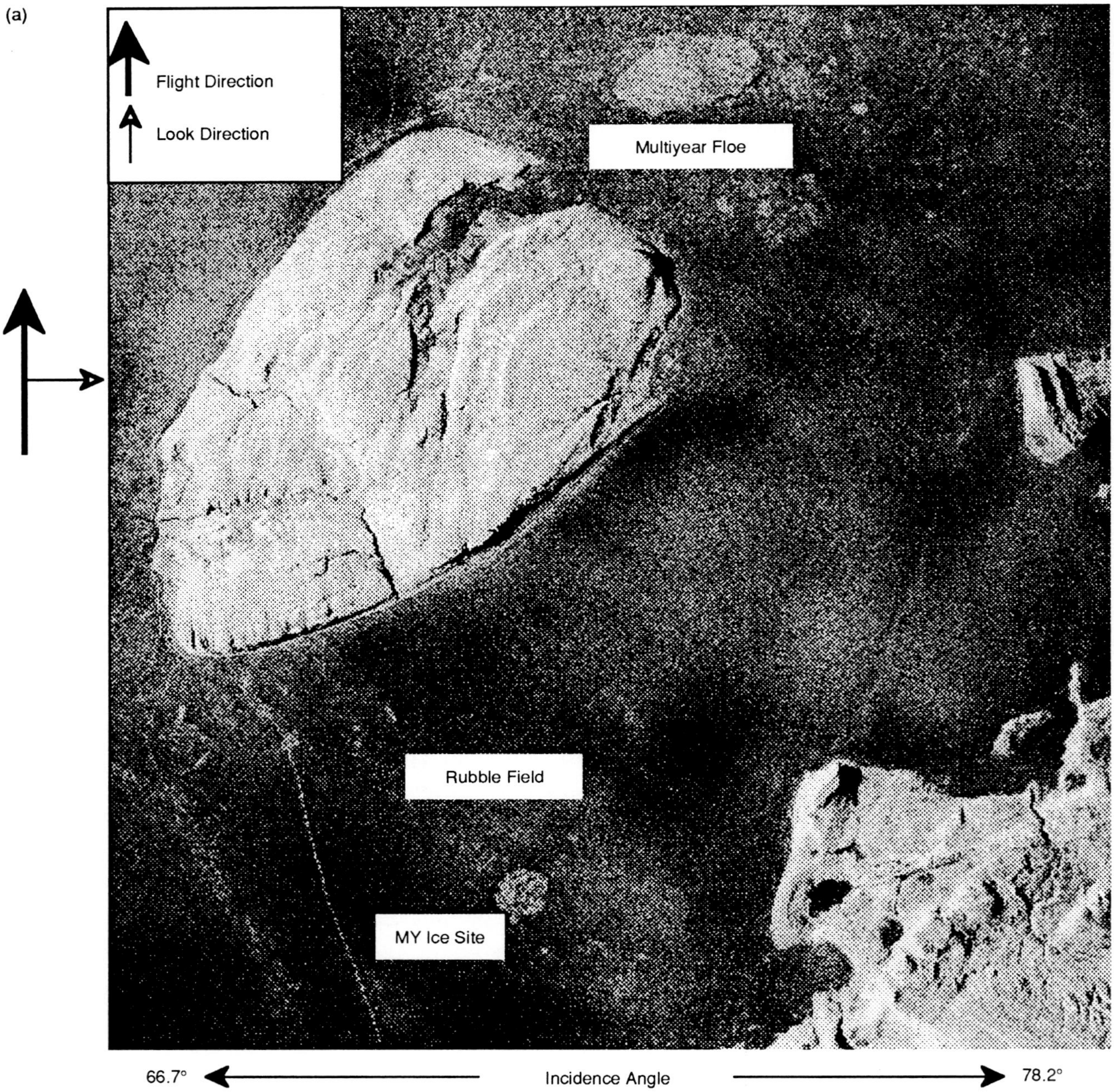

Fig. 17-9. SAR imagery of the SIMS'90 experiment site, taken June 13, 1990, at (a) 0818 hours local time and (b) 1100 hours local time.

and LeDrew, 1991] for details of the computation of this texture measure.

The MANOVA results (Table 17-1) show that both factor levels ("ice site" and "season") contain at least one GLCM contrast statistic that is significantly different from the others in each of the season and ice site factors. The difference in the ice site main effect (F-ratio = 1950.9) is larger than the difference in the season main effect (F-ratio = 590.91). This means that there is greater separation of at least one of the ice sites (i.e., ice types) than there is amongst the seasons when all ice sites are considered as one.

Examination of the pairwise main effects shows that the interaction of ice site and season is significant (F-ratio = 172.29). This means that season has an effect on at least one of the ice sites when they are considered over the three seasons. Figure 17-11 shows that there is good separation of the MY from the two FY sites (FY and home ice) in the winter season. The separation decreases in the early melt

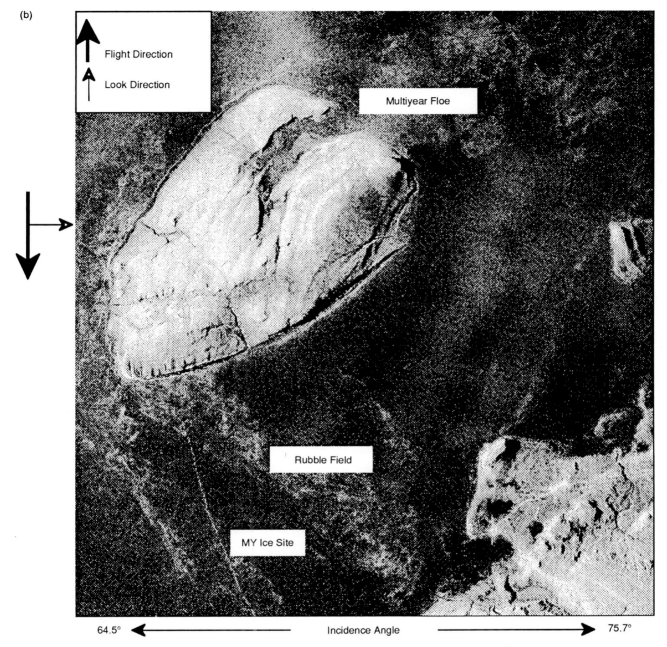

Fig. 17-9. SAR imagery of the SIMS'90 experiment site, taken June 13, 1990, at (a) 0818 hours local time and (b) 1100 hours local time. (Continued)

period and further decreases in the melt onset season. It is interesting to note that the separation of the FY and home sites increases in the melt onset season. These results have been used to suggest that MY ice types may act as an early indicator of the onset of melt in the Arctic and that FY ice types may be sensitive to the later stages of the seasonal evolution [Barber et al., 1992b].

In summary, the post hoc analysis shows that the three main ice types studied are separable over the three seasons

observed. In general, separation is best between MY ice and the two FY ice types. This separability between MY ice and the FY ice types degrades with the onset of melt. Separation, in general, is poorest between the two FY ice types and improves with the heterogeneity introduced by seasonal transition. When each ice site was examined for effects caused by seasonal change, the MY site showed considerable sensitivity to all seasonal variations. The home site showed poor sensitivity to the transition between winter

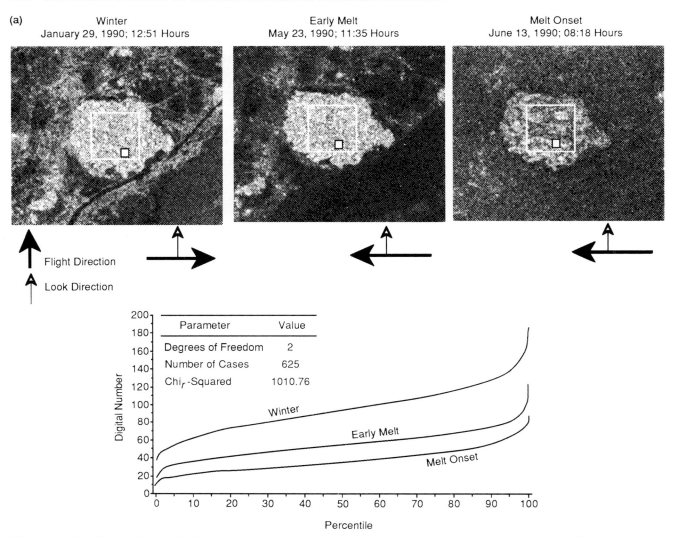

Fig. 17-10. (a) MY ice site X-band SAR imagery (256 by 256 pixels), and grey level distributions from the SIMS'90 field experiment sites (opaque white boxes) within each scene, (b) MY ice site X-band SAR imagery (256 by 256 pixels), and GLCM contrast statistic box plots from the SIMS'90 field experiment sites (opaque white boxes). Orientations used in computing the GLCM, and the flight direction and look direction of the SAR, are denoted.

and early melt, but significant sensitivities to winter versus melt onset and early melt versus melt onset. The FY ice site showed comparatively poor separation over the seasonal contrasts studied.

The geophysical and biophysical significance of these results is considerable and includes the potential to monitor the physical changes occurring in the snow and sea ice. These geophysical changes can then be used to infer the state and fluxes of the major energy-balance components of the snow-covered sea ice energy balance. These are the ongoing objectives of the SIMS'91 to SIMMS'95 (Seasonal sea Ice Monitoring and Modeling Site) field experiments.

17.3.1.4.   LIMEX. LIMEX provided surface measurements and X- and C-band airborne-SAR data for FY pack

ice; the surface measurements and the SAR data were each taken over a period of a few weeks. As shown by these experiments, surface conditions altered rapidly from early melt to midsummer as Arctic winter air was replaced by storms bringing moist, above-freezing air over the region.

According to the surface measurements, physical changes in snow and ice surfaces were similar to those observed in the high Arctic for FY ice in Mould Bay; however, the SAR return is remarkably different from the Mould Bay FY ice data because the large-scale morphology of the two areas is very different. In particular, surface flooding that can exist over many kilometers of FY ice in early summer to midsummer in the Arctic is not possible with Labrador Sea pack ice.

Examination of C- and X-band SAR data shows backscatter in Labrador Sea pack ice; the backscatter is a complex

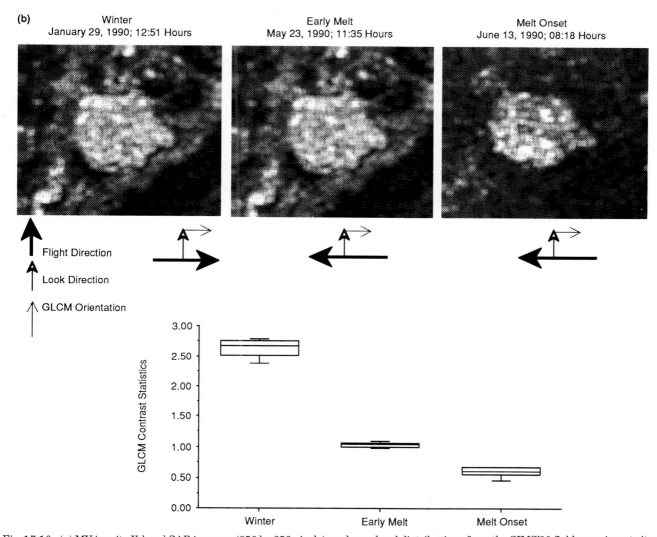

Fig. 17-10. (a) MY ice site X-band SAR imagery (256 by 256 pixels), and grey level distributions from the SIMS'90 field experiment sites (opaque white boxes) within each scene, (b) MY ice site X-band SAR imagery (256 by 256 pixels), and GLCM contrast statistic box plots from the SIMS'90 field experiment sites (opaque white boxes). Orientations used in computing the GLCM, and the flight direction and look direction of the SAR, are denoted. (Continued)

function of surface wetness, large-scale surface roughness [Drinkwater, 1989], floe size (a measure of surface roughness) [Carsey et al., 1989], and wave disturbance [Raney et al., 1989]. The percentage of the surface covered by frazil may also contribute to the overall backscatter. In contrast, melt season returns from the largely undeformed land-fast pack ice of the high Arctic are driven by surface-wetness characteristics.

17.3.1.5. Research summary. To summarize the results of the Mould Bay'82, MIZEX'84, and SIMS'90 experiments:

(1) In winter and spring, the return from MY ice is much stronger than that from FY ice.

(2) When the snow begins to melt in early summer, the contrast between MY and FY signatures vanishes.

(3) When no snow remains on the FY ice, the winter contrast is reversed, with the result that FY ice returns exceed MY ice returns by a few decibels. This shift can occur in less than a week.

(4) Melting of superimposed ice can cause the reversed contrast to vanish in less than a week.

(5) As melt ponds increase, the MY ice return once again exceeds FY ice return. This condition continues throughout the summer.

(6) The seasonal evolution of microwave scattering from snow-covered sea ice can be measured by using SAR image texture. MY ice types are more sensitive to the early changes (i.e., winter to early melt), and FY ice types in different consolidation states provide increased sensitivities for the latter seasonal transitions (i.e., early melt to melt onset).

TABLE 17-1. Two-way MANOVA of GLCM contrast statistics that were obtained for three levels of season (winter, early melt, and melt onset) and three levels of ice site (MY, home, and FY ice). The analysis is based on the square root transform of the GLCM contrast statistics. [Barber et al., 1992a].

Source	SSE	DF	MSE	F-ratio	P-value
Season	2.91	2	1.45	590.91	0.0001
Ice site	9.60	2	4.80	1950.09	0.0001
Interaction of season and ice type	1.70	4	0.42	172.29	0.0001
Residual error	0.20	81	0.002		

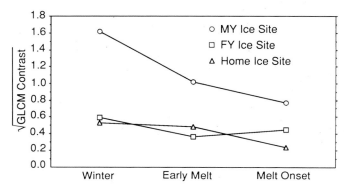

Fig. 17-11. Main-factor interaction effects as derived from the MANOVA of ice-site and season variances [Barber et al., 1992b, © 1992 IEEE].

(7) Seasonal evolution of σ^0 is geographically and seasonally dependent.

17.3.2 Radiometry

For surface-based observations, the observed "emissivity" for the ice has been calculated from the formula

$$\varepsilon_p(f,\phi) = \frac{T_B(p,f,\phi) - T_{\text{sky}}(f,\phi)}{T_{SI} - T_{\text{sky}}(f,\phi)}$$

where T_B is the brightness temperature of the surface, T_{sky} is the brightness temperature of the sky, f is the observational frequency, ϕ is the nadir angle (or corresponding zenith angle for the sky), and T_{SI} is the physical temperature at the snow–ice interface. While this formula is only an approximation for cold MY ice, it is quite accurate for cold FY ice and both FY and MY melting ice. Even for cold MY ice, it reduces the data scatter considerably as compared with using T_B.

Figure 17-12 shows cluster data for MY ice during fall freeze-up [Grenfell 1992]. These data were obtained at 18.7

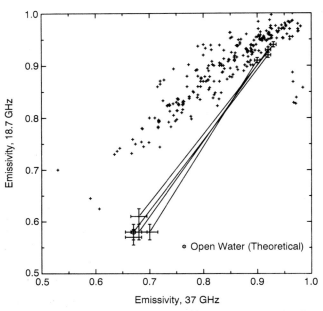

Fig. 17-12. Scatter plot of emissivities for sites on MY ice at 18 GHz and 37 GHz during the CEAREX'88 drift experiment [Grenfell, 1992]. The solid lines show the location of the cluster of points for melting ice measured during the ship's approach to the station floe. (Data = V-polarization; nadir angle = 50°.) The small cluster at about emissivity, 37 GHz = 0.97 and emissivity, 18 GHz = 0.85 is for partly frozen melt ponds.

GHz and 37 GHz and vertical polarization during the Coordinated Eastern Arctic Research Experiment (CEAREX). The individual points were collected over continuous MY ice floes in the observation area; the floes' identities were established from salinity profiles and from their total thicknesses, which were in excess of 3 meters. The angle of incidence for the observations (nadir angle) was 50°.

During late summer under melting conditions, the data were limited to a narrow strip indicated by the solid lines connected by the crosses, where the size of the crosses indicates the standard deviation at each end of the strip. As the surface began to freeze, the melting ice points shifted rapidly to another region defined by endpoints of approximately 0.97, 0.96 (for 18.7 and 37 GHz, respectively) and 0.725, 0.675. This occurred over an 8- to 10-day interval as the ship traveled northward through the ice pack, and was in part driven by the change in atmospheric conditions from the ice edge regime to the interior zone. By September 22, the transition to freezing conditions was well under way, and by September 30, the melt ponds were frozen solid. At this point, the MY emissivity distribution defined by the individual data points was established.

17.3.2.1. Melt ponds.
Since melt ponds cover a substantial fraction of the summer ice, the role of freezing and refrozen melt ponds can have a significant effect on the MY

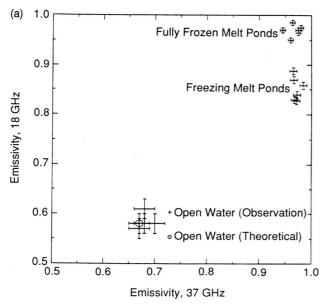

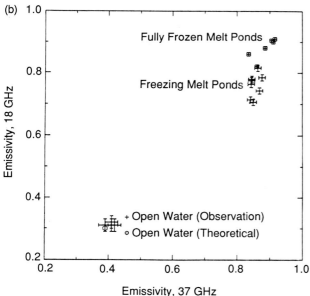

Fig. 17-13. (a) Scatter diagram for emissivities at 18 GHz versus 37 GHz at V-polarization and nadir angle of 50° for melt ponds on CEAREX'88. The three clusters are for open water, frozen-over ponds with underlying liquid water, and fully frozen ponds. The size of the crosses indicates the observational uncertainty for the measurements. (b) Corresponding data for H-polarization.

ice signature. Open melt ponds are indistinguishable from nearby leads or ocean water if either has surface roughness similar to that of the melt pond [Grenfell and Lohanick, 1985]. During freeze-up, however, melt pond signatures quickly become quite different from those of open water. Figures 17-13(a) and (b) present 18.7- versus 37-GHz scatter plots for melt ponds at both vertical and horizontal

polarizations. When the melt ponds are open, their emissivities cluster about the theoretical values calculated from Fresnel reflection theory. The melt pond represented by the "freezing" cluster had 25 to 30 cm of new fresh ice on the surface and 5 to 10 cm of liquid water remaining between the pond ice and the underlying sea ice. The emissivity reached saturation at 37 GHz because the fresh ice was optically thick. However, at 18.7 GHz, the effect of the higher reflectivity at the fresh ice–water boundary is still present, resulting in a higher net reflectivity for the freezing melt pond system. For the fully frozen ponds, the internal interface is no longer present, and the emissivity at 18 GHz also reaches saturation.

Virtually no scattering inhomogeneities are apparent down to at least the level where the brine volume in the ice rises significantly. This gives a flat spectrum very similar to that of FY ice. Since almost all Arctic MY ice is covered to some extent by melt ponds, it appears that even in winter the signature of MY ice depends on the degree of melt pond coverage. Estimates from surface observations [A. Hanson, personal communication, 1984] and from satellite data [Carsey, 1985] suggest wintertime coverages in the range of 15% to 30% at ice stations in the Arctic Basin.

These passive microwave results for melt ponds can be summarized:

(1) Spectra of open melt ponds were just like those of open leads.

(2) Spectra for freezing melt ponds closely resembled those of lake ice. They showed a positive spectral gradient and a rather high degree of polarization. The ice was not optically thick, and the high reflectivity of the underlying water affected the emissivity.

(3) As soon as the ponds were frozen through, their spectra became nearly the same as those for FY ice.

Passive microwave results can also be summarized for MY ice:

(1) For summer ice, the emissivities are determined by whether or not the surface is melting. If the upper layers have refrozen, the spectral gradient is negative with a slope that depends on the thickness and grain size of the refrozen layer. If the surface is melting, the spectrum is flat and the emissivity is close to 1. The slope of the spectrum depends only on the liquid-water content (LWC) of the surface, which takes up a very small range of the surface mass (0% to 4%), and the transition between melting and freezing is very fast. Liquid water between 4% and about 13% (which occurs during the melt season) increases the polarization ratio, especially at low frequencies, but does not affect the vertical-polarization-gradient ratio.

(2) For cold conditions, the emissivities for 100% MY ice cover a large range. The emissivities are stable over time, however. The range depends on the thickness of a near-surface decomposed and/or bubbly layer.

(3) In both summer and winter, the variance among emissivity spectra of MY ice is influenced primarily by the spatial distribution of melt ponds and by the presence of

significant amounts of scattering inhomogeneities that are above 0.5 mm in diameter in the snow and the upper 20 to 30 cm of the ice.

17.3.3 Satellite Data

Ketchum [1984] analyzed Sea Satellite (Seasat) SAR images collected for an area west of Banks Island. These images were acquired from summer melt through autumn freeze-up. He speculated that water-saturated slush layers would increase the backscatter of the signal and could be misinterpreted as rubble fields during the melt season. His analysis showed that the ambiguities introduced by the slush can be eliminated by utilizing both time-sequence data and information about floe sizes and shapes.

Drinkwater and Carsey [Observations of the later-summer to fall sea ice transition with 14.6-GHz Seasat scatterometer, submitted to *Journal of Geophysical Research,* 1992] used the Seasat satellite Ku-band (14.6-GHz) scatterometer (SASS) and collocated Scanning Multichannel Microwave Radiometer (SMMR) data to illustrate the pattern of signature transition between peak summer melt and fall freeze-up. These data were acquired between July 9 and October 8, 1978. A contiguous area of consolidated MY ice remaining in the Beaufort Sea at the end of summer was chosen because it had areas in which the 18-GHz SMMR brightness temperatures exceeded 210 K. Gridded data were averaged over five days to eliminate any change due to the short-term, passing, low-pressure systems that happen around that frequency [Carsey, 1985].

SASS data fall into the incidence angle range of 0 to 60°, and gates centered at 2.5° and 7.5° enable the near-normal incidence-response function to be evaluated. These two gates show the sensitivity of the near-normal backscatter coefficient to the presence of saturated snow or wet ice. At the time of melt pond maximum (about July 29), when surfaces are consistently wet, the near-nadir backscatter reaches a peak value of over 13 dB, as shown by Figure 17-14. This value is consistent with the reflectivity of a water-saturated snow surface with a relative permittivity of 4.0. After July 29, an abrupt drop in near-normal incidence backscatter (>4 dB) signals a rapid drainage event. The area-averaged signatures show this effect to be spread over a period of several days throughout the region of interest; sometimes they also show a time lag between drainage dates at different latitudes.

Meanwhile, values of backscatter in the high incidence-angle range, such as 40°-gate backscatter, begin to rise slowly. After the scattering coefficient reaches a minimum of −16 dB near peak melt (July 24) because of suppression of volume-scattering effects, ice-surface drainage also impacts the values by starting a rising trend. Cooling, drying, and drainage of the ice surface result in a 10-dB rise in 40° incidence backscatter over a two-month period. Values stabilize at around −6 dB by September 17, after widespread snowfall occurs. At this time, the variability in backscatter values also reduces dramatically. Figure 17-15 illustrates two contrasting mean H-polarized SASS signa-

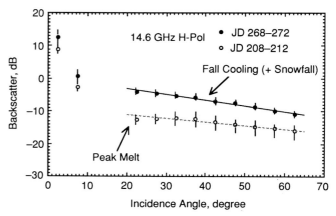

Fig. 17-14. Mean H-polarized SASS signatures for (a) peak summer melt and (b) fall freeze-up and snowfall.

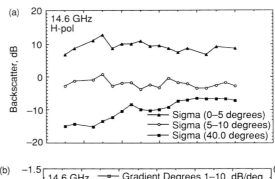

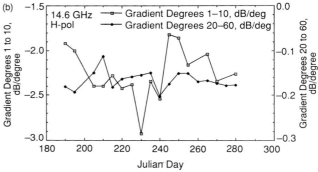

Fig. 17-15. (a) Summer–fall transition in backscatter, σ^0, at angles 0°–5°, 5°–10°, and 40°; (b) temporal change in the gradient of $\sigma^0(\theta)$ for the near and far swath, extracted by linear least-squares fits to the ice signatures.

tures, one from peak summer melt and one obtained after fall cooling and snowfall.

17.4 Conclusions

The seasonal variation is large for backscatter characteristics of FY and MY ice. During the early part of the melt season, a rough snow–ice interface develops on FY ice. This causes greater backscatter from FY ice at that time than in winter months. As summer progresses, the FY ice surface becomes heavily flooded from melting snow and eventually from the melting ice itself. Because of this, the scattering coefficient of FY ice at incidence angles well away from vertical decreases by many decibels. The scattering coefficient increases as ice refreezes, and eventually it attains a value similar to that of second-year ice.

Early in the melt period, MY ice returns are much lower than in winter. This results from blockage of the volume-scattering contribution by wet snow and wetness in the upper ice layers. The angular response of the σ^0 of MY ice also changes quite dramatically. For MY ice, the decrease of σ^0 with incidence angle is small in the winter, whereas MY ice σ^0 decreases very rapidly with incidence angle in summer.

Because of the rapid fluctuations in the backscatter response of FY and MY ice and reversals in the contrast between them, it is extremely difficult to classify C- and X-band radar images with intensity-based algorithms.

Summer passive-microwave signatures are also controlled by the upper layers of ice. The presence of moisture in snow forces MY ice brightness temperatures to become similar to those of FY ice. A passive sensor is much more sensitive to the presence of moisture than an active sensor and may be very useful in monitoring the onset of melt conditions. Once a significant amount of melting occurs, the passive sensor loses much of its ability to estimate ice concentration and to map ice types.

The available experimental data were taken during the melt-onset and peak-melt conditions. Although a few studies have been made of the microwave response of sea ice in the fall freeze-up, results are not yet available in the open literature. Data from the first European Remote Sensing Satellite (ERS-1) and the Japanese Earth Resources Satellite (JERS-1) sensors and from associated validation experiments should provide valuable information on monitoring sea ice in the melt season.

REFERENCES

Barber, D. G. and E. F. LeDrew, SAR Sea Ice Discrimination Using Texture Statistics: A Multivariate Approach, *Photogrammetric Engineering and Remote Sensing, 57(4)*, pp. 385–395, 1991.

Barber, D. G., E. F. LeDrew, D. G. Flett, M. Shokr, and J. Falkingham, Seasonal and Diurnal Variations in SAR Signatures of Sea Ice, *IEEE Transactions on Geoscience and Remote Sensing, 30(3)*, pp. 638–642, 1992a.

Barber, D. G., D. G. Flett, R. A. De Abreu, and E. F. LeDrew, Spatial and Temporal Variations in Sea Ice Geophysical Properties and Microwave Remote Sensing Observations: The SIMS'90 Experiment, *Arctic, 45(3)*, pp. 233–251, 1992b.

Carsey, F., Summer Arctic sea ice character from satellite microwave data, *Journal of Geophysical Research, 90(C3)*, pp. 5015–5034, 1985.

Carsey, F. D., S. A. Digby, S. Argus, M. J. Collins, B. Holt, C. E. Livingstone, and C. L. Tang, Overview of LIMEX'87 ice observations, *IEEE Transactions on Geoscience and Remote Sensing, 27(5)*, pp. 468–484, 1989.

Drinkwater, M. R., LIMEX'87 ice surface characteristics: Implications for C-band SAR backscatter signatures, *IEEE Transactions on Geoscience and Remote Sensing, GE-27(5)*, pp. 501–513, 1989.

Gogineni, S. P., *Radar Backscatter From Summer and Ridged Sea Ice, and the Design of Short-Range Radars*, Ph.D. dissertation, 128 pp., University of Kansas, Lawrence, Kansas, 1984.

Grenfell, T. C., Surface-based passive microwave studies of multiyear sea ice, *Journal of Geophysical Research, 97*, pp. 3485–3501, 1992.

Grenfell, T. C. and A. W. Lohanick, Temporal variations of the microwave signatures of sea ice during the late spring and early summer near Mould Bay, Northwest Territories, *Journal of Geophysical Research, 90(C3)*, pp. 5063–5074, 1985.

Haralick, R. M., K. S. Shanmugan, and I. Dinstein, Textural features for image classification, *IEEE Transactions on Systems, Man, and Cybernetics, SMC-3(6)*, pp. 610–621, 1973.

Holt, B. and S. A. Digby, Processes and imagery on fast first-year sea ice during the melt season, *Journal of Geophysical Research, 90(C3)*, pp. 5035–5044, 1985.

Ketchum, R. D., Jr., Seasat sea-ice imagery: Summer melt to autumn freeze-up, *International Journal of Remote Sensing, 5(3)*, pp. 533–544, 1984.

Livingstone, C. E., R. G. Onstott, L. D. Arsenault, A. L. Gray, and K. P. Singh, Microwave sea-ice signatures near the onset of melt, *IEEE Transactions on Geoscience and Remote Sensing, GE-25(2)*, pp. 174–187, 1987.

Onstott, R. G. and S. P. Gogineni, Active microwave measurements of Arctic sea ice under summer conditions, *Journal of Geophysical Research, 90(C3)*, pp. 5035–5044, 1985.

Onstott, R. G., T. C. Grenfell, C. Mätzler, C. A. Luther, and E. A. Svendsen, Evolution of microwave sea ice signatures during early summer and midsummer in the marginal ice zone, *Journal of Geophysical Research, 92(C7)*, pp. 6825–6835, 1987.

Raney, R. K., P. W. Vachon, R. A. DeAbreu, and A. S. Bhogal, Airborne SAR observations of ocean surface waves penetrating floating ice, *IEEE Transactions on Geoscience and Remote Sensing, GE-27(1)*, pp. 492–500, 1989.

Untersteiner, N., On the mass and heat budget of the Arctic sea ice, *Arch. Meteorol. Geophys. Bioklimatol., A(12)*, pp. 151–182, 1961.

Williams, T. H. L., G. M. Shirtliffe, R. K. Moore, and R. G. Onstott, *Active Microwave Spectrometry in the Marginal Ice Zone*, RSL Technical Report 3311-10, 25 pp., Radar Systems and Remote Sensing Laboratory, Lawrence,

Chapter 18. Determination of Sea Ice Motion From Satellite Images

BENJAMIN HOLT

Jet Propulsion Laboratory, California Institute of Technology, 4800 Oak Grove Drive, Pasadena, California 91109

D. ANDREW ROTHROCK

Polar Science Center, Applied Physics Laboratory, University of Washington, Seattle, Washington 98105

RONALD KWOK

Jet Propulsion Laboratory, California Institute of Technology, 4800 Oak Grove Drive, Pasadena, California 91109

18.1 INTRODUCTION

The motion of sea ice is important in a wide range of problems in polar oceanography. On scales larger than several hundred kilometers, there is a general circulation of the ice cover that determines the advective part of the ice mass balance and provides a velocity boundary condition on the ocean surface. Smaller scale processes involve the detailed motion of individual floes, aggregates of floes, and the formation of leads. Ice motion controls the abundance of thin ice and therefore many surface exchange processes dependent on thin ice, such as the turbulent heat flux to the atmosphere, ice production, and salinity forcing of the ocean. The field of motion is spatially discontinuous, so questions arise about the proper definition of spatial gradients. For examining ice motion, especially these smaller scale processes, images from satellite sensors that resolve floes and leads provide an attractive tool because they allow dense spatial sampling of the ice cover.

Sea ice motion is readily observable from satellite imagery. Of particular value has been imagery from the Synthetic Aperture Radar (SAR), the Advanced Very High Resolution Radiometer (AVHRR), and the Multispectral Scanner (MSS) on Landsat. A displacement measurement from imagery has two requirements: a means to recognize a common feature—a particular crook in a lead, a distinctive floe or ridge—in each of two sequential images, and a means to determine the geographic position of the feature in each image. The utility of the observation depends more on feature recognition, accurate positioning, and the geometric fidelity of the images than on radiometric accuracy or even on a thorough physical understanding of the ice signatures themselves.

Currently, two satellites, the First European Remote Sensing Satellite (ERS-1) and Japan's Earth Resources Satellite (JERS-1), are carrying SAR's. Extensive imagery from ERS-1 has been acquired in the polar regions. A third satellite is in preparation, Canada's Radarsat, which will be capable of essentially complete SAR coverage of the polar

regions every few days. Automated ice tracking processors are in place to provide scientists with routine motion data from this SAR imagery. A major advance in the observation of sea ice motion. These SAR observations will add to data being routinely acquired from drifting ice buoys and from visible and infrared satellite sensors, providing a rich new database for many areas of inquiry into sea ice geophysics.

18.2 ICE MOTION AND OBSERVATIONS

The fundamental concepts of ice motion are position, displacement, and velocity. Consider ice at a position $\mathbf{X}$ at time $t = 0$, at some later time the ice has moved to a new position $\mathbf{x}(t: \mathbf{X})$. A displacement is the difference in the positions of an ice particle at two different times

$$\mathbf{u} = \{\mathbf{x}(t_2) - \mathbf{x}(t_1)\}\mid_{\mathbf{X} = \text{constant}} \qquad (1)$$

The average velocity over the intervening time interval $T = t_2 - t_1$ is

$$\mathbf{v} = \mathbf{u}/T \qquad (2)$$

18.2.1 Observational Methods

Most of the existing data on ice motion have been provided by buoys drifting on the sea ice cover. Drifting ice stations and ships have also provided historical data. Satellite positioning systems such as Argos give several daily positions per buoy with an accuracy of about 300 meters. Buoys tend to be placed tens to hundreds of kilometers apart, and so provide data of low spatial density with good temporal resolution, about ten positions per day.

Satellite images provide displacements of as many ice features as can be identified and tracked between one pair of sequential images. The data are $(\mathbf{x}_j, \mathbf{u}_j, j = 1, ..., M)$ for M points or features, where $\mathbf{u}_j$ is the displacement from the original position $\mathbf{x}_j$ to the final position. Although the temporal resolution and spatial coverage are limited by sensor swath widths and orbit repeat intervals, the spatial resolution can be quite high, about one measurement every kilometer.

Microwave Remote Sensing of Sea Ice
Geophysical Monograph 68

Spaceborne SAR imagery has the advantage of having all-weather, day and night operational capability, fine ground resolution (generally about 25 m), good geometric accuracy, and sensitivity of the backscatter to the roughness, dielectric, and physical properties of both sea ice and open water. Thus SAR images show ridges, details of the variations in the surface and composition of ice floes, and floe boundaries. The radar imagery can also distinguish several ice types, even at reduced resolutions of about 300 m. The satellite SAR systems flown to date—Seasat (in 1978), ERS-1, and JERS-1—have approximately 100-km swath widths and 1 to 3 day near-repeat orbits. These swaths provide strips of ice motion data within a region. The ability to track ice within ERS-1 swaths has been estimated by Dade et al. [1991]. The wide swath (300 to 500 km) capabilities of Radarsat will enable complete and rapid mapping of the polar regions.

Optical and near-infrared imagery, such as from AVHRR and Landsat MSS, have a high contrast resulting from the albedo and temperature difference between the ice and ocean. The edges of floes are more clearly distinguished than surface features. The wide swath (2600 km) of AVHRR results in rapidly repeating coverage in the polar latitudes at a moderate resolution from 1.1 km at nadir to 4 km at the swath edges. Landsat MSS has a finer resolution (25 to 80 m) and a swath width of 185 km, but the repeat coverage has significant temporal gaps. In the Arctic, maximum cloud cover is from May to October [Herman, 1986], which precludes obtaining complete seasonal information from these sensors. Detection schemes used to separate clouds from the ice cover are imperfect since the ice cover and cloud often have similar radiances.

18.2.2 General Circulation

The pattern of ice transport across the ice-covered basins and oceans has been studied to date exclusively by using drifting buoys and stations. A collection of all Arctic Ocean buoy tracks from 1979 to 1982 is shown in Figure 18-1. Roughly an equal number of trajectories are known from earlier drifting stations. From these data have been computed the interpolated mean ice velocity field in the Arctic Ocean [Colony and Thorndike, 1984] and the mean vorticity. Estimates of shear and divergence have not yet been published. Although data from the seasonal ice zones are sparse, the circulation pattern also shows that the Laptev, Beaufort, and Chukchi Seas are likely source regions of ice and the Fram Strait and East Siberian Sea are likely sinks, because on average more ice moves into than out of these regions. Ice motion data from the Antarctic are even more sparse and localized. Buoy tracks in the western Weddell Sea show a northward drift of about 4 km/day [Limbert et al., 1989]. A few buoy trajectories have been obtained in the Ross Sea and the Southern Ocean off east Antarctica [Moritz, 1992; Allison, 1989], but currently there is no climatology of Antarctic ice motion.

Ice flux through major passages is an important element of the ice mass balance, and the global fresh water and salt

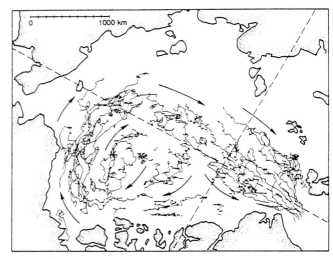

Fig. 18-1. Trajectories of automatic data buoys compiled from 1979 to 1982. The bold arrows provide the directional sense [Colony and Thorndike, 1984].

balances. An analysis of the ice velocity through Fram Strait from 20 years of buoy and station data shows considerable variation in velocity across the Strait, with the highest velocities being concentrated over the continental slope [Moritz, 1988]. Similar spatial structure has been measured from aircraft SAR imagery [Shuchman et al., 1987]. Zhang [1991] compares a one-month time series of ice motion and concentration derived from AVHRR with Moritz's results and finds similar mean velocities, but smaller velocity fluctuations, most likely because of the comparatively limited AVHRR data set (Figure 18-2). Both the mean and fluctuating components of ice motion in the Greenland Sea are dominated by currents, with the geostrophic wind having a secondary effect. Ice velocities and the strength of spatial gradients estimated from AVHRR imagery of the Fram Strait are greater than those predicted by a numerical circulation model [Emery et al., 1991]. Both discrepancies are probably due to the model's underestimation of the strength of the East Greenland Current.

18.2.3 Deformation and Spatial Statistical Structure

Spatial differences take on particular significance in ice motion for several reasons. The relative motion of floes causes leads to form, and these leads are the site of strong air–sea exchange and of ice production in winter and solar absorption in summer. The stresses by which ice resists motion are related to the strain rate, that is, to the spatial variation in ice velocity. There are several useful ways to describe this spatial structure.

The spatial autocorrelation function $\mathbf{R}$ of a homogeneous, isotropic ice velocity field is given by

$$
R(\mathbf{x}_1, \mathbf{x}_2) = q^{-2} E \mathbf{v}(\mathbf{x}_1) \mathbf{v}^T(\mathbf{x}_2)
$$

$$
= q^{-2} \begin{pmatrix} E u(\mathbf{x}_1) u(\mathbf{x}_2) & E u(\mathbf{x}_1) v(\mathbf{x}_2) \\ E u(\mathbf{x}_2) v(\mathbf{x}_1) & E v(\mathbf{x}_1) v(\mathbf{x}_2) \end{pmatrix} \quad (3a)
$$

(a)

(b)

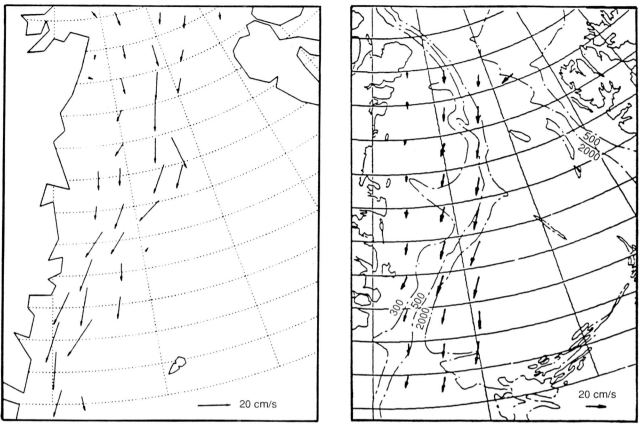

Fig. 18-2. Mean ice velocity vectors in Fram Strait. (a) Drifting buoy and station observations compiled by Moritz [1988]. (b) AVHRR observations from March 9–April 10, 1987, from Zhang [1991].

$$R\,(\mathbf{x}_1,\mathbf{x}_2) = \begin{pmatrix} \cos^2\theta\,B_p(r) & \cos\theta\,\sin\theta \\ +\sin^2\theta\,B_n(r) & \times(B_p(r)-B_n(r)) \\[1em] \cos\theta\,\sin\theta & \sin^2\theta\,B_p(r) \\ \times(B_p(r)-B_n(r)) & +\cos^2\theta\,B_n(r) \end{pmatrix} \qquad (3b)$$

where q^2 is the variance of velocity, E is the expected value, u and v are the components of $\mathbf{v}$, θ is the direction of the separation vector $(\mathbf{x}_2 - \mathbf{x}_1)$, r is the length of this vector, B_p is the autocorrelation function for the velocity component parallel to the separation vector, and B_n is the autocorrelation function for the velocity component normal to this vector [Thorndike, 1986a]. Estimates of these functions from Arctic Ocean buoy motions are shown in Figure 18-3.

To consider spatial correlation on shorter scales, it is useful to examine a related quantity that involves the relative velocities or the spatial increment in velocity between two ice particles $\mathbf{w} = \mathbf{v}(\mathbf{x}_2) - \mathbf{v}(\mathbf{x}_1)$. The relative velocity for a homogeneous, isotropic field depends only on the separation distance r

$$\omega(r) = u(x_1 + r, y_1) - u(x_1, y_1) \qquad (4a)$$

$$\upsilon(r) = v(x_1 + r, y_1) - v(x_1, y_1) \qquad (4b)$$

where the components of $\mathbf{w}$, $\mathbf{v}$, and $\mathbf{x}$ are (ω, υ), (u,v), and (x,y). The data in Figure 18-4, taken from a single pair of sequential SAR image strips, suggest that the variance of the increments $\omega(r)$ and $\upsilon(r)$ behaves as r^α for small separations r where α has a value of about unity. It can be shown that B_p and B_n behave as $1 - c\,r^\alpha$ for small values of r. This behavior is critical to the spatial differentiability of $\mathbf{v}$. Because α is only slightly greater than unity, the autocorrelation functions have a finite slope at $r = 0$, unlike the autocorrelation function of a differentiable field. The velocity of ice does not possess a spatial derivative in the mean square sense; α would require a value of 2 for the derivative to exist.

Although the spatial derivatives of velocity and displacement are not well defined in the limit $r \to 0$, we can compute the average velocity derivative over some finite region R as

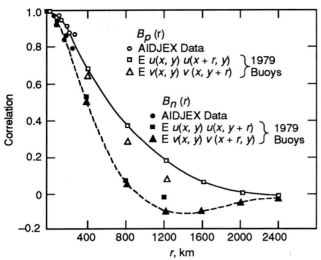

Fig. 18-3. Correlations for ice velocity components using buoy data from AIDJEX and First Global Atmospheric Research Program Experiment programs. The autocorrelation functions for the velocity components are (a) B_p parallel, and (b) B_n normal to the separation vector [Thorndike, 1986a].

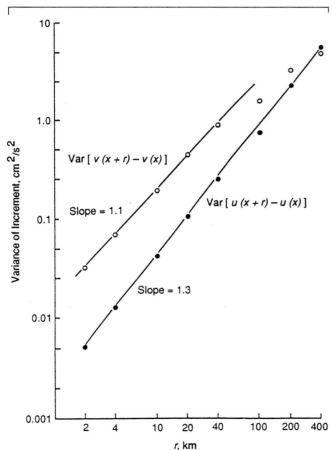

Fig. 18-4. The variance of the increments of ice velocity compared with the interval length computed from Seasat SAR data [Thorndike, 1986a].

$$\overline{\left.\frac{\partial u}{\partial x}\right|}_L = \frac{1}{L^2} \iint \frac{\partial u}{\partial x} \, da \tag{5}$$

where L^2 is the area of the region R, and the form of the other three derivatives is similar. The definition contains an implicit length scale L. The Green–Gauss theorem allows us to convert the area integral into a line integral around the boundary C of the region R

$$\overline{\left.\frac{\partial u}{\partial x}\right|}_L = \frac{1}{L^2} \oint u\mathbf{n} \cdot \mathbf{i} \, dl \tag{6}$$

where $\mathbf{n}$ is the outward normal to the boundary and $\mathbf{i}$ is the unit vector in the x direction. Satellite imagery can be analyzed to give velocities (or displacements) on a regular grid of points to which the discrete form of Equation (6) can be applied, ideally along four sides of a rectangle, but also around a more irregular polygon formed by points (x_j, y_j), $j = 1, ..., N$ (point $j = N + 1$ is the same point as $j = 1$). Then,

$$\overline{\left.\frac{\partial u}{\partial x}\right|}_L = \frac{1}{L^2} \sum_{j=1}^{N} \frac{u_j + u_{j+1}}{2} \left(y_{j+1} - y_j \right) \tag{7}$$

where L^2 is the area enclosed by the boundary. Note that Equations (5) through (7) apply equally to velocity and displacement.

An alternative estimate of deformation is obtained by finding the least squares fit of velocity measurements at N points to the expression

$$\mathbf{v}(\mathbf{x}) = \mathbf{v}_0 + \mathbf{D}_{0,L}\mathbf{x} + \mathbf{e} \tag{8}$$

where $\mathbf{D}_{0,L}$ is the 2×2 matrix of spatial velocity derivatives at the center of the N data points, $\mathbf{v}_0$ is the mean translation, and $\mathbf{e}$ is the residual in the least squares fit and gives the nonlinearity in the motion. Again, the deformation has an implicit length scale L whose square is roughly the area covered by the data. Figure 18-5 shows how the nonlinear contribution falls off as the time interval between sequential images lengthens.

18.2.4 Other Perspectives on Ice Motion

The discontinuous nature of ice motion within the Beaufort–Chukchi region of the Arctic Ocean has been observed with SAR (Figure 18-6) and AVHRR imagery [Hall and Rothrock, 1981; Curlander et al., 1985; Ninnis et al., 1986; Fily and Rothrock, 1987; Collins and Emery, 1988]. Spatial discontinuities are easily seen when the ice displacement field is represented as deformed grids, Figure 18-7. Large patches of floes, up to several hundred kilometers across, move as rigid pieces separated by narrow zones of intense deformation. A simulation of deformation by sequential

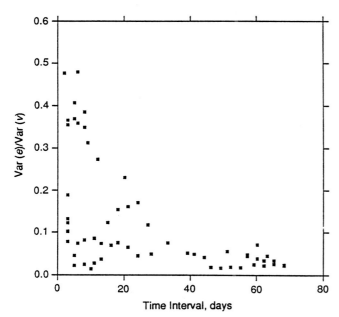

Fig. 18-5. The values of var (e)/var (v) as defined in Equation (8) from 59 pairs of images from the Beaufort Sea during the 100-day life of Seasat. The image pairs shared at least eight common features. The value of L is about 75 km.

shearing along random cracks shows how ice would slowly become mixed by diffusion [Thorndike, 1986b].

Ice motion in the Arctic can also be described as a random process in which a particle undergoes a daily displacement consisting of a mean displacement of roughly 2 km and a random isotropic displacement of about 7 km. Computed either forward or backward in time, this formalism allows one to investigate the probable origin or fate of ice occupying a particular location at time zero [Colony and Thorndike, 1985].

18.2.5 Processes Involving Motion and Deformation

A central question about sea ice is how it moves in response to forcing by wind and current. One approximation to the relation between ice velocity $\mathbf{v}$ and surface geostrophic wind $\mathbf{G}$ is that the current $\bar{\mathbf{c}}$ is steady, and that the varying part of the motion responds to the varying part of the wind

$$\bar{\mathbf{c}} = \bar{\mathbf{v}} - \mathbf{A}\,\overline{\mathbf{G}} \qquad (9)$$

$$\mathbf{v}' = \mathbf{A}\mathbf{G}' + \mathbf{e} \qquad (10)$$

where $\bar{\mathbf{e}}$ is zero. Here $\mathbf{e}$ contains all the errors, including any internal ice stress gradient or time dependence of the current. The matrix $\mathbf{A}$ can be considered as a scale factor $|\mathbf{A}|$ giving the ratio of ice speed to wind speed and a rotation through an angle β (positive counterclockwise) from the wind vector to the ice vector. An analysis of Arctic buoy data gives typical values of $(|\mathbf{A}|, \beta)$ from $(0.011, -18°)$ in summer to $(0.008, -5°)$ in the other seasons [Thorndike and Colony, 1982]. A similar analysis for the Weddell Sea shows that the ice drift with respect to the 15-m wind has a speed ratio of 0.03 and a direction $+23°$ to the left of the wind, as is appropriate for the Southern Hemisphere [Martinson and Wamser, 1990]. Ninnis et al. [1986] and Collins [1989] compare the ice velocity vectors from sequences of AVHRR image pairs with geostrophic winds in the Beaufort Sea, finding good correlations in response to high and low winds and changing wind directions. In the Greenland Sea, buoy data give a long-term mean $(|\mathbf{A}|, \beta)$ of $(0.012, -3°)$, although here the wind explains only 23% of the total variance of ice motion compared with 68% in the central Arctic [Moritz, 1988]. Overland et al. [1984] have examined bathymetry effects on the wind and current components of ice motion in the shallow seas of the Arctic.

The opening and closing of leads are parameterized in sea ice models as functions of the large-scale deformation. These functions can be tested with sequential SAR imagery by measuring both the mean deformation, Equation (7), and the opening and closing of individual leads. Measurement errors can be reduced to about 10% of the change in lead area [Fily and Rothrock, 1990]. The activity of individual leads may be better estimated by kinematic measures other than the mean strain, for instance, the statistics of the velocity increments, Equation (4), or the departures from rigid body motion [Thorndike, 1987].

Taken as an aggregate of floes, rigid pieces, leads, and deformation zones, sea ice behaves as a granular or fragmented material. Since it has been described equally as plastic, brittle, semibrittle, elastic, viscous, and inviscid, we must conclude that the material properties are not well understood. Investigations of its material behavior require some estimate of the loading placed on the ice by wind (and/or current), the resulting stress field set up in the ice, and of the response of the ice as evidenced in its deformation and failure patterns. These latter two quantities can be obtained from satellite imagery. The semibrittle hypothesis is supported by images showing leads intersecting at about 28°, the principal compressive stress lying in between at 14° from each lead [Marko and Thomson, 1977]. These angles may be deformation-rate dependent [Drinkwater and Squire, 1989]. The angle between fractures or leads has been related to the internal angle of friction for a granular material [Erlingsson, 1988]. The transient lead patterns north of the Bering Strait and intermittent arching and subsequent breakdown of the arches are compatible with plastic failure [Pritchard et al., 1979].

18.2.6 Mesoscale Kinematic Features

Structures in the field of ice motion indicate particular phenomena, such as a shear zone near a coast or underlying ocean eddies. Using a series of Landsat MSS images to examine deformation in the shear zone near Pt. Barrow, Hibler et al. [1974] found that the pack ice in the Beaufort

(a)

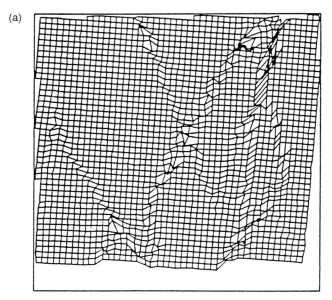

(b)

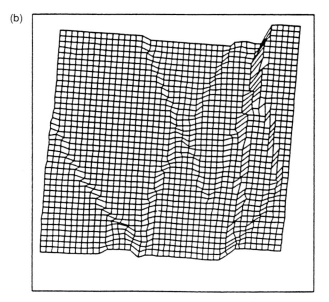

Fig. 18-6. Ice motion derived from a pair of Seasat SAR images of central pack ice in the Beaufort Sea, obtained October 5 and 8, 1978. Deformed grids of displacement field derived from imagery using (a) cross-correlation algorithm, and (b) manual tracking. Tie points are located at each grid intersection. (c) Difference between (a) automatically and (b) manually derived ice displacements [Fily and Rothrock, © 1987 IEEE].

(c)

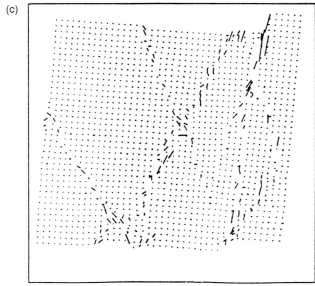

gyre moves essentially as a whole in a clockwise rotation, but that maximum shear (and counterclockwise rotation) occurs in a zone about 50-km wide between the shorefast and moving pack ice (Figure 18-7). Eddies are commonly observed in the Greenland Sea ice margins [Shuchman et al., 1987; Gascard et al., 1988]. Eddy structures in the ocean are resolved in the Greenland [Manley et al., 1987] and Labrador Seas [Ikeda et al., 1991] by removing the mean wind-driven component of the ice motion. Carsey and Holt [1987] showed examples of several image pairs from Seasat that resolved the appearance of an eddy embedded in the ice field near Barrow Canyon, and shear, divergence, and rotation in the coastal region near Banks Island.

Theory suggests that the increase in air–ocean stress coupling in the presence of an ice cover can cause either upwelling or downwelling along the ice edge (depending on wind direction), and that the resulting cross-ice-edge density gradients give rise to jets along the ice edge [Smedstad and Røed, 1985; Hakkinen, 1986]. There is evidence of these jets in AVHRR images of the Greenland Sea [Gascard et al., 1988]. Numerical experiments show that eddies interact with these jets, the more predominant cyclonic eddies being advected downstream and attracted toward the jet [Smith and Bird, 1991].

18.3 Ice Tracking Techniques Using Satellite Imagery

18.3.1 Ice Motion Algorithms

Early studies of ice motion relied on common ice features visually identified in sequences of aircraft and satellite photographs; examples are the shear zone study illustrated in Figure 18-7 and a comparison of modelled and observed motions in the Beaufort Sea by Hall [1980a, b]. The first SAR measurements of ice motion used optically correlated imagery strips acquired from aircraft during the Arctic Ice Dynamics Joint Experiment (AIDJEX) [Leberl et al., 1979] and from Seasat [Hall and Rothrock, 1981; Leberl et al.,

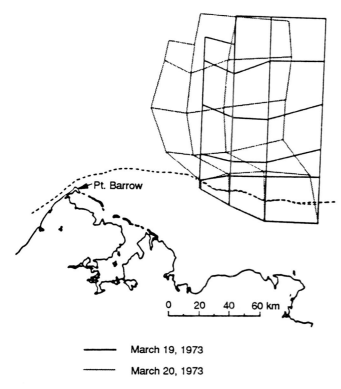

Pt. Barrow

0   20   40   60 km

——— March 19, 1973

——— March 20, 1973

Fig. 18-7. Deformation grid obtained from Landsat MSS imagery in the coastal region off Pt. Barrow indicating the spatial variation in the strain field [Hibler et al., 1974].

1983]. This form of radar imagery suffers from low dynamic range, geometric distortions (especially with aircraft imagery), and along-track location errors associated with timing errors.

More recently, the most common approach to automatic ice tracking has been cross-correlation between the two images. A common feature is likely when a small region ($n \times n$ pixels) of one image has a peak correlation with a small region of another image. The cross-correlation between the brightness a_{ij} of the first image and the brightness b_{kl} of the second image is

$$R(i,j;k,l) = \sum_{p=-t}^{+t} \sum_{q=-t}^{+t} \frac{\left(a_{(i+p,j+q)} - \overline{a}\right)\left(b_{(k+p,l+q)} - \overline{b}\right)}{n^2 \sigma_a \ \sigma_b} \quad (11)$$

where t is $(n-1)/2$ and $\overline{a}$ and σ_a are the mean and standard deviations, respectively, of a_{ij} over the $n \times n$ array. Several studies have used a pyramidal, or hierarchical, structure set of images. The original full-resolution images are first degraded by averaging pixels, and then some initial common features are found. More highly correlated features are then sought as the image resolution is increased through successive stages, until the displacement grid is filled in as completely as possible. These studies have used digital Seasat SAR [Fily and Rothrock, 1987; Collins and Emery,

1988; Vesecky et al., 1988] and AVHRR imagery [Emery et al., 1991; Zhang, 1991]. Examples of results from each sensor are shown in Figures 18-6 and 18-2(b), respectively. Ninnis et al. [1986] used fast Fourier transforms to compute the cross-correlation in AVHRR data. Lee and Yang [1987] and Zhang [1991] used a global motion predictor model initialized with a small set of visually identified features to guide the correlation matching. Kwok et al. [1990] incorporated a linear ice drift model, Equations (9) and (10), to confine the search for SAR image pairs before proceeding into the detailed correlation procedure.

Overall, cross-correlation is effective and accurate for deriving dense measurements of ice motion in the central Arctic ice pack where floe rotations over several days are small. Cross-correlation is less reliable in deformation zones and in the marginal ice zone, where significant floe rotations occur.

To add robustness for examining more active ice motion with SAR imagery, feature matching has been explored. The sharp contrast at lead–floe boundaries provides linear features that are generally stable over several days and can be successfully tracked. Vesecky et al. [1988] first isolated the darker leads from floes in SAR imagery by thresholding, then characterized the boundary segments as deviations from a straight line, and finally iteratively matched the segments. Rotation measurements are a by-product of the method since the orientations of the boundaries are found. Ridges, which appear as bright curvilinear features, are also useful for tracking, but have less distinct or sharp edges [Vesecky et al., 1990].

Another technique for tracking floe–lead boundaries describes each boundary curve in terms of its tangent direction ψ at each point along its arc length s measured from an arbitrary starting point [Kwok et al., 1990; McConnell et al., 1991]. Matching ψ-s curves are then sought between images. A dynamic programming procedure allows for small changes in the ψ-s curves [McConnell et al., 1991]. Banfield [1991] used an erosion-propagation algorithm to identify floes, then incorporated a probabilistic model to identify common features. This technique performs best in the marginal ice zone, where individual floes are more likely to be surrounded by open water and are therefore more distinct. Two techniques have been proposed that use invariant moments to optimally match sets of local image properties, such as brightness, feature shape, size, and orientation [Daida et al., 1991; Banfield, 1991; Zhang, 1991].

18.3.2 Displacement and Deformation Errors

There are two primary sources of error in measuring ice motion from satellite imagery: the absolute geographic position error e_g of each image pixel, and a tracking error e_f, which is the uncertainty in identifying common ice features in the second of two images. The position error applies independently to each position measurement in each image; i.e., a position estimate is the true position plus an error of $x + e_g$. The tracking error e_f applies to a displacement observed between two images. If we assume that e_g and e_f are each Gaussian with a zero mean, have standard deviations σ_g and σ_f, and are uncorrelated between images A and B, we can treat separately the errors of each scalar component of vectors or even matrix quantities. Including errors, an estimate of the displacement of an ice feature is given by

$$u = x_B + e_{gB} + e_f - x_A - e_{gA} \qquad (12)$$

The error in u has a zero mean and a variance of

$$\sigma_u{}^2 = 2\,\sigma_g{}^2 + \sigma_f{}^2 \qquad (13)$$

The error in velocity σ_v is σ_u divided by the time interval of the displacement; errors in the time interval are usually negligible.

Spatial differences in displacement for two features 1 and 2 are

$$\Delta u = [x_{B2} + e_{gB2} + e_{f2} - x_{A2} - e_{gA2}]$$
$$- [x_{B1} + e_{gB1} + e_{f1} - x_{A1} - e_{gA1}] \qquad (14)$$

The error in Δu has zero mean. Its variance contains a contribution from each of the tracking errors e_{f1} and e_{f2}, which are independent. If the geolocation errors are all independent, then their variances all add and the variance of Δu is $4\sigma_g{}^2 + 2\sigma_f{}^2$. This quantity is an upper bound on this error variance and is probably a good estimate of $\sigma_u{}^2$ when points 1 and 2 are separated by several hundred kilometers. However, if the two features are close (say, ten kilometers), the geolocation errors (e_{gB2} and e_{gB1}, for example) are no longer independent, and in fact tend to cancel; the error variance

in Δu tends towards the lower bound $2\sigma_f{}^2$. This means that if geolocation errors are large, differential motion or deformation can be well estimated, even if displacement cannot. To estimate a mean spatial gradient over the distance between two features, divide Equation (14) by $\Delta x = x_{A2} - x_{A1}$.

In the early studies that used aerial photography, measurement errors arose from geometric distortion due to the wide variation in incidence angle (0 to 55°) and from location and timing errors (Table 18-1). Leberl et al. [1979] required ground control points in the aircraft SAR data to reduce the geolocation error from ±3 km to a more acceptable ±0.2 km. These problems are largely reduced when using Seasat imagery [Leberl et al., 1983], but radiometric and geometric degradations are still present. Multiple land control points again are required to reduce the geolocation errors to ±0.3 km. Hall and Rothrock [1981] measured displacements of 10 to 25 km with errors of a few kilometers along-track due to film stretching and a few hundred meters across-track due to geometric distortions.

Errors in measurements using digital image data depend on the region (central pack or marginal ice zone), resolution, and, for AVHRR, land control points contained within the imagery (Table 18-1). Many of the SAR studies compare automatic algorithm results with manually derived ice motion measurements (Figure 18-6). Errors are higher in deformation zones because of the difficulty in locating and tracking ice features, but are reduced if the algorithm is designed with feature tracking capabilities. Curlander et al. [1985] assessed the end-to-end geometric error arising from the Seasat SAR system and processing and from errors in manual identification due to small floe size and indistinct surface areas within larger floes. Fily and Rothrock [1987] found mean measurement errors with various pixel sizes. Kwok et al. [1990] measured tracker displacement errors by

TABLE 18-1. Estimates of measurement errors

Data type	Processing optical/ digital	Geolocation error σ_g, km	Feature identification error σ_f, km	Displacement error σ_u, km	Pixel size/ resolution, km	Ground control points?	Reference
Aircraft SAR	O			0.2	$R - 0.025$	Y	Leberl et al. [1979]
Seasat SAR	O	0–3 (AT) 0.2 (CT)	0.04	0–4 (AT) 0.3 (CT)	$R - 0.05$	Y	Hall and Rothrock [1981]
Seasat SAR	O	0.3	0.15	0.45	$R - 0.05$	Y	Leberl et al. [1983]
Seasat SAR	D	0.1	0.1	0.17	$P - 0.025$	N	Curlander et al. [1985]
Seasat SAR	D	0 (assumed)	0.3	0.3	$P - 0.025 - 0.225$	N	Fily and Rothrock [1987]
Seasat SAR	D	0.1	0.3	0.33	$P - 0.1$	N	Kwok et al. [1990]
ERS-1 SAR	D	0.1	0.3	0.33	$P - 0.1$	N	Kwok et al. [1990]
AVHRR	D	1.1	0.9	1.8	$P - 1.1$	Y	Emery et al. [1991] and Zhang [1991]
AVHRR	D	1.7		>2.4	$P - 1.7$	Y	Ninnis et al. [1986]

AT = along track   CT = across track

using reduced-resolution Seasat SAR imagery. Land control points have been used to register AVHRR imagery and reduce the geolocation errors to a single pixel [Ninnis et al., 1986; Emery et al., 1991; Zhang, 1991].

18.3.3 Image Noise and Effects on Tracking

All tracking algorithms depend on the use of structure within an image. In feature-matching schemes, the structures are edges or boundaries. In intensity correlation schemes, the spatial patterns of pixel intensities in a subscene are important. For optical sensors, the assumed noise statistics for a homogeneous field are Gaussian with a fairly narrow distribution if the signal-to-noise ratio is quite high. For an active microwave sensor like SAR, speckle dominates if the number of *looks* (noncoherent summing of image intensities) is less than 10; the additive noise in the complex SAR data becomes speckle after the detection process (see Chapter 6). Since they are uncorrelated between images, additive noise and speckle affect the apparent location of common features because they appear slightly different, which reduces the correlation. For both feature- and correlation-based cases, the noise effect can be reduced by spatially averaging the image data, which consequently increases the spatial error. The decrease in the noise variance is then proportional to the number of independent samples taken.

18.3.4 Filtering Displacement Vectors

It is possible to set criteria for accepting a displacement vector and discarding any mismatched or otherwise poor quality vectors, thereby improving the accuracy of the resultant ice motion vector fields. If correlation is used to detect common features, the statistics of the correlation surface can be used to discard relatively poor matches. The magnitude of the correlation peak is used in most studies [Ninnis et al., 1986; Vesecky et al., 1988; Fily and Rothrock, 1987; Collins and Emery, 1988]. The use of the shape or sharpness of the correlation peaks was proposed by Vesecky et al. [1988]. Kwok et al. [1990] used a combination of the magnitude of the peak, the peak-to-background ratio, and peak-to-secondary-peak ratio as criteria for acceptance of a displacement vector.

The displacement vectors accepted by the tracking algorithm can also be filtered to discard outlying vectors that do not match well with surrounding vectors. The validity of a displacement vector can be determined by considering its correlation with the vectors in the neighborhood [Vesecky et al., 1988; Collins and Emery 1988]. Fily and Rothrock [1987] included a scheme to remove grid cells that were folded over themselves by deformation.

18.4 An Ice Motion Algorithm

The first operational algorithm for tracking ice motion has been implemented on the Geophysical Processor System (GPS) at the Alaska SAR Facility. This system is intended to use SAR imagery from ERS-1 and JERS-1 and eventually Radarsat. The design includes a comprehensive set of procedures and algorithms for locating image pairs using an ice motion estimator, for tracking ice in both the central pack and marginal ice zones using areal correlation and ψ-s feature matching, respectively, and finally for generating ice motion vectors displayed on an SSM/I polar stereographic projection mapped onto a 5×5 km grid system [Kwok et al., 1990; McConnell et al., 1991].

The ice motion estimator uses a linear ice drift model [Thorndike and Colony, 1982] for selection of images with potential overlap, with the geostrophic wind as input into the model. The system ingests geocoded imagery with a resolution (about 200 m) and a pixel size (100 m) that are reduced from the full-resolution data (25 m), and which theoretically has 64 looks. The system can automatically produce maps of ice motion for at least 20 pairs of images on a daily basis. Produced coincidentally with the ice motion maps are maps of sea ice types derived from a classification algorithm (Chapter 19). An example of a preliminary vector data set derived from ERS-1 SAR data is shown in Figure 18-8. Detailed assessments of these ice motion products will be available in the near future as the products are generated and begin to be used.

The GPS motion algorithm uses a combination of different filters at several stages of the tracking process to remove spurious or low-quality vectors, based on statistics of the correlation surfaces. Clustering of the motion vectors is used to identify the dominant modes of motion in the sampled field; the filtering process then discards the erroneous vectors by examining the cluster centroids that are inconsistent with the dominant modes. Also, a smoothness constraint is applied to ensure the spatial consistency of the displacement field. A quality factor is assigned to each vector to give a quantitative indication of the quality of the derived vector. The filtering process in the GPS attempts to optimize the ratio of good to bad vectors so that 95% of the motion vectors are accurate to less than 300 m displacement error.

18.5 Outlook

What is the present status of utilizing satellite observations for ice motion? The subject is in its adolescence. Over the last several years, studies have primarily focussed on examining limited data sets for their information content and measurement quality. However, as promising as the results are to date, they cannot be considered representative or conclusive; the results need to be reexamined using many more image scenes. The tools to utilize satellite imagery for routine motion observations are mature and ready for use, and fortunately adequate data are now being obtained to extend these observations to all scales, seasons, and regions.

In the Northern Hemisphere, buoys have provided a substantial base of knowledge of large-scale ice motion in the central Arctic Ocean, but gaps need to be filled, particularly in regions that have been sparsely sampled by buoys.

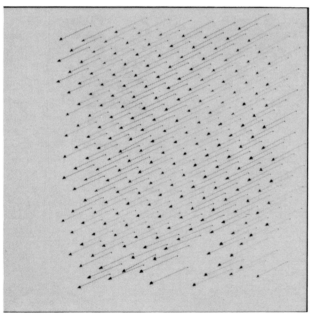

Fig. 18-8. A pair of ERS-1 SAR images obtained in the Beaufort Sea on November 27 and 30, 1991, and corresponding ice motion vectors derived by the ice motion algorithm at the Alaska SAR Facility Geophysical Processor System [Kwok et al., 1990; McConnell et al., 1991]. The images were processed at the Alaska SAR Facility and are copyrighted by the European Space Agency [1991].

0   50 km

Satellites should be used to intensely sample ice motion through the major straits—Fram, Denmark, and Bering—where the ice motion's vigor makes sampling by buoys inefficient and costly. The ice motion observations from these images will provide firm new data on the ice mass balance, which is now only generally known. Satellite coverage of the Southern Ocean is becoming available for most of the seasonal ice pack, and the combination of these data with data from buoys judiciously placed to improve wind estimates will provide our first climatology of the Antarctic ice mass and momentum balance.

Satellite imagery also has an important contribution to make in either confirming present theories or leading to new theories about the dynamics of leads in response to ice deformation. Are lead changes related tightly to the spatial variations in the large-scale field of motion, or are they best modeled as a motion or a Markov process with a large random component? Tests of stress/strain relationships require the spatial differences and even second spatial differences of displacement to be available from satellite images. Imagery needed to study these issues is pouring into archives. What for years we have viewed as an exciting future has become the present.

REFERENCES

Allison, I., Pack-ice drift off East Antarctica and some implications, *Annals of Glaciology, 12*, pp. 1–8, 1989.

Banfield, J., Automated tracking of ice floes: A stochastic approach, *IEEE Transactions on Geoscience and Remote Sensing, 29(6)*, pp. 905–911, 1991.

Carsey, F. and B. Holt, Beaufort–Chukchi ice margin data from Seasat: Ice motion, *Journal of Geophysical Research, 92(C7)*, pp. 7163–7172, 1987.

Collins, M. J., Synoptic ice motion from AVHRR imagery: Automated measurements versus wind-driven theory, *Remote Sensing of the Environment, 29*, pp. 79–85, 1989.

Collins, M. J. and W. J. Emery, A computational method for estimating sea ice motion in sequential Seasat synthetic aperture radar imagery by matched filtering, *Journal of Geophysical Research, 93(C8)*, pp. 9241–9251, 1988.

Colony, R. and A. S. Thorndike, An estimate of the mean field of Arctic Sea ice motion, *Journal of Geophysical Research, 89(C6)*, pp. 10,623–10,629, 1984.

Colony, R. and A. S. Thorndike, Sea ice motion as a drunkard's walk, *Journal of Geophysical Research, 90(C1)*, pp. 965–974, 1985.

Curlander, J. C., B. Holt, and K. J. Hussey, Determination of sea ice motion using digital SAR imagery, *IEEE Journal of Oceanic Engineering, OE-10(4)*, pp. 358–367, 1985.

Dade, E. F., D. A. Rothrock, R. Colony, and C. Olmsted, *Estimating Repeat Coverage of Arctic Sea Ice With ERS-1 SAR*, Technical Report 9114, 50 pp., Applied Physics Laboratory, University of Washington, Seattle, Washington, 1991.

Daida, J., R. Samadani, and J. F. Vesecky, Object-oriented feature-tracking algorithms for SAR images of the marginal ice zone, *IEEE Transactions on Geoscience and Remote Sensing, GE-28(4)*, pp. 573–589, 1990.

Drinkwater, M. R. and V. Squire, C-band SAR observations of marginal ice zone rheology in the Labrador Sea, *IEEE Transactions on Geoscience and Remote Sensing, 27(5)*, pp. 522–534, 1989.

Emery, W. J., C. W. Fowler, J. Hawkins, and R. H. Preller, Fram Strait satellite image-derived ice motions, *Journal of Geophysical Research, 96(C3)*, pp. 4751–4768, 1991. (Correction, *Journal of Geophysical Research, 96(C5)*, pp. 8917–8920, 1991.)

Erlingsson, B., Two-dimensional deformation patterns in sea ice, *Journal of Glaciology, 34(118)*, pp. 301–308, 1988.

Fily, M. and D. A. Rothrock, Sea ice tracking by nested correlations, *IEEE Transactions on Geoscience and Remote Sensing, GE-25(5)*, pp. 570–580, 1987.

Fily, M. and D. A. Rothrock, Opening and closing of sea ice leads: Digital measurements from synthetic aperture radar, *Journal of Geophysical Research, 95(C1)*, pp. 789–796, 1990.

Gascard, J. C., C. Kergomard, P.–F. Jeannin, and M. Fily, Diagnostic study of the Fram Strait marginal ice zone during summer from 1983 and 1984 Marginal Ice Zone Experiment Lagrangian observations, *Journal of Geophysical Research, 93(C4)*, pp. 3613–3641, 1988.

Hakkinen, S., Coupled ice-ocean dynamics in the marginal ice zones: Upwelling/downwelling and eddy generation, *Journal of Geophysical Research, 91(C1)*, pp. 819–832, 1986.

Hall, R. T., A test of the AIDJEX ice model using Landsat images, *Sea Ice Processes and Models*, edited by R. S. Pritchard, pp. 89–101, University of Washington, Seattle, Washington, 1980a.

Hall, R. T., AIDJEX modeling group studies involving remote sensing data, *Sea Ice Processes and Models*, edited by R. S. Pritchard, pp. 151–162, University of Washington, Seattle, Washington, 1980b.

Hall, R. T. and D. A. Rothrock, Sea ice displacement from Seasat synthetic aperture radar, *Journal of Geophysical Research, 86(C11)*, pp. 11,078–11,082, 1981.

Herman, G. F., Arctic stratus clouds, *Geophysics of Sea Ice*, edited by N. Untersteiner, pp. 465–488, Plenum Press, New York, 1986.

Hibler, W. D., III, S. F. Ackley, W. K. Crowder, H. L. McKim, and D. M. Anderson, Analysis of shear zone ice deformation in the Beaufort Sea using satellite imagery, *The Coast and Shelf of the Beaufort Sea*, edited by J. C. Reed and J. E. Slater, pp. 285–296, Arctic Institute of North America, Arlington, Virginia, 1974.

Ikeda, M., I. Peterson, and C. E. Livingstone, A mesoscale ocean feature study using synthetic aperture radar imagery in the Labrador Ice Margin Experiment: 1989, *Journal of Geophysical Research, 96(C6)*, pp. 10,593–10,602, 1991.

Kwok, R., J. C. Curlander, R. McConnell, and S. S. Pang, An ice-motion tracking system at the Alaska SAR Facility, *IEEE Journal of Oceanic Engineering, OE-15(1)*, pp. 44–54, 1990.

Leberl, F., M. L. Bryan, C. Elachi, T. Farr, and W. Campbell, Mapping of sea ice and measurement of its drift using aircraft synthetic aperture radar images, *Journal of Geophysical Research, 84(C4)*, pp. 1827–1835, 1979.

Leberl, F., J. Raggam, C. Elachi, and W. J. Campbell, Sea ice motion measurements from Seasat SAR images, *Journal of Geophysical Research, 88(C3)*, pp. 1915–1928, 1983.

Lee, M. and W.–L. Yang, Image-analysis techniques for determination of morphology and kinematics in Arctic sea ice, *Annals of Glaciology, 9*, pp. 92–96, 1987.

Limbert, D. W. S., S. J. Morrison, C. B. Sear, P. Wadhams, and M. A. Rowe, Pack-ice motion in the Weddell Sea in relation to weather systems and determination of a Weddell Sea sea-ice budget, *Annals of Glaciology, 12*, pp. 104–112, 1989.

Manley, T. O., R. A. Shuchman, and B. A. Burns, Use of synthetic aperture radar-derived kinematics in mapping mesoscale ocean structure within the interior marginal ice zone, *Journal of Geophysical Research, 92(C7)*, pp. 6837–6842, 1987.

Marko, J. R. and R. E. Thomson, Rectilinear leads and internal motions in the ice pack of the western Arctic Ocean, *Journal of Geophysical Research, 82(6)*, pp. 979–987, 1977.

Martinson, D. G. and C. Wamser, Ice drift and momentum exchange in winter Antarctic pack ice, *Journal of Geophysical Research, 95(C2)*, pp. 1741–1755, 1990.

McConnell, R., R. Kwok, J. C. Curlander, W. Kober, and S. S. Pang, ψ-S correlation and dynamic time warping: two methods for tracking ice floes in SAR images, *IEEE Transactions on Geoscience and Remote Sensing, GE-29(6)*, pp. 1004–1012, 1991.

Moritz, R. E., *The Ice Budget of the Greenland Sea*, Technical Report 8812, 116 pp., Applied Physics Laboratory, University of Washington, Seattle, Washington, 1988.

Moritz, R. E., Ice motion in the Ross Sea, *Journal of Geophysical Research*, in press, 1992.

Ninnis, R. M., W. J. Emery, and M. J. Collins, Automated extraction of pack ice motion from Advanced Very High Resolution Radiometer imagery, *Journal of Geophysical Research, 91(C9)*, pp. 10,725–10,734, 1986.

Overland, J. E., H. O. Mofjeld, and C. H. Pease, Wind-driven ice drift in a shallow sea, *Journal of Geophysical Research, 89(C4)*, pp. 6525–6531, 1984.

Pritchard, R. S., R. W. Reimer, and M. D. Coon, Ice flow through straits, *Fourth International Conference on Port and Oceanic Engineering Under Arctic Conditions, 3*, Trondheim, Norway, 3, pp. 61–74, 1979.

Shuchman, R. A., B. A. Burns, O. M. Johannessen, E. G. Josberger, W. J. Campbell, T. O. Manley, and N. Lannelongue, Remote sensing of the Fram Strait marginal ice zone, *Science, 236*, pp. 429–431, 1987.

Smedstad, O. M. and L. P. Røed, A coupled ice-ocean model of ice breakup and banding in the marginal ice zone, *Journal of Geophysical Research, 90(C1)*, pp. 876–882, 1985.

Smith, D. C. and A. A. Bird, The interaction of an ocean eddy with an ice edge ocean jet in a marginal ice zone, *Journal of Geophysical Research, 96(C3)*, pp. 4675–4689, 1991.

Thorndike, A. S., Kinematics of sea ice, *The Geophysics of Sea Ice*, edited by N. Untersteiner, pp. 489–550, NATO ASI Series B: Physics vol. 146, Plenum Press, New York, 1986a.

Thorndike, A. S., Diffusion of sea ice, *Journal of Geophysical Research, 91(C6)*, pp. 7691–7696, 1986b.

Thorndike, A. S., A random discontinuous model of sea ice motion, *Journal of Geophysical Research, 92(C6)*, pp. 6515–6520, 1987.

Thorndike, A. S. and R. Colony, Sea ice motion in response to geostrophic winds, *Journal of Geophysical Research, 87(C7)*, pp. 5845–5852, 1982.

Vesecky, J. F., R. Samadani, M. P. Smith, J. M. Daida, and R. N. Bracewell, Observation of sea-ice dynamics using synthetic aperture radar images: Automated analysis, *IEEE Transactions on Geoscience and Remote Sensing, GE-26(1)*, pp. 38–48, 1988.

Vesecky, J. F., M. P. Smith, and R. Samadani, Extraction of lead and ridge characteristics from SAR images of sea ice, *IEEE Transactions on Geoscience and Remote Sensing, GE-28(4)*, pp. 740–744, 1990.

Zhang, H., *Parameter Retrieval Algorithms and Data Analysis System for Sea Ice Remote Sensing*, Ph.D. dissertation, Publication LD 85, 231 pp., Technical University of Denmark, Lyngby, Denmark, 1991.

Chapter 19. An Approach to Identification of Sea Ice Types From Spaceborne SAR Data

RONALD KWOK, GLENN CUNNINGHAN, AND BENJAMIN HOLT

Jet Propulsion Laboratory, California Institute of Technology, 4800 Oak Grove Drive, Pasadena, California 91109

19.1 INTRODUCTION

During the 1990's, single-channel spaceborne SARs will be flown on the European ERS-1 (launched in July 1991), the Japanese ERS-1 (launched in February 1992), and the Canadian Radarsat (to be launched in 1994). These satellites will provide the first opportunities for extensive spaceborne monitoring of the polar regions with a SAR since Seasat, which operated for 3-1/2 months in 1978. To generate SAR data products to support polar climate analysis, algorithms have been developed that will reside in the Alaskan SAR Facility's (ASF) Geophysical Processor System (GPS) to automatically and routinely generate maps of ice motion (see Chapter 18), ocean surface wave direction and length [Holt et al., 1990a], and ice type and concentration [Kwok et al., 1992].

This chapter focuses on the ice classification algorithm implemented at ASF to generate maps of ice type and concentration, and on the evolution of such an approach using ancillary data sets (such as passive microwave observations). The ASF GPS ice-classification algorithm has been designed to efficiently segment the ice into major ice types, including multiyear and first-year ice, by using a clustering routine to isolate the SAR ice signatures. The clusters are labeled (as ice types) by comparison with tables of scatterometer-derived ice type signatures that have been compiled by season (see Chapter 5). The algorithm will account for these seasonal changes by utilizing separate lookup tables. The development status of this classification algorithm has been reported previously by Holt et al. [1989, 1990b].

The ice classification algorithm has been designed to be used initially with ERS-1 SAR, a single frequency (C-band, 5.3 GHz), single polarization (VV) radar with a fixed look angle of 23° off nadir. The satellite, launched on July 14, 1991, into a Sun-synchronous orbit at 98° inclination, will acquire extensive imagery of sea ice over the Arctic Ocean and peripheral seas.

The GPS will utilize reduced-resolution (approximately 200 m) SAR data processed at ASF. An advantage of using the lower resolution image products is the low level of speckle present in the imagery. The current algorithm is based entirely on the first-order statistics (mean and variance) of the backscatter intensity and does not attempt to exploit the two-dimensional or statistical structure (e.g., texture and higher order moments) of specific ice types. Although these features have been shown to be effective in some cases, there are insufficient observations at this time to support the implementation of such feature extraction schemes in an operational system.

The seasonal evolution of the active microwave responses of sea ice is described in Chapter 5. Preliminary assessment and comparison of SAR data with scatterometer data at C-band showed a consistent contrast of 5 to 6 dB between multiyear and first-year ice during winter conditions. This contrast serves as a primary discrimination feature. In the summer, a reduction and even reversals in this contrast have been observed during warm melt conditions due to changes in scattering characteristics caused by changes in surface conditions of the ice and snow cover (see Chapter 16). In this regime, the contrast between ice and water is still available for separation of the two classes for ice concentration calculations. Perhaps the most difficult time periods for ice classification are the seasonal transitions (spring and fall) when the most dramatic changes in the sea ice backscatter are expected.

The present plan is to implement a classification procedure for those seasons when the surface conditions remain fairly stable and a high confidence can be placed on the classification results. Validation and extension of the procedure to other seasons and conditions are dependent on routine SAR observations from ERS-1 as well as correlative data (e.g., surface measurements) for adapting the classifier to these data sets. A significant factor affecting the number of ice types that can be effectively discriminated is the performance of the radar sensor and processing system. The misclassification percentage is directly dependent on the relative, as well as absolute, calibration of the sensor. These issues are discussed in this chapter.

19.2 ICE CLASSIFICATION APPROACH

The major components and processes of the classification scheme, radar data calibration, image segmentation and labeling, and the lookup tables used to label the ice type of individual pixels, are described in the following subsections. A schematic diagram of the classification procedure is shown in Figure 19-1.

Microwave Remote Sensing of Sea Ice
Geophysical Monograph 68
©1992 American Geophysical Union

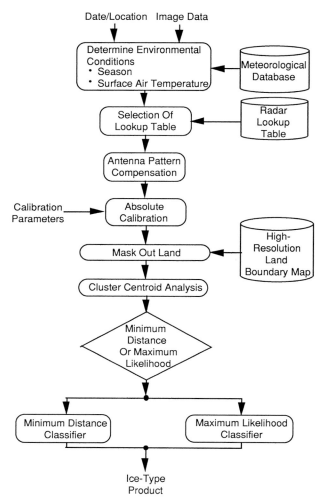

Fig. 19-1. Sea-ice classification approach.

19.2.1 Radar Data Calibration

Since this is an unsupervised procedure, the labeling of individual pixels is dependent on the overall system calibration. Thus, it is essential to monitor anticipated calibration changes. Incorrect compensation of the antenna pattern in SAR image data (causing range intensity gradients) is frequently a problem when there are uncertainties in antenna pointing and preflight pattern measurements. For the ERS-1 SAR, the residuals can be adequately modeled as a linear ramp. The backscatter from multiyear ice has very little incidence angle dependence (only 0.5 dB over the range of incidence angle within the ERS-1 swath) and is used in estimating the parameters of this linear model. Simulations show that the parameters of the linear error model can be estimated to approximately 0.5-dB accuracy. Thus, the relative calibration (consistency in the pixel value of the same ice type within an image frame) requirements of the pixel samples can be relaxed.

19.2.2 Image Segmentation and Labeling

19.2.2.1. Cluster analysis. The next step in the classification procedure identifies the dominant or reference cluster (the cluster with the largest sample population) within the image frame, with the goal of labeling this cluster as an ice type in the next step of the procedure. Clustering is performed in the intensity domain. The selection of the dominant cluster ensures that the mean backscatter of that ice type is reasonably estimated. The knowledge of the significant contrast between first-year and multiyear ice in winter C-band VV data (see Chapter 5) has been incorporated into our scheme for mechanizing this unsupervised clustering technique. The expected contrast will provide well-defined clusters in the data. Tests have shown that only a small subset of the image (3 to 5% in area) is needed to obtain stable clustering results, i.e., cluster centroids vary little after a sufficient number of pixels representative of the ice types in the image has been sampled.

19.2.2.2. Cluster labeling. The backscatter value of the dominant cluster's centroid does not necessarily correspond to a reference backscatter coefficient of an ice type stored in a lookup table (see Chapter 5) due to the variability of the backscatter or uncertainty in the absolute calibration. The cluster is labeled as belonging to that ice type with the mean backscatter closest (in absolute distance) to one listed in the lookup table. This is equivalent to identifying a radiometric tie-point where a radiometric misregistration exists between the measured and expected measurements.

19.2.2.3. Location of other ice types. The centroids of all the remaining ice types are located by using the expected contrast between different ice types stored in the lookup table. For example, if multiyear ice is identified as the reference cluster, then the relative locations of the other ice types, i, are

$$R^i_{MY} = \sigma_i / \sigma_{MY}$$

where σ_i is the backscatter of ice type, i. This step locates the small populations that were not resolved in the cluster analysis.

19.2.2.4. Pixel labeling. The final step is to give each image pixel an ice-type label. Two alternative methods for pixel classification are the minimum distance and Bayes' approach. The minimum distance method uses a simple one-dimensional threshold of the intensity values based on the cluster locations, C_i, where

$$x \sim C_i \quad \text{if } |x - C_i| \text{ is the minimum}$$

where x is the backscatter value and $\sim$ is the assignment operator. The Bayes approach is optimal if there are large differences in the expected variability of the different ice types. Simply

$$x \sim C_i \quad \text{if } P(C_i)p(x / C_i) \text{ is the maximum}$$

The advantage of this method is that it accounts for differences in the density functions. For example, if first-year ice has a larger variance compared to multiyear ice, then the error rate would be minimized if such an approach is taken. The disadvantage of the Bayes approach is that the distribution densities, $p(x / C_i)$, and a priori probabilities, $P(C_i)$, of the ice types, as well as parameters describing their distributions, have to be estimated. Also, the spatial and temporal dependence of these statistical parameters is probably quite high (due to snow cover, ice deformation, and surface temperature variations), and therefore the use of this technique is difficult to evaluate. Both techniques will be evaluated in the operational system.

The location uncertainty of the cluster centroid during cluster analysis (0.5 dB as pointed out earlier) obviously affects classification accuracy. The sensitivity of the classifier is analyzed by varying the centroid location about a reference point and assessing the percentage of misclassified pixels. Figure 19-2 shows the percentage of misclassified pixels as a function of the uncertainty in the location of the cluster centroid. The results based on evaluation of four aircraft images showed that we can expect a 5 to 10% error given the 0.5-dB uncertainty in the location of the cluster centroids. These errors are acceptable for many sea ice studies [ASF Prelaunch Science Working Team, 1989]. For example, the ambiguity between deformed first-year and multiyear ice in that (i.e., thickness) regime would have very little effect on regional flux calculations.

The lookup tables of ice backscatter are summarized in Chapter 5 and in Kwok et al. [1992]. Each individual ice type and its associated backscatter coefficient are included in each of these tables. The dependence of the backscatter signature of the principal ice types on season and environmental conditions are coded in Tables 19-1 and 19-2, and the improvement in the tables as more data become available is also accommodated.

The backscatter signatures in these lookup tables are based on scatterometer measurements. Only the best-available calibrated scatterometer measurements were included. The backscatter signature of the ice types in the lookup tables have been verified with in-situ measurements of the physical properties, and the environmental conditions during which the measurements were taken

have been well documented. Entries in the lookup tables represent years of observations and experience acquired during field campaigns. Because of the disparity in the resolution of scatterometer observations and SAR measurements, it is expected that the SAR backscatter measurements will be modulated to a higher degree by mixture distributions (of ice types) compared to scatterometer samples. The entries in the lookup tables represent the expected mean and variance of these ice types, and it should be noted that one could misclassify certain occurrences of grey ice, very rough first-year, or heavily ridged multiyear ice. Other limitations of our approach are listed below.

19.2.2.5. Effect of wind. The normalized radar cross section of open water is dependent on wind speed and direction. Currently, we do not have a strategy for classification of open water with its varying radar cross section. The limited fetch of leads may also result in significant backscatter variability within a single lead. The classifica-

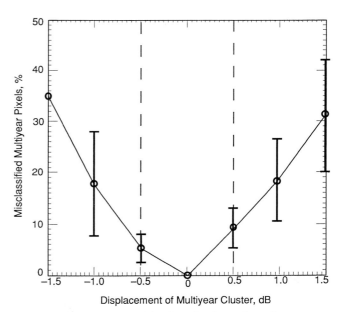

Fig. 19-2. The sensitivity of the classifier depends on the accuracy in the location of the cluster centroids.

TABLE 19-1. Seasonal and temperature dependence of active microwave sea ice signatures at C-band VV.

Season	General time span	Climatology ambient temperature	Principal ice types[a]
Winter to early spring	October–May	$T < -10°C$	MY, FY, NI/OW
Late spring	May and June	$-10°C < T < 0°C$	Ice, water
Early summer	June	$T \sim 0°C$	Ice, water
Midsummer	July	$T \geq 0°C$	Ice, water
Late summer	August	$T \geq 0°C$	Ice, water
Fall	September and October	$-10°C < T < 0°C$	MY, FY, NI/OW

[a]Ice types that are most likely to be separable in C-VV ERS-1 data. NI/OW = new ice/smooth open water.

TABLE 19-2. Seasonal dependent scatterometer-derived backscatter (C-band VV at 25° incidence) lookup tables for classification of sea types in SAR data.

Ice type	Ice thickness, cm	σ_0, dB	Standard backscatter deviation, dB	Expected incidence angle dependence of backscatter, dB
Winter to Early Spring				
MY	>220	−8.6	2.2	−0.08
FY	20–220	−14.0	2.1	−0.24
NI/OW	0–20	<−18.0	...	...
Late Spring				
MY	>220	−10.7	2.1	−0.27
FY	70–220	−13.2	1.1	−0.22
NI/OW	0–20	<−18.0	...	...
MY/FY	>20	>−16.0	...	...
NI/OW	0–20	<−18.0	...	...
Early Summer, Midsummer, Late Summer				
MY	>150	−10.5	1.7	−0.04
FY	30–120	−12.5	1.9	−0.21
NI/OW	0–30	<−18.0	...	...

There are gaps where no scatterometer measurements are available in certain thickness ranges.

tion products will indicate the wind velocity at the time of data acquisition. Wind velocity and air temperature fields used in the algorithm will be extracted from the numerical data products distributed by the National Meteorological Center (NMC).

19.2.2.6. Air temperature. The scattering characteristics of sea ice are highly dependent on processes controlled by air temperature. If the ice surface temperature (in response to air temperature) rises to a point where significant moisture in the ice increases the complex dielectric constant, the backscatter change will cause ambiguity in the classification. Other changes in the ice surface arise from vapor transport in warming air [Onstott and Gogineni, 1985]. It is obvious that the availability of air temperature measurements would help in identifying such conditions. The accuracy of the NMC temperature fields in the Arctic is currently being investigated.

Effectively, the lookup table selected to classify a specific image will be a function of the season and ambient air temperature. To account for all possible conditions, the decision table in Table 19-3 will be used. It is apparent that other factors (surface air pressure, geographic location, etc.) could be used in the decision table, and the system can easily accommodate future enhancement of this decision table based on available ancillary data. For example, passive microwave data could potentially be used synergistically with backscatter data to achieve more precise iden-

TABLE 19-3. Algorithm decision.

Ambient temperature	Winter	LSp	ES	MS	LS	Fall
T < −10°C	1	1	2	4	5	6
−10°C < T < 0°C	1	2	2	4	5	6
T ≥ 0°C	1	2	3	4	5	5

Numbers denote lookup tables as follows: 1. winter to spring; 2. late spring (LSp); 3. early summer (ES); 4. midsummer (MS); 5. late summer (LS); and 6. fall.

tification of seasonal change and, therefore, more accurate sea ice classification results from SAR imagery.

19.3 Data Set and Simulation

For realistic evaluation of the classification scheme before ERS-1 data became available, simulated ERS-1 image data approximating the expected quality of that processed at the ASF were generated. The NASA/JPL Airborne SAR (AIRSAR) C-band VV high-resolution data were used to generate four-look square-root intensity imagery of sea ice of sizes 1024 by 750 azimuth and range samples, respectively. The set of aircraft SAR imagery used in the algorithm evaluation was acquired during the March 1988 NASA DC-8 SMM/I algorithm validation program [Cavalieri et al., 1991]. Data from the Ka-band Radiometric Mapping System (KRMS) [Eppler et al., 1980], taken in March 1988 as well, were also used.

19.4 Results

Comparisons were made of aircraft SAR imagery with KRMS with the limited surface measurements obtained from a field camp overflown during March 1988. These results were used to formulate the ice-classification algorithm. The ice camp was located approximately 350 km north of Prudhoe Bay and was primarily established to support underwater acoustic propagation studies [Wen et al., 1989]. The camp was located on the edge of a multiyear floe, adjacent to a refrozen lead, which served as an aircraft landing strip.

The environmental and ice data that were collected provide the only known surface information obtained during the period of the aircraft campaign to confirm the winter conditions and identification of first-year and multiyear ice in the imagery. From March 11 to 19, the dates of the aircraft flights in the Beaufort and Chukchi Seas, the air temperatures were generally below −10°C, except on March 11 and 13 when they rose to about −7°C. Before and after these dates, the air temperatures were well below −20°C. The wind speeds were less than 7 m/s and often less than 3 m/s. The drift of the ice camp was determined by satellite positioning. On March 10 and 12, the drift speed rose rapidly to about 32 and 20 cm/s, respectively, due to winds of about 7 m/s. After March 14, the drift speed reduced to less than 5 cm/s as the wind decreased to less than 3 m/s. The day of the SAR overflight, March 19, fell during a period of quiescence with little or no ice motion and air temperatures below –20°C, reducing the likelihood of open water being present in any recently opened leads.

Measurements of the sea ice in the area [Wen et al., 1989] provided key information on radar signatures. Most of the measurements were made of the frozen lead rather than of the multiyear floes. The lead was approximately 1.4 m thick and had a snow cover of about 15 cm, although the snow cover was unevenly distributed into sastrugi. The lead was surrounded by multiyear floes with a recently formed rubble field to the north containing large ice chunks. Fissures several centimeters in width were contained within the lead. Two ice cores were taken from the lead from which salinity, temperature, and density were measured and brine volume calculated. The ice was observed to be largely columnar in structure. Salinity profiles have values and a C-shape characteristic of first-year ice of that thickness [Weeks and Ackley, 1986]. The temperature profiles are also characteristic of winter first-year ice. No measurements of surface roughness of the ice and variable snow cover depth were made on the lead, but it is presumed to be relatively undeformed since the lead was used as a landing strip and is referred to as flat. The large multiyear floe where the ice camp was located was 6 by 6 km in size and contained hummocks as high as 6 m. The ice camp was identified by knowledge of its location and by handheld photography and sketch maps of the camp itself [Wen et al., 1989].

19.4.1 Comparison With KRMS Data

During the March 1988 SMM/I validation, KRMS and AIRSAR data were acquired [Cavalieri, 1991; Eppler et al., 1986], and provided some coincident coverage of the same ice fields. The passive microwave data were spatially registered to the SAR data for comparison of the active and passive signatures of the sea ice to assess the performance of the classification scheme. The multiyear fraction of each classified image was calculated and compared, the difference in the multiyear fraction between the two classification maps being less than 2%. This was discussed more fully by Kwok et al. [1992]. Two other data sets comparing the classified KRMS data and the SAR data acquired in the Beaufort Sea have yielded comparable results.

The results are encouraging in that the ice-type signatures derived from the limited SAR data set are similar to those in the scatterometer winter lookup table. If the signatures remain stable as observed (under these winter conditions), then misclassification will be entirely driven by the computational aspects (segmentation) analyzed in Section 19.2. This gives us an expected misclassification accuracy of less than 10%. However, it is quite difficult, as in most classification procedures, to truly state the expected accuracy of the classification map. The most important factor that affects classification accuracy is still the predictability of the physical variability of sea ice signatures.

19.4.2 Preliminary ERS-1 Results

Figure 19-3 shows a geocoded image and the ice classification map produced by the GPS at the Alaska SAR Facility. This image was acquired in 1991 on day 331 with the image center located at approximately 72°N and 153°W. The air temperature was –23°C, and the winter lookup table was selected to classify the ice types in this image. Multiyear ice has an average backscatter between –9 and –10 dB, approximately 5 dB above the normalized cross section of first-year ice and 11 dB above that of thin ice. Within the calibration uncertainty, these values agree well with the expected behavior of the ice types during these winter conditions. Future efforts will focus on characterizing the stability of the signatures and extending the classifier to other seasons and regions.

19.5 Discussion

The classification procedure worked well with the aircraft data set acquired over the Beaufort and Chukchi Seas in March 1988. Comparison of the contrast observed in the SAR data and scatterometer measurements showed consistency in the separation between multiyear and first-year ice during this season. The discrimination of these two ice types is feasible under these winter Arctic conditions, although the detection of new ice and open water may be limited by system noise performance of the ERS-1 SAR. In the case of open water, the effect of wind on its scattering cross section may cause ambiguity in the intensity-based classification scheme suggested here. Also, the need for absolute and relative calibration of the radar data is demonstrated if our classification scheme is to be effectively mechanized. The effect of snow cover and surface melt conditions on scattering characteristics increases the possibility of classification error. Scatterometer measurements (see Chapter 5) have indicated that the contrast between first-year and multiyear ice remains fairly stable for the period between October and May. The expected scattering behavior of sea ice during the spring, summer, and fall has also been described. We suggest that even though there is very little contrast between ice types, it is possible to discriminate between ice and open water during the early summer and midsummer. During the late spring, late summer, and fall, it is possible to discriminate multiyear, first-year, and open water or new ice, although the contrast between these ice types will be slightly modulated by the moisture in the ice and snow. We thus anticipate that this classification algorithm will perform adequately during the winter, and that further enhancements to the algorithm will be required during the other seasons.

We emphasize that this is an intensity-based classification scheme and use of other features (e.g., texture and higher order moments) for ice classification have not been

(a)

(b)

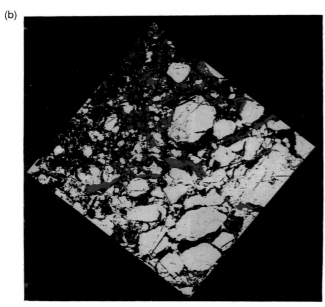

Fig. 19-3. Classification results from ERS-1 data: (a) SAR image geocoded at the Alaska SAR Facility, and (b) image classified into multiyear (white), first-year (blue), and thin ice (red).

explored fully due to lack of routine SAR observations required to understand and mechanize such schemes. Flexibility is designed into the classifier so that these additional features and other measurements (e.g., passive microwave data) can be easily incorporated into the classification scheme. It is expected that these sea-ice classification products from SAR imagery will increase in value and utility after thorough validation is accomplished. Design

changes and improvements will be required to extend the approach to the L-band data from JERS-1 and the C-band–HH data from Radarsat, with its wide-swath capability. Obtaining extended time series by region and season will be extremely valuable for monitoring seasonal and climate change, for calculating heat flux, and for determining the role of the polar regions in global climate.

REFERENCES

Alaska SAR Facility Prelaunch Science Working Team, *Science Plan for the Alaska SAR Facility Program*, JPL Publication 89-14, Jet Propulsion Laboratory, Pasadena, California, 1989.

Carsey, F., Review and status of remote sensing of sea ice, *IEEE Journal of Oceanic Engineering, 14(2)*, pp. 127–138, 1989.

Cavalieri, D. J., J. Crawford, M. R. Drinkwater, D. T. Eppler, L. D. Farmer, R. R. Jentz, and C. C. Wackerman, Aircraft active and passive microwave validation of sea ice concentration from the DMSP SMM/I, *Journal of Geophysical Research, 96(C12)*, pp. 21,989–22,008, 1991.

Eppler, D. T., L. D. Farmer, A. W. Lohanick, and M. Hoover, Classification of sea ice types with single band (33.6 GHz) airborne passive microwave imagery, *Journal of Geophysical Research, 91(C9)*, pp. 10,661–10,695, 1986.

Holt, B., R. Kwok, and E. Rignot, Ice classification algorithm development and verification for the Alaska SAR Facility using aircraft imagery, *Proceedings of the IGARSS'89*, European Space Agency, Vancouver, Canada, pp. 751–754, 1989.

Holt, B., R. Kwok, and J. Shimada, Ocean wave products from the Alaska SAR Facility Geophysical Processor System, *Proceedings of the IGARSS'90*, European Space Agency, University of Maryland, College Park, Maryland, pp. 1469–1472, 1990a.

Holt, B., R. Kwok, and E. Rignot, Status of the ice classification algorithm in the Alaska SAR Facility Geophysical Processor System, *Proceedings of the IGARSS'90*, European Space Agency, University of Maryland, College Park, Maryland, pp. 2221–2224, 1990b.

Kwok, R., E. Rignot, B. Holt, and R. Onstott, Identification of sea ice types in spaceborne SAR data, *Journal of Geophysical Research, 97(C2)*, pp. 2391–2402, 1992.

Onstott, R. G. and S. P. Gogineni, Active microwave measurements of Arctic sea ice under summer conditions, *Journal of Geophysical Research, 90(3)*, pp. 5035–5044, 1985.

Weeks, W. F. and S. F. Ackley, The growth, structure, and properties of sea ice, *The Geophysics of Sea Ice*, edited by N. Untersteiner, pp. 9–165, Plenum Press, New York, 1986.

Wen, T., W. J. Felton, J. C. Luby, W. L. J. Fox, and K. L. Kientz, *Environmental measurements in the Beaufort Sea, Spring 1988*, Applied Physics Laboratory Technical Report ALP-UW TR 8822, 34 pp., University of Washington, Seattle, Washington, 1989.

Chapter 20. Microwave Remote Sensing of Low-Salinity Sea Ice

MARTTI HALLIKAINEN

Laboratory of Space Technology, Helsinki University of Technology, Otakaari 5A, 02150, Espoo, Finland

20.1 INTRODUCTION

Microwave remote sensing of low-salinity sea ice in the Baltic Sea is discussed in this chapter. The Baltic Sea is a semienclosed brackish water basin in northern Europe. Only first-year low-salinity (brackish-water) sea ice exists in the Baltic. The ice starts to form in October or November, obtains its maximum areal extent in March, and completely melts by the end of May. The annual maximum relative ice-cover extent in the Baltic Sea varies from 12 to 100% with an average value of 50% since 1830 [Leppäranta and Seinä, 1985]. In most winters, the southern part of the Baltic Sea is ice free.

In the winter, icebreakers assist ships to Finnish and Swedish harbors. The icebreakers select their routes based on operationally provided sea-ice information, including visual observations from ships and aircraft. The results of these observations are published regularly in sea-ice charts, see Figure 20-1.

Ice charts contain information on the ice extent, ice types, and level ice thickness. For ship navigation, information on rafting, ridging, and open water is of primary importance. The height of ice ridges may reach 2 m above the water level and they may extend 10 to 15 m below the water level. Massive ice ridges may stop even icebreakers.

The feasibility of using optical satellite data for low-salinity sea-ice mapping in the Baltic Sea is limited by the average cloudiness of 60 to 70% and the shorter days during the early and mid-winter months. Satellite-borne and airborne microwave sensors appear to be the only realistic data source for monitoring the Baltic Sea ice.

20.2 CHARACTERISTICS OF LOW-SALINITY SEA ICE

The salinity of the Baltic Sea varies from 9‰ in the south to 2‰ or less in the north. The salinity of ice varies between 0.2 and 2‰ depending on the location, time, and weather history. Thus the salinity is much lower than that of the first-year Arctic sea ice (salinity is 4 to 10‰). The surface temperature of ice is usually above –5°C, due to the snow cover upon the ice. The density of Baltic ice is close to that of Arctic first-year ice, but substantially higher than that of multiyear ice. In a two-year dielectric study of sea ice, the mean density of samples taken from the Gulf of Finland

near Helsinki was $0.84\,g/cm^3$, and 95% of those samples had a density between 0.78 and $0.90\,g/cm^3$ [Hallikainen, 1983].

The vertical structure of Baltic Sea ice often consists of granular layers in the top, mixed layers of columnar and granular ice in the middle, and columnar ice in the bottom [Omstedt, 1985]. Columnar ice forms during undisturbed thermodynamic growth, whereas granular ice corresponds to frazil ice or snow ice, formed from slush. Granular ice may also result from rafting and ridging processes. In 1988, a detailed ice-structure investigation was carried out [Weeks, et al., 1990]. In several areas, the top layers were observed to contain snow ice or frazil ice, under which there was a transition layer, while the lower layers were congelation ice with columnar grains.

As discussed in Chapter 3, the complex permittivity of Baltic ice is lower than that of Arctic first-year ice because of its lower salinity. The $\varepsilon' - j\varepsilon''$ values of Baltic ice are close to those of Arctic multiyear ice (salinity is 1 to 2‰). The power absorption constant is about 20 dB/m (salinity is 1‰; temperature is –2°C) at 1 GHz, and is nearly 100 dB/m at 10 GHz (Chapter 3, Figure 3-7). The corresponding penetration depths are 40 cm and 10 cm (Chapter 3, Figure 3-8). At UHF, the penetration depth may be more than 1 m [Hallikainen, 1983], allowing backscatter and emission contributions from the ice bottom.

Since the physical characteristics of low-salinity ice are different from those of Arctic sea ice, the backscattering and emission behavior of Baltic Sea ice is likely to exhibit features different from those of Arctic sea ice.

20.3 MICROWAVE REMOTE SENSING EXPERIMENTS IN THE BALTIC SEA

Investigations on the feasibility of microwave sensors for Baltic Sea ice mapping were started in the early 1970s. The initial campaign was Sea Ice-75 [Blomquist et al., 1975], organized in March 1975 jointly by Sweden and Finland. It was followed in 1987 by the Bothnian Experiment in Preparation of ERS-1 (BEPERS) pilot study, in 1988 by the BEPERS-88 campaign, and in 1990 by the Surface and Atmospheric Airborne Microwave Experiment (SAAMEX'90) campaign. Information on these experiments is depicted in Table 20-1.

In most experiments, both airborne radars and radiometers were used to collect data, although, recently, the emphasis shifted to radar remote sensing. This is due to the poor spatial resolution of satellite-borne radiometers. Ad-

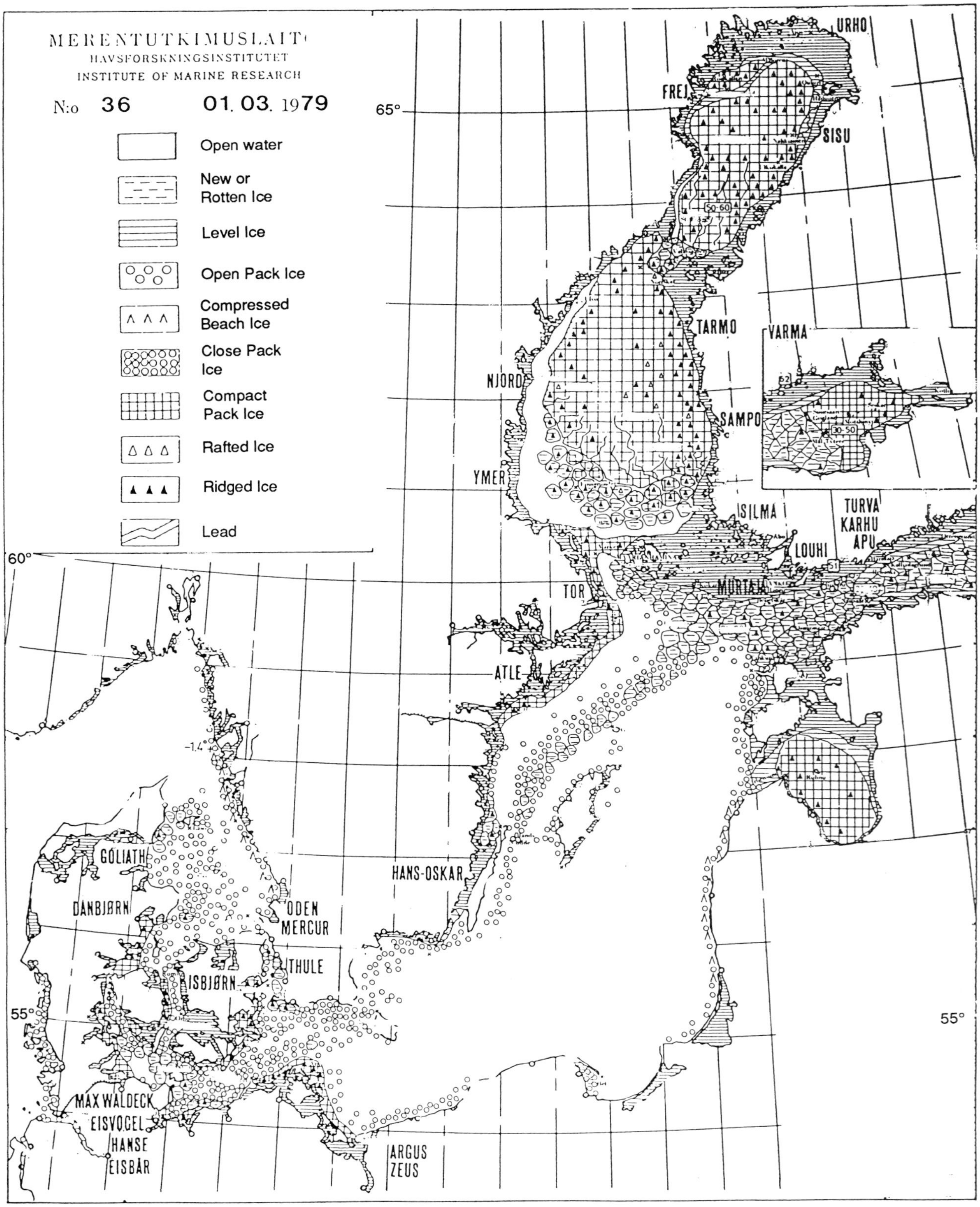

Fig. 20-1. Sea-ice chart for the Baltic Sea on March 1, 1979, by the Finnish Ice Service.

ditional data sources include airborne and spaceborne scanners and aerial photography.

The relatively small size of the Baltic Sea, especially the Gulf of Bothnia, provides easy access to test sites of remote sensing studies. Consequently, detailed surface information on snow and ice cover can be acquired, see Table 20-1.

Three test sites of different sizes were used in Sea Ice-75. The ice salinity ranged from 0.2 to 0.5‰, the snow thickness varied locally, and the air temperature was mostly below 0°C by night and slightly above 0°C by day [Tiuri et al., 1976].

Four test sites were used in the BEPERS pilot study, located in the Gulf of Bothnia between 65° N (Test Site I) and 59° N (Test Site IV) latitude. The air temperature

ranged from +1 to +3°C, ice thickness ranged from 10 to 80 cm, ice salinity from 0.4 to 1.2‰, snow thickness ranged from 0 cm to 30 cm, and snow wetness ranged from 0 to 5% depending on the test site [Leppäranta et al., 1991]. The main result of the experiment, in addition to ice-type classification studies [Kemppainen, 1989], was to provide synthetic aperture radar (SAR) data on Baltic Sea ice for training purposes.

In the BEPERS-88 campaign, low-salinity sea ice in the Gulf of Bothnia existed only north of latitude 63° N, due to the mild winter. Ice thickness ranged from 0 to 70 cm, the air temperature ranged between −7 and −2°C, and the ice salinity varied from 0.4 to 1.9‰ [Weeks, W. F. et al., Ice structure and salinity, *BEPERS-88: R/V Aranda Base Ice*

Table 20-1. Airborne microwave remote sensing campaigns of sea ice in the Baltic Sea.

Campaign date reference	Airborne radar	Airborne radiometer	Additional sensors	Sea-ice information	Snow cover information
Sea Ice-75, March 13–21, 975 [Blomquist et al., 1975]	Rijkswaterstaat Meetkundige Dienst 10-GHz SLAR; Saab–Scania Ltd. radar altimeter; National Defense Research Institute (FOA) FLAR, ODAR, and land-based radar	Helsinki University of Technology (HUT) 0.6/5 GHz	NOAA-4 VIS/IR; NOAA-4 VHRR; Landsat-2 MSS; Swedish Air Force high-altitude camera; Land Survey of Sweden multispectral camera; FOA infrared scanner	Thickness, temperature, roughness, and ice ridges	Thickness
BEPERS Pilot Study March 30–April 3, 1987 [Leppäranta et al., 1991]	Groupement pour le Developpement de la Teledetection Aerospatiale 9.4-GHz SAR; Swedish Coast Guard (SCG) SLAR		NOAA VIS/IR; SPOT VIS; SCG infrared scanner; aerial photography	Thickness, structure, salinity, temperature, and complex permittivity	Thickness, density, grain size, wetness, and surface temperature
BEPERS-88 March 2–15, 1988 [Askne et al., 1991]	Canada Centre for Remote Sensing 5/9-GHz SAR; HUT 5/10-GHz scatterometer; University of Bremen 5-GHz scatterometer	HUT 5/16 GHz	NOAA VIS/IR; SSM/I radiometer; Geosat radar altimeter; impulse radar; aerial photography	Thickness, temperature profile, salinity profile, density profile, and surface roughness	Thickness, temperature, wetness, density, and surface roughness
SAAMEX'90 March 14–23, 1990 [Uppala et al., 1991]	HUT 5/10-GHz scatterometer	British Meteorological Office 89/157 GHz; HUT 24/35/48/94 GHz		Thickness, salinity, surface temperature, and surface roughness	Thickness, wetness profile, and density profile

and Snow Geophysics Data Report, Internal Report 1990(4), Finnish Institute of Marine Research, Helsinki, Finland, pp. 18–47, 1990]. In a test site near the city of Umeå, the temperature at the snow-ice interface was about –1.6°C during the whole experiment. In some areas, the base of the snow was slightly wet. The surface roughness of snow and ice surfaces was measured in the BEPERS-88 both by ground-based laser equipment [Johansson, 1988] and by helicopter-borne equipment [Granberg, 1991].

The helicopter-borne eight-channel HUTSCAT (Helsinki University of Technology Scatterometer) was used in SAAMEX'90 to investigate ice-type classification under wet snow-ice conditions in March 1990 [Hallikainen et al., 1991]. The air temperature on March 17 and 18, 1990 was +1°C, and the snow wetness ranged from 0.5 to 6.5% on March 17, 1990, and from 4.3 to 11% on March 18, 1990. Based on the penetration depth for wet snow, Figure 3-14, the backscatter mainly came from the snowpack. Additionally, two microwave radiometer systems, covering a frequency range of 24 to 157 GHz, were flown over sea ice in the SAAMEX'90 [Uppala et al., 1991].

20.4 ACTIVE REMOTE SENSING OF SEA ICE

20.4.1 Basic Behavior of Backscattering Coefficient

The backscattering coefficient of snow-covered sea ice consists of surface and volume contributions, including the snow surface (σ^o_{ss}), snow volume (σ^o_{sv}), the ice surface (σ^o_{is}), the ice volume (σ^o_{iv}), and the water surface (σ^o_{ws}). Neglecting the surface-volume scattering interaction, the total backscattering coefficient [Kim et al., 1984] is given by

$$\sigma^o(\theta) = \sigma^o_{ss}(\theta) + t_s^2(\theta)\left\{\sigma^o_{sv}(\theta_s) + \frac{1}{L_s^2(\theta_s)}\left[\sigma_i^o(\theta_s) + \sigma_w(\theta_i)\right]\right\} \quad (1)$$

where $\sigma^o_i(\theta_s)$ and $\sigma^o_w(\theta_i)$ are the backscattering coefficients of ice and water, respectively,

$$\sigma_i^o(\theta_s) = \sigma_{is}^o(\theta_s) + t_i^2(\theta_s)\,\sigma_{iv}^o(\theta_i) \quad (2)$$

$$\sigma_w^o(\theta_i) = t_i^2(\theta_s)\frac{1}{L_i^2(\theta_i)}\,\sigma_{ws}^o(\theta_i) \quad (3)$$

The Fresnel power transmission coefficients at the air–snow and snow–ice interfaces are denoted by t_s and t_i, respectively. The radar local incidence angle is θ and the wave propagation angles in the snow and ice medium are θ_s and θ_i, respectively. The total attenuations of the snow and ice layers are L_s and L_i, respectively, in the direction of wave propagation.

Ulander et al. [1991] applied Equations (1) to (3) to Baltic Sea ice in order to compare theoretical values with BEPERS-88 SAR data. For ice surface scattering, they used either the Kirchhoff model with the scalar approximation or the small

perturbation model, depending on the observed values of the surface roughness parameters. The volume scattering contribution was computed according to Kim et al. [1985].

Figure 20-2 shows the backscattering components contributing to the total backscattering coefficient at 5.3 GHz under the conditions encountered in the BEPERS-88 campaign. The total backscattering coefficient is dominated by the snow-ice interface and has, additionally, a small contribution from the snow volume.

Manninen [1991] used a geometrical model for ice ridge sails in order to theoretically investigate discrimination of ridges from level ice. She assumed that the cross section of the sail is triangular and the individual ice blocks are rectangular polyhedrons described with three Euler angles and three dimensions. A first-order small-scale perturbation theory was used to account for surface scattering for like-polarizations and a second-order theory for cross-po-

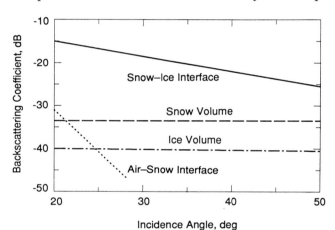

Fig. 20-2. Predicted backscattering coefficient at 5.3 GHz, HH polarization, for the BEPERS-88 Test Site 1 [Ulander et al., 1991]. Volume and surface scattering contributions from snow and ice are shown. The snow-ice ice surface rms height is 1.81 mm and the correlation length 22.1 mm.

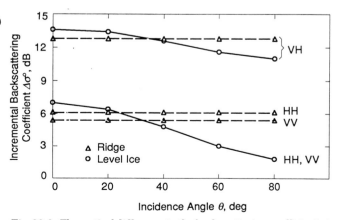

Fig. 20-3. Theoretical difference in the backscattering coefficient at 5.3 GHz for ice ridge and level ice due to different surface roughness parameters (rms height s and correlation length l) [Manninen, 1991]. $\Delta\sigma^o$ is defined as $\Delta\sigma^o = \sigma^o (ks = 0.3, kl = 1.5) - \sigma^o (ks = 0.2, kl = 1.0)$.

larizations. The results in Figure 20-3 show that the VH polarization (transmit vertical polarization, and receive horizontal polarization) is substantially more sensitive to the surface roughness than the like-polarizations.

The backscattering behavior of low-salinity sea ice at 5.4 and 9.8 GHz (VV, HH, HV, and VH polarization modes at each frequency) was investigated by Hyyppä and Hallikainen [1991] in BEPERS-88. At each frequency, the helicopter-borne HUTSCAT scatterometer employs four linear polarizations (VV, HH, HV, VH). The incidence angle was 23° off nadir, which is equal to that of the ERS-1 SAR, which was launched in July 1991. A review of the experimental results for different ice types is shown in Figure 20-4, where each data point represents the mean backscattering coefficient, obtained for a sea-ice type during the BEPERS-88 campaign. Ninety percent of the experimental results fall between the maximum and minimum values indicated by the vertical bar. Since the results for the VH and HV polarization modes agree within 1 dB, HV values are not shown. For comparison, results are shown for thick (1.5-m) snow-covered first-year Arctic sea ice at 5.2 GHz in Figure 20-4(a), and 9.0 GHz in Figure 20-4(b).

C-band radar signatures of Arctic sea ice and low-salinity Baltic Sea ice are compared in Figure 20-5 [Ulander, 1991]. Figures 20-4 and 20-5 show that the difference between the like-polarization backscattering coefficients measured in the two areas for an ice type is usually less than 5 dB. For the cross-polarization cases, the difference may be larger.

20.4.2 Retrieval of Ice Characteristics

20.4.2.1. Classification of ice types. The results in Figure 20-4 suggest that, using an incidence angle of 23°, the discrimination of ice types under the conditions encountered in BEPERS-88 is difficult if the backscattering coefficient from a single channel is used. Using like-polarization (either 5.4 or 9.8 GHz), only two broad categories can be distinguished: (a) open water and new ice, and (b) rubble field and ice ridges. The values for thick level ice overlap with those for the two categories. According to Figure 20-4, the discrimination capability of the cross-polarized channels (VH and HV) is substantially better than that of the like-polarized channels. The results suggest that three categories can be discriminated: (a) open water and new ice, (b) level ice, and (c) rubble field and ice ridges.

The best classification capability at 23° was obtained with a combination of VV and HV polarization modes at 5.4 GHz [Hyyppä and Hallikainen, 1991]. Based on Figure 20-6, the following categories can be distinguished: open water, new ice, thick level ice, and rubble field and ice ridges.

Discrimination of open water from new ice is affected by the wind speed and direction. The open water areas measured by Hyyppä and Hallikainen [1991] were small and, hence, the effects of wind speed on σ^0 are not evident.

The eight-channel HUTSCAT scatterometer data from the SAAMEX campaign indicated that the best classifica-

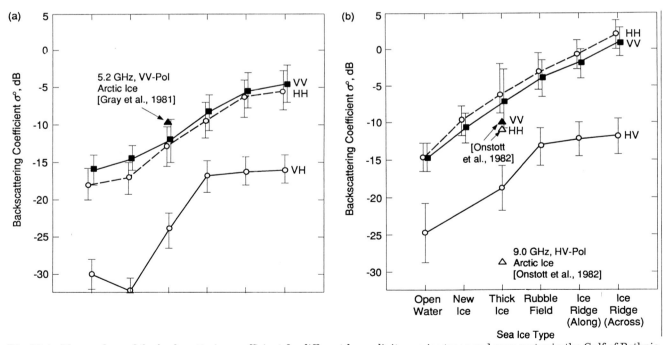

Fig. 20-4. Mean values of the backscattering coefficient for different low-salinity sea-ice types and open water in the Gulf of Bothnia, March 7–11, 1988, at an incidence angle of 23° and at (a) 5.4 and (b) 9.8 GHz [Hyyppä and Hallikainen, 1991]. Ninety percent of the results fall within the maximum and minimum values indicated by the vertical bar. For comparison, results for Arctic sea ice are included.

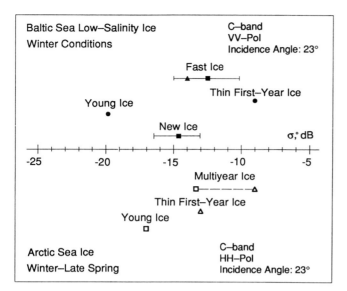

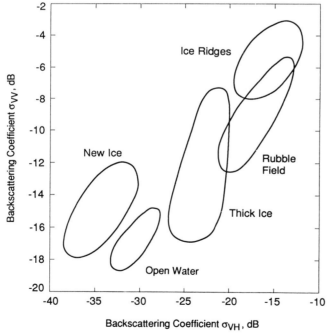

Fig. 20-5. Comparison of C-band sea-ice signatures in the Arctic and low-salinity ice in the Baltic Sea [Ulander, 1991].

Fig. 20-6. Discrimination of low-salinity sea-ice types by using the backscattering coefficient from VV and HV polarization modes at 5.4 GHz in BEPERS-88 [Hyyppä and Hallikainen, 1991]. Snow temperature is –1.5 to –4.1°C, base of snow in some locations is wet, snow-ice temperature is 0 to –2.5°C, and ice salinity is 1.3 (top) to 0.5‰ (bottom). Snow-free new-ice thickness ≤10 cm and level ice thickness is 40 to 60 cm.

tion capability using a single channel (incidence angle of 23°) under wet snow–ice conditions was obtained with the 5.4-GHz HV channel, in agreement with the results from BEPERS-88. Similarly, the VV and HV channels at 5.4 GHz provided the overall optimum channel combination. The classification capability under the wet snow-ice conditions was, however, substantially lower than that obtained in the dry snow-ice conditions encountered in BEPERS-88. The results in Figure 20-7 show that, basically, only snow-free new ice can be discriminated from the level ice and rubble field. This suggests that discrimination of ice types under typical spring conditions (wet snow–ice) may be much more difficult than under mid-winter conditions. As discussed in Chapter 3, the absorption coefficient of wet snow is so high that backscatter from the snow-ice interface may become negligible. Consequently, backscatter only from the snow medium contributes to the backscattering coefficient of the snow-ice-water system.

Ice classification was investigated in the BEPERS pilot study by Kemppainen [1989] using X-band SAR data. The mean and variance of backscatter, expressed in relative intensity values, were computed for eight ice categories, including open water. The results are based on the radar signatures from several small areas. The results for two test sites are shown in Figure 20-8. Basically, discrimination of open water and undeformed ice from deformed ice can be done in a reliable manner. The two main problems appear to be discrimination between open water and undeformed ice, and between brash ice and ridged ice. The open water signatures for the two test sites are different, due to different wind conditions. Brash ice areas (e.g., ship tracks) and ice ridges tend to have high intensities in both SAR images.

C-band SAR data from the BEPERS-88 were used by Ulander and Carlström [1991] to evaluate the feasibility of using data obtained at incidence angles between 27° and 68° for classification of ice types, including young ice (thickness of 15 cm), thin ice (thickness of 35 cm) and snow-covered fast ice (thickness of 40 cm of snow upon 60 cm of ice). Due to the limited SAR flight program, only thin ice was covered by both HH and VV polarization modes; for other ice types, VV data were available. The signature contrast between thin and young ice is about 10 dB independent of the incidence angle, see Figure 20-9. The fast ice exhibits a different incidence-angle dependency. Good agreement between the theoretical model and the experimental data was obtained by optimizing the rms (root mean square) ice surface height and the snow grain radius for each case. The optimum rms ice surface height varied from 0.15 to 0.60 cm and the correlation length from 3 to 5 cm, depending on the ice type.

Classification of ice types using Geosat Ku-band altimeter data was evaluated by Ulander and Carlström [1991]. The diameter of the Geosat pulse-limited footprint is 1.7 km. Based on interpretation results of the Canada Centre for Remote Sensing (CCRS) airborne SAR data, four ice types were considered: smooth and rough young ice, thin ice, and deformed ice. The values of the mean and standard

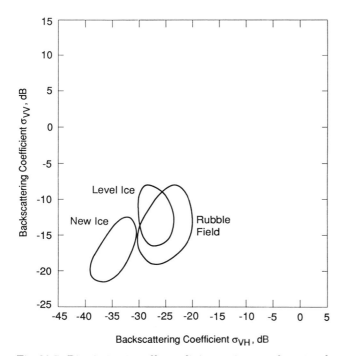

Fig. 20-7. Discrimination of low-salinity sea-ice types by using the backscattering coefficient from VV and HV polarization modes at 5.4 GHz in SAAMEX'90 [Hallikainen et al., 1991]. Ice and snow temperature is 0°C, new-ice snow-free snow depth is 10 to 20 cm upon level ice, snow wetness is 4.3 to 11.3%, and ice salinity is below 0.3‰.

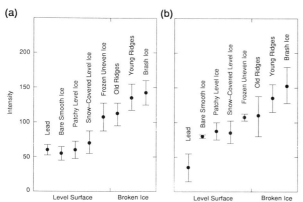

Fig. 20-8. Ice-type classification results based on X-band SAR data (HH polarization mode) for two test sites in the BEPERS pilot study [Kemppainen, 1989]. (a) Test Site I is near the city of Oulu (latitude 65° N), and (b) Test Site II is near the city of Vaasa (latitude 63° N).

deviation of the backscattering coefficient at normal incidence show that the four ice types are well separated, see Table 20-2.

20.4.2.2. Ridging intensity. In mapping sea-ice characteristics for ship navigation, one of the key tasks is to estimate ridging intensity. According to Similä [1991], ridges may account for up to 50% of the total ice mass. He

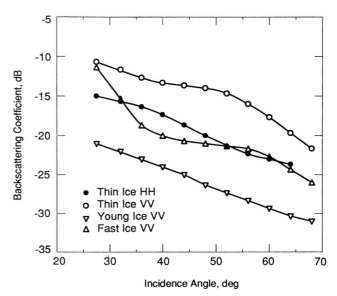

Fig. 20-9. Experimental SAR backscatter signatures for three ice types at 5.3 GHz as a function of incidence angle [Ulander and Carlström, 1991].

Table 20-2. Statistics of Geosat normal-incidence backscattering coefficient for four ice types in the Gulf of Bothnia, March 1988 [Ulander and Carlström, 1991].

Ice type	Mean, dB	Standard deviation, dB
Smooth young ice	31.6	1.9
Rough young ice	27.4	2.5
Thin ice	19.9	2.3
Deformed ice	11.8	1.5

investigated estimation of ridging intensity from SAR images by comparing BEPERS-88 SAR data (C-band, VV polarization) with ice surface profile measurements (elevation accuracy about 2 cm). Cutoff heights of 30 and 50 cm for ridges were used to distinguish ridges from other ice features in the analysis. Since information on ice ridges is contained in the upper tail of the SAR intensity distribution, he normalized the tail-to-mean ratio as

$$\tau_r = \frac{E(I \mid I > I_{\min})}{E(I)} \quad (4)$$

where E is the expected value, I is the intensity, and $I_{\min}$ is the lower limit of the tail. Based on experimental observations, the highest 10% of the intensity values were assumed to be related to ice ridges. Correlation analysis showed strong correlation between the tail-to-mean ratio and ridging intensity. For each of the two cutoff heights, the coefficient of determination was found to be about 0.80. Discrimination of brash ice from ridges was not successful.

20.4.2.3. Ice dynamics. Leppäranta and Yan [1991] determined ice displacement from three CCRS SAR images, acquired on March 6, 8, and 9, 1988 in the Gulf of Bothnia. The displacements were estimated from manually recognized ice features (e.g., floes, leads, etc.).

20.5 PASSIVE REMOTE SENSING

20.5.1 Basic Behavior of Brightness Temperature

Low-salinity sea ice is a nonscattering, low-loss medium at UHF and low microwave frequencies. Thus the phase coherence of the electromagnetic waves propagating in the sea-ice layer is preserved and the brightness temperature can be computed using a layered model that assumes the ice and snow properties to vary with depth only. For each infinitesimally thin layer, the ice and snow properties are assumed to remain constant. Based on extensive observations [Hallikainen, 1977, 1983], the computation can be made more realistic by assuming that (a) the ice salinity decreases with increasing depth and it includes random variations both vertically and horizontally, (b) the thickness of snow cover and ice cover varies within the antenna footprint [Uusitalo, 1957], (c) the ice surface temperature

varies according to ice and snow thickness variation, and (d) the effect of finite antenna beamwidth is accounted for.

The basic behavior of the brightness temperature in the UHF range has been investigated theoretically using the layered model and the above assumptions [Hallikainen, 1983]. The results are shown in Figure 20-10(a). Due to coherent reflections of the emitted electromagnetic field components within the low-loss ice layer, the brightness temperature oscillates strongly as a function of ice thickness. A local maximum in the brightness temperature is predicted for the nadir angle when the ice thickness is an odd number of quarter wavelengths:

$$L_i = n\frac{\lambda}{4}, \quad n = 1, 3, 5, 7, \cdots \tag{5}$$

where L_i is the ice thickness and λ is the wavelength in the ice medium. Correspondingly, a local minimum in the brightness temperature is predicted when the ice thickness is an integral number of half wavelengths

$$L_i = n\frac{\lambda}{2}, \quad n = 1, 2, 3, 4, \cdots \tag{6}$$

The average value of the brightness temperature approaches the value determined by the power transmission coefficients of the ice-snow and snow-air interfaces. For large ice thicknesses, brightness-temperature contributions from the

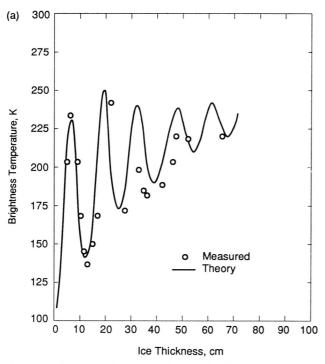

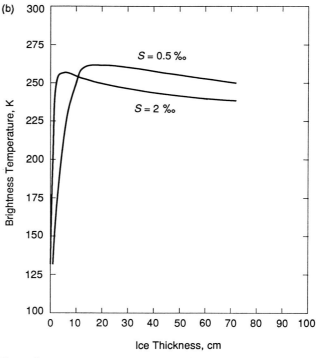

Fig. 20-10. Theoretical and experimental brightness temperatures of low-salinity sea ice at (a) 610 MHz and (b) 5 GHz as a function of ice thickness [Tiuri et al., 1978; Hallikainen, 1982, 1983]. Incidence angle is 0°: (a) at 610 MHz, and ice salinity is 0.7 to 1.0‰, ice surface temperature is −2.3 to −6.3°C, and snow depth is 0 to 15 cm (dry snow); and (b) at 5 GHz, and ice salinity is 0.5 and 2‰.

bottom of the ice layer are attenuated to a negligible level before reaching the air. The effects of the three main parameters—ice salinity, ice surface temperature, and snow depth—to the brightness temperature at 610 MHz were examined by Hallikainen [1983]. The brightness temperature has the highest sensitivity to the salinity of ice.

The measured brightness temperature at 610 MHz was observed to follow closely the theory, see Figure 20-10(a). The experimental results were obtained from a nonmoving platform (a hovering helicopter), so that no averaging due to platform movement is included.

At higher microwave frequencies, the incoherent approach, which ignores oscillations within the ice layer, is more realistic, see Figure 20-10(b). Due to the higher loss of sea ice at microwave frequencies (Chapter 3), the brightness temperature saturates for small ice thicknesses and it is determined primarily by the dielectric properties and surface roughness of the snow and sea-ice layers. For large ice thicknesses, the brightness temperature decreases with increasing ice thickness, due to assumed increase in the snow depth.

Available experimental airborne microwave radiometer data sets on Baltic Sea ice are limited to UHF and C band ranges [Tiuri et al., 1976, 1978; Hallikainen, 1983]. Additional data sets have been obtained in 1990–1992 by the Helsinki University of Technology (24, 35, 48, and 94 GHz) and in 1990 by the British Meteorological Office (89 and 157

GHz); the results have not been published yet. Nimbus-7 SMMR data have been applied to ice concentration studies by Hallikainen and Mikkonen [1991]. The feasibility of using 6.6 and 10.7 GHz data from Nimbus-7 in the Baltic Sea is limited, due to poor spatial resolution and antenna sidelobes over land.

20.5.2 Retrieval of Ice Characteristics

20.5.2.1. Ice thickness. The feasibility of using UHF and microwave radiometers for determining the thickness and other characteristics of sea ice in the Gulf of Bothnia has been investigated over a period of several winters [Tiuri et al., 1976, 1978; Hallikainen, 1983]. Based on Figure 20-10, at UHF and low microwave frequencies the brightness temperature of low-salinity sea ice depends on the ice thickness. The relationship is not unambiguous and, hence, retrieval of the ice thickness from its brightness temperature is not straightforward.

Figure 20-11 shows the 610-MHz and 5-GHz brightness temperatures measured along a test line that exhibits considerable ice thickness variations. As expected, the brightness temperature at 5 GHz is practically independent of the ice thickness. At 610 MHz, the brightness temperature follows well the smoothed ice thickness curve which takes into account the averaging due to the antenna beamwidth and the helicopter speed. Figure 20-11 suggests

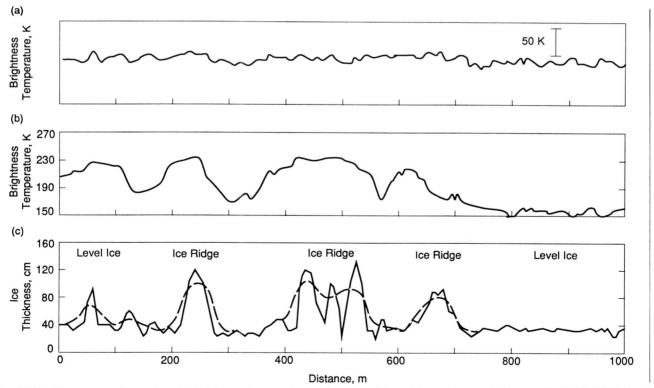

Fig. 20-11. Comparison of experimental brightness temperature and measured ice thickness along a test line in the Gulf of Bothnia at (a) 5 GHz and (b) 610 MHz and an incidence angle of 0° [Tiuri et al., 1976]. (c) The dotted line indicates smoothed ice thickness corresponding to a radiometer spatial resolution of 30 m. Ice salinity is 0.4‰, and ice surface temperature is –3°C.

that, due to the smoothing effects in the brightness temperature, ice thickness can be determined with a single-channel radiometer with reasonable accuracy.

Accuracy of the ice-thickness retrieval can be increased by using either a broadband radiometer or a multichannel radiometer. The use of a broadband radiometer at UHF and low microwave frequencies is not possible due to the presence of television transmitters and other man-made noise sources along the coast of the Gulf of Bothnia. The use of a multichannel radiometer for ice-thickness retrieval has been tested [Tiuri et al., 1976, 1978; Hallikainen, 1983]. The brightness temperature is computed as the mean value of the results at various channels,

$$T_B = \frac{1}{N} \sum_{i=1}^{N} T_{Bi} \qquad (7)$$

where T_{Bi} is the measured brightness temperature for channel i and N is the number of channels. In order to achieve an unambiguous relation between T_B and the ice thickness, the channel frequencies are selected properly. An example of the capability of a multichannel radiometer to retrieve ice thickness is shown in Figure 20-12. The channel frequencies are 530, 780, and 930 MHz. Theoretical and experimental values for the incidence angle of 0° agree well. An incidence angle of 45° off nadir and vertical polarization provide additional smoothing to the behavior of the brightness temperature. The effect of platform movement, which is not included in Figure 20-12, further linearizes the T_B versus ice-thickness relationship.

Accuracy of the ice-thickness retrieval is lowered by local variations in the ice salinity, surface temperature, and snow thickness. A detailed analysis performed by Hallikainen [1983] indicates that an accuracy of ±20 cm is possible if the above parameters are approximately known.

20.5.2.2. Ice concentration. Nimbus-7 SMMR data were used by Hallikainen and Mikkonen [1991] to investigate the

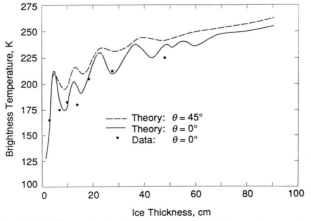

Fig. 20-12. Theoretical and experimental brightness temperature of sea ice using the mean value from frequencies of 530, 780, and 930 MHz at V polarization [Hallikainen, 1983].

effects of snow-cover variations on the retrieval of the ice concentration. Because of (a) the low spatial resolution and possible sidelobe contributions from land at 6.6 and 10.7 GHz, and (b) the atmospheric water vapor at 21 GHz, only the 18 and 37 GHz were used. For each frequency, the following algorithm was used,

$$C = A\,T_B - B \quad \text{(in percent)} \qquad (8\,a)$$

$$A = \frac{100}{T_{Bsi} - T_{Bsw}} \qquad (8\,b)$$

$$B = 100\,\frac{T_{Bsw}}{T_{Bsi} - T_{Bsw}} \qquad (8\,c)$$

where T_{Bsi} and T_{Bsw} are the brightness temperatures of sea ice and sea water, respectively. The values of coefficients A and B were determined from the observed brightness temperatures of open water (southern Baltic Sea) and ice.

To investigate the effect of the properties of the snow cover to retrieval of ice concentration, a date was selected when the snow cover upon the sea-ice layer was wet in the southern Baltic Sea and dry in the northern Baltic Sea. SMMR data from the morning pass of Nimbus-7 on March 1, 1979 was employed in the study.

Figure 20-13 suggests that the proper values of the tie points for the ice-concentration algorithm at 37 GHz strongly depend on the liquid water content of the snow cover. In Figure 20-13(a), the values of coefficients A and B were determined using the maximum value of T_{Bsi} occurring in the Baltic Sea on March 1, 1979. Since the maximum value is obtained for ice covered with wet snow, the ice concentration is underestimated for areas with a dry snow cover upon the ice layer. Due to increasing volume scattering in the snow medium with increasing frequency (Chapter 3), the ice concentration resulting from the 37-GHz algorithm is far too small for areas with dry snow cover. The ice chart for March 1, 1979, published by the Finnish Ice Service (based on visual observations from ships and aircraft), is depicted in Figure 20-1. It shows that the ice concentration in the northern Baltic Sea is 100%. The smallest satellite-derived value for that area is 58% from the 37-GHz algorithm, Figure 20-13(a). Using the 18-GHz algorithm, the smallest value is 86%.

The local variation in the snow properties can be accounted for by assuming that the brightness temperature of the snow-ice-water system is equal to that of the snow-soil system at the same latitude in Finland. The assumption holds reasonably well, due to the dielectric similarity of sea ice and soil [Hallikainen, 1983; Hallikainen et al., 1985]. The ice concentration obtained by this method using the 37-GHz algorithm is shown in Figure 20-13(b). It follows closely the reported ice concentration in Figure 20-1. Satellite-derived ice concentrations that are higher than 100% in Figure 20-13(b) are shown in order to exhibit the variation in satellite-derived ice concentrations.

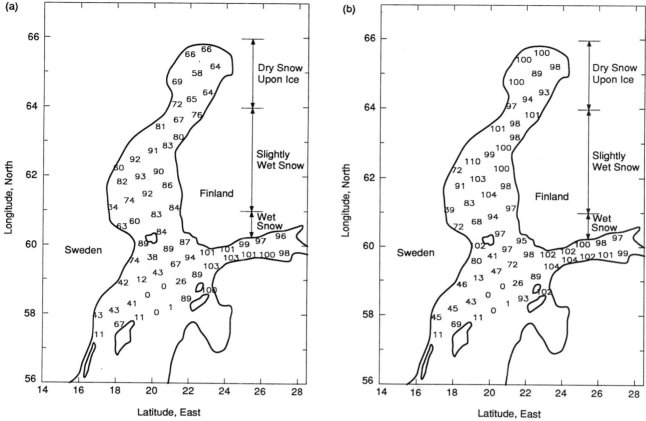

Fig. 20-13. Satellite-derived ice concentration (%) in the Baltic Sea for March 1, 1979, using SMMR 37-GHz data [Hallikainen and Mikkonen, 1991]. The brightness temperature of water is assumed to be 140 K. The sea-ice chart for March 1, 1979, by the Finnish Ice Service is given in Fig. 20-1. (a) The brightness temperature of ice is assumed to be 242 K. (b) The brightness temperature of ice is assumed to be equal to that of coastal snow-covered terrain at the same latitude in Finland.

The results indicate that the feasibility of frequencies higher than 15 GHz for remote sensing of snow-covered sea ice is limited, unless the effect of snow cover is accounted for. Because of volume scattering by snow particles, the brightness temperature of dry snow-covered ice may be tens of kelvins lower than that of bare ice. As discussed in Chapter 3, the penetration depth for dry snow at millimeter-wave frequencies is very small. Hence, high-frequency sensors may only provide information on the snow cover.

20.6 SUMMARY

The day/night and nearly all-weather capability and good spatial resolution of spaceborne and airborne SARs make them suitable instruments for sea-ice monitoring. The main disadvantage of SAR is the 100-km swath, as opposed to 1400 km for the SSM/I microwave radiometer. SAR data from several passes must be collected in order to cover the Baltic Sea.

Although the penetration depth of low-salinity ice in the Baltic Sea is much larger than that of Arctic sea ice, the main backscatter contribution at C-band under winter conditions comes from the snow-ice interface. Consequently,

the surface roughness is an important parameter in radar mapping of both Arctic and Baltic ice. Comparison between the experimental results for the two first-year ice types does not show any distinct difference, obviously due to similar surface roughness characteristics.

Based on the experimental results for low-salinity sea ice in Section 20.4, classification of the main categories (open water, level ice, and deformed ice) is possible under winter conditions using a single radar channel (one polarization mode, one frequency). However, retrieval accuracy depends on weather conditions, and discrimination of open water from level ice may be difficult, depending on wind speed and direction. The use of both a like-polarized (HH or VV) and cross-polarized (VH or HV) radar channel increases its classification capability. Under spring conditions (wet snow-ice surfaces), wet snow masks the underlying ice layer and only snow-free ice can be discriminated from snow-covered ice. The use of Geosat Ku-band altimeter data at normal incidence provided good classification results at BEPERS-88. All the above conclusions are based on limited data sets with good ground truth.

A microwave radiometer can distinguish between open water and ice much better than a radar. In the Arctic, ice concentration algorithms employing 6.6- and 10.7-GHz channels from the Nimbus-7 SMMR have been proved to work well [Cavalieri et al., 1984]. The poor spatial resolution of a spaceborne microwave radiometer excludes the use of low-frequency channels for ice mapping in the Baltic Sea. Only a 90-GHz channel provides an adequate spatial resolution. As discussed in Section 20.5, the brightness temperature of snow-covered ice may vary substantially at 37 GHz, depending on the liquid water content and other characteristics of the snow cover. Similarly, selection of the tie points for the ice-concentration algorithm at 90 GHz should be based on the brightness temperature of coastal snow-covered ground at the same latitude in Finland.

The feasibility of using a combination of ERS-1 SAR-type radar and SSM/I-type microwave radiometer for ice monitoring in the Baltic Sea has not been investigated yet. Based on the results obtained from the 37-GHz ice-concentration algorithm (Section 20.5), the ice concentration and ice extent can be derived from 85-GHz data (the effective field of view is 13 by 15 km) with a reasonable accuracy. Discrimination of deformed ice (e.g., ice ridges and rubble fields), new ice and, possibly, level ice can be done using SAR data (resolution of 30 m).

The swath of future spaceborne SAR's is planned to be about 500 km, allowing the Baltic Sea to be covered by data from a single pass. Accuracy of the ice type classification and ice extent determination with the proposed instrument combination will increase, when multipolarization SAR data and high-resolution microwave radiometer data are available. The footprint of the 89-GHz channel of the European Space Agency (ESA) Multifrequency Imaging Microwave Radiometer (MIMR) will be 4.8 km and that of the 36.5-GHz channel will be 11.6 km [Bernard et al., 1990].

The results in Section 20.5 suggest that, using an airborne multichannel UHF radiometer, the thickness of low-salinity ice can be determined with an adequate accuracy under winter conditions (when ice characteristics are known approximately). Similarly, an airborne P-band (400-MHz) SAR or SAR polarimeter should provide ice-thickness information.

References

Askne, J., M. Lepparanta, and T. Thompson, The Bothnian Experiment in Preparation of ERS-1, 1988 (BEPERS-88), an overview, *International Journal of Remote Sensing*, in press, 1991.

Bernard, R., M. Hallikainen, Y. Kerr, K. Künzi, C. Mätzler, P. Pampaloni, G. Duchossois, Y. Menard, and M. Rast, *The Multifrequency Imaging Microwave Radiometer— Instrument Panel Report*, ESA-SP 1138, European Space Agency, Paris, 39 pp., August 1990.

Blomquist, Å., C. Pilo, and T. Thompson, *Sea Ice–75 Programme*, Research Report No. 16:1, Winter Navigation Research Board, Stockholm, Sweden, 1975.

Cavalieri, D. J., P. Gloersen, and W. J. Campbell, Determination of sea ice parameters with the Nimbus 7 SMMR, *Journal of Geophysical Research, 89(D4)*, pp. 5355–5369, 1984.

Granberg, H. B., Interactive graphics methods for removal of helicopter motions from laser profile data, *Proceedings of the 11th Annual International Geoscience and Remote Sensing Symposium (IGARSS'91)*, vol. 3, Espoo, Finland, June 3–6, 1991, Institute of Electrical and Electronics Engineers, Inc., New York, pp. 1243–1246, 1991.

Gray, A. L., R. K. Hawkins, C. E. Livingstone, and L. D. Arsenault, Seasonal effects on the microwave signatures of Beaufort Sea ice, *Proceedings of the Fifteenth International Symposium on Remote Sensing of the Environment*, Ann Arbor, Michigan, May 11–15, 1981, vol. 1, Environmental Research Institute of Michigan, pp. 239–257, 1981.

Hallikainen, M., *Dielectric Properties of Sea Ice at Microwave Frequencies*, Report S 94, Helsinki University of Technology, Radio Laboratory, Espoo, Finland, 1977.

Hallikainen, M., The brightness temperature of sea ice and fresh-water ice in the frequency range 500 MHz to 37 GHz, *Digest 1982 International Geoscience and Remote Sensing Symposium (IGARSS'82)*, Munich, West Germany, June 1–4, 1982, vol. 2, Institute of Electrical and Electronics Engineers, Inc., New York, 6 pp., 1982.

Hallikainen, M., A new low-salinity sea ice model for UHF radiometry, *International Journal of Remote Sensing, 4*, pp. 655–681, 1983.

Hallikainen, M. and P.-V. Mikkonen, Retrieval of geophysical parameters from Nimbus-7 SMMR data in a semi-enclosed sea, Part II: Sea ice concentration, *IEEE Transactions on Geoscience and Remote Sensing*, in press, 1991.

Hallikainen, M., F. T. Ulaby, M. C. Dobson, and M. El-Rayes, Microwave dielectric behavior of wet soil—Part I: Empirical models and experimental observations, *IEEE Transactions on Geoscience and Remote Sensing, GE-23*, pp. 25–34, 1985.

Hallikainen, M., T. Tares, M. Toikka, and K. Heiska, Classification of low-salinity sea ice types by dual-frequency multipolarization scatterometer, *Proceedings of the 11th Annual International Geoscience and Remote Sensing Symposium (IGARSS'91)*, vol. 3, Espoo, Finland, June 3–6, 1991, Institute of Electrical and Electronics Engineers, Inc., New York, pp. 1211–1214, 1991.

Hyyppä, J. and M. Hallikainen, Classification of low-salinity sea ice types by ranging scatterometer, *International Journal of Remote Sensing*, in press, 1991.

Johansson, R., *Laser-Based Surface Roughness Measurements of Snow and Sea Ice on the Centimeter Scale*, Research Report 162, 34 pp. Department of Radio and Space Science, Chalmers University of Technology, Goteborg, Sweden, 1988.

Kemppainen, H., *Applicability of Synthetic Aperture Radar in Interpretation of Sea Ice in the Baltic Sea* (in Finnish), M.Sc. Thesis, Helsinki University of Technology, Department of Civil Engineering and Surveying, Espoo, Finland, 1989.

Kim, Y.-S., R. K. Moore, and R. G. Onstott, *Theoretical and Experimental Study of Radar Backscatter From Sea Ice*, Remote Sensing Laboratory Technical Report 331-37, University of Kansas, Lawrence, Kansas, 1984.

Kim, Y. S., R. K. Moore, R. G. Onstott, and S. Gogineni, Towards identification of optimum radar parameters for sea ice monitoring, *Journal of Glaciology, 31(109)*, pp. 214–219, 1985.

Leppäranta, M. and A. Seinä, Freezing, maximum annual ice thickness and breakup of ice on the Finnish coast during 1830–1984, *Geophysica, 21*, pp. 87–104, 1985.

Leppäranta, M. and S. Yan, Use of ice velocities from SAR imagery in numerical sea ice modeling, *Proceedings of the 11th Annual International Geoscience and Remote Sensing Symposium (IGARSS'91)*, vol. 3, Espoo, Finland, June 3–6, 1991, Institute of Electrical and Electronics Engineers, Inc., New York, pp. 1233–1237, 1991.

Leppäranta, M., R. Kuittinen, and J. Askne, BEPERS Pilot Study: An experiment with X-band synthetic aperture radar over Baltic Sea ice, *Journal of Glaciology*, in press, 1991.

Manninen, A. T., Surface backscattering dependence on various properties of ice ridges, *Proceedings of the 11th Annual International Geoscience and Remote Sensing Symposium (IGARSS'91)*, vol. 3, Espoo, Finland, June 3–6, 1991, Institute of Electrical and Electronics Engineers, Inc., New York, pp. 1219–1222, 1991.

Omstedt, A., *An Investigation of the Crystal Structure of Sea Ice in the Bothnian Bay*, Hydrology and Oceanography Report RHO-2, Swedish Meteorological and Hydrological Institute, Norrköping, 1985.

Onstott, R. G., T. C. Grenfell, C. Mätzler, C. A. Luther, and E. A. Svendsen, Evolution of microwave sea ice signatures during early summer and midsummer in the marginal ice zone, *Journal of Geophysical Research, 92(C7)*, pp. 6825–6835, 1987.

Onstott, R. G., R. K. Moore, S. Gogineni, and C. Delker, Four years of low-altitude sea ice broad-band backscatter measurements, *IEEE Journal of Oceanic Engineering, OE-7*, pp. 44–50, 1982.

Similä, M., Statistical description of sea ice ridging intensity by SAR imagery, *Proceedings of the 11th Annual International Geoscience and Remote Sensing Symposium (IGARSS'91)*, vol. 3, Espoo, Finland, June 3–6, 1991, Institute of Electrical and Electronics Engineers, Inc., New York, pp. 1223–1227, 1991.

Tiuri, M., M. Hallikainen, and A. Lääperi, *Microwave Radiometer Theory and Measurements of Sea Ice Characteristics*, Report S 89, Helsinki University of Technology, Radio Laboratory, Espoo, Finland, 1976.

Tiuri, M., M. Hallikainen, and A. Lääperi, Radiometer studies of low-salinity sea ice, *Boundary-Layer Meteorology, 3*, pp. 361–371, 1978.

Ulander, L., *Radar Remote Sensing of Sea Ice: Measurements and Theory*, Technical Report 212, Chalmers University of Technology, School of Electrical and Computer Engineering, Goteborg, Sweden, 1991.

Ulander, L. and A. Carlström, Radar backscatter signatures of Baltic sea ice, *Proceedings of the 11th Annual International Geoscience and Remote Sensing Symposium (IGARSS'91)*, vol. 3, Espoo, Finland, June 3–6, 1991, Institute of Electrical and Electronics Engineers, Inc., New York, pp. 1215–1218, 1991.

Ulander, L., R. Johansson, and J. Askne, C-band radar backscatter of Baltic sea ice: Theoretical predictions compared with calibrated SAR measurements, *International Journal of Remote Sensing*, in press, 1991.

Uppala, S., J. S. Foot, A. Chedin, N. A. Scott, C. Claud, F. Sirou, and M. Hallikainen, SAAMEX—Surface and Atmospheric Airborne Microwave Experiment, Finland 1990, *Proceedings of the 11th Annual International Geoscience and Remote Sensing Symposium (IGARSS'91)*, vol. 3, Espoo, Finland, June 3–6, 1991, Institute of Electrical and Electronics Engineers, Inc., New York, pp. 891–894, 1991.

Uusitalo, S., Beobachtungen mit Bezug auf das Meereseis, *Geophysica, 5*, pp. 139–146, 1957.

Weeks, W. F., A. J. Gow, P. Kosloff, and S. Digby-Argus, The internal structure, composition and properties of brackish ice from the Bay of Bothnia, *Sea Ice Properties and Processes*, edited by S. F. Ackley and W. F. Weeks, CRREL Monograph 90-1, Cold Regions Science and Engineering Laboratory, Dartmouth College, Hanover, New Hampshire, 1990.

Chapter 21. The Ice Thickness Distribution Inferred Using Remote Sensing Techniques

PETER WADHAMS

Scott Polar Research Institute, University of Cambridge, Lensfield Road, Cambridge, CB2 1ER, England

JOSEFINO C. COMISO

Laboratory for Hydrospheric Processes, Goddard Space Flight Center, Greenbelt, Maryland 20771

21.1 INTRODUCTION

The thickness distribution of sea ice is an important quantity that requires synoptic monitoring in both polar regions. Together with the ice extent, it defines the response of sea ice to climatic change. Together with the ice velocity, it defines the mass flux of sea ice, which in key regions such as the Fram Strait, is a major component of the overall energy and fresh water exchange between the polar oceans and subpolar seas. Together with the multiyear ice fraction, it defines the ice cover's strength and other mechanical properties that are important for ice-structure and ice-vessel interaction. By itself, the thickness distribution affects ocean–atmosphere heat exchange and its functional form is a measure of the degree of ice field deformation.

While it has been possible to design airborne radars that can survey the thickness of terrestrial ice sheets and obtain excellent echoes from the bedrock interface [Robin et al., 1977], only limited success has been achieved with direct radar sounding of sea ice [Rossiter and Bazeley, 1980; Kovacs and Morey, 1986]. Powerful nanosecond-pulsed systems, with large antennas mounted on helicopters that must be flown only a few meters above the ice at slow speeds, are required. This is clearly not an appropriate procedure for basin-wide synoptic monitoring.

The most successful means for measuring the ice thickness distribution is the synoptic technique of submarine sonar profiling. This technique has enabled accurate characterization of thickness distributions over some areas of the Arctic Basin [Lyon, 1961; Kozo and Tucker, 1974; Williams et al., 1975; Wadhams, 1981, 1983, 1989, 1990, 1992; Wadhams et al., 1985; McLaren, 1989; McLaren et al., 1984]. The obvious limitation of this technique is spatial coverage and temporal resolution. However, sufficient data have been collected to show a large variation in mean thickness over the Arctic, the highest values being found off the northern coast of the Greenland Sea and the Canadian Archipelago. Airborne operations involving passive and active microwave and laser profiling systems have also been carried out concurrently with submarine surveys, and results have suggested that data from such systems can be used for more comprehensive coverage of at least some aspects of the ice thickness distribution. In this chapter we discuss the current state of these surface-bottom relationships, and suggest some ways in which a reliable technique of ice thickness inference can be developed.

21.2 DIRECT TECHNIQUES FOR ESTIMATING ICE THICKNESS DISTRIBUTION

The ice thickness distribution is defined as the probability density function $g(h)$ of ice thickness h, where $g(h)dh$ is the probability that within a homogeneous region of ice cover the thickness lies between h and $(h + dh)$. A theoretical treatment of the formalism of $g(h)$ and its role in the continuity equations of ice was given by Thorndike et al. [1975] and Rothrock [1986]. A typical Arctic ice draft distribution $P(h)$ is shown in Figure 21-1(a). It is conventional to multiply drafts by a density factor of 1.11 to 1.12 to obtain the actual thickness distribution, $g(h)$, without otherwise altering the shape of the probability density function (PDF). The small peak in the range from 0 to 1 m corresponds to refrozen leads at different stages of development (e.g., nilas, young, and thin first-year ice). Typically the first-year ice peak occurs in the vicinity of 2 m and the multiyear ice peak is between 3 and 4 m. The tail represents ice that is thicker than that attained by thermodynamic growth (i.e., ridges). An alternative way of representing thickness distribution is by means of $G(h)$, the cumulative density function, such that the integral of $G(h)$ from 0 to h is the probability that the ice thickness within a region is less than h. Figure 21-1(b) shows the $G(h)$ that corresponds to the $P(h)$ in Figure 21-1(a). The $P(h)$ is more informative since it shows more clearly how the ice is distributed amongst different draft classes.

The techniques for measuring the ice thickness distribution fall into three classes: an upward-looking sonar method applied to submarine and other platforms; novel active electromagnetic techniques (impulse radar and eddy current sounding); and novel passive microwave techniques (long wavelength). A new, untested technique with some

Microwave Remote Sensing of Sea Ice
Geophysical Monograph 68
©1992 American Geophysical Union

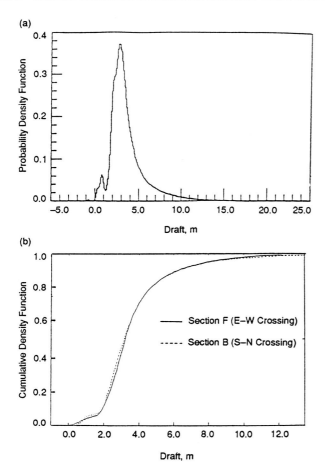

Fig. 21-1. (a) A probability density function $P(h)$ of ice draft from the Arctic Basin, and (b) the equivalent cumulative density function $G(h)$ (from [Wadhams and Horne, 1980]).

potential is acoustic tomography, by which modal thickness may be derived from travel time changes of acoustic pulses [Jin and Wadhams, 1989].

Our present knowledge of Arctic ice thickness distributions derives largely from the analysis and publication of sonar data from submarine cruises. Problems include the necessity of removing the effect of beamwidth where a wide-beam sonar has been employed [Wadhams, 1981], and the fact that the data are obtained during military operations, which necessitates restrictions on the publication of exact track lines. For the same reason, the data set is not systematic in time or space. The current state of knowledge about spatial variablity of ice thickness distribution in the Arctic for summer and winter is summarized by contour maps of the mean ice draft (Figure 21-2) as generated by Bourke and Garrett [1987].

Upward-looking sonars mounted on moorings have also been used to obtain $P(h)$ from a time series at a fixed location. Early experiments were carried out in shallow water in the Beaufort Sea [Hudson, 1990; Pilkington and Wright, 1991], but most recent studies have been in the Chukchi Sea

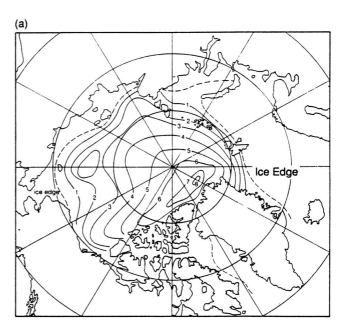

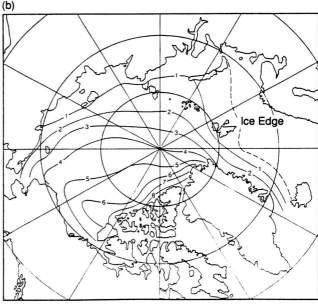

Fig. 21-2. Contour maps of estimated mean ice drafts derived from submarine data for (a) summer and (b) winter in the Arctic Basin.

[R. Moritz, personal communication, 1991], Fram Strait, and the southern Greenland Sea [Vinje, 1989]. More work of this kind can be expected, especially using lines of sonars that span key ice streams or choke points, such as the East Greenland current in Fram Strait and to the south. In conjunction with the use of AVHRR or ERS-1 SAR imagery to yield ice velocity vectors, this technique permits the time dependence of ice mass flux to be measured.

Upward-looking sonar systems mounted on nonmilitary mobile platforms would allow one to obtain systematic data sets along a repeatable grid of survey tracks, enabling interseasonal and interannual comparisons of identical geographical locations. Possible platforms with synoptic potential include: autonomous underwater vehicles, of which several are under development including long-range systems with basin-wide capability; long-range civilian manned submersibles, of which the Canadian–French Saga I is the only available vehicle [Grandvaux and Drogou, 1989]; and neutrally buoyant floats to accumulate and store a PDF that can then be transmitted acoustically to an Argos readout station on a floe [J. C. Gascard, personal communication, 1991].

The first active electromagnetic technique to be applied to sea ice was impulse radar [Kovacs and Morey, 1986], in which a nanosecond pulse (center frequency about 100 MHz) is applied to the ice via a paraboloidal antenna on a low- and slow-flying helicopter. Figure 21-3(a) shows a version of this radar constructed by Cambridge Consultants Ltd. mounted on a helicopter during a 1986 experiment in the Weddell Sea [Wadhams et al., 1987], while Figure 21-3(b) shows results obtained over multiyear Arctic ice [G. K. Oswald, personal communication, 1990]. The technique was found to have many limitations besides the slowness of data gathering. Superimposition of surface and bottom echoes means that this technique does not function well over ice thinner than 1 m, while absorption and scattering by brine cells means that better results are obtained over multiyear than first-year ice. In any case, a fading of the return signal occurs at depths beyond 10 to 12 m. Thus, the full profile of deeper pressure ridges is not obtained.

A more recent development is the use of electromagnetic induction. The technique was devised by Aerodat Ltd. of Toronto, and has subsequently been developed by CRREL [Kovacs and Holladay, 1989] and by Canpolar, Toronto [Holladay et al., 1990]. The technique involves towing a "bird" behind a helicopter flying at normal speeds. The bird contains a coil that emits an electromagnetic field in the frequency range 900 Hz to 33 kHz, inducing eddy currents in the water under the ice, which in turn generate secondary electromagnetic fields. The secondary fields are detected by a receiver in the bird; their strength depends on the depth of the ice–water interface below the bird. The bird also carries a laser profilometer to measure the depth of the ice–air interface below the bird, and the difference gives the absolute thickness of the ice. The method appears promising, but requires further validation. Figure 21-4(a) illustrates the principle, and Figure 21-4(b) shows some results obtained in the Labrador Sea during the Labrador Sea Ice Margin Experiment [Holladay et al., 1990].

The use of long-wavelength passive microwave (or UHF) systems to measure ice thickness has also been studied recently. A 611-MHz radiometer was mounted on board the RV Polarstern during the 1989 Winter Weddell Gyre

(a)

(b)

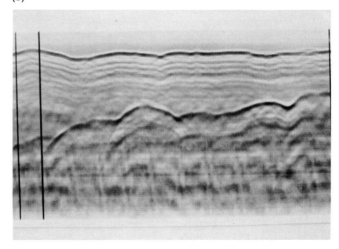

Fig. 21-3. (a) Impulse radar antennas mounted on a helicopter in the Weddell Sea. (b) An ice thickness profile obtained by impulse radar over multiyear ice in the Arctic (courtesy of Dr. Gordon Oswald, Cambridge, 1990).

Study. Theoretical considerations indicate that the system, which operates at about a 50-cm wavelength, is effective for ice thicknesses less than 70 cm (or 57 cm at 35° from nadir). The best data from the UHF measurements were those taken in the marginal ice zone. In this region, the ice cover was not so thick and it consisted of mixtures of different ice types with different thicknesses. Thicknesses were estimated from the brightness temperatures using a theoretical model [Menashi et al., 1992], and results showed qualitative consistency estimates derived from simultaneous video observations. While the potential is demonstrated, there are limitations for this system, such as weight (about half a ton), size (1.3 × 1.3 × 0.2 m), resolution, and the thickness of ice it can measure. Because of weight and size, it is platform-limited. With this system capable of measuring only about a 70-cm thickness of ice, it would

(a)

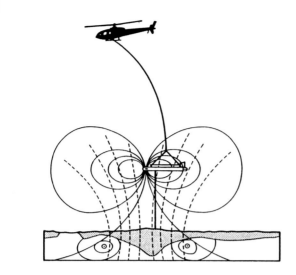

(b)

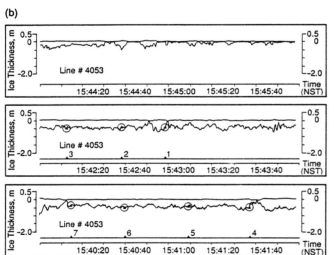

Fig. 21-4. (a) The electromagnetic induction system and ice sounding concept, and (b) results from electromagnetic induction sounding over first-year ice in the Labrador Sea.

be useful in only a very limited part of the Antarctic and Arctic regions.

21.3 INDIRECT REMOTE SENSING TECHNIQUES

21.3.1 Inferring Ice Thickness Distribution From Laser Profiling

Laser profiling of sea ice has been carried out extensively in the Arctic Ocean during the last two decades [e.g. Ketchum, 1971; Weeks et al., 1971; Wadhams, 1976; Tucker et al., 1979; Krabill et al., 1990], while limited studies have also been carried out in the Antarctic [Weeks et al., 1989]. The aim has been to delineate the frequency and height distri-

butions of pressure ridge sails and the spatial distribution of surface roughness. On only two occasions has it been possible to match a laser profile against a coincident profile of ice draft over substantial lengths of joint track. The first was a joint aircraft-submarine experiment [Lowry and Wadhams, 1979; Wadhams, 1980], and the second involved a NASA P-3A aircraft equipped with an Airborne Oceanographic Lidar (AOL) and a British submarine equipped with a narrow-beam upward-looking sonar [Wadhams, 1990; Comiso et al., 1991; Wadhams et al., 1992; Krabill et al., 1990].

The superior capability of the AOL over earlier lasers [Krabill et al., 1990] made it possible to make a direct comparison of the PDF's of draft and freeboard. Comiso et al. [1991] showed that over a 60-km sample of track, the overall PDF's of ice freeboard and draft match each other if a simple coordinate transform of the AOL, using a buoyancy factor r, is made. Specifically, if R is the ratio of the mean draft to mean freeboard, then the freeboard PDF matches the draft PDF if the elevation scale of the freeboard PDF is multiplied by a factor of R. This is equivalent to saying that if a fraction $F(h)$ of the ice cover has an elevation in the range h to $(h + dh)$, then the same fraction $F(h)$ will have a draft in the range $R(h)$ to $R(h + dh)$. Here R is related to the mean material density (ice plus snow) ρ_m and near-surface water density ρ_w by

$$R = \rho_m / (\rho_w - \rho_m) \qquad (1)$$

Figure 21-5 shows a semilogarithmic plot of sonar-draft and laser PDF after applying the coordinate transformation with $R = 7.91$. The two plots lie basically on top of each other except at high draft values where the number of data points is very limited.

The success of this correlation prompted an analysis of all good coincident-track data north of Greenland, within the zone 80.5° to 85° N and 2° to 35° W. The results of this analysis [Wadhams et al., 1992] indicated that, despite variations in mean draft from 3.6 to 6 m, the six values of R from each section all lay within a narrow range with a mean of 7.89 ±0.22. This corresponds to a mean material density of 908.8 ±2.3 kg/m^3, which is reasonable given that typical sea ice densities lie in the range 910 to 915 kg/m^3, with the mean density being reduced by the snow cover, which is sensed by the laser.

To study how R might vary with the time of year and location, a simple model for the seasonal variation of R, based on the best available data on seasonal snow thickness, the mean density of snow and ice, and the mean density of near-surface water displaced by the ice cover was developed [Wadhams et al., 1992]. The model indicated there was a large seasonal variation in R, mainly due to snow load. The value of R is at its highest value at the start of the snow season (with bare ice) on August 20. Then, R falls rapidly during September and October, when the surface is covered by snow, and diminishes slowly between November and the end of April, when little snow falls. A further onset of spring

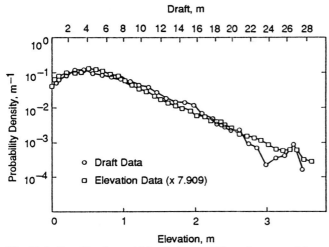

Fig. 21-5. Results of a matching between 60 km of sonar and laser profile in which freeboard PDF is stretched along the abscissa by a mean draft/freeboard ratio of 7.91 and reduced in magnitude by the same factor (from [Comiso et al., 1991]).

snow brought R to its lowest level at the beginning of June. As soon as surface snow melt began, R rises rapidly, until by the end of June it has risen again almost to its August value. The final slow drift is due to surface water dilution. In fact, there were one or more higher peaks for R during the summer due to ponding. Even in the absence of improved background data, these results suggest that in regions with mean ice thicknesses in the range from 4 to 6 m, R will lie in the vicinity of 7.9 in spring and can be estimated to an accuracy of ±2.4% in 300 km of track. This yields an accuracy of about ±12 cm in mean thickness over 300 km of track (30 cm in 50 km), neglecting other sources of error.

21.3.2 Relationships of SAR Backscatter to Ice Thickness Distribution

The 1987 Aircraft/Submarine Sea Ice Project described above included (among others) a joint survey between the submarine and an aircraft SAR system. The system operates at a 9.6 GHz and HH polarization, with a swath width of 63 km and resolution of 16.8 m. This provided an opportunity to examine correlations between ice draft as measured by the sonar, and backscatter level along the same track measured by the SAR.

A comparison of measurements from different sensors over the same 10 km track of sea ice is shown in Figure 12-6. A qualitative examination of the profile of SAR backscatter along the tracks of the submarine and P-3A, Figures 21-6(c) and (e), suggests that a positive correlation with both the draft and elevation profiles. This is to be expected since pressure ridges in particular give strong returns on account of their geometry [Onstott et al., 1987; Livingstone, 1989; Burns et al., 1987]. The correlation between SAR backscatter and sonar ice draft was studied over a 22-km section [Wadhams et al., 1991] where there was excellent matching

between the tracks. Figure 21-7 shows the scatter plot of draft versus SAR backscatter. It was found that at an average length of 252 m (15 SAR pixels), the correlation between draft and backscatter reached 0.68, which is better than the best correlation with elevation (about 0.51) and implies that 46% of the backscatter variance can be explained by draft differences. The higher correlation with draft may be because the draft distribution offers a greater range of depth discrimination for undeformed ice and a better representation of ridges, so a higher correlation can be developed over a shorter averaging length than for elevation. It is fortuitous that the SAR backscatter matches the draft distribution, which is close to the thickness distribution $g(h)$ since the mechanisms that govern backscatter are not related directly to thickness. However, it is clear that SAR alone cannot be used to infer the complete shape of the PDF since only 46% of the variance of the SAR can be explained by draft variations. Another utilization of SAR for ice thickness study is through ice-type classification [Holt et al., 1990; Kwok et al., 1992].

21.3.3 Relationship of Passive Microwave Signature to Thickness

There are presently no passive microwave satellite systems that can directly estimate the thickness of sea ice. The technology for doing this effectively does not even exist. However, some indirect information about thickness distribution may be obtainable from existing passive-microwave satellite systems. For example, the passive microwave signature of thick multiyear ice has been observed to be different from those of first-year ice during winter [Wilheit et al., 1972; Gloersen et al., 1973; Parkinson et al., 1987], while the signature of the latter is also known to be different from those of thinner new ice, Figure 12-6(h). It is, however, not clear that the signature of multiyear ice is entirely stable. Tooma et al. [1975] observed that second-year ice has signatures different from that of older ice. In a cluster analysis of multispectral emissivities of Arctic sea ice, Comiso [1983] observed nonlinearities in the distribution of consolidated ice, suggesting different signatures in different regimes of the Arctic Basin. The existence of considerable variability in the signatures for multiyear ice has been confirmed by in-situ measurements [Grenfell, 1992] and implied by high-resolution aircraft measurements [Comiso et al., 1991]. Similar variability is also observed for new and young ice [Comiso et al., 1989].

Based on the analysis of winter Arctic sea ice data, Comiso [1986, 1990] showed that the geographical locations of the different clusters in multichannel emissivity space are actually similar in pattern to that shown by Colony and Thorndike [1985] for statistical distribution of sea ice based on age. An example of a color-coded cluster map generated by using six SMMR channels is shown in Figure 21-8. The cluster map suggests the existence of different ice regimes. Such a region of lowest emissivity is located approximately where ice is oldest, as postulated by Colony and Thorndike

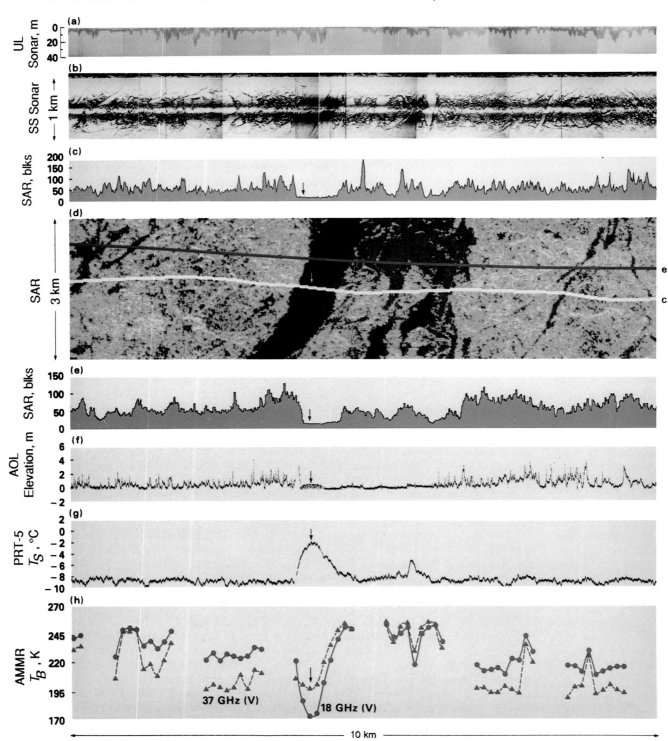

Fig. 21-6. A 10-km section of imagery and profiles from the joint aircraft-submarine experiment of May 1987 for the: (a) upward-looking sonar profile of ice draft; (b) sidescan sonar imagery of the ice underside, with a 1000-m swath width; (c) SAR backscatter values along the submarine track; (d) contrast-stretched SAR imagery with submarine and aircraft tracks overlaid; (e) SAR backscatter values along the track of the P-3A; (f) AOL laser profile of ice elevation; (g) PRT-5 infrared radiometer profile; and (h) microwave brightness temperatures at 18 and 37 GHz from the SMMR [Comiso et al., 1991].

[1985]. The region of the first-year thick ice (pink) and the marginal ice zone (dark blue), where new and young ice are dominant, are also identified by the cluster map.

Alternate information about multiyear ice coverage can be derived from studies of the summer ice cover. Since

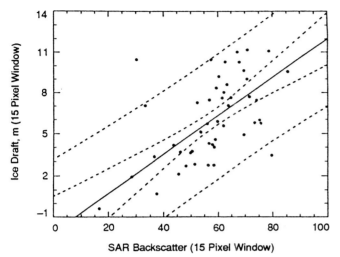

Fig. 21-7. Scatter diagram of SAR backscatter versus ice draft, 252-m window, with 9%5 and 99% confidence limits added [Comiso et al., 1991].

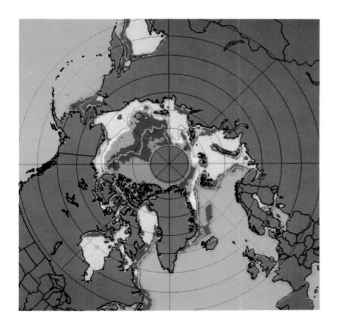

Fig. 21-8. Geographical map of radiometrically different regimes in the ice pack as derived from cluster analysis using six SMMR channels (37, 18, and 10 GHz at both horizontal and vertical polarizations). The color code from lowest to highest brightness temperatures in the consolidated ice region is light green, dark green, brown, orange, yellow, and pink, respectively. Red, dark blue, and blue correspond to young ice and new ice, while light blue represents open water.

multiyear ice corresponds to floes that survive the summer, the multiyear ice distribution can be independently inferred from the ice cover distribution during the summer [Comiso, 1990; Carsey, 1982]. While summer minimum may not occur simultaneously throughout the Arctic Basin, a good approximation of multiyear ice coverage can be inferred from time series data. Information about interannual changes in the perennial ice cover can be obtained by comparing summer data from different years. In combination with winter cluster maps, the summer ice data could provide improved interpretation of thickness distributions obtained by other means.

21.4 DISCUSSION AND CONCLUSIONS

The most reliable and comprehensive measurements of the thickness distributions of Arctic sea ice have come from submarine sonar studies. Attempts must be made to obtain and analyze sonar data accumulated during the past several years. While there might be a problem with consistency, this data set is currently the best source of thickness distribution and should be processed to supplement existing data so as to obtain a more complete impression of Arctic sea ice thickness. Other techniques, such as the use of upward-looking sonars from stationary platforms, are equally promising. Having fixed sonar systems in strategic locations in the Arctic Ocean would provide good time series information to complement submarine data. But monitoring the entire Arctic Ocean using these techniques may not be easy or practicable.

The SAR data provide very good spatial detail and yield thickness information about the ice as well since the backscatter levels and ice drafts are observed to be well correlated with correlation coefficients of about 0.68. However, this good correlation is due more to a relationship between the SAR backscatter and roughness, which is related to draft in a very indirect manner. High backscatter is associated with a rough ridged ice surface and old ice with a lot of air pockets that serve as scatterers, while low backscatter is associated with relatively thin new, young, or first-year ice with a smooth surface. Thus, the SAR data may not be accurate enough to detect interannual changes, especially for thick ice. Other factors, such as flooding or snow cover, could also hide the thickness information. However, an ability to track some of the large multiyear ice floes with SAR for a long time and to get thickness information from these same floes using other means in a periodic fashion are intriguing and would definitely remove the ambiguities mentioned earlier.

Satellite passive microwave sensors provide very detailed and comprehensive coverages, but direct estimates of thickness from the brightness temperature data are unlikely to be obtainable. Even very long wavelength radiometers may not be practical for a satellite system due to their weight and large size, as well as limited penetration depth and coarse resolution. The best use for existing passive microwave data for thickness distribution studies would be to utilize their ability to identify different types or different ice regimes. Since the average thickness is different in various ice regimes, multichannel cluster analysis could provide good complementary information about spatial distribution of ice thickness within the ice pack.

Other techniques for ice thickness determinations have been developed, including the electromagnetic induction technique and acoustic tomography, but they may not be practical for synoptic monitoring of the entire Arctic region. The most promising technique, other than submarine sonar, appears to be laser ranging altimeters. Using the basic laws of buoyancy, the draft distribution was found to be related quantitatively to elevation distribution. The virtue of this technique is that it is easier to plan long-term monitoring from the air than from a submarine platform and that much more area is covered during the same experimental period. When used in conjunction with SAR and passive microwave satellite data, this system could be a powerful tool for detecting ice thickness changes that may be related to climate change.

REFERENCES

Bourke, R. H. and R. P. Garrett, Sea ice thickness distribution in the Arctic Ocean, *Cold Regions Science Technology, 13*, pp. 259–280, 1987.

Burns, B. A., D. J. Cavalieri, M. R. Keller, W. J. Campbell, T. C. Grenfell, G. A. Maykut, and P. Gloersen, Multisensor comparison of ice concentration estimates in the marginal ice zone, *Journal of Geophysical Research, 92*, pp. 6843–6856, 1987.

Carsey, F. D., Arctic sea ice distribution at end of summer 1973–1976 from satellite microwave data, *Journal of Geophysical Research, 18*, pp. 5809–5835, 1982.

Colony, R. and A. Thorndike, Sea ice motion as a drunkard's walk, *Journal of Geophysical Research, 90(C1)*, pp. 965–974, 1985.

Comiso, J. C., Sea ice effective microwave emissivities from satellite passive microwave and infrared observations, *Journal of Geophysical Research, 88(C12)*, pp. 7686–7704, 1983.

Comiso, J. C., Characteristics of Arctic winter sea ice from satellite multispectral microwave observations, *Journal of Geophysical Research, 91(C1)*, pp. 975–994, 1986.

Comiso, J. C., Arctic multiyear ice classification and summer ice cover using passive microwave satellite data, *Journal of Geophysical Research, 95(C8)*, pp. 13,411–13,422, 1990.

Comiso, J. C., T. C. Grenfell, D. Bell, M. Lange, and S. Ackley, Passive microwave in situ observations of Weddell sea ice, *Journal of Geophysical Research, 88*, pp. 7686–7704, 1989.

Comiso, J. C., T. C. Grenfell, D. L. Bell, M. A. Lange, and S. F. Ackley, Passive microwave in situ observations of winter Weddell sea ice, *Journal of Geophysical Research, 94(C8)*, pp. 10,891–10,905, 1989.

Comiso, J. C., P. Wadhams, W. Krabill, R. Swift, J. Crawford, and W. Tucker, Top/Bottom multisensor remote sensing of Arctic sea ice, *Journal of Geophysical Research, 96(C2)*, pp. 2693–2711, 1991.

Gloersen, P., W. Nordberg, T. Schmugge, and T. Wilheit, Microwave signatures of first-year and multiyear ice, *Journal of Geophysical Research, 78*, pp. 3564–3572, 1973.

Grandvaux, B. and J. F. Drogou, Saga 1, une premiere etape vers les sousmarins autonomes d'intervention, *Arctic Technology and Economy—Present Situation and Problems, Future Issues*, Bureau Veritas, Paris, 1989.

Grenfell, T., Surface-based passive microwave studies of multiyear sea ice, *Journal of Geophysical Research, 97(C3)*, pp. 3485–3502, 1992.

Holladay, J. S., J. R. Rossiter, and A. Kovacs, Airborne measurement of sea ice thickness using electromagnetic induction sounding, *Proceedings of the Ninth International Conference on Offshore Mechanical and Arctic Engineering*, edited by O. A. Ayorind, N. K. Sinha, and D. S. Sodhi, pp. 309–315, American Society of Mechanical Engineers, 1990.

Holt, B., J. Crawford, and F. Carsey, Characteristics of sea ice during the Arctic winter using multifrequency aircraft radar imagery, *Sea Ice Properties and Processes*, edited by S. F. Ackley and W. F. Weeks, Monograph 90-1, 224 pp., U.S. Army Cold Regions Research and Engineering Laboratory, Hanover, New Hampshire, 1990.

Hudson, R., Annual measurement of sea-ice thickness using an upward-looking sonar, *Nature, London, 334*, pp. 135–137, 1990.

Ketchum, R. D., Airborne laser profiling of the Arctic pack ice, *Remote Sensing of the Environment, 2*, pp. 41–52, 1971.

Kovacs, A. and R. M. Morey, Electromagnetic measurements of multiyear sea ice using impulse radar, *Cold Regions Science and Technology, 12*, pp. 67–93, 1986.

Kozo, T. L. and W. B. Tucker III, Sea ice bottomside features in the Denmark Strait, *Journal of Geophysical Research, 79*, pp. 4505–4511, 1974.

Krabill, W. B., R. N. Swift, and W. B. Tucker III, Recent measurements of sea ice topography in the Eastern Arctic, *Sea Ice Properties and Processes*, edited by S. F. Ackley and W. F. Weeks, Monograph 90-1, pp. 132–136, U.S. Army Cold Regions Research and Engineering Laboratory, Hanover, New Hampshire, 1990.

Kwok, R., E. Rignot, B. Holt, and R. Onstott, Identification of sea ice types in spaceborne SAR data, *Journal of Geophysical Research, 97(C2)*, pp. 2391–2402, 1992.

Livingstone, C. E., Combined active/passive microwave classification of sea ice, *Proceedings of the IGARSS'89, 1*, pp. 376–380, 1989.

Lowry, R. T. and P. Wadhams, On the statistical distribution of pressure ridges in sea ice, *Journal of Geophysical Research, 84*, pp. 2487–2494, 1979.

Lyon, W. K., Ocean and sea-ice research in the Arctic Ocean via submarine, *Transactions of the New York Academy of Science, Series 2, 23*, pp. 662–674, 1961.

McLaren, A. S., The under-ice thickness distribution of the Arctic Basin as recorded in 1958 and 1970, *Journal of Geophysical Research, 94*, pp. 4971–4983, 1989.

McLaren, A. S., P. Wadhams, and R. Weintraub, The sea ice topography of M'Clure Strait in winter and summer of 1960 from submarine profiles, *Arctic, 37*, pp. 110–120, 1984.

Menashi, J., C. Swift, K. St. Germain, J. C. Comiso, and A. Lohanick, Passive microwave measurement of sea ice thickness, *Journal of Geophysical Research*, in press, 1992.

Onstott, R. G., T. C. Grenfell, C. Mätzler, C. A. Luther, and E. A. Svendsen, Evolution of microwave sea ice signatures during early and midsummer in the marginal ice zone, *Journal of Geophysical Research, 92(C7)*, pp. 6825–6837, 1987.

Parkinson, C. L., J. C. Comiso, H. J. Zwally, D. J. Cavalieri, P. Gloersen, and W. J. Campbell, *Arctic Sea Ice, 1973–1976: Satellite Passive-Microwave Observations*, NASA SP-489, 296 pp., National Aeronautics and Space Administration, Washington, DC, 1987.

Pilkington, G. R. and B. D. Wright, Beaufort Sea ice thickness measurements from an acoustic, under ice, upward looking ice keel profiler, *Proceedings of the First International Offshore and Polar Engineering Conference*, Edinburgh, August 11–16, 1991, in press, 1991.

Robin, G. de Q., D. J. Drewry, and D. T. Meldrum, International studies of ice sheet and bedrock, *Philatelic Transactions of the Royal Society of London, B279*, pp. 185–196, 1977.

Rossiter, J. R. and D. P. Bazeley, *Proceedings of the International Workshop on the Remote Estimation of Sea Ice Thickness*, Publication 80-5, Centre for Cold Ocean Resources Engineering, Memorial University, St. John's, Canada, 1980.

Rothrock, D. A., Ice thickness distribution—Measurement and theory, *The Geophysics of Sea Ice,* edited by N. Untersteiner, pp. 551–575, NATO ASI Series B: Physics vol. 146, Plenum Press, New York, 1986.

Thorndike, A. S., D. A. Rothrock, G. A. Maykut, and R. Colony, The thickness distribution of sea ice, *Journal of Geophysical Research, 80(33)*, pp. 4501–4513, 1975.

Tooma, S. G., R. A. Mannella, J. P. Hollinger, and R. D. Ketchum, Jr., Comparison of sea-ice type identification between airborne dual-frequency passive microwave radiometry and standard laser/infrared techniques, *Journal of Glaciology, 15*, pp. 225–239, 1975.

Tucker, W. B. III, W. F. Weeks, and M. Frank, Sea ice ridging over the Alaskan continental shelf, *Journal of Geophysical Research, 84*, pp. 4885–4897, 1979.

Vinje, T. E., An upward looking sonar ice draft series, *Proceedings of the Tenth International Conference on Port and Ocean Engineering Under Arctic Conditions, 1*, edited by K. B. E. Axelsson and L. A. Fransson, Lulea University of Technology, Lulea, pp. 178–187, 1989.

Wadhams, P., Sea ice topography in the Beaufort Sea and its effect on oil containment, *AIDJEX Bulletin, 33*, pp. 1–52, Division of Marine Resources, University of Washington, Seattle, Washington, 1976.

Wadhams, P., A comparison of sonar and laser profiles along corresponding tracks in the Arctic Ocean, *Sea Ice Processes and Models*, edited by R. S. Pritchard, pp. 283–299, University of Washington Press, Seattle, Washington, 1980.

Wadhams, P., Sea ice topography of the Arctic Ocean in the region 70°W to 25°E, *Philatelic Transactions of the Royal Society of London, A302(1464)*, pp. 45–85, 1981.

Wadhams, P., Sea ice thickness distribution in Fram Strait, *Nature, London, 305*, pp. 108–111, 1983.

Wadhams, P., Sea-ice thickness distribution in the Trans-Polar Drift Stream, *Rapp. P-v Reun Cons, International Explor. Mer., 188*, pp. 59–65, 1989.

Wadhams, P., Evidence for thinning of the Arctic ice cover north of Greenland, *Nature, London, 345*, pp. 795–797, 1990.

Wadhams, P., Sea ice thickness distribution in the Greenland Sea and Eurasian Basin, May 1987, *Journal of Geophysical Research, 97*, pp. 5331–5348, 1992.

Wadhams, P. and R. J. Horne, An analysis of ice profiles obtained by submarine sonar in the Beaufort Sea, *Journal of Glaciology, 25(93)*, pp. 401–424, 1980.

Wadhams, P., A. S. McLaren, and R. Weintraub, Ice thickness distribution in Davis Strait in February from submarine sonar profiles, *Journal of Geophysical Research, 90(C1)*, pp. 1069–1077, 1985.

Wadhams, P., M. A. Lange, and S. F. Ackley, The ice thickness distribution across the Atlantic sector of the Antarctic Ocean in midwinter, *Journal of Geophysical Research, 92(C13)*, pp. 14,535–14,552, 1987.

Wadhams, P., J. C. Comiso, J. Crawford, G. Jackson, W. Krabill, R. Kutz, C. B. Sear, R. Swift, W. B. Tucker, and N. R. Davis, Concurrent remote sensing of Arctic sea ice from submarine and aircraft, *International Journal of Remote Sensing, 12(9)*, pp. 1829–1840, 1991.

Wadhams, P., W. Tucker, W. Krabill, R. Swift, J. Comiso, and N. Davis, The relationship between sea ice freeboard and draft in the Arctic Basin, and applications for ice thickness monitoring, *Journal of Geophysical Research*, in press, 1992.

Weeks, W. F., A. Kovacs, and W. D. Hibler III, Pressure ridge characteristics in the Arctic coastal environment, *Proceedings of the First International Conference on Port and Ocean Engineering Under Arctic Conditions*, edited by S. S. Wetteland and P. Bruun, pp. 152–183, Technological University of Norway, Trondheim, 1971.

Weeks, W. F., S. F. Ackley, and J. W. Govoni, Sea ice ridging in the Ross Sea, Antarctica as compared with sites in the Arctic, *Journal of Geophysical Research, 94*, pp. 4984–4988, 1988.

Williams, E., C. W. M. Swithinbank, and G. de Q. Robin, A submarine sonar study of Arctic pack ice, *Journal of Glaciology, 15*, pp. 349–362, 1975.

Chapter 22. The Use of Satellite Observations in Ice Cover Simulations

RUTH H. PRELLER

Naval Research Laboratories, Stennis Space Center, Mississippi 39529-5004

JOHN E. WALSH

Department of Atmospheric Sciences, University of Illinois, 105 South Gregory Avenue, Urbana, Illinois 61801

JAMES A. MASLANIK

Cooperative Institute for Research in Environmental Sciences, University of Colorado, Boulder, Colorado 80309

22.1 INTRODUCTION

The combination of numerical models and observational data can provide a unique tool for studying the complex interactions of the atmosphere, the ice, and the ocean. The formulation of numerical ice and coupled ice–ocean–atmosphere models is based on our knowledge of dynamic and thermodynamic principles and how they relate to observed ice conditions. Field experiments such as the Arctic Ice Dynamics Experiment (AIDJEX) [Pritchard, 1980] and the Marginal Ice Zone Experiment (MIZEX) [*Journal of Geophysical Research Oceans, 88(C5), 92(C7), 96(C3)*] have provided observational data from which the basis of many of the formulations for ice drift, internal ice stresses, heat exchange, etc., have come. Numerical models, on the other hand, may be used to provide information on ice drift, ice thickness, and ice concentration in regions where observations are scarce or missing. In addition, numerical models may be used to forecast ice conditions.

Satellites have been able to provide observational data over larger areas and for longer periods of time than from conventional observations. Since 1972, passive microwave data from polar orbiting satellites have provided large-scale coverage of Arctic and Antarctic sea ice on a nearly continuous basis at resolutions as fine as 25 km. Visible and infrared imagery provide higher resolution data than the passive microwave, but do not have the all-weather capability that the passive microwave data have. A very promising source of high-resolution, all-weather data is the satellite-borne Synthetic Aperture Radar (SAR). Data from this instrument have recently become available via the launch of the ERS-1 satellite on July 17, 1991.

With the recent availability of larger amounts of satellite data in ice covered regions and the continuous development of more advanced numerical models for these regions, the logical next step is merging the data with the models. This merger will most likely occur through expanded use of data assimilation into the models. This combination should provide the best available analysis and forecast of ice conditions, as well as improve our understanding of dynamic and thermodynamic interactions in the ice. In addition, models are likely to be used more extensively to improve data processing algorithms and to help interpret complex responses recorded in the data.

22.2 NUMERICAL MODELS

22.2.1 Ice Models—Dynamic, Thermodynamic, and Dynamic-Thermodynamic

Numerical models of sea ice, developed over approximately the last twenty years, may be broken down into three categories: dynamic, thermodynamic, and dynamic-thermodynamic ice models.

Dynamic ice models use the various stresses on the ice to define the motion of the ice. The momentum balance used in dynamic ice models is

$$m \frac{\delta}{\delta t} \vec{u} = -m f \hat{k} \times \vec{u} + \vec{\tau}_a + \vec{\tau}_w - mg \, \mathrm{grad} H + \vec{F} \qquad (1)$$

where m is the ice mass per unit area and $\vec{u}$ is the ice drift velocity. The terms on the right side of the equation are the Coriolis term (where f is the Coriolis parameter), the wind stress on ice $\vec{\tau}_a$, the ocean stress on ice $\vec{\tau}_w$, the acceleration due to sea surface tilt $mg \, \mathrm{grad} H$ (where H is the sea surface dynamic height), and the internal ice stress term $\vec{F}$.

The simplest of the dynamic models, the free drift model, uses a balance between top and bottom surface stresses and the Coriolis force to determine ice motion. Free drift is often a good approximation away from boundaries and under divergent conditions. Removal of the last term in Equation (1), the effect of internal ice stress, results in the free drift balance. Ice drift may be significantly adjusted in both magnitude and direction by the internal ice stress that generally acts in a direction opposite to the resultant of the wind stress and Coriolis force. Models that include internal ice stress contain a constitutive law that treats ice as a viscous, elastic, viscous-plastic, or elastic-plastic medium.

Microwave Remote Sensing of Sea Ice
Geophysical Monograph 68

In a viscous rheology, stress can only be sustained through a nonrecoverable dissipation of energy by deformation. A plastic rheology allows stress to be sustained through a lack of deformation or elastic deformation in which the energy is recoverable. A linear viscous rheology in which stress depends on linear strain rates and a rigid plastic rheology in which the stress state is either dependent on the magnitude of the strain rates or indeterminate are two special case rheologies often used to describe sea ice. For a more detailed review of these constitutive laws, see Hibler [1980b, 1986].

An intermediate approach with considerable promise for climate simulations is the cavitating fluid approximation, which differs from free drift by allowing nonzero ice pressure under converging conditions, but, like free drift, offers no resistance to divergence or shear [Flato and Hibler, 1989]. In the corresponding numerical procedure, free-drift velocities are corrected iteratively in a momentum-conserving way.

Thermodynamic ice models take into account the interactions of both the atmospheric and oceanic heat fluxes with the ice to determine ice growth and decay. Many of the formulations of sea ice thermodynamics used in numerical ice models are based on the work of Maykut and Untersteiner [1971]. Figure 22-1 depicts the one-dimensional snow–ice–water system they use. The heat fluxes from the atmosphere and ocean are passed into the snow and ice by conduction.

At the snow–air surface, the balance of heat fluxes is given by

$$\left(1 - \alpha\right)F_r - I_0 + F_L - \varepsilon\sigma T^4 + F_s + F_e + K_0\left(\frac{\partial T}{\partial z}\right)_0 =$$

$$\begin{cases} 0 \text{ if } T \leq 273 \text{ K} \\ \\ -q\left(\frac{\partial}{\partial t}\right)(S + h) \text{ if } T \geq 273 \text{ K} \end{cases} \quad (2)$$

where α is the albedo, F_r is the solar (shortwave) radiation, and I_0 is the flux of shortwave radiation through the surface into the ice, assumed by Maykut and Untersteiner to be 17% of the net shortwave radiation at the surface. The term F_L is the incoming longwave radiation from the atmosphere (clouds) and $\varepsilon\sigma T^4$ is the outgoing longwave radiation based on the surface temperature, where ε is the longwave emissivity, σ is the Stefan–Boltzmann constant, and T is the surface temperature. Then, F_s is the sensible heat flux, F_e is the latent heat flux, K is the thermal conductivity, q is the latent heat of fusion, z is the depth within the ice–snow column, and $\partial(S + h)/\partial t$ is the ablation rate of snow and ice where S is the thickness of the snow and h is the thickness of the ice. The subscript 0 generically refers to the upper surface. At the snow–ice interface, the balance of fluxes is

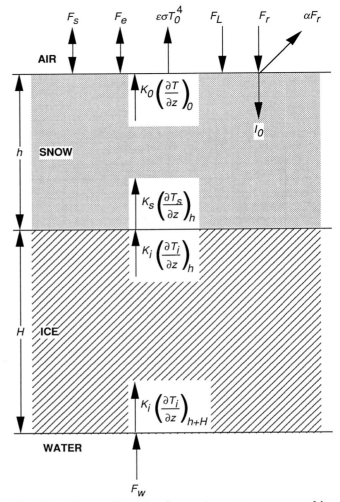

Fig. 22-1. The one-dimensional snow–ice–water system used by Maykut and Untersteiner [1971].

$$K_s\frac{\partial T_s}{\partial z} = K_i\frac{\partial T_i}{\partial z} \quad (3)$$

The subscripts s and i stand for snow and ice, respectively. At the ice–ocean interface the balance is

$$K_i\frac{\partial T_i}{\partial z} = F_w - q\frac{\partial}{\partial t}(S + h) \quad (4)$$

where F_w is the oceanic heat flux.

Thermal conductivity, specific heat, and ice density are all functions of salinity and temperature. This dependency is due in large part to brine pockets in the ice. Brine pockets can act as thermal reservoirs and slow down the heating or cooling of ice. Since brine has a small conductivity and greater specific heat than ice, the parameters also vary with temperature.

To calculate heat conduction, the diffusion equation for temperature is given by

$$(\rho c)_i \frac{\partial T_i}{\partial t} = K_i \frac{\partial^2 T_i}{\partial z^2} + k_i I_0 \exp(-k_i z) \qquad (5)$$

where k is the bulk extinction coefficient and c is the heat capacity. In snow, K and k are constant and I_0 is set equal to zero. In ice, K and k can vary due to the effects of brine pockets as noted above [Untersteiner, 1961].

Although the Maykut and Untersteiner model provides a thorough treatment of the snow–ice thermodynamic system, the complexity of the model can make it impractical for modeling large areas. Semtner [1976] simplified the Maykut and Untersteiner model by fixing the snow and ice conductivities. The salinity profile, required by the Maykut and Untersteiner model, does not have to be specified in Semtner's version. Internal melting is attributed to an amount of penetrating radiation stored in a heat reservoir without causing ablation. Energy from this reservoir keeps the temperature near the ice surface from dropping below freezing in the fall. Semtner's model was further simplified by assuming linear equilibrium temperature gradients in the snow and ice. The heat flux is uniform in both the ice and snow. No temperature flux exists at the snow–ice surface. The surface heat balance is set equal to zero and is used to solve for the surface temperature.

The third category, the dynamic-thermodynamic model, integrates ice motion and ice growth and decay effects into one model. Parkinson and Washington [1979] designed one of the first three-dimensional dynamic-thermodynamic models. Within each grid cell, the model was broken down vertically into four layers: a mixed layer ocean, an ice layer, a snow layer, and an atmospheric boundary layer. At the ice–ocean interface, geostrophic ocean currents were used to supply ocean stress to the ice. A temporarily invariant dynamic topography and a constant oceanic heat flux were proscribed. Atmospheric heat fluxes and wind stress were applied at the air–ice interface. The ice drift, derived from a free-drift formulation, was iteratively corrected to insure the existence of a minimum fraction of leads. This iteration does not conserve momentum and can damp the ice drift.

A more realistic treatment of ice dynamics was used by Hibler [1979]. This model uses a viscous-plastic constitutive law. A plastic rheology is used to relate the strength of ice interaction to a two-level ice thickness distribution. The ice strength P is defined by

$$P = P^* h \, e^{-C(1-A)} \qquad (6)$$

where A is the ice compactness or concentration and P^* and C are constants. Ice concentration is defined in these models as the fraction of a grid cell that is covered by thick ice. The area covered by thin ice or open water is $1 - A$. Equation (6) allows the ice strength to be greater in regions of ice inflow and weaker in regions of ice outflow.

The model also incorporates the equations of continuity for ice concentration and thickness defined as

$$\frac{\partial h}{\partial t} = -\frac{\partial(uh)}{\partial x} - \frac{\partial(vh)}{\partial y} + S_h + (\text{Diffusion}) \qquad (7)$$

$$\frac{\partial A}{\partial t} = -\frac{\partial(uA)}{\partial x} - \frac{\partial(vA)}{\partial y} + S_A + (\text{Diffusion}) \qquad (8)$$

where S_h and S_A are source and sink terms accounting for the growth and melt of ice. In later papers [e.g., Hibler, 1980a], these growth rates were calculated, as opposed to prescribed, from the heat budget balance similar to that used by Parkinson and Washington [1979].

Snow cover was not included explicitly in the Hibler ice model. Instead, based on the work of Bryan et al. [1975] and Manabe et al. [1979], the effects of snow cover were approximated by setting the ice surface albedo to that of snow when the surface temperature was below freezing and to that of snow-free ice when the surface was at the melting point. Walsh et al. [1985] were the first to include snow explicitly in a long-term simulation using daily winds and air temperature data from the National Center for Atmospheric Research (NCAR) for the years 1951 through 1980. In this model, both snow and ice were composed of seven thickness levels. The seven levels were defined by taking twice the average ice or snow thickness within a grid cell and dividing it into seven equal levels. The growth and melt of the ice and snow, based on the heat budget balance, was calculated at each level and then averaged back into one thick ice growth rate. Snowfall rates consisted of monthly varying climatological rates from Maykut and Untersteiner [1971]. In this system, heat from the heat budget balance is used to melt all the snow before it is used to melt ice. If a snow cover exists, the albedo is set to 0.80. When there is no snow, the albedo is 0.65. A version of the Hibler ice model, similar to that used by Walsh et al. [1985], is presently used as the basis for the Polar Ice Prediction System (PIPS), the U.S. Navy's numerical sea-ice forecast model.

22.2.2 Ice–Ocean Models

The development of models for ice covered regions was further expanded by including the temporal and spatial variability of the ocean. This was accomplished by fully coupling ice to ocean models. The first fully coupled, three-dimensional model was designed by Hibler and Bryan [1984, 1987]. The model coupled the Hibler ice model to the Bryan–Cox multilevel ocean model [Bryan, 1969] both dynamically and thermodynamically. The ocean model was initialized from climatological temperatures and salinities and weakly constrained to those values with a three-year relaxation period. Tests performed using this model highlighted the importance of the deep-ocean heat flux into the mixed layer. This heat is great enough to keep parts of the

Barents and Greenland Seas ice-free all winter. Semtner [1987] used a similar coupled ice–ocean model, but removed the constraint on the ocean temperature and salinity. He also used simplified ice dynamics that contained only bulk viscosity (no shear). Semtner's results verified the importance of the oceanic heat flux on keeping the marginal ice zone ice-free in winter. However, his ice thickness distribution contained thinner ice than found by Hibler and Bryan. Fleming and Semtner [1991] reduced the ice strength used by Semtner [1987] and obtained a much more realistic ice thickness distribution. Riedlinger and Preller [1991] used a model similar to the Hibler and Bryan model, but used daily varying forcing from the Navy's operational global atmospheric model. This study showed that the diagnostic ocean model could provide a useful tool for forecasting.

Additional ice–ocean models have been developed that show the importance of including a mixed layer in the ocean. One-dimensional models [Pollard et al., 1983; Lemke and Manley, 1984; Ikeda, 1985; Lemke, 1987; Bjork, 1989; Mellor and Kantha, 1989; Riedlinger and Warn–Varnas, 1990], two-dimensional models [Kantha and Mellor, 1989], and three-dimensional models [Piacsek et al., 1991] have all shown distinct changes to the heat and salt exchange (and therefore the ice growth and decay rates) when a mixed layer is included.

These dynamic-thermodynamic ice and ice–ocean models have mainly been applied to large-scale simulations using grid resolutions on the order of 100 km. The Hibler ice model has been applied, however, on regional scales to the Greenland Sea [Tucker, 1982; Preller et al., 1990] and to the Barents Sea [Preller et al., 1989] at grid resolutions of approximately 20 km.

Higher resolution models (of a few kilometers) have often been used to look at processes normally associated with the region near the ice edge. Roed and O'Brien [1983] and Hakkinen [1986] used one- and two-dimensional ice–ocean models, respectively, to study upwelling at the ice edge. Ikeda [1988] used a three-dimensional ice–ocean model that included thermodynamics to investigate upwelling at the ice edge. Tang [1991] used a two-dimensional thermodynamic ice–ocean model to study the advance and retreat of the ice edge near Newfoundland. Smith et al. [1988] used a three-dimensional, two-layer ocean model coupled to a Hibler-type ice model to investigate the behavior of an isolated ocean eddy within the marginal ice zone. Smith and Bird [1991] used the same model to study the interaction of an ocean eddy with a jet at the ice edge.

22.2.3 Ice–Atmosphere Models

Because ice–atmosphere interaction cannot be divorced from the ocean's influence, there have been relatively few studies utilizing interactive models of only the atmosphere and sea ice. The studies that fall into this category have generally been one- or two-dimensional experiments with energy balance models. Such studies have tended to focus on the effects of specific processes. For example, Ledley [1988, 1991] has used a coupled energy-balance sea-ice model to examine the effects of prescribed lead fractions, snowfall, and a highly parameterized transport of sea ice. Ledley's model is one-dimensional (latitudinal), but it does include a partitioning of each latitudinal zone into land and ocean. Harvey [1988] developed a more comprehensive sea ice model for use in zonally averaged energy-balance climate simulations. Harvey's model, which includes parameterizations of processes such as surface and lateral melting, leads, advection, and snow and ice thickness distributions, has been used to study various sea ice feedbacks (albedo, leads, etc.). Although models such as these are so highly idealized that their relevance to the actual climate system is unclear, they do identify high-leverage parameters and processes that merit further study with more sophisticated models.

Ice–atmosphere models have also been used to study atmospheric boundary layer processes and interactions at the air–ice interface. Overland [1985], for example, used an atmospheric boundary layer model to address momentum exchange and surface drag parameterizations at the air–ice interface. Bennett and Hunkins [1986] used a two-dimensional model to examine atmospheric boundary-layer modifications during advection over an inhomogeneous sea ice cover. Pease and Overland [1984] studied the drift and mass balance of sea ice in the Bering Sea by using an atmospherically driven sea ice model.

22.2.4 Linkage to General Circulation Models

A survey of current versions of the general circulation models (GCM's) used to simulate the global climate reveals a substantial gap between state-of-the-art sea ice models and the treatment of sea ice in GCM's. Most climate models treat sea ice as a motionless slab characterized by a single thickness in each cell. The simulated sea ice simply accretes or melts in response to a deficit or surplus in the surface energy budget. The treatments that permit sea ice motion do so with relatively crude parameterizations. For example, the model of Manabe and Stouffer [1988] retains the formulation of Bryan [1969], whereby ice moves in the direction of the surface ocean currents until the ice reaches a threshold thickness (4 m), after which ice motion ceases. Few models include leads and open water areas within the pack in spite of the importance of these elements to the atmospheric simulation, as shown in various sensitivity studies (e.g., [Simmonds and Budd, 1990]). The models that do include leads generally prescribe the lead fractions, as discussed below.

The inclusion of ice transport, leads, and open water in interactive models is essential to the realistic simulation of sea ice and climate for several reasons. Areas of thin ice and open water, continually created by deformation, are the sources of nearly all the new ice growth, salt rejection, and heat exchange with the atmosphere. The flux of sensible heat from the ocean to the atmosphere during winter is one to two orders of magnitude greater over thin ice and open water than over perennial ice. Maykut [1978] has shown that, in an areally averaged sense, a large fraction of the heat transfer from the polar oceans to the atmospheric boundary layer during winter takes place over young ice (thickness ≤ 0.5 m). The summer melt of ice and the heat storage in the ocean are accelerated dramatically by the absorption of solar radiation in leads and other low-albedo areas such as new ice.

With regard to ice transport, the climatological mean pattern of ice motion across the Arctic Basin and through the Fram Strait results in a net salinity flux to the continental shelves off Siberia. These shelves are regions of net ice growth. The Greenland, Iceland, and Norwegian Seas are regions of net ice melt. The net gain of salt by brine rejection in the shelf waters of the Arctic Ocean is thought to drive the thermohaline circulation of the Arctic Ocean [Aagaard et al., 1985]. The export of sea ice into the North Atlantic subpolar seas is a major source of fresh water for this region, in which deep water formation occurs intermittently and is highly sensitive to the salinity-determined stratification of the ocean surface layers [Aagaard and Carmack, 1989].

Although leads are not determined by ice dynamics in current versions of major GCM's, prescribed or simply parameterized leads have recently been included in several climate simulations. The United Kingdom Meteorological Office (UKMO) and Australian model, for example, have been run with seasonally varying prescribed lead fractions. Sensitivity experiments performed with the Australian model indicate that prescribed lead fractions of 50% in the Antarctic during winter cause the simulated circulation to resemble more closely the case of an ice-free southern ocean than the case of an ice-covered southern ocean [Simmonds and Budd, 1990]. Even when relatively modest lead fractions (several percent) are prescribed in accordance with observations, the atmospheric sea level pressure changes by 5 mb in some high-latitude regions. An alternative strategy has been followed in the Goddard Institute for Space Sciences (GISS) II model [Raymo et al., 1990], which computes a fraction of open water (leads) as $fr = 0.1/Z_{ice}$, where Z_{ice} is the ice thickness in meters. The latter is currently prescribed as 1 m at all times, resulting in a lead fraction of 10%; the use of computed values of Z_{ice} will permit more thorough tests of this parameterization. This parameterization of leads is mentioned here merely to indicate the relatively simplistic nature of such parameterizations in state-of-the-art global climate models.

It should be noted that the ice–ocean interactions involving salinity (freshwater) fluxes will not be reproduced correctly by coupled models until the models include ice transport. While ice transport is simulated well by atmospherically forced ice models (e.g., PIPS), a successful simulation of ice transport in coupled GCM's will require an adequate simulation of the surface wind field and, hence, of the sea level pressure field. There is no evidence yet that atmospheric GCM's reproduce the Arctic pressure patterns that will produce the key features of ice motion observed in the Beaufort Gyre, transpolar drift stream, and East Greenland drift.

22.2.5 Process Models

In addition to the model examples given in Sections 22.2.1 through 22.2.4, a variety of other physical models exist to represent the individual components of sea ice models, including detailed thermodynamics of heat transport, turbulent flux models at the air–ocean interface, atmospheric boundary layer processes, radiative models of longwave and shortwave radiation, and simplified oceanic mixed layer models.

As noted earlier, heat transport through the snowpack is typically treated in terms of conduction only, with one average thermal conductivity used for the entire snow column (e.g., [Parkinson and Washington, 1979; Semtner, 1976; Maykut, 1982]). Colbeck [1991] discusses some of the approaches used to address conduction as well as turbulent fluxes through homogeneous and layered snowpacks. Penetration of the snowpack at optical wavelengths is assumed to be negligible [Parkinson and Washington, 1979] or treated with a single transmittance value (e.g., [Gabison, 1987] and Section 22.2.1). More detailed treatments of thermal and radiative properties of the snowpack are described by Weller and Schwerdtfeger [1977] and Schwerdtfeger and Weller [1977]. Turbulent fluxes at the snow/ice–air surface and open-water–air surface are treated simplistically in three-dimensional ice models, with a single energy transfer coefficient for a variety of different surface types. Andreas and Murphy [1986], Andreas [1987], Morris [1989], and Smith et al. [1990] discuss more detailed treatments of turbulent transfer over ice surfaces and leads. Specific processes of ice growth and oceanic modifications in refreezing leads are discussed by Bauer and Martin [1983] and Kozo [1983].

In addition to the treatments of atmosphere–ocean–ice coupling presented earlier, simplified models exist with the potential to improve the performance of the existing ice models. Overland et al. [1983], Chu [1987], and Overland [1988] describe models of the atmospheric boundary layer over the ice pack. A simplified coupling of the ice cover to the boundary layer is described by Koch [1988]. Simple representations of the upper ocean applicable to ice modeling are described by Fichefet and Gaspar [1988] and Wood and Mysak [1989]. A variety of parameterizations to improve the accuracy of the longwave and shortwave fluxes that are input to the ice models are available, ranging from simplified approaches [Harvey, 1988; Shine and Henderson–Sellers, 1985; Ebert and Curry, 1990] to more complete transmittance models [Kneizys et al., 1988].

22.2.6 Model Parameterizations of Fields Available From Satellite Data

The previous section contained a summary of the types of numerical models that presently exist to describe the interaction of atmosphere, ice, and ocean. In order to better clarify the relationship between these numerical models and remotely sensed data, the following description of existing parameterizations of key model fields is presented.

In particular, the fields that are or will soon be available operationally from satellite data are ice extent, concentration, multiyear fraction, and velocity.

Ice extent is determined in a relatively straightforward manner from models containing sea ice thermodynamics and/or dynamics. When the temperature of the mixed layer (or upper layer of the ocean) falls below freezing (approximately $-1.8°C$) in a particular grid cell, sea ice forms in that grid cell. The ice is subsequently advected into other grid cells or it remains in its area of formation; melt may occur at any time. Ice extent can be defined as the equatorward limit of sea ice (of any concentration). Alternatively, ice extent is often expressed quantitatively as the ocean area poleward of this limit or, e.g., in the NASA sea ice atlases [Zwally et al., 1983; Parkinson et al., 1987] as the area with ice of concentration $\geq 15\%$. The areal summation is performed similarly whether one is evaluating ice extent from observational (satellite) data or from model output. Slight discrepancies may be introduced by the spatial resolution of the data or model output from which ice extent is evaluated.

Ice concentration, which provides direct information on the open water fractions within an area of sea ice, can change by several mechanisms in model simulations: advection, divergence or convergence, and freezing or melt. Small changes of ice concentration can also be caused by diffusion terms, which are included in some models in order to mimic the effects of subgrid-scale motions or to provide numerical stability. Melt can affect ice concentrations by changing the thickness (to zero) or by melting the ice laterally (e.g., in leads). In models containing only a single ice thickness, melting at the surface can change the concentration without melting all the ice in a grid cell if the single thickness is assumed to represent the mean of a range of thicknesses, e.g., a linear distribution of thicknesses between zero and twice the mean (e.g., [Hibler, 1979]). This strategy has not been adopted in GCM's, but it has been used in various stand-alone ice models [Preller and Posey, 1989; Walsh and Zwally, 1990]. Changes of ice concentration by freezing, on the other hand, are typically assumed to create ice of a uniform thickness over all open water in a grid cell unless the model formulation includes a prescribed minimum fraction of leads or open water.

The formulation of the multiyear ice fraction is based partially on the definition of multiyear ice: sea ice that has survived at least one summer's melt season. Consequently, models that carry multiyear ice as a variable impose an abrupt transition from first-year ice to multiyear ice. This transition can occur either at the time when the mixed layer temperature drops to freezing or at a prescribed date (e.g., September 1 in the Arctic). When multiyear ice is present, its concentration changes by the same processes that change total ice concentration: advection, convergence or divergence, diffusion, and melting. (Multiyear ice, by definition, cannot form instantaneously by freezing.) Surface heat budgets are computed separately for the first-year and multiyear ice within a grid cell. The ice rheology determines the extent to which convergence deforms first-year and

multiyear ice. According to present algorithms [Walsh and Zwally, 1990], convergence deforms solely first-year ice as long as the concentration of first-year ice exceeds 5%. The formulation of the ice strength, which has been made a rather ad hoc function of the amounts of first-year and multiyear ice, needs further attention in models that distinguish several types of ice. Hibler's [1980a] multiple-thickness formulation provides a possible point of departure for formulations of the strength of first-year and multiyear ice mixtures.

22.3 MODEL REQUIREMENTS OF SATELLITE DATA

Ice–ocean models require observational data for their initialization and validation. Moreover, the use of observational data for the forcing of ice–ocean models provides an objective means for model testing and assessment when the models are run in a decoupled (from the atmosphere) mode. The general lack of in-situ measurements from ice-covered waters makes satellite remote sensing the cornerstone of data assimilation schemes required for ice modeling.

The manner in which data are used for model initialization depends on which of three types of ice model simulation is being performed: short term (days), long term with periodic forcing, and long term with aperiodic forcing. An example of short-term simulation is the use of an operational model to forecast sea ice at ranges of several days (e.g., PIPS). For such simulations, an accurate initialization corresponding to the actual sea ice state is crucial to the success of the forecast. The second type of model integration, a long-term periodic simulation, spans at least several years with an annual cycle of forcing that repeats itself exactly. In such cases, the ice model should achieve equilibrium if run for a sufficiently long time. This type of simulation is useful for diagnostic studies of the ice–ocean system (e.g., [Hibler and Bryan, 1987; Semtner, 1987]) and for assuring that the ice model is free of long-term biases or model drift prior to interactive coupling (such as to an atmospheric GCM). Provided that sufficient computational resources are available, a realistic initialization is not essential for such simulations because the model should eventually achieve equilibrium with its forcing. The third type of simulation, the long-term aperiodic run, is used to obtain an interannually varying history tape of the sea ice state. Examples include the hindcast simulations of Rothrock and Thomas [1990], Walsh and Zwally [1990], and Riedlinger and Preller [1991]. As in the periodic simulations described above, the model can be spun up to the initial state of the desired date by integrating (periodically or aperiodically) through several annual cycles. For this reason and because the long-term statistics of the simulation are of primary interest, a precise initial state is not essential to the success of the simulation. Nevertheless, the initial state must conform reasonably closely to the corresponding actual state if the first several years are not to contaminate the verification statistics. Thus, the spin-up of such a simulation must be achieved in such a way that it ensures temporal homogeneity in the history tape's correspondence to observations. One may argue that the initialization strategy involves more subtleties in this type of simulation than in the other two.

The initialization and validation of ice models require observational data on several key sea ice variables: concentration, thickness and ice type, drift velocity and associated kinematic quantities, and surface properties (albedo, snow cover, temperature, etc.). While other properties such as ice salinity and temperature profiles may be important for diagnostic studies with ice–ocean models, we will focus primarily on the four variables listed above. These properties are also variables for which satellites provide useful information.

22.3.1   Key Variables Used for Model Initialization and Validation

22.3.1.1. Ice concentration. Ice concentration data implicitly include information on several quantities that are of central importance to studies of the large-scale distribution and variability of sea ice: fractional coverage of an area (ideally within different ice thickness ranges), ice extent, and the fraction of open water (e.g., leads) within the pack. For forecasting purposes, ice extent may be the primary concern. For coupled model studies of the Arctic or Antarctic surface energy balance, the lead fraction may be a primary concern because the small areas of leads and open water effectively govern the exchanges of heat and moisture between a pack-ice surface and the atmosphere. Different sensors provide the optimal information on ice concentration, extent, and lead fractions.

The most useful and homogeneous source of ice concentration data for large (50 km x 50 km) areas is the passive microwave data from polar orbiting satellites: the Electrically Scanning Microwave Radiometer (ESMR), the Special Sensor Microwave Imager (SSM/I), and the Scanning Multichannel Microwave Radiometer (SMMR). Since 1978, these satellites have provided nearly continuous coverage at 1- to 2-day intervals and 50-km resolution with more scattered coverage being available since 1972. The recent availability of gridded passive microwave data on CD-ROM (through the National Snow and Ice Data Center) has greatly facilitated the use of derived ice concentrations by the sea ice community.

The accuracy of any data used for model verification or data assimilation is a key consideration in determining its usefulness. Errors in the total ice concentration (first-year plus multiyear ice) and ice edge location as derived from passive microwave data are fully discussed in Chapters 4, 10, and 11. The estimated errors for total ice concentration generally range from 5% [Cavalieri et al., 1984; Swift and Cavalieri, 1985; Steffen and Maslanik, 1988] to about 15% [Steffen and Schweiger, 1990] using general versions of the NASA Team algorithm, with errors as low as 3% when the algorithm is tuned for local conditions. Absolute accuracies remain uncertain and somewhat controversial due to the

difficulties of comparing different data types and different scales. Total ice concentration estimates are relatively unaffected by surface melt on a regional basis, except when melt ponds are present (e.g., [Steffen and Maslanik, 1988]). Gloersen and Campbell [1988] describe some more specific effects of surface melt on retrievals of ice concentration and ice type. In addition to melt effects, sources of error include the effect of surface wind on open water, atmospheric water content, changes in the ice surface (snowpack, flooding, etc.), and the presence of radiometrically thin ice not accounted for by the algorithm.

For broad depictions of the total ice concentration fields in large-scale models (with resolutions on the order of 100 km), the accuracy and resolution of the passive microwave data are certainly adequate, except perhaps during late summer when substantial melt ponding is present. However, mesoscale models may require finer resolution and greater accuracy (for the determination of the location of leads and polynas in the pack) for initialization and validation at the 10-km scale.

The concentrations derived from satellite passive microwave data provide part of the input to the Navy and National Oceanic and Atmospheric Administration (NOAA) Joint Ice Center's (JIC) weekly analyses, which have been used in the initialization of the PIPS model and in the compilation of verification statistics from long-term simulations [Walsh et al., 1985]. The JIC depictions of the ice edge have also utilized information from the radar altimeter on the Geosat satellite. Because the shape of a radar altimeter's return pulse is significantly different over sea ice and open water, the altimeter permits the resolution of the ice edge to within 10 km, which is finer than the resolution of the digitized JIC ice charts and considerably finer than the resolution of the passive microwave concentration grids. For simulations with mesoscale models, radar altimeter data can serve a useful purpose in the initialization and validation of the ice edge.

Fine resolution imagery is needed to resolve the individual small areas of open water (leads and polynas) within the pack ice. Together with areas of thin ice, these open water areas are the primary determinants of the ice strength, the ocean–atmosphere heat exchange, and the evaporation from the ocean. Because the open water areas often have scales of tens to hundreds of meters, individual open water areas cannot be resolved by current passive microwave satellite imagery, although such subresolution features are included in the total open water area within the passive microwave field of view. Moreover, the uncertainties of several percent in the passive-microwave-derived concentrations are comparable to the variations in the small percentages of thin ice and open water that typically occur in pack ice (e.g., [Maykut, 1982]). Both visible and infrared imagery can provide fine-resolution (10 meters to 1 kilometer) observations of ice concentration, ice extent, lead locations, and floe characteristics. Visible and reflected infrared imagery are limited, however, by both clouds and darkness while thermal infrared imagery is limited by cloud cover. The most promising source of all-weather, fine-resolution data is the satellite-borne Synthetic Aperture Radar such as those on board ERS-1 (from which data are currently becoming available), JERS-1, and Radarsat. Nominal resolution from the JERS-1 SAR, for example, will be 30 m.

The accuracy of ice concentration from SAR data is discussed in Chapter 6. Generally, errors in SAR-derived ice concentration occur primarily when thin ice is present, since thin ice can have backscatter characteristics similar to that of open water. However, a comparison of ice concentrations calculated from visible-band aerial photographs, passive microwave acquired from aircraft, and SAR generally agreed to within 15%, with ice–water contrast and ice signature variability accounting for the error [Burns et al., 1987].

Considerable potential exists for the use of SAR data in model parameterization studies directed at the realistic simulation of the thin ice and open water areas within pack ice. Mesoscale models, in particular, will benefit from the order of magnitude improvement in resolution made possible by SAR. In addition to uses by the ongoing ice modeling efforts summarized here, such information may serve as a stimulus to the coupling of sea ice and atmospheric models through realistic parameterizations of surface–atmosphere heat exchange.

One additional use of satellite-derived ice concentrations is the specification of lead fractions in climatic models as noted in Section 22.2.4. In the Australian global climate model [W. Budd, personal communication, 1991] for example, the specified lead fractions are the seasonally varying fractions of open water derived from the microwave radiances of the ESMR and SMMR instruments.

22.3.1.2. Ice thickness and ice type. The spatial and temporal variations of ice thickness are notoriously difficult to quantify. The only data on sea ice thickness have been obtained from: (1) holes drilled through the ice, a logistically difficult and expensive procedure; (2) submarine sonar data, much of which is classified, providing snapshots for single year-months along specific cruise tracks; and (3) single-point time series obtained from moored acoustic sensors, which have come into use only during the last few years. Sources (1) and (2) have permitted the computation of a seasonal climatology of Arctic ice thickness (e.g., [Bourke and McLaren, 1992]) used to validate ice models in a general way, but they have provided little information on interannual variability.

There are no immediate prospects for the direct measurement of ice thickness by satellite, although relationships between ice thickness and passive microwave polarization ratios have been demonstrated (e.g., [Steffen and Maslanik, 1988]). The estimates of multiyear ice concentration obtained from passive microwave data can provide crude and quantitative indicators of ice thickness, but such indicators are inadequate for model initialization and verification. Thus, there will most likely be a continued reliance on the fragmentary thickness measurements obtained from the surface and subsurface sources.

Several models explicitly include the concentration and thickness of multiyear ice in their formulation [Rothrock and Thomas, 1990; Walsh and Zwally, 1990]. The results of the latter model show some agreement with the interannual fields of multiyear ice concentration derived from SMMR. However, uncertainties in the SMMR-derived concentrations of multiyear ice are substantial, and useful information on multiyear ice cannot be obtained during the spring and summer months due to changes in the dielectric properties of the ice pack with the onset of melt (see Chapter 4). During winter, average accuracy for ice type has been estimated to be within about 20% [Cavalieri et al., 1984], although the error may be considerably larger even in winter [Comiso, 1986]. For these reasons, the Kalman filter approach of Rothrock and Thomas provides an attractive means for assimilating the limited satellite data into a more temporally homogeneous framework. This framework offers the potential for a more meaningful validation of the time-varying fields of multiyear ice coverage (and ice thickness) simulated by the more traditional sea ice models.

22.3.1.3. Ice velocity and associated kinematic variables. Because ice velocity has a relatively short time scale (days), accurate initialization of ice motion is important primarily for short-term simulations. For long-term simulations, however, the mean patterns of ice drift determine the transports of mass and salinity. The validation of the simulated drift is therefore an important element of model simulations of long periods (seasons or longer).

The primary source of data on time-varying ice velocities has been the collection of position measurements of Arctic buoys and drift stations. The buoy data have been objectively analyzed on a daily basis since 1979 by the University of Washington's Polar Science Center. These velocities have been used to verify model simulations of ice velocity (e.g., [Serreze et al., 1989]). The satellite passive microwave data, although way too coarse in resolution to identify individual ice floes, nonetheless permit the derivation of approximate, generalized velocity vectors on the basis of changes of multiyear ice concentration over periods of several months to a season. Such derivations are based on the supposition that multiyear ice behaves somewhat like a tracer in subfreezing regions poleward of the ice edge. For example, velocities derived from changes of the position of the multiyear ice edge have been found to be spatially coherent, interannually variable, and consistent with the fields derived from a sea ice model [Zwally and Walsh, 1987].

The most promising source of information on individual floe ice velocities is SAR imagery as discussed in Chapter 18. Pattern recognition techniques can be applied to SAR images of the same region at intervals of several days, thereby providing kinematic vector images of the net ice motion during the intervening time period (e.g., [Curlander et al., 1985]). Emery et al. [1991] used a similar approach with Advanced Very High Resolution Radiometer (AVHRR) data to obtain ice velocity vectors for the Fram Strait region; these vectors were compared directly with model-derived ice velocities. The use of data in this manner is likely to provide the most precise means of initializing and validating ice velocities from mesoscale models and, perhaps, also of validating the integrated measure of the divergence and convergence over grid cell areas of larger scale models. It should be noted in this regard that the time differences of multiyear ice concentrations derived from satellite passive microwave data also provide potentially useful information on sea ice convergence and divergence over larger areas (hundreds of kilometers), as discussed in Section 22.4.2.

22.3.2 Additional Satellite Data for Future Use in Models

22.3.2.1. Albedo and ice surface temperature. One of the critical thermodynamic properties in a sea ice model simulation is the surface albedo. While the albedo's broad dependencies on the surface state (snow, bare ice, and puddled ice) are generally known, the spatial and temporal variations of surface albedo in ice-covered regions are poorly documented. In view of the high leverage of the surface albedo, especially in sea ice simulations that address climate change, the parameterization and validation of surface albedo merit increased attention.

Albedos over ice cover, on scales suitable for model input, have been both mapped from digital AVHRR imagery (e.g., [Rossow et al., 1989]) for short time periods and interpreted from Defense Meteorological Satellite Program (DMSP) Operational Line-Scan System (OLS) transparencies [Robinson et al., 1986; Scharfen et al., 1987] for about a ten-year period. These DMSP albedos were tested in an ice model by Ross and Walsh [1987]. Robinson et al.'s results cover only the central Arctic during the summer melt season (May through August), but do provide quantitative depictions of the interannual variations over areas of 1000 km x 1000 km. Unfortunately, the variation of regional surface albedo is intertwined with the distribution of open water within the region. Model biases regarding open water fractions may overwhelm the effects of variation in the snow–ice surface albedo. A need therefore exists for a more complete utilization of high-resolution satellite imagery to address the validity of model treatments of surface albedo within a grid cell. It is likely that combined usage of SAR and high-resolution visible imagery will be required for a meaningful assessment of the parametric needs relative to surface albedo.

The energy balance calculations necessary to estimate ice surface temperature for ice growth and decay in models are substantial sources of uncertainty. Direct measurement of ice surface temperatures would be particularly valuable for ice prediction models. Comiso [1983] used data from the temperature-humidity infrared radiometer (THIR) on board Nimbus 7 to estimate ice surface temperatures covering nine months in the Arctic. Schluessel and Grassl [1990] describe the use of AVHRR to retrieve water and ice surface temperatures in Antarctic polynyas. Key and Barry [1989] describe the retrieval of ice surface temperatures and albedos for a portion of the Arctic using AVHRR imagery. An example of the calculation of ice albedos and temperatures from AVHRR for operational ice forecasting is discussed by Condal and Le [1984]. With the exception of the albedos mapped from DMSP OLS data, no albedo or temperature data set has been prepared for the Arctic that covers more than several months. Henderson–Sellers and Wilson [1983] review some older albedo climatologies suitable for model input. Work is underway to refine the methodology to map ice surface temperatures (as well as cloud cover) from AVHRR as part of the Earth Observing System Polar Exchange at the Sea Surface (POLES) program.

22.3.2.2. Snow cover. Snow cover is clearly a key element of the albedo formulation discussed above. Snow depth also affects the wintertime energy budget over sea ice. While snowfall is an externally prescribed quantity in the context of ice modeling, there is nevertheless a need for more information on the distribution of snow depth within the grid-cell areas of models. The extent to which altimeters and/or SAR can provide such data is unclear. Attempts to use passive microwave (SMMR) data to estimate snow depth over land have, however, met with some success for specific regions [Chang et al., 1987]. A discussion of the use of passive microwave data to determine snow depths over sea ice is presented in Chapter 16. Advances in studies of snow depth in regions covered by sea ice are needed to permit modelers to determine whether biases in simulated sea ice (e.g., thickness) can be attributed to biases in the prescribed snowfall or snow depth.

22.3.2.3. Sea surface temperature. Sea surface temperatures suitable for ice modeling have been available for several years and are derived from ship observations (e.g., the comprehemsive ocean-atmosphere data set, COADS) and satellite imagery such as AVHRR, including the operational generation of sea surface temperatures by NOAA/NESDIS (National Environmental Satellite, Data and Information Service), the multichannel sea surface temperature fields (e.g., [McClain et al., 1985; Barton, 1985]) and from sounder data (e.g., [Susskind and Reuter, 1985]). Minnett [1990] reviewed some of the attributes of sea surface temperatures measured from space, including the problem of determining skin versus bulk temperatures (e.g., [Schluessel et al., 1987; Schluessel and Grassl, 1990]). While the noise level of individual temperature measurements from AVHRR is about 0.1 K, the unmodeled effects of the intervening atmosphere reduce this accuracy. The Along-Track Scanning Radiometer, carried on board ERS-1, is designed with both a lower noise level, as well as the ability to observe two paths through the atmosphere, to achieve more precise and accurate surface temperatures.

22.3.2.4. Cloud cover. Parameterizations of radiative fluxes in ice models can benefit from reliable cloud cover observations. Long-term cloud cover statistics derived from satellite data for the polar regions include the three-dimensional nephanalysis carried out by the U.S. Air Force [Fye, 1978; Tian and Curry, 1989], the International Satellite Cloud Climatology Program (ISCCP) [Rossow et al., 1985; Rossow and Schiffer, 1991], and analyses of THIR imagery [Stowe et al., 1988]. Henderson–Sellers [1984] describes some of these climatologies, as well as cloud climatologies, from surface observations. Some of the problems of comparing cloud climatologies based on satellite data versus surface observations are reviewed by Henderson–Sellers and McGuffie [1990], and different satellite-derived cloud statistics for the Arctic are discussed by McGuffie et al. [1988]. Serreze and Rehder [1990] describe a climatology for June spanning 12 years of DMSP interpretations for the western Arctic. Barry et al. [1987] give DMSP-derived cloud statistics for the Arctic for April through June 1979 and 1980. Key and Barry [1989] and Ebert [1989] describe some of the problems of automated mapping of polar clouds and AVHRR-based cloud mapping schemes designed specifically for the polar regions.

Outgoing longwave radiation fluxes are available directly from AVHRR data [Gruber and Jacobowitz, 1985], High-Resolution Infrared Radiation Sounder data [Ellingson et al., 1989], and the broadband channels of the Earth Radiation Budget Experiments (ERBE) (e.g., [Kopia, 1986; Jacobowitz et al., 1984; Barkstrom and Smith, 1986; Ramanathan et al., 1989]). Relatively long time series of these data are available at different spatial and temporal scales for the polar regions.

22.3.2.5. Atmospheric properties.
Atmospheric properties (water vapor, liquid water, and temperatures) are determined operationally using sounders as well as passive microwave imagery. Sounders provide information on the vertical distribution of atmospheric conditions, while passive microwave images provide integrated values for water content. Calculations are complicated over ice and snow due to the high and variable emissivity of the surface.

Surface air temperatures used in ice model simulations are generally prescribed, either from objective analyses of station data or from atmospheric model output. The effects of errors and biases in these temperatures need assessment in much the same way as do the effects of errors in the albedo and snow depth as discussed above. The use of satellite data for such assessments does not appear to have been fully exploited, although the products of the ERBE and ISCCP efforts may be useful in this regard.

Finally, wind speed over open ocean is routinely determined from passive microwave imagery and can be mapped from SAR data. Speed and direction can be measured by scatterometers. Over sea ice, no direct way of measuring surface winds from satellites is available, although cloud motions have been used to map upper-level winds over polar regions [Turner and Warren, 1989].

22.4 EXAMPLES OF THE INTERACTION OF MODELS WITH DATA

22.4.1 Model-to-Data Comparison Studies

Comparisons with satellite data have provided one of the major vehicles for recent assessments of sea-ice model performance. As discussed below, some model results have provided insights to the validity of satellite-derived fields of variables that are only indirectly measurable.

Comparisons between model output and satellite data generally require systematic compilations of satellite measurements of one or more of the variables simulated by a model. An outstanding example of such a compilation is the record of the sea ice extent and concentration derived from satellite passive microwave measurements (ESMR, SMMR, and SSM/I). The passive microwave data are also a primary input to the JIC's weekly ice charts, which have been used for a host of additional model-to-data comparisons. The availability of the passive microwave grids of ice concentration from 1973 onward provides a climatology as well as a 15- to 18-year sample of interannual fluctuations about the seasonal means.

Among the first model studies to have utilized data on ice extent derived from passive microwave imagery were those of Parkinson and Herman [1980] and Lemke et al. [1980]. In the latter study, a stochastic-dynamic model was calibrated using 10 years of monthly data on ice extent. Subsequently, the satellite-derived fields of ice extent have been used by Hibler and Walsh [1982], Hibler and Ackley [1983], Parkinson and Bindschadler [1984], Walsh et al. [1985], Preller [1985], Hibler and Bryan [1987], Semtner [1987], Stossel et al. [1989], Fleming and Semtner [1991], and Walsh and Chapman [1991]. Satellite-derived ice concentrations in the vicinity of the Weddell polynya stimulated the modeling studies of Martinson et al. [1981], Parkinson [1983], and Motoi et al. [1987].

While most model-to-data comparisons of ice extent have been intended to assess the climatologies simulated by models, several studies have used satellite-derived data to evaluate the ability of models to reproduce interannual fluctuations of ice cover. Walsh et al. [1985], for example, found that the Hibler [1979] two-level model reproduced approximately 50% of the year-to-year variance of ice coverage in 20 longitudinal sectors of the Arctic. Walsh and coworkers used interannually varying winds and air temperatures, but climatological ocean forcing to drive a stand-alone sea ice model. Fleming and Semtner [1991] subsequently showed that the verification statistics are improved somewhat when the ice model is coupled to an interactive ocean model.

More specific use of the ice concentrations derived from passive microwave measurements has been made by Serreze et al. [1990], who used an ice model and other tools to diagnose the origins of a large area of reduced ice concentrations within the Arctic pack ice during 1988. Serreze and coworkers supplemented the passive microwave data with visible-band imagery during the sunlit season.

Additional uses of satellite measurements for model-to-data comparisons have been made by Walsh and Zwally [1990], whose application of satellite-derived multiyear ice concentrations is discussed in Section 22.4.2, and by Emery et al. [1991], who used AVHRR imagery to obtain ice velocity vectors for the region east of Greenland. In the latter study, discrepancies between model- and satellite-derived velocities indicated that the strength of an ocean current had been underspecified in the model forcing. Ross and Walsh [1987] compared interannual fluctuations of Arctic Ocean surface albedo with those derived from satellite imagery by Scharfen et al. [1987].

22.4.2 Hindcast Studies: Walsh and Zwally [1990]

Hindcast studies of the interannual fluctuations of sea ice have been performed by using observational analyses of atmospheric forcing variables (air temperatures and pressure-derived winds) to drive the Hibler two-level model over time scales of several years to several decades [Walsh et al., 1985; Walsh and Zwally, 1990]. The inclusion of multiyear ice in the simulations was motivated by the availability of the multiyear ice concentrations derived from satellite passive microwave measurements. The oceanic component of the model consisted of only a constant-depth mixed layer and time-independent currents. The currents were obtained geostrophically from a prescribed dynamic topography.

As noted in the previous subsection, runs of the model in this mode explain about 50% of the interannual variability of satellite-derived coverage of total sea ice (first-year plus multiyear ice). The simulated values of multiyear ice concentration are generally larger than those derived from the SMMR data. Although several arbitrary elements of the model formulation may contribute to this discrepancy, Comiso [1990] has recently reported that SMMR-derived multiyear concentrations from the SMMR Team algorithm are also smaller than one would infer from the concentrations of ice surviving the summer melt in the Arctic. The interannual variations of SMMR- and model-derived multiyear ice coverage generally show somewhat ambiguous agreement in the Arctic Ocean (Figure 22-2).

The SMMR data also permit the estimation of generalized resultant velocity vectors on the basis of changes in the multiyear ice edge over periods of several weeks to a season. Such estimations are limited to the nonsummer months, when surface wetness does not contaminate the multiyear ice signature.

Instances of convergence/divergence in the numerical model results have been confirmed by corresponding changes in the multiyear ice concentrations derived from the SMMR data, although the model-derived changes are smaller than those obtained from SMMR. Larger samples of such comparisons, including buoy-derived velocity divergences, are needed to determine whether there is a bias in either the SMMR or the model depictions of short-term divergence events. The outcome of such comparisons has important

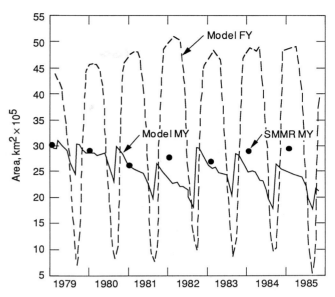

Fig. 22-2. Time series of model-derived multiyear and first-year ice coverage in the Arctic. Also shown are November–April averages from SMMR data, adjusted to include an estimated 8.2×10^5 km^2 in the unobserved area north of 84° N [Walsh and Zwally, 1990].

implications for the validity of the simulated mass budget, since most of the ice growth in the central Arctic occurs in areas of thin ice or open water (e.g., leads) that result from divergence of pack ice.

22.4.3 The Use of Models With Data in Forecasting

The Polar Ice Prediction System is a sea-ice forecasting system based on the Hibler ice model [Preller, 1985; Preller and Posey, 1989]. It is used for the daily prediction of ice drift, ice thickness, and ice concentration in the Arctic by the U.S. Navy (Figure 22-3). In this system, the Hibler ice model is driven by daily atmospheric stresses and heat fluxes from the Navy Operational Global Atmospheric Prediction System and by monthly mean ocean currents and heat fluxes derived from the Hibler and Bryan [1987] ice–ocean model. The model is run in an operational environment at the Fleet Numerical Oceanography Center (FNOC). A 120-hour forecast of several different ice fields is produced daily. Each day's forecast is initialized from the previous day's 24-hour forecast (Figure 22-4). Once each week, the model's ice concentration field is initialized from a digitized field of observed ice concentration made available by the Navy/NOAA Joint Ice Center (Figure 22-5). This field is a subjective analysis derived from multiple data sources [Wohl, 1991]. Satellite data used in the analysis come from the AVHRR, visible data from the DMSP OLS, and passive microwave data from the SSM/I. In addition to these satellite data, any available ice reconnaissance flights or ship observations are blended into the weekly analysis.

During model initialization, the observed ice concentration totally replaces the modeled ice concentration from the

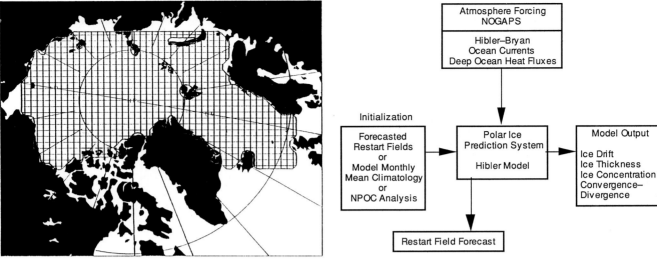

Fig. 22-3. Model domain and grid used by PIPS. Grid resolution is 127 km.

Fig. 22-4. The design of PIPS.

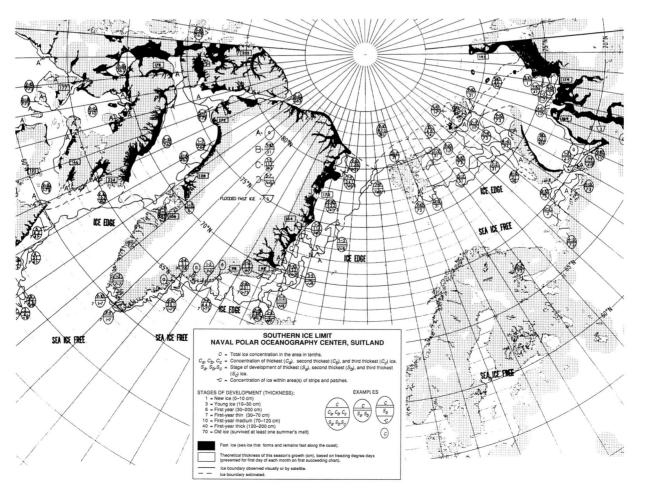

Fig. 22-5. Navy/NOAA Joint Ice Center weekly analysis of sea ice concentration for the eastern Arctic for June 5, 1991.

previous day's 24-hour forecast. Other variables from the previous day's forecast remain unchanged, with the exception of the ice thickness and the heat stored in the ocean's mixed layer. These fields are adjusted only near the ice edge to be consistent with the ice edge defined by the JIC data as shown in Figures 22-6(a) and (b). This adjustment involves removing ice from regions that should not be ice covered and adding heat to the mixed layer by raising its temperature. If the model forecasts no ice in a region where the data indicate it should exist, heat is removed from the mixed layer by setting the temperature back to freezing and either 0.5 or 1.0 m of ice is added to that grid cell depending on the

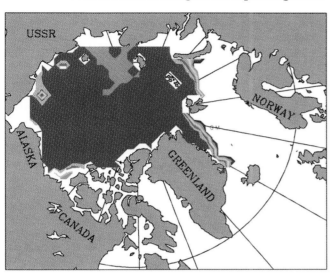

0.2 0.3 0.4 0.5 0.6 0.7 0.8 0.9 1.0

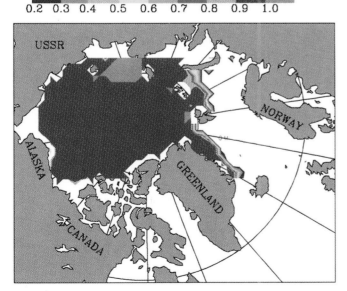

Fig. 22-6. PIPS 24-hour forecast ice concentration from (a) June 6, 1991, and (b) June 7, 1991, immediately following the initialization of PIPS by the JIC analysis. (The USSR is now referred to as the Commonwealth of Independent States.)

observed concentration [Preller and Posey, 1989]. Two similarly designed forecast systems, the Regional Polar Ice Prediction System Barents Sea (RPIPS-B) and the Regional Polar Ice Prediction System Greenland Sea (RPIPS-G) forecast over the Barents and Greenland Seas at a higher resolution than PIPS [Preller et al., 1989].

Data assimilation and model initialization in an operational forecasting system are dependent on a number of different factors. Availability of the data is the first important factor. The data must be available to the computer system used by the forecast model.

The second factor is the age of the data. The model uses data to provide the most realistic initial state for the forecast. If the data are old, they can provide incorrect information to the model. The JIC completes its analysis each Thursday; the analysis is ready for use in PIPS by the end of the day on Friday. The two-day lag in JIC data is not optimal, but is considered acceptable in the PIPS model. PIPS uses a resolution of 127 km. The ice edge, on average, does not move more than 100 km over 48 hours, so the observed changes may not be resolved by the model. There are certainly exceptions, such as the appearance of the Odden over a 24-hour period or the rapid movement of the ice edge brought about by extremes in the atmospheric forcing (strong winds). A 48-hour time delay in the data can, however, be a problem in the RPIPS-B and RPIPS-G models. Since these models use resolutions of 20 and 25 km, respectively, movement of the ice edge in these regions can be a grid distance or more over a 48-hour period. Ideally, the data used for initialization should be no more than 24 hours old. The importance of the age of the data also implies that data assimilation techniques such as Kalman filtering, which fits a time series of data into a predictive model, may not provide the most recent ice conditions for initializing a forecast. Also, the Kalman filter may not be the optimal choice in an operational system where computer time is limited (the Kalman filter is computer intensive) and where historical data are not often saved and available for use with the filter.

The third important factor is the frequency of the data. The JIC analysis is available only once per week. However, FNOC now has a digitized data set of ice concentration from the SSM/I available in real time each day (Figure 22-7). That data set is presently being tested as an initialization field for PIPS on those days that the JIC analysis is unavailable.

In addition, Cheng and Preller [personal communication, 1991] have designed a method of blending two or more concentration fields into one maximum likelihood estimate using nonlinear regression. The technique can be described by borrowing the terminology of the physical and measurement models from Thorndike [1988] and Thomas and Rothrock [1989] (see Chapter 23). The physical model describing pack ice evolution can be written as

$$\vec{X}_j^{t+1} = H_j\left(\vec{X}_j^t\right) + \vec{\varepsilon}_j^t \qquad (9)$$

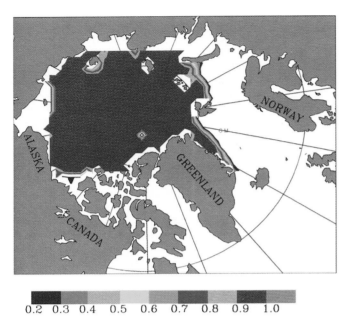

0.2 0.3 0.4 0.5 0.6 0.7 0.8 0.9 1.0

Fig. 22-7. SSM/I ice concentration from FNOC, derived using the Navy algorithm, for June 7, 1991. Values have been interpolated to and plotted on the PIPS grid.

where X is the ice concentration, H is the physical model PIPS, ε is the residual of the physical model, i and j are the spatial indices, and t is the time index. The physical model uses dynamic and thermodynamic equations along with atmospheric and oceanic forcing to predict $\vec{X}_j^{\,t+1}$. The measurement model is used to update the value of X for the physical model. Written in a general form,

$$\vec{Z}_m^t = G_m\left(\vec{X}_n^t\right) = \vec{e}_m^t \tag{10}$$

where Z represents the data (i.e., observations, measurements, constraints, and assumptions), G is the measurement model (in this case the continuity equation for ice concentration), e is the residual between the measurement and the corresponding model value, and m and n are spatial indices.

The JIC analysis and passive microwave data (other data may be included) provide field observations or measurements. The PIPS ice concentration provides a data set of historical information, while the continuity equation for ice concentration represents the constraint for these ice concentration fields. All of these data and the constraints imply that the measurement model is defined as an over-determined problem, which is particularly suited to nonlinear regression. The method also allows for emphasizing the reliability of the data in regression by applying variable weights to the data.

Remotely sensed data have also been used for validating the Navy's operational sea-ice forecast systems. Each U.S. Navy operational model is put through an operational evaluation, lasting anywhere from three months to a year. Since 1987, an operational forecast of at least 24 hours has been made daily and archived by the Naval Research Laboratory (NRL) (formerly the Naval Oceanographic and Atmospheric Research Laboratory). These forecasts are identical to those made by FNOC and are used by NRL for continuing model verification. During the operational evaluation of PIPS, the model-derived ice edge was compared qualitatively and quantitatively to the ice edge derived by JIC. This study showed that when initialized at least once each week, the mean error in the forecast ice edge location, after seven days, was less than or equal to one grid distance. This error increased if the initialization took place less frequently [Preller and Posey, 1989].

In addition to ice edge verification, the PIPS-derived ice drift is verified against Arctic buoy data on a regular basis. Ice drift derived from the change in buoy location is statistically compared to PIPS ice drift. The result of this verification has been the continued improvement in the accuracy of the PIPS ice drift field, mainly due to improvements in the treatment of surface wind stress on the ice [Preller and Posey, 1989]. During the operational evaluation of both regional models, satellite-derived ice drift data from successive AVHRR data were used for the validation of the model-derived ice drift, Figures 22-8(a) and (b). In the case of both regional models, the direction of the modeled ice drift agreed well with the satellite-derived ice drift, however, the magnitude of the modeled drift was weaker than that derived from the satellite data.

22.5 Summary and Conclusions

Numerical models and observational data have been used together in a number of ways. Data obtained from drifting ice station, aircraft, and ship measurements collected over many years provided highly localized point measurements that were used to define many characteristics of sea ice. A more synoptic view of sea ice was obtained from the data collected in field experiments such as AIDJEX and MIZEX. These data provided the basis for an improved understanding of the dynamic and thermodynamic processes that affect sea ice. As a result, more sophisticated formulations for ice motion, including the effect of internal ice stress and the growth and decay of sea ice, have been determined and incorporated into numerical models. The models, in turn, are used to help provide insight into the accuracy of these formulations by validating the model results against observations.

Satellites have been able to provide a source of observational data both over larger areas and for longer periods of time than the more conventional field observations. A combination of fine-resolution satellite imagery, such as visible or infrared imagery, and radar altimetry, along with coarser resolution data such as that from the passive microwave sources, have been used for verifying and validating many numerical models. A comparison of the ice edge derived from these sources with the modeled ice edge has been the favored method of model validation.

Satellite passive microwave data are particularly appealing to the large-scale models used to study interannual variability of sea ice. Passive-microwave total ice concentration has been used in the majority of model-to-data comparison studies. However, availability of multiyear ice concentration from the passive microwave data motivated Walsh and Zwally [1990] to incorporate multiyear ice into a version of the Hibler ice model. Using this formulation, they were able to show some compatibility between the interannual variations of the model- and SMMR-derived multiyear ice concentration.

A number of sources of remotely sensed data are used to create a weekly, subjective analysis of ice concentration by JIC. This analysis is used as a weekly initialization field for each of the three sea-ice forecast systems run by the U.S. Navy: PIPS, RPIPS-B, and RPIPS-G. Two upgrades are scheduled for the Navy forecast systems in the near future. The first upgrade involves the use of passive microwave data as an initialization field for the forecast systems. Passive microwave data, now available in real time from FNOC, are being tested as a data source that could provide more frequent initialization of these forecast systems. The second upgrade involves the replacement of PIPS with the PIPS 2.0 model. PIPS 2.0 will provide ice forecasts for most of the ice-covered regions in the Northern Hemisphere on a half-degree grid (Figure 22-9). The design of this system will be similar to that of PIPS, but will include a fully coupled ice–ocean model. Along with the JIC analysis, the SSM/I ice concentration will provide real-time initialization data over the entire model domain.

In addition to the passive microwave data, the recent launch of the ERS-1 satellite is now providing fine-resolu-

(a)

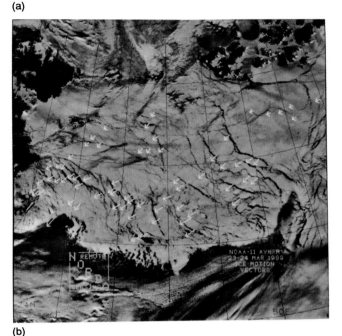

(b)

Fig. 22-8. Ice drift derived from (a) successive AVHRR images from March 23 and 24, 1989, and (b) the 24-hour forecast of RPIPS-B for the same time period.

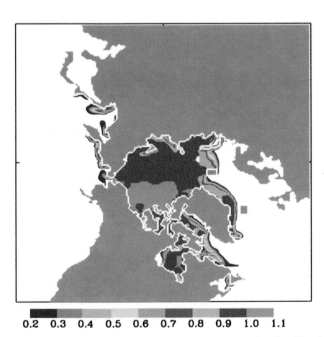

| 0.2 | 0.3 | 0.4 | 0.5 | 0.6 | 0.7 | 0.8 | 0.9 | 1.0 | 1.1 |

Fig. 22-9. PIPS 2.0 monthly mean ice concentration for March 1986.

tion coverage of sea ice conditions via the Synthetic Aperture Radar. Considerable potential exists for the use of SAR data in model studies of the parameterization of thin ice and open water regions. In addition, SAR may help provide more realistic parameterizations of the surface–atmosphere heat exchange, thereby serving as a stimulus to the coupling of ice and atmosphere models, as well as atmosphere–ice–ocean models.

Although large quantities of remotely sensed data already exist, there is great expectation for what will be available in the future. It is anticipated that remotely sensed data will provide additional important information to models on such variables as albedo, ice surface temperature, snow cover, ocean surface temperature, and atmospheric properties such as cloud cover, wind speed, and radiation budgets.

Although the interaction between data and models has been ongoing for many years, the future would suggest an even greater opportunity for interplay between the two. The combination of new sources of remotely sensed data and new products derived from these data should assist in the evolution of improved formulations for dynamic and thermodynamic processes affecting sea ice. These new formulations will be used to create more sophisticated models of the interaction between atmosphere, ice, and ocean. In addition, recent work by Thomas and Rothrock [1989], in Chapter 23, and the work presently being done by the U.S. Navy, suggests that the assimilation of data into models can provide both improved models as well as improved forecasts of sea ice conditions.

Acknowledgments.

John Walsh was supported by NASA's Interdisciplinary Research Program through grant IDP-88-009. James Maslanik was supported by NOAA's Climate and Global Change Program through grant NA85RAH05066. Ruth Preller was funded through the Navy Ocean Modeling and Prediction program by the Office of Naval Technology (program element 62435N) and by the U.S. Space and Naval Warfare Systems Command (program element 63207N). This book chapter, NRL contribution BC 001-92-322, is approved for public release; distribution unlimited.

REFERENCES

Aagaard, K. and E. C. Carmack, The role of sea ice and other fresh water in the Arctic circulation, *Journal of Geophysical Research, 94,* pp. 14,485–14,498, 1989.

Aagaard, K., J. H. Swift, and E. C. Carmack, Thermohaline circulation in the Arctic Mediterranean seas, *Journal of Geophysical Research, 90,* pp. 4833–4846, 1985.

Andreas, E. L., Comment on "Atmospheric boundary layer modification in the marginal ice zone" by T. J. Bennett, Jr., and K. Hunkins, *Journal of Geophysical Research, 92(C4),* pp. 3965–3968, 1987.

Andreas, E. L. and B. Murphy, Bulk transfer coefficients for heat and momentum over leads and polynyas, *Journal of Physical Oceanography, 16(11),* pp. 1875–1883, 1986.

Barkstrom, B. R. and G. L. Smith, The Earth Radiation Budget Experiment (ERBE): Science and implementation, *Review of Geophysics, 24,* pp. 379–390, 1986.

Barry, R. G., R. G. Crane, A. Schweiger, and J. Newell, Arctic cloudiness in spring from satellite imagery, *Journal of Climatology, 7,* pp. 423–451, 1987.

Barton, I. J., Transmission model and ground-truth investigation of satellite-derived sea surface temperatures, *Journal of Climate and Applied Meteorology, 24,* pp. 508–516, 1985.

Bauer, J. and S. Martin, A model of grease ice growth in small leads, *Journal of Geophysical Research, 88(C5),* pp. 2917–2925, 1983.

Bennett, T. J., Jr., and K. Hunkins, Atmospheric boundary layer modification in the marginal ice zone, *Journal of Geophysical Research, 91,* pp. 13,033–13,044, 1986.

Bjork, G., A one-dimensional time-dependent model for the vertical stratification of the upper Arctic Ocean, *Journal of Physical Oceanography, 19,* pp. 52–67, 1989.

Bourke, R. H. and A. S. McLaren, Contour mapping of Arctic Basin ice draft and roughness parameters, *Journal of Geophysical Research,* in press, 1992.

Bryan, K., Climate and the ocean circulation: III. The ocean model, *Monthly Weather Review, 97,* pp. 806–827, 1969.

Bryan, K., S. Manabe, and R. L. Pacanowski, A global ocean–atmosphere climate model. Part II. The ocean circulation, *Journal of Physical Oceanography, 5,* pp. 30–46, 1975.

Burns, B. A., D. J. Cavalieri, M. R. Keller, W. J. Campbell, T. C. Grenfell, G. A. Maykut, and P. Gloersen, Multisensor comparison of ice concentration estimates in the marginal ice zone, *Journal of Geophysical Research, 92,* pp. 6843–6856, 1987.

Cavalieri, D. J., P. Gloersen, and W. J. Campbell, Determination of sea ice parameters with the Nimbus 7 SMMR, *Journal of Geophysical Research, 89,* pp. 5355–5369, 1984.

Chang, A. T. C., J. L. Foster, and D. K. Hall, Nimbus-7 SMMR derived global snow cover parameters, *Annals of Glaciology, 9,* pp. 39–45, 1987.

Chu, P. C., An ice breeze mechanism for an ice divergence–convergence criterion in the marginal ice zone, *Journal of Physical Oceanography, 17(10),* pp. 1627–1632, 1987.

Colbeck, S. C., The layered character of snow covers, *Reviews of Geophysics, 29(1),* pp. 81–96, 1991.

Comiso, J. C., Sea ice effective microwave emissivities from satellite passive microwave and infrared observations, *Journal of Geophysical Research, 88,* pp. 7686–7704, 1983.

Comiso, J. C., Characteristics of Arctic winter sea ice from satellite multispectral microwave observations, *Journal of Geophysical Research, 91(C1),* pp. 975–994, 1986.

Comiso, J. C., Arctic multiyear ice classification and summer ice cover using passive microwave satellite data, *Journal of Geophysical Research, 95,* pp. 13,411–13,422, 1990.

Condal, A. R. and H. V. Le, Automated computer monitoring sea-ice temperature by use of NOAA satellite data, *Proceedings of the Eighth Canadian Symposium on Remote Sensing,* pp. 145–150, Montreal, Quebec, 1984.

Curlander, J. C., B. Holt, and K. J. Hussey, Determination of sea ice motion using digital SAR imagery, *IEEE Journal of Oceanic Engineering, OE-10(4),* pp. 358–367, 1985.

Ebert, E., Analysis of polar clouds from satellite imagery using pattern recognition and a statistical cloud analysis scheme, *Journal of Applied Meteorology, 28,* pp. 382–399, 1989.

Ebert, E. E. and J. A. Curry, Annual cycle of cloud radiative forcing over the Arctic Ocean, *Proceedings of the Seventh Conference on Atmospheric Radiation,* pp. 395–400, American Meteorological Society, Boston, Massachusetts, 1990.

Ellingson, R. G., D. J. Yanuk, H.-T. Lee, and A. Gruber, A technique for estimating outgoing long-wave radiation from HIRS radiance observations, *Journal of Atmospheric and Oceanic Technology, 6,* pp. 706–711, 1989.

Emery, W. J., C. W. Fowler, J. Hawkins, and R. H. Preller, Fram Strait satellite image-derived ice motions, *Journal of Geophysical Research, 96,* pp. 4751–4768, 1991.

Fichefet, T. and P. Gaspar, A model of upper ocean–sea ice interaction, *Journal of Physical Oceanography, 18(2),* pp. 181–195, 1988.

Flato, G. M. and W. D. Hibler, III, The effect of ice pressure on marginal ice zone dynamics, *IEEE Transactions on Geoscience and Remote Sensing, 27(5),,* pp. 514–521, 1989.

Fleming, G. H. and A. J. Semtner, Jr., A numerical study of interannual ocean forcing on Arctic sea ice, *Journal of Geophysical Research, 96,* pp. 4589–4604, 1991.

Fye, F. K., *The AFGWC automated cloud analysis model,* 97 pp., Technical Memorandum 78-002, United States Air Force, Offut Air Force Base, Nebraska, 1978.

Gabison, R., A thermodynamic model of the formation, growth, and decay of first-year sea ice, *Journal of Glaciology, 33(113),* pp. 105–119, 1987.

Gloersen, P. and W. J. Campbell, Variations in the Arctic, Antarctic, and global sea ice cover during 1978–1987 as observed with the Nimbus 7 Scanning Multichannel Microwave Radiometer, *Journal of Geophysical Research, 93,* pp. 10,666–10,674, 1988.

Gruber, A. and H. Jacobowitz, The long-wave radiation estimated from NOAA polar orbiting satellites: An update and comparison with Nimbus 7 ERB results, *Advanced in Space Research, 5(6),* pp. 111–120, 1985.

Hakkinen, S., Coupled ice–ocean dynamics in the marginal ice zones: Up/downwelling and eddy generation, *Journal of Geophysical Research, 91,* pp. 819–832, 1986.

Harvey, L. D. D., A semi-analytic energy balance climate model with explicit sea ice and snow physics, *Journal of Climatology, 1,* pp. 1065–1085, 1988.

Henderson–Sellers, A., *Satellite Sensing of a Cloudy Atmosphere: Observing the Third Planet,* 340 pp., Taylor & Francis Inc., Philadelphia, 1984.

Henderson–Sellers, A. and K. McGuffie, Are cloud amounts estimated from satellite sensor and conventional surface-based observations related?, *International Journal of Remote Sensing, 11(3),* pp. 543–550, 1990.

Henderson–Sellers, A. and M. F. Wilson, Surface albedo data for climate modeling, *Reviews of Geophysics and Space Physics, 21,* pp. 1743–1778, 1983.

Hibler, W. D., III, A dynamic-thermodynamic sea ice model, *Journal of Physical Oceanography, 9,* pp. 815–846, 1979.

Hibler, W. D., III, Modeling a variable thickness sea ice cover, *Monthly Weather Review, 108,* pp. 1943–1973, 1980a.

Hibler, W. D., III, Sea ice growth, drift and decay, in *Dynamics of Snow and Ice Masses,* edited by S. C. Colbeck, 468 pp., Academic Press, New York, 1980b.

Hibler, W. D., III, Ice dynamics, in *The Geophysics of Sea Ice,* edited by N. Untersteiner, pp. 577–640, NATO ASI Series B: Physics vol. 146, Plenum Press, New York, 1986.

Hibler, W. D., III and S. F. Ackley, Numerical simulation of the Weddell Sea pack ice, *Journal of Geophysical Research, 88,* pp. 2873–2887, 1983.

Hibler, W. D., III and K. Bryan, Ocean circulation: Its effects on seasonal sea-ice simulations, *Science, 224,* pp. 489–491, 1984.

Hibler, W. D., III and K. Bryan, A diagnostic ice–ocean model, *Journal of Physical Oceanography, 17,* pp. 987–1015, 1987.

Hibler, W. D., III and J. E. Walsh, On modeling seasonal and interannual fluctuations of Arctic sea ice, *Journal of Physical Oceanography, 12,* pp. 1514–1523, 1982.

Ikeda, M., A coupled ice–ocean model of a wind driven coastal flow, *Journal of Geophysical Research, 90,* pp. 9119–9128, 1985.

Ikeda, M., A three-dimensional coupled ice–ocean model of coastal circulation, *Journal of Geophysical Research, 93,* pp. 10,731–10,748, 1988.

Jacobowitz, H., H. V. Soule, H. L. Kyle, F. House, and the Nimbus-7 ERB Experiment Team, The Earth Radiation Budget (ERB) Experiment—An Overview, *Journal of Geophysical Research, 89(D4),* pp. 5021–5038, 1984.

Kantha, L. H. and G. L. Mellor, Application of a two-dimensional coupled ocean–ice model to the Bering Sea marginal ice zone, *Journal of Geophysical Research, 94,* pp. 10,921–10,935, 1989.

Key, J. and R. G. Barry, Cloud cover analysis with Arctic AVHRR data, 1. Cloud detection, *Journal of Geophysical Research, 94(D15),* pp. 18,521–18,535, 1989.

Kneizys, F. X., E. P. Shettle, L. W. Abreu, J. H. Chetwynd, G. P. Anderson, W. O. Gallery, J. E. A. Selby, and S. A. Clough, *Users Guide to LOWTRAN-7,* document AFGL-TR-88-00177, 138 pp., Air Force Geophysics Laboratory, Hanscom Air Force Base, Massachusetts, 1988.

Koch, C., A coupled sea ice atmospheric boundary layer model. Part 1: Description of the model and 1979 standard run, *Beitraege zur Physik der Atmosphaere, 61(4),* pp. 344–354, 1988.

Kopia, L. P., Earth Radiation Budget Experiment scanner instrument, *Review of Geophysics, 24,* pp. 400–406, 1986.

Kozo, T. L., Initial model results for Arctic mixed layer circulation under a refreezing lead, *Journal of Geophysical Research, 88(C5),* pp. 2926–2934, 1983.

Ledley, T. S., A coupled energy balance climate–sea ice model: Impact of sea ice and leads on climate, *Journal of Geophysical Research, 93,* pp. 15,915–15,932, 1988.

Ledley, T. S., The climatic response to meridional sea-ice transport, *Journal of Climate, 4,* pp. 147–163, 1991.

Lemke, P., A coupled one-dimensional sea ice–ocean model, *Journal of Geophysical Research, 92,* pp. 13,164–13,172, 1987.

Lemke, P. and T. O. Manley, The seasonal variation of the mixed layer and the pycnocline under polar sea ice, *Journal of Geophysical Research, 89(C4),* pp. 6494–6504, 1984.

Lemke, P., E. W. Trinkl, and K. Hasselmann, Stochastic dynamic analysis of polar sea ice variability, *Journal of Physical Oceanography, 10,* pp. 2100–2120, 1980.

Manabe, S. and R. J. Stouffer, Two stable equilibria of a coupled ocean–atmosphere model, *Journal of Climate, 1,* pp. 841–866, 1988.

Manabe, S., K. Bryan, and M. J. Spelman, A global ocean–atmosphere climate model with seasonal variation for future studies for climate sensitivity, *Dynamics of Atmospheres and Oceans, 3,* pp. 393–426, 1979.

Martinson, D. G., P. D. Killworth, and A. L. Gordon, A convective model for the Weddell polynya, *Journal of Physical Oceanography, 11,* pp. 466–488, 1981.

Maykut, G. A., Energy exchange over young sea ice in the Central Arctic, *Journal of Geophysical Research, 83,* pp. 3646–3658, 1978.

Maykut, G. A., Large-scale heat exchange and ice production in the Central Arctic, *Journal of Geophysical Research, 87,* pp. 7971–7984, 1982.

Maykut, G. A. and N. Untersteiner, Some results from a time-dependent thermodynamic model of sea ice, *Journal of Geophysical Research, 76,* pp. 1550–1575, 1971.

McClain, E. P., W. G. Pichel, and C. C. Walton, Comparative performance of AVHRR-based multichannel sea surface temperatures, *Journal of Geophysical Research, 90(C6),* pp. 11,587–11,601, 1985.

McGuffie, K., R. G. Barry, J. Newell, A. Schweiger, and D. A. Robinson, Intercomparison of satellite-derived cloud analyses for the Arctic Ocean in spring and summer, *International Journal of Remote Sensing, 9,* pp. 447–467, 1988.

Mellor, G. L. and L. Kantha, An ice–ocean coupled model, *Journal of Geophysical Research, 94,* pp. 10,937–10,954, 1989.

Minnett, P. J., The regional optimization of infrared measurements of sea surface temperature from space, *Journal of Geophysical Research, 95(C8),* pp. 13,497–13,510, 1990.

Morris, E. M., Turbulent transfer over snow and ice, *Journal of Hydrology, 105(314),* pp. 205–223, 1989.

Motoi, T., N. Ono, and M. Wakatsuchi, A mechanism for the formation of the Weddell polynya in 1974, *Journal of Physical Oceanography, 17,* pp. 2241–2247, 1987.

Overland, J. E., Atmospheric boundary layer structure and drag coefficients over sea ice, *Journal of Geophysical Research, 90,* pp. 9029–9049, 1985.

Overland, J. E., A model of the atmospheric boundary layer over sea ice during winter, *Second Conference on Polar Meteorology and Oceanography,* pp. 69–72, American Meteorological Society, Madison, Wisconsin, 1988.

Overland, J. E., R. M. Reynolds, and C. H. Pease, A model of the atmospheric boundary layer over the marginal ice zone, *Journal of Geophysical Research, 88(C5),* pp. 2836–2840, 1983.

Parkinson, C. L., On the development and cause of the Weddell polynya in a sea ice simulation, *Journal of Physical Oceanography, 13,* pp. 501–511, 1983.

Parkinson, C. L. and R. A. Bindschadler, Response of Antarctic sea ice to atmospheric temperature increases, in *Climate Processes and Climate Sensitivity, vol. 5,* edited by J. E. Hanson and T. Takahashi, pp. 254–264, American Geophysical Union, Washington, DC, 1984.

Parkinson, C. L. and G. F. Herman, Sea ice simulations based on fields generated by the GLAS GCM, *Monthly Weather Review, 108,* pp. 2080–2091, 1980.

Parkinson, C. L. and W. M. Washington, A large-scale numerical model of sea ice, *Journal of Geophysical Research, 84(C1),* pp. 311–337, 1979.

Parkinson, C. L., J. C. Comiso, H. J. Zwally, D. J. Cavalieri, P. Gloersen, and W. J. Campbell, *Arctic Sea Ice, 1973–1976: Satellite-Passive Microwave Observations,* NASA SP-489, 296 pp., National Aeronautics and Space Administration, Washington, DC, 1987.

Pease, C. H. and J. E. Overland, An atmospherically driven sea-ice drift model for the Bering Sea, *Annals of Glaciology, 5,* pp. 111–114, 1984.

Piacsek, S., R. Allard, and A. Warn–Varnas, Studies of the Arctic ice cover and upper ocean with a coupled ice–ocean model, *Journal of Geophysical Research, C3,* pp. 4631–4650, 1991.

Pollard, D., M. L. Batteen, and Y. Han, Development of a simple upper-ocean sea-ice model, *Journal of Physical Oceanography, 13,* pp. 754–768, 1983.

Preller, R. H., *The NORDA/FNOC Polar Ice Prediction System (PIPS)—Arctic: A Technical Description,* 60 pp., NORDA Report 108, Naval Oceanographic and Atmospheric Research Laboratory, Stennis Space Center, Mississippi, 1985.

Preller, R. H. and P. G. Posey, *The Polar Ice Prediction System—A Sea Ice Forecasting System,* NORDA Report 212, 45 pp., Naval Oceanographic and Atmospheric Research Laboratory, Stennis Space Center, Mississippi, 1989.

Preller, R. H., S. Riedlinger, and P. G. Posey, *The Regional Polar Ice Prediction Systems Barents Sea (RPIPS-B): A Technical Description*, NORDA Report 182, 38 pp., Naval Oceanographic and Atmospheric Research Laboratory, Stennis Space Center, Mississippi, 1989.

Preller, R. H., A. Cheng, and P. G. Posey, Preliminary testing of a sea ice model of the Greenland Sea, in *Proceedings of the W. F. Weeks Sea Ice Symposium,* CRREL Monograph 90-1, 19 pp., USA Cold Regions Research and Engineering Laboratory, Hanover, New Hampshire, 1990.

Pritchard, R. S., editor, *Sea Ice Processes and Models,* 474 pp., University of Washington Press, Seattle, Washington, 1980.

Ramanathan, V., R. D. Cess, E. F. Harrison, P. Minnis, B. R. Barkstrom, E. Ahmad, and D. Hartmann, Cloud-radiative forcing and climate: Results from the Earth Radiation Budget Experiment, *Science, 243,* pp. 57–63, January 6, 1989.

Raymo, R. E., D. Rind, and W. F. Ruddiman, Climatic effects of reduced Arctic sea ice limits in the GISS II general circulation model, *Paleoceanography, 5,* pp. 367–382, 1990.

Riedlinger, S. H. and R. H. Preller, The development of a coupled ice–ocean model for forecasting ice conditions in the Arctic, *Journal of Geophysical Research, 96,* pp. 16,955–16,978, 1991.

Riedlinger, S. H. and A. Warn–Varnas, Predictions and studies with a one-dimensional ice–ocean model, *Journal of Physical Oceanography, 20,* pp. 1545–1562, 1990.

Robinson, D. A., G. Scharfen, M. C. Serreze, G. Kukla, and R. G. Barry, Snow melt and surface albedo in the Arctic Basin, *Geophysical Research Letters, 13(9),* pp. 945–948, 1986.

Roed, L. P. and J. J. O'Brien, A coupled ice–ocean model of upwelling in the marginal ice zone, *Journal of Geophysical Research, 88,* pp. 2863–2872, 1983.

Ross, B. and J. E. Walsh, A comparison of simulated and observed fluctuations in Arctic surface albedo, *Journal of Geophysical Research, 92,* pp. 13,115–13,126, 1987.

Rossow, W. B. and R. A. Schiffer, ISCCP cloud data products, *Bulletin American Meteorological Society,* in press, 1991.

Rossow, W. B., F. Mosher, E. Kinsella, A. Arking, M. Desbois, E. Harrison, P. Minnis, E. Ruprecht, G. Seze, C. Simmer, and E. Smith, ISCCP cloud algorithm comparison, *Journal of Climate and Applied Meteorology, 24,* pp. 877–903, 1985.

Rossow, W. B., C. L. Brest, and L. C. Gardner, Global, seasonal and surface variations from satellite radiance measurements, *Journal of Climate, 2,* pp. 214–247, 1989.

Rothrock, D. A. and D. R. Thomas, The Arctic Ocean multiyear ice balance, 1979–1982, *Annals of Glaciology, 14,* pp. 252–255, 1990.

Scharfen, G., R. G. Barry, D. A. Robinson, G. Kukla, and M. C. Serreze, Large-scale patterns of snow melt on Arctic sea ice mapped from meteorological satellite imagery, *Annals of Glaciology, 9,* pp. 1–6, 1987.

Schluessel, P. and H. Grassl, SST in polynyas: A case study, *International Journal of Remote Sensing, 11(6),* pp. 933–945, 1990.

Schluessel, P., H.-Y. Shin, W. J. Emery, and H. Grassl, Comparison of satellite-derived sea surface temperatures with in situ skin temperatures, *Journal of Geophysical Research, 92(C3),* pp. 2859–2874, 1987.

Schwerdtfeger, P. and G. E. Weller, Radiative heat transfer processes in snow and ice, in *Meteorological Studies at Plateau Station, Antarctica,* edited by J. A. Businger, pp. 27–34, Antarctic Research Series 25, American Geophysical Union, Washington, DC, 1977.

Semtner, A. J., Jr., A model for the thermodynamic growth of sea ice in numerical investigations of climate, *Journal of Physical Oceanography, 6(3),* pp. 379–389, 1976.

Semtner, A. J., Jr., A numerical study of sea ice and ocean circulation in the Arctic, *Journal of Physical Oceanography, 17,* pp. 1077–1099, 1987.

Serreze, M. C. and M. C. Rehder, June cloud cover over the Arctic Ocean, *Geophysical Research Letters, 17(12),* pp. 2397–2400, 1990.

Serreze, M. C., R. G. Barry, and A. S. McLaren, Seasonal variations in the sea ice motion and effects on sea ice concentration in the Canada Basin, *Journal of Geophysical Research, 94,* pp. 10,955–10,970, 1989.

Serreze, M. C., J. A. Maslanik, R. H. Preller, and R. G. Barry, Sea ice concentrations in the Canada Basin during 1988: Comparisons with other years and evidence of multiple forcing mechanisms, *Journal of Geophysical Research, 95(C12),* pp. 22,253–22,268, 1990.

Shine, K. P. and A. Henderson–Sellers, The sensitivity of a thermodynamic sea ice model to changes in surface albedo parameterization, *Journal of Geophysical Research, 90(D1),* pp. 2243–2250, 1985.

Simmonds, I. and W. F. Budd, A simple parameterization of ice leads in a general circulation model and the sensitivity of climate to change in Antarctic ice concentration, *Annals of Glaciology, 14,* pp. 266–269, 1990.

Smith, D. C. and A. Bird, The interaction of an ocean eddy with an ice edge ocean jet in a marginal ice zone, *Journal of Geophysical Research, C3,* pp. 4675–4689, 1991.

Smith, D. C., A. Bird, and P. Budgell, A numerical study of mesoscale ocean eddy interaction with marginal ice zone, *Journal of Geophysical Research, 93,* pp. 12,461–12,473, 1988.

Smith, S. D., R. D. Muench, and C. H. Pease, Polynyas and leads: An overview of physical processes and environment, *Journal of Geophysical Research, 95(C6),* pp. 9461–9479, 1990.

Steffen, K. and J. A. Maslanik, Comparison of Nimbus 7 Scanning Multichannel Microwave Radiometer radiance and derived sea ice concentrations with Landsat imagery for the North Water area of Baffin Bay, *Journal of Geophysical Research, 93(C9)*, pp. 10,769–10,781, 1988.

Steffen, K. and A. J. Schweiger, A multisensor approach to sea ice classification for the validation of DMSP-SSM/I passive microwave derived sea ice products, *Photogrammetric Engineering and Remote Sensing, 56(1)*, pp. 75–82, 1990.

Stossel, A., P. Lemke, and B. Owens, *Coupled Sea Ice Mixed Layer Simulations for the Southern Ocean*, 29 pp., Report Number 30, Max-Planck-Institut fuer Meteorologie, Hamburg, Germany, 1989.

Stowe, L. L., C. G. Wellemeyer, T. F. Eck, H. Y. M. Yeh, and Nimbus-7 Cloud Data Processing Team, Nimbus-7 global cloud climatology. Part I: Algorithms and validations, *Journal of Climate, 1*, pp. 445–470, 1988.

Susskind, J. and D. Reuter, Retrieval of sea surface temperatures from HIRS2/MSU, *Journal of Geophysical Research, 90(C6)*, pp. 11,602–11,608, 1985.

Swift, C. T. and D. J. Cavalieri, Passive microwave remote sensing for sea ice research, *Eos, 66(49)*, pp. 1210–1212, 1985.

Tang, C. L., A two-dimensional thermodynamic model for sea ice advance and retreat in the Newfoundland marginal ice zone, *Journal of Geophysical Research, C3*, pp. 4723–4737, 1991.

Thomas, D. R. and D. A. Rothrock, Blending sequential scanning multichannel microwave radiometer and buoy data into a sea ice model, *Journal of Geophysical Research, 94*, pp. 10,907–10,920, 1989.

Thorndike, D. S., A naive zero-dimensional sea ice model, *Journal of Geophysical Research, 93*, pp. 5095–5099, 1988.

Tian, L. and J. A. Curry, Cloud overlap statistics, *Journal of Geophysical Research, 94(D7)*, pp. 9925–9935, 1989.

Tucker, W. B., *Application of a Numerical Sea Ice Model to the East Greenland Area*, CRREL Report 82-16, 40 pp., USA Cold Regions Research and Engineering Laboratory, Hanover, New Hampshire, 1982.

Turner, J. and D. E. Warren, Cloud track winds in the polar regions from sequences of AVHRR images, *International Journal of Remote Sensing, 10(4 and 5)*, pp. 695–703, 1989.

Untersteiner, N., On the mass and heat budget of the Arctic sea ice, *Archiv fuer Meteorologie, Geophysik, und Bioklimatologie, A(12)*, pp. 151–182, 1961.

Walsh, J. E. and W. L. Chapman, Short term climate variability of the Arctic, *Journal of Climate, 3*, pp. 237–250, 1991.

Walsh, J. E., W. D. Hibler, III, and B. Ross, Numerical simulation of northern hemisphere sea ice variability: 1951–1980, *Journal of Geophysical Research, 90*, pp. 4847–4865, 1985.

Walsh, J. E. and H. J. Zwally, Multiyear sea ice in the Arctic: Model- and satellite-derived, *Journal of Geophysical Research, 95*, pp. 11,613–11,628, 1990.

Weller, G. and P. Schwerdtfeger, Thermal properties and heat transfer processes of low-temperature snow, in *Meteorological Studies at Plateau Station, Antarctica*, edited by J. A. Businger, pp. 27–34, Antarctic Research Series 25, American Geophysical Union, Washington, DC, 1977.

Wohl, G., Sea ice edge forecast verification program for the Bering Sea, *National Weather Digest, 16(4)*, pp. 6–12, 1991.

Wood, R. G. and L. A. Mysak, A simple ice–ocean model for the Greenland Sea, *Journal of Physical Oceanography, 19(12)*, pp. 1865–1880, 1989.

Zwally, H. J. and J. E. Walsh, Comparison of observed and modeled ice motion in the Arctic Ocean, *Annals of Glaciology, 9*, pp. 136–144, 1987.

Zwally, H. J., J. C. Comiso, C. L. Parkinson, W. J. Campbell, F. D. Carsey, and P. Gloersen, *Antarctic Sea Ice, 1973–1976: Satellite Passive-Microwave Observations*, NASA SP-459, 206 pp., National Aeronautics and Space Administration, Washington, DC, 1983.

Chapter 23. Ice Modeling and Data Assimilation With the Kalman Smoother

D. Andrew Rothrock and Donald R. Thomas

Polar Science Center, Applied Physics Laboratory, University of Washington, 1013 NE 40th Street, Seattle, Washington 98105

The time series of sea ice passive microwave satellite data contains more information than is evident from the instantaneous data. One way to use this information is to assimilate the data into a sea ice model with a Kalman filter/smoother and make an optimal estimate of ice concentration and ice type. The Kalman algorithm smooths the observations to the modeled ice state rather than to some statistical behavior, as in other optimal estimation procedures. The estimated ice state is a compromise between our ability to interpret the passive microwave data and our understanding of the physical processes controlling concentration changes. The relative weights of the observations and the model prediction are determined by the relative errors in the measurement algorithm (the relation between the microwave data and ice concentration) and in the physical model. As part of the Kalman procedure, one computes the estimation error of the concentration by combining these measurement and physical model errors. The errors provide a tool for examining and tuning both the measurement model and the physical model.

Kalman smoothing has been applied to single points and to multiple cells in the Arctic Ocean. The physics includes changes to ice area by advection, ridging, thermodynamic growth and melt, and the aging of first-year into multiyear ice each autumn. Kalman smoother estimates of multiyear ice concentration for the Arctic Ocean vary from 0.66 in early winter to 0.59 at the onset of melt. Evaluating the individual terms in the model reveals that ridging and export through Fram Strait consume about 1.0×10^6 km²/yr and 0.8×10^6 km²/yr, and that these losses are balanced by the net thermal production.

The largest uncertainty in the application of Kalman smoothing, as well as other procedures for analyzing the satellite microwave record, is in the amount of seasonal, interannual, and regional variability of the signatures of first-year and multiyear ice.

23.1 Introduction

What are we to do with the long record of passive microwave satellite data of sea ice that we have accumulated from ESMR, SMMR, and SSM/I? The answer examined in this chapter is to analyze the temporal sequence of data with a Kalman smoother to provide an optimal estimate of the time series of ice concentration. The Kalman procedure uses knowledge of the measurements and their errors and knowledge of the physical system—in this instance, the conservation of area for each ice type—to estimate the ice concentration sequence most consistent with both the observational record and the physics. Kalman filtering uses the physical model to predict how the present ice state changes over a time step, compares the model prediction to the new observation, and weights the model prediction and the observation in inverse proportion to their respective error variances to make a filtered estimate of the true state. Hence the filtered estimate at any time is influenced by all past observations. Kalman smoothing is equivalent to filtering forward and backward through the data record and then optimally combining the two estimates. So the smoothed estimate at a point in time contains information from future as well as past data.

Why would one seek to interpret these data through the Kalman process, assimilating them into and compromising them with a physical model? There are alternative approaches. For example, one can simply use an algorithm to invert the passive microwave brightness temperatures at each time and position to concentrations of open water, first-year, and multiyear ice. Indeed, this is valuable and has been done (see other chapters in this volume). It is, in fact, the most straightforward approach to interpreting the data and is the first step in the usual paradigm in which one compares these observed concentration fields taken as "truth" with outputs of the same variable from some physical model and often "tunes" the model parameters to bring the model results closer to the observations. One has in the end two data sets: the modeled data and the observed data.

What if the concentration observations are themselves flawed—biased or noisy, or perhaps seasonally inaccurate? One is then uncertain how to use them to make the model better represent nature and may be unwilling to accept them as truer to nature than the modeled data. In this situation, the Kalman procedure offers a different approach. With it, one combines the observed data and the modeled data to compute a single data set that is the optimal estimate of the true values of the variable.

The notion of the true values of the variables, or true state, derives from statistical inference: The view is that one never knows the true state but has observations from which to estimate it. The Kalman filter/smoother output is

Microwave Remote Sensing of Sea Ice
Geophysical Monograph 68
©1992 American Geophysical Union

just such an estimate of the ice-concentration state at each moment. There are more familiar procedures used to estimate the state from noisy observations. For instance, in standard optimal interpolation one estimates the value of a variable at some point by assigning weights to data observed at other nearby points; one takes into account only the spatial or temporal statistics (the autocorrelation function) of the observed data in assigning the weights. The Kalman procedure is very much like optimal interpolation, except that instead of smoothing observations to some statistical behavior determined by an autocorrelation function, one smooths the observations to a physical behavior specified by a physical model.

Another motivation for Kalman filtering is that one can obtain estimates of variables for which there are no observations. For example, suppose that one has a mixture of open water, first-year and multiyear ice, and only single-channel ESMR data; in general the mixture fractions cannot be estimated. Even so, Kalman filtering can provide an estimate of, say, multiyear ice abundance, because it is related through the physical model and the temporal data record to a quantity that is observed—namely, all the ice at the end of summer [Thorndike, 1988]. This property is called "observability" and is discussed here in Section 23.2.7. By extension, the ice thickness distribution—or at least the thin ice—is observable if it can be related through the model to changes in observable quantities, such as first-year ice concentration. Not only does one obtain the desired estimate of the dependent variables, but also one can evaluate the terms in the physical model that govern these variables. This is an extremely good bargain: One begins by seeking optimal estimates of certain dependent variables and gets in addition estimates of the controlling physical processes.

Suppose one has a second observational data record, say, concentrations estimated from synthetic-aperture radar data or from visible or thermal satellite data. The Kalman procedure provides for the assimilation of more than one data source of the dependent variable. It also provides for the inclusion of kinematic data from buoys or ice tracking data from satellite images (Chapter 18) to specify the advective and deformational processes in the model.

We are aware of five papers that investigate or utilize the Kalman optimal estimation procedure for ice modeling: Thorndike [1988], Thomas and Rothrock [1989], Rothrock and Thomas [1990a, b], and Thomas and Rothrock [1992]. Our papers have focused on the formulation of the physical model for ice area conservation, the ice concentration measurements, and diagnosis of the Kalman smoother output. In this chapter, we pay more attention to a qualitative description of the Kalman procedure and the relevance of the three error terms that enter the formulation. A valuable resource on the Kalman procedure is Gelb [1974]. The topic of optimal estimation is vast and its applications in meteorology and oceanography are well established; see for example, Anderson and Willebrand [1989]. An equally broad application in studies of sea ice and ocean–atmosphere–ice interaction is foreseen.

In the next section of this chapter we review the fundamental concepts of the Kalman process; in Section 23.3 we consider some of its sea-ice applications; in Section 23.4 we close with a look to the future.

23.2 CONCEPTS

23.2.1 State Vector and State Space

Estimation theory is most conveniently applied with vector and matrix methods. Since passive microwave sensors discriminate among open water, first-year ice, and multiyear ice, it is natural to consider as dependent variables the concentrations of each of these surface types (X_{OW}, X_{FY}, X_{MY}). These three dependent variables at all p points or cells are stacked into a single vector X of length $3p$ that defines the state of the system at any instant.

Sea-ice applications have dealt with the discrete case, for which X_k is the state at time t_k, or time step k. A natural way to display the solutions is as time histories of one or several elements of X versus t. Because the concentrations of the three surface types sum to unity, there are two degrees of freedom at each point, and the vector to be found numerically has length $2p$. A second convenient way to think about the solution is as a trajectory in this $2p$-dimensional state space, or in some subspace.

23.2.2 The Measurement Model and Physical Model

The satellite data provide m observations at each of the p points or cells; these are stacked into a single measurement vector Z of length mp. The measurement process relates these mp elements of the observations Z to the $2p$ elements of the state vector X through the linear equation

$$Z_k = H_k X_k + W_k, \quad E(W_k W_j^T) = R_k \delta_{kj} \qquad (1)$$

where W_k is a random additive measurement error. Without the noise term W (discussed below), Equation (1) is just the familiar equation relating satellite brightness temperatures Z to ice type concentrations X. When H is invertible, Equation (1) is an algorithm for X in terms of Z. In summer, Equation (1) is not invertible because first-year and multiyear ice are not distinct. The Kalman filter uses Equation (1) without inverting it. We can take the brightness temperatures as our measurements Z, in which case H is just the matrix of brightness temperatures of the pure types contained in the X vector (as in all but one paper on the topic); or, we can take concentration estimates from some algorithm as our "measurements," in which case H is just the identity matrix [Thomas and Rothrock 1992]. In the sea ice applications described, it is assumed that the covariance matrix for W is diagonal (the errors in distinct elements of the Z vector are uncorrelated with each other), and that each element is zero mean, Gaussian, with known covariance R, and white (uncorrelated in time). We refer often to R, which is the (diagonal) covariance matrix of the measurement error, and has units of Z^2. Its elements must be specified from knowledge of the accuracy of the measurements; values of R that have been used in applications are given in Section 23.3.5.

The present state is assumed to be linearly related to the previous state X_{k-1} through a transition matrix Φ and an inhomogeneous term S:

$$X_k = \Phi_{k-1}X_{k-1} + S_{k-1} + V_{k-1}, E(V_k V_j^T) = Q_k \delta_{kj} \quad (2)$$

The matrix Φ describes known physical processes. The term S is not present when the formulation includes all three surface types, but when one surface type is eliminated (as discussed above), inhomogeneous terms arise. V is a random error and is assumed to be uncorrelated among elements of X, and zero mean, white, and Gaussian with known covariance Q. We interpret V as an error due to all the unmodeled or poorly modeled physics in Φ. Although V is conceptually part of the physical model, it is not specified, but estimated as part of the solution (see Section 23.3.2). We must, however, specify its covariance Q, as discussed in Section 23.3.5. Doing so is less precise than specifying the measurement error statistics R, because we have very limited ways to evaluate a model.

Without the noise term, Equation (2) is just a discrete form of the more familiar inhomogeneous, linear differential equation for the rate of change of concentrations

$$dX/dt = (\Phi - I) X + S$$

The physical processes include advection, ridging, divergence, thermodynamic growth and melt, and the aging of first-year into multiyear ice. We will not review their definitions here, but suggest that the reader find them in individual papers on the topic. It is useful, though, to imagine how these processes act on the trajectory of the solution. Consider the X_{FY}, X_{MY} plane of state space describing the state in a single cell or point. The arrows in Figure 23-1 show how each process moves the solution from the present state. Recall that $X_{OW} = 1 - X_{FY} - X_{MY}$, so the amount of open water can be read from the diagonal contours. Growth converts open water into first-year ice; melting of first-year ice does the opposite. Melting of multiyear ice reduces its area, producing an equal amount of open water. Divergence, or export, removes first-year and multiyear ice in proportion to their present concentrations, moving the state toward the origin. Convergence does the opposite. Aging converts first-year into multiyear ice. The annual trajectory for the entire Arctic Ocean, then, should roughly correspond to the solid curve in Figure 23-1. During winter, some export of ice and freezing of new first-year ice should slowly decrease the multiyear concentration (from state 1 to 2) near the no-open-water contour. A fine resolution of this process might show slight movement of the state toward the origin as some open water is formed, followed by refreezing of the water to ice, moving the state back to the no-open-water contour. During summer melt, mostly first-year but also some multiyear ice melts (from

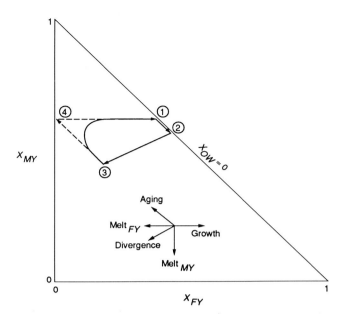

Fig. 23-1. State space for sea ice, showing the processes moving the state and a conceptual annual trajectory (solid curve). The circled numbers denote the state times of the year: (1) end of freezeup, (2) onset of melt, and (3) end of melt. The state at (4) denotes the end of summer, with all ice aging to appear as multiyear ice.

state 2 to 3). During freeze-up at the end of summer, all first-year ice should age into multiyear ice (from state 3 to 4) and open water freezes over to become first-year ice (from state 4 to 1). In practice, these last two steps are more or less simultaneous, but we expect the early winter state at 1 to be defined by the dashed lines in the figure. This is equivalent to saying that all ice remaining at the end of summer is multiyear ice the following winter.

23.2.3 Discrete Kalman Predictor, Filter, and Smoother Equations

The Kalman algorithm provides an estimate $\hat{X}$ of the true state X. We denote by $\hat{X}_{k|m}$ the estimate of X at time t_k given measurements $(Z_1, ..., Z_m)$ through time t_m. One strength of the Kalman procedure is that it tells us how good this estimate is. It does so by providing an estimate of the covariance of the estimation error (the discrepancy between our estimate $\hat{X}$ and the true X), defined as $P_{k|m} = E(\hat{X}_{k|m} - X_k)(\hat{X}_{k|m} - X_k)^T$, but estimated (since we do not know the true state) by other equations given below.

One must assume initial values $\hat{X}_{0|0}$ and $P_{0|0}$ for the state and the covariance of the estimation error. A reasonable initial condition for X is the initial measurement. In our application, the estimation error of the filtered total ice concentration is reduced when the next measurement is encountered, but remains large for ice type until after the first autumn's aging occurs; this is the time when summer's total ice is converted to the next winter's multiyear ice. After this, the estimation error remains fairly constant from year to year, no longer under the influence of the initial conditions. (One can run a preliminary Kalman smoother calculation just to acquire better initial conditions.)

For $k = 1$ to N (the duration of our entire record for Z), we solve the following five equations:

prediction:

$$\hat{X}_{k\,|k-1} = \Phi_{k-1}\hat{X}_{k-1\,|k-1} + S_{k-1} \qquad (3)$$

$$P_{k\,|k-1} = \Phi_{k-1}P_{k-1\,|k-1}\Phi_{k-1}^{T} + Q_{k-1} \qquad (4)$$

filtering:

$$K_k = P_{k\,|k-1}H_k^{T}(H_kP_{k\,|k-1}H_k^{T} + R_k)^{-1} \qquad (5)$$

$$\hat{X}_{k\,|k} = \hat{X}_{k\,|k-1} + K_k(Z_k - H_k\hat{X}_{k\,|k-1}) \qquad (6)$$

$$P_{k\,|k} = (I - K_kH_k)P_{k\,|k-1} \qquad (7)$$

Note that the filtered estimate is found in two steps. The first is a prediction, for which the estimate of X at time step k depends only on the model and the previous concentration estimate, which in turn depends on data up to time step $k-1$. In the second step, Equations (5) to (7), the filtered estimate $\hat{X}_{k\,|m}$ uses the prediction at time step k, but corrects it in proportion to the difference $(Z_k - H_k X_{k\,|k-1})$ between the actual observation and the predicted observation. These differences are called the "innovations"; their whiteness is discussed below. The sequence of steps by which the filtered estimates are made is pictured in state space in Figure 23-2.

The role of the gain matrix K in Equation (6) is to modify the predicted state $X_{k\,|k-1}$ in proportion to the accuracy of the new measurements at time step k. The gain measures the model error relative to the measurement error. Substituting Equation (4) into Equation (5), one has an expression for the gain K that has a term involving Q in the numerator and terms with Q and with R in the denominator. Hence, if the measurements are of poor quality and R is large compared with Q, the gain is small, and the predicted value is weakly affected by the measurements. Then the small arrows in Figure 23-2 showing the movement of the dots toward the open circles are very short. If the measurements are good and R is relatively small, the gain is then close to unity and the predicted value is corrected strongly toward the measurement. Then the small arrows in Figure 23-2 are longer, almost reaching the circles. This balance, contained in K, between model predictions and new measurements is optimal in the sense that it minimizes the scalar sum of diagonal elements of the estimation error covariance matrix $P_{k\,|k}$ [Gelb, 1974].

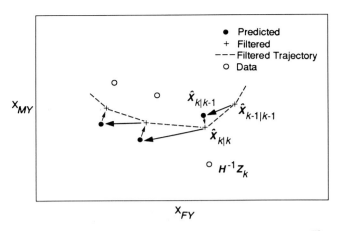

Fig. 23-2. State space showing the evolution of the solution. The smaller arrows show how much the predicted points (dots) are modified in the direction of the data (open circles) to obtain the filtered solution (+). The size of the gain matrix K determines the fraction of the distance the "+" is placed from the dot towards the circle. During summer when the data do not resolve ice type, each data point appears as a line $(X_{FY} + X_{MY} = \text{constant})$ in state space, and the filter moves the estimate towards that line.

Having stepped through the data in the forward sense, one then smooths the estimate by stepping backward through the sequence. For $k = N - 1$ to 1, we solve the following equations:

smoothing:

$$G_k = P_{k\,|k}\Phi_k^{T}P_{k+1\,|k}^{-1} \qquad (8)$$

$$\hat{X}_{k\,|N} = \hat{X}_{k\,|k} + G_k(\hat{X}_{k+1\,|N} - \hat{X}_{k+1\,|k}) \qquad (9)$$

$$P_{k\,|N} = P_{k\,|k} + G_k(P_{k+1\,|N} - P_{k+1\,|k})G_k^{T} \qquad (10)$$

The smoothed estimate $\hat{X}_{k\,|N}$ is found from the filtered estimate at time step k corrected by a term proportional to the difference between the smoothed and predicted estimates at time step $k + 1$. The result of this procedure is the same as filtering first forward through the data and then backward through the data and weighting the two solutions optimally to minimize the squared error.

23.2.4 The Relationship of P to Q and R

The Kalman process provides an estimate of the error in the solution. However, the estimation error covariance P depends on the model and measurement errors whose covariances Q and R are not well known but need to be specified a priori (as discussed in Section 23.3.5). Some of this dependence can be seen in the definitions of P. On the right-hand side of Equation (4) we see that the first term propagates the filter error covariance from Equation (7) at time $k-1$ to time k, and that the second term augments it by the model error covariance $\mathbf{Q}$. Similarly, in Equation (7), we see that because $(\mathbf{I}-\mathbf{KH})$ is positive and less than $\mathbf{I}$, the error covariance of the filtered estimate is less than that of the predicted estimate; this is expected because we add new information to obtain the filtered estimate. When one begins the Kalman process, one often has a fairly large estimation error (poorly known initial condition $\mathbf{X}_{0|0}$) that converges in a sawtooth fashion to some steady behavior. The sawtooth pattern reflects alternating $\mathbf{P}_{k|k-1}$ (larger prediction error covariance) and $\mathbf{P}_{k|k}$ (smaller filter error covariance) as k goes from 1 to N. What is the steady-state error covariance in filtering? It should not be larger than the measurement error covariance converted to units of X^2 (i.e., $\mathbf{HRH}^T$), but if the physical model is good, it can be lower. Combining Equations (4), (5), and (7) and ignoring the value of k, we can eliminate the prediction error covariance and solve for the filter error covariance $\mathbf{P}_{k|k} = \mathbf{P}$. We also consider the simple case in which $\mathbf{Q}$, $\mathbf{R}$, and $\mathbf{X}$ are scalars (a single variable), and the measurement matrix $\mathbf{H}$ is the identity and the transition matrix Φ is the identity (i.e., the state is a random walk and the measurement process merely adds a random error). Then we have $P = (1 - P/R)(P + Q)$ from which we solve for $1/P$

$$\frac{1}{P} = \frac{1}{2R} + \sqrt{\frac{1}{4R^2} + \frac{1}{QR}} \qquad (11)$$

When the model is weak compared with the measurements ($Q \gg R$), we see that $P \cong R$, and the estimate is only as good as the measurements. If the model is relatively strong ($Q \ll R$), then $P \cong (Q/R)^{1/2}$ or $R(Q/R)^{1/2}$, which is small compared with R; that is, the estimation error variance can be considerably less than R. One way to think of this is that the estimation process uses the physical model as a sort of filter window to integrate over some number n of independent measurements so that the error variance is reduced to $R\sqrt{n}$. This is the strength of Kalman filtering.

In complete analogy with the forward filter, one conceives of a backward filtering process for which the error covariance $\mathbf{P}_{\text{backward}}$ behaves as shown in Figure 23-3. The $\mathbf{P}_{\text{backward}}$ should level off at about the same level as that of the forward filter. The smoothed solution, being an optimal combination of the forward and the backward filtered solutions, has an error covariance given by

$$\mathbf{P}_{k|N}^{-1} = \mathbf{P}_{k|k}^{-1} + \mathbf{P}_{k|k}^{-1}(\text{backward})$$

that is less than the filter error covariance by a factor of about 1/2.

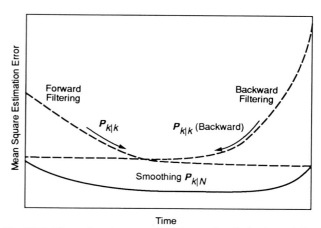

Fig. 23-3. The estimation error for a forward and a backward filter (dashed), and for the smoother (solid). The backward filter assumes an infinite estimation error for the first step [Gelb, 1974].

23.2.5 Consistency Between Model and Data

Thorndike [1988] has given a test for the consistency of the physical model with the observations. Compute the covariance of the innovations $(\mathbf{Z}_k - \hat{\mathbf{Z}}_k)$ as

$$\Gamma = \frac{1}{N} \sum_{k-1}^{N} (\mathbf{Z}_k - \hat{\mathbf{Z}}_k)(\mathbf{Z}_k - \hat{\mathbf{Z}}_k)^T \qquad (12)$$

where $\hat{\mathbf{Z}}_k = \mathbf{H}_k \hat{\mathbf{X}}_{k|k-1}$. It can also be shown that the covariance of the innovations at time k is

$$E(\mathbf{Z}_k - \hat{\mathbf{Z}}_k)(\mathbf{Z}_k - \hat{\mathbf{Z}}_k)^T = \mathbf{H}_k \Phi_{k-1} \mathbf{P}_{k-1|k-1} \Phi_{k-1}^T \mathbf{H}_k^T + \mathbf{H}_k \mathbf{Q}_k \mathbf{H}_k^T + \mathbf{R}_k \qquad (13)$$

(In Equation (13), typographical errors in Thorndike [1988] and in Thomas and Rothrock [1989] are corrected.) Let $\mathbf{C}$ be the average over k of the right-hand side of Equation (13). Then $\mathbf{C}$ is a function of all the uncertainties in the innovations (the measurement error covariance $\mathbf{R}$, the model error covariance $\mathbf{Q}$, and the estimation error covariance $\mathbf{P}$ for the previous state estimate), and the physical and measurement models Φ and $\mathbf{H}$, but it does not depend on the observations themselves. If the physical model Φ and error covariances $\mathbf{R}$ and $\mathbf{Q}$ are compatible with the measurements $\mathbf{Z}$, then $\Gamma \leq \mathbf{C}$. It is generally assumed that we know $\mathbf{R}$ better than we know $\mathbf{Q}$; in that case one can "tune" $\mathbf{Q}$ so that $\Gamma \cong \mathbf{C}$. (We actually compare trace $\mathbf{C}$ and trace Γ.) However, it should be noted that satisfying this compatibility criterion does not ensure that the model Φ is correct. If the measurement error covariance $\mathbf{R}$ is large, as we believe it is during summer and for the determination of ice type, then an incorrect model may be compatible with the measurements. Satisfaction of this criterion is a necessary but not sufficient condition for specifying Φ and $\mathbf{Q}$.

23.2.6 Adaptive Filtering

Another approach to using the filter results to improve knowledge of the model Φ, S, and H or the errors Q and R is called adaptive filtering. The whiteness of the innovations can be used to estimate Q or R when Φ, S, and H are known, but tuning the model to improve the whiteness of the innovations can yield nonunique solutions for Φ [Gelb, 1974]. Thomas and Rothrock [1989] and Thomas and Rothrock [1992] substitute the smoothed solution back into Equation (2) and examine the bias in the residual

$$\hat{V}_k = \hat{X}_k \mid N - \Phi_{k-1}\hat{X}_{k-1} \mid N \tag{14}$$

A sensitivity study [Rothrock and Thomas, 1990b] shows that the smoother estimates and the biases in $\hat{V}$ are only weakly sensitive to Q, R, and the parameters in the model (such as melt rates). The biases are strongly sensitive to the pure type signatures in the measurement model H. This is why the measurement model in Thomas and Rothrock [1992] is so different from earlier models; we felt the bias in our earlier pure-type signatures introduced more error into the system than any weakness in our physical model or uncertainty in its parameters. (See Section 23.3.4.)

23.2.7 Observability

One advantage of Kalman filtering and smoothing is that they allow estimation of variables not directly observed by passive microwave sensors. For instance, Thorndike [1988] used single-channel ESMR observations to estimate concentrations of three surface types. Another instance is the estimation of surface-type concentrations for the region north of 84°, which is not observed by SMMR [Rothrock and Thomas, 1990b]. Another possibility is to estimate concentrations of first-year ice of different ages (e.g., less than one month old and between one and two months old). That information, combined with a thermal model of growth in ice thickness, would provide an estimate of the ice thickness distribution.

It is the sequence of observations combined with the physical model that provides the additional information about the variables not directly observed. Suppose we have a sequence of measurements Z_0, Z_1, ..., Z_N for some region. Ignoring the inhomogeneous term S and the error term V, Equation (2) gives us

$$X_1 = \Phi_0 X_0$$

$$X_2 = \Phi_1 \Phi_0 X_0$$

$$\cdots \tag{15}$$

and from Equation (1) we have

$$Z_0 = H_0 X_0$$

$$Z_1 = H_1 X_1 = H_1 \Phi_0 X_0$$

$$\cdots \tag{16}$$

After $N + 1$ observations we have the system of equations

$$\begin{bmatrix} Z_0 \\ Z_1 \\ \cdots \\ Z_N \end{bmatrix} = \begin{bmatrix} H_0 \\ H_1\Phi_0 \\ \cdots \\ H_N\Phi_{N-1} \cdots \Phi_0 \end{bmatrix} X_0 = \Lambda X_0 \tag{17}$$

which can be solved for a unique X_0—provided Λ has rank equal to the number of variables in X_0. This is known as the observability condition. Thorndike [1988] gives an example showing how a sequence of two winter ESMR observations allows determination of open water, first-year ice, and multiyear ice concentrations, whereas a sequence of summer observations does not. For multichannel SMMR data, which is effectively two-dimensional [Rothrock et al., 1988], it is possible to determine no more than three surface-type concentrations—for example, open water, first-year ice, and multiyear ice (making use of the fact that the sum of the three concentrations must equal one). It is possible to determine the concentration of second-year ice with a data sequence that spans two winters and the intervening summer, if the second-year ice signature is known (even if it does not differ from the signature of older multiyear ice). If the second-year ice signature is unknown, it is theoretically possible to determine its concentration, and thus its signature, with a data sequence spanning three winters and two summers.

23.3 Results

23.3.1 Annual Cycle

In Section 23.2.2 and Figure 23-1, we discussed the idealized annual trajectory of the ice state. Some calculated state trajectories for small regions are shown in Figure 23-4. The results from Thorndike [1988] in Figure 23-4(a) are the equilibrium cycles calculated by his physical model, adjusting melt rate and divergence to minimize the squared differences between the observed and the modeled brightness temperatures for four years of ESMR data for each of two Eulerian regions. The optimal fit is for a summer melt rate (the same for first-year and multiyear ice) of 0.05/month and a divergence of 0.03/month in the central Arctic (80° N, 180° E) with corresponding values of 0.4/month and 0.06/month in the Beaufort Sea (75° N, 135° E).

Another regional result is for a Lagrangian region tracked by a drifting buoy. Figure 23-4(b) shows a two-year trajectory of the Kalman smoother estimate by Thomas and Rothrock [1989]. In this case a time history of neighboring buoy positions was used to specify divergence in the model, and a summer melt rate of 0.06/month was used for first-

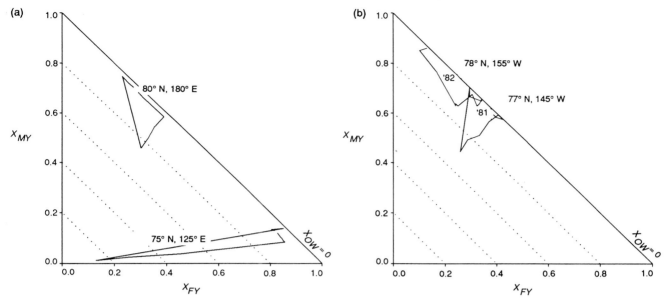

Fig. 23-4. Solution time histories as trajectories in state space: (a) annual cycles for two regions derived from ESMR data by Thorndike [1988], and (b) a 2-year cycle for a Lagrangian region (in the Beaufort Sea) tracked by a buoy, as estimated from SMMR data by Thomas and Rothrock [1989].

year ice; no multiyear melt was assumed to be zero. The microwave data show about half as much open water and melt ponds in the second summer (0.15) as in the first (0.30). The deformation data show about 20% divergence the first summer and no net divergence the second summer. This suggests that melting or ponding was heavier during the second summer. An attempt was made to include melt ponds as a fourth surface type, but without accurate information on either the signature of frozen melt ponds or the rate of pond formation, melt pond areas remain unobservable.

All four annual trajectories in Figure 23-4 obey the rule of thumb that early winter multiyear ice concentration equals the previous summer's minimum ice concentration. Thorndike's results (Figure 23-4(a)) show about 0.02 to 0.06 open water during the winter months; this is primarily due to the microwave data indicating some winter open water. Because of spatial variability of ice signatures, this amount of open water could easily be the result of a local signature bias, not true open water. Nevertheless, Thorndike adjusts the divergence, ending with quite high rates, to fit the model to the data. He also uses a growth rate of 0.5 per day, ensuring that some of the open water created by the divergence remains open. The results shown in Figure 23-4(b) are calculated by using buoy-derived velocity fields to estimate the divergences, which are considerably smaller than the constant values used by Thorndike. These results are computed by using time steps of a month, so a growth rate of 1 per month is used.

More recently, Kalman smoothing has been applied to determining ice concentrations over extended areas, in particular, the whole Arctic Ocean [Thomas and Rothrock,

1992]. Figure 23-5 shows the Arctic Ocean divided into seven computational cells. Advection of ice between cells is included in the physical model and is estimated from buoy motions. The vectors in Figure 23-5 represent the annual average area advection between cells. It is assumed that the only significant exchange of ice between the Arctic Ocean and surrounding seas takes place through Fram Strait.

Seven years (1979 to 1985) of Kalman smoother results for the Arctic Ocean are shown as an average annual cycle in Figure 23-6(a). The average winter ice concentration is greater than 0.99. The average minimum summer ice cover is about 0.70. At the start of winter, the multiyear ice concentration is about 0.66, decreasing to about 0.59 at the end of winter. The 0.04 difference between the minimum summer ice cover and the early winter multiyear ice concentration can be explained by freeze-up occurring at different times in the seven cells; after freeze-up in some cells, ice is still melting in other cells and continues to be exported out Fram Strait. This annual cycle differs significantly from the one calculated using the NASA algorithm (shown in Figure 23-6(b)). The NASA algorithm estimates, on average, a winter ice concentration of about 0.98 (less in spring and fall), a minimum summer ice cover of about 0.69 and an early winter multiyear ice concentration of about 0.37. Because of unaccounted variability in summer ice signatures, the algorithm's estimate of summer ice type is not very reliable; generally, only the total ice concentration (first-year plus multiyear) is used during summer. Processing the NASA algorithm concentrations with the Kalman smoother results in an annual cycle lying between those shown in Figure 23-6: roughly 0.5 summer ice concentration

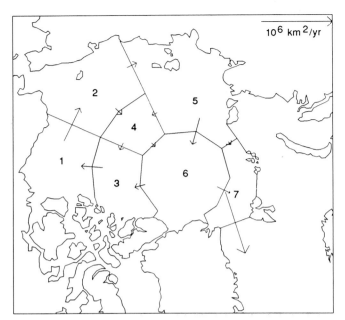

Fig. 23-5. The Arctic Ocean divided into seven computational cells. The average advection of area from cell to cell is shown in units of 10^6 km^2/yr (scale in upper right corner). The combined area of the seven cells is 6.85×10^6 km^2.

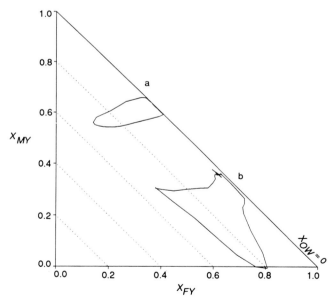

Fig. 23-6. The average annual solution (1979 to 1985) shown as a cycle in state space. The Kalman smoother solution is shown in (a) and the solution using the NASA algorithm and the pure type signatures given by Cavalieri et al. [1984] in (b).

and 0.5 winter multiyear ice. Most of the difference between the two solutions in Figure 23-6 is due to using different pure type signatures, as discussed in Section 23.3.4.

23.3.2 Strengths of Physical Processes

After obtaining the smoothed estimates of the surface-type concentrations, the physical model is used to decompose the results into area changes due to specific processes [Thomas and Rothrock, 1992]. Cumulative area changes (for the Arctic Ocean) resulting from each process are shown in Figure 23-7. The cumulative numbers are useful for evaluating the magnitudes of processes over seasons or years and biases in the error term. What Figure 23-7 does not show well is the size of individual events; with the exception of aging and growth at freeze-up, these are small. For the Arctic Ocean total, advection is just the ice exported through Fram Strait. Approximately equal amounts of first-year and multiyear ice are exported—about 0.43×10^6 km^2/yr of each. Ridging is roughly equal to advection (export) as a sink of total ice area—about 1.0×10^6 km^2/yr. Aging is a significant sink of first-year ice, about equal to ridging, and is the only source of multiyear ice. Thermal processes consist of summer melt and growth of new first-year ice. For first-year ice, growth is the only ice source. The last row in Figure 23-7, labeled error, is calculated as $\hat{V}_k = \hat{X}_{k|N} - \Phi_{k-1}\hat{X}_{k-1|N}$ and is our best estimate of the model error $V_k = X_k - \Phi_{k-1}X_{k-1}$ in Equation (2), which is assumed to be Gaussian, zero mean, and white, with covariance Q. These assumptions appear to be met for the Arctic Ocean as a whole, but the results shown in Figure 23-7 are actually weighted sums of the solution in each of the seven cells. For some of the individual cells, the error term has a small bias and is not white. This is a result of using the same pure-type signatures everywhere. Although our choice of signatures minimizes the Arctic-wide bias, regional biases cannot be eliminated without choosing different signatures for different regions.

23.3.3 Interannual Variability and Trends

Observing the year-to-year variability of the minimum and maximum ice extent and of the concentration of multiyear ice is one of the primary goals of passive microwave remote sensing of sea ice. One wants to know how much the concentrations vary and whether there is a long-term trend.

Figure 23-8(a) shows the mean annual cycle of multiyear and total ice concentration along with the range of concentrations for the years 1979 to 1985. These results are the Kalman smoother estimates, as discussed in Section 23.3.1 and Figure 23-6. The range of this basin-wide average is surprisingly small—plus or minus several percent. The larger range during freeze-up is due to variation in the timing of the onset of freeze-up and its duration. We see here the asymmetry between the onset of melt and of freeze-up. Melt begins a gradual decline of ice area, but freeze-up causes a rapid depletion of open water. Results for some

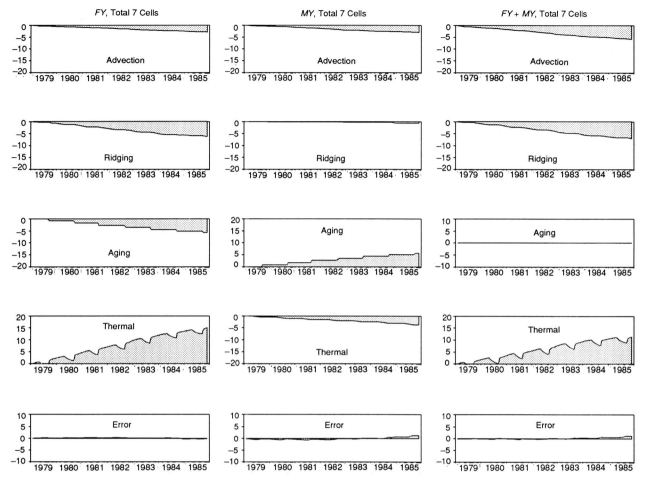

Fig. 23-7. Cumulative area changes of first-year, multiyear (center column), and total ice (right column) for the Arctic Ocean due to advection, ridging, aging, thermodynamics, and the estimated error in the physical model. Units are 10^6 km^2.

individual cells show much larger variation, particularly Cell 7 above Fram Strait.

Figure 23-8(b) shows the mean concentrations plus and minus one standard deviation of the estimation error. With just 7 years of results, all within about one standard deviation of the mean (except during the freeze-up period), we do not feel justified in making claims about interannual variability or the existence of trends.

Unfortunately, the interannual variability is probably confounded with unresolved spatial and temporal variability of pure-type signatures. The signatures obviously vary from region to region and throughout the year—for example, because of variations in summer melt pond coverage and early winter snowfall—so it seems likely that there is also variability from year to year. There are two ways we can better identify interannual variability in concentration and increase our confidence in estimates of trends. One way is to increase the number of years observed. The other is to improve our knowledge of signature variability, thus reducing the estimation error.

23.3.4 Ice-Type Signatures and Measurement Biases

Although the Kalman theory assumes that both the measurement error W and the model error V are unbiased, this is often not the case. These biases appear in the solution, and in particular in the error term $\hat{V}$. In our early Kalman calculations, for example, $\hat{V}$ was continually showing a surplus of multiyear ice that the physical model produced (by aging), but was not present according to the measurements. Because the solution is more sensitive to changes in the input measurements than to the model parameters, we have defined different pure-type signatures, attempting to reduce this bias.

The signatures used for the concentration measurements that are input for the Kalman smoother results shown here are defined [Thomas and Rothrock, 1992] by assuming that the cluster with the highest concentration of late summer ice evolves, in the absence of large divergence during freeze-up, into the cluster with the highest concentration of multiyear ice in early winter. The multiyear signature is defined as the brightness temperatures that give a cluster mean multiyear ice concentration in early winter equal to the same cluster's mean late summer ice concentration. The open water and first-year ice signatures are also redefined: Data from a wider area and from throughout the 1979 to 1985 period are included in the pure-type clusters, outliers are removed, and cluster *means* are used as the signatures. The signatures used for the NASA algorithm results (Figure 23-6(b)) are those given by Cavalieri et al. [1984] who use *extremes* of clusters of data from relatively small areas for a 5-day period. The two sets of signatures are given in Table 23-1. The new signatures still do not provide accurate estimates of summer ice type; the Kalman smoother estimate of ice type is extended through the summer by the physical model, subject to the constraint on total ice concentration imposed by the measurements.

23.3.5 The Sizes of the Measurement, Model, and Estimation Error Covariances

Having dealt with the issue of measurement biases, we turn to error covariances. The estimation error covariance for the smoothed solution is a function of error covariances in the measurements and the physical model. As we point out in Section 23.2.4 the covariance of the measurement error sets an upper limit to the estimation error covariance. The physical model can reduce the estimation error covariance, depending on the covariance of the error in the physical model.

There is some controversy regarding the magnitude of the measurement error variance. The accuracy of the microwave instrument is generally better than 1 K, which translates into concentration estimation errors of about 0.01. (We discuss errors in terms of standard deviations—square root of variance—rather than such parameters as variances, 90th or 95th percentiles, and ranges). The disagreement is over the variability of the "pure-type signatures" used in algorithms (measurement models) that interpret measured brightness temperatures as surface-type concentrations. For a variety of reasons (physical temperature, wind over open water, rain, clouds, amount and type of snow cover, melt ponds, and developmental history of the ice), the brightness temperature of open water, first-year ice, and multiyear ice vary in space and time. Commonly used measurement errors are 0.03 for open water or total ice concentration and 0.08 for ice type [Cavalieri et al., 1984]. These numbers are the standard deviations of the estimated concentrations for three regions assumed to contain just one of the surface types. Again, we take issue with the relatively small regions and the short (5-day) temporal sample from which these estimates are made.

Several studies compare ice concentrations computed from passive microwave data using the NASA algorithm and pure-type signatures with ice concentrations estimated from various other instruments (SAR, AVHRR, Landsat, aerial photos). The mean difference varies from a low of 0.02 between SMMR and Shuttle Imaging Radar B [Martin et al., 1987] to a high of 0.10 between SSM/I and AVHRR [Steffen and Schweiger, 1990]. The mean difference is an estimate of the accuracy of the concentration estimates, indicating that there is a bias in concentration estimates from one or both instruments. An estimate of the precision of the concentration estimates is the rms of the differences between sensor estimates, which range from 0.07 [Martin et al., 1987] to 0.13 [Burns et al., 1987]. While none of these numbers gives us the "true" measurement error, they suggest that it is larger than 0.03 for ice concentration.

Using the new pure-type signatures given in Table 23-1, new estimates of the precision of the NASA algorithm concentration estimates have been derived [Thomas and Rothrock, 1992]. The estimates are based on all the data from the Arctic Ocean and the Kara, Barents, Greenland, and Norwegian seas, and from throughout the winter seasons of the years 1979 to 1985. It is assumed that the three major peaks in the histograms of surface-type concentrations represent the pure types of open water, first-year, and (mostly) multiyear ice, and that the spread of the peaks is due to signature variability. Of course part of the spread, at least for multiyear ice, is actually due to concentration variability, so the precisions estimated from the peak widths are upper bounds to the true precisions. The precisions, or standard deviations of the pure-type peaks, are 0.05 for winter ice concentration, 0.08 for summer ice concentration, and 0.15 to 0.30 for ice type. The value of 0.30 for the multiyear ice concentration is, as just noted, probably partly due to concentration variability, so we use the first-year ice value, 0.15, for both ice types. We believe that these numbers are the appropriate estimation errors for use with

TABLE 23-1. Brightness temperatures used as pure-type signatures.

Reference	OW			FY			MY		
	18H	18V	37V	18H	18V	37V	18H	18V	37V
Cavalieri et al. [1984]	102.1	157.3	189.5	227.1	237.5	233.4	189.7	202.7	168.9
Thomas and Rothrock [1992]	108.1	159.6	194.3	229.6	240.6	239.5	205.5	218.5	195.6

"global" (Arctic-wide) pure-type signatures and have used them in our recent Kalman smoothing estimates of the Arctic Ocean ice balance [Thomas and Rothrock, 1992]. For regional studies using "local" pure-type signatures, the smaller errors given by Cavalieri et al. [1984] are probably more appropriate. For Cell 6, which is mostly unobserved by SMMR, we use the mean of the observations between 83°N and 84°N and double the measurement error variance. The measurement errors used in various papers are tabulated in Table 23-2. The measurements used in all the studies except Thomas and Rothrock [1992] consist of brightness temperatures or principal components of brightness temperatures. When brightness temperatures are used, the error in ice concentration depends on the uncertainty, or variation, of the pure-type signatures—the noise—relative to the distance—the signal—between these signatures in brightness temperature space. It is this noise-to-signal ratio that is given as the relative error in concentration in Table 23-2.

The error in the physical model must be estimated from what we know of the physical processes and parameter values. There are five parameters or sets of variables we must estimate or provide for the physical model: areal rates for aging first-year to multiyear ice, growth of first-year ice from open water, first-year and multiyear ice melt rates, and area fluxes across the cell boundaries (which we estimate from buoy motions). How well we know those parameters or input variables determines the estimation error for the model.

An aging rate of 1 at freezeup and 0 the rest of the year is true by definition; the only error that might occur is in specifying the time of freezeup. Experience has shown that changing the date of freezeup by one or two 10-day time steps has no significant effect on the concentration estimates. We assume that ice growth does not occur during the summer, or ice melt season, and that during the ice growth season the growth rate is 0.9. This means that if the creation of open water through divergence occurs uniformly in time, 90% of the open water created during a 10-day time step freezes over, leaving just the open water created on the 10th day. In midwinter, the growth rate is probably an underestimate—it takes less than 1 day for water to freeze over—while early or late in the ice growth season it may take longer than 1 day for open water to freeze over. During winter, the amount of open water created each time step is relatively small (it is traditionally assumed that the Arctic Ocean experiences about 1% per month of divergence) and varying the growth rate by ±0.1 has no significant effect on concentration estimates.

It is advection and summer melt that make the largest contribution to modeling errors. The velocity estimates for the buoys have an estimation error of about 0.002 m/s, but interpolation in space results in an estimation error of about 0.01 to 0.02 m/s [Thorndike and Colony, 1983]. Assuming a velocity error of 0.01 m/s perpendicular to a cell boundary and constant for 10 days, and a cell of the order of 1000 km on a side, the error in the area flux across one side of the cell is 8640 km^2, or about 0.01 of the cell area. During

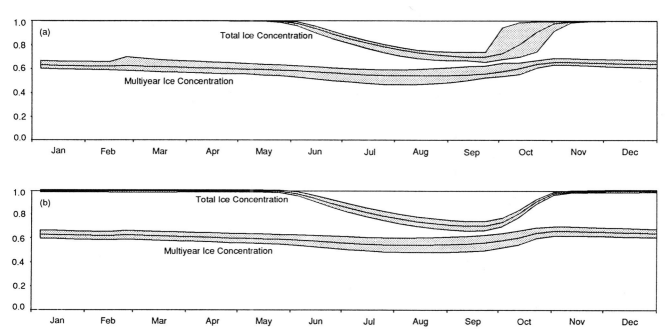

Fig. 23-8. The mean annual cycle of ice concentration: (a) the annual mean and interannual variability (1979–1985) in multiyear (lower curve) and total ice concentrations, and (b) the annual mean plus and minus one standard deviation of the estimation error [after Thomas and Rothrock, 1992].

TABLE 23-2. Standard deviations of measurement errors (in state space) used in various studies. Squares of these numbers are placed on the diagonal of R at the appropriate time step k.

Reference	Winter		Summer	
	Ice concentration	Ice type	Ice concentration	Ice type
Thorndike [1988]	0.06	0.17	0.06	—[a]
Thomas and Rothrock [1989]	0.02	0.04	0.08	—[a]
Rothrock and Thomas [1990a, b]	0.01	0.02	0.06	—[a]
Thomas and Rothrock [1992]	0.05	0.15	0.08	0.50

[a] The first-year and multiyear ice have the same summer signature, making the signal zero and the noise-to-signal ratio undefined.

winter, when the areal melt rates are assumed to be 0, the only significant error in the physical model is due to the error in advection, so we set $q = 0.01^2$ (where $Q = qI$).

Summer melt rates are more uncertain than winter growth and also have a stronger effect on the results. If we consider just the Arctic Ocean, winter melt rates of zero are a safe assumption, and contribute nothing to the modeling error. For summer, we assume that the error in the melt rates increases the error variance for the physical model to $q = 0.03^2$.

Using the above estimates of winter and summer R and Q, the Kalman smoother estimation error (which we extract from the P matrix found as part of the solution) is 0.01 for winter ice concentration, 0.04 for summer ice concentration, 0.05 for winter ice type, and 0.07 for summer ice type. Cell 6 always has a slightly larger estimation error due to the larger measurement error assumed for that cell, and the ice-type estimation error for the first winter is larger because of uncertainty in the initial conditions.

23.3.6 Comparisons With Other Estimates of Concentration

In Figure 23-6 we showed the annual average Kalman smoother estimate and the NASA algorithm estimate of the Arctic Ocean ice balance. The Kalman smoother estimate shows about 0.30 more multiyear ice, 0.05 less summer open water, and 0.02 or 0.03 less winter open water. These differences are due primarily to the different pure-type signatures (in Table 23-1) used in the two estimates.

Wittmann and Schule [1966] estimated various ice-type concentrations for the Arctic Ocean (excluding the Laptev Sea) from aerial surveys. Their winter ice concentration for the Arctic Ocean is only about 0.87; their comment is that open water "may be somewhat overestimated" and point out that submarine estimates of 0.98 winter ice concentration are more likely. Their winter (January through May) "polar pack" (assumed to be multiyear ice) concentration is about

0.7. Their August through October "polar pack" plus "thick winter" concentration is about 0.75 (thick winter ice is assumed to be the previous winter's first-year ice that did not melt during the summer). One expects a higher concentration of multiyear ice at the beginning of winter with export reducing it by about 1% per month through the winter. Including the Laptev Sea would probably lower the concentrations by a few percent, making them more comparable to our mean estimate of 0.66 in early winter and 0.59 in late winter. Spatially, their concentrations show about the same pattern as ours, with the highest multiyear concentrations (around 0.9) found in the Central Arctic and to the north of Greenland.

Carsey [1982] estimated the multiyear ice concentration in 200-km cells using ESMR data and the assumption that the summer ice in each cell becomes multiyear at freezeup. Carsey's Figure 23 gives the 4-year mean concentration of ice surviving the summer melt for each 200-km cell; summing over the Arctic Ocean, one computes an average multiyear concentration of 0.66. Parkinson et al. [1987] also use ESMR data to estimate multiyear ice concentration. This estimate, based on the minimum summer ice area for the Arctic Ocean as a whole, is from 0.54 to 0.60. Presumably this is an overestimate since no allowance is made for ice melt following the summer ice minimum, but Carsey's estimate [1982] which takes this into account is larger.

Another useful point of comparison is the estimates of ice export through Fram Strait, calculated as the product of ice concentration estimates north of the strait with the buoy-derived area flux through the strait. Vinje and Finnekasa [1986] have made independent estimates of the export for each individual year from 1976 through 1984, where ice drift speeds are estimated from buoy motions, and monthly average ice concentrations are derived from weather satellite images. Their average annual export of 1.25×10^6 km^2/yr is about 50% larger than our average annual export of 0.83×10^6 km^2/yr. Part of the difference can be explained by different choices for defining Fram Strait; our boundary lies about 150 km south of theirs, with the melt zone northwest of Spitsbergen lying in between. Vinje and Finnekasa estimate that about 0.1×10^6 km²/yr of (4-m thick) ice is melted between 81° N and 80° N, but this accounts for only about 25% of the difference in mean annual ice export. The remainder of the difference between the two estimates is probably due to different estimates of the ice velocity through Fram Strait. Vinje and Finnekasa base their velocity estimates on 45 buoys drifting across the 81st parallel; our velocity estimates are tuned to give the same average area flux through Fram Strait as Moritz [1988] calculated using a much larger set of buoy observations, 610 daily observations of ice velocity. The Kalman smoother estimates also show much more interannual variability than do Vinje and Finnekasa's results. The Kalman smoother estimate of average ice concentration in the vicinity of Fram Strait is 0.87; Vinje and Finnekasa's estimate is 0.88. Their estimate of summer multiyear ice fraction (0.85) is also near the Kalman smoother estimate (0.80 in late summer).

23.4 Conclusions and Outlook

The Kalman filtering and smoothing procedure provides a formalism for utilizing the temporal information in passive microwave data to improve our estimation of ice-type concentrations. It is a framework for incorporating independent information, such as ice motion, into the concentration estimates. It provides estimates of the changes in concentrations caused by the individual physical processes of advection, ridging, growth, melt, and aging. And, it gives an estimation error for each estimate.

Kalman smoothing has been used to estimate the first-year and multiyear ice balance of the Arctic Ocean for the years 1979 to 1985. The multiyear ice concentration, about 0.66 in early winter, is similar to other estimates of the Arctic Ocean ice cover. The estimation error is 0.01 for winter ice concentration and 0.05 for winter ice-type concentration. These are substantial reductions of the assumed measurement errors of 0.05 and 0.15. We note that the Kalman smoother estimation error corresponds to the precision of the measurements; biased measurements still lead to biased smoother results, although the bias is reduced. In fact, parameter studies [Thomas and Rothrock, 1992] show that the Kalman smoother multiyear ice concentration estimate is most sensitive by far—an order of magnitude or more—to biases in the concentration measurements, with summer melt rates a distant second. The Kalman smoother does not supersede the need for accurate (i.e., unbiased) measurements.

Analyzing all the physical processes that go into the Kalman smoothed solution provides a wealth of information beyond the concentration estimates. We see, for instance, that the export of ice from the Arctic and ridging of first-year ice consume roughly equal areas of ice (0.9×10^6 and 1.0×10^6 km²/yr) and balance the net thermal production of ice. The net error in the ice balance for the Arctic Ocean is quite small by comparison. We estimate that the annual export of ice area through Fram Strait is about one-third less and more variable than the independent estimate by Vinje and Finnekasa [1986]. We also find that the cell in the Chukchi and East Siberian seas is a "sink" of multiyear ice; more ice is advected into it than is advected out.

Kalman smoothing of passive microwave data has one overriding problem: We do not know how the signatures of first-year and multiyear ice vary throughout the year and from one region to another. The use of "global" signatures causes concentration estimates for some places and times to be biased. We have attempted to minimize the bias for the arctic as a whole, but subregions generally appear biased. There are two approaches to resolving this problem. First, one can try to reduce the model error covariance Q so that uncertainty in the data becomes less important. Better kinematic data with higher resolution, say, from SAR, will allow the model to better estimate advective and deformational changes in ice concentration. Better formulation of summer ice melt and melt pond formation will also help. Second, one can try to reduce the measurement error covariance R. The most direct improvement will come from better data on the evolution of ice signatures and of the natural variability caused by snow cover, temperature profiles, roughness, melt ponds, saltwater intrusions, growth rates, and salinity.

It should be noted that the signature issue is not unique to the Kalman procedure. Any algorithm that estimates concentrations from the microwave data is faced with the same uncertainty caused by signature variability. In fact, the Kalman procedure reduces signature-related errors in the concentration estimates, and, by making biases more apparent in the results, offers the possibility of improving our knowledge of signatures via adaptive filtering. Until these uncertainties and variabilities of ice-type signatures are better understood, we do not believe that one can discern long-term trends.

The potential of Kalman filtering and smoothing and other data assimilation schemes is great. Ice models far more rich than those used to date—richer both in dependent variables and in the formulation of the physical processes—could make good use of the information about the ice cover available from satellite data. The temporal record provides a great deal more information about the ice than an instantaneous snapshot will ever yield.

References

Anderson, D. L. and J. Willebrand, editors, *Oceanic Circulation Models: Combining Data and Dynamics*, NATO ASI Series, Kluwer Academic Publishers, Dordrecht, 1989.

Burns, B. A., D. J. Cavalieri, M. R. Keller, W. J. Campbell, T. C. Grenfell, G. A. Maykut, and P. Gloersen, Multisensor comparison of ice concentration estimates in the marginal ice zone, *Journal of Geophysical Research, 92*(C7), pp. 6843–6856, 1987.

Carsey, F. D., Arctic sea ice distribution at end of summer 1973–1976 from satellite microwave data, *Journal of Geophysical Research, 18*, pp. 5809–5835, 1982.

Cavalieri, D. J., P. Gloersen, and W. J. Campbell, Determination of sea ice parameters with the Nimbus 7 SMMR, *Journal of Geophysical Research, 89(D4)*, pp. 5355–5369, 1984.

Gelb, A., editor, *Applied Optimal Estimation*, M.I.T. Press, Cambridge, Massachusetts, 1974.

Martin, S., B. Holt, D. J. Cavalieri, and V. Squire, Shuttle Imaging Radar B (SIR-B) Weddell Sea ice observations: A comparison of SIR-B and Scanning Multichannel Microwave Radiometer ice concentrations, *Journal of Geophysical Research, 92(C7)*, pp. 7173–7179, 1987.

Moritz, R. E., *The Ice Budget of the Greenland Sea*, Technical Report 8812, 116 pp., Applied Physics Laboratory, University of Washington, Seattle, Washington, 1988.

Parkinson, C. L., J. C. Comiso, H. J. Zwally, D. J. Cavalieri, P. Gloersen, and W. J. Campbell, *Arctic sea ice, 1973–1976: Satellite passive-microwave observations*, NASA SP-489, 296 pp., National Aeronautics and Space Administration, Washington, DC, 1987.

Rothrock, D. A. and D. R. Thomas, An Arctic Ocean ice balance with assimilated satellite and buoy data, *Proceedings of the International Symposium on Assimilation of Observations in Meteorology and Oceanography*, Clermont–Ferrand, France, July 9–13, 1990a.

Rothrock, D. A. and D. R. Thomas, The Arctic Ocean multiyear ice balance, 1979–82, *Annals of Glaciology, 14*, pp. 252–255, 1990b.

Rothrock, D. A., D. R. Thomas, and A. S. Thorndike, Principal component analysis of satellite passive microwave data over sea ice, *Journal of Geophysical Research, 93(C3)*, pp. 2321–2332, 1988.

Steffen, K. and A. J. Schweiger, A multisensor approach to sea ice classification for the validation of DMSP-SSM/I passive microwave derived sea ice products, *Photogrammetric Engineering and Remote Sensing, 56(1)*, pp. 75–82, 1990.

Thomas, D. R. and D. A. Rothrock, Blending sequential Scanning Multichannel Microwave Radiometer and buoy data into a sea ice model, *Journal of Geophysical Research, 94(C8)*, pp. 10,907–10,920, 1989.

Thomas, D. R. and D. A. Rothrock, The Arctic Ocean ice balance: A Kalman smoother estimate, in press, *Journal of Geophysical Research*, 1992.

Thorndike, A. S., A naive zero-dimensional sea ice model, *Journal of Geophysical Research, 93(C5)*, pp. 5093–5099, 1988.

Thorndike, A. S. and R. Colony, Objective analysis of atmospheric pressure and sea ice motion over the Arctic Ocean, *Conference on Port and Ocean Engineering Under Arctic Conditions, 2*, pp. 1070–1079, Helsinki, Finland, 1983.

Vinje, T. and O. Finnekasa, The ice transport through the Fram Strait, *Norsk Polarinstitutt Skrifter, 186,* 39 pp., 1986.

Wittmann, W. I. and J. Schule, Jr., Comments on the mass budget of Arctic pack ice, *Proceedings of the Symposium on the Arctic Heat Budget and Atmospheric Circulation*, edited by J. O. Fletcher, pp. 215–246, Memorandum RM-5233-NSF, Rand Corporation, Santa Monica, California, 1966.

Chapter 24. Potential Applications of Polarimetry to the Classification of Sea Ice

MARK R. DRINKWATER, RONALD KWOK, AND ERIC RIGNOT

Jet Propulsion Laboratory, California Institute of Technology, 4800 Oak Grove Drive, Pasadena, California 91109

HANS ISRAELSSON

Department of Radio and Space Sciences, Chalmers University of Technology, S-41296, Göteborg, Sweden

ROBERT G. ONSTOTT

Environmental Research Institute of Michigan, P. O. Box 8618, Ann Arbor, Michigan 48107

DALE P. WINEBRENNER

Polar Science Center, Applied Physics Laboratory, University of Washington, 1013 NE 40th Street, Seattle, Washington 98105

24.1 INTRODUCTION

To date, conventional microwave radar studies of sea ice have enabled scattering characteristics of saline ice to be measured for particular sets of radar parameters, such as frequency, incidence angle (θ_i), and, foremost, polarization. Radar polarimetry employs a methodology that allows simultaneous measurement of the radar backscatter from a given surface at a number of different polarizations. The goals of polarimetry are to improve geophysical property retrievals and, in the near term, improve the estimate accuracy using single-channel radar techniques.

Technology has traditionally limited the number of variable parameters. Generally, a single antenna has been used for both signal transmission and reception (i.e., monostatic radar) with either vertical (v) or horizontal (h) fixed polarization. In most cases, the radar system was incoherent and did not preserve phase information (i.e., recording only the magnitude of the complex scattered field vector). With advances in available technology, practical airborne imaging with polarimetric synthetic aperture radar (SAR) is now being conducted [Zebker and van Zyl, 1991] and new airborne systems are under development [Livingstone et al., 1990]. Not only do these systems have frequency diversity (multiple channels at P-, L-, C-, and X-bands), but they also have polarization diversity (allowing reception of both the transmission polarization and its orthogonal) and polarimetric capability (where phase information is also recorded). Polarimetric radar selects h and v transmission and reception polarizations, simultaneously recording all combinations of linear polarizations (i.e., hh, hv, vh, and vv). Thus, the unknown complex scattered electromagnetic vector is sampled in two orthogonal directions so as to completely characterize the backscattered field.

The wave transmitted by a radar polarimeter can be considered a monochromatic and completely polarized plane wave. In contrast, a scattered signal is seldom completely polarized when observed as a function of time or space. Typically backscatter originates from a statistically random surface made of distributed scattering centers or facets. The resulting polarimetric response then comprises a superposition of a large number of waves with a variety of polarizations. For complete discussions on aspects of wave polarization, the reader is referred to Ulaby and Elachi [1990] and Kong [1990].

24.2 POLARIMETRIC DATA AND DEFINITIONS

For backscattering from a discrete target, the transmitted and received field vectors are uniquely related through the complex scattering matrix $\overline{\overline{S}}$, where

$$\overline{\overline{\mathbf{S}}} = \begin{bmatrix} |S_{hh}| \, e^{j\phi_{hh}} & |S_{hv}| \, e^{j\phi_{hv}} \\ |S_{vh}| \, e^{j\phi_{vh}} & |S_{vv}| \, e^{j\phi_{vv}} \end{bmatrix} \tag{1}$$

$|S_{rt}|$ is the magnitude of the scattering matrix element, and $e^{j\phi_{rt}}$ is the relative phase information for (t) transmit (r) receive polarization. Since in all operational imaging polarimeters the polarizations are linear and fixed by the antennas, the commonly used subscripts h and v refer to the horizontal and vertical transverse polarized components of the transmitted and received electric field. Each element of the scattering matrix is a function of the frequency and the scattering and illumination angles. A polarimetric radar essentially measures $\overline{\overline{S}}$ for every pixel or resolution element within an image

$$\mathbf{E}_s = e^{jkr}(kr)^{-1} \overline{\overline{\mathbf{S}}} \, \mathbf{E}_i \tag{2}$$

Microwave Remote Sensing of Sea Ice
Geophysical Monograph 68
©1992 American Geophysical Union

where k is the wave number, r is range, and $\mathbf{E}_i$ and $\mathbf{E}_s$ are the incident and scattered field vectors, respectively. The backscattered signal from sea ice is a vector sum of waves with a variety of polarizations from numerous randomly positioned scatterers. The SAR polarimeter provides a measure of the mean field components of the resulting partially polarized wave in each pixel. The effective backscattering coefficient σ^0_{rt} is then defined from elements of the scattering matrix

$$\sigma^o_{rt} = 4\pi \frac{\langle S_{rt} S^*_{rt}\rangle}{A} \qquad (3)$$

where A is the illuminated area, * denotes the complex conjugate, and $\langle\ \rangle$ denotes the ensemble averages of a number of pixels. Two equivalent forms for displaying the elements of $\overline{\mathbf{S}}$ in terms of the SAR-recorded complex polarimetric backscattering information are the Stokes matrix $\overline{\mathbf{M}}$ and the covariance matrix $\overline{\mathbf{C}}$—both representations consist of linear combinations of the cross products of the four elements of $\overline{\mathbf{S}}$. Importantly, each form explicitly characterizes the polarimetric scattering properties of the sea ice, thereby enabling synthesis of the backscattered power for arbitrary transmission and reception polarizations. Polarization synthesis expressions are given by Zebker and van Zyl [1991], and further discussion of the Stokes matrix for sea ice is contained in Drinkwater and Kwok [1991].

It is most convenient to convey the backscatter in terms of ensemble averaged properties of a group of pixels in the covariance matrix form

$$\overline{\overline{\mathbf{C}}} = \begin{bmatrix} \langle S_{hh} S^*_{hh}\rangle & \langle S_{hh} S^*_{hv}\rangle & \langle S_{hh} S^*_{vv}\rangle \\ \langle S_{hv} S^*_{hh}\rangle & \langle S_{hv} S^*_{hv}\rangle & \langle S_{hv} S^*_{vv}\rangle \\ \langle S_{vv} S^*_{hh}\rangle & \langle S_{vv} S^*_{hv}\rangle & \langle S_{vv} S^*_{vv}\rangle \end{bmatrix} \qquad (4)$$

This three-by-three Hermitian matrix fully characterizes the polarimetric scattering properties of a distributed scattering target such as sea ice. It is assumed that the surface satisfies the reciprocity relation where $S_{hv} = S_{vh}$. For practical purposes, one of the cross-polarized components of the scattering matrix is ignored and the scattering matrix is made symmetrical. In doing so, the mean phase imbalance between transmission and reception (in both h- and v-polarized antennas) is removed. This difference must be recorded for use during subsequent cross-channel phase balancing of images. Such a procedure is typical in generating the standard compressed data format for the National Aeronautics and Space Administration, Jet Propulsion Laboratory (NASA–JPL) synthetic aperture radar (SAR) polarimeter [Zebker and van Zyl, 1991].

24.3 Polarimetric Discriminants

Polarimetric parameters in this section are chosen to characterize the polarimetric signature of particular ice types. The term polarimetric discriminants here applies to combinations of elements of the matrix $\overline{\mathbf{C}}$ that describe the polarimetric response. In this section, several discriminants are introduced, namely: span, like-polarized (co-pol) and cross-polarized (cross-pol) power ratios, phase difference, correlation coefficient, and fractional polarization. Multifrequency data from JPL and the Environmental Research Institute of Michigan (ERIM) SAR polarimeters, acquired in March 1988 and March 1989, respectively, are used to illustrate examples of the variation in these discriminants for particular Arctic ice types.

Figure 24-1 shows a C-band sea ice image with five sample boxes. Contextual clues and the polarimetric response of this Beaufort Sea ice indicates that there is a mixture of multiyear (MY) and first-year (FY) ice. Boxes 1 and 3 are from large MY floes, while box 2 is from an area of ridged and rafted FY ice. The relatively dark response of boxes 4 and 5 indicates mechanically undisturbed and likely smoother FY ice that formed in cracks in the large MY floes. The brighter tones of the MY ice are explained by volume scatter [Drinkwater, 1990; Drinkwater et al., 1991a, 1991b, 1991c], as examples of subsequent polarimetric measures testify. All ice surfaces in this scene are likely covered by a veneer of dry snow [Wen et al., 1989].

Figure 24-2 is an image set illustrating polarization sensitivity at L-band. These images are 10×10 km with a resolution of about 2 m, and θ_i varies from 30° to 65°. Data shown in Figures 24-1 and 24-2 illustrate a range of polarimetric examples spanning frequencies from X-band to P-band, and provide the basis for the analyses contained in following sections.

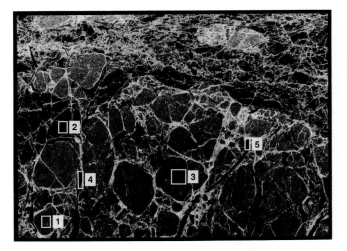

Fig. 24-1. Three-frequency total power image acquired by the JPL polarimeter at the location 73° 13.4′ N 142° 1.1′ W at 17:21:26 GMT (i.e., 08:21:26 local time) on March 11, 1988, in the Beaufort Sea [Drinkwater et al., 1991a]. These data are overlayed as a false-color image, with the C-band in blue, L-band in green, and P-band in red.

(a)

(b)

(c)

(d)

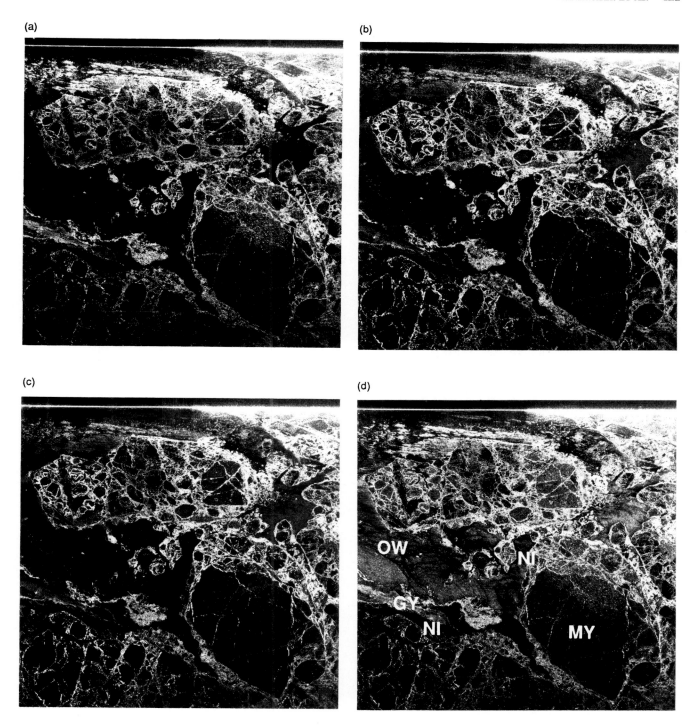

Fig. 24-2. ERIM L-band (1.25-GHz) SAR images of sea ice at (a) hh, (b) hv, (c) vh, and (d) vv polarizations from the Greenland Sea marginal ice zone during March 1989.   The surface categories indicated are open water (OW), nilas (NI), grey ice (GY), and multiyear ice (MY).

24.3.1 Total Power

Absolute calibration is not possible for many polarimetric data, owing to the absence of in-scene targets with known backscattering characteristics. Consequently, relative measures are derived using frequency and polarization. One such measure is the span, which represents the total power in the scattered field at a particular frequency

$$SPAN = \langle S_{hh} S_{hh}^* \rangle + \langle S_{vv} S_{vv}^* \rangle + 2 \langle S_{hv} S_{hv}^* \rangle \quad (5)$$

where $\langle . \rangle$ denotes ensemble averaging of a number of pixel values.

Israelsson and Askne [1991] choose examples taken from JPL polarimeter images of Beaufort Sea ice using 200-m^2 samples. They contrasted the $SPAN$ values for FY and MY ice at three frequencies and demonstrated the best discrimination at C-band with a separation of 10 dB. Typically, $SPAN$ is greater for MY than FY ice due to the fact that MY ice gives an additional depolarized component of backscatter along with the predominantly co-pol returns from FY ice of equivalent surface roughness (Chapter 8). The proportion of depolarized returns is significantly larger at C-band than at L- or P-band, leading to the hypothesis that some component is derived from multiple scattering by bubbles and inhomogeneities present in the upper layers of old ice in addition to second-order rough surface scattering and volume-surface interaction effects. Supporting evidence for volume scattering is that $SPAN$ is less sensitive to the incidence angle at C-band, in direct contrast to $SPAN$ values for L- and P-bands, each of which decrease with increasing θ_i. The decrease in L- and P-band $SPAN$'s for MY ice between 25° and 52° is between 10 and 15 dB, in contrast to the 5-dB decrease at C-band.

24.3.2 Power Ratios Between Channels

Polarization signatures are a convenient tool for three-dimensional viewing of the power recorded at each of the polarimeter's transmit and receive combinations [Evans et al., 1988]. However, polarization signatures in Drinkwater et al. [1991a] and Zebker and van Zyl [1991] indicate that sea ice backscattering does not appreciably transform the polarization of linearly polarized transmitted waves. Typically, the linear h- or v-polarization (hereafter abbreviated to h- or v-pol) of the transmitted waves is retained for a large fraction of the backscattered signal. Additionally, circularly polarized transmitted waves produce negligible co-pol returns [Drinkwater et al., 1991a]. For most geophysical media, circularly polarized waves are depolarized most effectively [Zebker et al., 1987]. Since there is some degree of redundancy in the information contained in nonlinear polarizations, linear h- and v- co- and cross-pol channels can be reasonably effective in characterizing the polarization signature of a given sea ice surface.

Heterogeneity of scattering mechanisms between pixels introduces a component of unpolarized or randomly polarized backscatter. This is manifested as a fraction of returns that exhibit diffuse scattering, characterized by rapidly varying polarizations from pixel to pixel (and thus widely differing covariance matrices in adjacent resolution elements). Generally, complex surfaces or anisotropic or multiple-scattering surface layers increase this unpolarized fraction of the backscatter. Another mechanism contributing to such an effect is the presence of system noise, which may occur at low backscatter. This circumstance is only really typical of situations of smooth young ice in the far range [Kwok et al., 1991a].

A more convenient method for using polarization combinations is by applying channel ratios. Ratios quantify the difference in power (in decibels) between specific polarizations and thus characterize the full polarization signature. The ratio between co-pol elements of $\overline{\overline{C}}$ is

$$R_{hh/vv} = \frac{\langle S_{hh} S_{hh}^* \rangle}{\langle S_{vv} S_{vv}^* \rangle} \quad (6a)$$

Similarly, the cross-pol ratio is

$$R_{hv/vv} = \frac{\langle S_{hv} S_{hv}^* \rangle}{\langle S_{hh} S_{hh}^* \rangle} \quad (6b)$$

The depolarization ratio is defined as the ratio of cross-pol to both co-pol channels

$$R_{depol} = \frac{2 \langle S_{hv} S_{hv}^* \rangle}{\langle S_{hh} S_{hh}^* \rangle + \langle S_{vv} S_{vv}^* \rangle} \quad (6c)$$

24.3.2.1 Co-pol ratio distributions.

The co-pol ratio distributions in Figure 24-3 illustrate how the hh- and vv-pol backscatter varies between each ice type with frequency. Individual pixel ratios are shown as a distribution, rather than by deriving a mean ratio using Equation (6a). The ratios are balanced by applying C- and L-band correction factors (–1.8 and 0.6 dB, respectively) based upon hh/vv imbalance calculations from radar calibrations. The resulting distributions are normalized with respect to the total number of pixels used in each sample box. Model predicted ratios, based upon first-order Bragg scattering [Winebrenner et al., 1989], are also made at L-band using typical measured properties of Beaufort Sea ice.

C-band consistently has the widest hh/vv ratio distribution, the lowest peak, and a tail extending to values above 1. This appears to be due to a combination of greater rough surface scattering as the wavelength becomes shorter and the apparent surface roughness increases. Figure 24-3 indicates the broadest spread of ratios in thick FY ice, and the narrowest distributions and lowest mean ratios for thin FY ice. MY ice exhibits an intermediate situation where C-band has the largest variance, but L- and P-bands show similar distributions. Based on this evidence, box 2 conforms to the characteristics of a rough surface scatterer at all

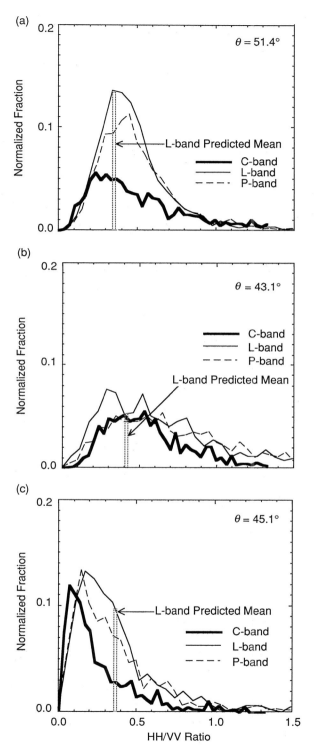

Fig. 24-3. Normalized distributions in the Beaufort Sea of individual hh/vv ratios at each frequency for (a) MY ice in box 1, (b) thick FY ice in box 2, and (c) thin FY ice in box 5. The measured L-band mean ratios are 0.44, 0.6, and 0.37 for Figure 24-1 boxes 1, 2 and 5, respectively; first-order model predictions are shown for comparison.

frequencies [Drinkwater et al., 1991a]. The mean C-band ratio falls close to –2.5 dB (0.6) for samples of this ice, but at L- and P-bands generally indicates slightly higher values. The $R_{hh/vv}$ at L-band is –2.2 dB, which is greater than the predicted value of –3.9 dB (0.41).

In contrast, box 5 (Figure 24-3) appears to be a smoother, higher salinity thin FY ice surface with a clustering of ratios less than 0.5 (–3 dB). Values are consistent with the small-perturbation scattering theory discussed in Chapter 8. First-order predictions suggest that vv-pol backscatter should exceed hh-pol backscatter for smooth surfaces at these incidence angles. The corresponding mean hh/vv ratios at each frequency become smaller as the wavelength or θ_i increases and as the ice salinity and permittivity increases [Winebrenner et al., 1989]. The $R_{hh/vv}$ falls close to predictions at L- and P-bands for smooth ice with a salinity of 15‰ and brine volume fraction close to 100‰.

Box 1 demonstrates an intermediate situation and the L-band model predicts lower mean values than those observed. Penetration depths are greater in MY ice and, consequently, the polarimetric characteristics are somewhat different than those observed for higher salinity surfaces. The most pervasive of all observations to date is that MY signatures are remarkably stable in their polarimetric characteristics at L- and P-bands: backscattered power varies by only a fraction of a decibel.

Figures 24-4(a) and (b) show the trend in $R_{hh/vv}$ for FY and MY ice as θ_i varies from 32° to 52°. In Figure 24-4(a), the C-band ratio rises with increasing θ_i, with the ratio tending to 1 (i.e., 0 dB) at higher angles in FY ice. This trend is consistent with geometric optics scattering from floe edges or slope facets of rough FY ice at higher incidence angles [Livingstone and Drinkwater, 1991]. In contrast, at L-band the ratio falls with θ_i to –4.0 dB at 50°. This is consistent with the predicted L-band ratio at 50° for thick FY ice [Onstott et al., 1991]. The MY ice plot in Figure 24-4(b) shows C-band ratios to have negligible gradient; L-band ratios for the same ice are slightly lower. Both sets of values fall close to the theoretical surface scattering predictions for MY ice. Volume scattering may account for the residual difference and that the C-band ratio is closer to 0 dB. Comparative X-band values of $R_{hh/vv}$ observed for MY by Onstott et al. [1991] indicate a mean ratio of 0 dB. This is also consistent with the general trend towards a mean ratio of 0 dB as the wavelength decreases and the volume scattering increases.

24.3.2.2 Cross-pol ratios. An increase in $R_{hv/hh}$ and second-order scattering effects generally implies anisotropy in the surface roughness or the dielectric structure of the FY ice (e.g., as a consequence of brine-inclusion size and orientation), or air bubbles and inhomogeneities in MY ice. In Figure 24-4(a) there is an increase in the FY ratio with θ_i. The C-band ratio rises from around –15 dB at 32° to –11.5 dB at 52°. Though the L-band ratio is more variable over this range, a similar trend is reproduced with a mean offset of –2 dB throughout the 20° range. Figure 24-4(b)

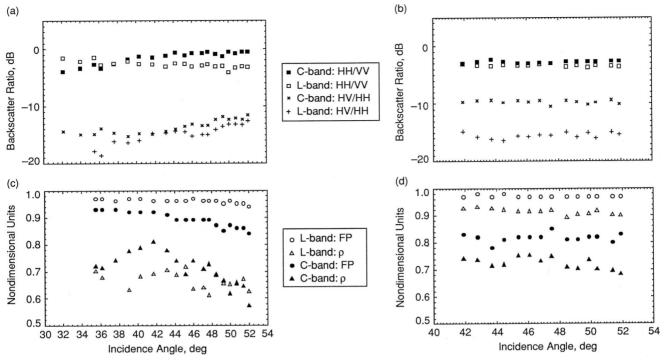

Fig. 24-4. Cross-swath C- and L-band polarimetric discriminant statistics. The upper panels show co- and cross-pol ratios in the Beaufort Sea for (a) FY and (b) MY ice. The lower panels show the correlation coefficient ρ between hh- and vv-pol returns and the corresponding fractional polarization (FP) for samples of (c) FY and (d) MY ice.

shows a contrasting MY situation with almost uniform mean ratios at each frequency: L-band has a mean $R_{hv/hh}$ value of around −15.0 dB, while the C-band mean is around −10.0 dB. Higher C-band MY ice cross-pol ratios of around −9.5 dB are recorded by Drinkwater et al. [1991a]. Notably, some of the highest cross-pol ratios have been observed for thin FY ice in the Beaufort and Bering Seas; certain L- and P-band examples result in values of $R_{hv/hh}$ between −6.9 dB and −9 dB [Drinkwater et al., 1991a]. Box 4 in Figure 24-1 is a particularly good example of high $R_{hv/hh}$, having L- and P-band values of −8.4 and −8.3 dB, respectively. The cross-pol isolation for the JPL radar system is around −27 and −17 dB for C- and L-band, respectively. Thus, these measurements are well within the sensitivity range of the JPL SAR.

At X-band, $R_{hv/hh}$ values are lowest for MY ice and increase as the surface salinity increases [Onstott et al., 1991]. In contrast, the above JPL SAR C-band results show that $R_{hv/hh}$ is highest for MY ice, generally decreasing the greater the salinity of the surface. Thick FY ice has a mean C-band ratio of −13.5 dB, while the lowest mean cross-pol ratios of −14.7 dB were observed for what is suspected to be thin frost-flowered ice [Drinkwater et al., 1991a]. Notably these latter examples also have the largest variability in the ratios. At L-band the reverse is true with MY ice exhibiting the smallest ratios.

24.3.3 Co-pol Correlation and Fractional Polarization

A condition for fully polarized radar backscattering from natural geophysical targets is that the cross-pol magnitude must be zero and hh and vv returns perfectly correlated (i.e., unity). The correlation coefficient between co-pol elements of $\overline{\overline{C}}$ is

$$\rho_{hhvv} = \left| \frac{\langle S_{hv} S_{vv}^* \rangle}{\sqrt{\langle S_{hh} S_{hh}^* \rangle \langle S_{vv} S_{vv}^* \rangle}} \right| \qquad (7)$$

At short wavelengths, FY sea ice typically has high correlation coefficients (tending to 1 at nadir). Figure 24-4(c) illustrates that the C-band correlation coefficient generally falls with θ_i in sea ice that is predominantly surface scattering, and at an increasing rate in higher salinity ice. In Figure 24-4(d), MY ice demonstrates slightly lower C-band values of ρ_{hhvv} than FY ice, due to the depolarization effects mentioned in Section 24.3.2. MY values indicate a much smoother and gradual monotonic decline with θ_i. Generally, the longer the wavelength, the lower the sensitivity of ρ_{hhvv} to surface or volume scattering effects. Figure 24-4(d) illustrates that the highest correlations ($0.89 \leq \rho_{hhvv} \leq 0.93$) are observed at L-band for MY ice.

The fractional polarization (FP) in Figure 24-4 is calculated from a sample by synthetically varying the polarization state and by recording the maximum (P_{max}) and minimum (P_{min}) intensities as both transmission and reception polarizations are varied [Zebker et al., 1987]. The value FP is then calculated from

$$FP = \frac{P_{max} - P_{min}}{P_{max} + P_{min}} \qquad (8)$$

where FP is a measure of the polarization purity of the return; thus, for lower values the unpolarized component is greater and ρ_{hhvv} is reduced. Figures 24-4(c) and (d) illustrate that FP is close to unity at L-band for both FY and MY ice, in direct contrast to other geophysical media such as forest vegetation where values of $FP < 0.5$ are typical. Wavelength effects are significant and as the wavelength is reduced to C-band FP is similarly reduced. The value of FP falls with θ_i in FY ice from above 90% to around 82%, while FP in MY ice is consistently lower and relatively insensitive to θ_i. The lowest C-band values of FP (around 0.8) were observed by Drinkwater et al. [1991a] in MY ice. Together with correlation coefficients as low as 0.7, this indicates that the largest fraction of unpolarized returns occurs in lower salinity ice, and is probably due to multiple scattering from bubbles or air-filled inhomogeneities in the upper ice sheet.

24.3.4 Phase Differencing

Relative differences in phase between channels are important because each scattering event, either at reflective horizons or from diffraction by particles in the medium, transforms the relative phase of co-pol waves. The mean hh–vv phase difference is

$$\phi_{hh-vv} = \tan^{-1}\left[\frac{\Im\left\langle S_{hh} S_{vv}^* \right\rangle}{\Re\left\langle S_{hh} S_{vv}^* \right\rangle}\right] \qquad (9)$$

where $\Re$ and $\Im$ indicate the real and imaginary parts, respectively. Normal incidence reflection from a highly conductive material such as sea water results in $\bar{\phi}_{hh-vv} = 0°$ and extremely low variability in the individual pixel phase difference. More complex distributed and layered targets such as sea ice often produce multiple reflections and sometimes nonzero values of $\bar{\phi}_{hh-vv}$. The hh–vv phase difference for a planar dielectric is also known to increase with θ_i. This arises from the sensitivity of a given linear polarization orientation to a particular scattering mechanism.

The co-pol phase difference ϕ_{hh-vv} can be estimated from the first-order Bragg scattering model. Assuming Fresnel reflection from a simple planar dielectric interface, Equation (9) reduces to

$$\phi_{hh-vv} = \tan^{-1}\left[\frac{\Im\left\langle R_h / R_v \right\rangle}{\Re\left\langle R_h / R_v \right\rangle}\right] \qquad (10)$$

where R is the power reflection coefficient at h and v polarizations. L-band results in Onstott et al. [1991] indicate that for MY ice ϕ_{hh-vv} is predicted to be zero and independent of θ_i. If the surface is lossy, the co-pol phase term becomes negative and with increasing negative phase difference with increasing θ_i. The higher the dielectric constant of the surface, the more rapid this increase becomes. Thus, the co-pol phase difference in the range of 45° to 50° is largest for open water and typically around −5° at L-band [Onstott et al., 1991].

Figures 24-5(a) and (b) contrast distributions of MY and FY ice single pixel values of ϕ_{hh-vv} at C-band, while Figures 24-5(c) and (d) show the same contrast at L-band. Results of comparing MY samples across the swath show that distributions of ϕ_{hh-vv} have zero means and are independent of θ_i, thus making MY a good target for relative phase calibration within a scene [Drinkwater et al., 1991a]. The spread in the MY ice distribution increases with decreasing wavelength, indicating that at sufficiently short wavelengths volume scatter from inhomogeneities introduces mixed phase differences.

Compared with MY ice examples, the FY ice in Figures 24-5(b) and (d) shows more variability in ϕ_{hh-vv}. Sample boxes 2, 4, and 5 from Figure 24-1 indicate means displaced less than 5° from zero, but which are within the expected phase accuracy of ±5°. At C-band the variance in ϕ_{hh-vv} differs negligibly between ice type, but at L-band the distribution is much broader for FY ice than for MY ice.

Values of ρ_{hhvv} shown in Figure 24-4(c) and (d) are closely related to the hh–vv phase distributions in Figure 24-5. Generally, the lower ρ_{hhvv} is, the greater the variance in the phase difference and the smaller FP becomes. This situation is characterized by the differences in variance in L- and C-band MY ice co-pol phase distributions shown in Figures 24-5(a) and (c). Figure 24-4(d) indicates that the higher L-band values of ρ_{hhvv} result in a much narrower distribution of phase difference, as in Figure 24-5(c), while the lower C-band correlation produces a much broader spread, as in Figure 24-5(a).

24.4 Geophysical Applications of Polarimetry

Multifrequency polarimetric SAR observations, together with passive microwave imagery, will ultimately yield retrievals of important geophysical ice information. In this section, preliminary results of geophysical importance are presented. Firstly, different types of thin ice and open water are identified using polarimetric observations. Secondly, images of sea-ice are classified using the polarimetric characteristics presented. Each result represents an improvement over single-frequency, single-polarization, incoherent radar observations of sea ice.

24.4.1 Thin Ice and Open Water Detection

One of the most significant problems facing single channel microwave techniques is their poor discrimination between

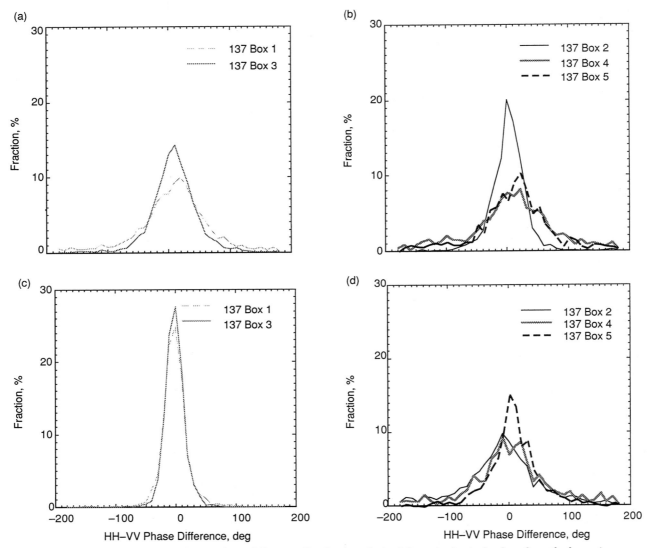

Fig. 24-5. Normalized C- and L-band hh–vv phase difference distributions derived from single pixel values from the boxes in Figure 24-1. Upper panels show C-band distributions for (a) the MY and (b) FY ice; the lower panels contrast L-band distributions of the same (c) MY and (d) FY examples.

open water and new ice. The early ice growth phase is of critical significance in controlling the vigor of heat, salt, and vapor fluxes taking place at the surface of the ocean. Thus, it is of particular importance to establish a technique for studying transformations in the ice during this period.

Polarization ratioing is employed as a means of discriminating between the ice and liquid phases of water [Winebrenner, 1990]. Providing that the sea ice is snow-free, as is typically the case for young or new ice, $R_{vv/hh}$ increases with θ_i at a rate dependent upon the complex dielectric constant. This relationship holds best for L- and P-bands in the range of roughness where the small-perturbation theory is valid (see Chapter 8). At C-band, the model is not valid, except for the smoothest surfaces encountered.

Thus, to first-order, the ratio is independent of roughness and is dependent only on θ_i and the brine volume fraction, and thus the thickness and age of the ice.

Winebrenner [1990] showed that single-pixel hh/vv ratios are noisy, requiring averaging of 100 pixels or more. Kwok et al. [1991a] extended this approach by averaging 20-by-20 pixel windows and using the mean ratio and the co-pol correlation to estimate and correct for additional system noise effects. Examples demonstrate L-band ratios between 2 and 10 dB in the range of 25° to 52°; these values are consistent with theoretical ratios for conditions between open water and thin ice. Confirmation for these findings is provided by collocated 33.6-GHz passive microwave brightness temperatures [Drinkwater et al., 1991b,

1991c]. The mean value of 150 K is well separated from the typical cold MY radiometric temperature (170 K) and is more consistent with the signature of new ice or open water. One example of ice–water discrimination is to use the calibrated L-band ERIM SAR data in Figure 24-2 from the Coordinated Eastern Arctic Experiment. Nilas and open water are each observed to produce weak backscatter at hh pol (–29.6 dB for nilas and –34.6 dB for water). In the vv-pol image where the water return is at its highest, the backscatter for nilas is weaker (–32.1 dB for nilas and –27.9 dB for water) and closer to that observed at hh pol.

Unique thin-ice signatures have been observed in many images acquired in the Beaufort Sea by the JPL SAR. This ice has characteristically bright C-band backscatter approaching that of MY ice, while having relatively low backscatter at L-band. In addition to the high vv/hh ratios noted above, a further effect has been recorded. Thin ice forming in leads is found to cause a significant phase shift between hh and vv returns, particularly at L-band, where values of –15° to –20° were noted by Kwok et al. [1991a]. The typical correlation coefficient for thin ice is also lower at L-band than for those examples observed for thick FY ice in Figure 24-3. At C-band though, ρ_{hhvv} is more consistent, with values for FY and MY ice ranging between 0.7 and 0.8. It is the latter clue that suggests that predominantly rough surface scattering is taking place at C-band while L-band penetrates the ice layer. Phase shifts can only be accounted for by a finite layer thickness effect that enables interference of top- and bottom-scattered signals. Ice salinities required to produce the observed hh/vv ratios and the observed phase effects are high and, since many of the observed ratios fall between the predictions for a thin ice layer and open water, this has led to speculation that the upper ice surface is roughened by frost flowers high in salinity.

24.4.2 Feature Use in Ice Classification

Polarimetric procedures for ice classification have been developed that utilize both magnitude and phase. In this final section, ice classification is broken into two stages: image segmentation using polarimetric image statistics, and labeling or tagging of the subdivided images into surface types.

24.4.2.1 Image segmentation. Polarimetric image segmentation requires no a priori information about sea ice and is simply based on the statistical distribution of the data. Several parameters need to be selected prior to segmentation, e.g., the number of image classes in relation to the number of sea-ice types that can be clearly identified, and the dimension and nature of the feature vector (combining different polarizations and frequencies).

Selection of image classes can be done either in a supervised or unsupervised manner. In the supervised polarimetric selection scheme used by van Zyl [1989], image classes, each corresponding to distinct polarimetric scattering behavior, were limited to single bounce, double bounce, and diffuse scattering. In Rignot and Drinkwater [1992], selection of the image classes was unsupervised, and was instead based on a multidimensional cluster analysis of the polarimetric covariance matrix data. This technique examines less than 10% of the pixels in the image, is robust to the presence of image speckle, and accounts for the distribution of each sample pixel into more than one image class .

Given distinct image classes and their polarimetric backscatter characteristics, various methods are then available to segment the entire polarimetric radar scene. Scattering behavior [van Zyl, 1989], single discriminants (such as the hh–vv phase difference or the hh/vv ratio), or complete polarimetry can be used. A fully polarimetric maximum likelihood (ML) Bayes classifier was developed by Kong et al. [1988] and Lim et al. [1988] on the basis that polarimetric data are characterized by a multivariate Gaussian distribution. Both show that from Monte Carlo simulations and various images the Bayers classifier produces better results than those obtained using single polarimetric discriminants. Van Zyl and Zebker [1990] used this method for ice in the Beaufort Sea. They found that C-band data alone give excellent discrimination between FY and MY sea ice, but that ridged FY ice is confused with the latter. Though P- and L-band classifications separate pressure ridges from MY and FY ice types better, the most accurate classification of the three surface types is achieved using a combination of all three frequencies. Results obtained from the ML classifier remain noisy, however, and unsatisfactory for most practical applications. Rignot and Chellappa [1992] extended the ML technique with what is termed the maximum a posteriori (MAP) classifier. In the MAP technique, image classes are not assumed to have equal a priori probabilities, and spatial context is used to improve the segmentation results. MAP segmentation accuracy is typically improved by 10 to 20% over the ML classifier. Using the MAP classifier, Rignot and Drinkwater [1992] extended the analysis of van Zyl and Zebker [1990] to the delineation of five surface types: compressed FY ice, MY floes, FY ridges and rubble, undeformed thick FY ice, and thin ice (see Figure 24-6). Compressed FY ice, which is the deformed FY matrix that holds together MY ice floes, and the hummocks present in the MY ice floes are clearly separated from other ice types based on polarimetric data and coincident passive microwave measurements. Such compressed ice is probably extremely thick, rafted FY ice that has deformed then desalinated due to its higher freeboard. The best overall classification results are in discriminating these five distinct ice types (>90% estimated accuracy) using combined L- and C-band fully polarimetric data. Notwithstanding this result, certain combinations of two frequencies at a single polarization, or of two polarizations at a single frequency, yield classification accuracies that are only slightly lower (84% classification accuracy using L-band hh and vv pol and 89% classification accuracy using L-band hh pol and C-band vv pol).

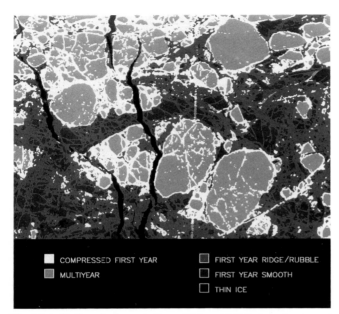

Fig. 24-6. Fully polarimetric C- and L-band MAP classification of five ice types [Rignot and Drinkwater, 1992].

24.4.2.2 Improvements in the labeling of segmented images. The final classification procedure requires labeling of segmented images. Identifying ice that demonstrates similar polarimetric characteristics requires a priori information about surface properties and scattering mechanisms. Most segmentations are labeled by deductive reasoning incorporating minimal surface data collected at ice camps, contextual information from patterns and the shape of features within the images, and experience from image analysis.

In the future, the required a priori ice information may be derived from two sources: surface measurements and model simulation data (i.e., a lookup table of the polarimetric backscatter characteristics of different ice types). Classifiers need to be constructed such that modeled polarimetric responses may be incorporated in the classification procedure. This will enable not only classification of broad classes of ice with different properties, but also extraction of geophysically important ice variables.

24.5 DISCUSSION

Knowledge of the Stokes or covariance matrices allows us to synthesize the scattering cross-section for all possible transmission and reception polarizations. Polarization signatures are not unique in that different combinations of scattering mechanisms may yield the same polarization response. However, the results from additional polarizations increase our present capability to solve for the geomet-

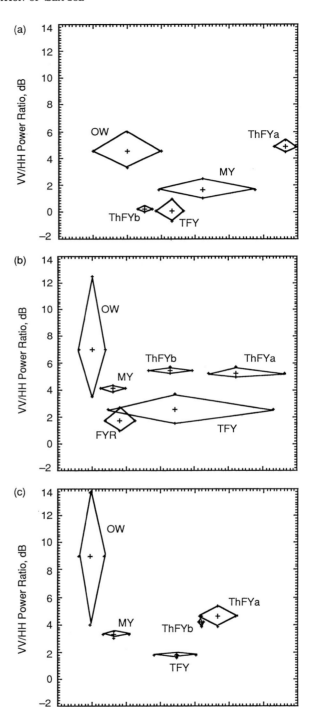

Fig. 24-7. Classification feature space for (a) C-band, (b) L-band, and (c) P-band, indicating the separability of ice types and open water using a polarimetric feature combining the spread in the hh–vv phase difference and the vv/hh power ratio. The categories are thin first-year ice (ThFYb), anomalous thin ice surfaces (ThFYa), thick first-year ice (TFY), rough Bering Sea first-year ice (FYR), multiyear ice (MY), and open water (OW).

ric structure of sea ice surfaces and dielectric properties of the main scatterers. This gives us a more flexible tool for identifying the dominant scattering mechanisms within an ice scene.

Measurement of the complete polarization response allows more rigorous testing of contemporary scattering models. These models must not only predict the backscatter coefficient as a function of θ_i or properties of the scatterer, but also as a function of the polarization of the transmitting and receiving antenna.

Results demonstrate that the essence of polarimetry is not simply in the added benefit of selectable polarization, so much as in the additional phase information embedded in correlations between differently polarized waves. The results of plotting the two single most effective discriminants are shown in Figure 24-7. This two-dimensional feature space incorporates the mean co-pol ratio $R_{vv/hh}$ and the standard deviation of the co-pol phase difference ϕ_{hh-vv}. Segmented clusters are labelled with the aid of Ka-band passive-microwave images and contextual information within radar images from the Beaufort and Bering Seas [Drinkwater et al., 1991a]. Discrimination of the three principal ice types (MY and thin and thick FY) is possible along with separation of all ice from open water. Clearly one of the the main obstacles of single channel radar is overcome, namely, separating ice from open water. One of the more interesting observations is that anomalous thin FY ice returns, supposedly from frost-flowered new ice, are very different from other thin ice surfaces at short wavelengths. These two signatures become similar at P-band. Multiyear ice provides a tight cluster at L- and P-bands, when volume scattering is negligible, while thick FY ice shows its greatest variability at L-band.

With a demonstrated capability, in particular for identifying the early stages of thin ice formation, monitoring new ice by high-resolution radar polarimeter will become of critical interest in those regions of intense air–ice–ocean interaction (Chapters 12, 13, 14, and 15). A future spaceborne polarimetric SAR, such as the proposed EOS–SAR, offers the observational capability at the necessary scale and resolution to provide information about the influence of such key regions upon global climate.

REFERENCES

Drinkwater, M. R., Multi-frequency imaging radar polarimetry of sea ice, Ice *Technology for Polar Operations,* edited by T. K. S. Murthy, J. G. Paren, W. M. Sackinger, and P. Wadhams, pp. 365–376, Computational Mechanics Publications, Southampton, UK, 1990.

Drinkwater, M. R. and R. Kwok, Stokes matrix statistics in sea ice polarimetric SAR images, *Proceedings of the IGARSS'91 Symposium, vol. 1,* pp. 99–102, IEEE catalog no. 91CH2971-0, IEEE, New York, 1991.

Drinkwater, M. R., R. Kwok, D. P. Winebrenner, and E. Rignot, Multifrequency polarimetric synthetic aperture radar observations of sea ice, *Journal of Geophysical Research, 96(C11),* pp. 20,679–20,698, 1991a.

Drinkwater, M. R., J. P. Crawford, D. J. Cavalieri, B. Holt, and F. D. Carsey, *Comparison of Active and Passive Microwave Signatures of Arctic Sea Ice,* JPL Publication 90-56, Jet Propulsion Laboratory, California Institute of Technology, Pasadena, California, pp. 29–36, 1991b.

Drinkwater, M. R., J. P. Crawford, and D. J. Cavalieri, Multi-frequency, multi-polarization SAR and radiometer sea ice classification, *Proceedings of the IGARSS'91 Symposium, vol. 1,* pp. 107–111, IIEEE catalog no. 91CH2971-0, IEEE, New York, 1991c.

Evans, D. L., T. G. Farr, J. J. van Zyl, and H. A. Zebker, Radar polarimetry: Analysis tools and applications, *IEEE Transactions on Geoscience and Remote Sensing, GE-26,* pp. 774–789, 1988.

Israelsson, H. and J. Askne, Analysis of polarimetric SAR observations of sea ice, *Proceedings of the IGARSS'91 Symposium, vol. 1,* IEEE catalog no. 91CH2971-0, pp. 89–92, IEEE, New York, 1991.

Kong, J. A. (editor), *Polarimetric Remote Sensing, Progress in Electromagnetic Research, vol. 3,* 520 pp., Elsevier Science Publishers, New York, 1990.

Kong, J. A., A. A. Swartz, H. A. Yueh, L. M. Novak, and R. T. Shin, Identification of terrain cover using the optimum polarimetric classifier, *Journal of Electromagnetic Wave Application, 2(2),* pp. 171–194, 1988.

Kwok, R., M. R. Drinkwater, A. Pang, and E. Rignot, Characterization and classification of sea ice in polarimetric SAR data, *Proceedings of the IGARSS'91 Symposium, vol. 1,* IEEE catalog no. 91CH2971-0, pp. 81–84, IEEE, New York, 1991a.

Lim, H. H., A. A. Swartz, H. A. Yueh, J. A. Kong, R. T. Shin, and J. J. van Zyl, Classification of Earth terrain using polarimetric synthetic aperture radar images, *Journal of Geophysical Research, 94,* pp. 7049–7057, 1989.

Livingstone, C. E. and M. R. Drinkwater, Springtime C-band SAR backscatter signatures of Labrador Sea marginal ice: Measurements versus modelling predictions, *IEEE Transactions on Geoscience and Remote Sensing, 29(1),* pp. 29–41, 1991. (Correction, *IEEE Transactions on Geoscience and Remote Sensing, 29(3),* p. 472, 1991.)

Livingstone, C. E., T. I. Lukowski, M. T. Rey, J. R. C. Lafontaine, J. W. Campbell, R. Saper, and R. Wintjes, CCRS/DREO synthetic aperture radar polarimetry—status report, *Proceedings of the IGARSS'90 Symposium, vol. 2,* IEEE catalog no. 90CH2825-8, pp. 1671–1674, IEEE, New York, 1990.

Onstott, R. G., T. C. Grenfell, C. Mätzler, C. A. Luther, and E. A. Svendsen, Evolution of microwave sea ice signatures during early summer and midsummer in the marginal ice zone, *Journal of Geophysical Research, 92(C7),* pp. 6825–6835, 1987.

Onstott, R. G., R. A. Shuchman, and C. C. Wackerman, Polarimetric radar measurements of Arctic sea ice during the Coordinated Eastern Arctic Experiment, *Proceedings of the IGARSS'91 Symposium, vol. 1*, pp. 93–97, IEEE catalog no. 91CH2971-0, IEEE, New York, 1991.

Rignot, E. J. M. and R. Chellappa, Segmentation of polarimetric synthetic aperture radar data, *IEEE Transactions on Geoscience and Remote Sensing*, in press, July 1992.

Rignot, E. J. M. and M. R. Drinkwater, On the application of polarimetric radar observations for sea ice classification, *Proceedings of the IGARSS'92 Symposium*, in press, Houston, 1992.

Ulaby, F. T. and C. Elachi, *Radar Polarimetry for Geoscience Applications*, 364 pp., Artech House, Dedham, Massachusetts, 1990.

van Zyl, J. J., Unsupervised classification of scattering behavior using radar polarimetry data, *IEEE Transactions on Geoscience and Remote Sensing, 27(1)*, pp. 36–45, 1989.

van Zyl, J. J. and H. A. Zebker, Imaging radar polarimetry, *Polarimetric Remote Sensing, Progress in Electromagnetics Research, vol. 3*, edited by J. A. Kong, 520 pp., Elsevier Science Publishers, New York, 1990.

Wen, T., W. J. Felton, J. C. Luby, W. L. J. Fox, and K. L. Kientz, *Environmental Measurements in the Beaufort Sea, Spring 1988*, Technical report APL-UW TR 8822, 34 pp., Applied Physics Laboratory, University of Washington, Seattle, 1989.

Winebrenner, D. P., Accuracy of thin ice/open water classification using multi-polarization SAR, *Proceedings of the IGARSS'90 Symposium, vol. 1*, IEEE catalog no. 90CH2825-8, pp. 2237–2240, IEEE, New York, 1990.

Winebrenner, D. P., L. Tsang, B. Wen, and R. West, Sea-ice characterization measurements needed for testing of microwave remote sensing models, *IEEE Journal of Oceanic Engineering, 14(2)*, pp. 149–158, 1989.

Zebker, H. and J. J. van Zyl, Imaging radar polarimetry: A review, *Proceedings of the IEEE, 79(11)*, pp. 1583–1606, 1991.

Zebker, H. A., J. J. van Zyl, and D. N. Held, Imaging radar polarimetry from wave synthesis, *Journal of Geophysical Research, 92*, pp. 683–701, 1987.

Chapter 25. Information Fusion in Sea Ice Remote Sensing

Michael J. Collins

Institute for Space and Terrestrial Science, Earth Observation Laboratory
4850 Keele Street, North York, Ontario, M3J 3K1, Canada

25.1 INTRODUCTION

The problem of inferring sea ice properties from remotely sensed data is extremely difficult, and can only be done reliably under certain very restricted circumstances. While it is generally held that combining data from multiple sources will improve this situation, it is not known how much improvement will be realized nor how to carry out the data integration. The former concern has no simple answer and will be revealed slowly as current investigations unfold. This chapter is a brief attempt to shed some light on the second problem.

Information fusion helps us solve the primary problem in terrestrial remote sensing—segmentation of the Earth's surface into known classes of materials (e.g., young sea ice). Information fusion is a numerical procedure combining multisource multitype data in some rational way, and helps us make classification decisions. Section 25.6 discusses the many good reasons why this problem should not be fully automated [Mackworth, 1990]. This chapter will introduce several information fusion theories and techniques that have shown potential. Very few of these methods have been applied to sea-ice remote-sensing problems, and many of them have not been used in any geophysical context. This fact necessarily makes the review appear rather short on examples and long on theory. Rather than providing a catalogue of examples, I have used several abstractions to help the reader think about the problem of data integration as a whole. The objective of the chapter is to provide motivated geoscientists with a review of literature that would normally be outside their reading list and hopefully to expand the dialogue between geoscientists, engineers, and computer scientists on the subject of information fusion.

25.2 THE NEED FOR MULTITYPE INFORMATION

To fully comprehend the complexity of the sea-ice remote-sensing problem, we must view the ice as being a component in the ocean–ice–snow–atmosphere system. This is a highly interactive system held in equilibrium through energy transfer over a wide variety of spatial and temporal scales. Sea ice hangs in a fragile balance within this system. The structural character of sea ice is the result of growth, ablation, transport, and deformation.

Microwave Remote Sensing of Sea Ice
Geophysical Monograph 68
©1992 American Geophysical Union

Given the complexity of this system, it is unlikely that we can look forward to deterministic solutions to sea-ice remote-sensing problems. The electromagnetic scattering models that predict backscatter and emission will often require more parameters than we can ever hope to provide. The utility of these theoretical models lies more in their ability to illuminate physical mechanisms, rather than act as operational geophysical algorithms. Our understanding of the physical mechanisms will help delineate the limitations of our information fusion techniques, improve empirical models (e.g., the passive microwave algorithms), and establish rules of thumb.

Notwithstanding any particular geophysical problem, the most important service that remote sensing can render to the sea ice community is ice type discrimination and the related problem of ice edge location. The way in which ice is categorized is, to a certain extent, driven by the motives of the user. Climatologists, biologists, and ship captains are each interested in different aspects of the ice and will necessarily use different taxonomies. The ice type categories may also be driven by the data at hand; the World Meteorological Organization (WMO) categories, for example, were determined from the appearance of ice in black-and-white photographs. However, we may categorize ice into four generally useful types: open water, young, first year, and old. Deformation features such as rubble fields, ridge lines, cracks, rafting, and the like, are often, of necessity, an ice type in themselves. This is because these features can confound the ice-classification ability of high-resolution sensors, especially radar (the effects of deformation on passive sensors are largely unknown) [Drinkwater et al., 1991]. Success in dealing with this problem will depend on our ability to utilize all the information at our disposal.

25.3 A SIMPLE MODEL OF INFORMATION FUSION

Information fusion has become an essential component of several fields, including computational vision [Clark and Yuille, 1990; Szeliski, 1989], military threat assessment [Waltz and Llinas, 1990; Camparato, 1988], and robotics [Brady, 1988], and in some remote sensing problems. In general, some form of information fusion must be used:
(1) When the problem to be solved is underconstrained, i.e., there are in infinite number of solutions
(2) When the relevant information seems to conflict
(3) When the relevant information is of a different type, e.g., terrain height, optical radiance, and backscatter coefficient

Almost all remote sensing problems fall into the first category, and the pressing need to incorporate all available information into the solution forces problems into the other two categories.

To facilitate an understanding of information fusion, we propose the general model shown in Figure 25-1. This model has been synthesized from several more specialized models [Clark and Yuille, 1990; Marr, 1982; Garvey, 1987a, b; Hanson et al., 1988]. It also has been suggested in a more philosophical context [Eells, 1982]. This model contains three levels of abstraction

(1) data includes the image pixels, the altimeter time series, and other forms of raw data

(2) information contains specific measurements made of the data that make important information explicit

(3) hypothesis is a qualified solution to the problem—it is the result of an inference procedure that utilizes all relevant information

Reducing the many complex methods of information fusion to such a simple framework focuses on the two most important processes in the system, namely representation and inference. While this review focuses on inference techniques, representation is of equal importance.

25.4 REPRESENTATION

As defined by Marr [1982], a representation is a system for making certain kinds of information explicit, together with a specification of how this is to be done. The result of applying a particular representation may be called a description or a token [Marr, 1982; Hanson et al., 1988]. It is at this level that we perform the fusion or integration, rather than at the sensed data level (although the sensed data comprise an information channel in themselves). There is a trade-off in applying a particular representation; while some information is made more explicit, other information becomes inaccessible. Hence, choosing a representation will greatly affect our ability to use the information. Most remote sensing problems call for several representations.

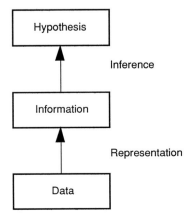

Fig. 25-1. A simple model for information fusion.

A familiar example of an important representation is the discrete Fourier transform that represents a signal by the superposition of a finite number of complex sinusoids and provides a mechanism for doing so. Another more general framework for representing information that has been extremely popular in remote sensing is the feature space. The axes of this multidimensional space correspond to individual measurements or derived features. This representation is the basis for the passive-microwave sea-ice algorithms [Cavalieri et al., 1984; Steffen and Schweiger, 1990] and has also been used in altimeter wave-form studies [Fetterer et al., 1991] and multisensor studies [Livingstone et al., 1987a, b; Drinkwater et al., 1991; Shuchman et al., 1991].

Unfortunately there have been very few published experiments that use derived features, as well as sensed data, as axes in a feature space. Features such as edges, lines, and texture primitives, which incorporate spatial context, should be incorporated into the feature space. If we have data from a well-known ice type, the feature space may be partitioned to emphasize certain aspects of the data. This is often called supervised classification, and is the first step towards inference.

There are several problems with the feature space representation that will ultimately limit its utility. The first is that it is difficult to incorporate certain vital types of information, such as season and location. The second problem is that most of the inference procedures that utilize the feature space are statistical in nature. This means that it is difficult to exploit any nonstatistical relationships between the axes, such as the reciprocity relation between active and passive microwave data.

25.5 INFERENCE

Inference systems are often theoretical models that treat information and knowledge in a formal way and use either deductive or inductive reasoning to reach a conclusion. While there may be practical difficulties in implementing a theoretical reasoning system, a major benefit of the theoretical over the heuristic methods is that the effects of computational compromises may be monitored [Pearl, 1988].

Just as information is a particular representation of data, an inference formalism is a way of representing the whole problem and, hence, its primary importance. Each formalism has its own semantic structure that imposes a meaning on all elements of the problem, both the solution (level 3) and the information (level 2). The actual information extracted from the data may be incompatible with this meaning. It is very important to investigate representation and inference in a concerted way.

If the system is to be useful, it is important that the solution be understandable, i.e., the system must be able to provide some rational explanation for its findings. This places special restrictions on the way inference may be performed. While inference systems may be categorized in a variety of ways [Pearl, 1988], we may classify them simply as logical and probabalistic [Garvey, 1987a]. Logical reasoning is deductive and is most useful in situations where the information is strictly Boolean (true or false). The basic rules of inference are the propositional and predicate calculus, but fundamental deficiencies with these systems have led some researchers to develop new logics and other researchers to reject the logic formalism altogether. Suffice to say that formal logic is used rarely in the solution of remote sensing problems.

25.5.1 Probabilistic Inference

Distinct from a logical approach, probabalistic inference recognizes that real world knowledge is uncertain. As stated by Pearl [1988], "The very act of preparing knowledge to support reasoning requires that we leave many facts unknown, unsaid, or crudely summarized." An important feature of the probabalistic approach is that it may operate in both an inductive and a deductive mode. The former mode is one of the fundamental methods of scientific investigation, the iterative process of hypothesis and experiment. However, probabalistic inference may also operate deductively, and the problem of ice type discrimination demands this mode of reasoning.

While many researchers believe that probability offers the ideal framework for reasoning under uncertainty, this view is not unanimous. A lively debate exists in the artificial intelligence community between logicists and probabalists, and even within the latter camp there is substantial disagreement [Cheeseman, 1983]. Part of the controversy centers on the nature of probability itself, for which there are at least three distinct definitions [Fischler and Firschein, 1987; Shafer, 1988]

(1) A measured frequency in situations that are like games of chance, i.e., repetition under fixed conditions is possible

(2) A subjective and numerical degree of belief that rational people assign to every proposition and event

(3) A numerical measure that describes the logical relationship between one set of propositions (evidence) and another (hypothesis)

These definitions are more than fodder for obscure philosophical discussions, they impact directly upon the ability of an inference model to integrate diverse forms of information. Remote sensing scientists have traditionally used definition (1). Until quite recently, remote sensing data have consisted of single sensor, often multispectral, visible or infrared imagery. The assumption was usually valid that radiance measurements from a pure sample of a particular target class could be modeled by a particular (invariably Gaussian) probability distribution function.

As more satellites are launched, a greater diversity of sensor data is becoming available over the same area. Remote sensing scientists have begun to look for methods of integrating multisensor imagery as well as image measurements and auxiliary, often nonnumeric, information. In addition, the past two decades have witnessed great advances in image processing, and many representations have been developed focusing on image information other than the simple numeric value of the pixel. It was discovered rather quickly that this additional information resists the more frequent, or multivariate, model, and alternatives have been sought.

25.5.1.1. Multivariate classification. In remote sensing, inference procedures are usually called classification or labeling. Before classification algorithms can be trusted in an operational setting (i.e., with predictable performance), they must be trained. This is the process by which the necessary parameters in the inference model are set to allow identification of a predetermined set of patterns or classes, $\vec{u}$. Training involves using subsets of data that are known to belong to a particular class and estimating parameters relating to a probability distribution. The data, $\vec{d}$, are generally treated as a vector, which may be a problem when dealing with multiple sensors and will almost certainly be a problem when dealing with diverse data sources.

The vector elements are now treated as statistically independent and assumed to be normally distributed. The labeling decisions are usually made with some form of Bayes' rule

$$p\left(\vec{u}\,|\,\vec{d}\right) = \frac{p\left(\vec{d}\,|\,\vec{u}\right)p(\vec{u})}{p(\vec{d})}$$

where, as usual, $p(\vec{d}\,|\,\vec{u})$ is the conditional probability of the data given the label set, u; $p(\vec{u})$ is the a priori probability of the label set elements; and $p(\vec{d})$ is the a priori probability of the data. The last term is a normalization that is a product of $p(\vec{d}\,|\,\vec{u})$ and $p(\vec{u})$ and is often omitted. The first term, also known as the likelihood function, characterizes the information by the observations, while the second term describes our state of knowledge before the observations were taken [Dempster, 1968]. Remote sensing applications of Bayes' rule have been interpreted strictly in a most frequent sense, leading to, for example, maximum likelihood (ML) classification.

Many researchers have found that traditional multivariate classification procedures do not lend themselves to multisource, and especially multitype, information. Several mechanisms have been developed by which nonspectral information may be integrated into the Bayesian classification framework. Richards et al. [1982] used an iterative supervised relaxation labeling procedure [Richards et al., 1981] to update a set of labels using the probability that a particular label is correct, as well as using spatial context to significantly improve classification accuracies of forest type using Skylab multispectral scanner data with elevation as auxiliary information.

Lee et al. [1987] noted that the data vector approach was problematic even when the sensor data are dissimilar, i.e., backscatter coefficients versus brightness temperatures. Their approach was to segment the individual data sources into clusters. These clusters were formed in a variety of ways, such as principal component or factor analysis, and represent the feature space information in the individual channels. Lee and his coworkers then defined a global membership function, Bayes' rule, which defines the relationship between the data clusters and a predefined set of labels. This relationship, as well as that between the data and the data clusters, was represented with conditional probabilities. The method was tested on Landsat Multispectral Scanner (MSS) data of an agricultural scene. Benediktsson and Swain [1989] improved the handling of the reliabilities and tested the system on a more diverse forestry data set, including Landsat MSS, elevation, slope, and aspect angle. The multisource information improved the classification accuracy over what was achieved with Landsat alone.

25.5.1.2. Subjective Bayesian analysis.

While Bayes' rule can be used with a frequentist interpretation of probability, its real power is unleashed only with the more flexible subjective perspective. (Details on the subjective probability and the subjectivist and most frequent controversy may be found in, for example, Weber [1973], Box and Tiao [1973], and Berger [1985].)

Unfortunately, to this author's knowledge, there have been no published remote sensing experiments using the subjective Bayesian paradigm. However, the method has been used very effectively in computational vision [Szeliski, 1989; Clark and Yuille, 1990], where regularization theory, Markov random fields, and Gibbs distributions have been used to facilitate the solution to low-level vision problems. The approach taken emphasizes that, given the raw sensor data, the number of possible solutions to the classification problem is large. A particular solution may only be found by constraining the search with extra information—a classical approach to solving inverse problems.

In this context, the important terms in Bayes' rule may be interpreted in a more general way. The conditional probability, $p(\vec{d} \mid u)$, corresponds to a model describing how a set of terrain classes, $\vec{u}$, could produce the sensor data, $\vec{d}$. This model is termed the image formation or measurement model. It contains any physical constraints that result from our knowledge of measurement theory (e.g., scattering models), as well as some representation of the uncertainties inherent in the measurement process. The a priori probability, $\vec{u}$, measures the likelihood of the set of terrain classes before the data were collected. As stated by Clark and Yuille [1990], "The role of the a priori constraints is to select a unique solution out of the possible infinite interpretations of the data." It is this prior knowledge (also known as the priors) that has driven the subjective perspective into disfavor due to its qualitative nature. The most popular forms of Bayesian inference are the maximum a posteriori (MAP) estimate, which maximizes $p(\vec{u} \mid \vec{d})$, and the maximum likelihood estimate (MLE), which is the MAP estimate with a uniform $p(\vec{u})$ (i.e., all solutions are equally likely).

Computational vision problems, like geophysical remote-sensing problems, are ill posed, i.e., the solution is not unique, and they are not stable with respect to small perturbations in the data. Solutions to ill-posed problems may be sought through regularization theory [Tikonov and Arsenin, 1977], which involves minimizing an energy functional of the form

$$E[u(x)] = \int \{u(x) - d(x)\}^2 \, dx + \lambda \int \{Lu(x)\}^2 \, dx$$

The first term, the data term, ensures that the solution is close to the data. The second term, the regularizer, imposes a smoothness constraint on the solution through the differential operator, L. The parameter, λ, determines the relative importance of the two terms.

Regularized Bayesian solutions are facilitated through the use of Markov random fields (MRF) and Gibbs distribution. These concepts have been borrowed from statistical physics and are most helpful in dealing with the priors. An MRF is a probability distribution defined over a discrete field such that the probability of an individual element, u_i, is dependent on only a small neighborhood

$$p(u_i \mid \vec{u}) = p(u_i \mid \{u_j\}), \quad j \in N$$

To illustrate the potential of this approach for solving remote sensing problems, we shall follow the simple example given by Szeliski [1989], who suggested that these conditional probabilities could be used to represent particular configurations of image labels. These conditional probabilities may be used to generate the priors through the use of a Gibbs distribution given by

$$p(\vec{u}) = Z_p^{-1} \exp\left\{ \frac{-E_p(\vec{u})}{T_p} \right\}$$

The important term in this equation is the energy function, E_p, which may be written as the sum of localized energy functions whose form we are at liberty to prescribe. A trivial example that simply counts distinct pixels is

$$E_p(u) = \sum_{i,j} \left(1 - \delta(u_{i,j}, u_{i+1,j})\right) + \left(1 - \delta(u_{i,j}, u_{i,j+1})\right)$$

The specification of the image formation model requires a careful consideration of the physics involved in the imaging process. For the purposes of our simple example, we shall use a truncated Gaussian distribution function

$$p(d_i \mid u_i = \alpha) = Z_\alpha^{-1} \exp\left\{ \frac{(d_i - u_\alpha)^2}{2\sigma_\alpha^2} \right\}$$

Further, let us associate the log probability of the data distribution with an energy function

$$E_d^*(u_i, d_i) = -\ln p\,(d_i, u_i)$$

hence,

$$p\left(\vec{d}\,|\,\vec{u}\right) = Z_d^{-1} \exp\left\{-E_d\left(\vec{u}, \vec{d}\right)\right\}$$

We may now substitute these expressions into an unnormalized Bayes' rule

$$p\left(\vec{u}\,|\,\vec{d}\right) = p\left(\vec{d}\,|\,\vec{u}\right) p\,(\vec{u}) = Z^{-1} \exp\left\{-E\,(u)\right\}$$

where

$$E\,(u) = \frac{E_p\,(\vec{u})}{T_p} + E_d\left(\vec{u}, \vec{d}\right)$$

From the last expression we can see that the product of probabilities in Bayes' rule has been converted into a sum of energies in the exponential. This sum bears a strong similarity to the regularization equation; we can see the role of the priors is to smooth or constrain the solution.

Time may be introduced into the Bayesian framework through the use of the Kalman filter, a Bayesian estimation technique that tracks a stochastic dynamic system being observed with noisy sensors (Chapter 23). Bayesian inference remains to be applied to multisource classification problems where the quantity of data is much greater and our understanding of the physics is more formative.

25.5.1.3. Theory of belief functions. The theory of belief functions, the Dempster–Shafer (D–S) theory, and evidential reasoning are methodologies that sidestep some of the difficulties of Bayesian reasoning by attempting to compute the probability of provability, rather than going straight for the "truth" [Pearl, 1988]. The approach views sensor data and all other relevant information as providing evidence for a set of predefined interpretations. This approach recognizes and tries to represent the fact that this evidence may be incomplete, uncertain, and inconsistent [Shafer, 1976; Pearl, 1988; Garvey, 1987b; Srinivasan and Richards, 1990].

Let us assume that a particular pixel can have one of four possible labels: water, young ice (YI), first year ice (FYI), and old ice (OI). The set of these labels $\Theta = \{$water, YI, FYI, OI$\}$ is called the frame of discernment. Any subset of Θ will contain a possible choice of labels for the pixel, e.g., {YI, OI}, indicating that the pixel could be either young ice or old ice. The belief that a pixel has a particular label or label set is indicated by a number between 0 and 1 and is called the mass, m. Hence $m(\{$YI, OI$\}) = 0.3$ indicates that the evidence at hand warrants a 30% belief that the pixel is either young ice or old ice. The belief in a set (denoted **Bel**) is the total mass committed to its subsets.

A prominent attribute of the theory, and one that makes it intuitively appealing as a reasoning system, is that the belief assigned to a group of labels is nonadditive, i.e., $\mathbf{Bel}(A) + \mathbf{Bel}(\neg A) \neq 1$, where $\mathbf{Bel}(\neg A)$ is the belief against A. Hence, the system allows a clear representation of ignorance: $1 - \{\mathbf{Bel}(A) + \mathbf{Bel}(\neg A)\}$. Another consequence of the nonadditivity is that belief is represented by an interval, rather than a single number (e.g., a point probability). This interval is bounded on one side by belief, $\mathbf{Bel}(A)$, and on the other side by plausibility, $1 - \mathbf{Bel}(\neg A)$.

The concept of a lack of belief (ignorance) as opposed to disbelief is an important one, and allows us to suspend judgment by assigning belief to a set of labels, rather than an individual label. Complete ignorance is indicated by assigned belief to the entire frame of discernment, rather than a subset. As more evidence becomes available, some portion of the suspended belief is assigned to a smaller subset or an individual label.

Two independent bodies of evidence may be combined via Dempster's rule of combination. Hence, we have

$$m\,(X) = K^{-1} \sum_{Y \cap Z = X} m_1(Y)\, m_2(Z)$$

where K is a normalizing constant given by

$$K = 1 - \sum_{Y \cap Z = \varnothing} m_1(Y)\, m_2(Z)$$

While the rule deals with only two bodies of evidence at a time, any number of beliefs may be integrated through repeated application, and the commutativity of multiplication assures that the order of integration does not matter.

The theory of belief functions has received little attention in remote sensing. Lee et al. [1987] experimented with it on a Landsat MSS data set and the results indicated that the performance was at least comparable to their modified multivariate method (see Section 25.5.1.1).

Kim [1990] implemented the theory more rigorously and applied it to a data set that included four channels of imagery from an airborne MSS, shallow and steep mode SAR (X- and L-band, hh, and hv pol, polarizations), and a digital elevation model (DEM). He found that the method gave superior performance to a MAP classifier.

Srinivasan and Richards [1990] implemented a modified form of the theory within a larger knowledge-based system. They recognized the problem of having an exhaustive and exclusive frame of discernment and the computational load implied by having to deal with all subsets. As their system is essentially rule based, we shall defer reporting their results to Section 25.6.

The theory of belief functions may be implemented in a variety of ways, many of which are arguably equivalent to an appropriately set-up Bayesian system [Buede, 1989; Cheeseman, 1985]. Many inference theorists are unhappy with the way in which ignorance is handled, the way Dempster's rule is normalized, and the way the belief is revised [Pearl, 1988]. While there are many subtleties that must be recognized before building such a system, belief functions offer a rational and defensible alternative for managing multisource information

25.5.2 Other Inference Formalisms

25.5.2.1. Fuzzy logic and fuzzy classification. Fuzzy classification is based on the concept of the fuzzy set whose development was initiated some 25 years ago most notably by L. Zadeh [Zadeh, 1965]. Fuzzy set theory is predicated on the notion of vagueness, which flows naturally from people's attempts to categorize the world. Whereas the classical set (or crisp set in the jargon of fuzzy set theory) demands that elements either belong to the set or not, fuzzy sets are defined by a list of elements and their associated degree of membership. As stated by Zadeh [Zadeh, 1965], the fuzzy set "provides a natural way of dealing with problems in which the source of imprecision is the absence of sharply defined criteria rather than the presence of random variables."

Fuzzy set theory lays down a rigorous mathematical framework for dealing with vague phenomena. A prominent application was the development of fuzzy logic and the associated possibility theory [Zadeh, 1988; de Mantaras, 1990]. These formalisms allow the representation of both the vagueness of terms such as "young ice" and the uncertainty inherent in our measurements. Fuzzy logic, like belief functions, is used in the development of general computational reasoning systems, as well as special purpose expert systems. However, unlike the D–S theory, which is an extension of probability theory, fuzzy logic is an extension of classical Boolean logic, where "a possibly imprecise conclusion is deduced from a collection of imprecise premises" [Dubois and Prade, 1980]. Recently, however, there has been some work to generalize the D–S theory to embrace the notion of vagueness [Yen, 1990]. There have been few applications of fuzzy logic to geophysical remote-sensing problems—an exception was the development of a fuzzy expert system for estimating rainfall rates from satellite-borne microwave radiometer data [Oh, 1988].

One of the first areas in which the power of fuzzy sets was harnessed was pattern classification [Bellman et al., 1966], and there have since been many applications to remote sensing. The majority of these applications have been unsupervised partitioning of the feature space, and several clustering algorithms have been generalized to handle the fuzzy set [Bezdek, 1981; Kandel, 1982]. An important advantage of fuzzy clustering algorithms is that pixels are not constrained to belong to a single class such as first-year ice. Instead, their membership in each class is computed; hence, the reality of mixed pixels is represented directly. The most popular fuzzy clustering method is the fuzzy c-means (FCM) algorithm. The applications have included Landsat MSS data collected over urban areas [Trevidi and Bezdek, 1986; Fisher and Pathirana, 1990], Advanced Very High-Resolution Radiometer (AVHRR) data collected over agricultural land [Cannon et al., 1986], and AVHRR data collected over polar regions [Key et al., 1989a]. Findings indicated that the FCM produces physically meaningful clusters and realistic estimates of class mixtures.

Wang [1990a, b] recently devised a supervised classification algorithm based on the fuzzy set. He found that his system provided enhanced information through pixel mixture estimates and improved classification accuracy when tested against conventional multivariate classifiers.

Unfortunately, fuzzy pattern recognition is prone to all the same problems that plague the multivariate techniques when dealing with multitype data. In spite of this, it has proven to be extremely effective in dealing with mixed pixel problems. However, it appears that the paradigm of reasoning, with its weighing of evidence, seems more appropriate as the overall framework for the problem. Pattern recognition is useful in empirical investigations into the information content of the data [Watanabe, 1985], but falls short in the more general context.

25.5.2.2. Neural networks. Neural networks represent a radical shift in their predication on an architectural, rather than an algorithmic paradigm. Such networks were originally focused on modeling the neural physiology of the brain. It was, and is, thought that the possibilities of a massively parallel distribution of simple processing elements can never be realized with conventional von Neuman computers. Excellent reviews of the technology may be found in Beale and Jackson [1990], Lippmann [1987, 1989], and Kohonen [1988, 1989].

The fundamental processing element in a neural network is a neuron (Figure 25-2). Neurons generally have several individually weighted inputs, a summation operator to combine the effects of the inputs, and a single output controlled by a nonlinear threshold. Neural nets are distinguished by the form of the input data (discrete or continuous), the type of nonlinear threshold performed, the overall topology, and the way in which the system learns to recognize new patterns through an iterative adjustment of its weights.

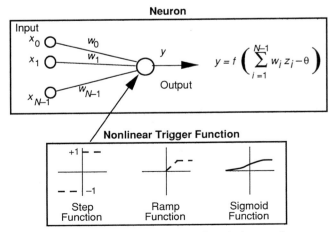

Fig. 25-2. The fundamental processing element in neural processing is the neuron, which performs the weighted sum of N inputs and passes the result through a nonlinearity. Three typical nonlinearities are shown (adapted from [Lippmann, 1987]).

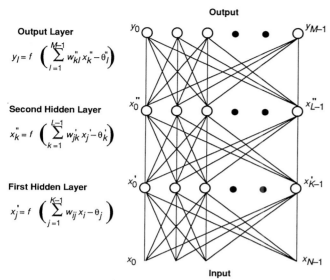

Output Layer

$$y_l = f\left(\sum_{l=1}^{M-1} w_{kl}'' x_k'' - \theta_l''\right)$$

Second Hidden Layer

$$x_k'' = f\left(\sum_{k=1}^{L-1} w_{jk}' x_j' - \theta_k'\right)$$

First Hidden Layer

$$x_j' = f\left(\sum_{j=1}^{K-1} w_{ij} x_i - \theta_j\right)$$

Fig. 25-3. A three layer perceptron with N input channels, M output channels, and two layers of hidden nodes. In the formulae, x_j' and x_j'' are the outputs of the nodes in the first and second hidden layers, Θ_j' and Θ_j'' are the internal offsets in those nodes, w_{ij} is the connection strength from the input to first hidden layer, and w_{ij}' and w_{ij}'' are the connection strengths between the first and second and between the second and output layers, respectively (adapted from [Lippmann, 1987]).

Neural networks have found great utility as pattern classifiers, where they can operate in either a supervised or an unsupervised mode (although particular network types cannot perform both). A distinctive characteristic of neural networks is that they do not attempt to model the data; hence, they are completely nonparametric. The most popular type of net used in remote sensing applications is called the multilayer perceptron with some back propagation training. A typical three-layer perceptron is shown in Figure 25-3. It has been shown that this network is capable of arbitrarily complex segmentation of a feature space [Beale and Jackson, 1990].

Recent studies have provided evidence that neural networks can outperform conventional classification techniques in remote sensing applications with less training data [Hepner et al., 1990; Benediktsson et al., 1990]. A study by Lee et al. [1990] used raw radiances and grey level co-occurrence (GLCM) texture measurements from Landsat MSS data to segment cloud types. They not only found that neural networks could outperform conventional nonparametric classifiers, but that the trained neural system could supply information relating to the relative significance of the input features. This point was echoed in Key et al. [1989b, 1990], who used a combined AVHRR and Special Sensor Microwave Imager data set to classify clouds and land cover types in polar regions.

The use of nonstatistical methods introduces a certain ambiguity in the interpretation of the neural network approach to unsupervised clustering. In spite of this, these

methods have been used successfully: to search for maximum-entropy solutions to ill-posed problems [Marrian and Peckerar, 1989]; to compute Hartley, Fourier, and Gabor transforms [Daugman, 1988; Culhane et al., 1989]; and to perform sensor fusion [Yuhas et al., 1989; Perlovsky and McManus, 1991]. Hence, neural nets offer great potential. They can be trained with minimal data, perform well in the face of missing and noisy input data, and accept features from disparate sources/sensors. However, as with the other methods, they rely on a careful consideration of representation and information for their success.

25.6 EXPERT SYSTEMS

An expert system is a computational system for representing the knowledge of human experts and emulating their problem solving ability. It is assumed from the outset that at least one human expert is capable of solving our problem. Expert systems conform to our simple model in that the two most important processes are knowledge representation and inference. Expert systems differ markedly from the other methods discussed in this section, however, in that the paradigm is a human expert. Expert systems make a point of separating the domain knowledge required to solve the problem (knowledge base) from the problem solving mechanism (inference engine). The former contains facts (data) and rules that make use of the facts. The latter interprets the facts and determines which rules to apply and in which order. This dichotomy differs from the conventional multivariate systems, which embed a good deal of domain knowledge in the models and assumptions underlying the inference procedure.

Before an expert system can be used, some minimal amount of domain-specific knowledge must be transferred to the system from the human expert(s). This preliminary step involves choosing the appropriate symbolic representation for the rules and facts required to solve the particular problem using the chosen inference procedure. The process by which the raw data, in our case consisting mainly of digital imagery, are converted to symbolic representations of relevant information, is a matter of preprocessing. Expert systems, like the other systems discussed in this chapter, occupy level two of our model. They distinguish themselves from other methodologies in that, when they are running, they are capable of adding new facts and rules to their knowledge base just as a human expert would. There are many ways an expert system can perform inference, and several of the methods discussed earlier have been used. These include Bayesian analysis [Cheeseman, 1983; Pearl, 1988], belief functions [Gordon and Shortliffe, 1985; Srinivasan and Richards, 1990] fuzzy logic [Oh, 1988; Graham, 1991], and even neural networks [Poli et al., 1991], as well as more ad hoc procedures [Waterman, 1986]. The hallmark of the expert system is that it is capable of providing the chain of reasoning behind its answers.

The success of an expert system relies heavily on its ability to represent the knowledge of a human expert.

When this knowledge involves complex sensory input such as vision, our ability to model the brain's representation of such data is rudimentary at best. Since all geophysical remote-sensing problems involve a high level of visual expertise, it seems ill-advised to take the human out of the loop. For such problems, Mackworth [1989] suggests, instead, that we build cooperative systems whose knowledge sources include the user. Such systems keep the user and the system in symmetrical roles where each can query the other for facts, rules, explanations, interpretations, and justifications. Such a system was developed by Mackworth and his colleagues as part of the Mapsee project [Mulder et al., 1987; Reiter and Mackworth, 1989]. The user provides the system with symbolic descriptions of aerial images of small towns in the form of vectors. The system interprets these "facts" using rules dictated by a formal logical theory of sketch maps. These interpretations then provide contextual constraints to a multivariate classifier.

Several expert systems have recently been developed to solve ice type classification problems. INTERA Technologies [McAvoy and Krakowski, 1989] developed a system that uses single-channel SAR data. It was quickly discovered during the process of knowledge transfer from a number of ice analysts, that not only did the experts have great difficulty articulating how they classify SAR imagery, but they disagreed with one another on the importance of the various image features. To ameliorate this situation, the experts were each presented with three randomly selected SAR images and asked how one of them differed from the other two, and how these two were similar. The expert was then asked to provide a name for the particular image features responsible for the discrimination. These names were then pooled and analyzed to determine which were most significant. Examples of names that are linguistic representations of complex image features include angular floes, fine texture, and homogeneous tone. The rule base consisted of affirmative, supportive, and precluding rules that confirm, support, or preclude the presence of an ice type based on certain facts. An example of an affirmative rule is

> if   there is evidence of floe
>   and floe.composition is "conglomerate"
>   and floe.shape is "round" or "round-angular"
>   and tone.inner is "grey" or "bright-grey" or "bright"
> then   old-ice-3 is confirmed

The author attempted to include many of the disparate knowledge sources that ice analysts commonly use, such as season and geographical region, and to use a commercial expert system shell to perform the inference. While the inference used is off the shelf and thus contains no physically based image formation model, the system demonstrates the appropriate paradigm for future systems—that of expert advisor rather than expert replacement.

A system developed at the University of Kansas uses two expert systems to classify ice in a data set consisting of both active and passive microwave imagery. This group recognizes the importance of integrating symbolic information, rather than raw data, and have implemented their system according to our simple model. The overall system is comprised of five steps

(1) Classification: Active and passive imagery are individually classified to produce two labeled images. The classifier will be multivariate if more than one frequency or polarization is available in each of the two streams (not yet implemented), or a histogram-based thresholding operation when the streams contain a single channel.

(2) Feature Detection: A type of region-growing procedure produces regions of a single ice type (label). These regions are the designated features and include a number of associated attributes such as area, perimeter, shape, and membership of adjacent regions.

(3) Classification Correction I: Errors in the initial labeling of the individual streams are corrected using two expert systems (one per stream) whose data consist of the regions in the individual streams and whose rules have been developed in association with certain experts.

(4) Feature Correspondence: Features in the two streams are matched and their attributes are pooled. The output of this step is a single set of features with both active and passive attributes, including two, possibly different, labels.

(5) Classification Correction II: A third expert system updates the labeling based on the pooled set of attributes.

The success of the system depends rather critically on the fact that there is a single symbolic element (the simply connected region) representing all the information in the image. In many geographic regions, such as the marginal ice zone, this representation would be inadequate and, even in areas dominated by discrete floes, there is a good deal of information left out. The system leaves the actual classification to a fairly unsophisticated procedure, and then uses sophisticated techniques to correct the errors. When more pathological image data are presented (and they will be), they may commit to particular labels too early, and may not recover from any initial mistakes. By referring to our model, we could group steps 1 and 2 together as a particular representation or feature-extraction module, and we could add other such modules leading to several feature types as suggested earlier in this chapter. These feature sets may be input as facts to an inference module consisting of steps 3, 4, and 5, and the actual labeling could be deferred until this stage. In any event, focus on symbolic features and the pooling of feature attributes reminiscent of Hanson et al. [1988] and the use of these as facts in the expert systems represents a significant innovation.

25.7 CONCLUSIONS

One aspect of the problem that has not been discussed is the temporal updating of a particular set of ice type labels.

As we are most often interested in the ongoing classification of ice in a particular region (e.g., the Beaufort Sea or the Grand Banks), it is desirable to develop a system capable of updating its labels as new remote-sensing data become available. These situations are most often handled with some form of Kalman filter (see Chapter 24) that interprets the data in a most frequent sense and blends them into an ice model in a rational way. More subjective interpretations of the data have also been used within the Kalman framework in the context of visual image-motion problems [Clark and Yuille, 1990]. A more ad hoc procedure was used for many years in the Beaufort Sea to manage ice information for oil exploration operations [Melrose et al., 1982].

This chapter has attempted to introduce some theories and techniques through which multisource and multitype information may be integrated or fused to provide a mutually consistent set of sea-ice-type labels. As many of these methods are active areas of research in themselves, there is no turn-key solution to our problems. However, the time is ripe to begin to tailor these theories to sea-ice remote-sensing problems. Fortunately, as many of the inference techniques discussed have been commercialized, it may not be necessary to develop software from scratch.

Acknowledgments. The writing of this chapter was funded through projects 97605 and 97504 of the Earth Observation Lab of the Institute for Space and Terrestrial Science. I would like to acknowledge the comments and suggestions of many colleagues during the initial stages of this review. I would also like to thank the anonymous reviewers for several important suggestions.

REFERENCES

Beale, R. and T. Jackson, *Neural Computing: An Introduction*, Hilger, Bristol, UK, 240 pp., 1990.

Bellman, R. E., R. Kalaba, and L. A. Zadeh, Abstraction and pattern classification, *Journal of Mathematical Analysis and Applications, 13*, pp. 1–7, 1966.

Benediktsson, J. A. and P. H. Swain, A method of statistical multisource classification with a mechanism to weight the influence of the data source, *Proceedings of the International Geoscience and Remote Sensing Society*, pp. 517–520, European Space Agency, Vancouver, Canada, 1989.

Benediktsson, J. A., P. H. Swain, and O. K. Ersoy, Neural network approaches versus statistical methods in classification of multisource remote sensing data, *IEEE Transactions on Geoscience and Remote Sensing, 28*, pp. 540–552, 1990.

Berger, J. O., *Statistical Decision Theory and Bayesian Analysis*, Springer–Verlag, Berlin, 617 pp., 1985.

Bezdek, J. C., *Pattern Recognition With Fuzzy Objective Function Algorithms*, Plenum Press, New York, 1981.

Box, G. E. P. and G. C. Tiao, *Bayesian Inference in Statistical Analysis*, Addison–Wesley Publishing Company, Reading, Massachusetts, 588 pp., 1973.

Brady, M., Foreword—special issue on sensor data fusion, *International Journal of Robotics Research, 7,* pp. 2–4, 1988.

Buede, D. M., Shafer–Dempster and Bayesian reasoning: A response to "Shafer–Dempster reasoning with applications to multisensor target classification systems," *IEEE Transactions on Systems, Man, and Cybernetics, 18*, pp. 1009–1011, 1989.

Camparato, V. G., Fusion—the key to tactical mission success, *Sensor Fusion, Proceedings of the SPIE, 931*, pp. 2–7, 1988.

Cannon, R. L., J. V. Dave, J. C. Bezdek, and M. M. Trivedi, Segmentation of a thematic mapper image using the fuzzy c-means algorithm, *IEEE Transactions on Geoscience and Remote Sensing, GE-24*, pp. 400–408, 1986.

Cavalieri, D. J., P. Gloersen, and W. J. Campbell, Determination of sea ice parameters with the Nimbus 7 SMMR, *Journal of Geophysical Research, 89(D4)*, pp. 5355–5369, 1984.

Cheeseman, P., A method of computing generalized Bayesian probability values for expert systems, *Proceedings of the International Joint Conference on Artificial Intelligence*, Karlsruhe, Germany, pp. 198–202, 1983.

Cheeseman, P., In defense of probability, *Proceedings of the International Joint Conference on Artificial Intelligence*, Los Angeles, pp. 1002–1009, 1985.

Clark, J. J. and A. L. Yuille, *Data Fusion for Sensory Information Processing Systems*, Kluwer Academic Publishers, Norwell, Massachusetts, 242 pp., 1990.

Culhane, A. D., M. C. Peckerar, and C. R. K. Marrian, A neural network approach to discrete Hartley and Fourier transforms, *IEEE Transactions on Circuits and Systems, 36*, pp. 695–703, 1989.

Daugman, J. G., Complete 2-D Gabor transforms by neural networks for image analysis and compression, *IEEE Transactions on Acoustics, Speech, and Signal Processing, 36*, pp. 1169–1179, 1988.

de Mantaras, R. L., *Approximate Reasoning Models*, Ellis–Horwood, Chichester, UK, 109 pp., 1990.

Dempster, A. P., A generalization of Bayesian inference, *Journal of the Royal Statistical Society, B30*, pp. 205–247, 1968.

Drinkwater, M. R., J. P. Crawford, and D. J. Cavalieri, Multifrequency, multipolarization SAR and radiometer sea ice classification, *Proceedings of the IGARSS '91 Symposium, vol. 2*, IEEE catalog no. 91CH2971-0, pp. 107–111, IEEE, New York, 1991.

Dubois, D. and H. Prade, *Fuzzy Sets and Systems: Theory and Applications*, Academic Press, New York, 1980.

Eells, E., *Rational Decision and Causality*, Cambridge University Press, Cambridge, UK, 234 pp., 1982.

Fetterer, F. M., S. Laxon, and D. R. Johnson, A comparison of Geosat altimeter and SAR measurements over east Greenland pack ice, *International Journal of Remote Sensing, 12(3)*, pp. 569–583, 1991.

Fischler, M. A. and O. Firschein, *Intelligence: The Eye, the Brain and the Computer*, Addison–Wesley Publishing Company, Reading, Massachusetts, 331 pp., 1987.

Fisher, P. F. and S. Pathirana, The evaluation of fuzzy membership of land cover classes in the suburban zone, *Remote Sensing of Environment, 34*, pp. 121–132, 1990.

Garvey, T. D., A survey of AI approaches to the integration of information, *Infrared Sensors and Sensor Fusion, Proceedings of the SPIE, 782*, pp. 68–82, 1987a.

Garvey, T. D., Evidential reasoning for geographic evaluation for helicopter route planning, *IEEE Transactions on Geoscience and Remote Sensing, 25*, pp. 294–304, 1987b.

Gordon, J. and E. H. Shortliffe, A method for managing evidential reasoning in a hierarchical hypothesis space, *Artificial Intelligence, 26*, pp. 323–377, 1985.

Graham, I., Fuzzy logic in commercial expert systems—results and prospects, *Fuzzy Sets and Systems, 40(3)*, pp. 451–472, 1991.

Hanson, A. R., E. M. Riseman, and T. D. Williams, Sensor and information fusion from knowledge-based constraints, *Sensor Fusion, Proceedings of the SPIE, 931*, pp. 186–196, 1988.

Hepner, G. F., T. Logan, N. Ritter, and N. Bryant, Artificial neural network classification using a minimal training set: Comparison to conventional supervised classification, *Photogrammetric Engineering and Remote Sensing, 56*, pp. 469–473, 1990.

Kandel, A., *Fuzzy Techniques in Pattern Recognition*, John Wiley and Sons, New York, 356 pp., 1982.

Key, J. R., J. A. Maslanik, and A. J. Schweiger, Classification of merged AVHRR and SMRR Arctic data with neural networks, *Photogrammetric Engineering and Remote Sensing, 55*, pp. 1331–1338, 1989a.

Key, J. R., J. A. Maslanik, and R. G. Barry, Cloud classification from satellite data using a fuzzy sets algorithm: A polar example, *International Journal of Remote Sensing, 10(12)*, pp. 1823–1842, 1989b.

Key, J. R., J. A. Maslanik, and A. J. Schweiger, Neural networks versus maximum likelihood classification of spectral and textural features in visible, thermal and passive microwave data, *Proceedings of the International Geoscience and Remote Sensing Society*, European Space Agency, Washington, DC, pp. 1277–1280, 1990.

Kim, H., *A Method of Classification for Multisource Data in Remote Sensing Based on Interval-Valued Probabilities*, doctoral dissertation, Purdue University, Department of Electrical Engineering, West Lafayette, Indiana, 120 pp., 1990.

Kohonen, T., An introduction to neural computing, *Neural Networks, 1*, pp. 3–16, 1988.

Kohonen, T., *Self Organization and Associate Memory*, Springer–Verlag, Berlin, 250 pp., 1989.

Lee, T., J. A. Richards, and P. H. Swain, Probabalistic and evidential approaches for multisensor data analysis, *IEEE Transactions on Geoscience and Remote Sensing, 25*, pp. 283–293, 1987.

Lee, J., R. C. Wegner, S. K. Sangupta, and R. M. Welch, A neural network approach to cloud classification, *IEEE Transactions on Geoscience and Remote Sensing, 28*, pp. 846–855, 1990.

Lippmann, R. P., An introduction to computing with neural nets, *IEEE ASSP Magazine, 4*, pp. 4–22, 1987.

Lippmann, R. P., Pattern classification using neural networks, *IEEE Communications Magazine, 27*, pp. 47–64, 1989.

Livingstone, C. E., R. G. Onstott, L. D. Arsenault, A. L. Gray, and K. P. Singh, Microwave sea ice signatures near the onset of melt, *IEEE Transactions on Geoscience and Remote Sensing, GE-25(2)*, pp. 174–187, 1987a.

Livingstone, C. E., K. P. Singh, and A. L. Gray, Seasonal and regional variations of active/passive microwave signatures of sea ice, *IEEE Transactions on Geoscience and Remote Sensing, GE-25*, pp. 159–173, 1987b.

Mackworth, A. K., Cooperative systems for perceptual tasks in a remote sensing environment, *Proceedings of the International Geoscience and Remote Sensing Society*, European Space Agency, Vancouver, Canada, pp. 819–822, 1990.

Marr, D., *Vision: A Computational Investigation Into the Human Representation and Processing of Visual Information*, W. H. Freeman and Company, New York, 397 pp., 1982.

Marrian, C. R. K. and M. C. Peckerar, Electronic neural net algorithm for maximum-entropy solutions of ill-posed problems, *IEEE Transactions on Circuits and Systems, 36*, pp. 288–294, 1989.

McAvoy, J. G. and E. M. Krakowski, A knowledge-based system for the interpretation of SAR images of sea ice, *Proceedings of the International Geoscience and Remote Sensing Society*, European Space Agency, Vancouver, Canada, pp. 844–847, 1989.

Melrose, S. K., K. W. Schuurman, C. S. Yow, J. B. Mercer, R. Routledge, and D. Trobak, A minicomputer system for ice management in support of Beaufort Sea drilling operations, paper presented at the Proceedings of the Offshore and Arctic Operations Symposium, 1982.

Mulder, J. A., A. K. Mackworth, and W. S. Havens, *Knowledge Structuring and Constraint Satisfaction: The Mapsee Approach*, Technical Report 87-21, University of British Columbia, Department of Computer Science, Vancouver, Canada, 1987.

Oh, K. W., *New Methodologies in the Design of a General Purpose Fuzzy Expert System: Applications With AI Based Precipitation Retrieval Designed for Satellite Microwave Measurements*, doctoral dissertation, Florida State University, Department of Computer Science, Tallahassee, 192 pp., 1988.

Pearl, J., *Probabalistic Reasoning in Intelligent Systems: Network of Plausible Inference*, Morgan Kaufman Publishers, San Mateo, California, 552 pp., 1988.

Perlovsky, L. I. and M. M. McManus, Maximum-likelihood, neural networks for sensor fusion and adaptive classification, *Neural Networks, 4*, pp. 89–102, 1991.

Poli, R., S. Cagnoni, R. Livi, G. Coppini, and G. Valli, A neural network expert system for diagnosing and treating hypertension, *Computer, 24*, pp. 64–71, 1991.

Reiter, R. and A. K. Mackworth, A logical framework for depiction and image interpretation, *Artificial Intelligence, 41*, pp. 125–155, 1989.

Richards, J. A., *Remote Sensing Digital Image Analysis: An Introduction*, Springer–Verlag, Berlin, 281 pp., 1986.

Richards, J. A., D. A. Landgrebe, and P. H. Swain, Pixel labelling by supervised probabalistic relaxation, *IEEE Transactions on Pattern Analysis and Machine Intelligence, 3*, pp. 188–191, 1981.

Richards, J. A., D. A. Landgrebe, and P. H. Swain, A means for utilizing ancillary information in multispectral classification, *Remote Sensing of Environment, 12*, pp. 463–477, 1982.

Shafer, G., *A Mathematical Theory of Evidence,* Princeton University Press, Princeton, 297 pp., 1976.

Shafer, G., Comments on "An inquiry into computer understanding" by P. Cheeseman, *Computer Intelligence, 4*, pp. 121–124, 1988.

Shafer, G. and T. Logan, Implementing Dempster's rule for hierarchical evidence, *Artificial Intelligence, 33*, pp. 271–298, 1987.

Shuchman, R. A., R. G. Onstott, C. C. Wackerman, C. A. Russel, L. L. Sutherland, O. M. Johannessen, J. A. Johannessen, S. Sandven, and P. Gloersen, Multifrequency SAR, SMM/I and AVHRR derived geophysical information in the marginal ice zone, *Proceedings of the International Geoscience and Remote Sensing Society*, European Space Agency, Espoo, Finland, 1991.

Srinivasan, A. and J. A. Richards, Knowledge-based techniques for multisource classification, *International Journal of Remote Sensing, 11*, pp. 505–525, 1990.

Steffen, K. and A. J. Schweiger, A multisensor approach to sea ice classification for the validation of DMSP-SMM/I passive microwave derived sea ice products, *Photogrammetric Engineering and Remote Sensing, 56(1)*, pp. 75–82, 1990.

Szeliski, R., *Bayesian Modelling of Uncertainty in Low-Level Vision*, Kluwer Academic Publishers, Norwell, Massachusetts, 198 pp., 1989.

Tikhonov, A. N. and V. Y. Arsenin, *Solution of Ill-Posed Problems*, Winston, Washington, DC, 1977.

Trevidi and Bezdek, Low-level segmentation of serial images with fuzzy clustering, *IEEE Transactions on Systems, Man, and Cybernetics, SMC-16(4)*, pp. 589–598, 1986.

Waltz, E. and J. Llinas, *Multisensor Data Fusion*, Artech House, Norwood, Massachusetts, 1990.

Wang, F., Fuzzy supervised classification of remote sensing images, *IEEE Transactions on Geoscience and Remote Sensing, 28*, pp. 194–201, 1990a.

Wang, F., Improving remote sensing image analysis through fuzzy information processing, *Photogrammetric Engineering and Remote Sensing, 56*, pp. 1163–1169, 1990b.

Watanabe, S., *Pattern Recognition: Human and Mechanical*, John Wiley and Sons, New York, 1985.

Waterman, D. A., *A Guide to Expert Systems*, Addison–Wesley Publishing, Reading, Massachusetts, 1986.

Weber, J. D., *Historical Aspects of the Bayesian Controversy*, University of Arizona, Tucson, 1973.

Yen, J., Generalizing the Dempster–Shafer theory to fuzzy sets, *IEEE Transactions on Systems, Man, and Cybernetics, 20*, pp. 559–570, 1990.

Yuhas, B. P., M. H. Goldstein, and T. J. Sejnowski, Integration of acoustic and visual signals using neural networks, *IEEE Communications Magazine, 27*, pp. 65–71, 1989.

Zadeh, L., Fuzzy sets, *Information and Control, 8*, pp. 338–353, 1965.

Zadeh, L., Fuzzy logic, *Computer, 21*, pp. 83–93, 1988.

Chapter 26. Status and Future Directions for Sea Ice Remote Sensing

FRANK D. CARSEY

Jet Propulsion Laboratory, California Institute of Technology, 4800 Oak Grove Drive, Pasadena, California 91109

ROGER G. BARRY

Cooperative Institute for Research in Environmental Sciences, University of Colorado, Boulder, Colorado 80309-0449

D. ANDREW ROTHROCK

Polar Science Center, Applied Physics Laboratory, University of Washington, 1013 NE 40th Street, Seattle, Washington 98105

WILFORD F. WEEKS

Alaska SAR Facility, Geophysical Institute, University of Alaska, Fairbanks, Alaska 99775-0800

26.1 SUMMARY

In Chapter 1 we set out with a rather short list of variables for our marching orders. We identified the need for access to information on the radiation balance, the vertical surface heat and brine fluxes, the horizontal fresh water fluxes, the processes at the ice margins, and the associated ice conditions. We also recognized the need to understand how ice conditions influence operations. We suggested that these needs could be satisfied by data on ice extent and thickness, snow depth, summer melt and melt pond coverage, ice motion and deformation, and the weather and oceanic state. What is our ability to supply this information?

• Ice Extent and Thickness Distribution. The determination of ice extent is the most important capability of microwave observations of sea ice (Chapters 1, 4, 10, 12, 13, 15, 19, and 23), and new ideas are being investigated to enhance this capability (Chapters 25 and 26). Although the rather small areas covered by frazil ice are not observed in some of the microwave data (Chapters 13, 14, and 15), in general the ice extent is well monitored. In view of the dynamic nature of sea ice, the coverage seems adequate in that the resolution of the data is comparable to a day's change in ice edge location. Information on ice thickness is less satisfactory in that it cannot, at present, be measured directly via microwave techniques (Chapter 22). Current research is under way to aid in more accurately resolving the distribution of thin ice types (Chapter 14). One problem is that the thin ice signal in the passive microwave data sets is rendered nearly unusable by the coarse resolution of the observations. New sensor technology will be needed to remedy this. The use of ice type as a thickness proxy, in substitution for the direct

resolution of the thicker ice classes, is at best a poor replacement for direct measurement (Chapters 2, 9, 10, 11, and 22). Unfortunately, we doubt that this problem will be resolved in the near future. In the marginal and seasonal ice zones, notably the Greenland and Weddell Seas, there are also numerous unresolved geophysical questions (Chapter 13). Although the quality of the data sets supporting the examination of these regions is improving, because these areas are quite complex and offer significant logistical difficulties for field operations, progress in identifying and monitoring the key surface processes is still limited (Chapters 12 through 15).

• Snow. We cannot presently effectively estimate snow depth on sea ice. There are carefully observed snow signals in the microwave data sets, and there are clear indications, even process models, of microwave sensitivity to the process of snow wetting in warm conditions (Chapters 3, 16, and 17). However, there have been insufficient opportunities for in-situ monitoring of the processes of snow cover formation, metamorphosis, and wetting with good time series of data. Snow thickness and density have great spatial variability, and the large-scale average information that would be useful, say, in thermodynamic ice models, has not been carefully formulated.

• Summer Melt and Pond Formation. The ice processes in summer, including snow melt and pond formation, have also been examined (Chapters 16 and 17), and the stages of snow wetting, snow and ice melt, and pond formation appear to be observable using microwave data. However, these results are tentative because logistical barriers have limited the studies to essentially pilot programs. Current results appear promising for the early melt and early fall, and new data from the first European Remote Sensing Satellite (ERS-1) and the Japanese Earth Resources Satellite (JERS-1) should fill in the midsummer period. Successful monitoring of the summer processes is a key to connecting data on sea ice to the current generation of global climate models, and it should have higher priority than it has received to date.

- Ice Motion and Deformation. The motion and deformation of the ice under wind and current stress are measurable in synthetic aperture radar (SAR) data, and in visible and infrared images such as those taken by the Advanced Very High-Resolution Radiometer (AVHRR). Given fine-scale ice motion data, other terms such as the estimation of freshwater fluxes due to ice melting can be made. Motion is sometimes difficult to determine in areas of high deformation, in areas of new ice, and perhaps in summer, but, in general, current capabilities for determining ice motion appear to be quite robust (Chapter 18). During the operational period for the Radar Satellite (Radarsat) (1995–2000), SAR coverage will be adequate for those needs. Finally, we need to test the geophysical validity of the ice motion data sets by determining whether the time series of the fine-scale ice motion field and estimates of thermodynamic flux fields will together yield an accurate thickness distribution, as they should.

- Weather and Oceanic State. A major gap exists with the absence of routine, high-quality information on the characteristics of the atmospheric and oceanic boundary layers in contact with the ice. The quality of the analyses for the weather over ice-covered seas is improving, but the accuracy of these analyses, in particular for the Southern Ocean, remains largely untested. Fortunately, ocean data sets are available from an expanding array of drifters and moorings, but these measurements cannot adequately monitor the seas, especially the ice-covered seas. For climatological studies, the lack of upper ocean data may prove to be very limiting.

26.2 FUTURE DIRECTIONS

It is clear that we require additional information on the microwave and physical properties of the ice, and additional development of algorithms for generating sea-ice geophysical data products, before we can adequately interpret satellite observations. This information is coming by way of in-situ observations, laboratory investigations, and theoretical modeling, with good progress noted in all these areas. Continued work in field, laboratory, and theoretical regimes is critically needed, especially on a better understanding of the processes of the summer season, the determination of snow cover properties, and the nature of the thin and thick ice species.

Sea ice investigators have, in the past, obtained useful in-situ data on various ice conditions by participating in field studies whose primary focus was not on remote-sensing technology development and testing; the microwave investigators have piggybacked on other programs. It is likely that this approach has inherent value and will continue. In fact, in areas of broad interest, such as the zones of oceanic convection, such deployments are in place now with more planned for the future. However, studies leading to microwave signature development and algorithm validation necessitate improved experimental designs. A better record and a more thorough understanding of the evolution of active and passive signatures throughout the year and at a variety of sites are needed.

The scales of variation originate variously in the ocean, the ice itself, the atmosphere, and the snow cover; together they range from submillimeter brine inclusions to storms covering thousands of kilometers, and they challenge our interpretations and confound our validation efforts. These variations are not only "noise" in a determination of a mean field, they are also the signals of processes. We have learned something about these variations, but there is much left to do in understanding them, in terms of both geophysics and algorithms accuracy.

Some clearly powerful applications are in their infancy. The use of the Kalman smoother (Chapter 24) is a prime example. With it, we can combine data sets with physical models that simulate the temporal behavior as aspects of the geophysical system; without some data assimilation techniques, we must treat each observation as an isolated event. Another research area with strong potential, not discussed in this book since adequate data are only now being collected, is the merging of active and passive microwave data sets. Given some necessary information about the microwave behavior of the thin ice species, the combination of the derived ice motion with the observed microwave radiance will make it possible to move beyond the use of simple, instantaneous ice concentration, a troublesome variable in the rapidly changing ice-covered seas. Other analyses involving the merging of data sets, including infrared and visible-light radiances, will surely evolve; the infrared data utilization is especially desirable in that infrared data can provide sea and ice surface temperatures [Key and Haefliger, 1992].

In the development of algorithms for generating geophysical data sets, there are areas, such as fluxes in polynyas, where good results have already been obtained, and there is promise in a number of additional areas, such as the evolution of thin ice and snow cover. However, a problem permeates all sea ice algorithm development—direct validation is essential, but has not as yet been accomplished. As a result, we have been forced to resort to unsatisfactory comparisons. In Chapter 11, validations consisting of two or more microwave data sets or a microwave and satellite visible (or infrared) image are discussed, and in neither case can ice type or concentration be properly validated. In the area of validation, we need the assistance of the traditional glaciological and oceanographic communities in generating data sets for comparison purposes. The sampling issue arising from comparisons of areal information and point observations must be examined. We can generate data sets that are purely measures of change in microwave properties, but they will not provide scientifically acceptable climate-change information until their relation to traditional geophysical variables is fully demonstrated.

26.3 Sea Ice in the Context of Global Change

An examination of the role of ice in the climate system, and the form that its investigation should take, is of considerable significance in any appraisal of current capabilities for monitoring the properties and processes of the ice cover.

There is currently considerable uncertainty in the resilience of sea ice cover to global warming conditions. Projections by global climate models of polar climate response to greenhouse gas-induced warming indicate the well-known increase in winter temperatures in high latitudes. The models suggest that warming may exceed 10°C for a CO_2 equivalent doubling as a result of the stability of the lower troposphere in high latitudes [Houghton et al., 1990]. In the transition seasons, the response is enhanced by the modeled temperature snow–ice albedo positive feedback. In the summer, however, the projected high latitude warming is less than the global average. Most model experiments show a substantial retreat and thinning of sea ice [Meehl and Washington, 1990], with some suggesting almost ice-free summers in the Arctic and Antarctic Oceans. However, a survey of 17 different climate models indicates that the simulated feedback of temperature snow–ice albedo effects ranges from weakly negative to strongly positive as a result of the complex interactions of shortwave and infrared radiation with snow–ice surfaces and cloud cover [Cess et al., 1991]. A further area of model shortcomings is in the treatment of snow on ice and leads. Given these uncertainties, additional data on the variability of sea ice extent, concentration, and thickness can assist in improving the parameterizations and validation of sea ice and coupled climate models [European Space Agency, 1985; Ledley, 1988; Morassutti, 1991].

Until now, only a limited quantity of ice thickness data has been available. Under-ice draft measurements have been obtained by upward-looking sonar on operational submarines and, more recently, by moored sensors. Changes have been reported in mean ice draft and open water coverage where submarine observations have been collected in the same area in different years [Wadhams, 1992]. However, even where the measurements were taken at the same season and along an identical transect, lack of knowledge of the inherent interyear variability precludes the accurate estimation of a trend. Monitoring of ice draft distribution by moored upward-looking sonars and by remotely controlled unmanned underwater vehicles is expected to provide answers to this question over the next decade. The application of accurate satellite ice algorithms incorporating improved weather filters and thin ice determinations should make significant contributions to this work.

Another prediction of the CO_2 doubling experiments is a general increase in precipitation in high latitudes, associated with increased open water and increased atmospheric vapor content. Since temperatures in winter will remain well below freezing, this may give rise to increased snowfall, as has been observed in southwestern Alaska during warm winters [Mayo and March, 1990]. Thicker snow cover may have the effect of decreasing the growth of land-fast ice, as noted at Alert, Northwest Territories in the 1970's–1980's [Brown, R. D. and P. Cole, Interannual variability of land-fast ice thickness in the Canadian High Arctic, 1950–1989, submitted to *Arctic*, 1992]. Conversely, Brown and Cole found a trend of thickening ice at Resolute associated with several years of decreased snow cover. Such local weather-induced complications show the necessity for a basin-wide analysis of changes in ice thickness, open water production, and snow properties.

26.4 Applications of Satellite Data Products

The availability of daily passive-microwave-derived ice extent and concentration data offers important opportunities for improving global weather analyses and forecasts. The provision of accurate all-weather maps of sea ice limits and concentrations within the pack ice is of direct importance for the operation of ships navigating in or near the ice [Brigham, 1991]. However, as operational weather forecasting models are upgraded to incorporate more elaborate parameterizations of surface processes, it will be necessary to incorporate the effects of turbulent fluxes of sensible and latent heat from polynyas and open leads on the atmospheric boundary layer. Data on ice concentration and roughness can also improve determinations of aerodynamic roughness lengths in the marginal ice zone and over the central pack.

The products derived from the Scanning Multichannel Microwave Radiometer and the Special Sensor Microwave/Imager are already widely distributed by the National Snow and Ice Data Center [Barry, 1991]. To these will be added the geophysical products from the Alaska SAR facility [Weeks et al., 1991]. These data sets are widely used for both analyses of ice–climate interactions [Cavalieri and Parkinson, 1981; Barry and Maslanik, 1989; Zwally and Walsh, 1987] and ice–ocean interactions [Jacobs and Comiso, 1989] to examine trends in ice extent and concentration [Parkinson and Cavalieri, 1989; Gloersen and Campbell, 1991] and to improve algorithm development [Comiso and Sullivan, 1985; Maslanik, 1992; Walters et al., 1987]. Ready availability and wide dissemination of such data sets are certain to enhance our knowledge and understanding of polar sea ice and its role in the climate system.

By the year 2000, there will be a nearly continuous record of global sea ice extent spanning more than 25 years and a time series of ice concentration data for almost that long. This will enable more definite assessments to be made concerning decadal-scale trends in ice area and should permit short-term variability and its regional expression to be defined more reliably. Eight years of SAR data will be available, which will provide a detailed look at the mechanistic interactions within the ice pack. The nature of atmospheric forcings on ice motion and the seasonal cycles of ice growth and decay will be clearer and their modeling with coupled atmosphere–ocean–ice models should be well in hand. The research community will thus be fully equipped with the knowledge and technical tools necessary to fully exploit the new multispectral data to be acquired by the Earth Observing System (EOS) sensors.

26.5 CONCLUSION

In applying remote sensing methods to observing sea ice for either research or operational purposes, one is faced with a reasonably clear-cut set of requirements and supplied with steadily improving data sets. While sea ice would indeed seem to be a reasonably simple and well-behaved material to monitor, it has proven to be far from simple in its microwave properties; the complexity of these properties is the theme throughout this book. The successful interplay of experimental and theoretical work is only just now maturing (e.g., in the examination of thin ice evolution), and this collaboration is essential to good progress. The analysis of satellite data has progressed well; there is much that we know how to do with this excellent data set, and significant augmentations of the archived sea ice data set will continue in the future. At the same time there are issues still to explore and many algorithms to develop, as we cannot yet determine many of the key state variables of the polar seas, nor can we properly validate much of our derived information. To address some variables, notably ice thickness, we clearly need additional data sets, such as the sonar ice-draft data. For new information on thin ice, summer processes, ice margin processes, and snow cover, we need continued theoretical progress, field and laboratory programs, and new data sets from improved sensors, such as the polarimetric SAR and the finer resolution microwave radiometers now in design. Work on improving our description of ice is proceeding with enthusiasm at a number of institutions, and highly interesting results have been presented at meetings in the brief period since these chapters were written. At this time, air–sea–ice interaction research is needed in the form of the generation and interpretation of accurate geophysical data sets that characterize those interactions within the global climate. This book has been written to assist that activity.

REFERENCES

Barry, R. G., Cryospheric products from the DMSP/SSM/I: Status and research applications, *Global and Planetary Change 90*, pp. 231–234, 1991.

Barry, R. G. and J. A. Maslanik, Arctic sea ice characteristics and associated atmosphere–ice interactions in summer inferred from SMMR data and drifting buoys: 1979–1986, *Geojournal, 18*, pp. 35–44, 1989.

Brigham, L. W., editor, *The Soviet Maritime Arctic* (WHOI Contribution 7609), 336 pp., Bellhaven Press, London, 1991.

Cavalieri, D. J. and C. L. Parkinson, Large-scale variations in observed Antarctic sea ice extent and associated atmospheric circulation, *Monthly Weather Review, 108*, pp. 2032–2044, 1981.

Cess, R. D., et al., Interpretation of snow-climate feedback as produced by 17 general circulation models, *Science, 253*, pp. 888–892, 1991.

Comiso, J. C. and C. W. Sullivan, Satellite microwave and in situ observations of the Weddell Sea ice cover and its marginal ice zone, *Journal of Geophysical Research, 91*, pp. 9663–9682, 1985.

European Space Agency, *A Programme for International Polar Ocean Research (PIPOR)*. ESA SP-1074, 47 pp., European Space Agency, Paris, 1990.

Gloersen, P. and W. J. Campbell, Recent variations in Arctic and Antarctic sea ice covers, *Nature, 352*, pp. 33–36, 1991.

Houghton, J. T., G. J. Jenkins, and J. J. Ephraums, editors, *Climate Change. The IPCC Scientific Assessment*, Intergovernmental Panel on Climate Change, WMP/UNEP, 365 pp., University Press, Cambridge, Massachusetts, 1990.

Jacobs, S. S. and J. C. Comiso, Sea ice and oceanic processes on the Ross Sea continental shelf, *Journal of Geophysical Research, 94*, pp. 195–211, 1989.

Key, J. and M. Haefliger, Arctic ice surface temperature retrieval from AVHRR thermal channels, *Journal of Geophysical Research, 97*, pp. 5885–5893, 1992.

Ledley, R. S., A coupled energy balance climate–sea model: Impact of sea ice and leads on climate, *Journal of Geophysical Research, 93*, pp. 15,919–15,932, 1988.

Maslanik, J. A., Effects of weather on the retrieval of sea ice concentration and ice type from passive microwave data, *International Journal of Remote Sensing, 13*, pp. 37–54, 1992.

Mayo, L. R. and R. S. March, Air temperature and precipitation at Wolverine Glacier, Alaska: Glacier growth in a warmer wetter climate, *Annals of Glaciology, 14*, pp. 191–194, 1990.

Meehl, G. A. and W. M. Washington, CO_2 climate sensitivity and snow–sea ice-albedo parameterization in an atmospheric GCM coupled to a mixed-layer ocean model, *Climatic Change, 16*, pp. 283–306, 1990.

Morassutti, M. P., Climate model sensitivity to sea ice albedo parameterization, *Theoretical and Applied Climatology, 44*, pp. 25–36, 1991.

Parkinson, C. L. and D. J. Cavalieri, Arctic sea ice 1973–1987: Seasonal, regional and interannual variability, *Journal of Geophysical Research, 94(14)*, pp. 499–523, 1989.

Wadhams, P., Sea ice distribution in the Greenland Sea and Eurasian Basin, *Journal of Geophysical Research, 97*, pp. 5331–5348, 1992.

Walters, J. M., C. Ruff, and C. T. Swift, A microwave radiometer weather-correcting sea ice algorithm, *Journal of Geophysical Research, 92*, pp. 6521–6534, 1987.

Weeks, W. F., G. Weller, and F. D. Carsey, The polar oceans program of the Alaska SAR Facility, *Arctic, 44, Supplement 1*, pp. 1–10, 1991.

Zwally, H. J. and J. E. Walsh, Comparison of observed and modeled ice motion in the Arctic Ocean, *Annals of Glaciology, 9*, pp. 136–139, 1987.

Appendix A. Glossary of Ice Terminology

BRASH ICE

Accumulation of floating ice made up of fragments not more than 2 m across (small ice cakes); the wreckage of other forms of ice.

BREAKUP

A general expression applied to the formation of a large number of fractures through a compact ice cover, followed by a rapid diverging motion of the separate fragments.

CANDLING

The separation of the elongate ice crystals in fresh and brackish-water ice into individual crystals (candles) as the result of differential melting along grain boundaries caused by the absorption of solar radiation.

CONCENTRATION

The ratio, in tenths, of the sea surface actually covered by ice to the total sea surface area, both ice-covered and ice-free at a specific location or over a defined area (cf. ice cover). May be expressed in the following terms:

• Compact pack ice—Concentration 10/10, no water visible.

• Consolidated pack ice—Concentration 10/10, floes frozen together.

• Very close pack ice—Concentration 9/10 to less than 10/10.

• Close pack ice—Concentration 7/10 to 8/10, floes mostly in contact.

• Open pack ice—Concentration 4/10 to 6/10, many leads and polynyas, floes generally not in contact.

• Very open pack ice—Concentration 1/10 to 3/10.

CRACK

Any fracture that has not yet parted.

DEFORMED ICE

A general term for ice that has been squeezed together and in places forced upward (and downward). Forms of deformation include rafting, ridging, and hummocking.

DIVERGENCE

Formally defined as div v_i = F($\partial U_x, \partial x$) + F($\partial V_y, \partial y$) where v_i is the ice drift velocity. The divergence can be considered as the change in area per unit area at a given point. This word is also used to indicate a general diverging motion in the ice.

DRAFT

The distance, measured normal to the sea surface, between the lower surface of the ice and the water level.

FAST ICE

Sea ice of any origin that remains fast (attached with little horizontal motion) along a coast or to some other fixed object.

FINGER RAFTING

A type of rafting whereby interlocking thrusts are formed, each floe thrusting "fingers" alternately over and under the other. Common in nilas and gray ice.

FIRST-YEAR ICE

Sea ice of not more than one winter's growth, developing from young ice and having a thickness of 30 cm to 3 m. May be subdivided into thin first-year ice or white ice (30–70 cm), medium first-year ice (70–120 cm), and thick first-year ice (over 120 cm).

FLAW

A narrow separation zone between pack ice and fast ice, where the pieces of ice are in a chaotic state that forms when pack ice shears under the effects of a strong wind or current along the fast ice boundary.

FLAW LEAD

A lead between pack ice and fast ice.

FLOE

Any relatively flat piece of sea ice 20 m or more across (cf. ice cake). Floes are subdivided according to horizontal extent:

Giant floe—more than 10 km across.
Vast floe—2–10 km across.
Big floe—500–2000 m across.
Medium floe—100–500 m across.
Small floe—20–100 m across.

According to some users, a floe is an area of the ice pack that moves as a rigid element.

FLOODED ICE

Sea ice that has been flooded by meltwater or river water and is heavily loaded with water and wet snow.

FRACTURE

Any break or rupture through very close, compact, or consolidated pack ice (see concentration), fast ice, or a single floe resulting from deformation processes (cf. lead). Fractures may contain brash ice and be covered with nilas or young ice. Their length may be a few meters or many kilometers.

447

FRAZIL ICE

Fine spicules or plates of ice suspended in water. It also sometimes forms at some depth, at an interface between water bodies of different physical characteristics, and floats to the surface. It may rapidly cover wide areas of water.

FREEBOARD

The distance, measured normal to the sea surface, between the upper surface of the ice and the water level.

GREASE ICE

A stage of freezing, later than that of frazil ice, in which the crystals have coagulated to form a soupy layer on the surface. Grease ice reflects little light, giving the sea a matte appearance.

GRAY ICE

Young ice, 10–15 cm thick. Less elastic than nilas, it breaks on swell. Usually rafts under pressure.

GRAY-WHITE ICE

Young ice, 15–30 cm thick. Under pressure, it is more likely to ridge than to raft.

GROUNDED ICE

Floating ice (e.g., ridge, hummock, or ice island) that is aground (stranded) in shoal water.

HUMMOCK

The raised area of multiyear ice formed by the ablation of the surrounding ice. Also, a hillock of broken ice that has been forced upward by pressure. May be fresh or weathered. The submerged volume of ice under the hummock is called a bummock.

ICEBERG

A massive piece of ice of greatly varying shape, with a freeboard of more than 5 m, which has broken away from a glacier and may be afloat or aground.

ICE CAKE

Any relatively flat piece of sea ice less than 20 m across (cf. floe). If less than 2 m across, it is a small ice cake.

ICE COVER

The ratio of an area of ice of any concentration to the total sea surface area within some large geographic locale; this locale may be global, hemispheric, or prescribed by a specific oceanographic entity, such as Baffin Bay or the Barents Sea.

ICE EDGE

The demarcation at any given time between the open sea and sea ice of any kind, whether fast or drifting.

ICE FIELD

Area of pack ice greater than 10 km across, consisting of floes of any size. Subdivided as:

Large ice field—more than 20 km across.
Medium ice field—15–20 km across.
Small ice field—10–15 km across.

ICE FREE

No sea ice present. There may, however, be some icebergs present (cf. open water).

ICE ISLAND

A large piece of floating ice, with a freeboard of approximately 5 m, which has broken away from an Arctic ice shelf. Ice islands usually have a thickness of 30–50 m, an area of from a few thousand square meters to several hundred square kilometers, and a regularly undulating upper surface.

ICE LIMIT

Climatological term referring to the extreme minimum or extreme maximum extent of the ice edge in any given month or period, based on observations over a number of years.

ICE RIND

A brittle shiny crust of ice formed on a quiet surface by direct freezing or from grease ice, usually in water of low salinity. This crust has a thickness to about 5 cm, and is easily broken by wind or swell, commonly breaking in rectangular pieces (cf. nilas).

ICE SHELF

A floating ice sheet of considerable thickness, showing 2–50 m or more above sea level, attached to the coast. Usually of great horizontal extent and with a level or gently undulating surface. Nourished by annual snow accumulation and often by the seaward extension of land glaciers. Parts of it may be aground. The seaward edge is called an ice front.

INTERNATIONAL ICE CODE

	Thickness
New ice, including frazil, grease, slush, shuga, and pancake	<10 cm
Nilas	<10 cm
Young ice	10–30 cm
Gray ice	10–15 cm
Gray-white ice	15–30 cm
First-year ice	30–200 cm
Thin first-year or white ice	30–70 cm
Medium first-year ice	70–120 cm
Thick first-year ice	>120 cm
Old ice—second-year or multiyear ice	
Second-year ice	
Multiyear ice	

KEEL

The underside of a ridge that projects downward below the lower surface of the surrounding sea ice.

LEAD

Any fracture or passage through sea ice that is generally too wide to jump across. A lead may contain open water (open lead) or be ice-covered (frozen lead).

LEVEL ICE

Sea ice that has been unaffected by deformation.

MELT POND

An accumulation of meltwater on the surface of sea ice that, because of appreciable melting of the ice surface, exceeds 20 cm in depth, is embedded in the ice (has distinct banks of ice), and may reach tens of meters in diameter.

MULTIYEAR ICE

Old ice 2 or more meters thick that has survived a summer's melt. The color, where bare, is usually blue. The melt pattern consists of interconnecting irregular melt ponds, with a well-developed drainage system. (In other terminology, multiyear ice is ice that has survived at least two summers, cf. old ice.)

NEW ICE

A general term for recently formed ice, which includes frazil ice, grease ice, slush, and shuga. These types of ice are composed of ice crystals that are only weakly frozen together (if at all) and have a definite form only while they are afloat.

NILAS

A thin elastic crust of ice up to 10 cm thick, with a matte surface. Bends easily under pressure, thrusting in a pattern of interlocking "fingers." Dark nilas, up to 5 cm thick, is very dark in color; light nilas, 5–10 cm thick, is lighter in color (cf. ice rind).

OLD ICE

Sea ice that has survived at least one summer's melt. Most topographic features are smoother than on first-year ice. May be subdivided into second-year ice and older multiyear ice.

OPEN WATER

A large area of freely navigable water in which sea ice is present in less than 1/10 concentration (cf. ice free).

PACK ICE

Any accumulation of sea ice, other than fast ice, no matter what form it takes or how it is disposed (cf. concentration).

PANCAKE ICE

Predominantly circular pieces of ice from 30 cm to 3 m in diameter, and up to about 10 cm in thickness, with raised rims due to the pieces striking against one another. It may be formed on a slight swell from grease ice, shuga, or slush, or as a result of the breaking up of ice rind, nilas, or, under severe conditions of swell or waves, gray ice.

POLYNYA

Any nonlinearly shaped opening enclosed in ice. Polynyas may contain brash ice or be covered with new ice, nilas, or young ice. If limited on one side by the coast, it is called shore polynya; if limited by fast ice, it is called a flaw polynya. If found in the same place every year, it is called a recurring polynya.

PRESSURE RIDGE

A general expression for any elongated (in plan view) ridgelike accumulation of broken ice caused by ice deformation.

RAFTING

Process whereby one piece of ice overrides another; most obvious in new and young ice (cf. finger rafting), but common in ice of all thicknesses.

RIDGING

The process whereby ice is deformed into ridges.

RUBBLE FIELD

An area of sea ice that has essentially all been deformed.

SASTRUGI

Sharp, irregular, parallel ridges formed on a snow surface by wind erosion and deposition. On mobile floating ice, the ridges are parallel to the direction of the prevailing wind at the time they were formed.

SECOND-YEAR ICE

Old ice that has survived only one summer's melt. Because it is thicker and less dense than first-year ice, it stands higher in the water. In contrast to multiyear ice, second-year ice during the summer melt shows a regular pattern of numerous small ponds. Bare patches and ponds are usually greenish blue.

SHEAR ZONE

An area in which a large amount of shearing deformation has been concentrated.

SHORE LEAD

A lead between pack ice and the shore, or between pack ice and an ice shelf or a glacier.

SHUGA

An accumulation of spongy white ice lumps a few centimeters across, formed from grease ice or slush and sometimes from ice formed at depth rising to the surface.

SLUSH

Snow that is saturated and mixed with water on land or ice surfaces, or forms as a viscous mass floating in water after a heavy snowfall.

SNOW ICE

The equigranular ice that is produced when slush freezes completely.

YOUNG ICE

Ice in the transition stage between nilas and first-year ice, 10–30 cm in thickness. May be subdivided into gray ice and gray-white ice. Young ice is also commonly used in a more general way to indicate the complete range of ice thickness between 0 and 30 cm (as in the formation and growth of young ice). Usually these differences in meaning are clear from the context of the discussion.

Appendix B. Archived Satellite Data

Sensor	Satellite systems	Duration	Wavelength	Spatial resolution	Swath width	Temporal resolution	Orbit	Data contact
ESMR	Nimbus-5	December 1972–December 1976	1.55 cm	25 x 25 km	3000 km	Repeat coverage every 3 days	Sun-synchronous, crosstrack	NSSDC NSIDC
	Nimbus-6 (technical problems)	June 1975–April 1977	0.81 cm	20 x 43 km	1270 km			NSSDC
SMMR	Seasat	June–October 1978	Channel 1: 4.55 cm	149 x 87 km	600 km	Repeat coverage every 36 hours	Near circular, not Sun-synchronous	NSSDC JPL–NODS
			Channel 2: 2.81 cm	89 x 58 km				
			Channel 3: 1.67 cm	53 x 31 km				
			Channel 4: 1.43 cm	42 x 27 km				
			Channel 5: 0.81 cm	27 x 16 km				
	Nimbus-7	October 1978–August 1987	Channel 1: 4.52 cm	136 x 89 km	783 km	Repeat period every 6 days	Sun-synchronous, equatorial crossings, noon, midnight	NSIDC NSSDC
			Channel 2: 2.81 cm	87 x 58 km				
			Channel 3: 1.67 cm	54 x 35 km				
			Channel 4: 1.43 cm	44 x 29 km				
			Channel 5: 0.81 cm	28 x 18 km				
SSM/I	DMSP Block 5D-2 Currently active: F8 F10 F11	June 1987– July 1987– December 1990– (nonstandard orbit) November 1991–	Channel 1: 1.55 cm	70 x 45 km	1394 km	Twice daily	Circular, Sun-synchronous, conically scanning	NSIDC SDSD RSS
			Channel 2: 1.35 cm	60 x 40 km				
			Channel 3: 0.81 cm	38 x 30 km				
			Channel 4: 0.35 cm	16 x 14 km				
SAR	Seasat	June–October 1978	L-band: 23.5 cm	25 m/4 looks (dependent upon processing)	107 km	Repeat coverage every 3 days	Near circular, not Sun-synchronous	JPL–NODS ASF
	ERS-1	May 1991– Life expectancy: 3 years	C-band: 5.7 cm	30 m/4 looks	80–100 km	Repeat coverage every 3 days	Sun-synchronous, exactly repeating	ESA ASF
	JERS-1	February 1992– Life expectancy: 2 years	L-band: 23.5 cm	18 x 30 m 4 looks	75 km	Repeat coverage every 41 days	Sun-synchronous	ASF
AVHRR	TIROS-N	October 1978–January 1980	Band 1/VIS: 0.58–0.68 µm	1.1 km	2580 km	Twice daily	Circular, Sun-synchronous	SDSD EDC ARC
			Band 2/NrIR: 0.725–1.1 µm					
			Band 3/IR: 3.55–3.93 µm					
			Band 4/TIR: 10.05–11.05 µm					
	NOAA-6, NOAA-8, NOAA-10, NOAA-12	June 1979–March 1983 July 1984– May 1983–June 1984 July–October 1985 November 1986– September 1991–	Band 1/VIS: 0.55–0.90 µm Band 2/NrIR: 0.725–1.1 µm Band 3/IR: 3.55–3.93 µm Band 4/TIR: 10.05–11.05 µm					
OLS	DMSP Block 5D-1, -2 Currently active: F9 F10 F11	OLS first launched in September 1976 February 1988– December 1990– (nonstandard orbit) November 1991–	Band 1/VIS: 0.4–1.1 µm	0.62 km for fine-resolution, 2.4 km for low-light, smoothed, stored (llss) data	3012 km	Twice daily	Circular, Sun-synchronous	NSIDC NGDC
			Band 2/TIR: 10.5–12.6 µm	0.56 km for fine-resolution, 2.8 km for llss data				
MSS	Landsat 1, Landsat 2, Landsat 3, Landsat 4, Landsat 5	July 1972–January 1978 July 1975–July 1983 (interrupted 1979–1980) March 1978–March 1983 July 1982– March 1984–	Band 4/VIS: 0.5–0.6 µm	79 m	185 km	Repeat coverage every 18 days	Circular, Sun-synchronous	EDC EOSAT
			Band 5/VIS: 0.6–0.7 µm					
			Band 6/NrIR: 0.7–0.8 µm					
			Band 7/TIR: 0.8–1.1 µm					
TM	Landsat 4, Landsat 5	July 1982– March 1984–	Band 1/VIS: 0.45–0.52 µm	30 m				
			Band 2/VIS: 0.52–0.60 µm					
			Band 3/VIS: 0.63–0.69 µm					
			Band 4/NrIR: 0.76–0.90 µm					
			Band 5/IR: 1.55–1.75 µm					
			Band 6/TIR: 10.40–12.50 µm	120 m				
			Band 7/IR: 2.08–2.35 µm	30 m				

Acronyms and Abbreviations

ADEOS	Advanced Earth Observing Satellite (Japan)		EOS	Earth Observing System
AES	Atmospheric Environment Service of Canada		EOSDIS	Earth Observing System Data and Information System
AGC	automatic gain control		ERB	Earth radiation budget
AGU	American Geophysical Union		ERIM	Environmental Research Institute of Michigan
AI	artificial intelligence		ERS-1	First European Remote Sensing Satellite
AIDJEX	Arctic Ice Dynamics Joint Experiment		ERTS	Earth Resources Technology Satellite
AIRS	Atmospheric Infrared Sounder		ESA	European Space Agency
AIRSAR	Airborne Synthetic Aperture Radar		ESMR	Electrically Scanning Microwave Radiometer
AMERIEZ	Antarctic Marine Ecosystem Research in the Ice Edge Zone		ESRIN	European Space Agency Research Institute
AMI	Active Microwave Instrument		ESSA	Environmental Sciences Service Administration (USA); Environmental Survey Satellite
AMS	American Meteorological Society			
AMSU	Advanced Microwave Sounding Unit		ESTEC	European Space Agency Technology Center
AOBP	Arctic Ocean Buoy Program			
APL	Applied Physics Laboratory			
ASF	Alaska Synthetic Aperture Radar Facility		FCM	fuzzy c-means algorithm
ATSR	Along-Track Scanning Radiometer		FLAR	forward-looking airborne radar
AVHRR	Advanced Very High-Resolution Radiometer		FNOC	Fleet Numerical Oceanography Center
			FNWC	Fleet Numerical Weather Central
			FOA	National Defense Research Institute (Sweden)
BEPERS	Bothnian Experiment in Preparation of ERS-1		FOV	field of view
BESEX	Bering Sea Experiment		FRAM	fine-resolution Antarctic model
			FY	first-year ice
CCRS	Canada Centre for Remote Sensing			
CCT	computer compatible tape		GAC	global area coverage
CEAREX	Coordinated Eastern Arctic Research Experiment		GCM	general circulation model
CIRES	Cooperative Institute for Research in Environmental Sciences		GDR	geophysical data record
			GEOS	Geodetic Earth Observing Satellite
COADS	Comprehensive Ocean-Atmosphere Data Set		GEOSAT	Geodynamics Experimental Ocean Satellite
CRREL	Cold Regions Research and Engineering Laboratory		GIS	Geographic Information System
			GISS	Goddard Institute for Space Science
CRRELEX	Cold Regions Research and Engineering Laboratory Experiment		GLCM	gray level co-occurrence
			GPS	Geophysical Processor System
CSA	Canadian Space Administration		GSFC	Goddard Space Flight Center
CZCS	Coastal Zone Color Scanner		GSP	Greenland Sea Project
DEM	digital elevation model		HCMM	Heat Capacity Mapping Mission
DMRT	dense median radiative transfer		HIRIS	High-Resolution Imaging Spectrometer
DMSP	Defense Meteorological Satellite Program		HRPT	high-resolution picture transmission
DN	digital number		HUTSCAT	Helsinski University of Technology Scatterometer
D–S	Dempster–Shafer			
ECMWF	European Centre for Medium-Range Weather Forecasting		IEEE	Institute for Electronics and Electrical Engineering
EMR	electromagnetic radiation		IFOV	instantaneous field of view

IR	infrared
ISCCP	International Satellite Cloud Climatology Program
ISTS	Institute for Space and Terrestrial Science (Canada)
JD	Julian day
JERS-1	Japanese Earth Resources Satellite
JIC	Navy–NOAA Joint Ice Center
JPL	Jet Propulsion Laboratory
JPL–NODS	JPL–NASA Ocean Data System, mail stop 300-323, 4800 Oak Grove Drive, Pasadena, CA 91109
KBS	knowledge-based systems
KRMS	Ka-band Radiometric Mapping System
LAS	Land Analysis System
LEADEX	Lead Experiment
LIMEX	Labrador Sea Ice Margin Experiment
MAP	maximum a posteriori
MESSR	Multispectral Electronic Self-Scanning Radiometer (Marine Observation Satellites, Japan)
MFMR	Multifrequency Microwave Radiometer
MFY	medium first-year
MIMR	Multifrequency Imaging Microwave Radiometer
MITI	Ministry of International Trade and Industry (Japan)
MIZ	marginal ice zone
MIZEX	Marginal Ice Zone Experiment
ML	maximum likelihood
MLE	maximum likelihood estimate
MOS	Marine Observation Satellites (Japan)
MRF	Markov random fields
MRT	modified radiative transfer
mse	minimum square error
MSR	Microwave Scanning Radiometer (Marine Observation Satellites, Japan)
MSS	multispectral scanner
MY	multiyear ice
NADC	Naval Air Development Center
NASA	National Aeronautics and Space Administration (USA)
NCAR	National Center for Atmospheric Research
NESDIS	National Environmental Satellite, Data, and Information Service
NIMBUS	NASA Meteorological Research and Development Satellite (USA)
NIPR	National Institute of Polar Research (Japan)
NMC	National Meteorological Center
NOAA	National Oceanic and Atmospheric Administration (USA)
NOARL	Naval Oceanographic and Atmospheric Research Laboratory
NODC	National Oceanographic Data Center (USA)
NORSEX	Norwegian Ocean Remote Sensing Experiment
NRL	Naval Research Laboratory
NSCAT	NASA Scatterometer (USA)
NSIDC	National Snow and Ice Data Center, University of Colorado, Boulder, CO 80309
NSSDC	National Space Science Data Center, Goddard Space Flight Center, Greenbelt, MD 20771
NWS	National Weather Service (USA)
OCTS	Ocean Color and Temperature Scanner
ODAR	omnidirectional airborne radar
OI	old ice
OLS	Optical Line-Scan System
OW	open water
PDF	probability density function
PIPOR	Program for International Polar Oceans Research
PIPS	Polar Ice Prediction System
PLF	pulse-limited footprint
PMEL	Pacific Marine Environmental Laboratory
POLES	Polar Exchange at the Sea Surface Program
RA	radar altimeter
Radarsat	Radar Satellite (Canada)
RAL	Rutherford Appleton Laboratory
rms	root mean square
RPIPS-B	Regional Polar Ice Prediction System Barents Sea
RPIPS-G	Regional Polar Ice Prediction System Greenland Sea
SAAMEX	Surface and Atmospheric Airborne Microwave Experiment
SAR	synthetic aperture radar
SCATT	Scatterometer
SCG	Swedish Coast Guard
SEASAT	Sea Satellite
SEAWIFS	Sea-Viewing Wide-Field Sensor
SFMR	Stepped Frequency Microwave Radiometer
SFT	strong fluctuation theory
SIR	Shuttle Imaging Radar
SIZEX	Seasonal Ice Zone Experiment
SLAR	side-looking airborne radar

SMHI	Swedish Meteorological and Hydrological Institute
SMLE	suboptimal maximum likelihood estimator
SMMR	Scanning Multichannel Microwave Radiometer
SMS	sea ice monitoring site
SPOT	Systeme Probatoire pour l'Observation de la Terre (France)
SSIZ	seasonal sea ice zones
SSM/I	Special Sensor Microwave/Imager
SST	sea surface temperature
STIKSCAT	Stick Scatterometer
SWH	significant wave height
SY	second-year ice

TBD	to be determined
TFY	thick first-year
ThFY	thin first-year
THIR	temperature-humidity infrared radiometer
TIR	thermal infrared
TIROS	Television and Infrared Observing Satellite
TM	Thematic Mapper
TOPEX	Ocean Topography Experiment
UKMO	United Kingdom Meteorological Office
USCG	U.S. Coast Guard
UW	University of Washington
VATT	voltage proportional to attitude
WMO	World Meteorological Organization
YI	young ice

461